characteristically produces an extensive longitudinal myelitis involving ≥3 cord levels. NMOSD is a serious disease and not infrequently may have a fatal outcome. The spectrum includes association of endocrinopathies as well as coexisting autoimmunity, such as Systemic Lupus Erythematous (SLE) and antiphospholipid and Sjögren's syndromes. Treatment with B-cell-depleting monoclonal antibodies and complementary activation cascade inhibitors, has shown a high degree of efficacy in reducing relapses and preventing disability.

Anti-MOG antibodies syndrome is also included by the editors in Part III, "Autoimmune Neurology." This recently identified disorder in which antibodies against membrane-embedded myelin oligodendrocyte glycoprotein (MOG) produce multifocal areas of demyelination, affecting optic nerves (simultaneous bilateral optic neuritis is common) and spinal cord as well, has provoked substantial interest in pediatric and adult neurology. The brain may also be involved. Almost 50% of patients with clinical diagnosis of NMO who test anti-NMO-IgG antibody negative are identified as belonging to the anti-MOG syndrome group. This disorder tends to be less aggressive clinically than the NMOSD. In children and adults there is a peculiar association with Acute Disseminated Encephalomyelitis (ADEM). High-dose steroids are typically used for the management of relapses while diverse immunosuppression protocols are proposed for long-term management. These aspects are updated and comprehensibly approached in the corresponding chapters of this section.

Outside the CNS, neuro-autoimmune disorders affecting the peripheral nervous system (Part IV) are historically indeed the "pioneers" of this recently established branch of neurology, even though in their very early origins the concept of autoimmunity did not exist yet. The part dedicated to acute and chronic immune neuropathies and radiculopathies reminds us of the classic collaboration as young French army officers during the First World War of Georg Guillain (1876–1961) and Jean A. Barré (1880–1967). They used the most advanced laboratory technology of that time: cerebrospinal fluid (CSF) analysis obtained through a lumbar puncture (first described and employed by Heinrich Quincke in 1891). Guillain and Barré described two soldiers with an acute symmetrical lower motor neuron picture and polyradiculopathy manifested by rapid ascending limb paralysis. CSF disclosed an acellular fluid with high protein content. The modern understanding of acute polyradiculopathies embraces the causal concept of autoimmune processes (triggered by a recent precedent infection?); hence these entities traditionally may respond clinically to acute and aggressive immunotherapy. Chronic peripheral neuropathies carry the necessity to activate an extensive differential diagnosis requiring the exclusion of metabolic and toxic causes, nutritional deficiencies, the previously noted paraneoplastic syndromes, genetic mutations, and even iatrogenic adverse effects. A comprehensive workup is discussed, including state-of-the-art imaging studies with muscle/nerve ultrasound techniques, designed electrophysiologic studies, and nerve and neuro-diagnostic skin biopsies utilizing advanced histopathological techniques. The long-noted response to immunomodulatory therapies such as intravenous humanized immunoglobulin-g (IVIg), plasma exchange, and particularly the exquisite sensitivity to many neuropathies to steroids poses a fundamental challenge for the clinician regarding the choice of initial and maintenance therapy. Satisfactory therapeutic response (improvement, remission) may be achieved utilizing oral steroid protocols in many patients with sensory/motor chronic inflammatory demyelinating neuropathies (CIDP). Some steroid-responsive neuropathies have shown demyelination and infiltration of immune cells in the endoneurium—predominantly natural killers (NK) lymphocytes, and other subsets with lytic capacity such as CD3−, CD56+, and other "small lymphocytes" (SLs).

Part IV of the book addressing autoimmune peripheral neuropathies includes the typically, polysymptomatic autoimmune autonomic disorders due to axonal small-fiber damage. This relatively recently identified type of neuropathy requires highly specialized laboratories for proper assessment and diagnosis, as commonly utilized tests in this field like QSART (Quantitative Sudomotor Axon Reflex Test) are not readily available to the community neurologist. Some researchers include complex regional pain syndromes in this group of neuropathies.

Chapter 19 addresses autoimmune neuromuscular junction diseases initiates the dissertation with Myasthenia Gravis (MG)—another "classic" example of a disorder eventually adjudicated to autoimmunity after a remarkable journey of scientific discoveries overcoming traditional non-evidenced thinking against scientifically based discoveries. In this case, even social and gender prejudices played a role in the modern development of the understanding of the MG pathogenesis and its effective management. Samuel Wilks, a British physician, reported the first case in 1877 as "bulbar paralysis," and Friedrich Jolly (Berlin, Germany) coined the term "myasthenia gravis pseudoparalytica." Nevertheless, the most compelling story of this saga is the utilization and demonstration of a symptomatic remedial therapy by Mary Broadfoot Walker (1888–1974), a Scottish physician, in 1934. Walker theorized MG resembled curare poisoning known to respond to physostigmine, an anticholinesterase agent. Dr. Walker showed a visually dramatic immediate response to a subcutaneous injection of this agent in a patient with MG manifesting bilateral ptosis and facial weakness. This report was received with tremendous skepticism by her British male peers at a clinical meeting of the Royal Society of Medicine gathering at the National Hospital Queen Square, London. Nevertheless, her report was initially published as a letter in the *Lancet* in 1934, initiating the epoch of effective symptomatic management of an otherwise totally untreatable disease. Notable response to removal of a diseased thymus in people with MG in the 1940s paved the road for consideration of MG as an autoimmune disorder.

MG is a relatively uncommon disease (global prevalence of 200–400/million) but it can be incapacitating irrespective of the age of the afflicted person, or the clinical variety. MG has been established as a multifactorial antibody-mediated disease directed to interfere with polysynaptic acetylcholine receptors (AchR) at the myoneural junction where removal of Ca+ channels involving transmitter release result in different clinical phenotypes. Detection of AchR antibody in serological assays is practically specific of MG. However, about 40% of individuals with MG disclose tyrosine muscle specific kinase protein (MuSK) receptor antibodies. This protein tends to cluster in AchRs at the motor-end plate. This immunologic development has not been completely elucidated but usually reflects clinically as aggressive clinical forms of MG in seronegative individuals. The section on autoimmune neuromuscular junction disorders discussing MG includes the current methods to determine a prompt clinical diagnosis and the utilization of present and future biomarkers expediting nosological identification.

The current international classification categorizing degrees of neuromuscular involvement (Class I, Class II, and Class IIa) is analyzed and criticized. Thymectomy and immunotherapy, including the novel utilization of monoclonal antibodies, and optimization of the diverse presentations of symptomatic medication (pyridostigmine), complement the current proposal of management of this potentially incapacitating disease.

In the same chapter, Lambert-Eaton myasthenic syndrome (LEMS) is discussed. This rare presynaptic motor-end plate autoimmune disorder is overwhelmingly associated with small-cell lung cancer (SCLC), but it has also been reported in conjunction with other malignancies, including malignant thymoma, lymphosarcoma, and even prostate cancer. While LEMS produces a myasthenic-like picture due to impaired release of Ach packages at the presynaptic space of the neuromuscular junction (producing a different electromyographic pattern than MG), and may be accompanied by autonomic dysfunction (the mechanism of this association is not completely understood). LEMS is typically refractory to immunomodulation, except for reports of some response to IVIg. A formulation manufactured by compounded pharmacies and used symptomatically off-label for years, Amifampridine, a chemical based on 3,4- diaminopyridine, a K+ channel blocker, has been approved by the FDA for symptomatic management of muscle weakness produced by LEMS. This chapter discusses may aspects on the pathophysiology that help with a prompt diagnosis and subsequent modern management of this syndrome.

Part IV of the book dedicated to peripheral nervous system disorders includes immune and inflammatory myopathies. These myopathic clinical pictures may produce profound proximal muscle weakness, tenderness to palpation of muscular groups, and eventually segmental atrophy. The typical example is Polymyositis, chronic or "active." Generally, if the picture presents as sporadic, isolated, and not associated with other disorders, the physical findings are purely myopathic without associated neurological deficits, i.e., primary sensory involvement or abnormal reflexes. The necessity for an extensive workup and differential diagnosis is inescapable. The chapter addresses the most updated data on electrophysiologic studies, muscle biopsy performed open or with needle utilization, including the proposal and indications for MRI-guided procedures. Advanced technology for histochemistry staining and immunologic management of the samples and the use of electronic microscopy are aspects the reader will find of interest.

Management of immune inflammatory myopathies tends to be confounding if myositis is part of the spectrum of other disorders such as dermatomyositis, or as a remote effect of cancer, or if this represents a case of sporadic gradually progressive inclusion body myositis (IBM), or the hereditary form (hIBM). Determination of autoantibodies and immune markers in a clinically suspected individual is assisted by the material discussed in the chapter.

The editors of *Neuroimmunology* include topics on neurorheumatology and neurosarcoidosis in Part V, delivered by experts on these commonly encountered clinical challenges in day-to-day practice. The elements and factors to be taken into consideration to reach the correct diagnosis in the extensive field of rheumatology (particularly with the advent of a host of available immunotherapies), and the adequate differential that sarcoid disease requires with the so many entities it may mimic, constitute more than intellectual exercises, but important clinical undertakings in view of the potential severity and complications these disorders may produce.

Chapter 24 on central nervous system vasculitis discusses one of the most serious disorders in the field of clinical neuroimmunology. While the autoimmune mechanism has not been elucidated, the clinical picture is usually one of severe acute or subacute encephalopathy with associated acute multi-ischemic state due to occlusion of arterial blood vessels. Involvement of the spinal cord may result in acute quadriplegia or paraplegia, depending on the affected level of the cord. The term "Primary Angiitis" is generally employed when the process is independent from other immune disorders; otherwise in association with other forms of vasculitis (i.e., granulomatosis with polyangiitis, SLE, Behçet's disease, rheumatoid arthritis, infectious processes), the preferred terminology is "secondary CNS vasculitis (or angiitis)." CNS vasculitis displays a typical angiographic pattern visualized by magnetic resonance angiography or standard arteriography. CSF analysis may show nonspecific inflammatory abnormalities, and serological blood tests usually do not contribute to diagnostic identification, particularly in primary angiitis. Frequently, invasive diagnostic procedures are required, including meningeal or cerebral artery biopsy. This chapter will be of invaluable assistance to the reader seeking the most comprehensive and updated information on this theme, including options for treatment.

The editors assigned Chapter 25 to review the infrequently discussed complications of immunotherapies. In the same section, infections and immunodeficiency are approached by known researchers in these areas. In contrast to paraneoplastic disease, Chapter 26 now redirects to oncological mimics of inflammatory CNS disease. The impact of this clinical situation in the real world requires effective differential diagnosis, and hence implications into specific managements. These aspects are emphasized.

The last part of the book (Part VI: Clinical Approaches in Neuroimmunology) is distributed into varied topics. Chapter 27 dedicated to autoimmune myelitis and myelopathic inflammatory disorders discusses spinal cord disease not associated with bona fide systemic autoimmune diseases or to demyelinating disorders such as MS, NMOSD, or anti-MOG syndrome. An episode of isolated transverse myelitis is usually characterized by an acute and devastating development of motor paraplegia showing a sensory dermatome level and sphincter disturbance. Most of these cases may exhibit a unique extensive lesion in the spinal cord by MRI images and generally have a discrete functional recovery despite steroids and other immunomodulatory modalities.

The underlying immunopathology has not been consistently reproduced in these cases. It is feasible, then, to conclude that mechanisms responsible for inflammatory myelopathies are compounded by heterogenous immunologic settings. The immunology mechanisms reported thus far and the current clinical management are presented.

Chapter 28 discussing the clinical approach to ophthalmic disease reminds us that the eye is a frequent target of autoimmune attack whether affecting the optic nerve as part of systemic disease (i.e., SLE and sarcoidosis) or as one of the manifestations of a more generalized inflammatory demyelinating disease (MS, NMOSD, anti-MOG syndrome, and ADEM). Optic neuritis can also develop as an isolated disorder. The typical example is the case of the formerly called clinically isolated syndrome (CIS) manifested as a pure case of optic neuritis. While the overwhelming majority of cases eventually will transform into clinically definite MS, just a very small proportion will remain as CIS. Another disorder specifically affecting the optic nerve is Chronic Recurrent Inflammatory Optic Neuropathy (CRION). This condition is a rare, poorly understood, and apparently heterogeneous autoimmune disorder. CRION produces recurrent attacks of optic neuritis involving one or both eyes. Other areas of the nervous system are not affected and tests conducive to diagnosis of the major demyelinating disease are negative. CRION attacks are highly sensitive to steroids but the patient usually requires long-term prophylactic use of immunosuppressants.

Neuro-ophthalmological conditions resulting from autoimmune processes are extremely diverse and varied. Uveitis may produce ocular pain, blurred vision, photophobia, and reddish inflammation of the eye secondary to reactive conjunctivitis. Uveitis is a common manifestation of sarcoidosis, Crohn's disease, Behçet's disease, and certain forms of ankylosing arthritis. Recent data suggest individuals carrying the HLA-B27 allele may experience uveitis associated with other disorders through the process of antigen mimicry triggering T-cell immunologic inflammatory attack. Other ophthalmological diseases intimately related to autoimmune mechanisms, including development of proptosis and extraocular myopathy, prominently manifest in cases of immune thyroiditis. The reader will find a rich collection of clinical examples and discussions on management in this section.

Chapter 29 addressing clinical approach to pediatric demyelinating disease discusses the formidable advances in this field. Monophasic disorders (mainly CIS and ADEM), and the multiphasic forms (relapsing) MS, NMOSD, anti-MOG syndrome, and even a type of ADEM, pose a diagnostic and management challenge to the clinicians in view of the particular neurological manifestations children with these disorders exhibit.

The editors include Chapter 30 to discuss "Stiff Person Syndrome" (SPS) assigned to experts on this unusual topic. This syndrome, formally known as "stiff man syndrome", is properly denominated at present since it affects woman in a larger proportion to men. The clinical picture is one of slow, gradual, insidious development of stiffness; pronounced rigidity; and spasms involving the trunk and extremities increasing in severity; and incapacitation with time. Advanced cases may develop joint deformities. Delay in diagnosis is usually due to the fact this disorder is not suspected or identified in its early stages, being frequently confused with a functional or conversion disorder. SPS results from autoimmune interference with GABAergic transmission in central motor pathways. The molecular abnormality has been adjudicated to elevated anti-GAD (glutamic acid decarboxylase) antibodies. Other immune-related disorders may be associated with SPS including vitiligo and thyroiditis. IVIg courses and Rituximab (anti CD20 monoclonal antibody) have been proposed as potential therapies along with a well-designed symptomatic program and physical therapy.

*Neuroimmunology* is, in fact, a sophisticated and yet practical scientific production serving as an extraordinary consultation source for students, practitioners, and scholars. One of the most meritorious aspects of this multi-authored work is precisely the stellar list of contributors that the editors constructed for this scientific communication artwork. The discussions from the authors displayed in each chapter, in each theme, provides the reader with *state-of-the-art* information and encourages thoughts on complex and fascinating areas of neuroimmunology that remain to be explored.

Victor M. Rivera, MD, FAAN
Distinguished Emeritus Professor
Baylor College of Medicine
Houston, TX, USA

# Preface

Neuroimmunology is a rapidly evolving field with significant diagnostic and treatment advancements over the last decade. Even basic concepts in clinical neuroimmunology have changed considerably over this time. New data has driven treatment paradigms for multiple sclerosis (MS) and increased clinical knowledge and antibody testing capability for a multitude of newly recognized autoimmune neurological syndromes. Treatment for neuromyelitis optica spectrum disorder (NMOSD), antibody-mediated autoimmune encephalitis, and stiff person syndrome (SPS), among others, has therefore dramatically improved. As the field of neuroimmunology evolves, the need for a comprehensive, up-to-date textbook has become apparent.

*Neuroimmunology* is a comprehensive handbook written with the clinician in mind and targets residents, fellows, advanced practice providers, general neurologists, and subspecialists. The aim of this book is to make recent developments in Neuroimmunology and Autoimmune Neurology accessible to clinicians as well as review basic concepts in Neuroimmunology. The chapters have been written by experts in their fields. *Part I* introduces concepts in Neuroimmunology, including an overview of the immune system, antibody testing, basics in neuroimaging, and concepts of immunotherapy. *Part II* consists of 6 chapters focused on MS, including diagnostic criteria, immunology, pathology, epidemiology, and genetics. *Part III* consists of 6 chapters focused on antibody-mediated diseases across a spectrum of various neurological subspecialties, including autoimmune and paraneoplastic encephalitis, epilepsy, movement disorders, and other demyelinating syndromes. This part includes a chapter discussing treatment approaches in Autoimmune Neurology that helps to highlight the successful completion of the first randomized trials in Autoimmune Neurology for NMOSD and the evolution of treatment algorithms in antibody-mediated autoimmune encephalitis. *Part IV* includes 4 chapters focused on immune disorders of the peripheral nervous system, a topic often overlooked in classic Neuroimmunology training programs, but includes relatively common conditions such as myasthenia gravis and Guillain-Barré syndrome along with newly described conditions. *Part V* includes 5 chapters of various systemic diseases with prominent neurological manifestations such as rheumatological disorders, sarcoidosis, vasculitis, immunodeficiencies, and oncological mimics. The final part, *Part VI*, includes 6 "clinical approach" chapters. These chapters, designed with the clinician in mind, use case-based examples to provide readers with a systematic approach to high-yield topics in Neuroimmunology, including clinical diseases (myelitis, ophthalmologic disease, pediatric demyelinating disease, and SPS) as well as treatment approaches with disease-modifying therapies and symptomatic therapies in MS.

We hope health professionals who are interested in neuroimmunological disorders will find this book useful. Finally, we would like to thank our contributing authors for all their hard work and dedication.

Aurora, CO, USA

Amanda L. Piquet, MD
Enrique Alvarez, MD, PhD

# Contents

# Contributors

**Enrique Alvarez, MD, PhD** Department of Neurology, Rocky Mountain Multiple Sclerosis Center at the University of Colorado, Aurora, CO, USA

**Christina J. Azevedo, MD, MPH** Department of Neurology, Saint Louis University School of Medicine, Saint Louis, MO, USA

**Laura E. Baldassari, MD, MHS** Mellen Center for Multiple Sclerosis Treatment and Research, Neurological Institute, Cleveland Clinic Foundation, Cleveland, OH, USA

**J. David Beckham, MD** Departments of Medicine, Neurology, and Immunology & Microbiology, Neuro-Infectious Diseases Group, University of Colorado School of Medicine, Rocky Mountain Regional VA Medical Center, Aurora, CO, USA

**Shamik Bhattacharyya, MD, MS** Department of Neurology, Brigham and Women's Hospital, Boston, MA, USA

**Kyle Blackburn, MD** Department of Neurology, University of Texas Southwestern Medical Center, Dallas, TX, USA

**Robert C. Bucelli, MD, PhD** Department of Neurology, Neuromuscular Section, Barnes Jewish Hospital, Washington University School of Medicine in Saint Louis, Saint Louis, MO, USA

**Tanuja Chitnis, MD** Pediatric Multiple Sclerosis and Demyelinating Disorders, Division of Pediatric Neurology, Department of Neurology, Massachusetts General Hospital, Department of Neurology, Harvard Medical School, Boston, MA, USA

**Tracey A. Cho, MD** Department of Neurology, University of Iowa Carver College of Medicine, University of Iowa Hospitals and Clinics, Iowa City, IA, USA

**Stacey L. Clardy, MD, PhD** Department of Neurology, University of Utah Health, Salt Lake City, UT, USA

Department of Neurology, George E. Wahlen Salt Lake City VHA, Salt Lake City, UT, USA

**Jeffrey A. Cohen, MD** Mellen Center, Neurologic Institute, Cleveland Clinic, Cleveland, OH, USA

**John R. Corboy, MD** Department of Neurology, Rocky Mountain Multiple Sclerosis Center at the University of Colorado, Aurora, CO, USA

**Anne Haney Cross, MD** Neuroimmunology Section, Department of Neurology, Washington University School of Medicine, Saint Louis, MO, USA

**Neda Dastgheyb, BA** Department of Neurosciences, University of California, San Diego, La Jolla, CA, USA

**Divyanshu Dubey, MD** Department of Laboratory Medicine & Pathology, Department of Neurology, Mayo Clinic, Rochester, MN, USA

**Jordon Dunham, MsC, PhD** Department of Neuroscience, Lerner Research Institute, Cleveland Clinic Foundation, Cleveland, OH, USA

**James E. Eaton, MD** Department of Neurology, Vanderbilt University Medical Center, Neurology Clinic, Nashville, TN, USA

**Megan Esch, MD** Department of Neurology, Geisinger Medical Center, Danville, PA, USA

**Michelle Fabian, MD** Department of Neurology, Mount Sinai Hospital, Icahn School of Medicine at Mount Sinai, New York, NY, USA

**Andrew R. Findlay, MD** Department of Neurology, Neuromuscular Section, Barnes Jewish Hospital, Washington University School of Medicine in Saint Louis, Saint Louis, MO, USA

**Eoin P. Flanagan, MD** Department of Neurology, and Laboratory Medicine and Pathology, Mayo Clinic, Rochester, MN, USA

**Jonathan R. Galli, MD** Department of Neurology, University of Utah Health, Salt Lake City, UT, USA

Department of Neurology, George E. Wahlen Salt Lake City VHA, Salt Lake City, UT, USA

**Rocio Garcia-Santibanez, MD** Department of Neurology, Emory University School of Medicine, Atlanta, GA, USA

**Christine M. Gill, MD** Department of Neurology, University of Iowa Carver College of Medicine, University of Iowa Hospitals and Clinics, Iowa City, IA, USA

**Elisabeth Golden, MD** Department of Neurology, University of Texas Southwestern Medical Center, Dallas, TX, USA

**Jennifer S. Graves, MD, PhD, MAS** Department of Neurosciences, Rady's Pediatric MS Center, UCSD Neuroimmunology Research Program, University of California, San Diego, ACTRI, La Jolla, CA, USA

**Benjamin Greenberg, MD, MHS** Transverse Myelitis Program, Department of Neurology and Neurotherapeutics, University of Texas Southwestern, Dallas, TX, USA

**Robert H. Gross, MD** Department of Neurology, Rocky Mountain Multiple Sclerosis Center at the University of Colorado, Aurora, CO, USA

**Thomas R. Haven, PhD** Department of Pathology, University of Utah School of Medicine, ARUP Institute for Clinical and Experimental Pathology, ARUP Laboratories, Salt Lake City, UT, USA

**Khalil Husari, MD** Department of Neurology, Johns Hopkins University, Johns Hopkins Hospital, Baltimore, MD, USA

**Jenny Joseph, MD** Department of Neurology, Emory University School of Medicine, Atlanta, GA, USA

**Stephen Krieger, MD, FAAN** Department of Neurology, Mount Sinai Hospital, Icahn School of Medicine at Mount Sinai, New York, NY, USA

**Christopher Langston, MD** Department of Neurology, Albert Einstein College of Medicine, Bronx, NY, USA

**A. Sebastian López-Chiriboga, MD** Department of Neurology, Division of Multiple Sclerosis and Autoimmune Neurology, Mayo Clinic, Florida, FL, USA

**Gabrielle Macaron, MD** Mellen Center for Multiple Sclerosis, Neurological Institute, Cleveland Clinic Foundation, Cleveland, OH, USA

**Kedar R. Mahajan, MD, PhD** Department of Neuroscience, Lerner Research Institute, Cleveland Clinic Foundation, Cleveland, OH, USA

Department of Neurology, Mellen Center for Multiple Sclerosis Treatment and Research, Neurological Institute, Cleveland Clinic Foundation, Cleveland, OH, USA

**Marisa P. McGinley, DO** Mellen Center, Neurologic Institute, Cleveland Clinic, Cleveland, OH, USA

**Kyla A. McKay, PhD** Department of Clinical Neuroscience, Karolinska Institutet, Stockholm, Sweden

**Andrew McKeon, MD** Department of Neurology, Department of Laboratory Medicine and Pathology, Division of Multiple Sclerosis and Autoimmune Neurology, Mayo Clinic, Florida, FL, USA

**Dery Miller, MD** Department of Neurology, Emory University School of Medicine, Atlanta, GA, USA

**Annalise E. Miner, BS** Department of Neurosciences, University of California, San Diego, La Jolla, CA, USA

**Alexander C. Mohler, MD, MS** Department of Neurology, Rhode Island Hospital and the Miriam Hospital, Warren Alpert Medical School of Brown University, Providence, RI, USA

**Nancy L. Monson, PhD** Department of Neurology, Department of Immunology, University of Texas Southwestern Medical Center, Dallas, TX, USA

**Douglas E. Ney, MD** Departments of Neurology and Neurosurgery, University of Colorado School of Medicine, Aurora, CO, USA

**Daniel Ontaneda, MD, PhD** Department of Neurology, Mellen Center for Multiple Sclerosis, Cleveland Clinic, Cleveland, OH, USA

**Miryam Palomino, BS** Department of Neurosciences, University of California, San Diego, La Jolla, CA, USA

**Daniel M. Pastula, MD, MHS** Department of Neurology, Medicine (Infectious Diseases), and Epidemiology, Neuro-Infectious Diseases Group, University of Colorado School of Medicine, Aurora, CO, USA

**Siddharama Pawate, MD** Department of Neurology, Vanderbilt University Medical Center, Nashville, TN, USA

**Lisa K. Peterson, PhD** Department of Pathology, University of Utah School of Medicine, ARUP Institute for Clinical and Experimental Pathology, ARUP Laboratories, Salt Lake City, UT, USA

**Amanda L. Piquet, MD** Department of Neurology, Neuroimmunology, Neuroinfectious disease and Neurohospitalist Sections, Neuro-Infectious Diseases Group, University of Colorado School of Medicine, Aurora, CO, USA

**David Prater III, MA** Department of Neurology, Rocky Mountain Multiple Sclerosis Center at the University of Colorado, Aurora, CO, USA

**Johannes Pulst-Korenberg, MD, MS** Department of Neurology, Keck School of Medicine, University of Southern California, Los Angeles, CA, USA

**Lokesh A. Rukmangadachar, MBBS, MD** Department of Neurology, Keck Hospital of USC, LAC+USC Medical Center, University of Southern California, Los Angeles, CA, USA

**Shiv Saidha, MBBCh, MD, MRCPI** Department of Neurology, Johns Hopkins University, Baltimore, MD, USA

**Jonathan Douglas Santoro, MD** Pediatric Multiple Sclerosis and Demyelinating Disorders, Division of Pediatric Neurology, Department of Neurology, Massachusetts General Hospital, Department of Neurology, Harvard Medical School, Boston, MA, USA

Department of Pediatrics, Keck School of Medicine at USC, Los Angeles, CA, USA

**Elia Sechi, MD** Department of Neurology, Mayo Clinic, Rochester, MN, USA

**M. Mateo Paz Soldán, MD, PhD** Neuroimmunology Division, Department of Neurology, University of Utah School of Medicine, Salt Lake City, UT, USA

**Marianna Spatola, MD, PhD** Massachusetts General Hospital Research Institute, Ragon Institute of Massachusetts General Hospital, Massachusetts

Institute of Technology, Harvard Medical School, Boston-Cambridge, MA, USA

**Maarten J. Titulaer, MD, PhD** Academic Center for Neuroinflammatory Disorders, Department of Neurology, Erasmus MC University Medical Center, Rotterdam, The Netherlands

**Helen Tremlett, PhD** Department of Medicine (Neurology), University of British Columbia, Vancouver, BC, Canada

**Kenneth L. Tyler, MD** Department of Medicine and Immunology-Microbiology, Department of Neurology, Neuro-Infectious Diseases Group, University of Colorado School of Medicine, University of Colorado Hospital and Rocky Mountain Regional VA Medical Center, Aurora, CO, USA

**Steven Vernino, MD, PhD** Department of Neurology, University of Texas Southwestern Medical Center, Dallas, TX, USA

**Juna M. de Vries, MD, PhD** Academic Center for Neuroinflammatory Disorders, Department of Neurology, Erasmus MC University Medical Center, Rotterdam, The Netherlands

**Cynthia Wang, MD** Department of Pediatrics and Department of Neurology and Neurotherapeutics, Children's Health Dallas, University of Texas Southwestern, Dallas, TX, USA

**Anson W. Wilks, MD** Department of Neurology, Washington University School of Medicine in Saint Louis, Saint Louis, MO, USA

**Mary Alissa Willis, MD** Department of Neurology, University of Mississippi Medical Center, Jackson, MS, USA

**Gregory F. Wu, MD, PhD** Neuroimmunology Section, Department of Neurology, Washington University School of Medicine, Saint Louis, MO, USA

**Anastasia Zekeridou, MD** Departments of Laboratory Medicine and Pathology and Neurology, Mayo Clinic College of Medicine, Rochester, MN, USA

# Part I

# Introduction: Immunology, Pathophysiology, and Immunotherapy

# 1 Introduction to Neuroimmunology: What the Cerebrospinal Fluid Teaches Us About Diseases of the Central Nervous System

Nancy L. Monson

**Key Points**

1. There are nonspecific (innate) and highly specific (adaptive) components to the immune system.
2. Lymphocytes (both B and T cells) are part of the adaptive immune system that are involved in immunosurveillance and are highly activated in many diseases of the central nervous system (CNS).
3. The innate immune system involvement in diseases of the CNS is largely unexplored, however our knowledge of its role continues to expand.
4. Our goal in focusing research efforts into these CNS diseases should be to identify cellular and humoral factors that are the best target candidates.

## Introduction to the Topic

There continues to be a rising interest in understanding the potential role of the immune response in diseases of the central nervous system (CNS). Neurologists, neuroscientists, and immunologists alike accept the challenge in learning each other's languages all for the good of a growing population of patients with diseases of the CNS who need better care. Several excellent reviews on the subject of neuroinflammation in neurodegenerative diseases of the brain are available [1–4], and so our task is instead to framework our subject matter in the context of the one compartment we can readily access in a living human subject that can provide important clues regarding the immunology of CNS disease: the cerebrospinal fluid. This is particularly important as CNS diseases that we once thought did not involve the immune system have now required a second consideration because of evidence acquired in the examination of their cerebral spinal fluid (CSF). Immunological factors that can influence the development or resolution of CNS disease are listed in Fig. 1.1. For example, memory B-cell frequencies in the CSF of patients at high risk to develop Alzheimer's disease correlate positively with amyloid burden in the brain [5]. Others have demonstrated that B cells from patients with multiple sclerosis (MS) secrete factors that can cause death to oligodendrocytes and neurons [6]. Generalized information regarding CSF (i.e., where it is made, how much of it is made, chemical composition, and the like) is available elsewhere [7].

N. L. Monson (✉)
Department of Neurology, Department of Immunology, University of Texas Southwestern Medical Center, Dallas, TX, USA
e-mail: nancy.monson@utsouthwestern.edu



A. L. Piquet, E. Alvarez (eds.), *Neuroimmunology*, https://doi.org/10.1007/978-3-030-61883-4_1

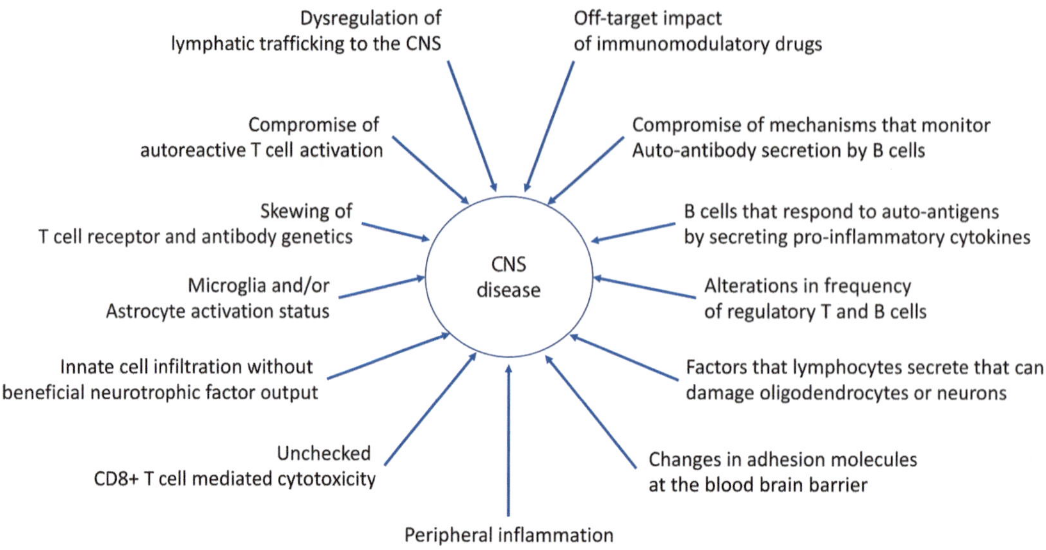


**Fig. 1.1** Immunological factors that may influence CNS disease

## An Introduction to the Immune System

The immune system is designed to protect the body from invasion by foreign agents such as viruses and bacteria (pathogenic agents) (reviewed in [8–11]). There are both nonspecific (innate) and highly specific (adaptive) components of the immune system. The innate immune response is a broadly reactive component of the immune system based on pattern recognition of pathogenic agents. The adaptive immune response is a specific reactive component of the immune system based on distinct proteins expressed by particular pathogenic agents. In general, the innate and adaptive immune responses work in concert to restrict the invasion space of the pathogenic agent (innate immunity) and either reduce its numbers or eradicate it altogether (innate and adaptive immunity). The adaptive immune response has a second feature which is to survey the body continually for possible invasion by the same pathogenic agent. Upon encounter with the same pathogenic agent a second time, the immune response occurs much more quickly since the adaptive immune response has already been established against that particular pathogenic agent.

Innate immunity is the first line of defense against pathogenic agents and is activated within hours or days of the invasion. The innate immune response is based on a barrier approach: anatomic barriers such as the skin prevent pathogenic agents from entering the body, physiological barriers such as fever prevent pathogenic agents from expansion, and phagocytic barriers include specialized cells that engulf the pathogenic agents and destroy them. The innate immune response also includes the secretion of inflammatory molecules that recruit cells of the adaptive immune response to more specifically target the invading pathogenic agent.

The adaptive immune response is particularly critical in those cases where innate immunity is unable to eradicate the pathogenic agent. It is composed of T cells and B cells that express receptors specific for distinct proteins expressed by particular pathogenic agents. T cells are activated by antigen-presenting cells (APC), which are part of the innate immune system that display peptides of the proteins expressed by foreign agents in the context of the MHC (major histocompatibility complex) to the T cell. If the T cell expresses a T-cell receptor that recognizes this peptide as presented by the APC, it will receive the necessary signals from the APC to become activated. Once activated, the T cell can participate in the immune response either by secreting cytokines that perpetuate the response by other immune cells to the pathogenic agent or by killing cells that have been invaded by it. There are also regulatory T cells that limit or suppress the immune response.

B cells express immunoglobulins on their surface that bind proteins directly. Once the membrane-bound immunoglobulin binds to its target, the complex is internalized which leads to the activation of the B cell. It is important to note that most B-cell responses require cytokines secreted by T cells for effective activation. The B cell undergoes a unique process of affinity maturation, which is designed to improve binding of the immunoglobulin to the specific protein. Those B cells that accomplish this goal secrete the immunoglobulin they express into the body. These secreted immunoglobulins circulate throughout the body and bind to the protein expressed by the pathogenic agent and are thus either neutralized or tagged for destruction by other immune cells.

This basic overview of the immune system was established from studies of blood and peripheral lymphoid organs of humans. The remaining chapter components focus on the historic context and current understanding of immunological elements within the central nervous system.

## There Are Lymphocytes in the CSF

The first indication of an immune response in a CNS disease was described by Kabat in 1942 [12]. Here, the data inferred that immunoglobulins were being synthesized in the cerebrospinal fluid of patients with either neurosyphilis or MS. This observation readily suggested that there must be cells in the CSF that could secrete immunoglobulin. In the following decade, scientists began to study the CSF for evidence of lymphocyte infiltration, and several scientists documented that lymphocyte numbers were increased in the CSF of patients with MS and other diseases [13–22]. Two laboratories linked immunoglobulin synthesis and lymphocytes in the CSF [23, 24] by demonstrating that CSF cells could make immunoglobulin in vitro without stimulation. This finding was confirmed by others over the next two decades [25–28].

The development of antibodies to identify T-cell subsets created a gateway for several studies that demonstrated T cells were present in the CSF of adults [17–22, 29–31] and pediatric [32] subjects. Prior to the use of antibodies, T-cell frequencies in the CSF were determined using rosetting techniques [29–31]. Cashman et al. (1982) used the then new OKT (Ortho-Kung T) antibodies to determine the frequency of OKT4 and OKT8 T cells in the CSF and blood of 40 patients with MS and 15 patients with other neurological diseases (OND) [33]. They found that OKT8+ T cells were reduced in the CSF of MS patients who had experienced an exacerbation within 14 days of CSF sampling compared to OND patients (14.5% vs. 28.4%; $p < 0.001$). The conclusion from these studies was that during an exacerbation, T cells leave the CSF and extravasate to the brain tissue where they can participate in the disease process.

## Lymphocytes Use the CSF as a Conduit to Get into the Brain

The journey toward understanding how lymphocytes extravasate from the CSF to the brain tissue has taken several decades. The first demonstration that lymphocytes travel from the periphery to the CNS was published by Hafler et al. in 1987 [34]. In this study, the authors introduced a monoclonal antibody recognizing the sheep red blood cell antigen on T cells by injecting it into the veins of four patients with progressive MS. Several hours to a few days later, they collected cerebrospinal fluid from each subject and looked for lymphocytes that were labeled with the antibody. By 3 days post-injection, 70% of the lymphocytes in the CSF were labeled with the antibody, indicating their ability to travel from the periphery to the CSF.

Other studies suggest that many of these cells do not survive the trip [35, 36]. Current understanding of immune cell trafficking from the CSF to the brain tissue is provided in excellent reviews [37, 38], including the use of lymphatics to enter the brain tissue [39, 40], the necessity of VCAM1 expression on the endothelial cells of the blood-brain barrier [41–43], and the influence of Th17 cells on blood-brain barrier breakdown [44].

## Lymphocytes Are Highly Activated in the CNS

Several studies have provided evidence that lymphocytes are present in brain tissue from patients with various diseases of the central nervous system. Henderson et al. demonstrated the presence of multiple inflammatory cell subtypes in the perivascular cuff of active lesions from patients with MS [45]. Lucchinetti et al. demonstrated that lesions could be stratified according to their immune cell component [46, 47]. Other excellent demonstrations of lymphocyte infiltration to the brain parenchyma have been reviewed elsewhere [48–53].

The brain offers an excellent environment for lymphocyte activation [54–56]. Thus, evidence of clonal expansion of T and B cells in brain tissue would be expected and has been confirmed [57–62]. Even activated astrocytes can support germinal center maintenance and activation of resident B cells in the CNS [63].

## The Lymphocyte Profile Varies Depending on Disease

While clonal expansion is strongly suggestive of activation, the first evidence that CSF cells are highly activated derived from a study by Noronha et al. in 1980 [64]. CSF was obtained from 17 MS patients and 21 controls. The cells were labeled to identify those in each phase of the cell cycle. From this analysis of CSF cells, it was evident that MS patients had more cells in G1 than control subjects (12.2 vs. 5.4, $p < 0.001$). Flow cytometric studies have been used to provide further confirmation that these activated CSF cells are subsets of T cells and B cells that can contribute to the inflammatory state of the MS brain [43, 65].

More recently, we have come to understand that other diseases of the CNS may also present a lymphocyte profile indicative of disease. For example, patients with autism demonstrate a pro-inflammatory profile, which may be related to disease worsening [66]. In other cases, an anti-inflammatory signature is identified with the disease rather than a pro-inflammatory signature. For example, the percentage of CD4+CD127(dim) regulatory T cells was significantly higher in patients with major psychiatric disorders (MPDs) in comparison to controls [67]. In contrast, the presence of regulatory B cells in the CSF may indicate improved prognosis in stroke [68]. Early detection of lymphoma in the CSF is also aided by flow cytometry [69], and the pro-inflammatory profile of patients with HIV can inform about cognitive decline [70]. Understanding the immune profile of atypical CNS disease is a growing need [71], but it is unlikely to be informative until we understand how the location of the CNS damage impacts the type of lymphocyte present and its influence on disease worsening or repair [72]. Such studies will be critical in understanding the underlying mechanism of the anti-MOG disorders [73–76].

## Adaptive Immune Cells in CNS Disease

An adaptive immune response requires the activation and differentiation of T cells [77]. There are several subtypes of T cells, all of which express the T-cell receptor (TCR). The interaction of a T cell with an antigen-presenting cell (APC) requires the TCR to bind the antigen presented by the APC in the context of the major histocompatibility complex (MHC) molecule. The source of the antigen is usually a complex protein that the APC internalized upon contact and processed into individual peptides that fit into the peptide groove of the MHC molecule. T cells that recognize this MHC-antigen complex are activated, undergo clonal expansion, and differentiate into one of several different T-cell subtypes that survey the body with the purpose of generating an immune response toward any tissue or cell expressing the antigen by which they were originally activated.

Helper T cells (identified by their expression of CD4) generally mediate the initiation of an immune response to targets expressing the activating antigen, while cytotoxic T cells (identified by their expression of CD8) generally mediate destruction of the targets expressing the activating antigen. Both T-cell subtypes secrete cytokines and chemokines that facilitate their optimal response, and further T-cell subtyping can be done using these secreted factor signatures, which facilitate an understanding of their ability to either antagonize or regulate an adaptive immune response [78]. For example, regulatory T cells are increased in the CSF of patients with MS [79, 80]. However, much work has yet to be done to determine the mechanism of autoimmune suppression by regulatory T cells in the CSF.

While T cells are the dominant lymphocyte subset in the CSF [81], there has been considerable debate regarding which T-cell subtype dominates in CNS diseases [82]. Studies in the model of MS, experimental autoimmune encephalomyelitis (EAE) perpetuated some of this debate. For example, CD4+ T cells secreting pro-inflammatory factors such as IFNgamma could transfer disease in EAE, but CD4+ T cells secreting anti-inflammatory factors such as IL-4 could not [83–86]. This finding was essential in understanding the impact of CD4+ T cells on the human MS disease, as early work had suggested that the frequency of myelin-reactive T cells was similar in both MS patients and healthy donors [87]. However, while the myelin-reactive T cells from the MS patients displayed a pro-inflammatory signature, the myelin-reactive T cells from the controls could not [88, 89]. Thus, model and human studies in aggregate led to a clearer understanding of how cytokine output by T-cell subsets can impact MS. Similar approaches should enhance our understanding of other CNS-related diseases.

Despite the intense study of CD4+ T cells in the pathogenesis of MS, the dominant T-cell subtype in human CSF and brain tissue by frequency are the cytotoxic CD8+ T cells [62, 90, 91]. CD8+ T cells undergo extensive clonal expansion in the CSF and brain tissue in comparison to CD4+ T cells [62, 92], and in one case, a single CD8+ T cell clone constituted 30% of all T cells in the sample [62]. Karandikar et al. demonstrated a high prevalence of CD8+ T cells in circulation of MS patients and not controls that respond to CNS autoantigens by proliferation using an in vitro flow cytometry assay [89]. In the EAE model, CD8+ cytotoxic T cells drive a progressive disease course [93, 94] and demonstrate the ability to transect axons [82]. Further understanding of the potential role of CD8+ T cells in MS and other CNS-related diseases will likely reveal important targets of therapy.

B cells are a central component of the humoral immune response within the adaptive immune system [95]. The primary purpose of a B cell is to produce antibodies that bind to any tissue or cell expressing the antigen against which they were originally produced to facilitate clearance of the target. B cells initially express their antibody on the cell surface in complex with signaling molecules. This receptor complex is called the B-cell receptor (BCR). Once the BCR binds the antigen, and the complex is internalized, the B cell is activated, undergoes clonal expansion, and finally differentiate to antibody-producing cells. Unlike T cells, B cells do not require presentation of the antigen by antigen-processing cells. Instead, B cells survey for antigen recognition directly through antigen interaction with the surface-bound antibody. Thus, B cells can recognize antigens in their native conformation rather than restricted to recognition of antigens processed as peptides.

Targeting B cells as a therapeutic strategy for MS was not considered until the late 1990s when case reports of B-cell presence and expansion in the CSF of MS patients were reported [96, 97]. The demonstration that B:T ratios correlate with disease progression [98] and that increases in B-cell subtypes correlate with larger numbers of

T2 lesions [99] also contributed to the pursuit of using B-cell depletion therapy in MS. Clear efficacy was demonstrated in the first B-cell depletion therapy trials in MS [100] resulting in FDA-approved use in the treatment of relapsing disease, and the first FDA-approved drug in the treatment of progressive disease. Mechanistic understanding of B cells as a therapeutic target emerged in tandem, demonstrating that B cells promote T-cell pro-inflammatory activity [101, 102], clonally expanded [57, 103–106], and often correlated with disease type [107]. Mouse models further demonstrated a prominent role of B cells in disease course culminating in the finding that increasing frequencies of plasmablasts that remain following B cell depletion drive residual disease [108, 109].

## Innate Cell Types in CNS Disease

Initiation of the immune response is mediated by cells of the innate immune subsystem, which include myeloid cell types (monocytes, macrophages, dendritic cells, and microglia in the CNS), NK cells, and other granulocytes [50, 110]. The study of these cell types remains largely unexplored, although both their potential role in CNS disease propagation and protection must be considered. For example, myeloid cells are able to produce neurotrophic factors as well as remove cellular debris [111]. In mouse models, in vivo activated microglia are also able to protect neurons from apoptosis by removing inhibitory synapses from neurons in the damaged area [112]. Innate cell studies are less frequent in human CNS disease. However, there have been recent demonstrations that innate cells are expanded in patients with stroke [113] and patients at high risk to develop Alzheimer's disease [5, 114]. Determining the impact these expanded innate cell subtypes on neurodegeneration and protection will certainly reveal a new depth in understanding the mechanism(s) of CNS disease. In this context, it is important to note that peripheral inflammation can impact the activation state and cytokine output of microglia in the brain [115].

## Other Inflammatory Factors in CNS Disease

The activation programming of immune cells includes secretion of pro-inflammatory cytokines [116] and other products such as neurotrophins [117]. For example, Bar-Or laboratory demonstrated that B cells from MS patients had an increased capacity to secrete pro-inflammatory cytokines such as lymphotoxin (LT) and tumor necrosis factor (TNF)-alpha, while their ability to secrete anti-inflammatory cytokines such as IL-10 was diminished [118]. Others have demonstrated that there is a relationship between cytokine profiles (particularly CXCL13) and the severity of cortical damage in MS [119]. Cytokine profiles in the CSF can also reveal mechanistic differences in the immune response of distinct CNS diseases [120].

## How This Information Changes the Way We Approach Diseases of the CNS

Aberrant immune responses are considered a significant contributor to the pathogenesis of MS, while the role of the immune system in other diseases of the CNS remain in the earliest stages of investigation. Observations made in the MS field have allowed the research community to expand its investigations in the neuroimmunology of other CNS diseases, resulting in some unexpected findings. Of particular interest here is the expansion of innate cells in the CSF of patients at increased risk for Alzheimer's disease [5, 114] and the possible egress of regulatory B cells in stroke [121]. Understanding the impact of biologicals in consideration for therapy on cells in the CSF [122, 123] can actually facilitate improved use and delivery. For example, the rise of B-cell depletion therapy in relapsing [124] and progressive [125] forms of MS is in part attributable to investigations demonstrating that B-cell depletion therapy impacts the targeted cells in the CSF [126, 127]. Understanding the impact of a drug on non-target cells in the CSF [43] is also an important consideration [128].

**Table 1.1** What we know about lymphocytes in the CSF

| |
|---|
| Mechanisms in place to enter the CNS |
| Highly activated and clonally expanded |
| Lymphocyte profile varies depending on disease |
| Produce both pro- and anti-inflammatory cytokines and chemokines |
| T cells dominate |
| Innate cell subtypes can impact disease |
| Cytokine profiles can reveal mechanistic differences in distinct CNS diseases |
| Impacted by immunomodulatory therapies delivered peripherally |

*CSF* cerebrospinal fluid, *CNS* central nervous system

## Conclusion

The CSF can provide important clues regarding the immune response in a variety of CNS diseases (Table 1.1). Our goal in focusing research efforts into these CNS diseases should be to identify cellular and humoral factors that are the best target candidates. These immune component targets are likely to vary depending on the immune response that either perpetuates or abrogates inflammation associated with each individual CNS disease. Study and translational application of relevant murine models is critical in this context. Finally, there is emerging evidence that mapping the neuro-immunological signature within the CSF is essential to effectively develop beneficial therapeutic strategies without causing undo harm.

## References


1. Chitnis T, Weiner HL. CNS inflammation and neurodegeneration. J Clin Invest. 2017;127(10):3577–87.
2. Wells E, Hacohen Y, Waldman A, Tillema JM, Soldatos A, Ances B, et al. Neuroimmune disorders of the central nervous system in children in the molecular era. Nat Rev Neurol. 2018;14(7):433–45.
3. Chabas D, Ness J, Belman A, Yeh EA, Kuntz N, Gorman MP, et al. Younger children with MS have a distinct CSF inflammatory profile at disease onset. Neurology. 2010;74(5):399–405.
4. Psimaras D, Carpentier AF, Rossi C, Euronetwork PNS. Cerebrospinal fluid study in paraneoplastic syndromes. J Neurol Neurosurg Psychiatry. 2010;81(1):42–5.
5. Stowe AM, Ireland SJ, Ortega SB, Chen D, Huebinger RM, Tarumi T, et al. Adaptive lymphocyte profiles correlate to brain Abeta burden in patients with mild cognitive impairment. J Neuroinflammation. 2017;14(1):149.
6. Lisak RP, Nedelkoska L, Benjamins JA, Schalk D, Bealmear B, Touil H, et al. B cells from patients with multiple sclerosis induce cell death via apoptosis in neurons in vitro. J Neuroimmunol. 2017;309:88–99.
7. Vecchio D. The history of cerebrospinal fluid analysis in multiple sclerosis: a great development over the last centuries. J Brain Disord. 2017;1:35–7.
8. Marshall JS, Warrington R, Watson W, Kim HL. An introduction to immunology and immunopathology. Allergy Asthma Clin Immunol. 2018;14(Suppl 2):49.
9. Warrington R, Watson W, Kim HL, Antonetti FR. An introduction to immunology and immunopathology. Allergy Asthma Clin Immunol. 2011;7(Suppl 1):S1.
10. Pariente D, Bihet MH, Tammam S, Riou JY, Bernard O, Devictor D, et al. Biliary complications after transplantation in children: role of imaging modalities. Pediatr Radiol. 1991;21(3):175–8.
11. Abbas AK, Lichtman A, Pillai S. Basic immunology: functions and disorders of the immune system. 5th ed. St. Louis: Elsevier; 2016. 332 p.
12. Kabat EA, Moore DH, Landow H. An electrophoretic study of the protein components in cerebrospinal fluid and their relationship to the serum proteins. J Clin Invest. 1942;21(5):571–7.
13. Freedman DA, Merritt HH. The cerebrospinal fluid in multiple sclerosis. Res Publ Assoc Res Nerv Ment Dis. 1950;28:428–39.
14. Greger J, Wieczorek V. [On the demonstration of plasma cells in the cerebrospinal fluid in neurological diseases]. Wien Z Nervenheilkd Grenzgeb. 1966;23(4):366–74.
15. Muller R. The correlation between the state of the cerebrospinal fluid and the clinical picture in disseminated sclerosis. Acta Med Scand. 1951;139(2):153–63.
16. Sornas R, Ostlund H, Muller R. Cerebrospinal fluid cytology after stroke. Arch Neurol. 1972;26(6):489–501.
17. Sandberg-Wollheim M, Turesson I. Lymphocyte subpopulations in the cerebrospinal fluid and peripheral blood in patients with multiple sclerosis. Scand J Immunol. 1975;4(8):831–6.
18. Allen JC, Sheremata W, Cosgrove JB, Osterland K, Shea M. Cerebrospinal fluid T and B lymphocyte kinetics related to exacerbations of multiple sclerosis. Neurology. 1976;26(6 PT 1):579–83.
19. Naess A. T lymphocytes in cerebrospinal fluid from patients with neurological diseases. Eur Neurol. 1979;18(3):183–8.
20. Sheremata W, Allen J, Sazant A, Cosgrove JB, Osterland K. Cerebrospinal fluid T & B lymphocyte responses in exacerbations of multiple sclerosis. Trans Am Neurol Assoc. 1976;101:40–5.

21. Traugott U. T and B lymphocytes in the cerebrospinal fluid of various neurological diseases. J Neurol. 1978;219(3):185–97.
22. Lisak RP, Zweiman B. In vitro cell-mediated immunity of cerebrospinal-fluid lymphocytes to myelin basic protein in primary demyelinating diseases. N Engl J Med. 1977;297(16):850–3.
23. Sandberg-Wollheim M. Immunoglobulin synthesis in vitro by cerebrospinal fluid cells in patients with multiple sclerosis. Scand J Immunol. 1974;3(6):717–30.
24. Cohen S, Bannister R. Immunoglobulin synthesis within the central nervous system in disseminated sclerosis. Lancet. 1967;1(7486):366–7.
25. Sandberg-Wollheim M, Zweiman B, Levinson AI, Lisak RP. Humoral immune responses within the human central nervous system following systemic immunization. J Neuroimmunol. 1986;11(3):205–14.
26. Levinson AI, Sandberg-Wollheim M, Lisak RP, Zweiman B, Sjogren K, Laramore C, et al. Analysis of B-cell activation of cerebrospinal fluid lymphocytes in multiple sclerosis. Neurology. 1983;33(10):1305–10.
27. Burns JB, Zweiman B, Lisak RP. Long-term growth in vitro of human cerebrospinal fluid T lymphocytes. J Clin Immunol. 1981;1(3):195–200.
28. Lisak RP, Zweiman B, Whitaker JN. Spinal fluid basic protein immunoreactive material and spinal fluid lymphocyte reactivity to basic protein. Neurology. 1981;31(2):180–2.
29. Naess A. Demonstration of T lymphocytes in cerebrospinal fluid. Scand J Immunol. 1976;5(1–2):165–8.
30. Manconi PE, Zaccheo D, Bugiani O, Fadda MF, Grifoni V, Mantovani G, et al. Letter: T and B lymphocytes in normal cerebrospinal fluid. N Engl J Med. 1976;294(1):49.
31. Kam-Hansen S, Fryden A, Link H. B and T lymphocytes in cerebrospinal fluid and blood in multiple sclerosis, optic neuritis and mumps meningitis. Acta Neurol Scand. 1978;58(2):95–103.
32. Hausler M, Sellhaus B, Schweizer K, Ramaekers VT, Opladen T, Kleines M. Flow cytometric cerebrospinal fluid analysis in children. Pathol Res Pract. 2003;199(10):667–75.
33. Cashman N, Martin C, Eizenbaum JF, Degos JD, Bach MA. Monoclonal antibody-defined immunoregulatory cells in multiple sclerosis cerebrospinal fluid. J Clin Invest. 1982;70(2):387–92.
34. Hafler DA, Weiner HL. In vivo labeling of blood T cells: rapid traffic into cerebrospinal fluid in multiple sclerosis. Ann Neurol. 1987;22(1):89–93.
35. Dux R, Kindler-Rohrborn A, Annas M, Faustmann P, Lennartz K, Zimmermann CW. A standardized protocol for flow cytometric analysis of cells isolated from cerebrospinal fluid. J Neurol Sci. 1994;121(1):74–8.
36. de Graaf MT, de Jongste AH, Kraan J, Boonstra JG, Sillevis Smitt PA, Gratama JW. Flow cytometric characterization of cerebrospinal fluid cells. Cytometry B Clin Cytom. 2011;80(5):271–81.
37. Ratnam NM, Gilbert MR, Giles AJ. Immunotherapy in CNS cancers: the role of immune cell trafficking. Neuro-Oncology. 2019;21(1):37–46.
38. Hawkins BT, Davis TP. The blood-brain barrier/neurovascular unit in health and disease. Pharmacol Rev. 2005;57(2):173–85.
39. Meyer C, Martin-Blondel G, Liblau RS. Endothelial cells and lymphatics at the interface between the immune and central nervous systems: implications for multiple sclerosis. Curr Opin Neurol. 2017;30(3):222–30.
40. Louveau A, Smirnov I, Keyes TJ, Eccles JD, Rouhani SJ, Peske JD, et al. Structural and functional features of central nervous system lymphatic vessels. Nature. 2015;523(7560):337–41.
41. Khatri BO, Man S, Giovannoni G, Koo AP, Lee JC, Tucky B, et al. Effect of plasma exchange in accelerating natalizumab clearance and restoring leukocyte function. Neurology. 2009;72(5):402–9.
42. Man S, Tucky B, Bagheri N, Li X, Kochar R, Ransohoff RM. alpha4 Integrin/FN-CS1 mediated leukocyte adhesion to brain microvascular endothelial cells under flow conditions. J Neuroimmunol. 2009;210(1–2):92–9.
43. Stuve O, Marra CM, Jerome KR, Cook L, Cravens PD, Cepok S, et al. Immune surveillance in multiple sclerosis patients treated with natalizumab. Ann Neurol. 2006;59(5):743–7.
44. Kebir H, Kreymborg K, Ifergan I, Dodelet-Devillers A, Cayrol R, Bernard M, et al. Human TH17 lymphocytes promote blood-brain barrier disruption and central nervous system inflammation. Nat Med. 2007;13(10):1173–5.
45. Henderson AP, Barnett MH, Parratt JD, Prineas JW. Multiple sclerosis: distribution of inflammatory cells in newly forming lesions. Ann Neurol. 2009;66(6):739–53.
46. Lucchinetti C, Bruck W, Parisi J, Scheithauer B, Rodriguez M, Lassmann H. Heterogeneity of multiple sclerosis lesions: implications for the pathogenesis of demyelination. Ann Neurol. 2000;47(6):707–17.
47. Ludwin SK. Understanding multiple sclerosis: lessons from pathology. Ann Neurol. 2000;47(6):691–3.
48. Thompson AJ, Baranzini SE, Geurts J, Hemmer B, Ciccarelli O. Multiple sclerosis. Lancet. 2018;391(10130):1622–36.
49. Hemmer B, Archelos JJ, Hartung HP. New concepts in the immunopathogenesis of multiple sclerosis. Nat Rev Neurosci. 2002;3(4):291–301.
50. Mishra MK, Yong VW. Myeloid cells - targets of medication in multiple sclerosis. Nat Rev Neurol. 2016;12(9):539–51.

51. Geurts JJ, Kooi EJ, Witte ME, van der Valk P. Multiple sclerosis as an "inside-out" disease. Ann Neurol. 2010;68(5):767–8; author reply 8.
52. Tsutsui S, Stys PK. Degeneration versus autoimmunity in multiple sclerosis. Ann Neurol. 2009;66(6):711–3.
53. Frohman EM, Racke MK, Raine CS. Multiple sclerosis – the plaque and its pathogenesis. N Engl J Med. 2006;354(9):942–55.
54. Corcione A, Casazza S, Ferretti E, Giunti D, Zappia E, Pistorio A, et al. Recapitulation of B cell differentiation in the central nervous system of patients with multiple sclerosis. Proc Natl Acad Sci U S A. 2004;101(30):11064–9.
55. Uccelli A, Aloisi F, Pistoia V. Unveiling the enigma of the CNS as a B-cell fostering environment. Trends Immunol. 2005;26(5):254–9.
56. Meinl E, Krumbholz M, Hohlfeld R. B lineage cells in the inflammatory central nervous system environment: migration, maintenance, local antibody production, and therapeutic modulation. Ann Neurol. 2006;59(6):880–92.
57. Qin Y, Duquette P, Zhang Y, Talbot P, Poole R, Antel J. Clonal expansion and somatic hypermutation of V(H) genes of B cells from cerebrospinal fluid in multiple sclerosis. J Clin Invest. 1998;102(5):1045–50.
58. Baranzini SE, Jeong MC, Butunoi C, Murray RS, Bernard CC, Oksenberg JR. B cell repertoire diversity and clonal expansion in multiple sclerosis brain lesions. J Immunol. 1999;163(9):5133–44.
59. Owens GP, Burgoon MP, Anthony J, Kleinschmidt-DeMasters BK, Gilden DH. The immunoglobulin G heavy chain repertoire in multiple sclerosis plaques is distinct from the heavy chain repertoire in peripheral blood lymphocytes. Clin Immunol. 2001;98(2):258–63.
60. Smith-Jensen T, Burgoon MP, Anthony J, Kraus H, Gilden DH, Owens GP. Comparison of immunoglobulin G heavy-chain sequences in MS and SSPE brains reveals an antigen-driven response. Neurology. 2000;54(6):1227–32.
61. Oksenberg JR, Stuart S, Begovich AB, Bell RB, Erlich HA, Steinman L, et al. Limited heterogeneity of rearranged T-cell receptor V alpha transcripts in brains of multiple sclerosis patients. Nature. 1990;345(6273):344–6.
62. Babbe H, Roers A, Waisman A, Lassmann H, Goebels N, Hohlfeld R, et al. Clonal expansions of CD8(+) T cells dominate the T cell infiltrate in active multiple sclerosis lesions as shown by micromanipulation and single cell polymerase chain reaction. J Exp Med. 2000;192(3):393–404.
63. Touil H, Kobert A, Lebeurrier N, Rieger A, Saikali P, Lambert C, et al. Human central nervous system astrocytes support survival and activation of B cells: implications for MS pathogenesis. J Neuroinflammation. 2018;15(1):114.
64. Noronha AB, Richman DP, Arnason BG. Detection of in vivo stimulated cerebrospinal-fluid lymphocytes by flow cytometry in patients with multiple sclerosis. N Engl J Med. 1980;303(13):713–7.
65. Cepok S, von Geldern G, Grummel V, Hochgesand S, Celik H, Hartung H, et al. Accumulation of class switched IgD-IgM- memory B cells in the cerebrospinal fluid during neuroinflammation. J Neuroimmunol. 2006;180(1–2):33–9.
66. Vargas DL, Nascimbene C, Krishnan C, Zimmerman AW, Pardo CA. Neuroglial activation and neuroinflammation in the brain of patients with autism. Ann Neurol. 2005;57(1):67–81.
67. Maxeiner HG, Rojewski MT, Schmitt A, Tumani H, Bechter K, Schmitt M. Flow cytometric analysis of T cell subsets in paired samples of cerebrospinal fluid and peripheral blood from patients with neurological and psychiatric disorders. Brain Behav Immun. 2009;23(1):134–42.
68. Selvaraj UM, Poinsatte K, Torres V, Ortega SB, Stowe AM. Heterogeneity of B cell functions in stroke-related risk, prevention, injury, and repair. Neurotherapeutics. 2016;13(4):729–47.
69. Di Noto R, Scalia G, Abate G, Gorrese M, Pascariello C, Raia M, et al. Critical role of multidimensional flow cytometry in detecting occult leptomeningeal disease in newly diagnosed aggressive B-cell lymphomas. Leuk Res. 2008;32(8):1196–9.
70. Hong S, Banks WA. Role of the immune system in HIV-associated neuroinflammation and neurocognitive implications. Brain Behav Immun. 2015;45:1–12.
71. Hardy TA, Reddel SW, Barnett MH, Palace J, Lucchinetti CF, Weinshenker BG. Atypical inflammatory demyelinating syndromes of the CNS. Lancet Neurol. 2016;15(9):967–81.
72. Reich DS, Lucchinetti CF, Calabresi PA. Multiple sclerosis. N Engl J Med. 2018;378(2):169–80.
73. Lopez-Chiriboga AS, Majed M, Fryer J, Dubey D, McKeon A, Flanagan EP, et al. Association of MOG-IgG serostatus with relapse after acute disseminated encephalomyelitis and proposed diagnostic criteria for MOG-IgG-associated disorders. JAMA Neurol. 2018;75:1355.
74. Di Pauli F, Reindl M, Berger T. New clinical implications of anti-myelin oligodendrocyte glycoprotein antibodies in children with CNS demyelinating diseases. Mult Scler Relat Disord. 2018;22:35–7.
75. Weber MS, Derfuss T, Metz I, Bruck W. Defining distinct features of anti-MOG antibody associated central nervous system demyelination. Ther Adv Neurol Disord. 2018;11:1756286418762083.
76. Spadaro M, Winklmeier S, Beltran E, Macrini C, Hoftberger R, Schuh E, et al. Pathogenicity of human antibodies against myelin oligodendrocyte glycoprotein. Ann Neurol. 2018;84(2):315–28.

77. Coico R, Sunshine G. Biology of the T lymphocyte. Immunology: a short course. Seventh ed. West Sussex: Wiley; 2015. p. 137–52.
78. Dominguez-Villar M, Hafler DA. Regulatory T cells in autoimmune disease. Nat Immunol. 2018;19(7):665–73.
79. Lowther DE, Hafler DA. Regulatory T cells in the central nervous system. Immunol Rev. 2012;248(1):156–69.
80. Feger U, Luther C, Poeschel S, Melms A, Tolosa E, Wiendl H. Increased frequency of CD4+ CD25+ regulatory T cells in the cerebrospinal fluid but not in the blood of multiple sclerosis patients. Clin Exp Immunol. 2007;147(3):412–8.
81. Svenningsson A, Andersen O, Edsbagge M, Stemme S. Lymphocyte phenotype and subset distribution in normal cerebrospinal fluid. J Neuroimmunol. 1995;63(1):39–46.
82. Medana I, Martinic MA, Wekerle H, Neumann H. Transection of major histocompatibility complex class I-induced neurites by cytotoxic T lymphocytes. Am J Pathol. 2001;159(3):809–15.
83. Kuchroo VK, Das MP, Brown JA, Ranger AM, Zamvil SS, Sobel RA, et al. B7-1 and B7-2 costimulatory molecules activate differentially the Th1/Th2 developmental pathways: application to autoimmune disease therapy. Cell. 1995;80(5):707–18.
84. Ando DG, Clayton J, Kono D, Urban JL, Sercarz EE. Encephalitogenic T cells in the B10.PL model of experimental allergic encephalomyelitis (EAE) are of the Th-1 lymphokine subtype. Cell Immunol. 1989;124(1):132–43.
85. Racke MK, Bonomo A, Scott DE, Cannella B, Levine A, Raine CS, et al. Cytokine-induced immune deviation as a therapy for inflammatory autoimmune disease. J Exp Med. 1994;180(5):1961–6.
86. Kennedy MK, Torrance DS, Picha KS, Mohler KM. Analysis of cytokine mRNA expression in the central nervous system of mice with experimental autoimmune encephalomyelitis reveals that IL-10 mRNA expression correlates with recovery. J Immunol. 1992;149(7):2496–505.
87. Jingwu Z, Medaer R, Hashim GA, Chin Y, van den Berg-Loonen E, Raus JC. Myelin basic protein-specific T lymphocytes in multiple sclerosis and controls: precursor frequency, fine specificity, and cytotoxicity. Ann Neurol. 1992;32(3):330–8.
88. Lovett-Racke AE, Trotter JL, Lauber J, Perrin PJ, June CH, Racke MK. Decreased dependence of myelin basic protein-reactive T cells on CD28-mediated costimulation in multiple sclerosis patients. A marker of activated/memory T cells. J Clin Invest. 1998;101(4):725–30.
89. Crawford MP, Yan SX, Ortega SB, Mehta RS, Hewitt RE, Price DA, et al. High prevalence of autoreactive, neuroantigen-specific CD8+ T cells in multiple sclerosis revealed by novel flow cytometric assay. Blood. 2004;103(11):4222–31.
90. Monteiro J, Hingorani R, Pergolizzi R, Apatoff B, Gregersen PK. Clonal dominance of CD8+ T-cell in multiple sclerosis. Ann N Y Acad Sci. 1995;756:310–2.
91. Monteiro J, Hingorani R, Peroglizzi R, Apatoff B, Gregersen PK. Oligoclonality of CD8+ T cells in multiple sclerosis. Autoimmunity. 1996;23(2):127–38.
92. Jacobsen M, Cepok S, Quak E, Happel M, Gaber R, Ziegler A, et al. Oligoclonal expansion of memory CD8+ T cells in cerebrospinal fluid from multiple sclerosis patients. Brain. 2002;125(Pt 3):538–50.
93. Sun D, Whitaker JN, Huang Z, Liu D, Coleclough C, Wekerle H, et al. Myelin antigen-specific CD8+ T cells are encephalitogenic and produce severe disease in C57BL/6 mice. J Immunol. 2001;166(12):7579–87.
94. Huseby ES, Liggitt D, Brabb T, Schnabel B, Ohlen C, Goverman J. A pathogenic role for myelin-specific CD8(+) T cells in a model for multiple sclerosis. J Exp Med. 2001;194(5):669–76.
95. Coico R, Sunshine G. Biology of the B lymphocyte. Seventh ed. West Sussex: Wiley; 2015.
96. Monson NL, Cravens PD, Frohman EM, Hawker K, Racke MK. Effect of rituximab on the peripheral blood and cerebrospinal fluid B cells in patients with primary progressive multiple sclerosis. Arch Neurol. 2005;62(2):258–64.
97. Stuve O, Cepok S, Elias B, Saleh A, Hartung HP, Hemmer B, Kieseier BC. Clinical stabilization and effective B cell depletion in the cerebrospinal fluid and peripheral blood in a patient with fulminant relapsing remitting multiple sclerosis. Arch Neurol. 2005;62:1620.
98. Cepok S, Jacobsen M, Schock S, Omer B, Jaekel S, Boddeker I, et al. Patterns of cerebrospinal fluid pathology correlate with disease progression in multiple sclerosis. Brain. 2001;124(Pt 11):2169–76.
99. Kuenz B, Lutterotti A, Ehling R, Gneiss C, Haemmerle M, Rainer C, et al. Cerebrospinal fluid B cells correlate with early brain inflammation in multiple sclerosis. PLoS One. 2008;3(7):e2559.
100. Hauser SL, Waubant E, Arnold DL, Vollmer T, Antel J, Fox RJ, et al. B-cell depletion with rituximab in relapsing-remitting multiple sclerosis. N Engl J Med. 2008;358(7):676–88.
101. Harp CT, Ireland S, Davis LS, Remington G, Cassidy B, Cravens PD, et al. Memory B cells from a subset of treatment-naive relapsing-remitting multiple sclerosis patients elicit CD4(+) T-cell proliferation and IFN-gamma production in response to myelin basic protein and myelin oligodendrocyte glycoprotein. Eur J Immunol. 2010;40(10):2942–56.
102. Harp CT, Lovett-Racke AE, Racke MK, Frohman EM, Monson NL. Impact of myelin-specific antigen presenting B cells on T cell activation in multiple sclerosis. Clin Immunol. 2008;128(3):382–91.
103. Columbo M, Dono M, Gazzola P, Roncella S, Valetto A, Chiorazzi N, et al. Accumulation of clonally related B lymphocytes in the cerebrospinal fluid of multiple sclerosis patients. J Immunol. 2000;164:2782–9.

104. Monson NL, Brezinschek HP, Brezinschek RI, Mobley A, Vaughan GK, Frohman EM, et al. Receptor revision and atypical mutational characteristics in clonally expanded B cells from the cerebrospinal fluid of recently diagnosed multiple sclerosis patients. J Neuroimmunol. 2005;158(1–2):170–81.
105. Owens G, Ritchie A, Burgoon M, Williamson R, Corboy J, Gilden D. Single cell repertoire analysis demonstrates clonal expansion is prominent feature of the B cell response in multiple sclerosis spinal fluid. J Immunol. 2003;171:2725–33.
106. Qin Y, Duquette P, Zhang Y, Olek M, Da RR, Richardson J, et al. Intrathecal B-cell clonal expansion, an early sign of humoral immunity, in the cerebrospinal fluid of patients with clinically isolated syndrome suggestive of multiple sclerosis. Lab Investig. 2003;83(7):1081–8.
107. Wurth S, Kuenz B, Bsteh G, Ehling R, Di Pauli F, Hegen H, et al. Cerebrospinal fluid B cells and disease progression in multiple sclerosis - a longitudinal prospective study. PLoS One. 2017;12(8):e0182462.
108. Chen D, Blazek M, Ireland S, Ortega S, Kong X, Meeuwissen A, et al. Single dose of glycoengineered anti-CD19 antibody (MEDI551) disrupts experimental autoimmune encephalomyelitis by inhibiting pathogenic adaptive immune responses in the bone marrow and spinal cord while preserving peripheral regulatory mechanisms. J Immunol. 2014;193(10):4823–32.
109. Chen D, Ireland SJ, Davis LS, Kong X, Stowe AM, Wang Y, et al. Autoreactive CD19+CD20- plasma cells contribute to disease severity of experimental autoimmune encephalomyelitis. J Immunol. 2016;196(4):1541–9.
110. Longhini AL, von Glehn F, Brandao CO, de Paula RF, Pradella F, Moraes AS, et al. Plasmacytoid dendritic cells are increased in cerebrospinal fluid of untreated patients during multiple sclerosis relapse. J Neuroinflammation. 2011;8(1):2.
111. Wee Yong V. Inflammation in neurological disorders: a help or a hindrance? Neuroscientist. 2010;16(4):408–20.
112. Chen Z, Jalabi W, Hu W, Park HJ, Gale JT, Kidd GJ, et al. Microglial displacement of inhibitory synapses provides neuroprotection in the adult brain. Nat Commun. 2014;5:4486.
113. Zrzavy T, Machado-Santos J, Christine S, Baumgartner C, Weiner HL, Butovsky O, et al. Dominant role of microglial and macrophage innate immune responses in human ischemic infarcts. Brain Pathol. 2018;28(6):791–805.
114. Monson NL, Ireland SJ, Ligocki AJ, Chen D, Rounds WH, Li M, et al. Elevated CNS inflammation in patients with preclinical Alzheimer's disease. J Cereb Blood Flow Metab. 2014;34(1):30–3.
115. Zrzavy T, Hoftberger R, Berger T, Rauschka H, Butovsky O, Weiner H, et al. Pro-inflammatory activation of microglia in the brain of patients with sepsis. Neuropathol Appl Neurobiol. 2019;45(3):278–90.
116. Brod SA, Benjamin D, Hafler DA. Restricted T cell expression of IL-2/IFN-gamma mRNA in human inflammatory disease. J Immunol. 1991;147(3):810–5.
117. Kerschensteiner M, Stadelmann C, Dechant G, Wekerle H, Hohlfeld R. Neurotrophic cross-talk between the nervous and immune systems: implications for neurological diseases. Ann Neurol. 2003;53(3):292–304.
118. Bar-Or A, Fawaz L, Fan B, Darlington PJ, Rieger A, Ghorayeb C, et al. Abnormal B-cell cytokine responses a trigger of T-cell-mediated disease in MS? Ann Neurol. 2010;67(4):452–61.
119. Magliozzi R, Howell OW, Nicholas R, Cruciani C, Castellaro M, Romualdi C, et al. Inflammatory intrathecal profiles and cortical damage in multiple sclerosis. Ann Neurol. 2018;83(4):739–55.
120. Bonin S, Zanotta N, Sartori A, Bratina A, Manganotti P, Trevisan G, et al. Cerebrospinal fluid cytokine expression profile in multiple sclerosis and chronic inflammatory demyelinating polyneuropathy. Immunol Investig. 2018;47(2):135–45.
121. Monson NL, Ortega SB, Ireland SJ, Meeuwissen AJ, Chen D, Plautz EJ, et al. Repetitive hypoxic preconditioning induces an immunosuppressed B cell phenotype during endogenous protection from stroke. J Neuroinflammation. 2014;11:22.
122. Calabresi PA. Advances in multiple sclerosis: from reduced relapses to remedies. Lancet Neurol. 2018;17(1):10–2.
123. Baecher-Allan C, Kaskow BJ, Weiner HL. Multiple sclerosis: mechanisms and immunotherapy. Neuron. 2018;97(4):742–68.
124. Hauser SL, Bar-Or A, Comi G, Giovannoni G, Hartung HP, Hemmer B, et al. Ocrelizumab versus interferon Beta-1a in relapsing multiple sclerosis. N Engl J Med. 2017;376(3):221–34.
125. Montalban X, Hauser SL, Kappos L, Arnold DL, Bar-Or A, Comi G, et al. Ocrelizumab versus placebo in primary progressive multiple sclerosis. N Engl J Med. 2017;376(3):209–20.
126. Cross AH, Stark JL, Lauber J, Ramsbottom MJ, Lyons JA. Rituximab reduces B cells and T cells in cerebrospinal fluid of multiple sclerosis patients. J Neuroimmunol. 2006;180(1–2):63–70.
127. Kowarik MC, Pellkofer HL, Cepok S, Korn T, Kumpfel T, Buck D, et al. Differential effects of fingolimod (FTY720) on immune cells in the CSF and blood of patients with MS. Neurology. 2011;76(14):1214–21.
128. Ransohoff RM. "Thinking without thinking" about natalizumab and PML. J Neurol Sci. 2007;259(1–2):50–2.

# Antibody Detection Methods for Neural Autoantibodies and Introduction to Antibody Pathogenesis

2

Thomas R. Haven and Lisa K. Peterson

**Key Points**

1. Neural autoantibodies are markers of autoimmune neurologic disorders, with only a few shown to be pathogenic.
2. Detection of neural autoantibodies can play an important role in the diagnosis, prognosis, and management of patients with autoimmune neurologic disorders.
3. Failure to detect a neural autoantibody does not rule out an autoimmune neurologic disorder.
4. Tests for detecting neural autoantibodies have complexities that must be considered, including the performance characteristics of the method used and the specimen type evaluated.
5. Results of neural antibody testing must be interpreted within the clinical context; taking them as conclusive evidence of autoimmune neurologic disorder could be a mistake.
6. The number of neural autoantibodies continues to grow, as does the number of specimens tested. This presents a challenge for both clinicians and laboratories in determining which autoantibodies to test, by which methods, and whether testing should be performed independently or in which combinations.

T. R. Haven (✉) · L. K. Peterson
Department of Pathology, University of Utah School of Medicine, ARUP Institute for Clinical and Experimental Pathology, ARUP Laboratories, Salt Lake City, UT, USA
e-mail: haventr@aruplab.com

## Introduction

Autoimmune neurology is a rapidly evolving field largely driven by the discovery of new autoantibodies (Table 2.1). Autoimmune neurologic disorders (ANDs) are a heterogeneous group of diseases thought to occur as a result of an aberrant immune response targeting the nervous system. Patients with these disorders are frequently identified by the detection of an autoantibody in their serum or cerebrospinal fluid (CSF), and thus the response is considered antigen specific. ANDs typically present with a subacute onset with rapid progression of symptoms that may affect any and often multiple parts of the nervous system. Thus, they can present with a wide array of symptoms ranging from nonspecific flu-like symptoms such as fever, headache, and pain to more specific neurologic symptoms including



A. L. Piquet, E. Alvarez (eds.), *Neuroimmunology*, https://doi.org/10.1007/978-3-030-61883-4_2

seizures, cognitive issues, movement disorders, dysautonomia, and psychiatric symptoms and can even result in loss of consciousness or death. Due to this wide array of symptoms, there are a number of other potential causes including infectious, metabolic, genetic, and toxic etiologies that need to be ruled out in order to diagnose a patient with an AND [1].

The workup for a suspected AND includes brain magnetic resonance imaging (MRI) and/or

**Table 2.1** Neural autoantibodies and methods for their detection in the clinical laboratory

| Category | Specific antibody | Detection methods[a] | | | | | |
|---|---|---|---|---|---|---|---|
| | | TBA | WB/LIA | RIA | ELISA | CBA/IFA | CBA/FACS |
| Intracellular antigens | AGNA-1 (Sox-1) | x | x | – | x | – | – |
| | Amphiphysin | x | x | – | x | – | – |
| | ANNA-1 (Hu) | x | x | – | x | – | – |
| | ANNA-2 (Ri) | x | x | – | x | – | – |
| | ANNA-3 | x | – | – | – | – | – |
| | CRMP-5 (CV2) (CV2) | x | x | – | x | x | – |
| | GAD65 | x | x | x | x | – | – |
| | Ma/Ta | x | x | – | – | – | – |
| | PCCA-1 (Yo) | x | x | – | x | – | – |
| | PCCA-2 | x | – | – | – | – | – |
| | PCCA-Tr (DNER) | x | x | – | x | – | – |
| | Recoverin | x | x | – | x | – | – |
| | Titin | – | – | – | x | – | – |
| | Zic4 | x | x | – | – | – | – |
| Neural cell-surface antigens | AMPAR | x | – | – | – | x | – |
| | AQP4 | x | x | – | x | x | x |
| | CASPR2 | x | – | – | – | x | – |
| | DPPX | x | – | x | – | x | – |
| | gACHR | – | – | x | – | – | – |
| | $GABA_BR$ | x | – | – | – | x | – |
| | LGI1 | x | – | – | – | x | – |
| | mGluR1 | x | – | – | – | x | – |
| | MOG | x | – | – | – | x | x |
| | Myelin | x | – | – | – | x | – |
| | NMDAR | x | – | – | x | x | – |
| Neuromuscular junction antigens | mACHRBIN | – | – | x | – | – | – |
| | MuSK | – | – | x | – | – | – |
| | N-VGCC | – | – | x | – | – | – |
| | PQ-VGCC | – | – | x | – | – | – |
| | STR | x | – | – | x | – | – |
| | VGKC | – | – | x | – | – | – |

*TBA* tissue-based assay, *WB* Western blot, *LIA* line immunoblot assay, *RIA* radioimmunoprecipitation assay, *ELISA* enzyme-linked immunosorbent assay, *CBA* cell-based assay, *IFA* indirect immunofluorescence assay, *FACS* fluorescence-activated cell sorting, *AGNA-1* anti-glial nuclear antibody, *ANNA* anti-neuronal nuclear antibody, *CRMP* collapsing response mediator protein, *GAD* glutamic acid decarboxylase, *PCCA* Purkinje cell cytoplasmic antibody, *AMPAR* alpha-amino-3-hydroxy-5-methyl-4-isoxazolepropionate receptor, *AQP* aquaporin, *CASPR* contactin-associated protein, *DPPX* dipeptidyl aminopeptidase-like protein, *gACHR* ganglionic acetylcholine receptor, $GABA_BR$ gamma-aminobutyric acid receptor, type B, *LGI1* leucine-rich glioma inactivated 1 protein, *mGluR1* metabotropic glutamate receptor 1, *MOG* myelin oligodendrocyte glycoprotein, *NMDAR* N-methyl-D-aspartate glutamate receptor, *mACHRBIN* muscle acetylcholine receptor binding antibody, *MuSK* muscle-specific tyrosine kinase, *VGCC* voltage-gated calcium channel, *STR* striated muscle, *VGKC* voltage-gated potassium channel

[a]Table properties limited to detection methods currently available for diagnostic testing at commercial laboratories (www.aruplab.com, www.mayocliniclaboratories.com and www.questdiagnostics.com; accessed January 1, 2019). Other autoantibodies have been identified but testing may only be available on a research basis (e.g., GlyR, DR2, $GABA_AR$, IgLON5, mGluR5, ARHGAP26) [3, 4].

positron-emission tomography (PET) to screen for hyperintensities or metabolic abnormalities, respectively; electroencephalography (EEG) to confirm or exclude seizures; CSF studies to evaluate for the presence of elevated levels of white blood cells, protein and/or immunoglobulin type G (IgG) and oligoclonal bands, as well as molecular methods or culture to explore infectious causes; and serum studies to evaluate for other potential autoimmune causes or indications of an autoimmune tendency and the presence of neural autoantibodies [1]. Depending on the results of these studies, additional testing may be performed to evaluate for malignancy. The diagnostic workup for various ANDs is discussed in detail in Part III of this book.

Neural autoantibodies are commonly divided into two categories based on the subcellular location of the antigens targeted [2]. One group of autoantibodies recognizes intracellular targets including RNA-binding proteins, transcription factors, and other nuclear and cytoplasmic proteins. Paraneoplastic syndromes (PNS), ANDs classically associated with malignancy, are most frequently associated with autoantibodies against intracellular targets (discussed in Chap. 16). The second group of autoantibodies recognizes cell-surface proteins including ion channels, water channels, and neurotransmitter receptors. Autoantibodies against cell-surface proteins have been associated with a variety of disorders, with two of the most common being autoimmune encephalitis (Chap. 12) and autoimmune neuromuscular junction disease (Chap. 19). Detection of any of these neural autoantibodies can play a significant role in the diagnosis, prognosis, and management of patients with ANDs.

## Methods for the Detection of Neural Autoantibodies

A variety of techniques are used to detect the presence of neural autoantibodies. These include the following: (1) tissue-based assays, (2) Western blot or line immunoblot assays, (3) immunoprecipitation assays, (4) cell-based assays, (5) enzyme-linked immunosorbent assays (ELISAs), and (6) primary culture-based immunofluorescence assays, with this last methodology primarily performed on a research basis (Table 2.1, Fig. 2.1) [3, 4].

The first autoantibodies associated with PNS were identified by incubating patient serum or CSF with brain tissue sections and observing autoantibodies binding to intracellular neural proteins [4]. The majority of neural autoantibodies can be screened for using this tissue-based assay (TBA) method on sections of the cerebellum and the hippocampus, with the exception of autoantibodies against neuromuscular junction antigens, since they are not present in these tissues. However, detection by TBA must be followed by testing using a different methodology in order to identify the specific antigen recognized by the autoantibody. Autoantibodies against intracellular neural antigens primarily recognize linear epitopes. Western blot or line immunoblot assays are frequently used to identify these autoantibodies. In contrast, autoantibodies against cell-surface or synaptic neural antigens primarily recognize conformational epitopes. Thus, different methodologies are preferred for the detection of these autoantibodies. Cell-based assays (CBAs) are the method of choice for autoantibodies against cell-surface receptors, and radioimmunoprecipitation assays (RIAs) are preferred for the detection of autoantibodies against many of the synaptic receptors.

### Tissue-Based Assays

TBAs for the detection of neural autoantibodies using indirect immunofluorescence assay (IFA) or immunohistochemistry (IHC) are performed by incubating patient serum or CSF on sections of primate or rodent neural tissue(s), bound autoantibodies are detected with a fluorescent- or enzyme-conjugated anti-human IgG secondary antibody, and the presence and pattern of bound autoantibodies are determined by microscopy (Fig. 2.1a). An important consideration for the optimal detection of neural autoantibodies is the region of the brain used and preparation of the tissue sections with regard to pretreatment and

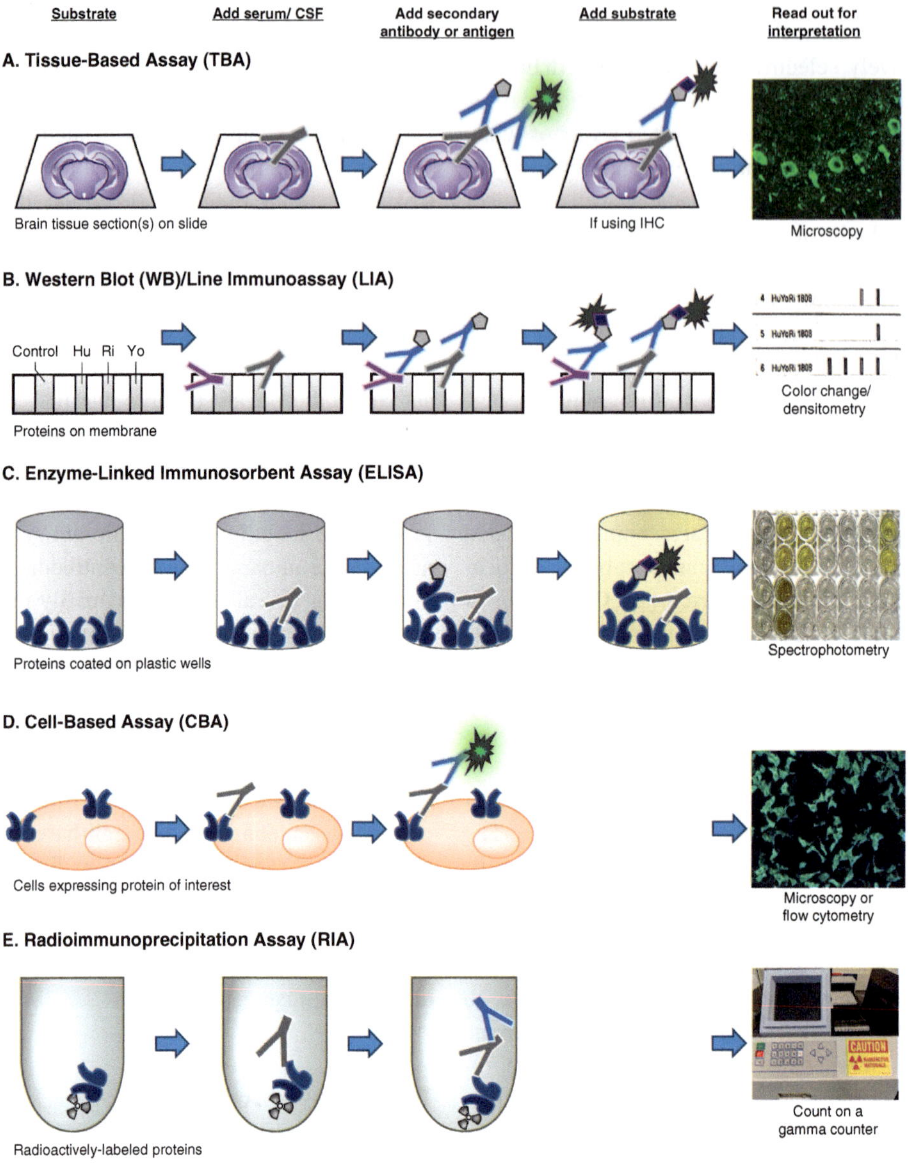


**Fig. 2.1** Overview of methods for the detection of neural autoantibodies. (**a**) Tissue-based assays are performed using sections of primate or rodent neural tissue(s), patient serum or CSF is added; bound autoantibodies are detected with a fluorescent- or enzyme-conjugated anti-human IgG secondary antibody; substrate is added to induce a color change when an enzyme-conjugated antibody is used; and the presence and pattern of bound autoantibodies is determined by microscopy. (**b**) Western blot or line immunoblot assays are performed using strips of membrane containing neural proteins, patient serum or CSF is added, bound autoantibodies are detected using an enzyme-conjugated antibody against human IgG, which after addition of the substrate are visualized as a change in color at a specific position on the strip. (**c**) Enzyme-linked immunosorbent assays are performed using plastic wells coated with neural proteins, patient serum or CSF is added, bound autoantibodies are detected by addition of biotin-conjugated protein of interest, which after addition of enzyme-conjugated streptavidin and substrate are visualized as a change in color measured by spectrophotometry. (**d**) Cell-based assays are performed using cells expressing the neural antigen and/or receptor of interest, patient serum or CSF is added, bound autoantibodies are detected using a fluorochrome-conjugated antibody against human IgG, which are visualized by either microscopy or flow cytometry. (**e**) Radioimmununoprecipitation assays are performed using radioactively labeled proteins, patient serum or CSF is added, bound autoantibodies are precipitated with an anti-human IgG secondary antibody, radioactivity in pelleted immune complexes is measured with a gamma counter

fixation, which differs between intracellular and cell-surface antigens [4, 5]. Primate cerebellum snap-frozen, sectioned using a cryostat and fixed with paraformaldehyde or acetone is the preferred substrate for the detection of autoantibodies against intracellular neural antigens. Whereas, rat hippocampus fixed with paraformaldehyde, cryoprotected in sucrose, snap-frozen, and sectioned using a cryostat is the preferred substrate for the detection of autoantibodies against cell-surface or synaptic neural antigens.

A major advantage of TBAs is that a large number of neural antigens are available and accessible in the tissue sections. Thus, TBAs can be used to screen for a wide variety of neural autoantibodies at the same time and to discover new autoantibodies. Indeed, many neural autoantibodies have been discovered using this methodology. A major disadvantage is that it takes significant training to become proficient at identifying all of the possible patterns [5]. Additional disadvantages include the fact that autoantibodies against different antigens can produce similar patterns of staining, so additional testing must be performed to confirm the specificity of the autoantibodies. It can also be difficult to identify coexisting autoantibodies using this method. Many of these autoantibodies are very rare making it difficult to obtain positive specimens for validating assays, functioning as controls for the assay, training new staff, and maintaining competency and proficiency. TBAs are also time consuming, labor intensive, lack standardization, and can be subjective [4].

Detection of autoantibodies using TBAs can be performed individually or using mosaics of biochips containing various brain or other tissue sections [6–8]. This technology consolidates the ability to screen for multiple neural autoantibodies and identification/confirmation of some of their specific targets into a single assay. An important consideration when using this approach is whether positive controls for all autoantibodies to be reported are tested on every run [4].

## Western Blot or Line Immunoblot Assays

Western blot (WB) or line immunoblot assays (LIAs) are the preferred method for confirming the presence of autoantibodies against intracellular targets. These methods are performed using lysates or proteins purified from extracts of brain tissue or cells expressing the proteins of interest, which are either run on a polyacrylamide gel and transferred to a membrane in the case of WBs or printed directly on a membrane in the case of LIAs. The membranes are cut into strips, incubated with patient serum or CSF and bound autoantibodies are detected using an enzyme-conjugated antibody against human IgG, which after addition of the substrate are visualized as a change in color at a specific position on the strip (Fig. 2.1b). Advantages of this methodology are that multiple autoantibodies can be tested for simultaneously, the testing can be automated and the results are more specific than those obtained by TBAs because specific antigens are present at particular locations on the membrane. Disadvantages of this method are that purification of the proteins often affects their conformation and/or interactions with other proteins, which can lead to false-negative results if the autoantibodies in the patient serum recognize conformational epitopes. This method also suffers from the same problem as TBAs with regard to difficulty in obtaining samples containing rare autoantibodies in order to validate the assay, serve as controls for performance of the assay, train laboratory staff, and maintain competency and proficiency. In addition, clinical significance of an immunoblot positive but TBA negative result is uncertain.

WB of brain tissue extracts allows for the detection of multiple autoantibodies. However, this advantage is off-set by the possibility of more than one antigen occupying the same location on the membrane. This problem is solved using LIAs where antigens are placed at specific locations. Thus, LIA does not offer the advantage of examining the entire repertoire of proteins observed by WB, as it is limited to the number of proteins selected for inclusion on the membrane.

## Enzyme-Linked Immunosorbent Assays

ELISAs can be used to identify autoantibodies against intracellular antigens as well as select cell-surface or synaptic receptors. Similar to immunoblots, this method is performed using protein extracts but instead of using a membrane, the antigens are coated on the wells of a 96-well plate, and bound autoantibodies are detected using spectrophotometry. Several ELISAs used to detect neural autoantibodies use a variation of the technique, where autoantibodies present in patient serum or CSF form a bridge between antigen coated on the plate and biotinylated-antigen, which after addition of streptavidin peroxidase and substrate are detected using a spectrophotometer (Fig. 2.1c). In bridge ELISA testing, the detection method is not a secondary antigen against human IgG. Therefore, these assays are not antibody isotype specific. This can result in detection of autoantibodies of the IgA and/or IgM isotypes in addition to IgG autoantibodies. The clinical relevance of IgA or IgM autoantibodies is currently uncertain [9, 10]. In contrast to an immunoblot, ELISAs typically only test for autoantibodies against one target at a time, which can be considered a disadvantage of this method. Advantages of ELISA include increased sensitivity and specificity, decreased subjectivity compared to TBAs, and it is a high-throughput method that can be performed in many laboratories and can be automated. ELISAs can suffer a similar disadvantage to immunoblots in that the antigens may not be in their native conformation as a result of the purification process, which can lead to false-negative results. ELISAs can also yield false-positive results as a result of nonspecific binding due to the antibodies themselves binding to the plate or to the presence of heterophile antibodies [3]. In addition, some ELISAs, such as those used for the detection of autoantibodies against aquaporin 4 (AQP4) and glutamic acid decarboxylase (GAD65) antibodies, evaluate serum directly, whereas most methodologies dilute serum prior to testing in order to reduce background signal. Taken together, the lack of isotype specificity and the use of undiluted serum may explain differences in correlation with disease and/or other methodologies, especially for sera found to have low positive results by ELISA.

## Cell-Based Assays

CBAs are the preferred method for detecting autoantibodies against cell-surface antigens and some synaptic receptors. They are performed using cells transfected with the antigen and/or receptor of interest. Transfected cells are incubated with patient serum or CSF, and bound autoantibodies are detected using a fluorescently conjugated antibody against human IgG and evaluated either by microscopy or flow cytometry (Fig. 2.1d). Advantages of this method include that the antigens are in their native conformation and that the results are more specific than TBAs because the cells are transfected with a single antigen of interest. Thus, interpretation is less subjective than TBAs and requires less training to become proficient. Disadvantages include that only autoantibodies against the antigen expressed by the cells are detected. Thus, this method cannot be used for the discovery of new autoantibodies.

Both live and fixed CBAs have been used for the detection of autoantibodies against cell-surface antigens. Commercially available CBAs use fixed cells out of necessity. Use of live cells requires continuous culturing of cells and the generation and maintenance of transfected cell lines. An important difference between assays using live cells instead of fixed cells is that antibodies only have access to targets on the surface of the cells. Fixation of cells can lead to permeabilization of the cell membrane, which can allow antibodies access to antigens inside the cell in addition to those on the cell surface. Fixation can also alter the presentation or accessibility of antigens, so the antigens present on live cells may be present in a more native form than those of fixed cells. Difference in performance between fixed and live CBAs varies among antigens, with live cells showing slightly better sensitivity for some autoantibodies, whereas fixed cells demonstrate higher sensitivity for others such as N-methyl-D-aspartate glutamate receptor (NMDAR) [11]. However, it is important to consider the number of clinically defined patient specimens used to make these comparisons. Differences in the results for a single specimen can appear to have a considerable effect on sensitivity or specificity when few clinically defined specimens are included in the analysis, as is frequently the case for these rare autoantibodies.

Detection of autoantibodies using fixed CBAs evaluated by IFA can be performed individually or using mosaics of biochips containing various transfected cells expressing different neural antigens, brain and/or other tissue sections [6–8]. This technology consolidates the ability to screen for multiple neural autoantibodies and identification/confirmation of some of their specific targets into a single assay. An important consideration when using this approach is whether positive controls for all autoantibodies to be reported are tested on every run. Multiplexing of CBA using mosaics can present a challenge with regard to manual reading and interpretation. As the number of biochips included in the mosaic increases so does the risk of confusing which biochip is being observed. Automation can aid in this process, but additional process controls should be incorporated to ensure the accuracy of results.

Detection of autoantibodies using live cell CBAs evaluated by flow cytometry decreases the subjectivity in visual interpretation commonly observed in CBA/IFA assays [5]. This method is gaining in popularity, but it is currently only available for diagnostic testing of a few neural autoantibodies (Table 2.1).

## Immunoprecipitation Assays

RIAs are the preferred method for detecting many of the autoantibodies against synaptic targets. RIA is performed using either iodine-125 radioactively labeled recombinant proteins or lysates of brain, muscle or cells expressing the antigen of interest that have been incubated with radioactively labeled toxins with high affinity for specific synaptic receptors. Patient serum or CSF is incubated with the radioactively labeled proteins or lysates containing the radioactively labeled receptors, unbound antibodies are washed away, and then anti-human IgG or protein A or G sepharose is added to form immune complexes and facilitate precipitation of the antigen/autoantibody complexes. Presence of autoantibodies is measured by the detection of radioactivity using a gamma counter, and quantitation is based on comparison to a standard curve or directly based on the specific activity of the radioactive ligand (Fig. 2.1e). Advantages of this method include that the antigens are in their native conformation and that the method is very sensitive due to the use of radioactivity. However, use of radioactivity is also a disadvantage because it poses a health hazard to laboratory personnel and requires a license for use and proper disposal [4]. Another disadvantage of this method is that it is possible to precipitate entire immune complexes, which then requires additional testing to confirm the specificity of the autoantibody [5]. This is the case for immunoprecipitation of the voltage-gated potassium channel (VGKC) complex, which is then followed by CBA to evaluate whether the reactivity is specific for leucine-rich glioma inactivated 1 protein (LGI1), contactin-associated protein 2 (CASPR2), or other proteins in the complex.

## Primary Neuronal Cell Cultures

Primary cell culture-based IFA is performed by isolating cells from specific parts of the brain and culturing them for 2–3 weeks before using them to detect autoantibodies. Thus, this method is primarily used on a research basis. Patient serum or CSF is incubated with live neurons, and bound autoantibodies are detected using a fluorescently conjugated antibody against human IgG and visualized by microscopy. Autoantibodies to a variety of extracellular antigens can be detected using this method, but the staining patterns produced are often indistinguishable requiring the use of additional methods to determine the specific targets. Important considerations when interpreting the results are the presence and level of expression of the antigen of interest at the time of testing and that the binding of antibodies against some cell-surface proteins can alter their localization. This second point is relevant for both autoantibodies that may be present in patient serum or CSF and purified antigen-specific antibodies used as controls or in co-localization studies performed to evaluate the specificity of the autoantibodies [4, 5].

## Specimen Type (Serum or CSF)

In addition to considering which methodology to use when testing for neural autoantibodies, another important consideration is specimen type. Some autoantibodies may only be present in one body fluid due to the location of their antigenic target, for example, autoantibodies against the muscle acetylcholine receptor (mAChR) are primarily only detected in serum. Alternatively, intrathecally synthesized autoantibodies may only be detectable in CSF (e.g., NMDAR autoantibodies) [4, 12]. Advantages of using serum include that its collection is less invasive making it more suitable for serial testing in monitoring response to treatment, and autoantibodies are often present at higher titers in serum. However, serum contains other proteins and non-neural autoantibodies that can cause high background or nonspecific binding resulting in false-positive results. Advantages of testing CSF include that it contains less extraneous proteins, so fewer false positives are observed due to nonspecific binding. CSF can be more sensitive and specific for the detection of neural cell-surface autoantibodies, and if the neural autoantibodies are being produced intrathecally, serum may be negative. Disadvantages to testing CSF include that its collection is more invasive and that autoantibodies are often present at low titers, if at all, which can lead to false-negative results. For example, CSF has been reported to be less sensitive than serum for the detection of AQP4 and LGI1 [4, 7]. Thus in some cases it may be important to test both serum and CSF, specifically in the setting of suspected LGI1 autoimmune encephalitis [13].

## Sensitivity and Specificity

Widely divergent figures of the combined sensitivity for the known neural autoantibodies have been published [14, 15]. Discussion of this large group of heterogeneous surrogate biomarkers of disease and/or pathogenic autoantibodies is complicated by autoantibody presence being the defining characteristic in certain ANDs. Inconsistent laboratory findings in a setting of multiple autoantibody-associated disorders, with similar clinical presentation, add to the confusion. In addition, autoantibodies are not detected in all patients with clinically defined encephalitis suggestive of an autoimmune etiology [3]. This is in part due to the ongoing identification of additional antigenic targets and differing composition of autoantibody panels performed at different reference laboratories. In one single-center 1-year retrospective cohort, a combined sensitivity for paraneoplastic autoantibodies of 34% was estimated [14]. Whereas, in another multiyear retrospective study an estimated combined sensitivity between 60% and 80% was reported for clinically defined autoimmune encephalitis patients [15]. An important consideration when evaluating sensitivity of specific methods are the species and/or the region(s) of the brain from which the tissues or proteins are derived. Lack of detection of neural autoantibodies may be due to absence of epitopes/antigenic targets due interspecies differences [5].

Neural autoantibodies are rarely detected in serum from non-encephalitis disease control or healthy individuals. False-positive rates vary by methodology, isotype of secondary detection antibody used (polyclonal, IgG only, IgG1), and autoantibody of interest (NMDAR IgA, IgM, and IgG polyclonal detection by CBA has approximately a 10% false-positive rate) [6, 14, 16]. Interestingly, autoantibodies associated with classical PNS can be detected at an elevated frequency in patients with particular malignancies (approximately 20% of small cell lung cancer patients have Hu autoantibodies), yet very few of these patients develop neurologic symptoms (<0.01%) [3, 17].

Important considerations when evaluating sensitivity and specificity include the cut-offs used, how they were generated, and whether the results reported are qualitative or quantitative. Interpretation of low-titer antibody results can be challenging since some autoantibodies such as those against the VGKC complex and GAD65 have been found at low levels in patients without neurologic disease, but these have also been shown to be clinically relevant as is the case for patients with low level VGKC complex results but high LGI1 or CASPR2 results [1]. Low titer results can also be seen in the setting of immunotherapy as some antibody levels may decrease in response to therapy. It is important to note that the majority of information regarding sensitivity and specificity of assays to detect neural autoantibodies is based on testing of serum. Data are lacking for sensitivity and specificity for detection of neural autoantibodies in CSF.

## Challenges Related to Detection of Neural Autoantibodies

Current challenges for the detection of autoantibodies associated with ANDs include that testing is very segmented in some countries with only certain labs able to perform testing for certain autoantibodies due to intellectual property restrictions. Many neural autoantibodies are very rare creating difficulty in obtaining positive samples for validation, training, competency, proficiency, and to function as controls when performing the assays. This is complicated by the fact that detection of staining patterns associated with neural autoantibodies requires significant training to become proficient and that overlap of symptoms between patients with multiple autoantibodies makes it difficult to determine the sensitivity of an assay since the diseases are defined by the presence of the autoantibody. Another challenge is that the number of autoantibodies associated with ANDs is continuing to grow, as is the availability of commercial assays to detect them, but the thorough characterization required to establish clinical context and prevalence often lags due to the rarity of positive patients.

## Introduction to the Role of Neural Autoantibodies in Pathogenesis

Although clear associations between autoantibodies and ANDs have been demonstrated, it is less clear whether these autoantibodies play a role in the pathogenesis of the diseases or are simply markers of the disease process, since only a few have actually been shown to cause disease. Distinction between neural autoantibodies based on the subcellular location of their antigenic targets is also relevant in discussions on the pathogenic role of these autoantibodies. The three main groups include autoantibodies that target intracellular nuclear and cytoplasmic antigens, autoantibodies that target intracellular synaptic antigens, and autoantibodies that target cell-surface and synaptic antigens (Fig. 2.2). Evidence for the pathogenicity of the autoantibodies is based on data from in vitro studies, animal models, and biopsy and autopsy tissue studies, as well as the responsiveness of patients positive for these autoantibodies to immunotherapy [18].

Autoantibodies to intracellular nuclear or cytoplasmic targets have limited access to their target antigens and are therefore considered not to be directly pathogenic. Instead, they are thought to be biomarkers of a T-cell-mediated response against their corresponding neuronal target antigen [2]. Evidence exists that autoantibodies to intracellular cytoplasmic or nuclear targets present in the CNS, may be synthesized intrathecally, can be taken up by neurons and in some cases lead to neuronal cell death *in vitro* [19–21]. However, animal models involving passive transfer of these autoantibodies or

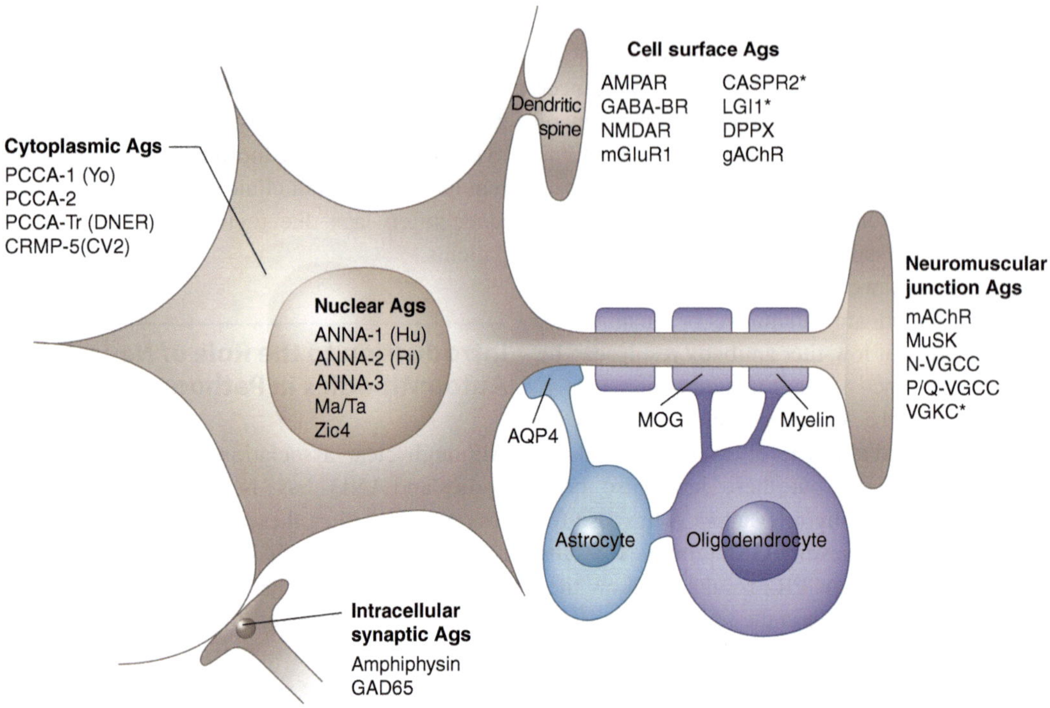


**Fig. 2.2** Depiction of the subcellular location of the antigenic targets of neural autoantibodies and the relationship between location and the pathogenic role of these autoantibodies. Three main groups of neural autoantibodies include autoantibodies that target intracellular nuclear and cytoplasmic antigens, autoantibodies that target intracellular synaptic antigens, and autoantibodies that target cell-surface and synaptic antigens. Autoantibodies to intracellular nuclear or cytoplasmic targets have limited access to their target antigens and are therefore not considered directly pathogenic. In contrast, autoantibodies to intracellular synaptic targets are proposed to have access to their target antigens during fusion and reuptake of synaptic vesicles, and thus maybe directly pathogenic. Autoantibodies against cell surface and neuromuscular junction have direct access to their antigenic targets and are considered to be directly pathogenic. The majority of the targets of neural autoantibodies are expressed in neurons, but AQP4 and MOG are expressed on the cell surface of astrocytes and oligodendrocytes, respectively. * = antigens expressed both on the neuronal cell surface and in the neuromuscular junction. AGNA-1 anti-glial nuclear antibody, AMPAR alpha-amino-3-hydroxy-5-methyl-4-isoxazolepropionate receptor, ANNA anti-neuronal nuclear antibody, AQP4 aquaporin 4, CASPR2 contactin-associated protein 2, CRMP5 collapsing response mediator protein 5, DPPX dipeptidyl aminopeptidase-like protein, $GABA_BR$ gamma-aminobutyric acid receptor, type B, GAD glutamic acid decarboxylase, gAChR ganglionic acetylcholine receptor, LGI1 leucine-rich glioma inactivated 1 protein, mGluR metabotropic glutamate receptor, MOG myelin oligodendrocyte glycoprotein, mAChR muscle acetylcholine receptor, MuSK muscle-specific tyrosine kinase, NMDAR N-methyl-D-aspartate glutamate receptor, PCCA Purkinje cell cytoplasmic antibody, VGCC voltage-gated calcium channel, VGKC voltage-gated potassium channel

immunization with their corresponding target antigens have failed to confirm a pathogenic role for these autoantibodies *in vivo* [22, 23]. Evidence for a T-cell-mediated response against intracellular nuclear or cytoplasmic neuronal antigens includes the detection of neuronal antigen-specific T-cell responses in patients with paraneoplastic encephalomyelitis, the presence of more T cells than B cells in their brain and peripheral nerve tissues, expression of a marker of cytotoxic effector T-cell function, granzyme B, in close proximity to neurons in areas with evidence of neuronal cell loss [19, 24, 25].

Intracellular synaptic targets such as GAD65 and amphiphysin may be targeted by both T cells and autoantibodies. In contrast to the antigenic targets discussed in the previous section, autoantibodies have access to these antigens during fusion and reuptake of synaptic vesicles [16]. A direct pathogenic role for this group of autoantibodies was demonstrated by intrathecal injection of anti-amphiphysin into rats resulting in stiff person syndrome-like symptoms [26]. Evidence for a T-cell-mediated response against intracellular synaptic antigens includes development of encephalomyelitis in immunized mice producing GAD65-specific T cells and development of neurologic symptoms upon transfer of these GAD65-specific T cells to naïve mice or mice lacking B cells [2, 27].

Autoantibodies against cell-surface receptors are thought to play a more direct role in pathogenesis through agonistic or antagonistic effects on the receptors, disrupting the function of the receptors either by causing them to be internalized or preventing their ligands from binding to them, or potentially leading to cytotoxicity due to antibody- and/or complement-mediated mechanisms [18, 25]. These pathogenic effects of cell-surface neural autoantibodies have been demonstrated both *in vitro* and *in vivo* using passive transfer of patient IgG into mice [28, 29]. Reports that this neural dysfunction is frequently reversible upon removal of the autoantibodies and many patients often experience complete recovery in response to immunotherapy suggests that autoantibodies play a direct pathogenic role [18]. Additional support for a pathogenic role is that autoantibodies against some neural cell-surface receptors produce effects similar to genetic or pharmacologic disruption of the receptors [2].

In PNS, pathogenesis is thought to be due to an immune response against a neural protein that is aberrantly expressed on a tumor, leading to activation and expansion of autoreactive T and B cells, and production of autoantibodies [17]. When these immune agents gain access to the nervous system, they can cause damage leading to neurologic symptoms. However, tumors are detected in less than a third of patients with ANDs at the onset of neurologic symptoms and autoantibody detection, which begs the question of what triggers autoantibody production in these patients. Current hypotheses include an infectious trigger and multiple recent publications have drawn a link between herpes simplex encephalitis and NMDA receptor encephalitis (discussed further in Chap. 25) [9]. Another possibility is that a tumor is not detected because the immune response is effective in fighting the malignancy [17]. It also remains to be determined, how chronicity is established or how relapses are triggered in the case of idiopathic AND [29].

Additional questions about the role of autoantibodies in the pathogenesis of ANDs include the following: why there is diversity in clinical presentations associated with a particular autoantibody and how specificity of symptoms occurs despite widespread expression? Potential explanations include heterogeneity in the antibody response with respect to subtype of IgG and epitope(s), post-translational modifications or conformational changes specific to particular regions of the brain, and the presence of co-existing autoantibodies [2, 17, 18].

## Conclusion

The number of autoantibodies associated with ANDs continues to grow, as does the number of specimens being tested. Clinical laboratories are faced with the challenge of determining how many and which assays to offer, by which methods and in which combinations. Clinicians are faced with deciding which tests or panels of tests to use in the assessment for neural autoantibodies in their patients. Despite uncertainty about the role of autoantibodies in the pathogenesis of ANDs, they can play a significant role in diagnosis and treatment. Thus, it is important for clinicians to be aware of the limitations of the various methods for detecting autoantibodies and to note that failure to detect a neural autoantibody does not rule out diagnosis of an AND.

## References


1. Linnoila J, Pittock SJ. Autoantibody-associated central nervous system neurologic disorders. Semin Neurol. 2016;36:382–96.
2. Lancaster E, Dalmau J. Neuronal autoantigens – pathogenesis, associated disorders and antibody testing. Nat Rev Neurol. 2012;8:380–90.
3. Tebo AE, Haven TR, Jackson BR. Autoantibody diversity in paraneoplastic syndromes and related disorders: the need for a more guided screening approach. Clin Chim Acta. 2016;459:162–9.
4. Waters P, Pettingill P, Lang B. Detection methods for neural autoantibodies. Handb Clin Neurol. 2016;133:147–63.https://doi.org/10.1016/B978-0-444-63432-0.00009-8. PMID: 27112676.
5. van Coevorden-Hameete MH, Titulaer MJ, Schreurs MW, de Graaff E, Sillevis Smitt PA, Hoogenraad CC. Detection and characterization of autoantibodies to neuronal cell-surface antigens in the central nervous system. Front Mol Neurosci. 2016;9:37. https://doi.org/10.3389/fnmol.2016.00037.
6. Dahm L, Ott C, Steiner J, Stepniak B, Teegen B, Saschenbrecker S, et al. Seroprevalence of autoantibodies against brain antigens in health and disease. Ann Neurol. 2014;76:82–94.
7. McCracken L, Zhang J, Greene M, Crivaro A, Gonzalez J, Kamoun M, et al. Improving the antibody-based evaluation of autoimmune encephalitis. Neurol Neuroimmunol Neuroinflamm. 2017;4:e404.
8. Naides SJ. The role of the laboratory in the expanding field of neuroimmunology: autoantibodies to neural targets. J Immunol Methods. 2018;463:1–20.
9. Prüss H, Höltje M, Maier N, Gomez A, Buchert R, Harms L, et al. IgA NMDA receptor antibodies are markers of synaptic immunity in slow cognitive impairment. Neurology. 2012;78:1743–53.
10. Hara M, Martinez-Hernandez E, Ariño H, Armangue T, Spatola M, Petit-Pedrol M, et al. Clinical and pathogenic significance of IgG, IgA, and IgM antibodies against the NMDA receptor. Neurology. 2018;90:e1386–94.
11. Ramberger M, Peschl P, Schanda K, Irschick R, Höftberger R, Deisenhammer F, et al. Comparison of diagnostic accuracy of microscopy and flow cytometry in evaluating N-methyl-D-aspartate receptor antibodies in serum using a live cell-based assay. PLoS One. 2015;10:e0122037.
12. Suh-Lailam BB, Haven TR, Copple SS, Knapp D, Jaskowski TD, Tebo AE. Anti-NMDA-receptor antibody encephalitis: performance evaluation and laboratory experience with the anti-NMDA-receptor IgG assay. Clin Chim Acta. 2013;421:1–6.
13. Graus F, Titulaer MJ, Balu R, Benseler S, Bien CG, Cellucci T, et al. A clinical diagnostic approach to diagnosis of autoimmune encephalitis. Lancet Neurol. 2016;15:391–404.
14. Albadareen R, Gronseth G, Goeden G, Sharrock M, Lechtenberg C, Yang Y. Paraneoplastic autoantibody panrels: sensitivity and specificity, a retrospective cohort. Int J Neurosci. 2017;127:531–8.
15. Wandinger KP, Leypold F, Jumker R. Autoantibody-mediated encephalitis - differential diagnosis in patients with impaired consciousness of unclear origin. Dtsch Arztebl Int. 2018;115:666–73.
16. Lang K, Prüss H. Frequencies of neuronal autoantibodies in healthy controls: estimation of disease specificity. Neurol Neuroimmunol Neuroinflamm. 2017;4:e386.
17. Darnell RB, Posner JB. Paraneoplastic syndromes involving the nervous system. N Engl J Med. 2003;349:1543–54.
18. van Coevorden-Hameete MH, de Graaff E, Titulaer MJ, Hoogenraad CC, Sillevis Smitt PA. Molecular and cellular mechanisms underlying anti-neuronal antibody mediated disorders of the central nervous system. Autoimmun Rev. 2014;13:299–312.
19. Dalmau J, Furneaux HM, Rosenblum MK, Graus F, Posner JB. Detection of the anti-Hu antibody in specific regions of the nervous system and tumor from patients with paraneoplastic encephalomyelitis/sensory neuronopathy. Neurology. 1991;41:1757–64.
20. Greenlee JE, Clawson SA, Hill KE, Wood B, Clardy SL, Tsunoda I, et al. Neuronal uptake of anti-Hu antibody, but not anti-Ri antibody, leads to cell death in brain slice cultures. J Neuroinflammation. 2014;11:160.
21. Greenlee JE, Clawson SA, Hill KE, Wood B, Clardy SL, Tsunoda I, et al. Anti-Yo antibody uptake and interaction with its intracellular target antigen causes Purkinje cell death in rat cerebellar slice cultures: a possible mechanism for paraneoplastic cerebellar degeneration in humans with gynecological or breast cancers. PLoS One. 2015;10:e0123446.
22. Sillevis Smitt PA, Manley GT, Posner JB. Immunization with the paraneoplastic encephalomyelitis antigen HuD does not cause neurologic disease in mice. Neurology. 1995;45:1873–8.
23. Graus F, Illa I, Agusti M, Ribalta T, Cruz-Sanchez F, Juarez C. Effect of intraventricular injection of an anti-Purkinje cell antibody (anti-Yo) in a guinea pig model. J Neurol Sci. 1991;106:82–7.
24. Rousseau A, Benyahia B, Dalmau J, Connan F, Guillet JG, Delattre JY, et al. T cell response to Hu-D peptides in patients with anti-Hu syndrome. J Neuro-Oncol. 2005;71:231–6.
25. Bien CG, Vincent A, Barnett MH, Becker AJ, Blümcke I, Graus F, et al. Immunopathology of autoantibody-associated encephalitides: clues for pathogenesis. Brain. 2012;135(Pt 5):1622–38.
26. Geis C, Weishaupt A, Hallermann S, Grünewald B, Wessig C, Wultsch T, et al. Stiff person syndrome-associated autoantibodies to amphiphysin mediate reduced GABAergic inhibition. Brain. 2010;133:3166–80.

27. Burton AR, Baquet Z, Eisenbarth GS, Tisch R, Smeyne R, Workman CJ, et al. Central nervous system destruction mediated by glutamic acid decarboxylase-specific CD4+ T cells. J Immunol. 2010;184:4863–70.
28. Williams JP, Carlson NG, Greenlee JE. Antibodies in autoimmune human neurological disease: pathogenesis and immunopathology. Semin Neurol. 2018;38:267–77.
29. Bradl M, Lassmann H. Neurologic autoimmunity: mechanisms revealed by animal models. Handb Clin Neurol. 2016;133:121–43.

# 3 Neuroimaging Modalities in Neuroimmunology

Lokesh A. Rukmangadachar
and Christina J. Azevedo

**Key Points**

1. Because of its high sensitivity to detect white matter hyperintensities on T2-weighted images, magnetic resonance imaging (MRI) plays a central role in diagnosing and monitoring disease activity in multiple sclerosis (MS). MS plaques have a typical appearance on MRI (i.e., location, size, and morphology), which helps distinguish MS from other conditions. MRI can facilitate an earlier diagnosis of MS, often at first clinical presentation, by demonstrating dissemination in space and time within the central nervous system (CNS), the central tenet of MS Diagnostic Criteria.
2. MS misdiagnosis is common when MRI criteria are not applied in the appropriate clinical context, particularly when the clinical presentation and/or MRI appearance is atypical or nonspecific. Detection of central veins within white matter lesions and/or demonstration of cortical lesions may increase the specificity of MRI for MS and are active areas of research currently.
3. Neurodegeneration is a fundamental component of MS pathology. Whole brain volume decline on MRI can be measured reproducibly, is clinically relevant and modifiable with several available disease-modifying therapies, and has good face validity as a surrogate marker of neurodegeneration. There is also a high degree of interest in several regional volumes as MRI surrogate markers of neurodegeneration, including thalamus, other deep gray matter structures, and spinal cord.
4. Positron emission tomography (PET) is a promising clinical research tool that can offer a high degree of pathologic specificity and can complement MRI to better understand MS pathology. PET can study aspects of the disease which are "MRI invisible," such as microglial activation, or to which MRI is relatively insensitive, such as remyelination.

L. A. Rukmangadachar
Saint Louis University School of Medicine, Saint Louis, MO, USA

C. J. Azevedo (✉)
Keck School of Medicine, University of Southern California, Los Angeles, CA, USA
e-mail: cazevedo@usc.ed



A. L. Piquet, E. Alvarez (eds.), *Neuroimmunology*, https://doi.org/10.1007/978-3-030-61883-4_3

## Introduction

Over the past few decades, magnetic resonance imaging (MRI) has evolved to play an essential role in the diagnosis and management of neuroimmunological diseases, including multiple sclerosis (MS), and has furthered our basic understanding of MS substantially. This chapter discusses MRI as a tool to diagnose and monitor MS, and to distinguish MS from other neurologic conditions. Volumetric MRI and positron emission tomography (PET) imaging are highlighted as specific areas of interest that are currently within the realm of MS clinical research, but may have applications for clinical care in the future.

## The Role of MRI in the Diagnosis of MS

Despite having undergone multiple revisions, the central tenet of the diagnosis of MS has remained unchanged, that is, dissemination in space and time (DIS and DIT, respectively) within the central nervous system (CNS). Each iteration of the Diagnostic Criteria for MS, including the current 2010 Criteria [1] and the recently proposed 2017 revisions [2], has allowed the diagnosis of clinically definite MS to be made following two separate clinical events that demonstrate both DIS and DIT. MRI was first incorporated into the Diagnostic Criteria in 2001 [3] to facilitate earlier diagnosis at the time of the initial clinical presentation, that is, clinically isolated syndrome (CIS).

MRI is highly sensitive to visualize white matter (WM) signal abnormalities on T2-weighted images, and the characteristic appearance of demyelinating lesions has become well recognized. Typical demyelinating lesions are hyperintense on T2-weighted images, are round or ovoid in shape, are at least 3 mm in size (in-plane), and occur in characteristic locations within the CNS [4]. Because of the location of venules within the brain, classic periventricular demyelinating lesions are oriented perpendicularly to the long axis of the ventricles (so-called Dawson's fingers; Fig. 3.1a, b). If demyelination involves the subcortical U-fibers, the lesions appear juxtacortical on MRI, which abut the cortex but respect the gray-white border (Fig. 3.1c). Demyelination often occurs in the infratentorium, commonly in the spinal cord, brainstem, middle cerebellar peduncle, or the deep WM of the cerebellum (Fig. 3.1d, f–h).

Contrast-enhanced MRI can be useful in differentiating acute or "active" lesions from chronic or "inactive" lesions [5] (Fig. 3.1e, h). Acute inflammation leads to breakdown of the blood-brain barrier, allowing leakage of the paramagnetic contrast agent gadolinium into the brain parenchyma. Most acute MS lesions enhance for 2–4 weeks [6, 7], though this time window is dependent on the dose and timing of contrast administration, particularly the delay in acquisition of the post-contrast images following gadolinium injection. Most MS lesions enhance in either a nodular or ring pattern, though dynamic imaging has shown that both patterns can be seen in the same lesion depending on when the image is acquired [8]. Initially, contrast extravasates from the inflamed central vein, spreads centrifugally outward, and will be seen as nodular enhancement. However, in more established acute lesions, the leakage of contrast occurs at the inflamed margins of the lesions, resulting in ring enhancement, which spreads centripetally over minutes [9]. Periventricular or juxtacortical white matter lesions may enhance in an "open ring" fashion, with the open portion of the ring facing either the ventricle or the cortex [9]. Persistent enhancement beyond 6 weeks is uncommon in MS and should prompt consideration of other etiologies, such as malignancy or sarcoidosis.

In the 2010 Criteria [1], DIS can be demonstrated by >1 T2 lesion in ≥2 of the four typical locations for MS (periventricular, juxtacortical, infratentorial, and spinal cord; Fig. 3.1). DIT can be demonstrated by a new T2 lesion on any follow-up scan, irrespective of its timing, or by the simultaneous presence of gadolinium-enhancing and non-enhancing lesions on a single scan. From an MRI perspective, the proposed 2017 revisions to the Diagnostic Criteria [2] are largely similar, but with some important modifications. Whereas the 2010 Criteria did not allow the symptomatic

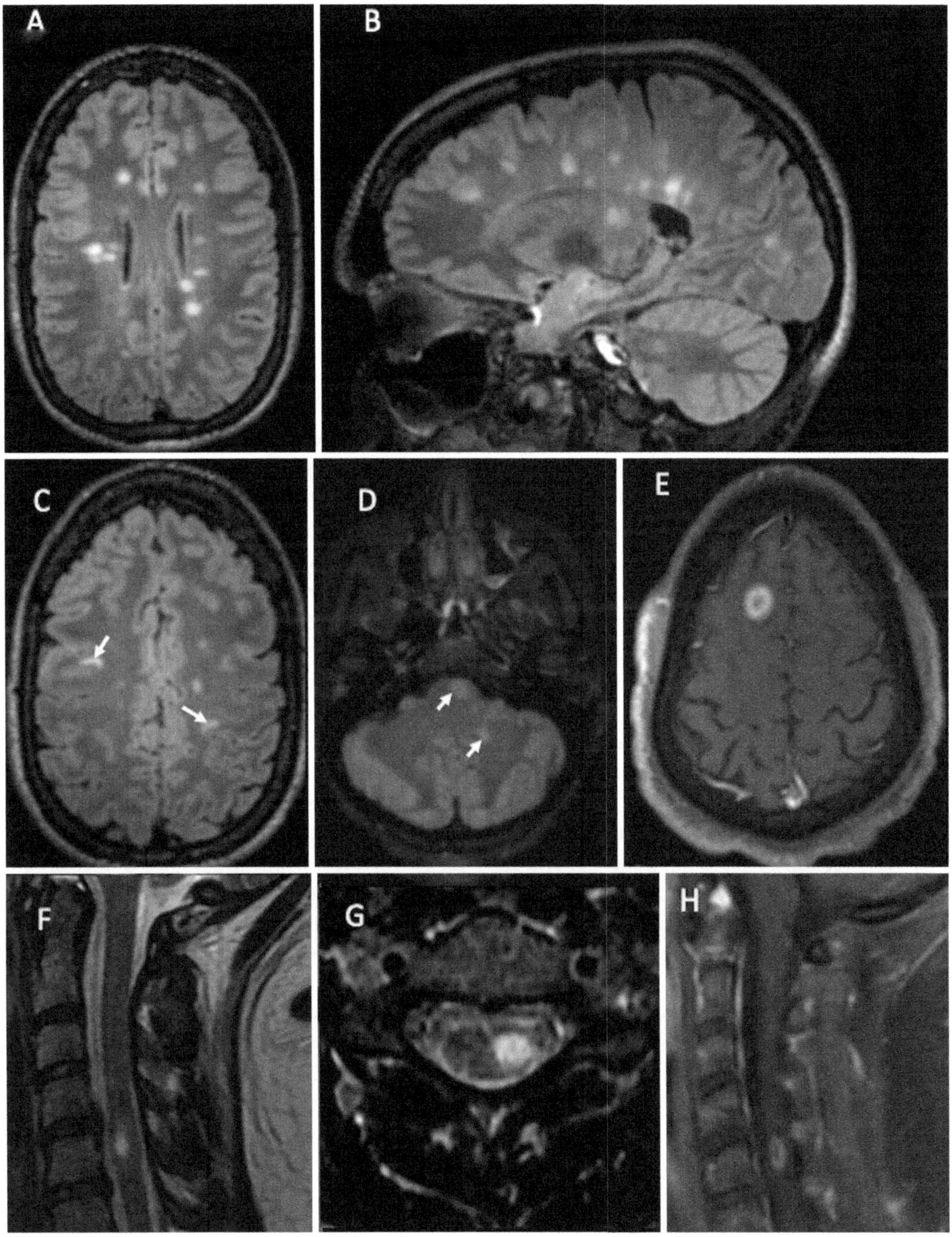


**Fig. 3.1** Brain and spinal cord MRIs demonstrating typical demyelinating lesions from a patient with MS. (**a**) Ovoid, periventricular lesions oriented perpendicularly to the long axis of the lateral ventricles on axial FLAIR. (**b**) The same periventricular lesions from Panel A are seen in the sagittal plane, oriented perpendicularly to the ventricle and involving the corpus callosum, classically described as "Dawson's fingers." (**c**) Typical juxtacortical lesions (arrows) involving the subcortical U-fibers, abutting the cortex and respecting the gray-white border. (**d**) Demyelinating lesions in the pons and middle cerebellar peduncle, typical for infratentorial location. (**e**) An acute demyelinating lesion showing ring enhancement on post-contrast T1-weighted image. (**f**) Demyelinating lesion in the cervical spinal cord on sagittal T2-weighted image, which is located, (**g**) in a typical dorsolateral position in the spinal cord on axial T2. (**h**) After gadolinium administration, this lesion demonstrates enhancement on post-contrast T1-weighted sagittal image

lesion to be included in the lesion count to satisfy DIS, the 2017 revisions no longer distinguish between symptomatic and asymptomatic lesions. Perhaps the biggest proposed revision is that, for the first time, lesions in the optic nerve and/or cortical lesions, if present, can be used to demonstrate DIS. As with previous revisions, the 2017 Criteria aim to facilitate an earlier diagnosis of MS, which they may achieve with increased sensitivity, but reduced specificity for a second attack [10]. However, there are some important caveats in the real-world application of the proposed 2017 MRI Criteria, with particular caution advised regarding cortical lesions. Cortical lesions are not routinely seen using standard clinical MRI protocols, and early attempts to detect cortical lesions such as double inversion recovery (DIR) are highly artifact-prone and suffer from poor sensitivity, specificity, and interobserver agreement [11–13]. Efforts to visualize cortical lesions are ongoing in the field and will be of interest (see “Cortical Lesion Detection”).

## Increasing the Specificity of MRI for CNS Demyelination

It is of fundamental importance to understand that each version of the Diagnostic Criteria for MS, including the MRI Criteria, has been devised to facilitate an early MS diagnosis by demonstrating DIS and DIT in patients who present with a clinical syndrome that is typical of CNS demyelination. Current and historic MRI criteria were not created to differentiate MS from other (non-MS) conditions, nor were they designed to be applied in clinical scenarios in which the symptoms and/or MRI lesions are atypical for CNS demyelinating disease. Although MRI is highly sensitive to detect white matter abnormalities, it is inherently nonspecific pathologically. T2-hyperintensity can result from any process that increases the water content of the tissue and therefore may reflect not only demyelination, but also inflammation, edema, gliosis, or any combination thereof. In recent years, an emerging literature focused on MS misdiagnosis has found that the misinterpretation of MRI findings is a common cause of misdiagnosis, particularly when the clinical presentation is nonspecific or atypical [14]. Entities commonly misdiagnosed as MS include migraine, nonspecific symptoms with abnormal MRI, fibromyalgia, and conversion or psychogenic disorders [15]. This underscores the need to develop and incorporate MRI techniques with improved specificity for demyelination into routine clinical care, which is an area of high interest in the field currently.

## Radiologically Isolated Syndrome

Occasionally, typical demyelinating lesions may be demonstrated incidentally on MRI obtained for an unrelated indication, such as headache or trauma. When there are no clinical symptoms or signs of MS but MRI demonstrates demyelinating-appearing lesions that meet 2005 DIS Criteria without alternate explanation, radiologically isolated syndrome (RIS) may be diagnosed [16]. Observational studies suggest that over a period of 5 years, roughly 1/3 of RIS patients will develop clinical symptoms and therefore fulfill Criteria for CIS/RRMS or PPMS [17, 18] suggesting that RIS is a pre-symptomatic stage of MS for many patients. However, because of the lack of typical clinical symptoms that would ordinarily provide specificity for CNS demyelination (e.g., optic neuritis, partial myelitis, or brainstem syndrome), diagnosing RIS requires extreme caution. Nonspecific T2 WM changes are common, and RIS may be misdiagnosed if MRI criteria for DIS are inappropriately applied, which is especially concerning if treatment is initiated. RIS is a poignant example of the need to develop techniques that can increase the specificity of MRI for CNS demyelination.

## Central Vein Imaging

Central vein imaging is a promising technique to increase the specificity of MRI and was the topic of a recent Consensus Statement by the North American Imaging in MS (NAIMS) Cooperative [19]. That WM MS plaques form around venules was described pathologically over 100 years ago [20]. Susceptibility-based MR techniques (e.g., phase imaging, T2*, quantitative susceptibility mapping) are highly sensitive to iron and can be used to demonstrate the presence of a central vein within a T2-hyperintense lesion (Fig. 3.2a, b). One method with particular promise is FLAIR* [23], which combines 3D FLAIR and 3D T2*-weighted images in the post-processing setting (after they have been acquired). FLAIR* leverages the high sensitivity of FLAIR to demonstrate WM lesions, combined with the ability of T2* to detect blood vessels. Currently, FLAIR* is available on some commercial scanners, with increased availability across multiple MR manufacturers expected in the near future.

Further work is needed in order to implement central vein imaging into routine clinical care. The optimal MR technique to identify central veins is not known, and there is no standardized definition of the "central vein sign." Relatively small studies using variable susceptibility-based MRI techniques and field strengths have demonstrated central veins in the majority (67–80%) of demyelinating lesions and the minority (20–30%) of WM lesions from other causes such as small vessel disease and vasculopathies [24–26]. Many of these studies have relied on manually counting the proportion of WM lesions that surround a central vessel, which is time-consuming and likely not feasible in clinical care. As such, simple, practical rules have been proposed to define and implement the central vein sign [19]. Once a standardized definition of the central vein sign is adopted, its operating characteristics (sensitivity, specificity, positive predictive value, and negative

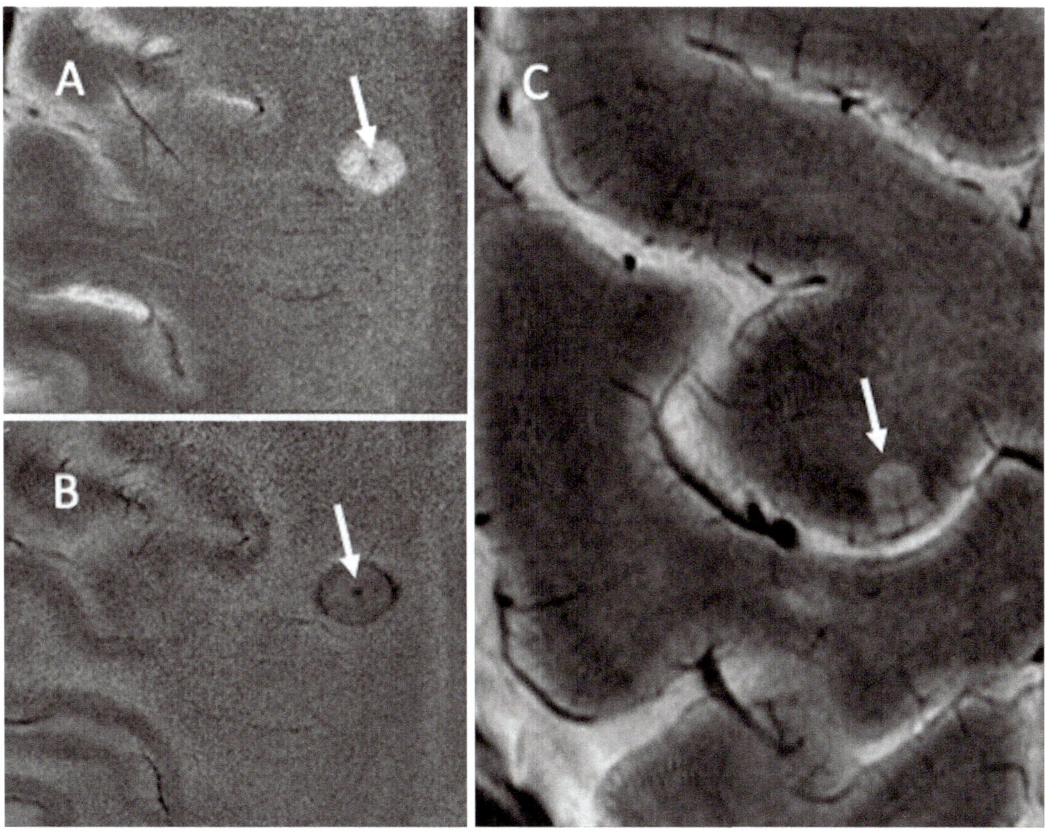


**Fig. 3.2** T2*-weighted image acquired at 7T demonstrating a hypointense central vein (arrow) on magnitude (**a**) and phase images (**b**). On phase images, a hypointense rim around the periphery of the lesion is also seen, which may suggest the presence of iron-laden microglia at the periphery of the demyelinating plaque. (**c**) High-resolution T2*-weighted image acquired at 7 T in axial plane showing an intracortical (Type 2) lesion (arrow). Source images in this figure were previously published and [21 (Panels A and B), 22 (Panel C)] and are used with permission

predictive value) will need to be determined prior to implementation into clinical care, not only in patients who present with typical demyelinating syndromes, but also in clinical scenarios with nonspecific/atypical symptoms and/or MRI findings, where it is perhaps needed most.

### Cortical Lesion Detection

Cortical lesion detection may offer another approach to increase the specificity of MRI for CNS demyelination. Cortical lesions were described on histopathology several decades ago [27], but their clinical significance is not known due to difficulty visualizing them with current MR technology. Histopathologically, they are classified into Types 1, 2, and 3 (leukocortical, intracortical, and subpial, respectively) [28]. Subpial lesions are thought to be the most common, may span several gyri, and may be extensive, particularly in progressive MS [29]. Visualization of all three subtypes has been challenging, but particularly subpial lesions, which remain essentially undetected at conventional field strengths (1.5 and 3T). Multiple reasons for these challenges exist, including partial voluming from adjacent CSF, the small size of cortical lesions within an already thin cortex of 2–3 mm (which is often below the resolution of typical clinical images), and their relative lack of MR contrast due to their paucity of inflammation. Emerging MR techniques at higher resolutions and field strengths (7T) can generate better tissue contrast and may offer better visualization of these lesions (Fig. 3.2c). Recent work at 7T has described subpial lesions in the postmortem setting [30, 31]. However, translating this in vivo will require a clinically feasible scan time, which remains a challenge. Like central vein imaging, once cortical lesions can be reliably detected, further studies will be needed prior to implementation into routine clinical care to determine sensitivity and specificity prospectively in patients who present with typical demyelinating syndromes, and in clinical scenarios with nonspecific/atypical symptoms and/or MRI findings.

## Role of MRI in Monitoring Patients with Established MS

### New White Matter Lesion Formation

In addition to its utility in establishing the diagnosis of MS, MRI plays a central role in monitoring MS longitudinally. Formation of new lesions over time is one of the main hallmarks of MS, and detecting new lesions is one of the primary roles of MRI. The clinical relevance of new MRI lesion formation has been demonstrated conclusively as a predictor of clinical relapse in the short term [32] and of disability accrual in the longer term [33, 34]. Importantly, MRI lesions satisfy the stringent Prentice criteria as a statistically valid surrogate marker of clinical relapse both at the group level [35] and the individual level [36]. As such, in relapsing MS, it has become standard to use new MRI lesions as the primary endpoint in Phase 2 clinical trials to screen candidate drugs in the developmental pipeline. If a drug effectively prevents new MRI lesion formation in Phase 2, it has historically worked when tested in Phase 3, where clinical relapses are the primary endpoint.

Despite the clear implications of new MRI lesion formation, monitoring MS patients in clinical practice remains somewhat challenging. Although several national and international consortia have published recommendations for standardized MRI protocols by which to monitor MS patients in clinical practice [37, 38], these guidelines have not been adopted in the real world. Clinicians routinely face the challenge of manually comparing MRI scans that have been obtained on different MRI scanners with heterogeneous acquisition protocols, often with varying pulse sequences and tissue contrasts, and gaps between slices. Images are generally not realigned to facilitate lesion-by-lesion comparison. Manually determining new lesion formation can be particularly difficult in patients with a high lesion burden. Given these challenges, there has been interest in automated lesion detection algorithms, which may provide better power to detect new lesions [39, 40]. The output of these algorithms could be presented to clinicians to assist a manual reading and overall interpretation of the scan. Automated lesion detection remains an area of active interest.

In addition to new/enlarging T2 lesions, the detection of contrast-enhancing lesions is standard in clinical practice to monitor disease activity. Although there are certain clinical scenarios in which the detection of gadolinium enhancement is very useful (e.g., upon diagnosis, to exclude alternate diagnoses, or during a clinical relapse), there are limitations of relying on gadolinium enhancement to detect new MS disease activity on routine follow-up MRIs, which are often obtained annually. Because new MS lesions enhance with gadolinium for an average of 4 weeks [6, 7], detecting gadolinium enhancement on an annual MRI scan is essentially random, and new lesion formation can typically be identified on T2-weighted images without contrast. Moreover, recent descriptions of gadolinium accumulation in the brain [41] have raised concerns about administering repeated dosages of gadolinium over time. Whether gadolinium accumulation has long-term clinical effects is unknown; nonetheless, the North American Consortium of Multiple Sclerosis Centers (CMSC) MRI working group revised their guidelines to address this concern [38]. The new guidelines recommend judicious use of gadolinium, recognizing that gadolinium is essential when monitoring a patient with highly active disease, especially in the first few years, when there is an unexpected decline in the patient's clinical status, upon first clinical presentation (i.e., at CIS), and when there is question of an alternative diagnosis.

## Recommended MRI Protocol and Clinical Guidelines in Diagnosing and Monitoring MS

The reader is referred to published CMSC recommendations for standardized MRI protocols and clinical guidelines, most recently revised in 2018 [38]. Scans should be of good quality, with adequate signal-noise ratio (SNR) and spatial resolution (in slice pixel resolution of $\leq$1 mm × 1 mm), cover the whole brain, and have $\leq$3 mm slice thickness without gaps for 2D acquisition or 3D reconstruction. 3D acquisitions ($\leq$1 mm × 1 mm × 1 mm isotropic voxel size) are generally recommended, but options for 2D acquisitions are also provided. Recommended core sequences include 2D/3D sagittal and axial FLAIR, 2D/3D axial T2, axial 2D DWI, 3D gradient echo T1, and post gadolinium 2D/3D axial T1 as required. The CMSC recommends a baseline scan and at 6 months after initiation of a DMT. Thereafter, a periodic brain MRI, typically annually, should be performed to assess subclinical disease activity. Interim imaging is indicated when there is unexpected decline or suspicion for a relapse or PML. Cervical spinal cord imaging is recommended at the time of diagnosis and if new symptoms develop that are referable to the spinal cord. Spinal cord MRI may also be useful to increase specificity while establishing the diagnosis in atypical presentations.

## Volumetric MRI

Neurodegeneration is a fundamental component of MS pathology and probably results from multiple mechanisms, including axonal transection within WM lesions [42] causing downstream degeneration, as well as glutamate excitotoxicity, iron accumulation, mitochondrial dysfunction, and microglial activation. These mechanisms may lead to a final common pathway of oxidative stress that eventually overwhelms cellular compensatory mechanisms and results in neuronal cell death [29]. Whole brain volume decline on MRI likely represents the net accumulation of tissue damage in MS, including neuroaxonal, myelin, and glial cell loss, and reduced synaptic density. Neuroaxonal loss is thought to be the major pathologic substrate of irreversible clinical disability in MS [43]; as such, understanding and preventing neuroaxonal loss has become a major focus in the field.

Whole brain volume has been studied extensively in MS [44, 45]. Most studies suggest that the rate of whole brain atrophy in MS averages

−0.7% per year, which is about 3x the rate of healthy controls [46]. In MS, whole brain volume loss correlates with several clinical endpoints, including ambulation [47], cognition [48], and quality of life [49]. Several available DMTs can slow the rate of whole brain atrophy significantly [44], and it has now become standard to include whole brain volume decline as a secondary or tertiary outcome in clinical trials testing primarily anti-inflammatory agents, and/or to use whole brain volume as the primary endpoint in phase 2 trials testing primarily neuroprotective agents [50].

Despite its many practical advantages, there are limitations of whole brain volume measurements. Whole brain volume may lack sensitivity in early phases of MS, such as CIS and RIS, though data in these phases are mixed [45]. Whole brain volume measurements, particularly at a single time point, can be confounded by several other factors that reduce the specificity for neurodegeneration, such as diurnal fluctuations [51], patient hydration status [52], or corticosteroid administration and/or newly initiated DMT (so-called pseudoatrophy [53]). Because of these limitations, interest has emerged in regional brain atrophy metrics, which may be more sensitive in earlier phases of MS and should be less confounded by tissue fluid dynamics. Thalamic volume loss, for example, is an early occurrence in MS and has been documented in RIS, CIS, and pediatric MS [54–57]. Thalamic volume declines persistently throughout the MS disease duration [58], correlates with clinical endpoints including cognition [58–60], and appears to provide feasible sample sizes as a primary MRI endpoint [58]. For these reasons, regional gray matter metrics, such as thalamic and other deep gray matter volumes, cortical thickness, and spinal cord volumes, are an active area of ongoing research in the field.

Despite a high degree of interest in volumetrics and their face validity as measures of neurodegeneration, several barriers exist in incorporating brain volume measurements into MS clinical practice, a topic which has been recently reviewed [61, 62]. The heterogeneous images that are collected in clinical practice present major challenges to current image processing software. No "gold standard" software to measure brain volumes has been identified. Most published work has assessed brain volume loss at the group level; the optimal statistical methods to translate brain volumes into a clinically meaningful metric at the individual level remain to be defined. Efforts have been made to define pathologic "cutoff" values for whole brain volume decline [63], but further work is needed to refine these values, including an adjustment for age. Because of the aforementioned fluctuations, brain volume decline over multiple time points may be more useful than single time point measurements, but this is difficult to implement in a clinical setting. Robust data from a large reference population that include both MS patients and healthy controls will be needed to develop an individual level metric. Statistical methods to adjust for variation from scan parameters, tissue fluid status, etc., could be developed, but these will need validation before implementing in the clinic [62].

## Positron Emission Tomography (PET)

Although MRI is an invaluable tool in MS clinical practice and clinical research, it has limitations. MRI may not detect "other" types of inflammation besides WM lesions that are present in MS, such as microglial activation, and MRI does not reliably detect diffuse pathology in the NAWM and/or cortical GM. In addition, because MRI is pathologically nonspecific, its ability to measure remyelination is limited. Positron emission tomography (PET) can investigate aspects of MS not visualized by MRI and can provide a higher degree of specificity depending on the ligand used. PET measures radiation (positrons) emitted by specific radioisotopes tagged to a specific ligand, which binds to a target of interest. The signal measured by PET can provide in vivo quantitative information about the concentration of these target molecules. The use of PET in MS has been reviewed elsewhere [64, 65]. Herein, we focus on current PET approaches to measure microglial activation and remyelination.

## Activated Microglia

Microglia are the key component of the innate CNS immune system and are activated in the setting of chronic inflammation [66]. Although activated microglia are an area of high interest in MS, their role in disease pathophysiology is not entirely clear. Histopathologically, activated microglia are seen at the edge of chronically active and expanding (aka "smoldering") MS lesions, but not at chronic inactive lesions [67]. Both blood-derived monocytes and resident CNS microglia may contribute to neuronal damage by releasing pro-inflammatory cytokines and reactive oxygen species, leading to oxidative stress [68], which is thought to be an important mechanism in the pathophysiology of progressive MS [29]. Chronic, "smoldering" inflammation occurs behind an intact blood-brain barrier and is MRI-invisible, but it can be detected by PET strategies targeting activated microglia. This could enhance our fundamental understanding of MS and ultimately lead to the development of new therapeutic targets for progressive MS, which are urgently needed in the field.

Translocator protein (TSPO) is an 18 KDa protein that is expressed on the outer mitochondrial membrane of activated microglia and is the most commonly studied radiotracer target to visualize microglial activation in MS. Formerly known as the peripheral benzodiazepine receptor, TSPO is expressed predominantly on activated CNS resident microglia, but it is also found on blood-derived macrophages, reactive astrocytes, and vascular endothelial and smooth muscle cells [66]. TSPO is also expressed in the normal human brain, mainly neurons [69]. The first-generation radioligand, [$^{11}$C]PK11195, has high specificity for TSPO, but binds to multiple cell types that express TSPO including reactive astrocytes, endothelial cells, and plasma proteins, in addition to activated microglia. [$^{11}$C]PK11195 has a short half-life of about 20 minutes and a low signal-to-noise ratio, and signal quantification can be difficult. Typically, [$^{11}$C]PK11195 binding quantification uses a normal reference region, but because such a region does not exist in MS, relatively complex mathematical modeling is required. The second-generation TSPO radioligands, such as [$^{11}$C]PBR28 and [$^{18}$F]PBR111, have higher binding affinity and better signal-to-noise ratio, but their binding affinity is affected by TSPO gene polymorphisms, making genetic testing mandatory for proper interpretation of the PET signal [70]. Like [$^{11}$C]PK11195, the second-generation TSPO ligands also bind to activated astrocytes and endothelial cells [71, 72].

TSPO uptake is increased in MS plaques during relapse [73] (Fig. 3.3) and in some but not all chronic lesions [70, 75], potentially consistent with the "smoldering" inflammation described pathologically in chronic active lesions. Diffuse TSPO binding in NAWM has been shown in RRMS patients compared to healthy controls [76], and may be more pronounced in SPMS compared to RRMS [77]. TSPO uptake in the cortex, cortical lesions, deep GM, and NAWM in MS patients is also associated with worse clinical disability, cognitive function, and more cortical thinning on MRI [77].

It has been thought that TSPO expression on the surface of microglia is upregulated upon exposure to pro-inflammatory stimuli. However, TSPO was recently shown to be downregulated on macrophages exposed to pro-inflammatory stimuli and unchanged upon exposure to anti-inflammatory stimuli [78]. In another study, TSPO was upregulated upon exposure to pro-inflammatory stimuli in rodents, but did not change in human microglia exposed to the same stimulus [79]. The implications of these findings are not clear in humans with MS, but further study is needed to understand TSPO localization in MS brain tissue in situ [66] and the correct interpretation of TSPO binding in vivo in MS. It is possible that TSPO expression measured by PET may reflect microglial/macrophage density rather than activation status [79].

## Myelination

Imaging myelin content is perhaps one of the most promising applications of PET imaging in MS. Developing remyelinating agents, which are

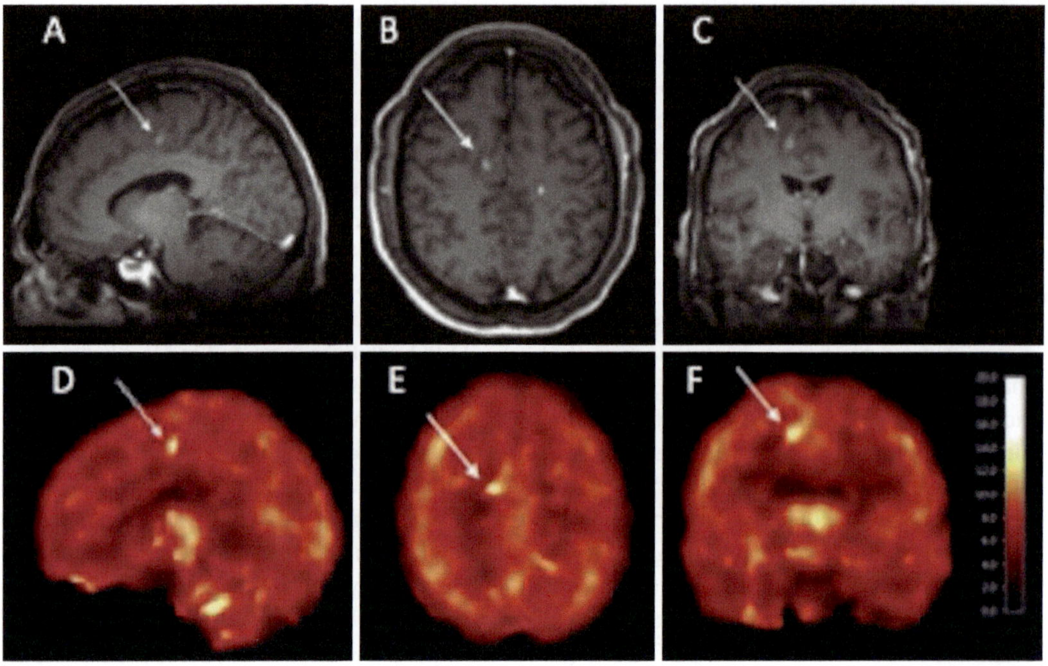


**Fig. 3.3** MRI and $^{11}$C-PBR28 PET images from a patient with relapsing multiple sclerosis. Panels **a–c** show post-contrast T1-weighted MR images demonstrating an acute lesion that enhances following gadolinium administration (arrows) in sagittal (**a**), axial (**b**), and coronal (**c**) plane. Panels **d–f** show the corresponding $V_T$ (volume of distribution) parametric map demonstrating a focal increase (arrows) in uptake of $^{11}$C-PBR28, a second-generation TSPO ligand, perhaps suggesting the presence of activated microglia within these acute lesions. Source images in this figure were previously published and are used with permission [74]

currently lacking in the field, has become a major focus in MS clinical research. Remyelination should not only improve functional recovery after demyelinating injury in the short term, but should also protect axons in the long term, as chronically denuded axons are more susceptible to inflammatory insults that eventually lead to neurodegeneration [29]. One major barrier to developing remyelinating agents has been evaluating their efficacy in vivo. Compared to MR-based methods that have been used for this purpose (e.g., magnetization transfer ratio, diffusion tensor imaging, myelin water imaging), PET offers a more direct and specific measure of myelin and is thus a promising approach.

Stilbene derivative radiotracers bind to intact myelin sheath proteins such as PLP, MBP, or the sites of interactions between myelin lipids and these proteins [80]. Amyloid tracers such as Pittsburgh Compound B (PiB) also bind to myelin and may be useful. PiB has been shown to be highly sensitive to detect myelin loss in EAE [81] and has been used in humans longitudinally to create a global index of myelin content change in a voxel-wise, individual-level analysis [82]. However, PiB must be mixed in an on-site cyclotron prior to administration to the patient, which many centers do not have; newer fluorinated amyloid tracers such as florbetapir, flutemetamol, and florbetaben may improve availability because they are more stable and do not require a cyclotron.

In addition to its low availability, several limitations of PET exist. PET has a low spatial resolution (typically 2–3 mm at best), which can make changes in small lesions difficult to detect due to partial voluming. PET is an expensive technique that requires a high degree of on-site expertise, and data analysis/signal quantification can be challenging. Finally, radiation exposure is a concern with PET, particularly in longitudinal studies.

## Conclusion

MRI plays a central role in the diagnosis and clinical management of MS due to its high sensitivity, but approaches such as central vein imaging and cortical lesion detection are needed to increase specificity for CNS demyelination. MRI is a useful tool to monitor disease activity and assess efficacy of DMTs both in clinical trials and in clinical practice. Brain volume decline is clinically relevant, but more work is needed before incorporation of volumetrics into routine clinical care. Finally, PET may further our understanding of the disease biology by its ability to study specific aspects of the disease not visible with MRI.

## References

1. Polman CH, Reingold SC, Banwell B, Clanet M, Cohen JA, Filippi M, et al. Diagnostic criteria for multiple sclerosis: 2010 revisions to the McDonald criteria. Ann Neurol. 2011;69(2):292–302.
2. Thompson AJ, Banwell BL, Barkhof F, Carroll WM, Coetzee T, Comi G, et al. Diagnosis of multiple sclerosis: 2017 revisions of the McDonald criteria. Lancet Neurol. 2018;17(2):162–73.
3. McDonald WI, Compston A, Edan G, Goodkin D, Hartung HP, Lublin FD, et al. Recommended diagnostic criteria for multiple sclerosis: guidelines from the International Panel on the diagnosis of multiple sclerosis. Ann Neurol. 2001;50(1):121–7.
4. Fazekas F, Barkhof F, Filippi M, Grossman RI, Li DKB, McDonald WI, et al. The contribution of magnetic resonance imaging to the diagnosis of multiple sclerosis. Neurology. 1999;53(3):448–56.
5. Filippi M, Rocca MA, Barkhof F, Brück W, Chen JT, Comi G, et al. Association between pathological and MRI findings in multiple sclerosis. Lancet Neurol. 2012;11(4):349–60.
6. Filippi M. Enhanced magnetic resonance imaging in multiple sclerosis. Mult Scler. 2000;6(5):320–6.
7. Cotton F, Weiner HL, Jolesz FA, Guttmann CRG. MRI contrast uptake in new lesions in relapsing-remitting MS followed at weekly intervals. Neurology. 2003;60(4):640–6.
8. Lassmann H. A dynamic view of the blood-brain barrier in active multiple sclerosis lesions. Ann Neurol. 2011;70(1):1–2.
9. Gaitán MI, Shea CD, Evangelou IE, Stone RD, Fenton KM, Bielekova B, et al. Evolution of the blood-brain barrier in newly forming multiple sclerosis lesions. Ann Neurol. 2011;70(1):22–9.
10. Van Der Vuurst De Vries RM, Mescheriakova JY, YYM W, Runia TF, Jafari N, Samijn JP, et al. Application of the 2017 revised McDonald criteria for multiple sclerosis to patients with a typical clinically isolated syndrome. JAMA Neurol. 2018;75(11):1392–8.
11. Geurts JJG, Roosendaal SD, Calabrese M, Ciccarelli O, Agosta F, Chard DT, et al. Consensus recommendations for MS cortical lesion scoring using double inversion recovery MRI. Neurology. 2011;76(5):418–24.
12. Seewann A, Kooi EJ, Roosendaal SD, Pouwels PJW, Wattjes MP, Van Der Valk P, et al. Postmortem verification of MS cortical lesion detection with 3D DIR. Neurology. 2012;78(5):302–8.
13. Sethi V, Muhlert N, Ron M, Golay X, Wheeler-Kingshott CA, Miller DH, et al. MS cortical lesions on DIR: not quite what they seem? PLoS One. 2013;8(11):8–12.
14. Solomon AJ, Bourdette DN, Cross AH, Applebee A, Skidd PM, Howard DB, et al. The contemporary spectrum of multiple sclerosis misdiagnosis. Neurology. 2016;87(13):1393–9.
15. Solomon AJ, Corboy JR. The tension between early diagnosis and misdiagnosis of multiple sclerosis. Nat Rev Neurol. 2017;13(9):567–72.
16. Okuda DT, Mowry EM, Beheshtian A, Waubant E, Baranzini SE, Goodin DS, et al. Incidental MRI anomalies suggestive of multiple sclerosis: the radiologically isolated syndrome. Neurology. 2009;72(9):800–5.
17. Okuda DT, Siva A, Kantarci O, Inglese M, Katz I, Tutuncu M, et al. Radiologically isolated syndrome: 5-year risk for an initial clinical event. PLoS One. 2014;9(3):e90509.
18. Kantarci OH, Lebrun C, Siva A, Keegan MB, Azevedo CJ, Inglese M, et al. Primary progressive multiple sclerosis evolving from radiologically isolated syndrome. Ann Neurol. 2016;79(2):288–94.
19. Sati P, Oh J, Todd Constable R, Evangelou N, Guttmann CRG, Henry RG, et al. The central vein sign and its clinical evaluation for the diagnosis of multiple sclerosis: a consensus statement from the north American Imaging in Multiple Sclerosis Cooperative. Nat Rev Neurol. 2016;12(12):714–22.
20. Dawson JW. The histology of disseminated sclerosis. Trans R Soc Edinburgh. 1916;50(03):517–740.
21. Hammond KE, Metcalf M, Carvajal L, Okuda DT, Srinivasan R, Vigneron D, et al. Quantitative in vivo magnetic resonance imaging of multiple sclerosis at 7 Tesla with sensitivity to iron. Ann Neurol 2008;64:707–13.
22. Metcalf M, Xu D, Okuda DT, Carvajal L, Srinivasan R, Kelley DAC, et al. High-resolution phased-array MRI of the human brain at 7 Tesla: Initial experience in multiple sclerosis patients. J Neuroimaging 2010;20:141–47.
23. Sati P, George IC, Shea CD, Gaitán MI, Reich DS. FLAIR*: a combined MR contrast technique for visualizing white matter lesions and parenchymal veins. Radiology. 2012;265(3):926–32.

24. Tallantyre EC, Brookes MJ, Dixon JE, Morgan PS, Evangelou N, Morris PG. Demonstrating the perivascular distribution of MS lesions in vivo with 7-tesla MRI. Neurology. 2008;70(22):2076–9.
25. Mistry N, Dixon J, Tallantyre E, Tench C, Abdel-Fahim R, Jaspan T, et al. Central veins in brain lesions visualized with high-field magnetic resonance imaging: a pathologically specific diagnostic biomarker for inflammatory demyelination in the brain. JAMA Neurol. 2013;70(5):623–8.
26. Sparacia G, Agnello F, Gambino A, Sciortino M, Midiri M. Multiple sclerosis: high prevalence of the 'central vein' sign in white matter lesions on susceptibility-weighted images. Neuroradiol J. 2018;31(4):356–61.
27. Lumsden C. The neuropathology of multiple sclerosis. In: Vinken PJ, Bruyn G, editors. Handbook of clinical neurology. Amsterdam: Elsevier; 1970. p. 217–309.
28. Dutta R, Trapp BD. Pathogenesis of axonal and neuronal damage in multiple sclerosis. Neurology. 2007;68(22 Supp):S22–31.
29. Mahad DH, Trapp BD, Lassmann H. Pathological mechanisms in progressive multiple sclerosis. Lancet Neurol. 2015;14(2):183–93.
30. Pitt D, Boster A, Pei W, Wohleb E, Jasne A, Zachariah CR, et al. Imaging cortical lesions in multiple sclerosis with ultra–high-field magnetic resonance imaging. Arch Neurol. 2010;67(7):812–8.
31. Kilsdonk ID, Jonkman LE, Klaver R, Van Veluw SJ, Zwanenburg JJM, Kuijer JPA, et al. Increased cortical grey matter lesion detection in multiple sclerosis with 7 T MRI: a post-mortem verification study. Brain. 2016;139(5):1472–81.
32. Sormani MP, Bruzzi P. MRI lesions as a surrogate for relapses in multiple sclerosis: a meta-analysis of randomised trials. Lancet Neurol. 2013;12(7):669–76.
33. Fisniku LK, Brex PA, Altmann DR, Miszkiel KA, Benton CE, Lanyon R, et al. Disability and T2 MRI lesions: a 20-year follow-up of patients with relapse onset of multiple sclerosis. Brain. 2008;131(3):808–17.
34. Rudick RA, Lee JC, Simon J, Fisher E. Significance of T2 lesions in multiple sclerosis: a 13-year longitudinal study. Ann Neurol. 2006;60(2):236–42.
35. Sormani MP, Bonzano L, Roccatagliata L, Cutter GR, Mancardi GL, Bruzzi P. Magnetic resonance imaging as a potential surrogate for relapses in multiple sclerosis: a meta-analytic approach. Ann Neurol. 2009;65(3):268–75.
36. Sormani MP, Stubinski B, Cornelisse P, Rocak S, Li D, De Stefano N. Magnetic resonance active lesions as individual-level surrogate for relapses in multiple sclerosis. Mult Scler J 2011;17(5):541–549.
37. Traboulsee A, Létourneau-Guillon L, Freedman MS, O'Connor PW, Bharatha A, Chakraborty S, et al. Canadian expert panel recommendations for MRI use in MS diagnosis and monitoring. Can J Neurol Sci. 2015;42(3):159–67.
38. Consortium of MS centers MRI protocol and clinical guidelines for the diagnosis and follow-up of MS: 2018 Revised Guidelines [Internet]. Available from: www.mscare.org/mri
39. Moraal B, Wattjes MP, Geurts JJG, Knol DL, van Schijndel RA, Pouwels PJW, et al. Improved detection of active multiple sclerosis lesions: 3D subtraction imaging. Radiology. 2010 Apr;255(1):154–63.
40. Moraal B, Van Den Elskamp IJ, Knol DL, Uitdehaag BMJ, Geurts JJG, Vrenken H, et al. Long-interval T2-weighted subtraction magnetic resonance imaging: a powerful new outcome measure in multiple sclerosis trials. Ann Neurol. 2010;67(5):667–75.
41. McDonald RJ, McDonald JS, Kallmes DF, Jentoft ME, Murray DL, Thielen KR, et al. Intracranial gadolinium deposition after contrast-enhanced MR imaging. Radiology. 2015;275:150025.
42. Trapp BD, Peterson J, Ransohoff RM, Rudick R, Mörk S, Bö L. Axonal transection in the lesions of multiple sclerosis. N Engl J Med. 1998;338(5):278–85.
43. Bjartmar C, Wujek JR, Trapp BD. Axonal loss in the pathology of MS: consequences for understanding the progressive phase of the disease. J Neurol Sci. 2003;206(2):165–71.
44. De Stefano N, Airas L, Grigoriadis N, Mattle HP, O'Riordan J, Oreja-Guevara C, et al. Clinical relevance of brain volume measures in multiple sclerosis. CNS Drugs. 2014;28(2):147–56.
45. Bermel RA, Bakshi R. The measurement and clinical relevance of brain atrophy in multiple sclerosis. Lancet Neurol. 2006;5(2):158–70.
46. Vollmer T, Signorovitch J, Huynh L, Galebach P, Kelley C, DiBernardo A, et al. The natural history of brain volume loss among patients with multiple sclerosis: a systematic literature review and meta-analysis. J Neurol Sci. 2015;357(1–2):8–18.
47. Fisher E, Rudick RA, Simon JH, Cutter G, Baier M, Lee JC, et al. Eight-year follow-up study of brain atrophy in patients with MS. Neurology. 2002;59(9):1412–20.
48. Rocca MA, Amato MP, De Stefano N, Enzinger C, Geurts JJ, Penner IK, et al. Clinical and imaging assessment of cognitive dysfunction in multiple sclerosis. Lancet Neurol. 2015;14(3):302–417.
49. Mowry EM, Beheshtian A, Waubant E, Goodin DS, Cree BA, Qualley P, et al. Quality of life in multiple sclerosis is associated with lesion burden and brain volume measures. Neurology. 2009;72(20):1760–5.
50. Fox RJ, Coffey CS, Conwit R, Cudkowicz ME, Gleason T, Goodman A, et al. Phase 2 trial of Ibudilast in progressive multiple sclerosis. N Engl J Med. 2018;379(9):846–55.
51. Nakamura K, Brown RA, Narayanan S, Collins DL, Arnold DL. Diurnal fluctuations in brain volume: statistical analyses of MRI from large populations. NeuroImage. 2015;118:126–32.
52. Nakamura K, Brown RA, Araujo D, Narayanan S, Arnold DL. Correlation between brain volume change and T2 relaxation time induced by dehydration and

rehydration: implications for monitoring atrophy in clinical studies. NeuroImage Clin. 2014;6:166–70.
53. Miller DH, Soon D, Fernando KT, MacManus DG, Barker GJ, Yousry TA, et al. MRI outcomes in a placebo-controlled trial of natalizumab in relapsing MS. Neurology. 2007;68(17):1390–401.
54. Azevedo CJ, Overton E, Khadka S, Buckley J, Liu S, Sampat M, et al. Early CNS neurodegeneration in radiologically isolated syndrome. Neurol Neuroimmunol NeuroInflamm. 2015;2(3):e102.
55. Henry RG, Shieh M, Okuda DT, Evangelista A, Gorno-Tempini ML, Pelletier D. Regional grey matter atrophy in clinically isolated syndromes at presentation. J Neurol Neurosurg Psychiatry. 2008;79(11):1236–44.
56. Aubert-Broche B, Fonov V, Ghassemi R, Narayanan S, Arnold DL, Banwell B, et al. Regional brain atrophy in children with multiple sclerosis. NeuroImage. 2011;58(2):409–15.
57. Henry RG, Shieh M, Amirbekian B, Chung SW, Okuda DT, Pelletier D. Connecting white matter injury and thalamic atrophy in clinically isolated syndromes. J Neurol Sci. 2009;282(1–2):61–6.
58. Azevedo CJ, Cen SY, Khadka S, Liu S, Kornak J, Shi Y, et al. Thalamic atrophy in multiple sclerosis: a magnetic resonance imaging marker of neurodegeneration throughout disease. Ann Neurol. 2018;83(2):223–34.
59. Zivadinov R, Bergsland N, Dolezal O, Hussein S, Seidl Z, Dwyer MG, et al. Evolution of cortical and thalamus atrophy and disability progression in early relapsing-remitting MS during 5 years. Am J Neuroradiol. 2013;34(10):1931–9.
60. Houtchens MK, Benedict RHB, Killiany R, Sharma J, Jaisani Z, Singh B, et al. Thalamic atrophy and cognition in multiple sclerosis. Neurology. 2007;69(12):1213–23.
61. Rocca MA, Battaglini M, Benedict RHB, De Stefano N, Geurts JJG, Henry RG, et al. Brain MRI atrophy quantification in MS. Neurology. 2017;88(4):403–13.
62. Azevedo CJ, Pelletier D. Whole-brain atrophy: ready for implementation into clinical decision-making in multiple sclerosis? Curr Opin Neurol. 2016;29(3):237–42.
63. De Stefano N, Stromillo ML, Giorgio A, Bartolozzi ML, Battaglini M, Baldini M, et al. Establishing pathological cut-offs of brain atrophy rates in multiple sclerosis. J Neurol Neurosurg Psychiatry. 2016;87(1):93–9.
64. Moccia M, Ciccarelli O. Molecular and metabolic imaging in multiple sclerosis. Neuroimaging Clin N Am. 2017;27(2):343–56.
65. De Paula Faria D, Copray S, Buchpiguel C, Dierckx R, De Vries E. PET imaging in multiple sclerosis. J Neuroimmune Pharmacol. 2014;9(4):468–82.
66. Cosenza-Nashat M, Zhao ML, Suh HS, Morgan J, Natividad R, Morgello S, et al. Expression of the translocator protein of 18 kDa by microglia, macrophages and astrocytes based on immunohistochemical localization in abnormal human brain. Neuropathol Appl Neurobiol. 2009;35(3):306–28.
67. Frischer JM, Bramow S, Dal-Bianco A, Lucchinetti CF, Rauschka H, Schmidbauer M, et al. The relation between inflammation and neurodegeneration in multiple sclerosis brains. Brain. 2009;132(5):1175–89.
68. Fischer MT, Sharma R, Lim JL, Haider L, Frischer JM, Drexhage J, et al. NADPH oxidase expression in active multiple sclerosis lesions in relation to oxidative tissue damage and mitochondrial injury. Brain. 2012;135(3):886–99.
69. Doble A, Malgouris C, Daniel M, Daniel N, Imbault F, Basbaum A, et al. Labelling of peripheral-type benzodiazepine binding sites in human brain with [3H] PK 11195: anatomical and subcellular distribution. Brain Res Bull. 1987;18(1):49–61.
70. Airas L, Nylund M, Rissanen E. Evaluation of microglial activation in multiple sclerosis patients using positron emission tomography. Front Neurol. 2018;9:181.
71. Dickens AM, Vainio S, Marjamaki P, Johansson J, Lehtiniemi P, Rokka J, et al. Detection of microglial activation in an acute model of neuroinflammation using PET and radiotracers 11C-(R)-PK11195 and 18F-GE-180. J Nucl Med. 2014;55(3):466–72.
72. Rizzo G, Veronese M, Tonietto M, Zanotti-Fregonara P, Turkheimer FE, Bertoldo A. Kinetic modeling without accounting for the vascular component impairs the quantification of [11C]PBR28 brain PET data. J Cereb Blood Flow Metab. 2014;34(6):1060–9.
73. Debruyne JC, Versijpt J, Van Laere KJ, De Vos F, Keppens J, Strijckmans K, et al. PET visualization of microglia in multiple sclerosis patients using [11C] PK11195. Eur J Neurol. 2003;10(3):257–64.
74. Park E, Gallezot JD, Delgadillo A, Liu S, Planeta B, Lin SF, et al. 11C–PBR28 imaging in multiple sclerosis patients and healthy conrols: test-retest reproducibility and focual visualization of active white matter areas. Eur J Nucl Med Mol Imaging 2015;42:1081–92.
75. Rissanen E, Tuisku J, Rokka J, Paavilainen T, Parkkola R, Rinne JO, et al. In vivo detection of diffuse inflammation in secondary progressive multiple sclerosis using PET imaging and the radioligand 11C-PK11195. J Nucl Med. 2014;55(6):939–44.
76. Sucksdorff M, Rissanen E, Tuisku J, Nuutinen S, Paavilainen T, Rokka J, et al. Evaluation of the effect of fingolimod treatment on microglial activation using serial PET imaging in multiple sclerosis. J Nucl Med. 2017;58(10):1646–51.
77. Herranz E, Giannì C, Louapre C, Treaba CA, Govindarajan ST, Ouellette R, et al. Neuroinflammatory component of gray matter pathology in multiple sclerosis. Ann Neurol. 2016;80(5):776–90.
78. Narayan N, Mandhair H, Smyth E, Dakin SG, Kiriakidis S, Wells L, et al. The macrophage marker translocator protein (TSPO) is down-regulated on pro-inflammatory 'M1' human macrophages. PLoS One. 2017;12(10):1–19.
79. Owen DR, Narayan N, Wells L, Healy L, Smyth E, Rabiner EA, et al. Pro-inflammatory activation of primary microglia and macrophages increases

18 kDa translocator protein expression in rodents but not humans. J Cereb Blood Flow Metab. 2017;37(8):2679–90.
80. Wu C, Tian D, Feng Y, Polak P, Wei J, Sharp A, et al. A novel fluorescent probe that is brain permeable and selectively binds to myelin. J Histochem Cytochem. 2011;54(9):997–1004.
81. De Paula Faria D, Copray S, Sijbesma JW, Willemsen AT, Buchpiguel CA, Dierckx RA, et al. PET imaging of focal demyelination and remyelination in a rat model of multiple sclerosis: comparison of [11C] MeDAS, [11C]CIC and [11C]PIB. Eur J Nucl Med Mol Imaging. 2014;41(5):995–1003.
82. Bodini B, Veronese M, García-Lorenzo D, Battaglini M, Poirion E, Chardain A, et al. Dynamic imaging of individual remyelination profiles in multiple sclerosis. Ann Neurol. 2016;79(5):726–38.

# 4 Clinical Trial Design in Neuroimmunology

Enrique Alvarez, David Prater III, and Daniel Ontaneda

**Key Points**

1. Clinical trial design in neuroimmunology has evolved as we learn about these diseases and has led to the approval of treatments in multiple sclerosis (MS) and neuromyelitis optica spectrum disease (NMOSD).
2. Randomized controlled trials remain the gold standard trial design but newer studies are increasingly being used.
3. The process of developing and conducting a clinical trial starts with a well-thought-out question that can then be tested and analyzed following an approved protocol.
4. Treatments must follow a series of phases in drug development before they are approved by regulatory agencies.

## Introduction

Clinical trial design in neuroimmunology has evolved over time, which reflects our increased knowledge of these diseases and they have provided great insights into these conditions. Clinical studies have resulted in the approval of disease-modifying therapies (DMTs) for multiple sclerosis (MS) and neuromyelitis optica spectrum disease (NMOSD) along with symptomatic therapies such as dalfampridine. Treatment of relapsing-remitting MS (RRMS) has seen dramatic changes with therapies that can stabilize the vast majority of patients with more than 20 drugs available including generics. Progressive MS has proven to be a more challenging frontier, although treatments are beginning to become available. As of 2019, the first three clinical trials were successfully conducted in NMOSD, paving the way for new therapies in this rare disease. Additionally, clinical trials are also helping us understand how to use these DMTs. This chapter describes how treatments are approved, frameworks for evaluating clinical trials, and the evolution of clinical trial design including newer trial design concepts to help the reader better interpret these studies.

E. Alvarez (✉) · D. Prater III
Department of Neurology, Rocky Mountain Multiple Sclerosis Center at the University of Colorado, Aurora, CO, USA
e-mail: Enrique.alvarez@cuanschutz.edu

D. Ontaneda
Department of Neurology, Mellen Center for Multiple Sclerosis, Cleveland Clinic, Cleveland, OH, USA



A. L. Piquet, E. Alvarez (eds.), *Neuroimmunology*, https://doi.org/10.1007/978-3-030-61883-4_4

## Types of Clinical Studies

Clinical trials are important as they translate our knowledge of basic scientific research into treatments that help prevent, diagnose, or treat a disease. Clinical trials are considered the "gold standard" in clinical research. A clinical trial is a type of research study that is defined by the National Institutes of Health (NIH) as a study where "one or more human subjects are prospectively assigned to one or more interventions to evaluate the effects of those interventions on health-related biomedical or behavioral outcomes (https://grants.nih.gov/policy/clinical-trials/definition.htm)." Their design aims to reduce bias and variability in our understanding of therapies and provide information on the magnitude of the efficacy and their safety in humans. They often can help our understanding of the pathology of the underlying disease by answering other research questions.

Case reports and observational studies are complementary to clinical trials. Case reports, for instance, can provide examples of drugs that could be explored in larger clinical trials such as a description of rituximab in a patient with MS who was repeatedly failing interferons and mitoxantrone [1] or tocilizumab for patients with NMOSD who were failing rituximab [2]. Observational studies can provide information where a clinical trial is not appropriate such as due to a lack of clinical equipoise, unfeasible due to lack of cost to compare multiple drugs, or difficulty in studying certain populations such as patients with rare autoantibodies. Observational studies are also developing rapidly with improved statistical techniques such as the use of propensity-based matching or weighing methods.

## Clinical Trial Overview

Designing a clinical trial requires significant planning in order to ensure the quality of data obtained and its generalizability while reducing biases and maximizing the amount of data obtained. The planning stages can help reduce amendments later and increase the speed at which a question can be answered by remaining focused yet allow for the flexibility to adapt to potentially changing conditions. The general overview of a study is outlined in Fig. 4.1.

## Defining the Research Question

A clinical trial needs to start with a good question, which can be evaluated with the FINER criteria (Table 4.1) [3]. Once a question is identified, a study plan can be developed around it. Defining the question does require one to start developing the trial design, but a good question makes this process much easier.

Feasibility includes many aspects and has posed challenges in the design of clinical trials evaluating antibody-mediated autoimmune neurological conditions where few patients exist. Additionally, feasibility for these conditions can be limiting because of the increasing costs

**Table 4.1** FINER criteria for evaluating a research question

| |
|---|
| **F**easible (answerable due to recruitment, expertise, tools, and costs) |
| **I**nteresting |
| **N**ovel |
| **E**thical |
| **R**elevant |

**Fig. 4.1** Clinical trial overview. A finalized protocol often requires several iterative cycles of developing the research question and designing the trial to help develop the best protocol. IRB Institutional Review Board

needed to bring in more sites to study enough patients with diseases that have low prevalence rates. The expertise and resources can also limit what questions are feasible by identifying sites that can conduct a study. It is important to have the tools needed to study certain conditions. The Schumacher criteria in 1965 provided the diagnostic criteria that provided the backbone to identify patients with multiple sclerosis that could be included in clinical trials [4]. Having laboratory and clinical tools led to the refinement of these criteria. The Poser criteria in 1965 included paraclinical information such as oligoclonal banding in cerebrospinal fluid and visual evoked potentials [5]. These criteria were subsequently replaced in 2001 by the McDonald criteria with the advent of magnetic resonance imaging (MRI) [6].

In addition to Feasibility, the FINER criteria also help formulate a research question that is Interesting, Novel, Ethical, and Relevant. The effort and resources needed to conduct a clinical trial can be tremendous, and it can be difficult to drive a study forward unless the study is interesting or novel. Ethical considerations in conducting a clinical trial will be expanded upon below, but the question has to meet the *principle of equipoise* in which there is a genuine uncertainty about which treatment is most beneficial and provides the ethical basis for assigning patients to the different treatment arms in a clinical trial. This term was first used by Benjamin Freedman in 1987 [7]. Finally, the question has to be relevant to the field that is going to use the information learned from the study.

## Developing the Trial Design

Once an appropriate question is identified, a trial can be designed to help answer it. The plan for the trial can be designed and refined using the PICO or PICOT criteria (Table 4.2) [8]. These two steps often form an iterative cycle that together can result in identifying the best study plan. The steps for developing the trial design are described below.

**Table 4.2** PICO(T) criteria for developing research criteria

| |
|---|
| Population (patients) – Disease and its characteristics along with demographics of subjects |
| Intervention – Dose, duration, location, route of administration, study design, etc. |
| Comparison – Placebo or other treatment if any |
| Outcome – Measure of benefit/risk, at what time/frequency, analyzed as dichotomous/continuous |
| Time – Follow-up time |

## Selecting the Patient Population

Defining the population to study can greatly influence clinical trials in patients with MS and NMOSD, and it is generally defined by a set of inclusion and exclusion criteria. Despite these criteria, there are inevitable differences among studies, some planned while others are unplanned. For example, in RRMS trials, it has been noted that the patients entering studies tend to have more benign disease than they did in earlier studies and have less disability [9]. This possibly reflects a reluctance to place patients with very active disease into a trial whereby a patient could receive placebo, or a drug deemed to have a low therapeutic effect as more drugs become available. This changing demographic makes comparing newer trials with older trials challenging. Additionally, it can make recruitment difficult, and therefore some studies rely heavily on recruiting in countries where there are fewer therapeutic options. However, most regulatory agencies require that studies be done in patients from their region or country, which can sometimes delay entry into those markets but also increases the generalizability of that study to that country.

In progressive MS trials, the age of the population and the percent of patients with contrast enhancing lesions (CELs) can greatly affect a study. This was seen in the European and North American interferonβ(beta)-1b studies in patients with secondary progressive MS where a treatment effect on delaying disability progression was seen compared to placebo in the European study but not in the North American study. It is worth pointing out that the North American study had a mean age of 46.8 years at study entry compared to 41.0 years in the European study, while the mean number of CEL was 1.5 and 2.6, respectively [10]. Since younger patients tend to have more active disease, this comparison suggested that interferonβ(beta)-1b had its effects by treating this active component of MS. Similarly, the OLYMPUS (A Study to Evaluate the Safety and Efficacy of Rituximab in Adults with Primary Progressive MS) study in primary progressive MS did not find a reduction in the 12-week confirmed disability progression (CDP) by rituximab versus placebo [11]. However, the subpopulations that were younger (<51 year old) or had CEL showed a reduction in CDP of 48% and 59%, respectively. This likely played a role in changing the age limit from 65 years in this study to 55 years for the ocrelizumab primary progressive MS trial (ORATORIO) to 55 years [12]. This resulted in a reduction in the mean age of the B-cell depleting therapy groups from 50.1 years to 44.7 years. ORATORIO showed a reduction of 24% in the 12-week CDP, which led to the approval of ocrelizumab as the first treatment for primary progressive MS.

Similarly in NMOSD, the decision of including patients who were aquaporin-4 (AQP4) antibody negative was variable and had dramatic differences in the NMOSD trials, including for eculizumab (PREVENT) [13], satralizumab (in 2 trials, SAkuraSky as an add-on therapy [14] and SAkuraStar as monotherapy [15]), and inebiluzimab (N-Momentum) [16]. PREVENT included only AQP4 antibody positive patients and saw a reduction in the risk of relapses of 94.2%. In SAkuraSky, the reduction was 79.2% in AQP4 antibody positive patients but only 33.7% if seronegative. Similarly in N-Momentum, the overall reduction in the study was 72.8%, but increased to 77.3% in AQP4 antibody positive patients. Other differences in study length and relapse ascertainment made comparison across these trials difficult, but this difference in inclusion criteria in NMOSD clearly affected the study results.

Another aspect of identifying the study population includes the decision to include special populations, which can sometimes pose challenges but is essential for conclusions that benefit all populations. Studies in MS tend to have poor representation of minorities and often exclude patients with significant medical comorbidities and other vulnerable populations such as children or pregnant or lactating women.

The importance of studying DMTs in these other populations has been highlighted by the recent studies conducted in pediatric populations. The PARADIGMS study of fingolimod versus interferon beta-1a in patients younger than 18 years old showed a reduction in the annualized relapse rate of 82% (0.12 vs. 0.67; $p < 0.001$) [17]. However, the phase 3 trial in adults (TRANSFORMS) that similarly studied fingolimod and interferon beta-1a showed a reduction in the annualized relapse rate of 52% (0.16 vs. 0.33; $p < 0.0001$) [18]. One of the likely differences in these studies is that younger patients as described above are more likely to experience more relapses allowing fingolimod to demonstrate a larger effect.

## Defining the Intervention

The intervention in the clinical trial is informed by prior studies. The phases of clinical trials will be described below; these gradually help to define the dose, frequency, duration, and route of administration of drugs, procedure, or interventions in a study. The pharmacokinetics and pharmacodynamics greatly influence these decisions. Doses are gradually narrowed down to one or two options in most phase 3 studies.

Other aspects of study design can also greatly impact a study. Designing the study well is quite understandably crucial and includes several factors:

- *Drug washouts.* Some immunosuppressive medications can greatly affect the immune system for long periods of time. In order to reduce the effects of prior medications, it is important to define periods of time to let prior medications wash out from a patient before entering a study. This also needs to be considered in studies that employ a crossover study design including going into extension studies. These washout periods can be reduced using activated charcoal or cholestyramine (i.e., for rapid elimination of teriflunomide).

- *Randomization*. Multiple methods can be used to create random groups to help reduce the possibility of bias in a clinical trial. The ease of use across multiple sites needs to be considered in larger studies. Examples include simple randomization (e.g., using sealed envelopes with an equal number of envelopes with control vs. treatment group), random allocation (e.g., using a random number generator), blocking designed to allow randomization into equal-sized groups ensuring balance across groups over time, and stratification to allow randomization within a categorical covariate such as gender or age. Although this can be done with the best of intentions, some studies can end up with unbalanced groups such as was seen in the HERMES (B-Cell Depletion with Rituximab in Relapsing MS) where the rituximab group had a mean number of CEL of 2.1 vs. 0.3 in the placebo arm [19]. Consider having a preplanned method of analysis for correcting these imbalances should they occur in a study.
- *Allocation* (1:1, 2:1, 3:1, etc.). For several reasons, including to help with recruitment into a trial by making it more likely for participants to receive treatment or to allow for stratification, studies may recruit more heavily into certain groups.
- *Blinding*. This can help reduce bias by having either the subject, the person administering the treatment, and/or the person evaluating the response blind to the treatment that the subject received. Single blinding is when only one party is blinded, usually the subject, whereas double blind is when both the subject and study staff are blinded. Triple binding extends blinding to the data analysis. In open-label studies, no blinding is used.
- *Duration of the study*. The TENERE study compared teriflunomide to interferon beta-1a and failed to show a difference between annualized relapse rate between teriflunomide 14 mg and interferon beta-1a [20]. The primary outcome of this study was time to failure, although teriflunomide is known to have a significant lag to therapeutic efficacy while interferons have shown a quick onset of action. A better comparison would have been likely achieved by examining the relapses over a fixed period of time or resetting a baseline after several months when both treatments have reached steady state.
- *Study design*. Selection of the primary study design is key, and most times, it relates to the type of therapeutic approach (Fig. 4.2). *Crossover designs* have been used to test symptomatic medications, especially when the

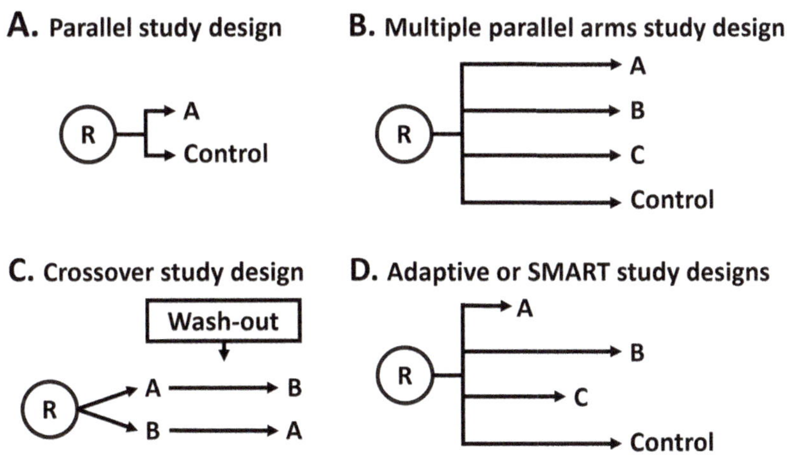


**Fig. 4.2** Randomized clinical trial designs. Parallel arm studies are the classic randomized clinical trials, which can compare one treatment (**a**) or multiple treatments (**b**) to either a control (shown here) or an active comparator. Crossover studies have some statistical advantages by studying multiple treatments in the same patient population and are shown here with a washout between treatments (**c**). Adaptive or SMART (Sequential Multiple Assignment Randomized Trials) are more complicated but can help reduce the exposure to less promising treatment by eliminating certain arms early during an interim analysis (**d**)

effect of the intervention is expected to wash-out over a short period of time. This allows studying treatments within one subject, which can reduce the variability observed and can increase the statistical power of the study. For DMTs in RRMS, double-blind randomized active comparator studies are preferred with *parallel treatment arms*. *Multiple parallel arms* have been used in progressive studies to help minimize the number of patients in placebo arms should each of these treatments required its own placebo control group. The MS-SMART study examined riluzole, fluoxetine, and amiloride in a 1:1:1:1 ratio vs. placebo, which helped effectively eliminate two placebo arms in studying these three medications [21]. An additional layer of complexity is added to these multiple arm studies in *adaptive or SMART* (Sequential Multiple Assignment Randomized Trials) clinical trial designs, which are gaining popularity. These studies are designed to adapt after initial in-trial observations by dropping lower performing treatments and reducing exposure to ineffective therapies. Additionally, *pragmatic trials* are designed to evaluate the effectiveness of interventions in real-life routine practice conditions by reducing the inclusion and exclusion criteria so that the results can be more broadly applied in routine clinical practice.

## Comparison

Deciding whether the study drug or intervention should be against placebo or an active comparator can determine whether differences can be observed and whether the treatment might be approved by regulatory agencies. The comparison drug can affect recruitment and/or blinding (e.g., the comparator drug has notable side effects). The decision to use an active comparator depends on whether other treatment options are available. For example, it is much more acceptable to have a placebo arm in a progressive trial in MS compared to RRMS.

## Outcomes and Endpoints

Outcome selection is a cornerstone of good clinical trial design. Outcomes should be selected to optimally demonstrate the effects of the therapeutic intervention. Selection of the primary outcome will vary based on the ultimate objective of the study. For phase 1 trials, outcomes will center on safety. In phase 2 trials, outcomes will focus on demonstrating a therapeutic effect across a range of doses and ideally will be biologically closer to the mechanism of action. For phase 3 trials, outcomes will need to satisfy the requirements of regulators and reflect the function of patients. The phases of clinical trials are described in more detail below. Secondary or even tertiary outcomes support effectiveness of the primary outcome, while exploratory outcomes investigate novel effects of the intervention or mechanisms. Clinical trials are powered to demonstrate an effect on the primary outcome, and statistical power is concentrated on that outcome. For secondary outcomes, hierarchical testing is prespecified for a rational spending of alpha function in the setting of multiple comparisons. Exploratory outcomes are considered hypothesis generating and typically do not require corrections for multiple comparisons. In these sections, we will review the most commonly employed outcomes in the context of phase 3 clinical trials focusing first on measures that are generally used as primary or secondary outcomes:

- *Annualized Relapse Rate (ARR):* The ARR is the most commonly used outcome measure in RRMS clinical trials [22], and it formed the basis of the approval of more than a dozen MS DMTs. The ARR is calculated from the number of relapses annualized over a 12-month in study follow-up. The ARR many times is expressed as a relative reduction between the active and comparator arm. The ARR, when presented as a percentage reduction is immediately interpretable both for neurologists and people with MS. Relapses have the advantage that they are easily discernable and generally require confirmation of neurological function on the neurological exam. In addition to this, relapses in phase 3 trials are highly predicted by development of new MRI lesions in phase 2 trials, which provides reasonable estimates of efficacy and sample sizes for phase 3 trials

[23]. There is evidence that the ARR has decreased over time in MS trials [24]. This finding and the fact that placebo comparators are likely no longer ethical in RRMS trials has prompted the use of alternative methods of relapse analysis, including time to first relapse [25]. Relapses and ARR will continue to be the primary outcome for ongoing anti-inflammatory drug programs in relapsing MS.

- *Disability Progression.* A second and perhaps more meaningful outcome measure in clinical trials has included assessment of neurological disability. Disability in MS has traditionally been measured using the Expanded Disability Status Scale (EDSS). The EDSS has several advantages as it is well known to neurologists, is relatively easy to perform, has been used in most clinical trials to date, and provides a simple 0–10 determination of neurological disability that is easy to understand and compare among patients/groups of patients [26]. The EDSS, however, also has several limitations including low inter-rater reliability, low intra-rater reliability, ordinal nature of the scale, roof and ceiling effects with plateaus at certain levels of disability, overreliance on lower extremity function, noise, relative underrepresentation of cognition, and limited use in general clinical practice [27]. Nonetheless, the EDSS has been the most widely used disability measure in clinical trials. Worsening of disability progression in trials typically must be sustained over a given time period, with 3 or 6 months being typically used. Specific thresholds for change including a half a point, a whole point, and greater than a point have been used in the past. The rationale for requiring a confirmed worsening at a second time point is that the EDSS may have some noise. Additionally, ascertaining worsening outside of relapses or daily fluctuation is desirable. EDSS progression has been used both in relapsing trials to document progression of disability that may occur through incomplete recovery of relapses, and perhaps more importantly, in progressive MS trials where disability is accrued slowly and independent of relapses. In the progressive MS trials, EDSS progression is normally the primary outcome, while in RRMS trials, it is usually a secondary outcome.
- *Composite Disability Outcomes.* Due to the multiple known limitations of the EDSS, the Multiple Sclerosis Functional Composite (MSFC) was developed as a quantitative neuroperformance test, initially designed to capture disability in three realms: upper extremity function (9 hole peg test), lower extremity function (timed 25 foot walk), and cognition (Paced Serial Addition Test [PASAT]) [28]. Newer iteration of the MSFC now includes measures that capture vision (low contrast letter acuity) and the PASAT was replaced with the Symbol Digit Modalities Test (SDMT) [29]. The MSFC components can have Z-scores derived, which improves psychometric properties, and the composite is formed by the addition of the Z-Scores. Although the MSFC is quite widely used, it has still not been fully endorsed by regulatory agencies; this, however, is likely to change in the near future. The MSFC generally is reserved as a secondary outcome, but it has been used in combination with the EDSS and alone as a primary outcome. Specific thresholds of meaningful change in the components has been established, and disability worsening can be defined as worsening of one or more components along with changes in the EDSS.
- *MRI Measures.* MRI outcomes are commonly used in clinical trials as secondary outcomes and are often primary outcomes in phase 2 studies. Typical measures include number of new T2, new gadolinium-enhancing lesions, new combined unique lesions (T2 or gadolinium without double counting) [30]. New T1-hypointense lesions and conversion to T1 black holes (persistent T1-hypointense lesions) have also been used. Semi-automated and fully automated methods also can derive new T2, gadolinium, and T1-hypointense lesion volumes. Volumetric measures of whole brain have also been employed and are highly predictive of future clinical disability, making them a particularly interesting secondary outcome measure. Indeed, several progressive

MS trials have successfully used brain volume loss as phase 2 outcomes, and ongoing studies will determine if atrophy will translate into effects on disability progression in phase 3 studies [31, 32]. Segmenting brain volume further including measurement of deep gray matter, cortical gray matter, or specific structures such as thalamus might have some value, but how much noise is generated with segmentation of smaller structures has to be weighed against improved interrogation along a specific biological pathway [33].

- *Patient Experience*. Patient-reported outcomes (PROs) are of interest as they interrogate the patient directly on clinically meaningful constructs. Several general neurological PROs have been validated in MS and have been used in the clinical trial setting. One such measure, Neurological quality of life (Neuro-QoL), has been validated in the MS population and has both long forms and more efficient computer adaptive testing [34]. More specific MS scales have also been developed. Perhaps one of the most used scales in MS is the MS Impact Scale (MSIS-29) and the Multiple Sclerosis Quality of Life-54. These scales have the advantage that they interrogate MS-related disability better, but perhaps fall short of dedicated instruments for specific symptoms, such as the Modified Fatigue Impact Scale, which has also been used in clinical trials. One overriding problem with PROs is that differences in clinical trials, although many times statistically significant, do not reach the threshold of clinically meaningful differences.
- *Biomarkers*. Blood biomarker for MS remains elusive; however, significant progress has been made with one candidate, neurofilament light chain (NFL). Although NFL is readily measured in cerebrospinal fluid, it has not been until recently, with the use of new technologies, that it can be measured in blood due to its low concentration. NFL is an axonal cytoskeletal protein that has been shown to increase in the setting of focal lesion development in MS. NFL has also shown to have treatment effects in clinical trials [35]. While NFL seems most promising as a measure of focal inflammation, and an outcome for anti-inflammatory DMT trials, the use in progressive disease has also been proposed. A key unknown for implementation in progressive MS has been defining normal levels on NFL. NFL levels increase in the blood as we age, along with other factors including disability level and comorbidities that also affect the central and peripheral nerves; moreover, there is potential decrease in levels in obese patients because of dilution.

## Institutional Review Board Approval and Study Registration

Clinical trials are closely monitored to maintain patient safety. They must be approved by supervising ethics committees before a trial can start. In the United States, these committees are called Institutional Review Boards (IRBs), and in the European Union, they are called ethics committees. Although most ethics committees are located at the investigator's institutions, some independent central IRBs are available for investigators at smaller institutions and are sometimes preferred due to quicker reviews of studies. Additionally, local IRBs must often certify researchers and their staff in order to conduct clinical studies by understanding patient privacy law (HIPAA) and good clinical practice principles set forth by the International Conference of Harmonisation Guidelines for Good Clinical Practice. The ongoing safety of subjects in a study is often done by a Data and Safety Monitoring Board (DSMB), which is an independent committee or an Independent Safety Officer.

The ethical principles involved in clinical trials are often extracted from the following documents:

1. *Nuremburg Code* (1947) is a 10-point statement meant to prevent future abuse of human subjects by emphasizing that consent for research must be voluntary, avoid unnecessary suffering, and participants are free to withdraw from the research [36]. It was

developed in response to Nazi atrocities of using concentration camp prisoners for human experiments.

2. *Declaration of Helsinki* (1964) is a code of medical ethics developed by the World Medical Association (WMA) and expands informed consent protections by describing that study subjects should be fully informed of the study procedures including risks and benefits, goals of the study, and sources of funding including potential conflicts of interest (https://www.wma.net/what-we-do/medical-ethics/declaration-of-helsinki/doh-jun1964/).
3. *Belmont Report* (1979) that identified the primary principles underlying ethical research with human beings as respect for persons (informed consent with subjects able to comprehend the information given), beneficence (benefits and risks of a study), and justice (the fair selection of subjects) (https://www.hhs.gov/ohrp/regulations-and-policy/belmont-report/index.html).

Once a clinical trial protocol has been developed and approved, it is important to consider registering the study in a public registry. This is increasingly becoming a requirement by both publishers and funders with the goal of reducing publication bias and selective reporting. The International Committee of Medical Journal Editors announced in 2004 a policy that as a condition of publication, clinical trials would be required to be registered [37]. Additionally, regulatory authorities started requiring clinical trial information and, in some cases, a summary of results to a publicly accessible registry. The World Health Organization subsequently specified elements that clinic trials should contain and maintains an international registry portal at http://apps.who.int/trialsearch/ [38]. The seventh revision of the Declaration of Helsinki in 2008 stated, “Every clinical trial must be registered in a publicly accessible database before recruitment of the first subject” [39]. Many registries are country specific with the two most common registries being ClinicalTrials.gov (https://www.clinicaltrials.gov/ct2/home) run by the United States National Library of Medicine in the United States and the European Union Clinical Trials registry (https://www.clinicaltrialsregister.eu/).

## Data Collection and Analysis

After developing and getting approval for a clinical study, it is helpful, if not imperative, to develop good tools for data collection. This is often in the form of a case report form (CRF) to help guarantee that all of the data are collected in a way that can be entered into a database easily. During this time, it is helpful to work directly with a statistician or the person who is going to have to analyze the data by setting up the database in such a way to easily extract the data and create reports or enter into statistical analysis software for analysis. Evaluation of data integrity and protection of privacy is crucial. These processes are often overlooked in the excitement to start the study and can create a lot of work later in having to transcribe data or reclassify data, which is prone to errors and can create delays in the analysis. A full description of the statistics is out of scope for this chapter but working with a statistician early and often cannot be emphasized enough. Preplanned interim analysis can be helpful to make sure that enough events are occurring in the study to analyze and assess feasibility to continue the study. These interim analyses can also include safety evaluations by a DSMB or safety officer as described above.

Data from a trial are often analyzed with an *intent-to-treat (ITT) analysis* in which the data from subjects are analyzed according to the group to which they were assigned, even if they did not receive or adhere to the intended treatment. ITT compares intervention strategy and not the intervention. Treatments or interventions that have tolerability issues will result in poor adherence, and an ITT may not provide useful information for subjects who adhered to the treatment or intervention. A *per-protocol analysis* can then be instructive in identifying the treatment or intervention effect in patients who are fully compliant.

## Reporting and Publishing the Clinical Study

Disseminating the information gathered in the study ensures that the information learned can be used by others. It also helps to reduce the risks to subjects by reducing duplication of studies especially if studies were not published because of harmful side effects. This process can begin early even before the data collection begins by publishing the clinical trial protocol especially in more complicated trials. Additionally, interim analyses can be presented at meetings or published to help introduce the study to the scientific community as well as helping guide our thinking about what the results will look like and what secondary or exploratory analysis might be more interesting.

The reporting of clinical trials is increasingly following the CONsolidated Standards of Reporting Trials or CONSORT statement, which was last revised in 2010 [40]. This is an evidence-based set of minimum recommendations for reporting randomized trials. This statement consists of 25 items consisting of how the trial was designed, analyzed, and interpreted (Fig. 4.3). Additionally, a participant flow diagram follows the flow of participants through the study (Fig. 4.4). These documents can be found and downloaded at http://www.consort-statement.org/. Most journals have endorsed the CONSORT guidelines and require that manuscripts follow them before consideration of any manuscript reporting a clinical trial.

## Phases of Clinical Research

The process of getting treatments approved involves a series of steps including clinical trials that eventually can lead to approved treatments. These steps have been described as phases of clinical research with each step having different goals. Here we will provide an overview of these phases (Fig. 4.5):

### Preclinical

Before drugs are tried in humans, they are often tested in the laboratory in cell or tissue cultures. This process involves screening many compounds to identify those that are most promising. Sometimes these drugs can then be modified or designed to improve some of their chemical properties, which may improve the efficacy or decrease their side effect profiles. The drug discovery often then proceeds to animal testing depending on whether good models of the disease being studied exist.

### Phase 1

These studies are often the first studies involving human participants. Phase 0 studies are sometimes done, which are sometimes called human microdosing studies, and they provide no safety or efficacy data as the dose is too low to cause a therapeutic effect. Phase 1 studies are often carried out in healthy volunteers, although they are sometimes carried out in patients, which are then designated as phase 1/2 studies. This occurs if the test drug is too toxic in healthy volunteers, the dose needed is higher than healthy volunteers can tolerate, or the therapeutic range is very narrow.

Phase 1 studies are used to assess the *pharmacokinetics* (PK) and *pharmacodynamics* (PD) of a drug. PK is how our bodies affect a drug and involve absorption, distribution, metabolism, and excretion (ADME) of a drug. PD is how a drug affects our bodies and evaluates the drug activity on at its target and the therapeutic window including the onset and duration of action.

Phase 1 studies are often carried out in ascending dose studies. These can either be single ascending dose studies (phase 1a) where a small group of people are given a single dose of a drug and observed before moving to larger doses until the maximum tolerated dose is achieved or multiple ascending dose studies (phase 1b) where several groups are given different doses and followed.

a

**CONSORT 2010 checklist of information to include when reporting a randomised trial***

| Section/Topic | Item No | Checklist item | Reported on page No |
|---|---|---|---|
| **Title and abstract** | | | |
| | 1a | Identification as a randomised trial in the title | |
| | 1b | Structured summary of trial design, methods, results, and conclusions (for specific guidance see CONSORT for abstracts) | |
| **Introduction** | | | |
| Background and objectives | 2a | Scientific background and explanation of rationale | |
| | 2b | Specific objectives or hypotheses | |
| **Methods** | | | |
| Trial design | 3a | Description of trial design (such as parallel, factorial) including allocation ratio | |
| | 3b | Important changes to methods after trial commencement (such as eligibility criteria), with reasons | |
| Participants | 4a | Eligibility criteria for participants | |
| | 4b | Settings and locations where the data were collected | |
| Interventions | 5 | The interventions for each group with sufficient details to allow replication, including how and when they were actually administered | |
| Outcomes | 6a | Completely defined pre-specified primary and secondary outcome measures, including how and when they were assessed | |
| | 6b | Any changes to trial outcomes after the trial commenced, with reasons | |
| Sample size | 7a | How sample size was determined | |
| | 7b | When applicable, explanation of any interim analyses and stopping guidelines | |
| Randomisation: | | | |
| Sequence generation | 8a | Method used to generate the random allocation sequence | |
| | 8b | Type of randomisation; details of any restriction (such as blocking and block size) | |
| Allocation concealment mechanism | 9 | Mechanism used to implement the random allocation sequence (such as sequentially numbered containers), describing any steps taken to conceal the sequence until interventions were assigned | |
| Implementation | 10 | Who generated the random allocation sequence, who enrolled participants, and who assigned participants to interventions | |
| Blinding | 11a | If done, who was blinded after assignment to interventions (for example, participants, care providers, those assessing outcomes) and how | |
| | 11b | If relevant, description of the similarity of interventions | |
| Statistical methods | 12a | Statistical methods used to compare groups for primary and secondary outcomes | |
| | 12b | Methods for additional analyses, such as subgroup analyses and adjusted analyses | |
| **Results** | | | |
| Participant flow (a diagram is strongly recommended) | 13a | For each group, the numbers of participants who were randomly assigned, received intended treatment, and were analysed for the primary outcome | |
| | 13b | For each group, losses and exclusions after randomisation, together with reasons | |
| Recruitment | 14a | Dates defining the periods of recruitment and follow-up | |
| | 14b | Why the trial ended or was stopped | |
| Baseline data | 15 | A table showing baseline demographic and clinical characteristics for each group | |
| Numbers analysed | 16 | For each group, number of participants (denominator) included in each analysis and whether the analysis was by original assigned groups | |
| Outcomes and estimation | 17a | For each primary and secondary outcome, results for each group, and the estimated effect size and its precision (such as 95% confidence interval) | |
| | 17b | For binary outcomes, presentation of both absolute and relative effect sizes is recommended | |
| Ancillary analyses | 18 | Results of any other analyses performed, including subgroup analyses and adjusted analyses, distinguishing pre-specified from exploratory | |
| Harms | 19 | All important harms or unintended effects in each group (for specific guidance see CONSORT for harms) | |
| **Discussion** | | | |
| Limitations | 20 | Trial limitations, addressing sources of potential bias, imprecision, and, if relevant, multiplicity of analyses | |
| Generalisability | 21 | Generalisability (external validity, applicability) of the trial findings | |
| Interpretation | 22 | Interpretation consistent with results, balancing benefits and harms, and considering other relevant evidence | |
| **Other information** | | | |
| Registration | 23 | Registration number and name of trial registry | |
| Protocol | 24 | Where the full trial protocol can be accessed, if available | |
| Funding | 25 | Sources of funding and other support (such as supply of drugs), role of funders | |

b

*We strongly recommend reading this statement in conjunction with the CONSORT 2010 Explanation and Elaboration for important clarifications on all the items. If relevant, we also recommend reading CONSORT extensions for cluster randomised trials, non-inferiority and equivalence trials, non-pharmacological treatments, herbal interventions, and pragmatic trials. Additional extensions are forthcoming: for those and for up to date references relevant to this checklist, see www.consort-statement.org.

**Fig. 4.3** (**a**, **b**) CONSORT 2010 checklist. These 25 items form the minimum list of items to include when reporting a randomized trial. http://www.consort-statement.org/consort-2010. (Reprinted under terms of Creative Commons Attribution CC BY 2.0 from Schulz KF, Altman DG, Moher D, for the CONSORT Group. CONSORT 2010 Statement: updated guidelines for reporting parallel group randomised trials. BMC Medicine 2010, 8:18. (24 March 2010))

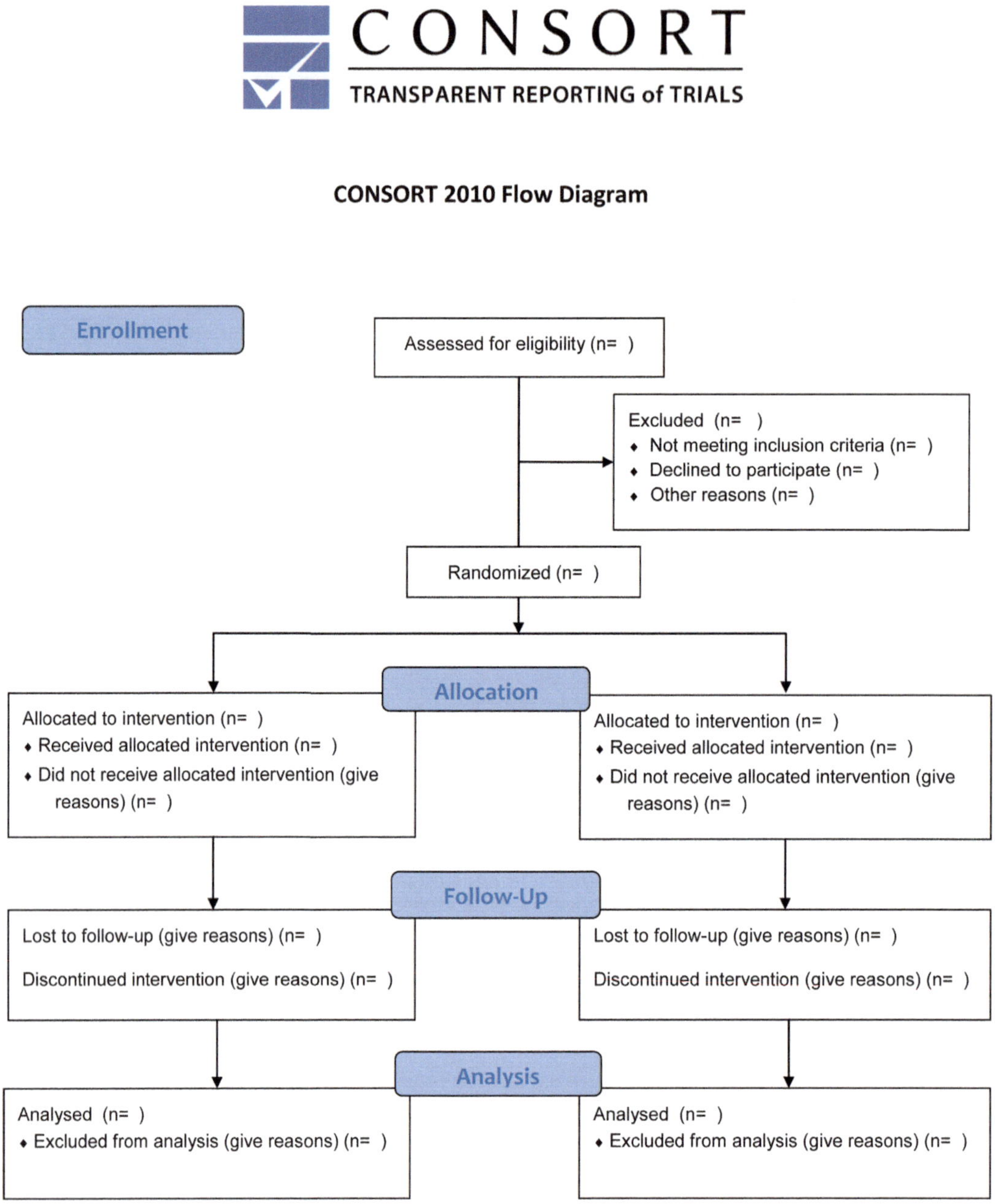


**Fig. 4.4** CONSORT 2010 Participant flow diagram. This diagram describes how subjects dropped out from a study and how many were finally included in the final analysis. http://www.consort-statement.org/consort-2010. (Reprinted under terms of Creative Commons Attribution CC BY 2.0 from Schulz KF, Altman DG, Moher D, for the CONSORT Group. CONSORT 2010 Statement: updated guidelines for reporting parallel group randomised trials. BMC Medicine 2010, 8:18. (24 March 2010))

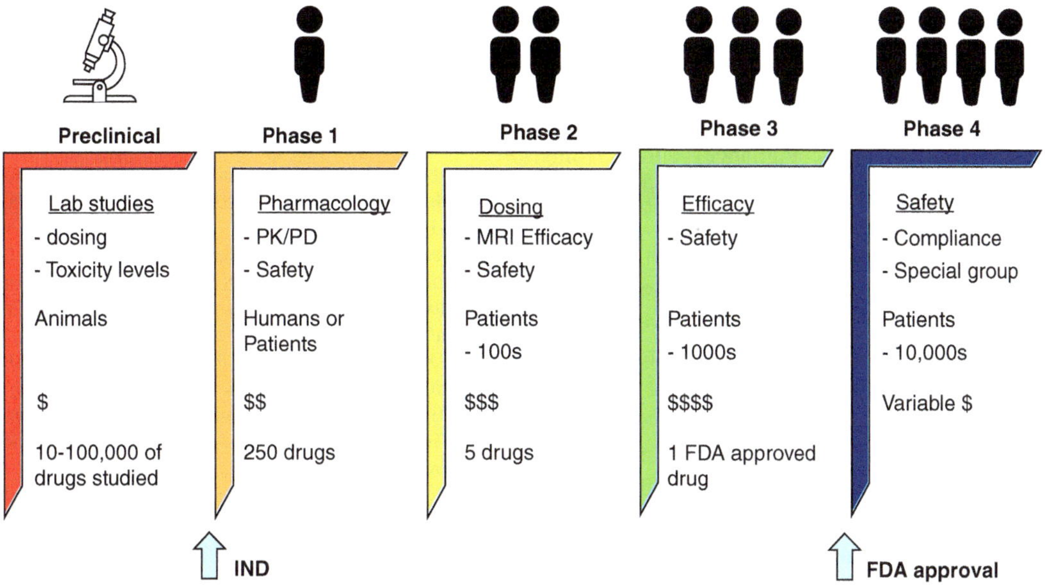


**Fig. 4.5** Phases of clinical research. Drug development follows a series of steps that can lead to drug approval. The goals of the study evolve through these steps, which involve an increasing number of patients and costs

## Phase 2

Once a dose range is determined, the next goal becomes evaluating if the drug has any biological effect. These are conducted in larger populations often involving hundreds of patients and often involve several doses. Surrogate endpoints are often evaluated and the drug effect on biomarkers is explored. Preliminary data are obtained on clinical outcomes to help determine the size of phase 3 studies. Patients who respond better to treatment are identified, which can influence the design and size of phase 3 studies.

These studies assess *efficacy* which is the effect of a drug under ideal and controlled settings. *Effectiveness* refers to the performance of a drug under real-world conditions. As studies move on to later phases, the drug's effectiveness is increasingly studied.

## Phase 3

Phase 3 studies are the cornerstone studies designed to assess the effectiveness of a new treatment or intervention that helps to assess their value in clinical practice. These studies require a clinical endpoint and involve multiple centers to recruit the large number of patients needed in these studies. They are also often multinational in order to help get approval in a variety of countries as there is often a requirement that a certain number of patients or subjects come from the country in which the study will be approved. The inclusion and exclusion criteria are more relaxed, although they may still not represent the full spectrum of disease in which a study will be approved. These studies are also longer than prior studies helping to evaluate adverse event with longer drug exposures. For all of these reasons, these studies become the gold standard assessment for a treatment. It is typical that two phase 3 studies be required for approval by the FDA (Food and Drug Administration) in the United States and the EMA (European Medicine Agency) in the European Union. Once a drug has shown satisfactory results, all of the manufacturing, preclinical, and trial data accumulated to this point are combined into a large document that is submitted to the regulatory agencies as a regulatory submission for drug approval though an NDA (New Drug Application). The process for approval of a drug has averaged 10–15 years and cost 2.6 billion dollars in the early 2010s [41].

## Phase 4

Phase 4 studies are also known as post marketing surveillance trials and provide longer-term safety data (pharmacovigilance) and ongoing studies, which may be required by the regulatory agencies at the time of approval. The inclusion and exclusion criteria are more relaxed and can include studies in special populations. These studies provide a more real-world experience on the effectiveness of a treatment. The longer period of evaluation allows for the study of rare events, including interactions with other drugs. These studies often involve reporting databases, registries, and monitoring health records. Pharmaeconomic studies help differentiate drugs of equal efficacy and safety.

## Conclusion

The treatment of patients with neuroimmune conditions will require combining the information gleaned from all of these methods and evaluating that they yield similar results. Comparing across clinical trials should be undertaken with caution due to differences in baseline demographics, how outcomes are defined, underlying biases, etc. However, when studies are dissimilar, studying why they are different can provide invaluable information into a disease process. Similarly, studying subpopulations within a clinical trial can provide information of who is responding better or not at all and help apply this information into clinical practice.

## References

1. Stuve O, Cepok S, Elias B, Saleh A, Hartung HP, Hemmer B, et al. Clinical stabilization and effective B-lymphocyte depletion in the cerebrospinal fluid and peripheral blood of a patient with fulminant relapsing-remitting multiple sclerosis. Arch Neurol. 2005;62(10):1620–3.
2. Ayzenberg I, Kleiter I, Schroder A, Hellwig K, Chan A, Yamamura T, et al. Interleukin 6 receptor blockade in patients with neuromyelitis optica nonresponsive to anti-CD20 therapy. JAMA Neurol. 2013;70(3):394–7.
3. Hulley SB. Designing clinical research. 4th ed. Philadelphia: Wolters Kluwer/Lippincott Williams & Wilkins; 2013.
4. Schumacher GA, Beebe G, Kibler RF, Kurland LT, Kurtzke JF, McDowell F, et al. Problems of experimental trials of therapy in multiple sclerosis: report by the panel on the evaluation of experimental trials of therapy in multiple sclerosis. Ann N Y Acad Sci. 1965;122:552–68.
5. Poser CM, Paty DW, Scheinberg L, McDonald WI, Davis FA, Ebers GC, et al. New diagnostic criteria for multiple sclerosis: guidelines for research protocols. Ann Neurol. 1983;13(3):227–31.
6. McDonald WI, Compston A, Edan G, Goodkin D, Hartung HP, Lublin FD, et al. Recommended diagnostic criteria for multiple sclerosis: guidelines from the international panel on the diagnosis of multiple sclerosis. Ann Neurol. 2001;50(1):121–7.
7. Freedman B. Equipoise and the ethics of clinical research. N Engl J Med. 1987;317(3):141–5.
8. Brian Haynes R. Forming research questions. J Clin Epidemiol. 2006;59(9):881–6.
9. Klawiter EC, Cross AH, Naismith RT. The present efficacy of multiple sclerosis therapeutics: is the new 66% just the old 33%? Neurology. 2009;73(12):984–90.
10. Kappos L, Weinshenker B, Pozzilli C, Thompson AJ, Dahlke F, Beckmann K, et al. Interferon beta-1b in secondary progressive MS: a combined analysis of the two trials. Neurology. 2004;63(10):1779–87.
11. Hawker K, O'Connor P, Freedman MS, Calabresi PA, Antel J, Simon J, et al. Rituximab in patients with primary progressive multiple sclerosis: results of a randomized double-blind placebo-controlled multicenter trial. Ann Neurol. 2009;66(4):460–71.
12. Montalban X, Hauser SL, Kappos L, Arnold DL, Bar-Or A, Comi G, et al. Ocrelizumab versus placebo in primary progressive multiple sclerosis. N Engl J Med. 2017;376(3):209–20.
13. Pittock SJ, Berthele A, Fujihara K, Kim HJ, Levy M, Palace J, et al. Eculizumab in aquaporin-4-positive neuromyelitis optica spectrum disorder. N Engl J Med. 2019;381(7):614–25.
14. Yamamura T, Kleiter I, Fujihara K, Palace J, Greenberg B, Zakrzewska-Pniewska B, et al. Trial of satralizumab in neuromyelitis optica spectrum disorder. N Engl J Med. 2019;381(22):2114–24.
15. Traboulsee A, Greenberg BM, Bennett JL, Szczechowski L, Fox E, Shkrobot S, et al. Safety and efficacy of satralizumab monotherapy in neuromyelitis optica spectrum disorder: a randomised, double-blind, multicentre, placebo-controlled phase 3 trial. The Lancet Neurology. 2020;19(5):402–12. https://doi.org/10.1016/S1474-4422(20)30078-8.
16. Cree BAC, Bennett JL, Kim HJ, Weinshenker BG, Pittock SJ, Wingerchuk DM, et al. Inebilizumab for the treatment of neuromyelitis optica spectrum disorder (N-MOmentum): a double-blind, randomised placebo-controlled phase 2/3 trial. Lancet. 2019;394(10206):1352–63.

17. Chitnis T, Arnold DL, Banwell B, Bruck W, Ghezzi A, Giovannoni G, et al. Trial of fingolimod versus interferon beta-1a in pediatric multiple sclerosis. N Engl J Med. 2018;379(11):1017–27.
18. Cohen JA, Barkhof F, Comi G, Hartung HP, Khatri BO, Montalban X, et al. Oral fingolimod or intramuscular interferon for relapsing multiple sclerosis. N Engl J Med. 2010;362(5):402–15.
19. Hauser SL, Waubant E, Arnold DL, Vollmer T, Antel J, Fox RJ, et al. B-cell depletion with rituximab in relapsing-remitting multiple sclerosis. N Engl J Med. 2008;358(7):676–88.
20. Vermersch P, Czlonkowska A, Grimaldi LM, Confavreux C, Comi G, Kappos L, et al. Teriflunomide versus subcutaneous interferon beta-1a in patients with relapsing multiple sclerosis: a randomised, controlled phase 3 trial. Mult Scler. 2014;20(6):705–16.
21. Chataway J, De Angelis F, Connick P, Parker RA, Plantone D, Doshi A, et al. Efficacy of three neuroprotective drugs in secondary progressive multiple sclerosis (MS-SMART): a phase 2b, multiarm, double-blind, randomised placebo-controlled trial. Lancet Neurol. 2020;19(3):214–25.
22. Nicholas R, Straube S, Schmidli H, Schneider S, Friede T. Trends in annualized relapse rates in relapsing-remitting multiple sclerosis and consequences for clinical trial design. Mult Scler. 2011;17(10):1211–7.
23. Sormani MP, Stubinski B, Cornelisse P, Rocak S, Li D, De Stefano N. Magnetic resonance active lesions as individual-level surrogate for relapses in multiple sclerosis. Mult Scler. 2011;17(5):541–9.
24. Nicholas R, Straube S, Schmidli H, Pfeiffer S, Friede T. Time-patterns of annualized relapse rates in randomized placebo-controlled clinical trials in relapsing multiple sclerosis: a systematic review and meta-analysis. Mult Scler. 2012;18(9):1290–6.
25. Sormani MP, Signori A, Siri P, De Stefano N. Time to first relapse as an endpoint in multiple sclerosis clinical trials. Mult Scler. 2013;19(4):466–74.
26. Meyer-Moock S, Feng YS, Maeurer M, Dippel FW, Kohlmann T. Systematic literature review and validity evaluation of the Expanded Disability Status Scale (EDSS) and the Multiple Sclerosis Functional Composite (MSFC) in patients with multiple sclerosis. BMC Neurol. 2014;14:58.
27. Goldman MD, Motl RW, Rudick RA. Possible clinical outcome measures for clinical trials in patients with multiple sclerosis. Ther Adv Neurol Disord. 2010;3(4):229–39.
28. Cutter GR, Baier ML, Rudick RA, Cookfair DL, Fischer JS, Petkau J, et al. Development of a multiple sclerosis functional composite as a clinical trial outcome measure. Brain. 1999;122(Pt 5):871–82.
29. Ontaneda D, LaRocca N, Coetzee T, Rudick R, Force NMT. Revisiting the multiple sclerosis functional composite: proceedings from the National Multiple Sclerosis Society (NMSS) Task Force on Clinical Disability Measures. Mult Scler. 2012;18(8):1074–80.
30. Daumer M, Neuhaus A, Morrissey S, Hintzen R, Ebers GC. MRI as an outcome in multiple sclerosis clinical trials. Neurology. 2009;72(8):705–11.
31. Fox RJ, Coffey CS, Conwit R, Cudkowicz ME, Gleason T, Goodman A, et al. Phase 2 trial of ibudilast in progressive multiple sclerosis. N Engl J Med. 2018;379(9):846–55.
32. Chataway J, Schuerer N, Alsanousi A, Chan D, MacManus D, Hunter K, et al. Effect of high-dose simvastatin on brain atrophy and disability in secondary progressive multiple sclerosis (MS-STAT): a randomised, placebo-controlled, phase 2 trial. Lancet. 2014;383(9936):2213–21.
33. Oh J, Ontaneda D, Azevedo C, Klawiter EC, Absinta M, Arnold DL, et al. Imaging outcome measures of neuroprotection and repair in MS: a consensus statement from NAIMS. Neurology. 2019;92(11):519–33.
34. Miller DM, Bethoux F, Victorson D, Nowinski CJ, Buono S, Lai JS, et al. Validating neuro-QoL short forms and targeted scales with people who have multiple sclerosis. Mult Scler. 2016;22(6):830–41.
35. Kuhle J, Kropshofer H, Haering DA, Kundu U, Meinert R, Barro C, et al. Blood neurofilament light chain as a biomarker of MS disease activity and treatment response. Neurology. 2019;92(10):e1007–e15.
36. Annas GJ, Grodin MA. The Nazi doctors and the Nuremberg Code: human rights in human experimentation. New York: Oxford University Press; 1992. xxii, 371 p. p.
37. De Angelis C, Drazen JM, Frizelle FA, Haug C, Hoey J, Horton R, et al. Clinical trial registration: a statement from the International Committee of Medical Journal Editors. Stroke. 2005;36(4):924–5.
38. Gulmezoglu AM, Pang T, Horton R, Dickersin K. WHO facilitates international collaboration in setting standards for clinical trial registration. Lancet. 2005;365(9474):1829–31.
39. Krleza-Jeric K, Lemmens T. 7th revision of the declaration of Helsinki: good news for the transparency of clinical trials. Croat Med J. 2009;50(2):105–10.
40. Schulz KF, Altman DG, Moher D, Group C. CONSORT 2010 statement: updated guidelines for reporting parallel group randomised trials. BMJ. 2010;340:c332.
41. DiMasi JA, Grabowski HG, Hansen RW. Innovation in the pharmaceutical industry: new estimates of R&D costs. J Health Econ. 2016;47:20–33.

# 5 Concepts of Immune Therapy and Disease Management

Gabrielle Macaron and Mary Alissa Willis

**Key Points**

1. Early identification and treatment of clinical, biological, radiological, or electrophysiological activity in neurological autoimmune disorders is key in the prevention of irreversible disability.
2. Differentiating between pseudorelapses, fluctuating symptoms, and symptoms due to comorbid conditions from actual relapses is needed to avoid unnecessary treatment escalation.
3. Treatment of acute relapses of neurological autoimmune disorders includes corticosteroids, intravenous immunoglobulin, and plasmapheresis, depending on the condition and the severity of symptoms.
4. Long-term immunosuppressive therapy is needed to control most neurological autoimmune disorders.
5. Optimal symptom management is essential to improve the quality of life of patients and to complement the beneficial effect of long-term maintenance therapies.
6. With the exception of multiple sclerosis (MS), high-quality trials studying the efficacy of acute and chronic treatment approaches and strong evidence on optimal disease monitoring strategies are lacking in the field of autoimmune neurological disorders.

## Introduction

Autoimmune inflammatory disorders constitute an important proportion of neurological diseases of the central, peripheral, and, to a lesser extent, autonomic nervous system. Their course is often marked by subacute development of new symptoms or worsening of previous symptoms. These relapses reflect active inflammation and should be promptly recognized. The importance of identifying clinical, biological, radiological, or electrophysiological activity is crucial to (1) treat the acute episode timely and appropriately to avoid irreversible damage to neural tissues and accumulation of disability over time and (2) escalate maintenance immunosuppressive therapy to better control the disease and avoid further recrudescence of inflammation. On the other hand, the evolution of these disorders often involves fluctuation of established symptoms, rather than

G. Macaron
Mellen Center for Multiple Sclerosis, Neurological Institute, Cleveland Clinic Foundation, Cleveland, OH, USA

M. A. Willis (✉)
Department of Neurology, University of Mississippi Medical Center, Jackson, MS, USA
e-mail: mwillis@umc.edu



A. L. Piquet, E. Alvarez (eds.), *Neuroimmunology*, https://doi.org/10.1007/978-3-030-61883-4_5

bona fide relapses. Distinguishing these episodes of intermittent worsening from actual disease activity is key. Hence, a major part of disease management includes symptomatic treatments, complementing the long-term benefit of immunosuppressive medications.

Autoimmune neurology is a complex field. In order to understand the mechanism of action and potential therapeutic advantage of immunomodulating therapies, one should keep in mind the pathophysiology of each group of diseases. For some disorders, development of symptoms is thought to be due to direct antibody-mediated effect on a specific antigen in the nervous system. Examples include autoimmune neuromuscular disease (e.g., antibodies directed against acetylcholine receptors, muscle-specific tyrosine kinase [MuSK], lipoprotein receptor-related protein 4 [LRP4] in myasthenia gravis), [1] limbic encephalitis with antibodies directed toward extracellular antigens (e.g., antibodies directed against the anti-N-methyl-D-aspartate [NMDA] receptor in NMDA encephalitis) [1], and neuromyelitis optica spectrum disorders (NMOSD) [2]. In the latter, antibody binding to aquaporin-4 (AQP4) receptors induces demyelination and neuronal death by antibody-mediated cytotoxicity, whereas in limbic encephalitis and myasthenia gravis, antibodies modulate reversibly the effect of surface or synaptic proteins [2, 3]. For other disorders, cytotoxic T-cells are the major culprit in disease development, like the classical paraneoplastic disorders associated with antibodies directed toward intracellular antigens (e.g., anti-Hu also known as anti-neuronal nuclear antibody type 1 [ANNA-1]) [1]. Antibody pathogenesis is further discussed in Chap. 2. Finally, many disorders have a more complex pathophysiology, with incompletely understood dysregulation of the immune system. These include multiple sclerosis (MS) [4], chronic inflammatory demyelinating polyneuropathy (CIDP) [5], and systemic diseases with CNS involvement such as sarcoidosis [6].

In this chapter, we will review the general concepts of monitoring and treatment of disease activity as well as the management of common symptoms. Detailed treatment strategies are reserved for subsequent chapters dedicated for each disorder.

## Diagnosis of Disease Activity

### Identifying Disease Activity and Recognizing Pseudorelapses or Comorbid Syndromes

A relapse or exacerbation is broadly defined as an immune-mediated insult to the nervous system resulting in new or worsening neurological symptoms and ongoing neuronal damage or modification of neural cell function. The demyelinating CNS diseases (MS and NMOSD) are representative examples, where a relapse can result in irreversible damage and disability accumulation [7, 8]. Other autoimmune disease can also present with a relapsing course, including autoimmune encephalitis (e.g., NMDA-R encephalitis [9, 10], LGI-1 encephalitis [3]), neurosarcoidosis [11], and CIDP [12], among others. Diagnosis of a relapse is largely based on clinical history and detection of objective changes on neurological examination. Validated neuroperformance assessments—measurements of gait, manual dexterity, cognition, and quality of life—have been incorporated into clinical practice in some MS centers [13] with the hope of detecting clinical change more sensitively. In neurosarcoidosis, diseases of the neuromuscular junction, and autoimmune polyneuropathies, neurologic symptoms are often chronic and treatment-dependent rather than distinct and well-defined [12, 14, 15]. Worsening when weaning immune suppression may be the clearest indicator that ongoing symptoms are related to an active immune process.

The term pseudorelapse is mostly used in MS and NMOSD. It is defined as worsening of preexisting neurological symptoms in the context of systemic metabolic derangement. The presence of typical and previously unexperienced symptoms is highly suspicious of a bona fide relapse. Conversely, worsening or reoccurring neurological symptoms should raise concern for a pseudorelapse induced by infection, heat, metabolic abnormality, initiation of a new medication, or emotional stress. For example, new vision loss in a patient with NMOSD would be strongly suggestive of a true relapse. Symptoms of increased leg spasticity in a patient with a previous myelitis could be a pseudorelapse [16]. In addition to evaluation for common causes of pseudorelapse, imaging can help differentiate true from pseudorelapses in MS and NMOSD. Pseudorelapses should generally not be treated with corticosteroids or a change in disease-modifying therapy.

## Monitoring of Disease Activity

Regular clinical assessments remain the foundation of disease activity monitoring. For some disorders, clinical judgment is sufficient to initiate treatment and modify long-term management. Two examples are recurrent seizure activity and behavioral symptoms in a patient with NMDA encephalitis or new-onset unilateral painful visual impairment in a patient with NMOSD [10]. Disease activity monitoring in MS includes evaluation of subclinical activity with serial magnetic resonance imaging (MRI) studies in addition to identification of relapses and objective clinical changes. Most MS patients develop asymptomatic lesions more commonly than symptomatic lesions [14], and detection of the subclinical lesions would warrant consideration of a change in disease-modifying therapy [17, 18]. Serial MRIs of the neuroaxis are also helpful in assessing disease activity and treatment response in neurosarcoidosis with CNS involvement [19, 20].

Biological markers can help identify inflammatory activity in some cases. The presence of CSF NMDA-R antibodies confirms a diagnosis of NMDA encephalitis, and changes in titers after treatment may be helpful for prognostication, although strong evidence to support following titers is lacking [15, 16]. AQP4 antibodies increase from their baseline weeks before the occurrence of a relapse in NMOSD [21]; however, clinical meaningfulness of the absolute value is difficult to interpret, and serial titer measurements are not routinely done in clinical practice to monitor disease activity. In other cases—such as anti-GAD antibodies in stiff person syndrome (SPS) [22]—antibody titers are not correlated to disease course.

Though MS has been studied more extensively than most other neuroinflammatory disorders, there are currently no biomarkers for monitoring of disease activity and treatment response.

The interest in finding such biomarkers remains high, with hope for markers such as neurofilament light chain detectable in both serum and CSF [17].

## Acute Management of Relapses

### Corticosteroids

The anti-inflammatory action of glucocorticoids is complex, pleiotropic, dose-dependent, and incompletely elucidated. Glucocorticoids are thought to inhibit initial events triggering inflammation as well as the processes maintaining an inflammatory response [18]. After being exposed to glucocorticoids, the glucocorticoid receptor subunit translocates in the nucleus and binds response elements that modulate the transcription of genes controlling leukocyte function to glucocorticoid response elements on various genes, leading to up- or downregulation of these genes depending on their function [18–20]. Glucocorticoids also act on non-genomic pathways, which ultimately modulate transcription of pro- and anti-inflammatory genes [19]. The net effects of glucocorticoids in the CNS are suppression of inflammation (decreasing the production of pro-inflammatory cytokines and growth factors), modulation of the immune repertoire in the blood and CSF (stimulating lymphocyte apoptosis through complex mechanisms, redistribution of T-cells, decreased memory T-cells, decreased Fc receptor expression by macrophage), and restoration of the blood-brain barrier function (by regulating expression of adhesion molecules on the blood-brain barrier and decreasing leukocyte migration in the injured CNS) [19]. Examples on the use of steroids in the acute management of autoimmune diseases of the nervous system are summarized in Table 5.1.

**Table 5.1** Overview of the use of corticosteroids in the treatment of exacerbations in neurological autoimmune diseases

| Disease | Comment |
|---|---|
| Multiple sclerosis | First-line therapy: 1000 mg IV MP for 3–5 d<br>Evidence supports beneficial effect of high-dose steroids on relapse outcome<br>High-dose oral prednisone (1250 mg) or MP (1000 mg) for 3–5 d is equivalent to IV administration<br>No evidence supporting the use of an oral taper<br>IV course can be repeated for up to 10 d in severe relapses with incomplete improvement after 5 days |
| NMOSD/ anti-MOG disease | First-line therapy: 1000 mg IV MP for 3–5 d<br>Strong evidence for specific regimen is lacking<br>Prompt use of plasmapheresis is advocated if minimal or no response is observed |
| Idiopathic acute transverse myelitis and optic neuritis | First-line therapy: 1000 mg IV MP for 3–7 d<br>Strong evidence based on the large ONTT |
| Autoimmune encephalitis | First-line therapy: 1000 mg IV MP for 3–5 d<br>Strong evidence for specific regimen is lacking<br>Followed by a maintenance dose of 1000 mg IV MP/w for 4–8 w with a gradual decrease in dose frequency over months, depending on response<br>Prompt use of plasmapheresis in severe/refractory cases<br>Response to treatment may be limited in disorders with antibodies directed toward intracellular antigens |
| Neurosarcoidosis | High-dose bolus (500–1000 mg IV MP for 3–5 d) reserved to severe cases<br>Followed by a maintenance dose of oral prednisone, typically 60 mg/d with a slow taper over 6 to 12 months depending on clinical and radiological response<br>Mild to moderate neurological involvement can be initially treated with low-dose oral prednisone (20 to 40 mg/d) |
| CIDP | IVIg advocated as first-line therapy, but corticosteroids are overall equally effective<br>Multiple corticosteroid regimens have been used, including monthly 1000 mg IV MP, 80–100 mg oral prednisone daily for at least 1 month followed by a slow taper, pulse 500 mg oral MP/week for at least 3 months followed by a taper, and pulse 40 mg/d oral dexamethasone for 4 days every month for 6 months, followed by a taper |
| Inflammatory myopathies[a] | First-line therapy<br>Prednisone 1 mg/kg/d, to a maximum daily dose of 80 mg/d for 4–6 w followed by dose adjustment depending on response<br>1000 mg IV MP for 3 d can be used initially in severely affected patients |

*Abbreviations: CIDP* chronic inflammatory demyelinating polyneuropathy, *d* day, *IV* intravenous, *IVIg* intravenous immunoglobulin, *MP* methylprednisolone, *MOG* myelin oligodendrocyte glycoprotein, *NMO* neuromyelitis optica, *NMO-SD* neuromyelitis optica spectrum disorder, *ONTT* optic neuritis treatment trial, *w* week(s)
[a]Inflammatory myopathies include polymyositis, dermatomyositis, and inclusion body myositis

In the 1980s, high-dose methylprednisolone (HDMP) gradually became the standard of care in MS relapses following three randomized trials showing its non-inferiority to intramuscular (IM) or intravenous (IV) adrenocorticotropic hormone (ACTH), which was previously the gold standard approach [23, 24]. The high cost of ACTH and the accessibility of HDMP have limited the use of ACTH (HP Acthar® gel, repository corticotropin injection) though it remains available to treat MS relapses for those patients who do not tolerate or do not respond to HDMP [22]. Numerous randomized controlled trials subsequently evaluated the benefit of various doses and routes of HDMP in the treatment of MS relapses [22, 25, 26]. A methylprednisolone dose of 500 to 1000 mg/day for 3 to 5 days is most common, though variations may be considered in some scenarios. Several recent studies in MS patients showing comparable effectiveness of high-dose oral steroids [22–26] led to increased adoption of oral methylprednisolone 1000 mg daily or prednisone 1250 mg daily as an alternative to IV administration. There is no strong evidence supporting the use of tapering regimen after a short course of high-dose steroids [24]; however, this approach is often advocated in clinical practice, particularly in patients with a high number/volume of active lesions.

While acute management is not as well-studied in other disorders, HDMP is typically initiated when confronting a patient with a subacute neurological presentation suggestive of an inflammatory monophasic or relapsing disease, and after ruling out an infectious process, HDMP remains the first-line therapy in suspected or relapsing autoimmune encephalitis [3], NMOSD [27, 28], idiopathic acute transverse myelitis [29], isolated optic neuritis [30], inflammatory myopathies [31], and to a lesser extent autoimmune peripheral neuropathies [5, 32]. High-dose steroids can be detrimental in some neuroinflammatory disorders. For example, HDMP may precipitate a myasthenic crisis or may worsen symptoms in multifocal motor neuropathy; the neuropathy is steroid-unresponsive and can even be worsened by their use [33].

The use of high-dose steroids is associated with a number of potential side effects (Table 5.2). Patients should be assessed for preexisting conditions that may be associated with higher risk of complications. Side effects of short-term pulse steroids are typically mild and manageable with the use of antacids, proton-pump inhibitors, anxiolytics, and/or a short-acting sedative-hypnotic. Blood glucose and blood pressure should be monitored in patients with diabetes and hypertension during therapy. Inpatient administration should be considered for patients with severe psychiatric disorders.

## Intravenous Immunoglobulins (IVIg)

Pooled polyclonal immunoglobulins from the blood of thousands of donors are delivered intravenously at high dose to provide an immunomodulatory effect in autoimmune disease [34]. The immunomodulatory and anti-inflammatory mechanism of action of intravenous immuno-

**Table 5.2** Side effects of corticosteroid use

| Onset | Adverse effect |
|---|---|
| Occurring during high-dose short-term use | Gastrointestinal side effects:<br>Abdominal pain<br>Nausea and vomiting<br>Peptic ulcer[a]<br>Intractable hiccups<br>Pancreatitis<br>Hepatitis |
| | Neuropsychiatric:<br>Insomnia<br>Euphoria<br>Mania<br>Anxiety<br>Psychosis[a]<br>Depression[a] |
| | Cutaneous:<br>Flushing and diaphoresis<br>Easy bruising<br>Acne[a] |
| | Other:<br>Hypertension<br>Diabetes[a]<br>Cardiac arrhythmia<br>Metallic taste<br>Headache<br>Myalgia<br>Increased appetite<br>Glaucoma |
| Occurring with long-term use | Musculoskeletal:<br>Osteopenia and osteoporosis<br>Avascular osteonecrosis<br>Myopathy |
| | General appearance:<br>Cushingoid features<br>Weight gain |
| | Cardiovascular:<br>Premature atherosclerotic disease<br>Hypertension<br>Diabetes |
| | Ophthalmologic:<br>Cataract<br>Glaucoma<br>Exophthalmos<br>Central serous chorioretinopathy |
| | Other:<br>Skin thinning and ecchymosis<br>Pseudotumor cerebri<br>Hypothalamic-pituitary-adrenal axis suppression<br>Immunosuppression |

[a]Side effects occurring more frequently in people with predisposing conditions

globulin (IVIg) is thought to be mediated by neutralizing and eliminating autoantibodies, blocking cellular receptors and hence, pro-inflammatory cell interactions, neutralizing complement proteins, inhibiting pro-inflammatory cytokine production (interleukin 1 [$IL_1$] and tumor necrosis factor-alpha [TNF-α]), blocking effector cells by saturating Fc receptors, inhibiting antibody-dependent cellular toxicity, and modulating T-cell function and antigen recognition [34, 35].

Table 5.3 lists neuroinflammatory disorders for which IVIg is used in either the acute or chronic setting. The role of IVIg is best established in PNS disorders such as Guillain-Barré syndrome [36, 37], CIDP [32, 36, 38], and multifocal motor neuropathy [36, 39]. The efficacy of high-dose IVIg has been reported in other peripheral neuropathies including painful sensory neuropathy associated to Sjögren's syndrome [40] and sarcoidosis [41]. IVIg may be used for treatment of worsening myasthenia gravis or myasthenic crisis [36, 42], although its use as a maintenance therapy in this disease is controversial. There is some evidence to support the use of short- and long-term therapy in Lambert-Eaton myasthenic syndrome [36, 43] as well as severe or steroid-refractory inflammatory myopathy [44].

IVIg is also commonly used as a first-line therapy in patients with suspected autoimmune

**Table 5.3** Overview of the use of intravenous immunoglobulin in the treatment of neurological autoimmune diseases

| Disease | Comment |
|---|---|
| CIDP and GBS | Recommendation based on strong evidence<br>For CIDP: 2 g/kg infused over 2 to 5 d, followed by maintenance therapy of 1–2 g/kg infused over 2–5 d every 3 w for a few months until improvement stabilizes<br>For GBS: 2 g/kg infused over 2 to 5 d |
| Multifocal motor neuropathy | Recommendation based on strong evidence<br>2 g/kg infused over 2 to 5 d followed by 1–2 g/kg every 2–8 w, depending on response<br>Corticosteroids and plasma exchange are ineffective |
| Myasthenia gravis and Lambert-Eaton myasthenic syndrome | Equally effective as plasma exchange in myasthenia crisis<br>Efficacy of IVIg less certain in patients with mild disease and purely ocular form<br>Chronic use of IVIg is less certain in myasthenia gravis but can be reserved for patients with refractory disease or contraindication to immunosuppressants<br>In myasthenia crisis/acute worsening: 2 g/kg infused over 2 to 5 d<br>In Lambert-Eaton: 2 g/kg infused over 2 to 5 d followed by maintenance therapy with repeat infusions at 4- to 12-w intervals in initial responders |
| Small-fiber neuropathy in neurosarcoidosis and Sjögren's disease | Evidence based on small case series<br>Efficacy on neuropathic pain<br>2 g/kg infused over 2 to 5 d<br>Repeat administration depending on clinical response |
| Inflammatory myopathies[a] | Most convincing data for dermatomyositis, less established for polymyositis<br>Benefit in inclusion body myositis is controversial<br>2 g/kg per month for 3 months based on one RCT in dermatomyositis, no established regimen available |
| Autoimmune encephalitis | Efficacy based on retrospective studies and rare randomized trials<br>Often used as first-line therapy<br>2 g/kg infused over 2 to 5 d |
| MS | Insufficient evidence for recommending its use in MS relapses<br>Positive results in reversing deficits after steroid-refractory MS-related optic neuritis in one uncontrolled trial<br>Can be used in patients with severe refractory relapses and contraindication to plasmapheresis<br>2 g/kg infused over 2 to 5 d |
| NMOSD | Insufficient evidence for recommending its use in NMOSD relapses<br>Can be used in steroid-refractory cases with contraindication to plasmapheresis<br>2 g/kg infused over 2 to 5 d |

*Abbreviations: CIDP* chronic inflammatory demyelinating polyneuropathy, *d* day, *GBS* Guillain-Barré syndrome, *HDMP* high-dose methylprednisolone, *IV* Intravenous, IVIg intravenous immunoglobulin, *MS* multiple sclerosis, *NMOSD* neuromyelitis optica spectrum disorder, *w* week(s)

[a]Inflammatory myopathies include polymyositis, dermatomyositis, and inclusion body myositis

encephalitis, often after HDMP fail to improve symptoms [45]. The efficacy of IVIg in this setting is limited to retrospective observations and small randomized trials. For example, a randomized trial of IVIg versus placebo in SPS with anti-GAD antibodies showed a significant decrease in stiffness scores and heightened sensitivity scores in patients treated with IVIg [46]. Controlled studies are needed to formally establish the efficacy of IVIg in autoimmune encephalitis.

Trials of IVIg for treatment of MS relapses have shown little benefit [47–49]. A small open-label nonrandomized study evaluating the efficacy of IVIg in steroid-refractory optic neuritis within 60 to 90 days after onset showed a significant benefit on visual outcome for patients on IVIg versus those without additional treatment [50]. Though this result was not confirmed in a larger randomized controlled trial, it suggests a potential benefit of IVIg for acute treatment of some MS patients. The efficacy of IVIg in NMOSD relapses was observed in case series and reports [51] and is based on the rationale that IVIg should be effective in humoral-mediated disorders.

Potential side effects of IVIg are summarized in Table 5.4 [42].

**Table 5.4** Side effects of intravenous immunoglobulin

| | |
|---|---|
| *Immediate reactions* | Infusion site pain and erythema |
| | Headache |
| | Myalgia, arthralgia |
| | Nausea, vomiting |
| | Phlogistic reactions[a]<br>*Generalized inflammatory reaction specifically in patients with active infections* |
| | Anaphylaxis-like reaction [a]<br>*Non-Ig-E-mediated reaction*<br>*Patients present with urticaria, flushing, tachycardia, hypertension, shortness of breath, chest pain, anxiety* |
| | Transfusion-related acute lung injury |
| | Transfusional volume overload[b, c] |
| | Anaphylaxis in IgA-deficient patients |
| *Delayed reactions* | Persistent headache |
| | Aseptic meningitis |
| | Thromboembolic events |
| | Acute kidney injury[c] |
| | Hyponatremia[c] |
| | Transient hemolytic anemia (positive Coombs) or neutropenia |
| | Enterocolitis |
| *Late reactions* | Eczematous dermatitis |
| | Impaired immune response to vaccination |
| | Interference with immunodiagnosis |
| | Blood-borne infections |

Generalized inflammatory reaction specifically in patients with active infections
[a]Rate-related adverse event
[b]In patients with preexisting cardiac insufficiency
[c]In patients with preexisting renal insufficiency

## Plasmapheresis

Plasmapheresis is used in numerous autoimmune disorders in the acute setting, particularly in severely ill patients. See Table 5.5 for overview. Different apheresis techniques are currently available. Plasma exchange (PLEX) can be performed using a centrifugation (separates plasma from cellular components based on density) or a filtration technique (separates plasma from cellular components using a filtration membrane), both of which require replacement of the removed fluid [52]. Other techniques include double filtration plasmapheresis and immunoadsorption where IgG and other humoral factors are selectively removed, and both do not require replacement fluid [52]. The evident mechanism of action of PLEX is through the removal of pathogenic autoantibodies. Other postulated actions of PLEX include increased sensitivity of antibody-producing cells to immunosuppressants following plasma exchange, removal of immune complexes and, hence, improvement of monocyte and macrophage function, removal of inflammatory cytokines, and increase in T suppressor and decreased natural killer cell function [52].

The use of PLEX is usually reserved for the acute setting as a second-line therapy after

**Table 5.5** Overview of the use of plasmapheresis in the treatment of neurological autoimmune diseases (non-exhaustive list)

| Disease | Comment |
|---|---|
| GBS and CIDP | Recommendation based on strong evidence<br>For GBS: five plasma exchanges, approximately every other day (7–14 d)<br>For CIDP: four to six plasma exchanges every other day (8–10 d) followed by one exchange every 3–4 w and depending on clinical response<br>Equally effective as IVIg for both disorders<br>In GBS: early administration after onset of symptom associated with improved efficacy of plasmapheresis; beneficial effect can be observed for up to 4 weeks |
| Myasthenia gravis | Established as an effective treatment based on numerous trials<br>Used in myasthenia crisis and in the perioperative period in patients undergoing major surgery<br>Five plasma exchanges, approximately every other d (7–14 d)<br>Equally effective as immunoglobulin in myasthenia crisis<br>Plasmapheresis could be more effective in MUSK-positive patients |
| Lambert-Eaton myasthenic syndrome | Small trials suggest short-term benefits |
| Autoimmune encephalitis | Efficacy based on retrospective studies<br>Can be used as first-line therapy, specifically in severely affected patients<br>Five plasma exchanges, approximately every other d (7–14 d)<br>Response to treatment is deceiving in disorders with antibodies directed toward intracellular antigens |
| MS | Efficacy based on few small randomized controlled trials and observational studies<br>Reserved as a rescue therapy in severe relapses unresponsive to HDMP<br>Five plasma exchanges, approximately every other d (7–14 d) |
| NMO/NMO-SD | Recommendation based on numerous positive retrospective trials<br>Typically used in severe attacks as an add-on therapy to HDMP early in the course of the attack<br>Early administration after onset of symptom associated with improved efficacy of plasmapheresis<br>Five plasma exchanges, approximately every other d (7–14 d) |

*Abbreviations: CIDP* chronic inflammatory demyelinating polyneuropathy, *d* day, *GBS* Guillain-Barré syndrome, *HDMP* high-dose methylprednisolone, *IV* intravenous, *IVIg* intravenous immunoglobulin, *MS* multiple sclerosis, *MUSK* muscle-specific kinase, *NMO* neuromyelitis optica, *NMO-SD* neuromyelitis optica spectrum disorder, *RCT* randomized controlled trial, *w* week

HDMP or first-line therapy in severely ill patients. PLEX is equally effective as IVIg as first-line therapy for Guillain-Barré Syndrome, CIDP, and myasthenia gravis [53–56]. Short-term benefits of PLEX have been reported in Lambert-Eaton myasthenic syndrome [57]. PLEX use in autoimmune encephalitis is based on numerous retrospective observations [45]. Small randomized controlled trials and retrospective studies [58] have shown benefits of PLEX for MS patients with severe attacks unresponsive to repeat courses of HDMP. Several studies suggest high efficacy of PLEX when used early in NMOSD [27, 47].

Potential complications with plasmapheresis are summarized in Table 5.6 [48].

## Long-Term Management

Long-term immunosuppressive therapy is needed to control most neurological autoimmune disorders. These therapies will be detailed in each disease-specific chapter; however, it is important to highlight the concepts of long-term therapy to prevent disability accumulation. An illustrative example is NMOSD, where disability is relapse-driven, relapses are typically severe, poor recovery from relapses is frequent, and the risk of having a relapse after an initial event is high [49, 59]. Hence, and in the absence of predictive biomarkers of worse prognosis, the risk of discontinuing therapy in NMOSD (and particularly

**Table 5.6** Complications of plasmapheresis

| | |
|---|---|
| *General reactions* | Hypotension |
| | Anaphylaxis |
| | Catheter-related complications<br>*Hematoma at puncture site*<br>*Pneumothorax*<br>*Infection of catheter* |
| | Bleeding<br>*Due to coagulation factor depletion* |
| *Pulmonary complications* | Transfusion-related acute lung injury (TRALI) |
| | Transfusional volume overload (TACO) |
| | Bronchospasm<br>*With or without an anaphylactic reaction* |
| | Dyspnea, wheezing, and chest pain<br>*Rarely in the context of a complement-mediated membrane incompatibility* |
| *Ion metabolism disorder* | Citrate-induced hypocalcemia |
| | Citrate-induced metabolic alkalosis |
| | Hypokalemia |
| | Hypocalcemia |

seropositive cases) appears to outweigh the benefit, and this approach is not recommended [59].

In MS, there are also many arguments advocating for the long-term and early use of disease-modifying therapies (DMT) (see Chap. 31). Evidence show that the probability of achieving long-term NEDA (no evidence of disease activity) is low even with higher-efficacy DMT, ranging from 13% to 46% after 2 years and 7.9% after 7 years of treatment [60]. Autologous hematopoietic stem cell transplantation yields a higher probability of achieving NEDA (78–83% at year 2 and 60–68% at year 5) [61]. Recent data also suggest that early use of higher-efficacy DMT is associated with a lower risk of secondary conversion to progressive MS [62]. The question of DMT discontinuation frequently arises in older individuals often facing increased financial burden of medication use or higher risk of adverse effects in the context of an increasing burden of comorbid medical issues. The Discontinuation of Disease Modifying Therapies in Multiple Sclerosis (DISCOMS) study (clinicaltrial.gov identifier: NCT03073603) aims to define characteristics of patients who may benefit from DMT discontinuation.

Long-term use of immunosuppressive therapies is increasingly common in other neuroinflammatory disorders. For example, in NMDA encephalitis, the absence of use of immunotherapy is associated with a worse prognosis [16]. In CIDP and MMN, long-term maintenance treatment (typically with IVIg or corticosteroids) is standard of care, although some patients might remain stable off-therapy [63]. Dosing algorithms to optimize the use of long-term therapies in inflammatory neuropathies have been proposed in an effort to avoid unnecessary overdosing and costs [64]. More details will be provided in individual chapters.

## Management of Symptoms

Optimal symptom management is essential to improve quality of life of patients and to complement the beneficial effect of long-term maintenance therapies. Chronic symptoms may occur early in the disease and might even precede the diagnosis by several years as in the case of many MS patients [65]. Many patients will have more than one symptom and these may be interrelated [66]. Routine evaluations should include screening for persistent symptoms, preferably using validated scales (i.e., Symbol Digit Modalities Test, Fatigue Impact Scale, Patient Health Questionnaire-9) [67, 68].

Table 5.7 summarizes the main symptoms encountered and their treatment options. Detailed symptom management will be discussed in Chap. 32. Many of these issues are best addressed by a multidisciplinary care team.

## Future Perspectives in Autoimmune Neurology

Neuroimmunology is a rapidly evolving field with many immune-mediated disorders yet to be identified and best practices in monitoring and treatment yet to be defined. With the exception of

**Table 5.7** Overview of the management of the most frequent symptoms in neurological autoimmune disorders

| Symptom | Management |
|---|---|
| Fatigue | Address primary cause of fatigue if any (depression, sleep disturbance, physical deconditioning, chronic pain, medication adverse effect)<br>Behavioral changes: physical activity, regular sleep, stress management, cognitive-behavioral therapy<br>The use of oral amantadine (up to 100 mg bid), modafinil (200 to 400 mg/d), and armodafinil (150 to 250 mg/d) has been particularly studied in MS |
| Sleep disturbance | Address primary cause of sleep disorder if any (depression, anxiety, neuropathic pain, spasticity, nocturia, medication adverse effect)<br>Consider polysomnography if a sleep disorder is suspected or insomnia is treatment-resistant<br>Sleep hygiene<br>Cognitive-behavioral therapy<br>Reserve small doses of benzodiazepine, sedating antidepressants, or hypnotics for short-term management or failure of non-pharmacologic approaches |
| Cognitive impairment | Evaluate cognitive function with objective measures<br>Address fatigue and mood disorders (major contributors of patient-perceived cognitive dysfunction)<br>Cognitive rehabilitation |
| Spasticity | Rehabilitation, physical therapy, stretching exercises<br>Oral therapy with baclofen (30 to 90 mg per day divided in three to four doses), tizanidine (2 to 8 mg at bedtime), and rarely dantrolene and benzodiazepines<br>Botulinum toxin injection for focal spasticity<br>Intrathecal baclofen pump for severe and diffuse spasticity |
| Neuropathic pain | Pharmacologic management consists of antiepileptics, SNRIs, TCAs, topical lidocaine, capsaicin |
| Mood disorders | Psychotherapy<br>Pharmacologic management with SSRI, SNRI, mirtazapine, bupropion, or TCA<br>Atypical antipsychotic medications when needed for agitation, delirium, and euphoria refractory to behavioral interventions |
| Neurogenic bladder | Reduced fluid intake at bedtime, decreased bladder irritants, scheduled voiding, bladder retraining<br>Oral anticholinergic drugs (for overactive bladder), alpha-adrenergic blockers (for obstructive symptoms)<br>Botulinum toxin injection (in the detrusor)<br>Intermittent or permanent urinary catheterization<br>Sacral neuromodulation |

*Abbreviations: MS* multiple sclerosis, *SNRI* serotonin-norepinephrine reuptake inhibitor, *SSRI* selective serotonin reuptake inhibitor, *TCAs* tricyclic antidepressants

MS, there is little consensus on how to monitor disease activity. For example, questions remain about the utility of serial antibody titers in autoimmune encephalitis, the need for repeat imaging in NMOSD, and the benefits of serial electrophysiologic studies in MMN.

Biomarkers of inflammation or axonal damage—such as neurofilament light chain—are needed to enhance monitoring of disease activity and evaluation of treatment response [69, 70].

With the exception of MS and more recently NMOSD, high-quality trials studying the efficacy of acute and chronic treatment approaches are lacking. Randomized controlled trials are challenging to undertake in most neuroinflammatory disorders due to the rarity of these diagnoses. Thus, development of clinical management guidelines will need to come from data captured in pragmatic trials, large cohorts, and disease registries [66]. Further treatment approaches for autoimmune encephalitis and a discussion of the barriers to clinical trials in autoimmune encephalitis are discussed in Chap. 17.

## References

1. Lancaster E. Paraneoplastic disorders. Contin Lifelong Learn Neurol. 2017;23(6):1653–79.
2. Papadopoulos MC, Bennett JL, Verkman AS. Treatment of neuromyelitis optica: State-of-the-art and emerging therapies. Nat Rev Neurol. 2014;10(9):493–506.

3. Lancaster E. The diagnosis and treatment of autoimmune encephalitis. J Clin Neurol. 2016;12(1):1–13.
4. Reich DS, Lucchinetti CF, Calabresi PA. Multiple Sclerosis. N Engl J Med. 2018;378(2):169–80.
5. Dalakas MC. Advances in the diagnosis, pathogenesis and treatment of CIDP. Nat Rev Neurol. 2011;7(9):507–17.
6. Valeyre D, Prasse A, Nunes H, Uzunhan Y, Brillet PY, Müller-Quernheim J. Sarcoidosis. Lancet. 2014;383(9923):1155–67.
7. Wingerchuk DM, Frcp C, Pittock SJ, Lucchinetti CF, Lennon VA, Weinshenker BG. A secondary progressive clinical course is uncommon in neuromyelitis optica. Neurology. 2007;68:603–5.
8. Lublin FD, Baier M, Cutter G. Effect of relapses on development of residual deficit in multiple sclerosis. Neurology. 2003;61:1528–32.
9. Dalmau J, Gleichman AJ, Hughes EG, Rossi JE, Peng X, Lai M, et al. Anti-NMDA-receptor encephalitis: case series and analysis of the effects of antibodies. Lancet Neurol. 2008;7(12):1091–8.
10. Florance NR, Davis RL, Lam C, Szperka C, Zhou L, Ahmad S, et al. Anti-N-methyl-D-aspartate receptor (NMDAR) encephalitis in children and adolescents. Ann Neurol. 2009;66(1):11–8.
11. Joubert B, Chapelon-Abric C, Biard L, Saadoun D, Demeret S, Dormont D, et al. Association of prognostic factors and immunosuppressive treatment with long-term outcomes in neurosarcoidosis. JAMA Neurol. 2017;74(11):1336–44.
12. Chiò A, Cocito D, Bottacchi E, Buffa C, Leone M, Plano F, et al. Idiopathic chronic inflammatory demyelinating polyneuropathy: an epidemiological study in Italy. J Neurol Neurosurg Psychiatry. 2007;78(12):1349–53.
13. Cutter G, Baier M, Rudick R, Cookfair D, Fischer J, John Petkau J, et al. Development of a multiple sclerosis functional composite as a clinical trial outcome measure. Brain. 1999;122:871–82.
14. Isaac C, Li DK, Genton M, Jardine C, Grochowski E, Palmer M, et al. Multiple sclerosis: a serial study using MRI in relapsing patients. Neurology. 1988;38(10):1511–5.
15. Vincent A, Bien CG, Irani SR, Waters P. Autoantibodies associated with diseases of the CNS: new developments and future challenges. Lancet Neurol. 2011;10(8):759–72.
16. Broadley J, Seneviratne U, Beech P, Buzzard K, Butzkueven H, O'Brien T, et al. Prognosticating autoimmune encephalitis: a systematic review. J Autoimmun. 2019;96:24–34.
17. Hakansson I, Tisell A, Cassel P, Blennow K, Zetterberg H, Lundberg P, et al. Neurofilament light chain in cerebrospinal fluid and prediction of disease activity in clinically isolated syndrome and relapsing – remitting multiple sclerosis. Eur J Neurol. 2017;24:703–12.
18. Coutinho AE, Chapman KE. The anti-inflammatory and immunosuppressive effects of glucocorticoids, recent developments and mechanistic insights. Mol Cell Endocrinol. 2011;335(1):2–13.
19. Fox RJ, Kinkel RP. High-dose methylprednisolone to treat multiple sclerosis. In: Cohen JA, Rudick RA, editors. Multiple sclerosis therapeutics. Cambridge: Cambridge University Press; 2011. p. 418–35.
20. Smoak KA, Cidlowski JA. Mechanisms of glucocorticoid receptor signaling during inflammation. Mech Ageing Dev. 2004;125(10–11 SPEC. ISS):697–706.
21. Jarius S, Aboul-Enein F, Waters P, Kuenz B, Hauser A, Berger T, et al. Antibody to aquaporin-4 in the long-term course of neuromyelitis optica. Brain. 2008;131(11):3072–80.
22. Citterio A, LaMantia L, Ciucci G, Candelise L, Brusaferri F, Midgard R, et al. Corticosteroids or ACTH for acute exacerbations in multiple sclerosis. Cochrane Database Syst Rev. 2013;(4):CD001331.
23. Barnes D, Hughes RAC, Morris RW, Wade-Jones O, Brown P, Britton T, et al. Randomised trial of oral and intravenous methylprednisolone in acute relapses of multiple sclerosis. Lancet. 1997;349(9056):902–6.
24. Thomson A, Kennard C, Swash M, Summers B, Yuill G, Sheperd D, et al. Relative efficacy of intravenous methylprednisolone and ACTH in the treatment of acute relapse in MS. Neurology. 1989;39(July):969–72.
25. Milligan NM, Newcombe R, Compston DA. A double-blind controlled trial of high dose methylprednisolone in patients with multiple sclerosis: 1. Clinical effects. J Neurol Neurosurg Psychiatry. 1987;50(5):511–6.
26. Sellebjerg F, Frederiksen JL, Nielsen PM, Olesen J. Double-blind, randomized, placebo-controlled study of oral, high-dose methylprednisolone in attacks of MS. Neurology. 1998;51(2):529–34.
27. Abboud H, Petrak A, Mealy M, Sasidharan S, Siddique L, Levy M. Treatment of acute relapses in neuromyelitis optica: steroids alone versus steroids plus plasma exchange. Mult Scler. 2016;22(2):185–92.
28. Sherman E, Han MH. Acute and chronic management of neuromyelitis optica spectrum disorder. Curr Treat Options Neurol. 2015;17:48.
29. Greenberg BM, Thomas KP, Krishnan C, Kaplin AI, Calabresi PA, Kerr DA. Idiopathic transverse myelitis: corticosteroids, plasma exchange, or cyclophosphamide. Neurology. 2007;68(19):1614–7.
30. Beck RW, Cleary PA, Anderson MMJ, Keltner JL, Shults WT, Kaufman DI, et al. A randomized, controlled trial of corticosteroids in the treatment of acute optic neuritis. The Optic Neuritis Study Group. N Engl J Med. 1992;326(9):581–8.
31. Clark KEN, Isenberg DA. A review of inflammatory idiopathic myopathy focusing on polymyositis. Eur J Neurol. 2018;25(1):13–23.
32. Hughes R, Bensa S, Willison H, Van den Bergh P, Comi G, Illa I, et al. Randomized controlled trial of intravenous immunoglobulin versus oral prednisolone in chronic inflammatory demyelinating polyradiculoneuropathy. Ann Neurol. 2001;50(2):195–201.

33. Jinka M, Chaudhry V. Treatment of multifocal motor neuropathy. Curr Treat Options Neurol. 2014;16(2):269.
34. Luemann JD, Nimmerjahn F, Dalakas MC. Intravenous immunoglobulin in neurology-mode of action and clinical efficacy. Nat Rev Neurol. 2015;11(2):80–9.
35. Dalakas MC. Mechanism of action of intravenous immunoglobulin and therapeutic considerations in the treatment of autoimmune neurologic diseases. Neurology. 1998;51(6 Suppl 5):S2–8.
36. Patwa HS, Chaudhry V, Katzberg H, Rae-Grant AD, So YT. Evidence-based guideline: intravenous immunoglobulin in the treatment of neuromuscular disorders. Neurology. 2012;78:1009–15.
37. Kleyweg RP, van der Meche FG, Meulstee J. Treatment of Guillain-Barre syndrome with high-dose gammaglobulin. Neurology. 1988;38(10):1639–41.
38. Hughes RAC, Donofrio P, Bril V, Dalakas MC, Deng C, Hanna K, et al. Intravenous immune globulin (10% caprylate-chromatography purified) for the treatment of chronic inflammatory demyelinating polyradiculoneuropathy (ICE study): a randomized placebo-controlled trial. Lancet Neurol. 2008;7(2):136–44.
39. Leger JM, Chassande B, Musset L, Meininger V, Bouche P, Baumann N. Intravenous immunoglobulin therapy in multifocal motor neuropathy: a double-blind, placebo-controlled study. Brain. 2001;124(Pt 1):145–53.
40. Morozumi S, Kawagashira Y, Iijima M, Koike H, Hattori N, Katsuno M, et al. Intravenous immunoglobulin treatment for painful sensory neuropathy associated with Sjogren's syndrome. J Neurol Sci. 2009;279(1–2):57–61.
41. Tavee JO, Karwa K, Ahmed Z, Thompson N, Parambil J, Culver DA. Sarcoidosis-associated small fiber neuropathy in a large cohort: clinical aspects and response to IVIG and anti-TNF alpha treatment. Respir Med. 2017;126:135–8.
42. Gajdos P, Chevret S, Toyka KV. Intravenous immunoglobulin for myasthenia gravis. Cochrane Database Syst Rev. 2012;(12):Art. No.: CD002277.
43. Keogh M, Sedehizadeh S, Maddison P. Treatment for Lambert-Eaton myasthenic syndrome. Cochrane Database Syst Rev. 2011;(2):Art. No.: CD003279.
44. Dalakas MC, Illa I, Dambrosia JM, Soueidan SA, Stein DP, Otero C, et al. A controlled trial of high-dose intravenous immune globulin infusions as treatment for dermatomyositis. N Engl J Med. 1993;329(27):1993–2000.
45. Nosadini M, Mohammad SS, Ramanathan S, Brilot F, Dale RC. Immune therapy in autoimmune encephalitis: a systematic review. Expert Rev Neurother. 2015;15(12):1391–419.
46. Dalakas MC, Fujii M, Li M, Lutfi B, Kyhos J, McElroy B. High-dose intravenous immune globulin for stiff-person syndrome. N Engl J Med. 2001;345(26):1870–6.
47. Bonnan M, Valentino R, Debeugny S, Merle H, Ferge J-L, Mehdaoui H, et al. Short delay to initiate plasma exchange is the strongest predictor of outcome in severe attacks of NMO spectrum disorders. J Neurol Neurosurg Psychiatry. 2018;89(4):346–51.
48. Basic-Jukic N, Kes P, Glavas-Boras S, Brunetta B, Bubic-Filipi L, Puretic Z. Complications of therapeutic plasma exchange: experience with 4857 treatments. Ther Apher Dial. 2005;9(5):391–5.
49. Wingerchuk DM, Hogancamp WF, O'Brien PC, Weinshenker BG. The clinical course of neuromyelitis optica (Devic's syndrome). Neurology. 1999 Sep;53(5):1107–14.
50. Tselis A, Perumal J, Caon C, Hreha S, Ching W, Din M, et al. Treatment of corticosteroid refractory optic neuritis in multiple sclerosis patients with intravenous immunoglobulin. Eur J Neurol. 2008;15(11):1163–7.
51. Elsone L, Panicker J, Mutch K, Boggild M, Appleton R, Jacob A. Role of intravenous immunoglobulin in the treatment of acute relapses of neuromyelitis optica: experience in 10 patients. Mult Scler. 2014;20(4):501–4.
52. Reeves HM, Winters JL. The mechanisms of action of plasma exchange. Br J Haematol. 2014;164(3):342–51.
53. Hughes RAC, Swan AV, van Doorn PA. Intravenous immunoglobulin for Guillain-Barré syndrome. Cochrane Database Syst Rev. 2014;(9):Art. No.: CD002063.
54. Dyck PJ, Litchy WJ, Kratz KM, Suarez GA, Low PA, Pineda AA, et al. A plasma exchange versus immune globulin infusion trial in chronic inflammatory demyelinating polyradiculoneuropathy. Ann Neurol. 1994;36(6):838–45.
55. Barth D, Nabavi Nouri M, Ng E, Nwe P, Bril V. Comparison of IVIg and PLEX in patients with myasthenia gravis. Neurology. 2011;76(23):2017–23.
56. Gwathmey K, Balogun R, Burns T. Neurologic indications for therapeutic plasma exchange: 2013 update. J Clin Apher. 2013;29:211–9.
57. Newsom-Davis J. Therapy in myasthenia gravis and Lambert-Eaton myasthenic syndrome. Semin Neurol. 2003;23(2):191–8.
58. Weinshenker BG, O'Brien PC, Petterson TM, Noseworthy JH, Lucchinetti CF, Dodick DW, et al. A randomized trial of plasma exchange in acute central nervous system inflammatory demyelinating disease. Ann Neurol. 1999;46(6):878–86.
59. Weinshenker BG, Wingerchuk DM, Vukusic S, Linbo L, Pittock SJ, Lucchinetti CF, et al. Neuromyelitis optica IgG predicts relapse after longitudinally extensive transverse myelitis. Ann Neurol. 2006 Mar;59(3):566–9.
60. Rotstein DL, Healy BC, Malik MT, Chitnis T, Weiner HL. Evaluation of no evidence of disease activity in a 7-year longitudinal multiple sclerosis cohort. JAMA Neurol. 2015 Feb;72(2):152–8.
61. Sormani MP, Muraro PA, Saccardi R, Mancardi G. NEDA status in highly active MS can be more easily obtained with autologous hematopoietic stem cell transplantation than other drugs. Mult Scler. 2017;23(2):201–4.

62. Brown JWL, Coles A, Horakova D, Havrdova E, Izquierdo G, Prat A, et al. Association of Initial Disease-Modifying Therapy with later conversion to secondary progressive multiple sclerosis. JAMA. 2019;321(2):175–87.
63. Adrichem ME, Eftimov F, van Schaik IN. Intravenous immunoglobulin treatment in chronic inflammatory demyelinating polyradiculoneuropathy, a time to start and a time to stop. J Peripher Nerv Syst. 2016;21(3):121–7.
64. Lunn M, Ellis L, Hadden R, Rajabally Y, Winer J, Reilly MM. A proposed dosing algorithm for the individualized dosing of human immunoglobulin in chronic inflammatory neuropathies. J Peripher Nerv Syst. 2016;21(1):33–7.
65. Disanto G, Zecca C, MacLachlan S, Sacco R, Handunnetthi L, Meier UC, et al. Prodromal symptoms of multiple sclerosis in primary care. Ann Neurol. 2018;83(6):1162–73.
66. Macaron G, Cohen JA. Integrating multiple sclerosis guidelines into practice. Lancet Neurol. 2018;17(8):658–60.
67. Learmonth YC, Dlugonski D, Pilutti LA, Sandroff BM, Klaren R, Motl RW. Psychometric properties of the fatigue severity scale and the modified fatigue impact scale. J Neurol Sci. 2013;331(1–2):102–7.
68. Kroenke K, Spitzer RL, Williams JBW. The PHQ-9: validity of a brief depression severity measure. J Gen Intern Med. 2001;16(9):606–13.
69. Novakova L, Zetterberg H, Sundstrom P, Axelsson M, Khademi M, Gunnarsson M, et al. Monitoring disease activity in multiple sclerosis using serum neurofilament light protein. Neurology. 2017;89(22):2230–7.
70. Mariotto S, Farinazzo A, Magliozzi R, Alberti D, Monaco S, Ferrari S. Serum and cerebrospinal neurofilament light chain levels in patients with acquired peripheral neuropathies. J Peripher Nerv Syst. 2018;23(3):174–7.

# Part II

# Multiple Sclerosis

# Evolution of the Diagnostic Criteria in Multiple Sclerosis

# 6

Marisa P. McGinley and Jeffrey A. Cohen

**Key Points**

1. The goal of the diagnostic criteria is to facilitate accurate and early diagnosis of multiple sclerosis (MS).
2. The key features of the diagnostic criteria are clinical presentation, establishing dissemination in time and space, role of magnetic resonance imaging (MRI), and utility of cerebrospinal fluid (CSF) and other paraclinical tests.
3. More research is needed to understand the utility and applicability in diverse populations and nonclassical presentations.
4. The diagnostic criteria will continue to evolve, and there is a need for development of unique biomarkers and advanced imaging techniques to improve diagnostic accuracy.

## Introduction

Diagnostic criteria in multiple sclerosis (MS) provide an important framework for how we define the disease and the requirements for making a formal diagnosis. The criteria have evolved over time from a purely clinical definition to incorporation of imaging and laboratory evidence. As our understanding of MS has expanded, the need to evaluate how we distinguish MS from other conditions, including other autoimmune and inflammatory disorders, has become more important. Similarly, harmonization of the criteria used to diagnose MS in diverse populations is needed to promote consistency. Although there have been substantial improvements to the criteria over time, there remain areas that need refinement including nonclassical presentations, (e.g., radiologically isolated syndrome, solitary sclerosis, etc.), the need for nonimaging diagnostic biomarkers, better utilization of advanced magnetic resonance imaging (MRI) measures, and the role of ancillary testing. The purpose of the criteria is to provide formal definitions of several key themes that create a useful framework to aid with early and accurate diagnosis (Table 6.1).

M. P. McGinley (✉) · J. A. Cohen
Mellen Center, Neurologic Institute, Cleveland Clinic, Cleveland, OH, USA
e-mail: mcginlm@ccf.org



A. L. Piquet, E. Alvarez (eds.), *Neuroimmunology*, https://doi.org/10.1007/978-3-030-61883-4_6

**Table 6.1** Purposes of diagnostic criteria

| |
|---|
| Define diagnostic categories (indicating certainty of the diagnosis) |
| Define relapse |
| Define progression |
| Provide criteria for dissemination in space and time |
| Define the role of laboratory and ancillary tests |
| Define the role of imaging |

## Clinical Criteria

### Clinical Features and Courses

MS was first described by Jean-Martin Charcot in 1868 based on the clinical triad of nystagmus, intention tremor, and scanning speech [1]. This description was the initial step to defining the clinical characteristics of the disease, which still form the foundation for diagnosis. An early attempt to define formal criteria and course was made by Allison and Millar in 1954 [2]. In this paper, four distinct classification schemes were proposed: early disseminated sclerosis, probable disseminated sclerosis, possible disseminated sclerosis, and discarded cases. This breakdown was based on clinical presentations and examination findings. The distinction between a relapsing and progressive course was also alluded to and was the main distinguishing feature between probable and possible cases. Another key concept was the need to exclude all other potential diagnoses. The concepts of core clinical criteria and courses coupled with exclusion of other potential etiologies have been retained in subsequent criteria.

Schumacher expanded on the clinical courses by acknowledging two distinct courses: relapsing episodes or slow/stepwise progression [3]. McDonald and Halliday combined the classification schemes and concept of different courses in 1977 [4]. They defined five diagnostic categories: clinically definite, early probably or latent, progressive probable, progressive possible, and suspected. These criteria were similar to Schumacher with further refinement of the distinct relapsing remitting and progressive courses. This criterion was created to provide a more reliable and structured framework to the diagnosis of MS to facilitate the development of therapeutic agents. Definite MS was defined by six criteria: (1) objective neurological abnormalities, (2) involvement of two or more separate parts of the central nervous system (CNS) based on history or exam, (3) objective evidence that the process predominately reflects white matter involvement, (4) evolution over time, (5) onset at age 10–50 years, and (6) the manifestations not better explained by another disease.

The Poser criteria in 1983 proposed two distinct categories: definite and probable MS [5]. The clinically definite MS was defined as clinical evidence of two separate attacks that localized to two anatomic sites typical of MS with no other explanation. The novel revision was the addition of definite MS by laboratory evidence, discussed more below. Probable MS was defined as two attacks with evidence of only one lesion. The purpose of incorporating probable MS was to allow prospective evaluation of new diagnostic methods.

New criteria were developed nearly 20 years later by McDonald et al. in 2001 [6]. Unlike many of the prior criteria, the so-called McDonald or international criteria represented an attempt to create a scheme intended not only for research but also for clinical use. In addition to integrating MRI, the McDonald criteria also aimed to simplify the scheme and formally incorporate primary progressive disease. Definite MS could be defined through five different presentations based on a combination of clinical events and paraclinical data. Diagnostic options were simplified to only three categories: MS, possible MS, and not MS. This simplification has continued throughout the McDonald revisions to make it easier to use in clinical practice, and the incorporation of MRI has helped facilitate earlier diagnosis (Table 6.2).

### Relapse Definition

Although it was recognized early that MS is associated with periods of acute clinical worsening, the first formal definition of relapse was by Schumacher [3]. He described the major prob-

**Table 6.2** Evolution of the McDonald criteria

| Clinical presentation | McDonald 2001 [6] and 2005 [15] | McDonald 2010 [7] and 2017 [9] |
|---|---|---|
| Two or more relapses *and* objective evidence of two or more lesions | No additional testing needed | No additional testing needed |
| Two or more relapses *and* objective evidence of one lesion | Dissemination in space by MRI *or* 2+MRI lesions plus positive CSF *or* second clinic relapse in different location | Dissemination in space proven by MRI *or* wait for the relapse with different symptomatology |
| One relapse *and* objective evidence of two lesions | Dissemination in time by MRI *or* two or more MRI lesions plus positive CSF and dissemination in time on MRI *or* second clinical relapse | Dissemination in time proven by MRI *or* wait for the second relapse (Note: 2017 allows for OCBs to substitute for DIT criteria) |
| One relapse and objective evidence of one lesion (clinically isolated syndrome) | Dissemination in space by MRI or two or more MRI lesions plus positive CSF and dissemination in time on MRI or second clinical relapse | Dissemination in time and space by MRI *or* wait for the second relapse *and* dissemination in space proven by MRI scan *or* wait for the second relapse with different symptomatology |

*CSF* cerebrospinal fluid, *DIT* dissemination in time, *MRI* magnetic resonance imaging, *OCBs* oligoclonal bands

lems associated with utilizing data around relapses including uncertain number of patients with this type of disease activity, wide disagreement among observers regarding the frequency, questionable validity of retrospective data, and lack of a uniform definition. The proposed definition stated the symptoms had to be new or aggravated for at least 24 hours with associated objective change on exam and preceded by at least a month of stability or improvement. This definition was continued to be used by McDonald and Halliday in 1977 [4]. Poser also utilized this basic definition but acknowledged that some symptoms may last for considerably less time, notably Lhermitte's sign or vertigo [5]. In the Poser criteria, these short-lasting symptoms were not defined as a relapse. Similarly, the 2001 McDonald criteria used the same definition but stated multiple paroxysmal events recurring over at least 24 hours could constitute a relapse [6]. In 2010, the panel clarified that before a definitive diagnosis of MS can be made, at least one relapse must be corroborated by findings on neurological exam, visual evoked potential (VEP), or MRI consistent with the reported symptoms [7]. When the clinical course of MS was redefined in 2013, the concept of "activity" was also introduced [8]. Activity in a relapsing remitting MS patient is indicated by a clinical relapse and/or MRI activity (contrast enhancing lesion or new or enlarging T2 lesion). Finally, the most recent 2017 McDonald criteria clarified that relapse, attack, and exacerbation are synonyms [9]. In addition to requiring the symptom to last 24 hours, it was also stipulated that it must occur in the absence of fever or infection. The refinement of the definition provides better understanding of what constitutes a relapse clinically and aids with distinguishing relapse from fluctuating symptoms, pseudo-relapses, or progression.

## Progression Definition

Although progression was first mentioned in the 1954 Millar and Allison criteria, it has been more difficult to define and utilize compared to relapse. Schumacher discussed progression as a potential clinical course and discussed quantitative methods to monitor neurological function. It was at this time that Kurtzke first developed a scale for evaluating disability in MS, which ultimately led to the expanded disability status scale (EDSS) [10, 11]. In the multiple iterations of the diagnostic criteria, the definition of progression offered is simply disability worsening that is sustained over at least 6 months, but an objective method to quantify this concept has not been agreed upon for the purposes of the diagnostic criteria. In the first iteration of the primary progressive MS diagnostic criteria, it was acknowledged that objective documentation of progression was desirable, but not always available [12]. The concept of progression was also incorporated in the definition of MS phenotypes. In these definitions, it was acknowledged that progression does not occur in a uniform fashion and should be evaluated by history or objective measures [8]. In the field, efforts have focused on the development of quantitative neurological testing, most notably the multiple sclerosis functional composite (MSFC) [13]. The most common way to demonstrate disability accrual in MS clinical trials and observational studies has been worsening on the EDSS or MSFC independent of relapses, but these rating scales or other methods have not been formally incorporated into diagnosis criteria.

## Evolution of the Concept of Dissemination in Time

The concept of dissemination in time (DIT) refers to identifying development of new CNS lesions over time. This concept was first suggested by McAlpine [14]. He detailed several typical clinical symptoms including acute retrobulbar neuritis, paraesthesias, motor weakness, sphincter disturbance, and brainstem signs. He emphasized the importance of obtaining a detailed history to elucidate prior episodes that suggested distinct CNS events. Although not explicitly stated, this description was the first attempt at incorporating DIT into diagnosis. It was not until the Schumacher criteria in 1965 that a formal requirement was stated [3]. This initial iteration refined the distinct clinical scenarios and clinical features and added the need for DIT and dissemination in space (DIS). Schumacher's description of temporal dissemination could be by either two or more episodes of worsening separated by a period of 1 month or more and lasting at least 24 hours (relapses) or slow/stepwise progression of signs and symptoms over a period of at least 6 months (progression). The goal of the time parameters was to prevent labeling a new clinical event that was a fluctuation attributed to another cause or acute nonrecurrent neurological disease. All of these features were used again by McDonald and Halliday in 1977 for the definition of "clinically definite" MS with the additional requirement that symptoms or signs were present for at least 1 year [4]. The Poser criteria utilized a similar concept.

**Table 6.3** Demonstration of dissemination in time by MRI in the McDonald criteria

| McDonald 2001 [6] | McDonald 2005 [15] | McDonald 2010 and 2017 [7, 9]* |
|---|---|---|
| (1) One Gd-enhancing lesion at least 3 months after the onset of first clinical symptoms. This lesion should not be the one responsible for the clinical symptoms<br>(2) Presence of a new T2 hyperintense lesion or a Gd-enhancing lesion on the second MRI scan performed not sooner than 3 months after the first scan | (1) Unchanged or<br>(2) Detection of a new T2 lesion if it appears at any time compared with a reference scan done at least *30 days* after the onset of the initial clinical event | (1) A new T2 and/or Gd-enhancing lesion(s) on follow-up MRI, with reference to a baseline scan, irrespective of the timing of the baseline MRI<br>(2) Simultaneous presence of asymptomatic Gd-enhancing and nonenhancing lesions at the any time<br>*2017 criteria make no distinction between symptomatic and asymptomatic lesions |

*Gd* gadolinium

The next shift in defining DIT occurred with the 2001 McDonald criteria [6]. Prior to this criterion, DIT had always been defined by clinical events, but with the advent of MRI, DIT could now be defined by either clinical events or MRI findings (Table 6.3). The goal of these definitions was to ensure any new lesion was unrelated to previous clinical events, intentionally conservative to maximize specificity of the criteria. In 2005 the definition changed to allow for a T2 lesion detected 30 days from the baseline MRI to fulfill DIT [15]. This change was made to allow for more rapid diagnosis. Further refinements in 2010 and 2017 removed the repeat scan time restriction and allowed for the presence of simultaneous Gd-enhancing and nonenhancing lesions to fulfill DIT (Fig. 6.1) [7, 9].

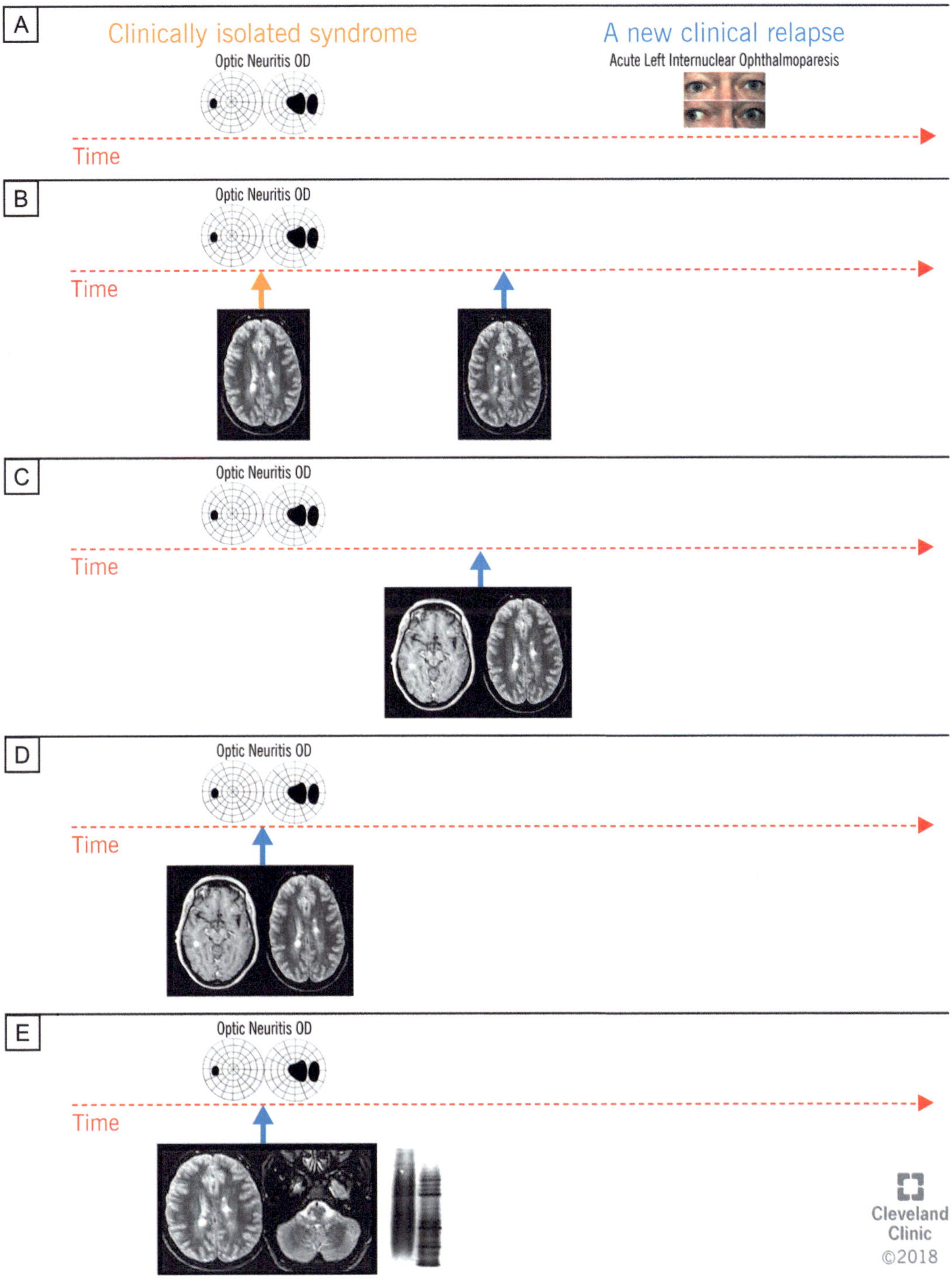


**Fig. 6.1** 2017 McDonald Criteria for Dissemination in Time. (**a**) Two clinical events. (**b**) Clinical event and a new T2 lesion on follow-up scan. (**c**) Clinical event and simultaneous gadolinium enhancing (GdE) and non-GdE lesion at any time. (**d**) Clinical event and simultaneous GdE and non-GdE lesions. (**e**) Clinical event and MRI demonstrating dissemination in space plus CSF specific oligoclonal bands. (Reprinted with permission, Cleveland Clinic Center for Medical Art & Photography © 2018-2020. All Rights Reserved)

## Evolution of the Concept of Dissemination in Space

The Schumacher criteria first required involvement of two or more distinct neuroanatomic sites in the CNS, which was the origin of the DIS concept [3]. He made the important point that the presence of multiple neurological symptoms is not sufficient, because a solitary lesion can produce a variety of symptoms. Therefore, careful examination to confirm true "multiplicity" of lesions is needed. The definition remained primarily a clinical definition until the Poser criteria, which allowed cerebrospinal fluid (CSF) (presence of unique CSF oligoclonal bands or elevated immunoglobulin G [IgG]) to be substituted for demonstration of a second lesion [5].

Another shift in this concept occurred with the incorporation of MRI in the McDonald criteria. As with DIT, MRI could now fulfill DIS (Table 6.4). The several most recent iteration of the McDonald criteria has required T2 lesions in two or more characteristic areas of the CNS: periventricular, juxtacortical/cortical, and infratentorial brain regions and spinal cord.

## Laboratory Evidence

The Schumacher criteria explicated stated that "no laboratory test pathognomonic of multiple sclerosis useful in the selection of cases for an experimental study has been discovered" [3]. This statement remains true today, but both CSF and other laboratory tests can contribute to diagnostic certainty. The Poser criteria were the first to include a diagnostic category that allowed laboratory evidence to be utilized in place of a clinical evidence to support the diagnosis [5].

## Cerebrospinal Fluid

Throughout the iterations of the diagnostic criteria, the role of CSF has fluctuated in importance. Starting with McAlpine, it was acknowledged that CSF was often abnormal in MS [14]. At that time, it was thought CSF was abnormal in ~70% of cases. The most commonly seen abnormalities were mild pleocytosis, increased protein, and abnormal large colloidal gold curve, which was related to the gamma-globulin content. The utility of CSF was discussed by Schumacher to rule out other conditions and support an MS diagnosis based on the presence of abnormalities that could be attributed to MS [3]. Recommended testing included manometrics, gross appearance, cell count, protein content, serological reaction for syphilis, and colloidal gold curve. There was growing evidence at this time that the test with the greatest diagnostic utility for MS was the quantification of the gamma-globulin content.

In the Poser criteria, "laboratory-supported definite MS" consisted of demonstration of CSF-

**Table 6.4** Demonstration of dissemination in space by MRI in the McDonald criteria

| McDonald 2001 [6] | McDonald 2005 [15] | McDonald 2010 [7] | McDonald 2017 [9] |
|---|---|---|---|
| Three of the four following conditions:<br>(1) At least one Gd-enhancing lesion or at least nine T2 hyperintense lesions<br>(2) At least one infratentorial lesion must be present<br>(3) At least one subcortical lesion must be present<br>(4) At least three periventricular lesions must be present<br>*Spinal cord lesion can substitute for a brain lesion | 2001 definitions *plus*<br>A spinal cord lesion can be considered equivalent to a brain infratentorial lesion<br>A Gd-enhancing spinal cord lesion is considered to be equivalent to an enhancing brain lesion, and individual spinal cord lesions can contribute together with individual brain lesions to reach the required number of T2 lesions | ≥1 T2 lesion in at least two of four areas: periventricular, juxtacortical, infratentorial, or spinal cord<br>*A Gd-enhancing lesion not required for DIS<br>*If subject has a brainstem or spinal cord lesion that is the symptomatic lesion, it is excluded | 2010 criteria *plus*<br>Cortical lesions can be used in fulfilling MRI criteria for dissemination in space<br>No distinction between symptomatic and asymptomatic MRI lesions is required |

*DIS* dissemination in space, *Gd* gadolinium

specific IgG oligoclonal bands or increased IgG synthesis rate [5]. It was stipulated that the oligoclonal bands must not be present in the patient's serum and the serum IgG level had to be normal. CSF abnormalities suggestive of another diagnosis must also be absent. In this criterion, one event could be clinical and the second "event" could be paraclinical CSF evidence. The incorporation of CSF for diagnosis had a goal of allowing for an earlier diagnosis.

CSF continued to be utilized in the McDonald 2001 criteria to satisfy DIS coupled with MRI definitions [6]. When the MRI criteria for DIS were liberalized in 2010, the role of CSF was only utilized to demonstrate the inflammatory demyelinating nature of the condition and rule out alternative diagnoses. CSF could no longer be utilized to fulfill DIS [7]. In the most recent McDonald criteria (2017), the utility of CSF was reemphasized, and now oligoclonal bands can be utilized in place of the requirement for demonstration of DIT [9]. A subsequent retrospective study demonstrated that oligoclonal bands increased the specificity for diagnosis (80.6 vs. 88.1) for clinically isolated syndrome patients that had DIS on MRI [16]. This finding further supports the use of oligoclonal bands in lieu of demonstration of DIT in the 2017 McDonald criteria, specifically the presence of oligoclonal bands that are CSF specific [9].

## Other Laboratory Testing

The minimum routine laboratory investigations initially proposed by Schumacher were intended to rule out other potential diagnoses [3]. Core testing included hematocrit, hemoglobin, leukocyte count, urinalysis, serological reaction for syphilis, and fasting glucose. Also, depending on the patient's history, further testing may include blood urea nitrogen, glucose tolerance test, serum electrophoresis, and gastric analysis. Over time, the differential diagnosis of MS and potential testing have evolved. The differential diagnosis can be extensive; therefore, the workup should be tailored to the individual patient. Vascular, autoimmune, infection, genetic, neurodegenerative, metabolic, and neoplastic etiologies should be evaluated under the appropriate clinical contexts. The potential for misdiagnosis of MS is still a concern; therefore, differentials should always be considered, and the clinician should remain vigilant for atypical features that suggest an alternative diagnosis. None of the diagnostic criteria were created to distinguish MS from other disorders, and a detailed discussion is beyond the scope of this chapter.

With the development of the aquaporin-4 (AQP4) serum antibody testing for neuromyelitis optica spectrum disorder (NMOSD), it was first recommended to be checked in the 2005 McDonald criteria, specifically in cases of recurrent optic neuritis and transverse myelitis [15]. In the 2010 McDonald criteria, consideration of NMOSD was again emphasized with special attention added to the occurrence of area postrema syndromes or periaqueductal medullary lesion on MRI [7]. Additionally, patients with antibodies to myelin-oligodendrocyte glycoprotein (MOG) have a clinical picture that overlaps that of anti-AQP4-positive NMOSD patients. As more information has accumulated about anti-AQP4 and anti-MOG diseases, the 2017 McDonald criteria recommended that the possibility of NMOSD be considered in all cases being evaluated for MS and tested for when there are manifestations suggestive of NMOSD [9]. The diagnostic criteria for NMOSD are further discussed in Chap. 15.

## Magnetic Resonance Imaging Evidence

Before the advent of MRI, imaging included roentgenography of the skull or spine and myelography, but these were considered ancillary testing if the clinical history warranted more detailed testing [3]. The Poser criteria also allowed for the use of computed tomography (CT) and nuclear magnetic resonance scans as paraclinical evidence to fulfill diagnostic criteria of definite MS, but it was not until MRI became clinically useful and generally available that imaging became a key factor in diagnostic criteria.

MRI findings were formally incorporated for the first time in the 2001 McDonald criteria (Table 6.3), based on the criteria proposed by Barkhof and Tintore [6, 17, 18]. In addition to specifying characteristic locations of the lesions, the criteria also stated the lesions should be at least 3 mm in cross section. The goal of these criteria was to create enough sensitivity while maintaining rigor to keep the criteria accurate. The 2005 McDonald criteria made no changes to location and number of lesion requirements except to bring more clarity to the utility of a spinal cord lesion for diagnosis [15]. The 2010 and 2017 McDonald criteria further simplified the criteria based on the number of lesions required and role of Gd-enhancing lesions [7, 9]. Both 2010 and 2017 McDonald criteria adopted the European Magnetic Resonance Imaging in Multiple Sclerosis (MAGNIMS) network recommendations, except that the 2017 criteria did not incorporate optic nerve as a fifth anatomic location and maintained that only one periventricular lesion was required [19–22]. MAGNIMS criteria had proposed that since a solitary periventricular lesion can occur in other disease states (migraine, vascular), a minimum of three periventricular lesions should be utilized. A recent comparison of the 2016 MAGNIMS criteria to the 2010 McDonald criteria reported similar accuracy for predicting the development of definite MS [23]. The removal of the distinction between symptomatic and asymptomatic lesions was intended to simplify use in clinical practice. Future efforts are needed to determine optimal number of lesions in certain locations—e.g., periventricular lesions and optic nerve—to find the ideal balance between diagnostic accuracy and sensitivity to allow early diagnosis.

## Primary Progressive MS

The McDonald criteria were primarily developed for “attack onset” MS, i.e., to make the diagnosis of MS beginning with a clinically isolated syndrome. Throughout the evolution of the diagnostic criteria, the existence of MS with progression from onset was acknowledged, but formal diagnostic criteria for primary progressive (PP) MS were not devised until 2000 [12]. To make the diagnosis of PPMS, a patient needed to have 1 year of progression and positive CSF and meet DIS criteria. Over time, the requirement for positive CSF was removed, and the MRI criteria for DIS reflected the changes in the overall diagnostic criteria (Table 6.5). A recent comparison of

**Table 6.5** Diagnostic criteria for primary progressive MS

| Thompson 2000 [13] and McDonald 2001 [6] | McDonald 2005 [7] | McDonald 2010 and 2017 [8, 9] |
|---|---|---|
| Clinical progression for 1 year<br>*and*<br>Positive CSF<br>*and*<br>Dissemination in space demonstrated by:<br>(1) Nine or more T2 lesions in brain, *or*<br>(2) Two or more lesions in the spinal cord, *or*<br>(3) Four to eight brain plus one spinal cord lesion *or* abnormal VEP associated with four to eight brain lesions *or* with fewer than four brain lesions plus one spinal cord lesion demonstrated by MRI<br>*and*<br>Dissemination in time, demonstrated by MRI, *or*<br>Continued progression for 1 year | (1) One year of disease progression (retrospectively or prospectively determined)<br>(2) Plus two of the following:<br>(a) Positive brain MRI (nine T2 lesions or four or more T2 lesions with positive VEP)<br>(b) Positive spinal cord MRI (two focal T2 lesions)<br>(c) Positive CSF (isoelectric focusing evidence of oligoclonal IgG bands or increased IgG index or both) | One year of disease progression (retrospectively or prospectively determined) plus two of three of the following criteria:<br>(1) Evidence for DIS in the brain based on one T2 lesions in the MS characteristic (periventricular, juxtacortical, or infratentorial regions)<br>(2) Evidence for DIS in the spinal cord based on two T2 lesions in the cord<br>(3) Positive CSF (isoelectric focusing evidence of oligoclonal bands and/or elevated IgG index) |

*CSF* cerebrospinal fluid, *DIS* dissemination in space, *IgG* immunoglobulin G, *MRI* magnetic resonance imaging, *VEP* visual evoked potentials the 2010 McDonald criteria and MAGNIMS 2016 criteria indicated that 2016 MAGNIMS criteria for DIS, incorporation of CSF, and at least 1 year of progression had the highest sensitivity and specificity. The 2013 revised MS phenotypes also introduced the concept of "activity" to progressive disease. In addition to the formal diagnostic criteria, progressive patients could further be defined by the presence or absence of activity (clinical relapse and/or MRI activity) [8].

## Diagnostic Criteria in Diverse Populations

The McDonald criteria were originally developed for and validated in adult Caucasian cohorts of Western European ethnic origin. There have been several small studies that have indicated that the criteria are valid in diverse populations including Canada [24], Italy [25], the Netherlands [26], Spain [27], Russia [28], Asia [29–31], Middle East [32, 33], and Latin American [34, 35]. Care is needed when making the diagnosis of MS outside of Europe and North America, particularly to address potential mimics, including infectious diseases and NMOSD.

MS typically presents between the ages of 20 and 40, but both pediatric-onset MS (POMS) and later-onset MS are well-recognized variants. POMS typically presents with a relapsing-remitting course, but there are several unique features that can complicate diagnosis. The 2010 McDonald criteria were evaluated in postpubertal patients (≥12 years) with a reported positive predictive value of 76% and negative predictive value of 100% [36]. For younger, prepubertal children (<12 years), the positive predictive value was only 55%. In order to address some of the unique challenges in POMS diagnosis, the International Pediatric MS Study Group (IPMSSG) revised the proposed POMS diagnostic to allow acute disseminated encephalomyelitis to be the first clinical presentation of POMS as long as it is followed by a clinical event more typical of an MS relapse 3 or more months after symptom onset and is associated with new MRI lesions fulfilling the 2010 McDonald criteria [37]. A recent study of 2017 McDonald criteria in POMS indicated that the inclusion of CSF improved performance in POMS [38]. This study also looked at the utility of testing anti-MOG serology and found it did not improve performance of the diagnostic criteria. As in adults, it should be tested in patients with atypical features.

Later-onset MS is defined as MS initially appearing after age 50. Recent reports indicate it remains rare (3–4%) and most often presents with a progressive course [39, 40]. This population has more comorbidities, which need to be considered when interpreting symptoms and test results. Since late onset is atypical of MS, this is another situation in which CSF testing should be considered to increase diagnostic certainty. It is thought that the McDonald criteria are applicable in this age group, but further research is needed to validate this assumption.

## Balancing Early Diagnosis and Avoiding Misdiagnosis

The desire to diagnose early while avoiding misdiagnosis has been a concern for decades [41–44]. McAlpine first discussed this issue in 1957 when he noted that most individuals are not diagnosed until the fourth or fifth year of disease [14]. In his description of MS, he pleaded for early diagnosis. Since that time, there has been an effort to simplify the diagnostic criteria for utilization in clinical practice and facilitate earlier diagnosis in patients likely to have MS, particularly as effective therapies emerged. As the criteria have been simplified, there has been an increasing concern they are too liberal, potentially resulting in more misdiagnosis. In addition to the changes in criteria, the widespread use of MRI has resulted in identification of nonspecific abnormalities that can lead to inappropriate diagnoses without careful consideration of the entire clinical picture. This topic was discussed in the 2017 McDonald criteria revision [9]. Previous reports indicated that a sizable number of people diagnosed with MS are misdiagnosed [45]. This is a significant concern because many of these patients are started on disease-modifying therapy (DMT), which is not risk-free. One report of 110 misdiagnosed patients found that 70% of these patients received DMT and 31% experienced unnecessary morbidity because of the misdiagnosis [45]. The most common disorders that were misdiagnosed as MS were migraine, fibromyalgia, nonspecific symptoms with abnormal MRI, and NMOSD. Misdiagnosis is, unfortunately, partly a consequence of the lack of any unique diagnostic marker for MS; the diagnosis is based on a combination of clinical, laboratory, and imaging findings. It remains crucial for clinicians to remain vigilant for atypical features, because ultimately every diagnostic criterion requires that there is not a more appropriate diagnosis. There will always be the trade-off of sensitivity and specificity for the MS diagnostic criteria. Therefore, future efforts should focus on development of laboratory, imaging, and clinical measures that aid the diagnostic process. Future research should also be done to validate the 2017 McDonald criteria in a routine clinical practice setting with an unselected population.

## Areas That Need Further Refinement

### Nonclassical Situations: Radiologically Isolated Syndrome, Solitary Sclerosis, Etc.

As MRI has advanced and became readily available, people are sometimes identified with imaging abnormalities characteristic of MS but without any definite clinical manifestations, termed radiologically isolated syndrome (RIS) [46]. Although the situation is rare (0.8 cases per 100,000), approximately one-third of RIS patients are diagnosed with MS within 5 years [47]. When a patient meets radiological criteria for DIS and DIT and has positive CSF, some studies have demonstrated subtle sequelae [48]. These findings have led to the recommendation that these patients should be treated as MS. However, making the erroneous diagnosis of MS in patients with absent or nonspecific clinical manifestations and nonspecific MRI is a common source of misdiagnosis [45]. Therefore, the 2017 McDonald criteria maintained the requirement for a definitive clinically isolated syndrome or progression to make an MS diagnosis.

Similarly, there are patients who have an isolated lesion with a progressive course, termed solitary sclerosis [49]. Again, these patients do not meet the diagnostic criteria for primary progressive MS due to insufficient MRI findings to fulfill DIS.

### Visual System

The visual system is commonly affected in MS, but there has not been a consistent approach to how to incorporate this manifestation into diagnostic criteria. Optic neuritis is a classic clinical isolated syndrome presentation, but the utility of visual system objective testing is unclear. In the 2001 McDonald criteria, there was an attempt made by allowing abnormal VEP to be used as supplemental information to provide objective evidence of a second lesion [6]. The 2016 MAGNIMS criteria went a step further by proposing the optic nerve as a fifth anatomical location to fulfill MRI criteria for DIS, in patients without a clear-cut history of optic neuritis [22]. This recommendation was not adopted in the 2017 McDonald criteria, because this change only minimally improved sensitivity and lowered the specificity. Optical coherence tomography (OCT) also has demonstrated some potential as a meaningful assessment in MS. A recent study that investigated the OCT-derived ganglion cell and inner plexiform layer thickness in patients with clinically isolated syndrome indicated that patients with lower thickness of the retinal nerve fiber layer were more likely to have recurrent disease activity [50]. The existing data for VEP and OCT suggest they may be more useful than MRI to confirm a prior optic neuritis, but the current research is inadequate to justify their ability to fulfill criteria for DIS. There is a need for better evidence supporting and validating the objective visual system techniques, to refine their value for MS diagnosis.

## Advanced MRI Measures

MRI has become an important component to the diagnostic criteria, but since the initial incorporation into the 2001 McDonald criteria, only T2 lesions and Gd lesions have been utilized. The 2017 McDonald criteria were the first to incorporate cortical lesions but acknowledged that currently there is limited ability to identify these lesions with available MRI techniques [23, 51, 52]. A recent study found that less than half of cortical lesions were reliably detected with 3T double inversion recovery sequences, thus emphasizing the need to develop and validate techniques to identify these lesions in clinical practice [53].

The McDonald MRI criteria have focused on T2 lesion location and number, but there is a need to develop techniques that are capable of differentiating MS lesions from other etiologies. Potential novel techniques include the central vein sign and thalamic volume and warrant further validation [54–56]. MRI criteria are further complicated by the recent discovery of the so-called myelocortical MS (MCMS) [57], in which patients had typical T2 white matter lesions on MRI but on pathological examination had demyelinating lesions only in the cortex and spinal cord but not in the cerebral white matter. These patients illustrate the lack of specificity of currently used MRI markers of disease.

## Nonimaging Biomarkers

A significant obstacle in the diagnosis of MS is the lack of a diagnostic biomarker. With the exception of CSF oligoclonal bands that support the diagnosis of MS in the appropriate setting but are not specific for MS, all of the other laboratory testing sent as part of the diagnostic process is primarily intended to rule out other conditions. When a biomarker is identified, the accurate identification of a disease state dramatically improves. This was well illustrated with NMOSD, which has a very sensitive and specific biomarker. Diagnostic biomarkers have been proposed for MS, but none has been validated [58].

## Conclusion

The evolution of diagnostic criteria for MS reflects our improved understanding of the disease process and advances in medical technology. From the initial clinical description by Charcot, the clinical criteria have been refined to include better defined syndromes and improved definitions of dissemination in space and time. The addition of MRI dramatically changed the diagnostic process, allowing for early and more accurate identification of MS. Nevertheless, there are several areas that need additional studies to allow for better clinical and research outcomes.

## References


1. Charcot JM. Lectures on the diseases of the nervous system: delivered at La Salpêtrière. Lea, 1879, 1889.
2. Allison RS, Millar JH. Prevalence of disseminated sclerosis in Northern Ireland. Ulster Med J. 1954;23:1–27.
3. Schumacher GA, Beebe G, Kibler RF, Kurland LT, Kurtzke JF, McDowell F, et al. Problems of experimental trials of therapy in multiple sclerosis: report by the panel on the evalaution of experimental trials of therapy in multiple sclerosis. Ann N Y Acad Sci. 1965;122:552–68.
4. McDonald WI, Halliday AM. Diagnosis and classification of multiple sclerosis. Br Med Bull. 1977;33:4–9.
5. Poser CM, Paty DW, Scheinberg L, McDonald WI, Davis FA, Ebers GC, et al. New diagnostic criteria for multiple sclerosis: guidelines for research protocols. Ann Neurol. 1983;13:227–31.
6. McDonald WI, Compston A, Edan G, Goodkin D, Hartung HP, Lublin FD, et al. Recommended diagnostic criteria for multiple sclerosis: guidelines from the International Panel on the diagnosis of multiple sclerosis. Ann Neurol. 2001;50:121–7.
7. Polman CH, Reingold SC, Banwell B, Clanet M, Cohen JA, Filippi M, et al. Diagnostic criteria for multiple sclerosis: 2010 revisions to the McDonald criteria. Ann Neurol. 2011;69:292–302.
8. Lublin FD, Reingold SC, Cohen JA, Cutter GR, Sørensen PS, Thompson AJ, et al. Defining the clinical course of multiple sclerosis: the 2013 revisions. Neurology. 2014;83:278–86.
9. Thompson AJ, Banwell BL, Barkhof F, Carroll WM, Coetzee T, Comi G, et al. Diagnosis of multiple sclerosis: 2017 revisions of the McDonald criteria. Lancet Neurol. 2018 Feb;17(2):162–73.
10. Kurtzke JF. On the evaluation of disability in multiple sclerosis. Neurology. 1961;11:686–94.

11. Kurtzke JF. Rating neurologic impairment in multiple sclerosis: an expanded disability status scale (EDSS). Neurology. 1983;33:1444–52.
12. Thompson AJ, Montalban X, Barkhof F, Brochet B, Filippi M, Miller DH, et al. Diagnostic criteria for primary progressive multiple sclerosis: a position paper. Ann Neurol. 2000;47:831–5.
13. Rudick R, Antel J, Confavreux C, Cutter G, Ellison G, Fischer J, et al. Recommendations from the national multiple sclerosis society clinical outcomes assessment task force. Ann Neurol. 1997;42:379–82.
14. Mcalpine D. Multiple sclerosis: a plea for a fresh outlook. Br Med J. 1957;1:475–80.
15. Polman CH, Reingold SC, Edan G, Filippi M, Hartung HP, Kappos L, et al. Diagnostic criteria for multiple sclerosis: 2005 revisions to the "McDonald Criteria". Ann Neurol. 2005;58:840–6.
16. Arrambide G, Tintore M, Espejo C, Auger C, Castillo M, Río J, et al. The value of oligoclonal bands in the multiple sclerosis diagnostic criteria. Brain. 2018;141:1075–84.
17. Barkhof F, Filippi M, Miller DH, Scheltens P, Campi A, Polman CH, et al. Comparison of MRI criteria at first presentation to predict conversion to clinically definite multiple sclerosis. Brain. 1997;120(Pt 1):2059–69.
18. Tintoré M, Rovira A, Martínez MJ, Rio J, Díaz-Villoslada P, Brieva L, et al. Isolated demyelinating syndromes: comparison of different MR imaging criteria to predict conversion to clinically definite multiple sclerosis. AJNR Am J Neuroradiol. 2000;21:702–6.
19. Swanton JK, Rovira A, Tintore M, Altmann DR, Barkhof F, Filippi M, et al. MRI criteria for multiple sclerosis in patients presenting with clinically isolated syndromes: a multicentre retrospective study. Lancet Neurol. 2007;6:677–86.
20. Rovira A, Swanton J, Tintoré M, Huerga E, Barkhof F, Filippi M, et al. A single, early magnetic resonance imaging study in the diagnosis of multiple sclerosis. Arch Neurol. 2009;66:587–92.
21. Montalban X, Tintore M, Swanton J, Barkhof F, Fazekas F, Filippi M, et al. MRI criteria for MS in patients with clinically isolated syndromes. Neurology. 2010;74:427–34.
22. Filippi M, Rocca MA, Ciccarelli O, De Stefano N, Evangelou N, Kappos L, et al. MAGNIMS Study Group. MRI criteria for the diagnosis of multiple sclerosis: MAGNIMS consensus guidelines. Lancet Neurol. 2016;15:292–303.
23. Filippi M, Preziosa P, Meani A, Ciccarelli O, Mesaros S, Rovira A, et al. Prediction of a multiple sclerosis diagnosis in patients with clinically isolated syndrome using the 2016 MAGNIMS and 2010 McDonald criteria: a retrospective study. Lancet Neurol. 2018;17:133–42.
24. Kang H, Metz LM, Traboulsee AL, Eliasziw M, Zhao GJ, Cheng Y, et al. Application and a proposed modification of the 2010 McDonald criteria for the diagnosis of multiple sclerosis in a Canadian cohort of patients with clinically isolated syndromes. Mult Scler. 2014;20:458–63.
25. D'Alessandro R, Vignatelli L, Lugaresi A, Baldin E, Granella F, Tola MR, et al. Risk of multiple sclerosis following clinically isolated syndrome: a 4-year prospective study. J Neurol. 2013;260:1583–93.
26. Runia TF, Jafari N, Hintzen RQ. Application of the 2010 revised criteria for the diagnosis of multiple sclerosis to patients with clinically isolated syndromes. Eur J Neurol. 2013;20:1510–6.
27. Gómez-Moreno M, Díaz-Sánchez M, Ramos-González A. Application of the 2010 McDonald criteria for the diagnosis of multiple sclerosis in a Spanish cohort of patients with clinically isolated syndromes. Mult Scler. 2012;18:39–44.
28. Belova AN, Shalenkov IV, Shakurova DN, Boyko AN. Revised McDonald criteria for multiple sclerosis diagnostics in central Russia: sensitivity and specificity. Mult Scler. 2014;20:1896–9.
29. Hsueh CJ, Kao HW, Chen SY, Lo CP, Hsu CC, Liu DW, et al. Comparison of the 2010 and 2005 versions of the McDonald MRI criteria for dissemination-in-time in Taiwanese patients with classic multiple sclerosis. J Neurol Sci. 2013;329:51–4.
30. Piccolo L, Kumar G, Nakashima I, Misu T, Kong Y, Wakerley B, et al. Multiple sclerosis in Japan appears to be a milder disease compared to the UK. J Neurol. 2015;262:831–6.
31. Huh SY, Kim SH, Kim W, Lee SH, Park MS, Ahn SW, et al. Evaluation of McDonald MRI criteria for dissemination in space in Korean patients with clinically isolated syndromes. Mult Scler. 2014;20:492–5.
32. Yamout B, Alroughani R, Al-Jumah M, Khoury S, Abouzeid N, Dahdaleh M, et al. Consensus guidelines for the diagnosis and treatment of multiple sclerosis. Curr Med Res Opin. 2013;29:611–21.
33. Alroughani R, Al Hashel J, Lamdhade S, Ahmed SF. Predictors of conversion to multiple sclerosis in patients with clinical isolated syndrome using the 2010 revised McDonald criteria. ISRN Neurol. 2012;2012:792192.
34. Patrucco L, Rojas JI, Miguez JS, Cristiano E. Application of the McDonald 2010 criteria for the diagnosis of multiple sclerosis in an Argentinean cohort of patients with clinically isolated syndromes. Mult Scler. 2013;19:1297–301.
35. Patrucco L. Application of the McDonald criteria in Latin America. Mult Scler J - Exp Transl Clin. 2017;3:2055217317721943.
36. Sadaka Y, Verhey LH, Shroff MM, Branson HM, Arnold DL, Narayanan S, et al. Canadian Pediatric Demyelinating Disease Network. 2010 McDonald criteria for diagnosing pediatric multiple sclerosis. Ann Neurol. 2012;72:211–23.
37. Tardieu M, Banwell B, Wolinsky JS, Pohl D, Krupp LB. Consensus definitions for pediatric MS and other demyelinating disorders in childhood. Neurology. 2016;87:S8–S11.
38. Fadda G, Brown RA, Longoni G, Castro DA, O'Mahony J, Verhey LH, et al. Canadian Pediatric

Demyelinating Disease Network. MRI and laboratory features and the performance of international criteria in the diagnosis of multiple sclerosis in children and adolescents: a prospective cohort study. Lancet Child Adolesc Health. 2018;2:191–204.

39. Roohani P, Emiru T, Carpenter A, Luzzio C, Freeman J, Scarberry S, et al. Late onset multiple sclerosis: is it really late onset? Mult Scler Relat Disord. 2014;3:444–9.
40. Lotti CB de C, Oliveira ASB, Bichuetti DB, Castro I, Oliveira EML. Late onset multiple sclerosis: concerns in aging patients. Arq Neuropsiquiatr. 2017;75:451–6.
41. Freedman MS, Francis GS, Sanders EA, Rice GP, O'Connor P, Comi G, et al. Once weekly interferon beta-1alpha for Multiple Sclerosis Study Group; University of British Columbia MS/MRI Research Group. Randomized study of once-weekly interferon beta-1la therapy in relapsing multiple sclerosis: three-year data from the OWIMS study. Mult Scler. 2005;11:41–5.
42. Murray TJ, Murray SJ. Characteristics of patients found not to have multiple sclerosis. Can Med Assoc J. 1984;131:336–7.
43. Poser CM. Misdiagnosis of multiple sclerosis and β-interferon. Lancet. 1997;349:1916.
44. Carmosino MJ, Brousseau KM, Arciniegas DB, Corboy JR. Initial evaluations for multiple sclerosis in a university multiple sclerosis center: outcomes and role of magnetic resonance imaging in referral. Arch Neurol. 2005;62:585–90.
45. Solomon AJ, Bourdette DN, Cross AH, Applebee A, Skidd PM, Howard DB, et al. The contemporary spectrum of multiple sclerosis misdiagnosis. Neurology. 2016;87:1393–9.
46. Okuda DT, Mowry EM, Beheshtian A, Waubant E, Baranzini SE, Goodin DS, et al. Incidental MRI anomalies suggestive of multiple sclerosis: the radiologically isolated syndrome. Neurology. 2009;72:800–5.
47. Okuda DT, Siva A, Kantarci O, Inglese M, Katz I, Tutuncu M, et al. Radiologically Isolated Syndrome Consortium (RISC); Club Francophone de la Sclérose en Plaques (CFSEP). Radiologically isolated syndrome: 5-year risk for an initial clinical event. PLoS One. 2014;9:e90509.
48. Amato MP, Hakiki B, Goretti B, Rossi F, Stromillo ML, Giorgio A, et al. Italian RIS/MS Study Group. Association of MRI metrics and cognitive impairment in radiologically isolated syndromes. Neurology. 2012;78:309–14.
49. Keegan BM, Kaufmann TJ, Weinshenker BG, Kantarci OH, Schmalstieg WF, Paz Soldan MM, et al. Progressive solitary sclerosis: gradual motor impairment from a single CNS demyelinating lesion. Neurology. 2016;87:1713–9.
50. Zimmermann HG, Knier B, Oberwahrenbrock T, Behrens J, Pfuhl C, Aly L, et al. Association of Retinal Ganglion Cell Layer Thickness with future disease activity in patients with clinically isolated syndrome. JAMA Neurol. 2018;75:1071–9.
51. Filippi M, Rocca MA, Calabrese M, Sormani MP, Rinaldi F, Perini P, et al. Intracortical lesions: relevance for new MRI diagnostic criteria for multiple sclerosis. Neurology. 2010;75:1988–94.
52. Preziosa P, Rocca MA, Mesaros S, Meani A, Montalban X, Drulovic J, et al. Diagnosis of multiple sclerosis: a multicentre study to compare revised McDonald-2010 and Filippi-2010 criteria. J Neurol Neurosurg Psychiatry. 2017;89:jnnp-2017-315863.
53. Faizy TD, Thaler C, Ceyrowski T, Broocks G, Treffler N, Sedlacik J, et al. Reliability of cortical lesion detection on double inversion recovery MRI applying the MAGNIMS-criteria in multiple sclerosis patients within a 16-months period. PLoS One. 2017;12:e0172923.
54. Sati P, Oh J, Constable RT, Evangelou N, Guttmann CR, Henry RG, et al. NAIMS Cooperative. The central vein sign and its clinical evaluation for the diagnosis of multiple sclerosis: a consensus statement from the North American Imaging in Multiple Sclerosis Cooperative. Nat Rev Neurol. 2016;12:714–22.
55. Azevedo CJ, Overton E, Khadka S, Buckley J, Liu S, Sampat M, et al. Early CNS neurodegeneration in radiologically isolated syndrome. Neurol Neuroimmunol Neuroinflammation. 2015;2:e102.
56. Zivadinov R, Havrdová E, Bergsland N, Tyblova M, Hagemeier J, Seidl Z, et al. Thalamic atrophy is associated with development of clinically definite multiple sclerosis. Radiology. 2013;268:831–41.
57. Trapp BD, Vignos M, Dudman J, Chang A, Fisher E, Staugaitis SM, et al. Cortical neuronal densities and cerebral white matter demyelination in multiple sclerosis: a retrospective study. Lancet Neurol. 2018;17:870–84.
58. Thouvenot É. Multiple sclerosis biomarkers: helping the diagnosis? Rev Neurol (Paris). 2018;174:364–71.

# Natural History of Multiple Sclerosis

7

Laura E. Baldassari and M. Mateo Paz Soldán

**Key Points**

1. The clinical presentation of multiple sclerosis (MS) can vary among patients in terms of age at onset, rate of disability worsening, and clinical symptoms.
2. Different phenotypes and presentations along the spectrum of central nervous system (CNS) demyelinating disease, including clinically and radiologically isolated syndromes, have heterogeneous clinical features that have variable prognostic utility.
3. Relapsing-remitting MS (RRMS) is the most common MS phenotype and is characterized by relapses consistent with CNS demyelination, from which there is a spectrum of recovery.
4. Over time, the majority of patients with RRMS transition from predominantly inflammatory to more neurodegenerative driving forces of the disease, with age-dependent development of secondary progressive (SP) MS. Use of disease modifying therapies (DMTs) has reduced the incidence of conversion to SPMS and ultimate level of disability.
5. Clinical and imaging features in MS that aid in prognostication include age at disease onset, gender, race, relapse rate, recovery from relapses, gadolinium-enhancing lesions, infratentorial and brainstem lesions, tobacco use, and the presence of comorbidities.

## Introduction

The clinical presentation of multiple sclerosis (MS) can vary significantly among patients in terms of age at onset, rate of disability worsening, and clinical symptoms. MS phenotypes have been described and refined over time [1], but even within subtypes of the disease, presentations can be quite heterogeneous.

In this chapter, we review the literature regarding the natural history of multiple sclerosis by phenotype, known factors that can influence its clinical course, and the impact of disease modifying therapies (DMTs) on MS. Several historic observational studies of population-, clinic- and hospital-based MS patient cohorts have made important contributions to the understanding of MS and will be discussed as well. Although some aspects of each topic will be briefly mentioned in

L. E. Baldassari
Mellen Center for Multiple Sclerosis Treatment and Research, Neurological Institute, Cleveland Clinic Foundation, Cleveland, OH, USA

M. M. Paz Soldán (✉)
Neuroimmunology Division, Department of Neurology, University of Utah School of Medicine, Salt Lake City, UT, USA
e-mail: mateo.pazsoldan@hsc.utah.edu



A. L. Piquet, E. Alvarez (eds.), *Neuroimmunology*, https://doi.org/10.1007/978-3-030-61883-4_7

this chapter, detailed discussions regarding MS diagnostic criteria, epidemiology, and genetics can be found in Chaps. 6, 10, and 11, respectively.

Disability in MS has historically been tracked using the Expanded Disability Status Scale (EDSS) [2], in which patients receive a score ranging from 0 to 10 based on seven functional systems and ambulation scores. In the EDSS, 0 indicates no disability, 6.0 indicates the need for a unilateral walking aid, and 9.0 indicates that the patient is bedbound from MS (Fig. 7.1). Several studies discussed in this chapter utilize the EDSS as a benchmark for disability development.

## Multiple Sclerosis Phenotypes

Defining the course of MS is an evolving area that, similar to diagnostic criteria, is periodically updated. This section will discuss MS phenotypes according to the Lublin-Reingold MS phenotypes described in 2014 [1]. In these criteria, patients with MS are characterized as having either relapsing-remitting (RRMS), secondary progressive (SPMS), or primary progressive multiple sclerosis (PPMS) (Fig. 7.2). Each of these three phenotypes is further qualified as having activity in the form of gadolinium-enhancing or new T2 lesions on magnetic resonance imaging (MRI) or in the form of clinical relapses. Additionally, the progressive MS phenotypes are further qualified as to whether they have evident clinical progression independent from relapses.

Different phenotypes and presentations along the spectrum of CNS demyelinating disease, including clinically and radiologically isolated syndromes (CIS and RIS, respectively), have heterogeneous clinical and prognostic features of which a clinician should be aware (Tables 7.1 and 7.2).

### Radiologically Isolated Syndrome

The term radiologically isolated syndrome (RIS) is applied when a patient has MRI findings typical of MS, but without any clinical symptoms [3]. Such incidental findings are often revealed when a patient undergoes MRI for other indications, such as headache. Given the absence of clinical symptoms, patients with RIS are not considered

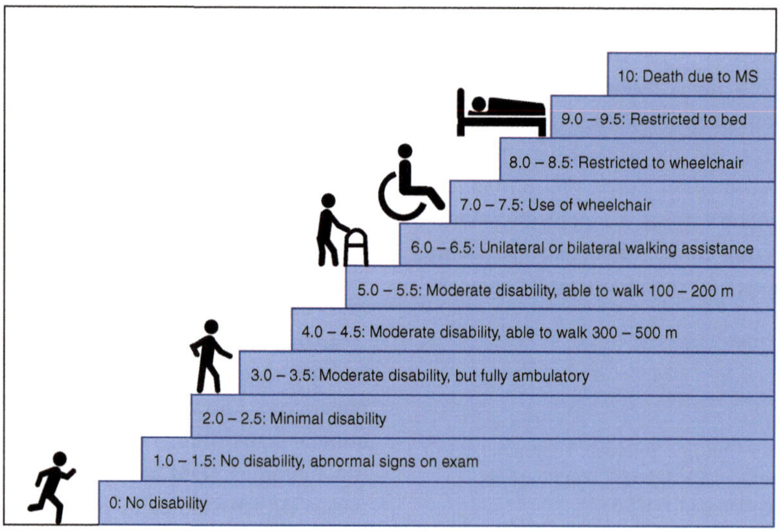


**Fig. 7.1** The Expanded Disability Status Scale (EDSS) is considered the standard disability measure in multiple sclerosis clinical trials. Patients are assigned an overall EDSS score based on their disability within seven functional systems (vision, brainstem, pyramidal, cerebellar, sensory, bowel/bladder, cerebral) and ambulation ability. In the EDSS, 0 indicates no disability, 4.0 indicates ambulation limitations, 6.0 indicates the need for a unilateral walking aid, and 9.0 indicates that the patient is essentially bedbound from MS

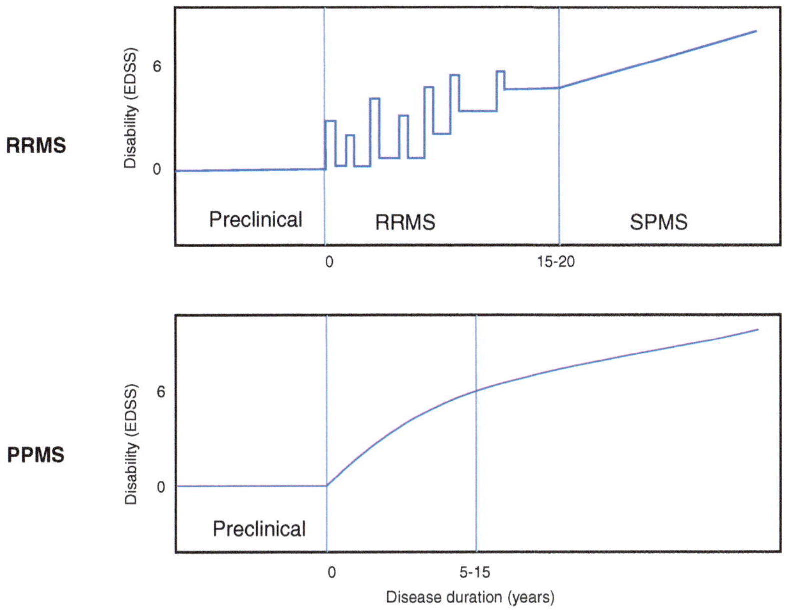


**Fig. 7.2** Comparison of multiple sclerosis phenotypes. Patients with relapsing-remitting multiple sclerosis (RRMS) present initially with relapses from which there is variable recovery. After approximately 15 to 20 years, patients generally transition to secondary progressive multiple sclerosis (SPMS), which is characterized by gradual accumulation of disability with or without superimposed relapses. Generally, patients reach an Expanded Disability Status Scale (EDSS) score of 6 within 15 to 20 years after RRMS onset. Patients with primary progressive multiple sclerosis (PPMS) have a gradual accumulation of disability from onset, with or without superimposed relapses

**Table 7.1** Predictors of conversion to multiple sclerosis from radiologically and clinically isolated syndromes

| | Radiologically isolated syndrome | Clinically isolated syndrome |
|---|---|---|
| Demographic characteristics | Younger age<br>Male gender (possible) | Younger age<br>Non-Caucasian race |
| Clinical characteristics | N/A | Myelitis or brainstem presentation (possible) |
| Imaging characteristics | Gadolinium-enhancing lesions on MRI<br>Posterior fossa involvement<br>Spinal cord lesions | Abnormal brain MRI |
| Ancillary testing | Abnormal visual evoked potentials | Abnormal cerebrospinal fluid (either positive oligoclonal bands or increased intrathecal IgG synthesis) |

*MRI* magnetic resonance imaging, *IgG* immunoglobulin G

to have MS. However, patients with RIS can transition to either RRMS or PPMS.

Several observational studies have followed cohorts of RIS patients over time to determine its natural history [3–8]. It is estimated that between 27% and 91% of patients with RIS experience radiographic dissemination in time (DIT), with a mean time to DIT of 0.6 to 2.7 years. Estimates for the overall percentage of patients who have conversion from RIS to either CIS or definite MS range from 15% to 50% across studies, with a mean time to clinical conversion ranging between 2.3 and 6 years [4]. In terms of CIS alone, 31% to 45% of patients converted to CIS over a mean period of 2.3 to 4.1 years. In terms of clinically definite MS, studies estimate that 3% to 36% of patients with RIS will convert to definite MS over a mean of 2.4 to 8.5 years. Although the majority of patients who convert from RIS to clinically definite MS develop RRMS, there is a subset of patients with RIS who develop PPMS [4, 6].

**Table 7.2** Summary of prognostic factors in multiple sclerosis

| | Factors associated with better prognosis | Factors associated with worse prognosis |
|---|---|---|
| Demographic characteristics | | Male gender<br>Older age at onset<br>African-American race |
| Clinical characteristics | Use of disease modifying therapies<br>Isolated optic neuritis | Early, frequent, severe relapses<br>Incomplete relapse recovery<br>Rapid conversion to SPMS<br>Multifocal presentations involving motor function, bowel/bladder function, cerebellum, cognition |
| Imaging characteristics | Minimal lesion burden at presentation | Multiple gadolinium-enhancing lesions<br>Infratentorial lesions<br>Spinal cord involvement<br>Early brain atrophy<br>T1 hypointensities |
| Comorbidities and health behaviors | | Tobacco use<br>Presence of medical comorbidities, including vitamin D deficiency<br>Presence of mental health comorbidities |

*SPMS* secondary progressive multiple sclerosis

Additionally, studies have investigated potential characteristics associated with RIS that are predictive of development of CIS or definite MS (Table 7.1) [4]. Clinical and demographic characteristics associated with development of CIS or MS include younger age (less than 35 or 37 years) and male gender in some studies [6–8] but not in others [3]. In terms of MRI characteristics, the presence of gadolinium-enhancing lesions [3, 4, 8], spinal cord involvement [5–7], and posterior fossa involvement [6] are more predictive of developing CIS or clinically definite MS. The number of T2 lesions at presentation has not been predictive of CIS or MS [8]. Abnormal cerebrospinal fluid (CSF) findings have not consistently been associated with increased risk of CIS or MS in the setting of RIS [8]. Abnormal visual evoked potential testing [8] and retinal nerve fiber layer thickness on optical coherence tomography [9] have also been identified as predictors of conversion to CIS or MS.

## Clinically Isolated Syndrome

Clinically isolated syndrome (CIS) refers to a single demyelinating event, such as optic neuritis or partial myelitis, that is often the first clinical manifestation of MS. Certain patients can meet diagnostic criteria for MS at the time of CIS; it is important to recognize that criteria have changed over time, which has resulted in newer CIS cohorts averaging less disease burden than older CIS cohorts. Per 2017 McDonald Diagnostic Criteria, patients can be diagnosed with MS at the time of CIS if there are both enhancing and non-enhancing lesions characteristic of MS on MRI that would meet criteria for dissemination in time and space or if dissemination in space criteria are met and there are CSF-specific oligoclonal bands (OCBs) present [10]. In patients who do not meet MS diagnostic criteria at the time of CIS, certain characteristics are associated with a higher risk of developing MS in the future and therefore warrant closer clinical and radiographic monitoring in these patients (Table 7.1). Of note, changes in diagnostic criteria have resulted in more patients with CIS meeting MS criteria, so studies of CIS cohorts per older criteria may overestimate the risk of conversion among patients with CIS per 2017 McDonald Diagnostic Criteria. From a pragmatic perspective, however, CIS studies utilizing either older or newer diagnostic criteria remain the best source of information by which to estimate conversion risk.

In terms of CIS presentations, the most common syndromes are optic neuritis (approximately 20 to 30%), myelitis (approximately 30%), and brainstem syndromes (approximately 20%) [11, 12]. Some studies report that patients with optic neuritis are less likely to develop clinically definite MS compared to those with myelitis or brainstem syndromes [11, 13], but others have reported equal incidence of MS following each CIS subtype [14]. The Optic Neuritis Treatment Trial (ONTT) reported the risk of MS following CIS with optic neuritis at 30% at 5 years [15] and 50% at 15 years [16], with higher incidence of conversion to MS in patients with abnormal brain MRI at baseline in both analyses.

Age at CIS presentation, race/ethnicity, and gender are also reported as determinants of MS risk following CIS. The risk of MS following CIS by gender has been variably reported, with no clear gender predominance due to conflicting results. However, a meta-analysis suggested a nonstatistically significant increased risk of MS in women with CIS [17]. Overall, younger age is associated with a higher risk of conversion to MS following CIS [18, 19]. One observational study of a CIS cohort found patients with non-white race/ethnicity, younger age, and a lower number of functional systems affected by CIS were more likely to have a second demyelinating event within 1 year [18].

Regarding imaging characteristics at the time of CIS, brain and cervical spine MRI are crucial for both initial evaluation and ongoing surveillance for development of demyelinating disease activity. Studies of CIS patients demonstrate that increasing abnormalities characteristic of MS on baseline MRI as defined by the Barkhof-Tintoré criteria [19–22] are highly predictive of the risk of subsequent definite MS and also correlate with disability at 5 years as measured by the EDSS [2, 23]. Similarly, several studies report increased risk of conversion to MS in patients with abnormal brain MRI at CIS presentation [16, 24, 25]. Overall, patients with normal brain MRI at the time of CIS are thought to have approximately a 20% risk of MS development, compared to 60% to 80% in patients with abnormal brain MRI at the time of presentation [26].

Patients with CIS who have CSF abnormalities (defined as either CSF-specific OCB or increased intrathecal immunoglobulin G [IgG] synthesis) at presentation have increased risk of conversion to MS [19, 27, 28]. A meta-analysis of 71 studies estimated that CIS patients with CSF-specific OCBs had a 9.88 (95% confidence interval, 6.54 to 15.27) times increased odds of conversion to MS compared to those who did not have this finding [27].

## Relapsing-Remitting and Secondary Progressive Multiple Sclerosis

Relapsing-remitting MS is the most common MS phenotype, as it accounts for approximately 85% of MS cases at symptomatic onset [29]. Patients with RRMS present with relapses consistent with demyelination, from which there is a spectrum of recovery. The median age of RRMS onset is approximately 28 to 30 years [30–32]. Certain patients can present with a highly aggressive course characterized by frequent relapse despite DMT. In general, the frequency of relapse decreases over time [33, 34], with an estimate of relapse decrease by 17% for every 5 years of disease duration [33]. The observed decrease in relapse risk in this study is more pronounced in patients with older age at onset. However, not all studies have demonstrated this finding.

Several large studies of the natural history of MS have reported extensively on RRMS and subsequent SPMS development [29–32, 35–43]. These heterogeneous studies in terms of geographic location, setting, timing in relation to DMT availability, and patient populations have provided various estimates regarding time to certain disability milestones, conversion to SPMS, and other elements of the natural history of MS. A summary of various elements of the natural history of RRMS and PPMS is presented in Table 7.3 [39].

In terms of development of disability with relapsing-onset MS, the median time to EDSS of 3, indicating relatively mild to moderate levels of disability, has been reported as ranging from 8 to 30 years from diagnosis [32, 44–46]. The median time to EDSS of 6 has been estimated to be between 15.0 and 23.1 years [31, 32, 35, 40]. Overall, the median age at which patients with RRMS reach an EDSS of 3 is reported as 43 years [47], EDSS of 4 as

**Table 7.3** Natural history of relapsing-onset MS, primary progressive MS, and pediatric-onset MS

| | Relapsing-onset MS[a] | Primary progressive MS | Pediatric-onset MS |
|---|---|---|---|
| *Median age at:* | | | |
| Onset of symptoms | 28 to 30 years | 35 to 42 years | 15 years |
| Onset of progression | 38 to 49 years | 35 to 42 years | 41 years |
| EDSS 4 | 44 to 46 years[b] | 40 to 44 years[b] | 34 to 40 years |
| EDSS 6 | 50 to 56 years | 48 to 58 years | 42 to 45 years |
| *Time from symptom onset to:* | | | |
| Onset of progression (SPMS) | 18.9 to 20 years | N/A | 28 years |
| EDSS 6 | 15 to 23 years | 3 to 14 years | 28 to 31 years |

*CI* confidence interval, *EDSS* Expanded Disability Status Scale, *MS* multiple sclerosis, *SPMS* secondary progressive MS

[a]Some estimates for relapsing-onset MS disability milestones were derived from cohort studies that included either only SPMS patients or a mixed population of those with SPMS and those that did not transition to SPMS. Patients with SPMS have earlier age at disability milestones compared to overall relapsing-onset MS patients

[b]Estimate obtained from one natural history cohort [39]

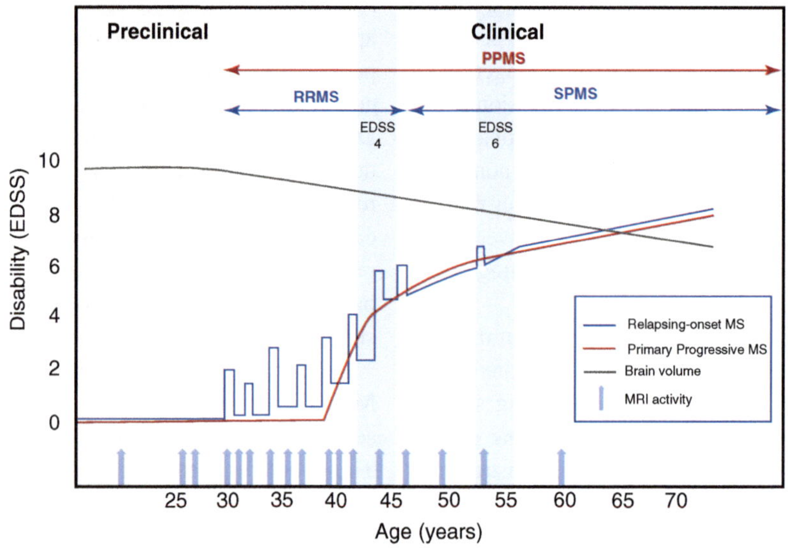


**Fig. 7.3** Natural history of multiple sclerosis by age. Patients with relapsing-remitting multiple sclerosis (RRMS) generally develop their first demyelinating event between the ages of 28 to 30 years, with subsequent relapses for the next 10 to 15 years. Between the ages of 38 to 49 years, RRMS patients generally develop secondary progressive multiple sclerosis (SPMS), characterized by a gradual accumulation of disability. RRMS patients generally reach an Expanded Disability Status Scale (EDSS) of 4 between the ages of 44 and 46 years and EDSS of 6 between 50 and 56 years. Primary progressive multiple sclerosis (PPMS) patients have onset of disease and therefore progression between the ages of 35 and 42, which is approximately the same age of SPMS onset in patients with RRMS. Despite later age of onset, PPMS patients reach disability milestones earlier than patients with relapsing-onset MS, with EDSS of 4 between 40 and 44 years and EDSS of 6 between 50 and 56 years. MRI disease activity occurs in the preclinical phase and continues over the course of disease. Brain atrophy also occurs over time, as brain volume begins to decrease even early in the disease process

44.8 years (95% CI, 43.8 to 45.9) [39], and EDSS of 6 between 50 and 56 years (Fig. 7.3 and Table 7.3) [39, 47–49]. It should be noted, however, that some of these median age estimates are increased by the proportion of RRMS patients who do not transition to SPMS [39]. The ages at which these disability milestones were reached were higher in patients with older age of MS onset [39, 41]. Additionally, it has been shown that the number and severity of relapses are indicative of both short- and long-term disability [34, 42, 43].

Over time, the majority of patients with RRMS transition from predominantly inflammatory to more neurodegenerative driving forces of the disease, with development of SPMS [49]. The transition to SPMS is an age-dependent process [39, 41, 50], and it is estimated that approximately 30% to 40% of patients with SPMS have relapses superimposed upon this typical gradual decline [31, 43]. The time from RRMS diagnosis to SPMS conversion is age of onset-dependent and ranges from 18.9 to 20 years for all patients per several of the natural history studies [33, 39, 46, 51]. Generally, the younger a patient is at the time of RRMS onset, the longer the disease duration prior to SPMS conversion. The time to SPMS is shorter for males at approximately 15 years versus females at approximately 20 years; this difference is also confounded by the gender difference in the mean age of RRMS onset [33, 46, 51]. The overall median age of SPMS onset is approximately 40 to 49 years but is generally younger in men and those with motor symptoms at onset [35, 39, 41, 47, 51–53]. In one study, the median age at disability milestones was younger for RRMS patients who transitioned to SPMS than for the entire relapsing-onset cohort (37.6 versus 44.8 years for EDSS 4 and 45.5 versus 55.3 years for EDSS 6, respectively) and was also younger than PPMS patients (Table 7.3) [39].

## Primary Progressive Multiple Sclerosis

Patients with MS can present with a gradual accumulation of neurological symptoms from onset, which is referred to as primary progressive MS (PPMS). Generally, in PPMS, patients have a more neurodegenerative presentation, but superimposed relapses or inflammatory MRI activity can exist. It is estimated that between 3% and 39% of patients with PPMS have superimposed relapses [31, 43]. A 5-year follow-up study of patients with RIS estimated that approximately 3% of patients converted to PPMS, indicating the presence of inflammatory activity preceding clinical onset [5].

The large natural history studies in MS have also provided information regarding PPMS, though the proportions of PPMS patients within each cohort were relatively low [31, 35, 43, 54–56]. The median age of onset of PPMS is approximately 39 to 41 years of age [35, 55]. These studies have demonstrated that the accrual of disability in PPMS can be quite varied across patient populations, but it is important to note the heterogeneity in PPMS definition across studies. In terms of development of disability with PPMS, the median time to EDSS of 6 has been reported as between 3 and 14 years [32, 40, 44, 46, 54, 55], which is lower than that of RRMS (Fig. 7.3 and Table 7.3). Overall, the median age at which patients with PPMS reach an EDSS of 4 was reported as 42.1 years (95% CI, 40.2 to 44.0) [39] and EDSS of 6 as approximately 48 to 58 years [47–49, 55]. As noted above, these median ages are older than for relapsing-onset patients who transition to SPMS.

## Fulminant Onset Multiple Sclerosis

Patients can present with fulminant or aggressive disease, requiring high potency medications to rapidly control the inflammatory disease activity. Proposed criteria for aggressive MS suggest that patients with RRMS have one or more of the following: EDSS of 4 within 5 years, at least two relapses with incomplete resolution in the previous year, more than two MRIs demonstrating new or enlarging T2 or gadolinium-enhancing lesions despite treatment, and no response to one or more DMTs for a year [57]. In these patients, the severity of disease often warrants potent immunosuppression as first-line treatment [58]. Entities such as Balo's concentric sclerosis [59], Marburg variant MS [60], tumefactive MS, and acute disseminated encephalomyelitis (ADEM)-like presentations of MS are often included in the category of fulminant onset MS.

## Benign Multiple Sclerosis

Given the heterogeneity of clinical presentation in MS, the concept of "benign MS" has emerged, where patients can have relatively infrequent relapses and slow or no disability accrual over time. Although certain prognostic factors for MS are known, identification of patients who will have a benign course at onset is unreliable [61–63]. It is also difficult to discern an individual patient's clinical course on DMT from the actual natural history if he or she was not on DMT. Additionally, patients who have a milder MS course can experience potentially disabling cognitive impairment in the absence of physical disability [62, 63]. Therefore, the diagnosis of "benign MS" should only be made in hindsight.

## Pediatric-Onset Multiple Sclerosis

Pediatric-onset MS (POMS) refers to patients with MS who have their first clinical event prior to the age of 18 years, usually in childhood and early adolescence, and comprises approximately 10% of adults with MS [64]. The median age of onset of POMS is estimated to be approximately 14 to 16 years [65, 66]. Although patients with POMS generally present with similar syndromes as adults with MS, ataxia, encephalopathy, seizures, and brainstem syndromes are more common in children below the age of 10 [64]. Studies estimate that between 6% and 18% of patients who present with acute disseminated encephalomyelitis (ADEM) will go on to meet diagnostic criteria for MS [67]. The vast majority of patients with POMS have a relapsing-remitting course and tend to have good recovery from relapses despite relatively high relapse frequency early in the disease course.

In terms of longer-term prognosis in the setting of POMS, median time to SPMS from disease onset was 28 years, with a median age at conversion of 41 years [66]. Although patients with POMS have longer time to SPMS conversion from disease onset (estimates ranging from 28 to 32 years, versus approximately 20 years in adult-onset MS patients [65, 66]), the median age at onset of SPMS has been reported as 41.4 years (95% CI, 37.8 to 45.7), approximately 5 years younger than patients with adult-onset MS (Table 7.3) [66]. Similarly, patients with POMS have longer time to disability milestones but younger median age at these milestones compared to those with adult-onset MS; age at EDSS 4 has been reported to range between 34 and 40 years (at disease duration 20 to 24 years) and EDSS 6 between 42 and 45 years (at disease duration 28 to 31 years) (Table 7.3) [65, 66]. Another important consideration for patients with POMS is that of early cognitive impairment and severe fatigue, with resultant impairment in quality of life [67].

## Prognostic Factors in Multiple Sclerosis

Through various studies, it has become apparent over time that certain clinical and other characteristics portend different prognoses in MS. However, it remains difficult to predict prognosis of MS in an individual patient. Prognostic factors in MS are summarized in Table 7.2 and are discussed further in this section.

### Age

In MS, younger age at onset tends to be associated with better prognosis in terms of time to disability milestones [31]. A study involving a large cohort of 1023 patients with MS indicated that older age at RRMS onset was associated with higher risk of reaching disability milestones, independent of disease duration and relapse frequency [47]. This observation was thought to be driven by the increased risk of SPMS conversion, which portends poor prognosis in terms of reaching disability milestones. A European Database for MS study (n = 957) also demonstrated that the risk of PPMS presentation increased with age [50]. This study also demonstrated that with increased age of onset, the number of relapses between onset and SPMS conversion, as well as median time to SPMS conversion, decreased.

Though not always explicitly considered, each of these observations reflects the age-dependent nature of progressive MS onset. Specifically, older age at RRMS onset equates to fewer years until reaching the median age of progressive MS onset.

## Gender

Men with MS tend to have worse prognosis than women in terms of having shorter time to disability milestones and SPMS from an initially RRMS course [31]. In the Lyon MS Cohort Study, the time to each disability milestone of EDSS 4, 6, and 7 was significantly higher in women [31]. Women had median time to onset of EDSS 4, 6, and 7 of 9.6 years, 23.1 years, and 30.4 years, respectively. In contrast, men had median time to these disability milestones of 7.2 years, 17.2 years, and 25.1 years, respectively. It should be considered, though, that some of this difference may be accounted for by men having an average older age of RRMS onset compared to women.

## Race

Race is another prognostic factor in patients with MS. Although MS is less common in non-Caucasian races, studies indicate that MS tends to be more aggressive in African Americans [68–70]. African-American patients with MS are more likely to have multifocal signs and symptoms or transverse myelitis at presentation. Additionally, African-American patients with MS had higher risk for development of disability compared to Caucasian patients, even with adjustment for other baseline characteristics and DMT use; this observation was partially attributed to the older age of onset seen in African-American patients [68]. An observational study of a CIS cohort found patients with non-white race/ethnicity were more likely to have a second demyelinating event within 1 year [18]. However, the underrepresentation of minorities in MS clinical trials and other socioeconomic barriers potentially limit the understanding of treatment response and natural history of MS in non-Caucasians with MS [71].

## Imaging Characteristics

Certain imaging characteristics can also be useful in determining an MS patient's prognosis. Patients who have a high burden of T2 lesions early in the disease course tend to have worse prognosis, as demonstrated by a 13-year longitudinal study of 30 patients [72]. Baseline T2 lesion volume correlated with normalized whole-brain volume after 13 years. These results were also corroborated in a study of patients with CIS, where the MRI T2 lesion volume at 5 years and the increase in lesion volume over the first 5 years of follow-up were both moderately correlated with the median EDSS at 14 years of follow-up [73].

Accumulation of new T2 lesions early in the course of disease also correlates with long-term imaging and clinical outcomes. A 13-year longitudinal study of 30 patients demonstrated that increase in T2 lesions over the first 2 years was moderately correlated with normalized whole-brain volume, Multiple Sclerosis Functional Composite scores, and Paced Auditory Serial Addition Test scores at last follow-up [72].

Multiple gadolinium-enhancing lesions at baseline MRI and at follow-up are also shown to predict poor prognosis. In a study of 307 patients, the mean number of gadolinium-enhancing lesions on MRIs over the first 6 months following presentation was associated with the relapse rate in the first year following diagnosis [74]. Additionally, the mean number of gadolinium-enhancing lesions over the first 6 monthly MRIs was weakly predictive of EDSS change after 2 years.

Other imaging markers are also thought to indicate long-term prognosis. In terms of axonal injury, T1 hypointensities, otherwise known as "black holes" on MRI, have been shown to predict disability worsening over time [75]. Neurodegeneration as captured by brain atrophy can occur early in some patients with MS, and rate of ventricular enlargement indicating central atrophy was predictive of disability worsening in patients with MS at 5.5 years of follow-up [76]. Patients with infratentorial lesions have also been shown to have worse prognosis [57, 77].

## Clinical Characteristics

Several aspects regarding MS clinical presentation are also known to influence natural history and prognosis of MS. Patients who present with RRMS have longer time to disability milestones compared to those with PPMS, with estimates for a median time of 23.1 years to EDSS of 6 in patients with RRMS compared to 7.1 years in patients with PPMS [31]. Some studies have demonstrated that the presence of relapses in the setting of PPMS or SPMS does not affect time to disability milestones [31, 40]. However, one study reported that superimposed relapses did result in shorter time to EDSS 6 in a cohort that included both PPMS and SPMS patients [43]. Another study indicated that in patients with PPMS and superimposed relapses, treatment of inflammatory disease activity (and therefore prevention of relapses) via DMT was associated with lower likelihood of confirmed disability worsening [78].

The initial clinical presentation of MS is also thought to affect time to disability milestones. In the Lyon MS Cohort Study, patients who presented with isolated optic neuritis had longer time to disability milestones, with a median time to EDSS 4 of 14.1 years, compared to 10.5 years and 6.0 years in patients with a brainstem or long tract dysfunction presentation, respectively [31]. Patients with multifocal presentations also are thought to have poorer prognoses [57]. The severity of MS relapses is also indicative of prognosis, as patients with severe relapses are at higher risk of experiencing subsequent severe events [18]. However, relapses are thought to indicate more short-term rather than long-term disability but do likely accelerate conversion to SPMS [34]. The majority of patients with frequent early relapses in RRMS convert to SPMS, but some did not convert to SPMS in a large cohort study [79]. Overall, frequent early relapses portend a more aggressive MS course [37, 57].

Recovery from early relapses is also an important prognostic indicator in MS, along with relapse frequency [11, 80]. In particular, patients with poor recovery from early relapses are at significant risk for shorter duration to conversion to SPMS [42]. Clinical relapses can occur in any MS phenotype and have been shown to result in worsening of 0.24 to 0.57 points on the EDSS that is sustained at least to some degree in the majority of patients [81]. Studies have demonstrated that patients with complete early relapse recovery had a longer time interval to reach disability milestones compared to patients with incomplete recovery [31, 42]. Time to disability milestones was also increased in patients who had longer intervals between the first and second demyelinating event [31]. Poor initial relapse recovery also has been shown to predict poor subsequent relapse recovery [18, 42].

## Genetics

Although several genes that confer risk for MS have been identified, the etiology of MS remains multifactorial, with both genetic susceptibility and environmental components [82]. Genome-wide association studies have identified genes involved in a broad range of cellular functions that confer certain MS characteristics [83]. However, these are generally not utilized in clinical practice due to their unknown predictive value regarding disease course, as well as high cost. Patients with MS who had a first-, second- or third-degree relative with MS were followed in the Ontario Natural History of MS Study [84]. Although age of MS was lower in those with a family history of MS, time to disability milestones did not differ compared to those without relatives with MS. The genetics of MS is further discussed in Chap. 11.

## Impact of Disease Modifying Therapies on the Natural History of MS

Over the past 20 years, over a dozen medications have received regulatory approval for treatment of MS. Although these immunotherapies have diverse mechanisms of action and properties, the clinical course of MS has been positively impacted by the advent of the DMT era. Most patients with MS accumulate disability and evolve to a SPMS course, but multiple studies demonstrate that patients treated with DMT have substantially decreased risk of and delayed conversion to SPMS compared to patients in the pre-DMT era [85–91] (Fig. 7.4). Additionally, DMT use has been shown to delay reaching disability milestones [85, 88–90] (Fig. 7.4).

One study investigated changes in overall disability worsening among all MS phenotypes from the pre- to post-DMT eras in Nova Scotia [90]. Prior to DMT availability, annual EDSS increases were 0.1 for RRMS, 0.31 for SPMS, and 0.16 for relapsing-onset (combined RRMS and SPMS) MS patients. The study concluded that the EDSS increases avoided per year on first DMT were statistically significant in the RRMS, SPMS, and relapsing-onset groups, with relative treatment effect estimates of 112%, 21%, and 105%.

Despite these improvements in overall MS prognosis with initiation of DMTs, several questions remain regarding management of MS. First, the ideal overall treatment strategy for MS is unclear; the two major considerations are escalation versus early treatment with high potency medication. Escalation is the classic treatment paradigm in MS and involves initiating a safer, but generally less efficacious, DMT. Patients then transition to more potent and potentially less safe medications with breakthrough disease activity. Although this approach is appealing from a safety perspective, it can delay time to optimal control of inflammatory disease activity, therefore increasing the risk of ongoing relapses, tissue injury, and disability worsening. On the other hand, early treatment with high efficacy medications involves utilizing more potent medications initially. Although safety concerns may arise with the use of these medications, this strategy is gaining popularity due to increasing focus on early, effective control of inflammatory disease [92]. Overall, there is currently a state of clinical equipoise between escalation and early, highly efficacious treatment strategies, and large pragmatic clinical trials are underway to compare both short- and long-term outcomes of each strategy.

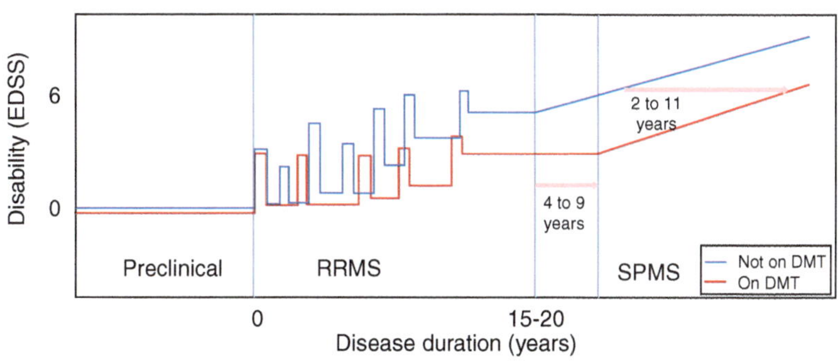


**Fig. 7.4** Effect of disease modifying therapies (DMTs) on the natural history of multiple sclerosis (MS). DMTs for MS have several beneficial effects on the natural history of the disease. First, patients started on DMT, particularly earlier in their disease course, have reduction in clinical relapses and MRI disease activity, with resultant decreased disability accumulation during the relapsing phase of the disease. Second, DMTs have been shown to decrease the risk of conversion to SPMS [85]. Third, DMTs have been shown to delay the time to conversion to SPMS, by approximately 4 to 9 years [87]. Finally, DMTs overall delay time to disability milestones based on the Expanded Disability Status Scale (EDSS). In particular, time to EDSS of 6.0 was reported to be delayed by approximately 2 to 11 years in patients on DMT [87]

Another currently unanswered question regarding DMTs is that of discontinuation timing [93]. As patients age, the inflammatory activity of MS tends to become quiescent, and DMTs are less likely to have a positive impact in that stage [49, 94]. Currently, DMT discontinuation is considered in older patients with clinical and radiographic stability over several years [95]. Several studies indicate that it is appropriate to consider DMT discontinuation in older patients (generally >60 years) who have not had evidence of active inflammatory disease for several years, but no widely accepted guidelines exist for this scenario [93, 95, 96]. Clinical trials are underway at the time of this publication to investigate appropriate timing of DMT discontinuation in older patients, as well as associated short- and long-term outcomes. The clinical approach to selection and discontinuation of DMT in MS is further discussed in Chap. 31.

## Comorbidities and Health Behaviors

Overall health and wellness have been increasingly emphasized as part of MS multidisciplinary care [97]. A growing body of literature is indicative of the fact that medical comorbidities portend worse overall prognosis in patients with MS [98]. In addition to higher levels of disability [99], patients with comorbidities have delays in diagnosis [100], worse cognition, increased mortality [101], increased disease activity [102], and worse health-related quality of life [98, 103]. Vascular risk factors and obesity are thought to be important determinants of overall health and the aging process of the central nervous system, which is accelerated in the setting of MS [97]. Management of comorbidities in the setting of MS should be an important part of multidisciplinary care with a patient's primary care physician in order to minimize the risk of these adverse outcomes.

Mental health comorbidities are also known to impact prognosis and mortality in patients with immune-mediated inflammatory diseases, including MS [104]. A study of 28,384 patients with immune-mediated inflammatory disease, 5496 of whom had MS, indicated that having depression in particular was associated with a greater than additive interaction with disease status in terms of increased mortality [104]. The authors also found that having either depression, anxiety, or bipolar disorder in the setting of MS and other immune-mediated inflammatory diseases was associated with increased risk of suicide.

Vitamin D deficiency is also recognized as an important determinant of MS risk and disease activity. Studies have suggested that low vitamin D is associated with MS risk, future clinical and MRI disease activity [105], and greater disability [97, 106].

Adverse health behaviors have been associated with worse outcomes in the setting of MS, particularly tobacco use. Smoking has been associated with worse disease at baseline in terms of EDSS, MS Severity Score, and brain parenchymal fraction [107]. Smokers were more likely to have PPMS and had faster conversion from RRMS to SPMS compared to nonsmokers [107]. Smokers with MS have also been noted to have a more sustained malignant course overall [108]. MRI metrics of disease activity were also worse in smokers, as T2 lesion volume increased and brain parenchymal fraction decreased more quickly in a longitudinal analysis [107].

Given this knowledge, it is important for clinicians to inform their patients about the role of wellness, regular medical care, and avoiding adverse health behaviors such as tobacco use in the setting of MS.

## Conclusions

MS is a heterogeneous disease with varying phenotypes and clinical characteristics. In general, patients presenting with CIS are diagnosed with RRMS with a second demyelinating event or supportive paraclinical data at the time of CIS, and many evolve to develop SPMS over time in an age-dependent process (Fig. 7.3). Development of SPMS and disability in general has been reduced by the advent of more than a dozen DMTs for MS that have positively impacted the natural history of the disease. A smaller percentage of patients develop progressive disease from onset and are referred to as PPMS. RRMS/CIS patients who present with relapses affecting the posterior fossa, enhancing lesions, high spinal cord lesion burden, and incomplete recovery from relapses tend to have worse long-term prognosis. It is becoming increasingly recognized that health behaviors, particularly tobacco use, and comorbidities affect the natural history of MS. Clinicians should be mindful of these prognostic and some potentially modifiable risk factors for MS-related disability and, in addition to treatment with DMTs, should encourage patients to optimize their general health as part of comprehensive MS management.

## References


1. Lublin FD, Reingold SC, Cohen JA, Cutter GR, Sorensen PS, Thompson AJ, et al. Defining the clinical course of multiple sclerosis: the 2013 revisions. Neurology. 2014;83(3):278–86.
2. Kurtzke JF. Rating neurologic impairment in multiple sclerosis: an expanded disability status scale (EDSS). Neurology. 1983;33(11):1444–52.
3. Okuda DT, Mowry EM, Beheshtian A, Waubant E, Baranzini SE, Goodin DS, et al. Incidental MRI anomalies suggestive of multiple sclerosis: the radiologically isolated syndrome. Neurology. 2009;72(9):800–5.
4. Granberg T, Martola J, Kristoffersen-Wiberg M, Aspelin P, Fredrikson S. Radiologically isolated syndrome--incidental magnetic resonance imaging findings suggestive of multiple sclerosis, a systematic review. Mult Scler (Houndmills, Basingstoke, England). 2013;19(3):271–80.
5. Okuda DT, Siva A, Kantarci O, Inglese M, Katz I, Tutuncu M, et al. Radiologically isolated syndrome: 5-year risk for an initial clinical event. PLoS One. 2014;9(3):e90509.
6. Okuda DT, Mowry EM, Cree BA, Crabtree EC, Goodin DS, Waubant E, et al. Asymptomatic spinal cord lesions predict disease progression in radiologically isolated syndrome. Neurology. 2011;76(8):686–92.
7. Yamout B, Al Khawajah M. Radiologically isolated syndrome and multiple sclerosis. Mult Scler Relat Disord. 2017;17:234–7.
8. Lebrun C, Bensa C, Debouverie M, Wiertlevski S, Brassat D, de Seze J, et al. Association between clinical conversion to multiple sclerosis in radiologically isolated syndrome and magnetic resonance imaging, cerebrospinal fluid, and visual evoked potential: follow-up of 70 patients. Arch Neurol. 2009;66(7):841–6.
9. Vural A, Okar S, Kurne A, Sayat-Gürel G, Acar NP, Karabulut E, et al. Retinal degeneration is associated with brain volume reduction and prognosis in radiologically isolated syndrome. Mult scler (Houndmills, Basingstoke, England). 2018; https://doi.org/10.1177/1352458518817987.
10. Thompson AJ, Banwell BL, Barkhof F, Carroll WM, Coetzee T, Comi G, et al. Diagnosis of multiple sclerosis: 2017 revisions of the McDonald criteria. Lancet Neurol. 2018;17(2):162–73.
11. Eriksson M, Andersen O, Runmarker B. Long-term follow up of patients with clinically isolated syndromes, relapsing-remitting and secondary progressive multiple sclerosis. Mult Scler (Houndmills, Basingstoke, England). 2003;9(3):260–74.
12. Brownlee WJ, Miller DH. Clinically isolated syndromes and the relationship to multiple sclerosis. J Clin Neurosci. 2014;21(12):2065–71.
13. Tintore M, Rovira A, Rio J, Nos C, Grive E, Tellez N, et al. Is optic neuritis more benign than other first attacks in multiple sclerosis? Ann Neurol. 2005;57(2):210–5.
14. Fisniku LK, Brex PA, Altmann DR, Miszkiel KA, Benton CE, Lanyon R, et al. Disability and T2 MRI lesions: a 20-year follow-up of patients with relapse onset of multiple sclerosis. Brain J Neurol. 2008;131(Pt 3):808–17.
15. Optic Neuritis Study Group. The 5-year risk of MS after optic neuritis. Experience of the optic neuritis treatment trial. Neurology. 1997;49(5):1404–13.
16. Group ONS. Multiple sclerosis risk after optic neuritis: final optic neuritis treatment trial follow-up. Arch Neurol. 2008;65(6):727–32.
17. Dobson R, Ramagopalan S, Giovannoni G. The effect of gender in clinically isolated syndrome (CIS): a meta-analysis. Mult Scler (Houndmills, Basingstoke, England). 2012;18(5):600–4.
18. Mowry EM, Pesic M, Grimes B, Deen SR, Bacchetti P, Waubant E. Clinical predictors of early second event in patients with clinically isolated syndrome. J Neurol. 2009;256:1061–6.
19. Polman C, Kappos L, Freedman MS, Edan G, Hartung HP, Miller DH, et al. Subgroups of the BENEFIT study: risk of developing MS and treatment effect of interferon beta-1b. J Neurol. 2008;255(4):480–7.
20. Barkhof F, Filippi M, Miller DH, Scheltens P, Campi A, Polman CH, et al. Comparison of MRI criteria at first presentation to predict conversion to clinically definite multiple sclerosis. Brain J Neurol. 1997;120(Pt 11):2059–69.
21. CHAMPS Study Group. MRI predictors of early conversion to clinically definite MS in the CHAMPS placebo group. Neurology. 2002;59(7):998–1005.
22. Tintore M, Rovira A, Martinez MJ, Rio J, Diaz-Villoslada P, Brieva L, et al. Isolated demyelinating syndromes: comparison of different MR imaging criteria to predict conversion to clinically definite multiple sclerosis. AJNR Am J Neuroradiol. 2000;21(4):702–6.
23. Tintore M, Rovira A, Rio J, Nos C, Grive E, Tellez N, et al. Baseline MRI predicts future attacks and dis-

ability in clinically isolated syndromes. Neurology. 2006;67(6):968–72.
24. Filippi M, Horsfield MA, Morrissey SP, MacManus DG, Rudge P, McDonald WI, et al. Quantitative brain MRI lesion load predicts the course of clinically isolated syndromes suggestive of multiple sclerosis. Neurology. 1994;44(4):635–41.
25. O'Riordan JI, Thompson AJ, Kingsley DP, MacManus DG, Kendall BE, Rudge P, et al. The prognostic value of brain MRI in clinically isolated syndromes of the CNS. A 10-year follow-up. Brain J Neurol. 1998;121(Pt 3):495–503.
26. Miller DH, Chard DT, Ciccarelli O. Clinically isolated syndromes. Lancet Neurol. 2012;11(2):157–69.
27. Dobson R, Ramagopalan S, Davis A, Giovannoni G. Cerebrospinal fluid oligoclonal bands in multiple sclerosis and clinically isolated syndromes: a meta-analysis of prevalence, prognosis and effect of latitude. J Neurol Neurosurg Psychiatry. 2013;84(8):909–14.
28. Cole SR, Beck RW, Moke PS, Kaufman DI, Tourtellotte WW. The predictive value of CSF oligoclonal banding for MS 5 years after optic neuritis. Optic Neuritis Study Group. Neurology. 1998;51(3):885–7.
29. Deshpande R, Kremenchutzky M, Rice GP. The natural history of multiple sclerosis. Adv Neurol. 2006;98:1–15.
30. Confavreux C, Vukusic S. Natural history of multiple sclerosis: a unifying concept. Brain J Neurol. 2006;129(Pt 3):606–16.
31. Confavreux C, Vukusic S, Adeleine P. Early clinical predictors and progression of irreversible disability in multiple sclerosis: an amnesic process. Brain J Neurol. 2003;126(Pt 4):770–82.
32. Weinshenker BG, Bass B, Rice GP, Noseworthy J, Carriere W, Baskerville J, et al. The natural history of multiple sclerosis: a geographically based study. I. Clinical course and disability. Brain J Neurol. 1989;112(Pt 1):133–46.
33. Tremlett H, Zhao Y, Joseph J, Devonshire V. Relapses in multiple sclerosis are age- and time-dependent. J Neurol Neurosurg Psychiatry. 2008;79(12):1368–74.
34. Tremlett H, Yousefi M, Devonshire V, Rieckmann P, Zhao Y. Impact of multiple sclerosis relapses on progression diminishes with time. Neurology. 2009;73(20):1616–23.
35. Kremenchutzky M, Rice GP, Baskerville J, Wingerchuk DM, Ebers GC. The natural history of multiple sclerosis: a geographically based study 9: observations on the progressive phase of the disease. Brain J Neurol. 2006;129(Pt 3):584–94.
36. Scalfari A, Neuhaus A, Degenhardt A, Rice GP, Muraro PA, Daumer M, et al. The natural history of multiple sclerosis: a geographically based study 10: relapses and long-term disability. Brain J Neurol. 2010;133(Pt 7):1914–29.
37. Weinshenker BG, Bass B, Rice GP, Noseworthy J, Carriere W, Baskerville J, et al. The natural history of multiple sclerosis: a geographically based study. 2. Predictive value of the early clinical course. Brain J Neurol. 1989;112(Pt 6):1419–28.
38. Weinshenker BG, Rice GP, Noseworthy JH, Carriere W, Baskerville J, Ebers GC. The natural history of multiple sclerosis: a geographically based study. 3. Multivariate analysis of predictive factors and models of outcome. Brain J Neurol. 1991;114(Pt 2):1045–56.
39. Confavreux C, Vukusic S. Age at disability milestones in multiple sclerosis. Brain J Neurol. 2006;129(Pt 3):595–605.
40. Confavreux C, Vukusic S, Moreau T, Adeleine P. Relapses and progression of disability in multiple sclerosis. N Engl J Med. 2000;343(20):1430–8.
41. Tutuncu M, Tang J, Zeid NA, Kale N, Crusan DJ, Atkinson EJ, et al. Onset of progressive phase is an age-dependent clinical milestone in multiple sclerosis. Mult Scler (Houndmills, Basingstoke, England). 2013;19(2):188–98.
42. Novotna M, Paz Soldan MM, Abou Zeid N, Kale N, Tutuncu M, Crusan DJ, et al. Poor early relapse recovery affects onset of progressive disease course in multiple sclerosis. Neurology. 2015;85(8):722–9.
43. Paz Soldan MM, Novotna M, Abou Zeid N, Kale N, Tutuncu M, Crusan DJ, et al. Relapses and disability accumulation in progressive multiple sclerosis. Neurology. 2015;84(1):81–8.
44. Pittock SJ, Mayr WT, McClelland RL, Jorgensen NW, Weigand SD, Noseworthy JH, et al. Disability profile of MS did not change over 10 years in a population-based prevalence cohort. Neurology. 2004;62(4):601–6.
45. Tremlett H, Paty D, Devonshire V. Disability progression in multiple sclerosis is slower than previously reported. Neurology. 2006;66(2):172–7.
46. Debouverie M, Pittion-Vouyovitch S, Louis S, Guillemin F. Natural history of multiple sclerosis in a population-based cohort. Eur J Neurol. 2008;15(9):916–21.
47. Scalfari A, Neuhaus A, Daumer M, Ebers GC, Muraro PA. Age and disability accumulation in multiple sclerosis. Neurology. 2011;77(13):1246–52.
48. Tremlett H, Zhao Y, Devonshire V. Natural history comparisons of primary and secondary progressive multiple sclerosis reveals differences and similarities. J Neurol. 2009;256(3):374–81.
49. Leray E, Yaouanq J, Le Page E, Coustans M, Laplaud D, Oger J, et al. Evidence for a two-stage disability progression in multiple sclerosis. Brain J Neurol. 2010;133(Pt 7):1900–13.
50. Stankoff B, Mrejen S, Tourbah A, Fontaine B, Lyon-Caen O, Lubetzki C, et al. Age at onset determines the occurrence of the progressive phase of multiple sclerosis. Neurology. 2007;68(10):779–81.
51. Tremlett H, Yinshan Z, Devonshire V. Natural history of secondary-progressive multiple sclerosis. Mult Scler (Houndmills, Basingstoke, England). 2008;14(3):314–24.
52. Minderhoud JM, van der Hoeven JH, Prange AJ. Course and prognosis of chronic progressive

multiple sclerosis. Results of an epidemiological study. Acta Neurol Scand. 1988;78(1):10–5.
53. Koch M, Mostert J, Heersema D, De Keyser J. Progression in multiple sclerosis: further evidence of an age dependent process. J Neurol Sci. 2007;255(1–2):35–41.
54. Cottrell DA, Kremenchutzky M, Rice GP, Koopman WJ, Hader W, Baskerville J, et al. The natural history of multiple sclerosis: a geographically based study. 5. The clinical features and natural history of primary progressive multiple sclerosis. Brain J Neurol. 1999;122(Pt 4):625–39.
55. Koch M, Kingwell E, Rieckmann P, Tremlett H. The natural history of primary progressive multiple sclerosis. Neurology. 2009;73(23):1996–2002.
56. Kremenchutzky M, Cottrell D, Rice G, Hader W, Baskerville J, Koopman W, et al. The natural history of multiple sclerosis: a geographically based study. 7. Progressive-relapsing and relapsing-progressive multiple sclerosis: a re-evaluation. Brain J Neurol. 1999;122(Pt 10):1941–50.
57. Rush CA, MacLean HJ, Freedman MS. Aggressive multiple sclerosis: proposed definition and treatment algorithm. Nat Rev Neurol. 2015;11(7):379–89.
58. Freedman MS, Rush CA. Severe, highly active, or aggressive multiple sclerosis. Continuum (Minneapolis, Minn). 2016;22(3):761–84.
59. Amini Harandi A, Esfandani A, Pakdaman H, Abbasi M, Sahraian MA. Balo's concentric sclerosis: an update and comprehensive literature review. Rev Neurosci. 2018;29:873.
60. Nunes JC, Radbruch H, Walz R, Lin K, Stenzel W, Prokop S, et al. The most fulminant course of the Marburg variant of multiple sclerosis-autopsy findings. Mult Scler (Houndmills, Basingstoke, England). 2015;21(4):485–7.
61. Pittock SJ, McClelland RL, Mayr WT, Jorgensen NW, Weinshenker BG, Noseworthy J, et al. Clinical implications of benign multiple sclerosis: a 20-year population-based follow-up study. Ann Neurol. 2004;56(2):303–6.
62. Ramsaransing GS, De Keyser J. Predictive value of clinical characteristics for 'benign' multiple sclerosis. Eur J Neurol. 2007;14(8):885–9.
63. Sayao AL, Devonshire V, Tremlett H. Longitudinal follow-up of "benign" multiple sclerosis at 20 years. Neurology. 2007;68(7):496–500.
64. Lee JY, Chitnis T. Pediatric Multiple Sclerosis. Semin Neurol. 2016;36(2):148–53.
65. Harding KE, Liang K, Cossburn MD, Ingram G, Hirst CL, Pickersgill TP, et al. Long-term outcome of paediatric-onset multiple sclerosis: a population-based study. J Neurol Neurosurg Psychiatry. 2013;84(2):141–7.
66. Renoux C, Vukusic S, Mikaeloff Y, Edan G, Clanet M, Dubois B, et al. Natural history of multiple sclerosis with childhood onset. N Engl J Med. 2007;356(25):2603–13.
67. Narula S, Banwell B. Pediatric demyelination. Continuum (Minneapolis, Minn). 2016;22(3):897–915.
68. Cree BA, Khan O, Bourdette D, Goodin DS, Cohen JA, Marrie RA, et al. Clinical characteristics of African Americans vs Caucasian Americans with multiple sclerosis. Neurology. 2004;63(11):2039–45.
69. Kister I, Chamot E, Bacon JH, Niewczyk PM, De Guzman RA, Apatoff B, et al. Rapid disease course in African Americans with multiple sclerosis. Neurology. 2010;75(3):217–23.
70. Marrie RA, Cutter G, Tyry T, Vollmer T, Campagnolo D. Does multiple sclerosis-associated disability differ between races? Neurology. 2006;66(8):1235–40.
71. Khan O, Williams MJ, Amezcua L, Javed A, Larsen KE, Smrtka JM. Multiple sclerosis in US minority populations: clinical practice insights. Neurol Clin Pract. 2015;5(2):132–42.
72. Rudick R, Lee J, Simon J, Fisher E. Significance of T2 lesions in multiple sclerosis: a 13-year longitudinal study. Ann Neurol. 2006;60(2):236–42.
73. Brex PA, Ciccarelli O, O'Riordan JI, Sailer M, Thompson AJ. Miller DH. A longitudinal study of abnormalities on MRI and disability from multiple sclerosis. N Engl J Med. 2002;346(3):158–64.
74. Kappos L, Moeri D, Radue EW, Schoetzau A, Schweikert K, Barkhof F, et al. Predictive value of gadolinium-enhanced magnetic resonance imaging for relapse rate and changes in disability or impairment in multiple sclerosis: a meta-analysis. Gadolinium MRI Meta-analysis Group. Lancet (London, England). 1999;353(9157):964–9.
75. Tomassini V, Paolillo A, Russo P, Giugni E, Prosperini L, Gasperini C, et al. Predictors of long-term clinical response to interferon beta therapy in relapsing multiple sclerosis. J Neurol. 2006;253(3):287–93.
76. Lukas C, Minneboo A, de Groot V, Moraal B, Knol DL, Polman CH, et al. Early central atrophy rate predicts 5 year clinical outcome in multiple sclerosis. J Neurol Neurosurg Psychiatry. 2010;81(12):1351–6.
77. Rahn AC, Kopke S, Stellmann JP, Schiffmann I, Lukas C, Chard D, et al. Magnetic resonance imaging as a prognostic disability marker in clinically isolated syndrome: a systematic review. Acta Neurol Scand. 2018;139:18.
78. Hughes J, Jokubaitis V, Lugaresi A, Hupperts R, Izquierdo G, Prat A, et al. Association of Inflammation and Disability Accrual in patients with progressive-onset multiple sclerosis. JAMA Neurol. 2018;75(11):1407–15.
79. Scalfari A, Neuhaus A, Daumer M, Deluca GC, Muraro PA, Ebers GC. Early relapses, onset of progression, and late outcome in multiple sclerosis. JAMA Neurol. 2013;70(2):214–22.
80. Scott TF, Schramke CJ. Poor recovery after the first two attacks of multiple sclerosis is associated with poor outcome five years later. J Neurol Sci. 2010;292(1–2):52–6.
81. Lublin FD, Baier M, Cutter G. Effect of relapses on development of residual deficit in multiple sclerosis. Neurology. 2003;61(11):1528–32.
82. International Multiple Sclerosis Genetics Consortium (IMSGC) ea. Analysis of immune-related loci iden-

tifies 48 new susceptibility variants for multiple sclerosis. Nat Genet. 2013;45(11):1353–60.
83. Baranzini SE, Wang J, Gibson RA, Galwey N, Naegelin Y, Barkhof F, et al. Genome-wide association analysis of susceptibility and clinical phenotype in multiple sclerosis. Hum Mol Genet. 2009;18(4):767–78.
84. Ebers GC, Koopman WJ, Hader W, Sadovnick AD, Kremenchutzky M, Mandalfino P, et al. The natural history of multiple sclerosis: a geographically based study: 8: familial multiple sclerosis. Brain J Neurol. 2000;123(Pt 3):641–9.
85. Bergamaschi R, Quaglini S, Tavazzi E, Amato MP, Paolicelli D, Zipoli V, et al. Immunomodulatory therapies delay disease progression in multiple sclerosis. Mult Scler (Houndmills, Basingstoke, England). 2016;22(13):1732–40.
86. Tedeholm H, Lycke J, Skoog B, Lisovskaja V, Hillert J, Dahle C, et al. Time to secondary progression in patients with multiple sclerosis who were treated with first generation immunomodulating drugs. Mult Scler (Houndmills, Basingstoke, England). 2013;19(6):765–74.
87. Trojano M, Pellegrini F, Fuiani A, Paolicelli D, Zipoli V, Zimatore GB, et al. New natural history of interferon-beta-treated relapsing multiple sclerosis. Ann Neurol. 2007;61(4):300–6.
88. Cree BA, Gourraud PA, Oksenberg JR, Bevan C, Crabtree-Hartman E, Gelfand JM, et al. Long-term evolution of multiple sclerosis disability in the treatment era. Ann Neurol. 2016;80(4):499–510.
89. Palace J, Duddy M, Lawton M, Bregenzer T, Zhu F, Boggild M, et al. Assessing the long-term effectiveness of interferon-beta and glatiramer acetate in multiple sclerosis: final 10-year results from the UK multiple sclerosis risk-sharing scheme. J Neurol Neurosurg Psychiatry. 2018;90:251.
90. Brown MG, Kirby S, Skedgel C, Fisk JD, Murray TJ, Bhan V, et al. How effective are disease-modifying drugs in delaying progression in relapsing-onset MS? Neurology. 2007;69(15):1498–507.
91. Brown JWL, Coles A, Horakova D, Havrdova E, Izquierdo G, Prat A, et al. Association of Initial Disease-Modifying Therapy with later conversion to secondary progressive multiple sclerosis. JAMA. 2019;321(2):175–87.
92. Harding K, Williams O, Willis M, Hrastelj J, Rimmer A, Joseph F, et al. Clinical outcomes of escalation vs early intensive disease-modifying therapy in patients with multiple sclerosis. JAMA Neurol. 2019;76:536.
93. Tobin WO, Weinshenker BG. Stopping immunomodulatory medications in MS: frequency, reasons and consequences. Mult Scler Relat Disord. 2015;4(5):437–43.
94. Weideman AM, Tapia-Maltos MA, Johnson K, Greenwood M, Bielekova B. Meta-analysis of the age-dependent efficacy of multiple sclerosis treatments. Front Neurol. 2017;8:577.
95. Birnbaum G. Stopping disease-modifying therapy in nonrelapsing multiple sclerosis: experience from a clinical practice. Int J MS Care. 2017;19(1):11–4.
96. Hua LH, Fan TH, Conway D, Thompson N, Kinzy TG. Discontinuation of disease-modifying therapy in patients with multiple sclerosis over age 60. Mult Scler (Houndmills, Basingstoke, England). 2018; https://doi.org/10.1177/1352458518765656.
97. Coyle PK. Symptom management and lifestyle modifications in multiple sclerosis. Continuum (Minneapolis, Minn). 2016;22(3):815–36.
98. Marrie RA. Comorbidity in multiple sclerosis: implications for patient care. Nat Rev Neurol. 2017;13(6):375–82.
99. Zhang T, Tremlett H, Zhu F, Kingwell E, Fisk JD, Bhan V, et al. Effects of physical comorbidities on disability progression in multiple sclerosis. Neurology. 2018;90(5):e419–e27.
100. Marrie RA, Horwitz R, Cutter G, Tyry T, Campagnolo D, Vollmer T. Comorbidity delays diagnosis and increases disability at diagnosis in MS. Neurology. 2009;72(2):117–24.
101. Marrie RA, Elliott L, Marriott J, Cossoy M, Blanchard J, Leung S, et al. Effect of comorbidity on mortality in multiple sclerosis. Neurology. 2015;85(3):240–7.
102. Marck CH, Neate SL, Taylor KL, Weiland TJ, Jelinek GA. Prevalence of comorbidities, overweight and obesity in an international sample of people with multiple sclerosis and associations with modifiable lifestyle factors. PLoS One. 2016;11(2):e0148573.
103. Mitchell AJ, Benito-Leon J, Gonzalez JM, Rivera-Navarro J. Quality of life and its assessment in multiple sclerosis: integrating physical and psychological components of wellbeing. Lancet Neurol. 2005;4(9):556–66.
104. Marrie RA, Walld R, Bolton JM, Sareen J, Patten SB, Singer A, et al. Psychiatric comorbidity increases mortality in immune-mediated inflammatory diseases. Gen Hosp Psychiatry. 2018;53:65–72.
105. Ascherio A, Munger KL, White R, Kochert K, Simon KC, Polman CH, et al. Vitamin D as an early predictor of multiple sclerosis activity and progression. JAMA Neurol. 2014;71(3):306–14.
106. Thouvenot E, Orsini M, Daures JP, Camu W. Vitamin D is associated with degree of disability in patients with fully ambulatory relapsing-remitting multiple sclerosis. Eur J Neurol. 2015;22(3):564–9.
107. Healy BC, Ali EN, Guttmann CR, Chitnis T, Glanz BI, Buckle G, et al. Smoking and disease progression in multiple sclerosis. Arch Neurol. 2009;66(7):858–64.
108. Gholipour T, Healy B, Baruch NF, Weiner HL, Chitnis T. Demographic and clinical characteristics of malignant multiple sclerosis. Neurology. 2011;76(23):1996–2001.

# Pathology of Multiple Sclerosis

# 8

Jordon Dunham and Kedar R. Mahajan

**Key Points**

1. Multifocal demyelinated lesions occur in both the white and gray matter of the central nervous system (CNS), are present early in the disease course, and accumulate over time.
2. Active lesions are associated with demyelination, axonal injury, gliosis, oligodendrocyte destruction, microglial/macrophage activation, and infiltration of immune cells including macrophages and T- and B-cells.
3. Axonal transection has historically been described as "relative sparing," although in some chronic lesions up to 80% of axons are lost.
4. Secondary pathogenic mechanisms including oxidative energy and mitochondrial deficits may drive neurodegeneration.
5. Although magnetic resonance imaging (MRI) is a useful tool in the diagnosis of multiple sclerosis (MS), limitations exist in discriminating the underlying pathology of lesions, e.g., only approximately 55% of T2-only lesions are demyelinated, subpial and intracortical lesions are poorly visualized on MRI, and individuals with myelocortical MS have cerebral white matter lesions on MRI without significant cerebral white matter demyelination (but do have cortical and spinal cord demyelination).

J. Dunham
Department of Neuroscience, Lerner Research Institute, Cleveland Clinic Foundation, Cleveland, OH, USA

K. R. Mahajan (✉)
Department of Neuroscience, Lerner Research Institute, Cleveland Clinic Foundation, Cleveland, OH, USA

Department of Neurology, Mellen Center for Multiple Sclerosis Treatment and Research, Neurological Institute, Cleveland Clinic Foundation, Cleveland, OH, USA
e-mail: mahajak@ccf.org

## Introduction

Pathological characterization of multiple sclerosis (MS) postmortem and biopsy tissue has led to an improved understanding of inflammatory lesions and neurodegeneration and shed light on potential therapeutic targets. Since Charcot's recognition of MS as a distinct disease in the nineteenth century [1, 2], several seminal concepts have emerged. Demyelination is a key pathological feature of MS associated with inflammation that is widespread, involving the white matter (WM), cortical and deep gray matter (GM), and spinal cord [3]. Focal inflammatory lesions are associated with



A. L. Piquet, E. Alvarez (eds.), *Neuroimmunology*, https://doi.org/10.1007/978-3-030-61883-4_8

neuronal injury and axon/neurite transection [4, 5]. Neurodegeneration, either as a secondary consequence to focal inflammatory lesions, mediated by oxidative injury and energy failure [4], or perhaps independent of demyelination, contributes to clinical disability more significantly than WM lesions. Magnetic resonance imaging (MRI), while an invaluable tool, may not reflect pathology as accurately as we presume; contrary to popular belief, only 55% of T2-hyperintense lesions in patients with MS are demyelinated [6]. Here, we detail our current understanding of MS pathology, correlations with MRI where applicable, its clinical relevance, and gaps in our understanding.

## Lesions

A hallmark of MS is the presence of focal lesions characterized by demyelination and neuronal/axonal injury in the brain and spinal cord [7]. Myelin sheaths, produced by oligodendrocytes enwrapping axons to promote rapid saltatory conduction and trophic support to axons, can be extensively lost in demyelinated lesions. A higher proportion of myelin is composed of lipids (70–85%weight/weight), with the remainder being proteins—mainly myelin basic protein (MBP), proteolipid protein (PLP), and myelin oligodendrocyte glycoprotein (MOG). Myelin loss is readily detected by immunohistochemistry using monoclonal antibodies to these proteins. Chemical staining techniques such as Luxol fast blue are limited in that they are poorly able to visualize GM demyelination and are not specific to demyelination as they can also reflect lipid perturbations. Focal demyelinating lesions have clearly demarcated margins [8] and occur either in perivenular or subependymal/subpial areas. Areas with a greater predilection for lesions include the hippocampus, lateral ventricles, and superficial cortical layers [8].

A second prominent feature of MS lesions is neurodegeneration associated with neuronal injury (dystrophic neurons and reduced neuronal size), axonal pathology (transection and swelling), and synaptic/dendritic changes. "Relative sparing" of axons can be misleading as axonal loss in chronically demyelinated lesions can reach up to 80% [9]. The extent of neuronal loss associated with demyelination has been mixed [10–12], but neuronal pathology in lesions includes neurite transection, shrinking of the neuronal soma, and synaptic stripping [5, 12, 13].

Axonal injury, which has been recognized since the early 1900s, is associated with clinical disability worsening and is noted in both demyelinated and remyelinating lesions with high variability between individuals. Smaller caliber axons are more vulnerable [14], and pathology includes transection and proximal spheroids [9, 15]. Direct mechanisms of injury may include proteases and reactive oxygen/nitrogen species (ROS/RNS) from activated macrophages/microglia (mØ) in close contact and granules released by cytotoxic T-cells. Secondary means of injury could be by impaired axoplasmic membrane permeability and intra-axonal energy failure due to oxidative tissue injury and accumulation of respiratory-deficient neurons [16–20]. Axonal loss in eloquent areas such as the pyramidal tract can cause significant clinical disability [21]. Together these processes comprise mechanisms of tissue injury that contribute to physical disability.

## White Matter Lesions

White matter lesions (WMLs) gather more attention as they are readily identifiable as hyperpigmented plaques on gross pathology and associated with demyelination, inflammation, and axonal loss with immunohistochemistry [5, 15]. They are commonly located in corpus callosum, centrum semiovale, periventricularly, juxtacortically/cortically, and in the spinal cord [7, 8]. The extent of inflammatory cell infiltration with mØ and T-cells, demyelination, and axonal loss varies with the stage of the lesion noted below [22].

Histologically, earlier WMLs are more inflammatory than later stage lesions. The majority of CD3+ cells (76%) are major histocompatibility complex class I (MHC-I) restricted CD8+ T-cells and express markers associated with tissue-resident memory cells [23]. In active lesions, up to a third of CD8+ T-cells are granzyme B positive, suggesting cytotoxic activity against a yet unidentified self-antigen [23, 24]. B-cells (CD20+), in contrast, are predominantly in perivascular spaces with a lower number infiltrating the lesion parenchyma [23]. These observations are common in either relapsing or progressive disease courses and even in acute fulminant lesions. MHC-I is expressed on inflammatory cells and neuroglia [25], and MHC-II is observed in activated mØ [26, 27]. The inflammatory milieu of the active lesion changes when chronic active, characterized by a hypocellular lesion center with a rim of activated mØ [23].

Axonal pathology in the WML includes disruption of axonal transport with spheroids and transection [9]. Proposed mechanisms of acute transection include oxidative injury, cytokines, and proteolytic enzymes [4, 28]. Chronically demyelinated axons likely degenerate due to intra-axonal energy failure, altered sodium-potassium ($Na^+/K^+$) ATPase function, and accumulation of intra-axonal calcium levels [6, 16, 29].

While MRI has been paramount in the diagnosis of MS and monitoring for disease activity, there are clear limitations in detecting and staging lesions. Lesions suggestive of MS on MRI are defined by location (periventricular, juxtacortical/cortical, infratentorial, and spinal cord) and the presence of gadolinium enhancement (GdE) associated with a breakdown in the blood-brain barrier seen at the onset of lesion formation. Both the presence of GdE lesions and new/enlarging T2 lesions are useful in establishing the diagnosis of MS and surveillance of disease activity. GdE aids in the identification of new lesions on conventional T1 post-contrast images despite being relatively insensitive to blood-brain barrier disruption [30]. It is important to recognize that the presence of an MRI lesion does not necessarily equate to pathologically identifiable demyelination. In one analysis, 45% of T2-only (T2-hyperintense) and 17% of T2T1MTR (T2-hyperintense, T1-hypointense, and reduced magnetization transfer ratio) were not demyelinated with postmortem validation [29]. T2T1MTR lesions are more likely to be chronic inactive (68%) and have reduced axonal density [29]. T2-only changes do not always reflect demyelination, as they can be due to mØ activation and non-demyelinating axonal pathology [31]. T2 lesions that are not apparent on gross pathology can include preactive and active demyelinating lesions, while grossly visible lesions are often chronic inactive [32]. Remyelinated lesions cannot be discriminated by T2, but using T1 and MTR may improve sensitivity [33]. As lesions track venous vasculature, a "central vein sign" (imaged using T2*-echo planar imaging [34, 35]) can be useful to discriminate MS lesions from mimics such as microvascular disease, migraine, vasculitis, and aquaporin 4-positive neuromyelitis optica spectrum disorder (NMOSD) [36–38].

## White Matter Lesion Classification/ Staging

Histologically, focal lesions can be classified by the presence and distribution of inflammation (mØ) and the temporal pattern of demyelination. Staging lesions with in situ MRI-histopathology postmortem correlations provides insight into the efficacy of therapeutics—as one would expect minimal active lesions in individuals on highly effective therapies. More importantly, validating conventional and advanced imaging modalities with pathology lends to better clinical translation with in vivo imaging of myelin and neurodegeneration [39]. The lexicon for delineating lesion types has evolved over time to be more descriptive and is summarized in Table 8.1 [9, 22, 27, 40–42]. The use of "early" and "late" activity can be noted by mØ-containing myelin degradation products: cyclic nucleotide phosphodiesterase (CNP), myelin oligodendrocyte glycoprotein (MOG), or myelin-associated protein (MAG) in early and MBP and PLP in late [22]. Features of more recent lesion characterization include a description of the extent of inflammation, typically with mØ, and the presence of ongoing demyelination as evidenced by myelin degradation product inclusions in mØ.

The simplest classification (Bö/Trapp) characterizes lesions as active, chronic active, and inactive based on cellularity, predominantly by mØ [27]. The Lucchinetti/Brück/Lassmann classifica-

**Table 8.1** Lesion types characterized by immunohistochemistry for the presence of inflammation and myelin protein inclusions

| Classification and histological MS lesion staging proposed by the respective groups | | | | |
|---|---|---|---|---|
| Bö/Trapp 1994 [9, 27] | Vienna consensus 1997 [40] | De Groot/van der Valk 1999 [41] | Lucchinetti/Brück/ Lassmann 2000 [42] | Kuhlmann/Brück/ Lassman 2017 [22] |
| Active—hypercellular<br>Chronic active—hypocellular center with hypercellular rim<br>Chronic inactive—hypocellular | Inflammatory; demyelinating<br>Inflammatory; not demyelinating<br>Inflammatory rim with hypocellular center; not demyelinating<br>No inflammation; demyelinating(not yet observed)<br>No inflammation; not demyelinating | Preactive —abnormal WM, HLA-DR+/ CD45+ microglial clusters, few perivascular inflammatory cells, no demyelination<br>Active demyelinating—macrophages with myelin inclusions<br>Active non-demyelinating—lacking myelin inclusions<br>Chronic active—hypocellular center with hypercellular rim<br>Chronic inactive—hypocellular | Early active—mØ with myelin protein (MOG+/ MRP14$^+$)/lipids<br>Late active—mØ (MRP14$^-$, 27E10$^+$) with myelin debris (MBP$^+$, PLP$^+$, MOG$^-$)<br>Inactive—mØ (MRP14$^-$, 27E10$^-$) with glycolipids (PAS+), without myelin breakdown products<br>Early remyelinating—lymphocytes, mØ, thinly myelinated axons<br>Late remyelinating—mØ, astrogliosis, thinly myelinated axons | Active and early demyelinating—mØ throughout containing myelin degradation products (CNP, MAG, MOG, MBP, PLP)<br>Active and late demyelinating—mØ-containing MBP and PLP<br>Active and post-demyelinating—foamy lipid laden mØ lacking myelin proteins or LFB<br>Mixed active/inactive and demyelinating—mØ rim with hypocellular center and mØ-containing myelin degradation products<br>Mixed active/inactive and post-demyelinating—mØ devoid of myelin degradation products<br>Inactive—sharply demarcated hypocellular lesion with sparse mØ |

*Abbreviations: LFB* Luxol fast blue, *mØ* microglia/macrophages, *WM* white matter

tion, derived from biopsy and autopsy samples, includes myelin degradation products to discriminate early and late active lesions [26, 42]. Recently, Kuhlmann proposed including activity without demyelination (myelin degradation products) and "mixed active/inactive" ± demyelination; smoldering/slowly expanding lesions may be included in the latter category [22]. The De Groot/van der Valk modification introduces the concept of a preactive/early active lesion type with mØ clusters in normal-appearing regions without demyelination [39]. The Vienna consensus identifies six types based on inflammation as well as demyelination by the presence of myelin degradation products but is uncommonly utilized. For most purposes, the Bö/Trapp and Lucchinetti/Brück/Lassmann methods are sufficient to interpret pathological findings. See Fig. 8.1.

## Gray Matter Lesions

GM pathology correlates with physical disability, disability worsening, and cognitive impairment more significantly than WM lesion burden but is greatly underappreciated due to limitations with gross examination and MRI [43–46]. GM demyelination involves the cortex, deep GM (hippocampus, thalamus, caudate, putamen, amygdala, hypothalamus, and substantia nigra), and the spinal cord [47–50]. Grossly, cortical GM demyelination does not exhibit characteristic hyperpigmentation from marked myelin loss as in WM lesions. GM atrophy correlates with clinical disability, can precede WM atrophy, is associated with GM lesions, and is detectable in individuals even with low cerebral WM lesion burden [51–54].

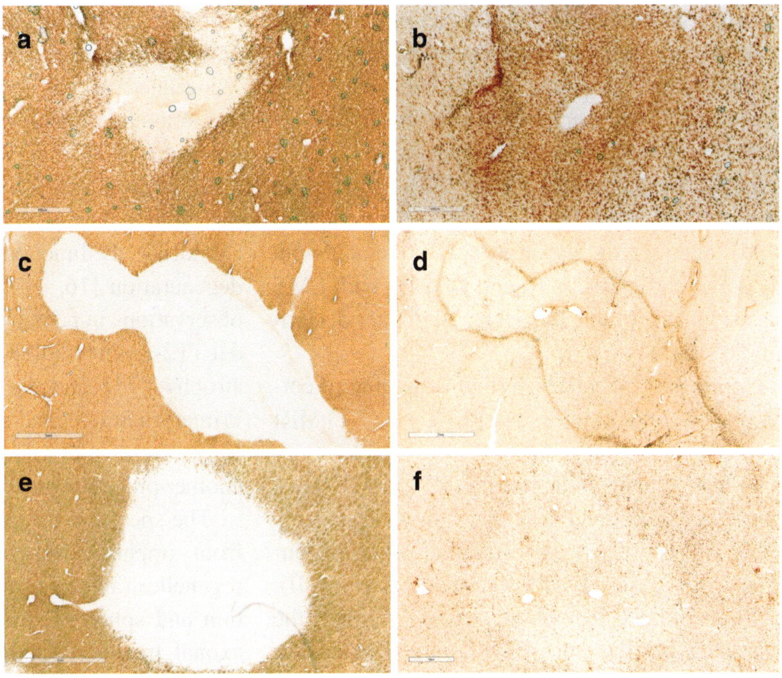


**Fig. 8.1** Classification of white matter demyelinating lesions using 1994 Bö/Trapp staging (see Table 8.1). Areas of perivascular demyelination were identified with proteolipid protein (PLP) staining (**a**, **c**, **e**) in 30 μm free floating sections of MS thalamic tissue. Staining for activated microglia/macrophages (MHC class II; **b**, **d**, **f**) is useful for identifying cellularity. Active lesions (**a**, **b**) are hypercellular, chronic active lesions (**c**, **d**) have a rim of hypercellularity (**d**), and chronic inactive lesions are hypocellular (**e**, **f**)

Cortical GM lesions are classified by their anatomic location: leukocortical (type I) lesions span the GM/WM interface, intracortical (type II) lesions are limited to the cortical GM, subpial (type III) lesions are on the surface of the cortex and extend to layers III/IV, and "type IV" lesions are sometimes used to describe subpial lesions spanning the entire width of the cortex and are relatively infrequent compared to other lesion types [5]. The degree and nature of inflammation and injury is variable between these lesion types. In early lesions characterized by biopsy specimens, cortical lesions are highly inflamed and suggested to precede WM demyelination [55]. Leukocortical lesions are seen earlier in the disease and typically start from subcortical WM (with greater inflammation) and extend into the cortex (with lesser inflammation). Intracortical lesions project radially from vessels and have less inflammation than leukocortical and WMLs [56]. Subpial lesions are the most common type of cortical lesion and are unique to MS. These lesions can be extensive, involving entire gyri, frequently occur in deep sulci regions, and often have clearly demarcated demyelinated lesion borders that sharply halt at layer III/IV [9, 50]. Similar to WML, subpial lesion borders frequently contain a line of activated mØ.

Subpial lesions in deep sulci and WMLs adjacent to the lateral and third ventricle suggest a role of cerebrospinal fluid (CSF) factors mediating demyelination. Consistent with this hypothesis are findings of subpial cortical demyelination and neurodegeneration correlating to the presence of meningeal follicle-like structures in a cohort of postmortem tissue from the UK Multiple Sclerosis Tissue Bank from donors with an aggressive disease course [57, 58]. In patients with a secondary progressive (SP) course, a gradient of neuron, astrocyte, and oligodendrocyte loss associated with follicle-like structures in low-flow sulci regions was observed, suggesting the role of yet unidentified soluble factors in mediating neurodegeneration [11]. Additionally, subpial demyelination was noted in association with persistent leptomeningeal enhancement (LME) in biopsy specimens from two subjects with MS [59]. Others have failed to replicate these findings, however, and found no correlation between meningeal inflammation and subpial lesions, nor cortical demyelination and neuronal loss [44, 60, 61]. In one report, meningeal inflammation was similar across myelinated and demyelinated associated subpial regions [60]. There is considerable debate as to the cellular organization of leptomeningeal inflamed regions and what degree aggregates of immune cells constitute a "follicle-like" structure. It has been suggested that such discrepancies could be due to cohort differences and methods of analysis [62]. LME on MRI was noted in 17% (40/240) of individuals with MS [59]. Furthermore, the presence of LME in a longitudinal study was associated with GM and cortical atrophy, and subjects with LME had a longer disease duration and greater disability [63, 64].

A clear unmet need is reliable imaging of cortical pathology. Ultrahigh field 7 Tesla (7 T) MRI and sequences such as 3D T1 magnetization-prepared 2 rapid gradient echo (MP2RAGE) have improved detection of cortical lesions. Using 7 T MRI with postmortem validation, 100% of leukocortical (type I), 11% intracortical (type II), 32% of subpial extending partially through the cortex (type III), and 68% of subpial extending the length of the cortex (type IV) lesions are detected [65].

## Atypical Demyelinating Lesions

Some atypical lesions in patients with MS include Balo's concentric sclerosis and tumefactive lesions. Balo's is characterized by WMLs with alternating concentric rings of demyelination and non-demyelinated regions associated with T-cell and mØ inflammation and hypoxia-like tissue injury [8, 66, 67]; no GM lesions have been noted [68]. Tumefactive lesions are greater than 2 cm with perilesional edema and/or ring enhancement mimicking a malignant glioma or cerebral abscess [69] and are hypercellular with atypical reactive astrocytes and mitotic figures.

## Neurodegeneration

Clinical symptoms such as cognitive impairment and fatigue as well as the accumulation of disability, particularly in progressive courses, are frequently attributed to neurodegeneration including neuronal injury and axonal transection. For example, deep GM atrophy, particularly thalamic, occurs at a greater rate attributable to MS earlier in the disease course, and later the contribution of age-related atrophy matches or exceeds MS-related atrophy [70]. While its precise etiology remains elusive, neurodegeneration (neuronal and axonal injury) can be directly mediated by cytokines and ROS/RNS or indirect consequences of chronic demyelination such as mitochondrial dysfunction and antero-/retrograde degeneration [16, 71, 72]. This is suggested by observations in pediatric-onset MS of WML burden (T2-lesion volume) correlating with thalamic atrophy [73] and limited brain growth [74]. Primary neurodegeneration with loss of neurons and axons independent of demyelination is another proposed mechanism.

The spectrum of axonopathy in WML ranges from impaired axonal transport of proteins/organelles (amyloid precursor protein accumulation and spheroid structures) and impaired fast axonal transport (nonphosphorylated neurofilament heavy chain/SMI-32 immunoreactivity) to irreparable axonal transection and the presence of terminal axonal ovoids distal to site of injury [9, 75]. The degree of axon loss in WMLs correlates with the extent of peripheral immune or resident glial inflammation [9, 15]. Anterograde axonal loss by Wallerian degeneration can occur in normal-appearing WM either adjacent or distal to the lesion and occurs early in MS as evidenced by pathologic and MRI studies [76–79].

GM normal-appearing and demyelinated areas are associated with neurite transection, synaptic and dendritic loss, and reduced neuron size without clear evidence of neuronal loss [13]. Up to 79% of hippocampi in cohort exhibited demyelinating lesions with reduced synaptic density, neuronal proteins essential for synaptic plasticity, and reduced glutamate neurotransmission, without a significant decrease in neuronal count [80, 81]. These observations are consistent with altered hippocampal MRI measures correlating with memory dysfunction [82].

## Myelocortical MS

Contrary to dogma, *both* WM and GM exhibit areas of demyelination [71, 83], and only approximately *half* of MRI T2-only lesions in patients with MS are demyelinated [29]. Recently, a new subtype of MS, termed myelocortical MS (MCMS), has been identified with the application of a rapid postmortem in situ MRI with immunohistochemistry [61]. Both patients with typical MS (TMS) and MCMS had similar appearing cerebral WM T2-weighted fluid-attenuated inversion recovery (FLAIR)-lesion burden on MRI (Fig. 8.2a–d) and had severe disability (mean expanded disability status scale [EDSS] > 8.0). However, subjects only had demyelination of the spinal cord and cerebral

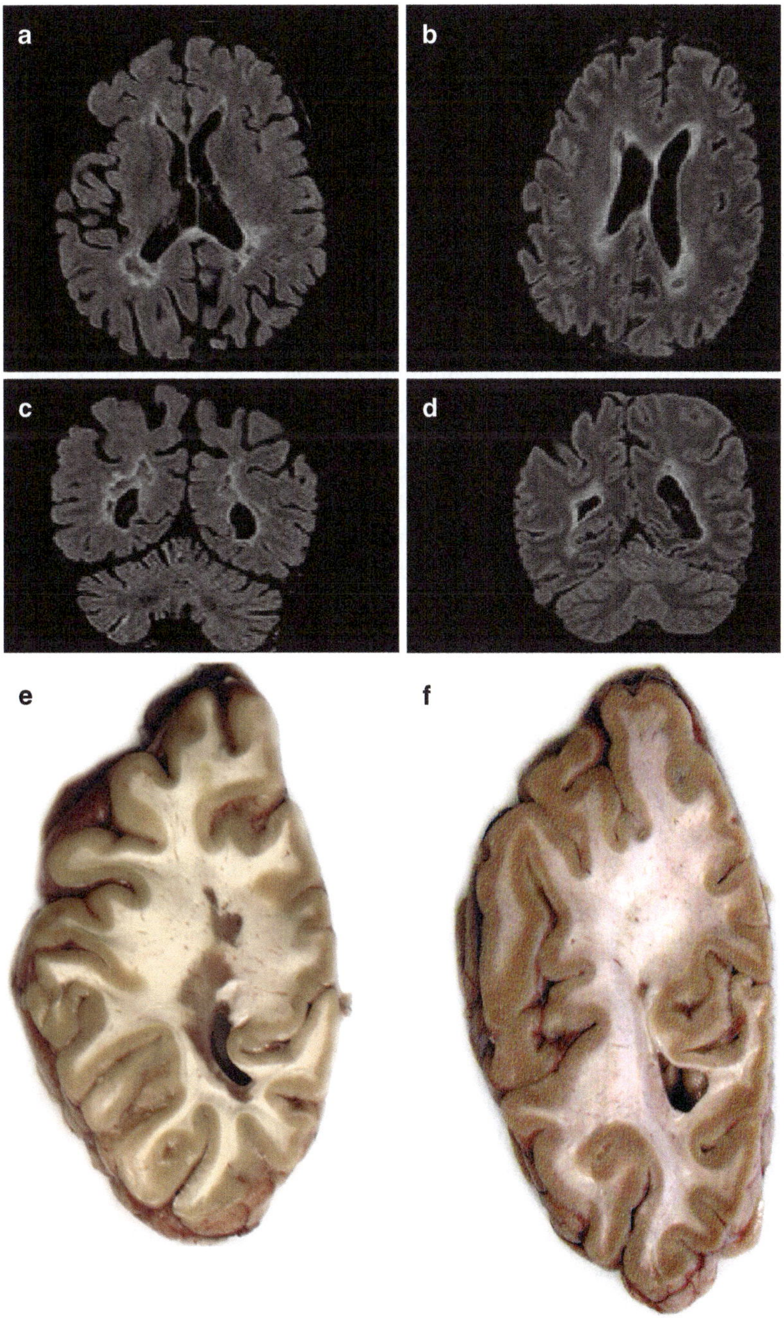


**Fig. 8.2** A novel subtype of multiple sclerosis (MS), termed "myelocortical," is characterized by cortical and spinal cord demyelination without significant cerebral white matter demyelination. Distribution of MRI FLAIR lesions is comparable between typical (**a**, **c**) and myelocortical (**b**, **d**) MS, but no apparent demyelination is noted on gross examination (**e** and **f**, respectively) or with immunohistochemistry for myelin using PLP (not shown). MRI images are shown in radiological convention (left side of screen represents right hemisphere); both axial (**a**, **b**) and coronal (**c**, **d**) orientations for the subjects shown in gross images are depicted to appreciate lesion distribution. (**e**, **f**: Reprinted with permission from Trapp et al. [61])

cortex without any significant cerebral WM demyelination observed grossly (Fig. 8.2f) and histologically. MRI regions of interest were selected for greater demyelination and axonal/neuronal injury—T2T1MTR rather than T2-only areas [29]. These T2T1MTR regions in patients with MCMS however lacked significant demyelination and did not have any T- or B-cell infiltration but had swollen myelinated axons.

Compared to MCMS, TMS had more WM and GM atrophy and spinal cord demyelination. A striking observation from this study was the reduction in neuronal densities in five cortical regions not directly connected to the spinal cord in layers III, V, and VI in MCMS compared to control; significant differences were only observed between TMS and control in layer V. In all layers, neuronal densities trended lower in MCMS compared to TMS. Confirming previous work, cortical neuronal loss did not correlate with subpial cortical demyelination in either MCMS or TMS [12]. Neuronal loss in MCMS was not explained by cerebral WM or cortical demyelination, supporting the concept that cortical neurodegeneration can occur independent of cerebral WM demyelination [29].

## Oxidative Tissue Injury and Mitochondrial Deficits

Oxidative damage and mitochondrial dysfunction can mediate axonal pathology and neurodegeneration (reviewed in [16, 84]). Oxidative injury, iron accumulation, energy failure, and impaired ion channels are all potential mechanisms of axonal and neuronal injury, which could occur independently or due to inflammatory lesions.

An example of oxidative damage is the production of ROS by upregulation of the nicotinamide adenine dinucleotide phosphate (NADPH) oxidase 2 catalytic subunit (NOX2). NOX2 upregulation is observed on mØ in pre-active lesions, diffusely in active demyelinating lesions, and in the rim of chronic active lesions [85, 86]. The release of iron from demyelinating axons, demonstrated by iron deposition and changes in iron homeostasis, is another example [87, 88]. Oxidative injury can be amplified by liberation of ferric iron ($Fe^{3+}$) from myelin loss and oligodendrocyte death [87], reduction to ferrous iron ($Fe^{2+}$) by a superoxide anion, and production of highly reactive hydroxyl radicals after combination with hydrogen peroxide (Fenton reaction) [89].

Collectively, ROS/RNS cause metabolic stress, deoxyribonucleic acid (DNA) alkylation, and peroxidation of phospholipids and proteins [90]. Oxidative injury is extensively observed in oligodendrocytes, neurons, myelin, and axons [4], and markers of oxidative insult coincide with upregulation of antioxidant enzymes [4, 28].

A functional consequence of oxidative stress and tissue injury is mitochondrial dysfunction and consequent disease progression [6, 91–94]. Reduced energy production occurs by disruption of the mitochondrial respiratory chain which involves oxidative phosphorylation of adenosine diphosphate to adenosine triphosphate (ATP) [95]. Due to the lack of protective histones, mitochondrial DNA (mtDNA) is particularly susceptible to oxidative stress leading to deletions and point mutations that may be propagated during clonal expansion. Lesions have neurons functionally deficient in mitochondrial respiratory chain complex IV and mtDNA deletions throughout the GM [6, 91]. Furthermore, deceased ATP production causes dysfunction of the $Na^+/K^+$ ATPase, which leads to a reversal of the axolemmal $Na^+/Ca^{2+}$ exchanger, increased intracellular calcium, and consequently axonal degeneration. Demyelinated axons in chronic lesions have a 50% reduction in $Na^+/K^+$ ATPase expression [16, 96]. Histotoxic hypoxia, an impaired ability to uptake oxygen, has also been described in lesions as a consequence of mitochondrial injury and a potential contributor to neurodegeneration [96].

## Distinctions Between Relapsing and Progressive MS

The clinical course of MS is heterogeneous, and there are several patterns of disease phenotypes characterized based on progression of disability [97]. Of MS patients, 85% start with a relapsing remitting (RR) disease course, and half of those individuals will transition to a SP course after 10–15 years. From onset, 15% of MS patients have a primary progressive (PP) course. Postmortem analysis of 185 subjects with PP and SP with 3188 tissue blocks with 7562 lesions shows similar lesion activity and lesion load [98]. While no differences in lesion pathology were noted, a greater lesion burden and a greater proportion of mixed active/inactive lesions at time of death was associated with a faster clinical disability worsening (time to EDSS 6). Consistent with these findings, subjects with RR had lower lesion load and a greater proportion of lesions with remyelination.

## Conclusion

Pathological analysis of postmortem and biopsy tissue has fundamentally shaped our understanding of MS and has provided valuable insight to advance therapeutic discovery and disease monitoring. A postmortem donation program for the procurement of tissue with in situ and post-fixed co-registered MRI, as well as clinical characterization, is vital in advancing our knowledge of MS pathology and clinical disability worsening. Postmortem validation of advanced MRI sequences striving for quantification of myelin and axonal perturbations will be necessary as biomarkers of neurodegeneration for clinical in vivo applications and outcome measures in clinical trials targeting inflammation and neurodegeneration.

**Acknowledgments** KRM has received funding from the National MS Society Clinician Scientist Development Award (FAN-1507-05605) and is currently funded by NIH/NINDS K23NS109328. Figures 8.1 and 8.2 were obtained by KRM in the laboratories of Dr. Bruce D. Trapp (Department of Neuroscience) and Dr. Kunio Nakamura (Department of Biomedial Engineering) at the Cleveland Clinic – Lerner Research Institute. The rapid MS postmortem autopsy program has received funding from Sanofi Genzyme.

## References

1. Pearce JM. Historical descriptions of multiple sclerosis. Eur Neurol. 2005;54(1):49–53.
2. Charcot M. Histologie de la sclerose en plaques. Gaz Hosp. 1868;141:554–5; 7–8.
3. Frischer JM, Bramow S, Dal-Bianco A, Lucchinetti CF, Rauschka H, Schmidbauer M, et al. The relation between inflammation and neurodegeneration in multiple sclerosis brains. Brain. 2009;132(Pt 5):1175–89.
4. Haider L, Fischer MT, Frischer JM, Bauer J, Hoftberger R, Botond G, et al. Oxidative damage in multiple sclerosis lesions. Brain. 2011;134(Pt 7):1914–24.
5. Peterson JW, Bo L, Mork S, Chang A, Trapp BD. Transected neurites, apoptotic neurons, and reduced inflammation in cortical multiple sclerosis lesions. Ann Neurol. 2001;50(3):389–400.
6. Campbell GR, Ziabreva I, Reeve AK, Krishnan KJ, Reynolds R, Howell O, et al. Mitochondrial DNA deletions and neurodegeneration in multiple sclerosis. Ann Neurol. 2011;69(3):481–92.
7. Popescu BF, Pirko I, Lucchinetti CF. Pathology of multiple sclerosis: where do we stand? Continuum (Minneap Minn). 2013;19(4 Multiple Sclerosis):901–21.
8. Popescu BF, Lucchinetti CF. Pathology of demyelinating diseases. Annu Rev Pathol. 2012;7:185–217.
9. Trapp BD, Peterson J, Ransohoff RM, Rudick R, Mork S, Bo L. Axonal transection in the lesions of multiple sclerosis. N Engl J Med. 1998;338(5):278–85.
10. Carassiti D, Altmann DR, Petrova N, Pakkenberg B, Scaravilli F, Schmierer K. Neuronal loss, demyelination and volume change in the multiple sclerosis neocortex. Neuropathol Appl Neurobiol. 2018;44(4):377–90.
11. Magliozzi R, Howell OW, Reeves C, Roncaroli F, Nicholas R, Serafini B, et al. A Gradient of neuronal loss and meningeal inflammation in multiple sclerosis. Ann Neurol. 2010;68(4):477–93.
12. Klaver R, Popescu V, Voorn P, Galis-de Graaf Y, van der Valk P, de Vries HE, et al. Neuronal and axonal loss in normal-appearing gray matter and subpial lesions in multiple sclerosis. J Neuropathol Exp Neurol. 2015;74(5):453–8.
13. Jurgens T, Jafari M, Kreutzfeldt M, Bahn E, Bruck W, Kerschensteiner M, et al. Reconstruction of single cortical projection neurons reveals primary spine loss in multiple sclerosis. Brain. 2016;139(Pt 1):39–46.
14. Evangelou N, Konz D, Esiri MM, Smith S, Palace J, Matthews PM. Size-selective neuronal changes in the anterior optic pathways suggest a differential susceptibility to injury in multiple sclerosis. Brain. 2001;124(Pt 9):1813–20.
15. Bitsch A, Schuchardt J, Bunkowski S, Kuhlmann T, Bruck W. Acute axonal injury in multiple sclerosis. Correlation with demyelination and inflammation. Brain. 2000;123(Pt 6):1174–83.

16. Witte ME, Mahad DJ, Lassmann H, van Horssen J. Mitochondrial dysfunction contributes to neurodegeneration in multiple sclerosis. Trends Mol Med. 2014;20(3):179–87.
17. Anthony DC, Miller KM, Fearn S, Townsend MJ, Opdenakker G, Wells GM, et al. Matrix metalloproteinase expression in an experimentally-induced DTH model of multiple sclerosis in the rat CNS. J Neuroimmunol. 1998;87(1–2):62–72.
18. Smith KJ, Lassmann H. The role of nitric oxide in multiple sclerosis. Lancet Neurol. 2002;1(4):232–41.
19. Neumann H, Medana IM, Bauer J, Lassmann H. Cytotoxic T lymphocytes in autoimmune and degenerative CNS diseases. Trends Neurosci. 2002;25(6):313–9.
20. Smith KJ, Kapoor R, Hall SM, Davies M. Electrically active axons degenerate when exposed to nitric oxide. Ann Neurol. 2001;49(4):470–6.
21. Bjartmar C, Kidd G, Mork S, Rudick R, Trapp BD. Neurological disability correlates with spinal cord axonal loss and reduced N-acetyl aspartate in chronic multiple sclerosis patients. Ann Neurol. 2000;48(6):893–901.
22. Kuhlmann T, Ludwin S, Prat A, Antel J, Bruck W, Lassmann H. An updated histological classification system for multiple sclerosis lesions. Acta Neuropathol. 2017;133(1):13–24.
23. Machado-Santos J, Saji E, Troscher AR, Paunovic M, Liblau R, Gabriely G, et al. The compartmentalized inflammatory response in the multiple sclerosis brain is composed of tissue-resident CD8+ T lymphocytes and B cells. Brain. 2018;141(7):2066–82.
24. Babbe H, Roers A, Waisman A, Lassmann H, Goebels N, Hohlfeld R, et al. Clonal expansions of CD8(+) T cells dominate the T cell infiltrate in active multiple sclerosis lesions as shown by micromanipulation and single cell polymerase chain reaction. J Exp Med. 2000;192(3):393–404.
25. Hoftberger R, Aboul-Enein F, Brueck W, Lucchinetti C, Rodriguez M, Schmidbauer M, et al. Expression of major histocompatibility complex class I molecules on the different cell types in multiple sclerosis lesions. Brain Pathol. 2004;14(1):43–50.
26. Bruck W, Porada P, Poser S, Rieckmann P, Hanefeld F, Kretzschmar HA, et al. Monocyte/macrophage differentiation in early multiple sclerosis lesions. Ann Neurol. 1995;38(5):788–96.
27. Bo L, Mork S, Kong PA, Nyland H, Pardo CA, Trapp BD. Detection of MHC class II-antigens on macrophages and microglia, but not on astrocytes and endothelia in active multiple sclerosis lesions. J Neuroimmunol. 1994;51(2):135–46.
28. van Horssen J, Schreibelt G, Drexhage J, Hazes T, Dijkstra CD, van der Valk P, et al. Severe oxidative damage in multiple sclerosis lesions coincides with enhanced antioxidant enzyme expression. Free Radic Biol Med. 2008;45(12):1729–37.
29. Fisher E, Chang A, Fox RJ, Tkach JA, Svarovsky T, Nakamura K, et al. Imaging correlates of axonal swelling in chronic multiple sclerosis brains. Ann Neurol. 2007;62(3):219–28.
30. Saade C, Bou-Fakhredin R, Yousem DM, Asmar K, Naffaa L, El-Merhi F. Gadolinium and multiple sclerosis: vessels, barriers of the brain, and glymphatics. AJNR Am J Neuroradiol. 2018;39(12):2168–76.
31. Moll NM, Rietsch AM, Thomas S, Ransohoff AJ, Lee JC, Fox R, et al. Multiple sclerosis normal-appearing white matter: pathology-imaging correlations. Ann Neurol. 2011;70(5):764–73.
32. De Groot CJ, Bergers E, Kamphorst W, Ravid R, Polman CH, Barkhof F, et al. Post-mortem MRI-guided sampling of multiple sclerosis brain lesions: increased yield of active demyelinating and (p)reactive lesions. Brain. 2001;124(Pt 8):1635–45.
33. Barkhof F, Bruck W, De Groot CJ, Bergers E, Hulshof S, Geurts J, et al. Remyelinated lesions in multiple sclerosis: magnetic resonance image appearance. Arch Neurol. 2003;60(8):1073–81.
34. Samaraweera AP, Clarke MA, Whitehead A, Falah Y, Driver ID, Dineen RA, et al. The central vein sign in multiple sclerosis lesions is present irrespective of the T2* sequence at 3 T. J Neuroimaging. 2017;27(1):114–21.
35. Campion T, Smith RJP, Altmann DR, Brito GC, Turner BP, Evanson J, et al. FLAIR* to visualize veins in white matter lesions: a new tool for the diagnosis of multiple sclerosis? Eur Radiol. 2017;27(10):4257–63.
36. Solomon AJ, Watts R, Ontaneda D, Absinta M, Sati P, Reich DS. Diagnostic performance of central vein sign for multiple sclerosis with a simplified three-lesion algorithm. Mult Scler. 2018;24(6):750–7.
37. Sati P, Oh J, Constable RT, Evangelou N, Guttmann CR, Henry RG, et al. The central vein sign and its clinical evaluation for the diagnosis of multiple sclerosis: a consensus statement from the North American Imaging in Multiple Sclerosis Cooperative. Nat Rev Neurol. 2016;12(12):714–22.
38. Cortese R, Magnollay L, Tur C, Abdel-Aziz K, Jacob A, De Angelis F, et al. Value of the central vein sign at 3T to differentiate MS from seropositive NMOSD. Neurology. 2018;90(14):e1183–e90.
39. van der Valk P, De Groot CJ. Staging of multiple sclerosis (MS) lesions: pathology of the time frame of MS. Neuropathol Appl Neurobiol. 2000;26(1):2–10.
40. Lassmann H, Raine CS, Antel J, Prineas JW. Immunopathology of multiple sclerosis: report on an international meeting held at the Institute of Neurology of the University of Vienna. J Neuroimmunol. 1998;86(2):213–7.
41. van Waesberghe JH, Kamphorst W, De Groot CJ, van Walderveen MA, Castelijns JA, Ravid R, et al. Axonal loss in multiple sclerosis lesions: magnetic resonance imaging insights into substrates of disability. Ann Neurol. 1999;46(5):747–54.
42. Lucchinetti C, Bruck W, Parisi J, Scheithauer B, Rodriguez M, Lassmann H. Heterogeneity of multiple sclerosis lesions: implications for the pathogenesis of demyelination. Ann Neurol. 2000;47(6):707–17.

43. Calabrese M, Magliozzi R, Ciccarelli O, Geurts JJ, Reynolds R, Martin R. Exploring the origins of grey matter damage in multiple sclerosis. Nat Rev Neurosci. 2015;16(3):147–58.
44. Klaver R, De Vries HE, Schenk GJ, Geurts JJ. Grey matter damage in multiple sclerosis: a pathology perspective. Prion. 2013;7(1):66–75.
45. Rocca MA, Amato MP, De Stefano N, Enzinger C, Geurts JJ, Penner IK, et al. Clinical and imaging assessment of cognitive dysfunction in multiple sclerosis. Lancet Neurol. 2015;14(3):302–17.
46. Steenwijk MD, Geurts JJ, Daams M, Tijms BM, Wink AM, Balk LJ, et al. Cortical atrophy patterns in multiple sclerosis are non-random and clinically relevant. Brain. 2016;139(Pt 1):115–26.
47. Geurts JJ, Bo L, Roosendaal SD, Hazes T, Daniels R, Barkhof F, et al. Extensive hippocampal demyelination in multiple sclerosis. J Neuropathol Exp Neurol. 2007;66(9):819–27.
48. Vercellino M, Masera S, Lorenzatti M, Condello C, Merola A, Mattioda A, et al. Demyelination, inflammation, and neurodegeneration in multiple sclerosis deep gray matter. J Neuropathol Exp Neurol. 2009;68(5):489–502.
49. Papadopoulos D, Dukes S, Patel R, Nicholas R, Vora A, Reynolds R. Substantial archaeocortical atrophy and neuronal loss in multiple sclerosis. Brain Pathol. 2009;19(2):238–53.
50. Bo L, Vedeler CA, Nyland HI, Trapp BD, Mork SJ. Subpial demyelination in the cerebral cortex of multiple sclerosis patients. J Neuropathol Exp Neurol. 2003;62(7):723–32.
51. Tedeschi G, Lavorgna L, Russo P, Prinster A, Dinacci D, Savettieri G, et al. Brain atrophy and lesion load in a large population of patients with multiple sclerosis. Neurology. 2005;65(2):280–5.
52. Fisniku LK, Chard DT, Jackson JS, Anderson VM, Altmann DR, Miszkiel KA, et al. Gray matter atrophy is related to long-term disability in multiple sclerosis. Ann Neurol. 2008;64(3):247–54.
53. Fisher E, Lee JC, Nakamura K, Rudick RA. Gray matter atrophy in multiple sclerosis: a longitudinal study. Ann Neurol. 2008;64(3):255–65.
54. Dalton CM, Chard DT, Davies GR, Miszkiel KA, Altmann DR, Fernando K, et al. Early development of multiple sclerosis is associated with progressive grey matter atrophy in patients presenting with clinically isolated syndromes. Brain. 2004;127(Pt 5):1101–7.
55. Lucchinetti CF, Popescu BF, Bunyan RF, Moll NM, Roemer SF, Lassmann H, et al. Inflammatory cortical demyelination in early multiple sclerosis. N Engl J Med. 2011;365(23):2188–97.
56. Bo L, Vedeler CA, Nyland H, Trapp BD, Mork SJ. Intracortical multiple sclerosis lesions are not associated with increased lymphocyte infiltration. Mult Scler. 2003;9(4):323–31.
57. Choi SR, Howell OW, Carassiti D, Magliozzi R, Gveric D, Muraro PA, et al. Meningeal inflammation plays a role in the pathology of primary progressive multiple sclerosis. Brain. 2012;135(Pt 10):2925–37.
58. Howell OW, Reeves CA, Nicholas R, Carassiti D, Radotra B, Gentleman SM, et al. Meningeal inflammation is widespread and linked to cortical pathology in multiple sclerosis. Brain. 2011;134(Pt 9):2755–71.
59. Absinta M, Vuolo L, Rao A, Nair G, Sati P, Cortese IC, et al. Gadolinium-based MRI characterization of leptomeningeal inflammation in multiple sclerosis. Neurology. 2015;85(1):18–28.
60. Kooi EJ, Geurts JJ, van Horssen J, Bo L, van der Valk P. Meningeal inflammation is not associated with cortical demyelination in chronic multiple sclerosis. J Neuropathol Exp Neurol. 2009;68(9):1021–8.
61. Trapp BD, Vignos M, Dudman J, Chang A, Fisher E, Staugaitis SM, et al. Cortical neuronal densities and cerebral white matter demyelination in multiple sclerosis: a retrospective study. Lancet Neurol. 2018;17(10):870–84.
62. Wicken C, Nguyen J, Karna R, Bhargava P. Leptomeningeal inflammation in multiple sclerosis: insights from animal and human studies. Mult Scler Relat Disord. 2018;26:173–82.
63. Zivadinov R, Ramasamy DP, Vaneckova M, Gandhi S, Chandra A, Hagemeier J, et al. Leptomeningeal contrast enhancement is associated with progression of cortical atrophy in MS: a retrospective, pilot, observational longitudinal study. Mult Scler. 2017;23(10):1336–45.
64. Makshakov G, Magonov E, Totolyan N, Nazarov V, Lapin S, Mazing A, et al. Leptomeningeal contrast enhancement is associated with disability progression and grey matter atrophy in multiple sclerosis. Neurol Res Int. 2017;2017:8652463.
65. Kilsdonk ID, Jonkman LE, Klaver R, van Veluw SJ, Zwanenburg JJ, Kuijer JP, et al. Increased cortical grey matter lesion detection in multiple sclerosis with 7 T MRI: a post-mortem verification study. Brain. 2016;139(Pt 5):1472–81.
66. Balo J. Encephalitis periaxialis concentrica. Arch Neurol. 1928;19:242–64.
67. Stadelmann C, Ludwin S, Tabira T, Guseo A, Lucchinetti CF, Leel-Ossy L, et al. Tissue preconditioning may explain concentric lesions in Balo's type of multiple sclerosis. Brain. 2005;128(Pt 5):979–87.
68. Behrens JR, Wanner J, Kuchling J, Ostendorf L, Harms L, Ruprecht K, et al. 7 Tesla MRI of Balo's concentric sclerosis versus multiple sclerosis lesions. Ann Clin Transl Neurol. 2018;5(8):900–12.
69. Kaeser MA, Scali F, Lanzisera FP, Bub GA, Kettner NW. Tumefactive multiple sclerosis: an uncommon diagnostic challenge. J Chiropr Med. 2011;10(1):29–35.
70. Azevedo CJ, Cen SY, Khadka S, Liu S, Kornak J, Shi Y, et al. Thalamic atrophy in multiple sclerosis: a magnetic resonance imaging marker of neurodegeneration throughout disease. Ann Neurol. 2018;83(2):223–34.
71. Lassmann H, Bruck W, Lucchinetti CF. The immunopathology of multiple sclerosis: an overview. Brain Pathol. 2007;17(2):210–8.
72. Kutzelnigg A, Lucchinetti CF, Stadelmann C, Bruck W, Rauschka H, Bergmann M, et al. Cortical demy-

elination and diffuse white matter injury in multiple sclerosis. Brain. 2005;128(Pt 11):2705–12.
73. Mesaros S, Rocca MA, Absinta M, Ghezzi A, Milani N, Moiola L, et al. Evidence of thalamic gray matter loss in pediatric multiple sclerosis. Neurology. 2008;70(13 Pt 2):1107–12.
74. Aubert-Broche B, Fonov V, Narayanan S, Arnold DL, Araujo D, Fetco D, et al. Onset of multiple sclerosis before adulthood leads to failure of age-expected brain growth. Neurology. 2014;83(23):2140–6.
75. Wegner C, Esiri MM, Chance SA, Palace J, Matthews PM. Neocortical neuronal, synaptic, and glial loss in multiple sclerosis. Neurology. 2006;67(6):960–7.
76. Bjartmar C, Kinkel RP, Kidd G, Rudick RA, Trapp BD. Axonal loss in normal-appearing white matter in a patient with acute MS. Neurology. 2001;57(7):1248–52.
77. Simon JH, Kinkel RP, Jacobs L, Bub L, Simonian N. A Wallerian degeneration pattern in patients at risk for MS. Neurology. 2000;54(5):1155–60.
78. Dziedzic T, Metz I, Dallenga T, Konig FB, Muller S, Stadelmann C, et al. Wallerian degeneration: a major component of early axonal pathology in multiple sclerosis. Brain Pathol. 2010;20(5):976–85.
79. Casanova B, Martinez-Bisbal MC, Valero C, Celda B, Marti-Bonmati L, Pascual A, et al. Evidence of Wallerian degeneration in normal appearing white matter in the early stages of relapsing-remitting multiple sclerosis: a HMRS study. J Neurol. 2003;250(1):22–8.
80. Sicotte NL, Kern KC, Giesser BS, Arshanapalli A, Schultz A, Montag M, et al. Regional hippocampal atrophy in multiple sclerosis. Brain. 2008;131(Pt 4):1134–41.
81. Dutta R, Chang A, Doud MK, Kidd GJ, Ribaudo MV, Young EA, et al. Demyelination causes synaptic alterations in hippocampi from multiple sclerosis patients. Ann Neurol. 2011;69(3):445–54.
82. Roosendaal SD, Moraal B, Vrenken H, Castelijns JA, Pouwels PJ, Barkhof F, et al. In vivo MR imaging of hippocampal lesions in multiple sclerosis. J Magn Reson Imaging. 2008;27(4):726–31.
83. Lassmann H, Bruck W, Lucchinetti C. Heterogeneity of multiple sclerosis pathogenesis: implications for diagnosis and therapy. Trends Mol Med. 2001;7(3):115–21.
84. Stephenson E, Nathoo N, Mahjoub Y, Dunn JF, Yong VW. Iron in multiple sclerosis: roles in neurodegeneration and repair. Nat Rev Neurol. 2014;10(8):459–68.
85. Fischer MT, Sharma R, Lim JL, Haider L, Frischer JM, Drexhage J, et al. NADPH oxidase expression in active multiple sclerosis lesions in relation to oxidative tissue damage and mitochondrial injury. Brain. 2012;135(Pt 3):886–99.
86. van Horssen J, Singh S, van der Pol S, Kipp M, Lim JL, Peferoen L, et al. Clusters of activated microglia in normal-appearing white matter show signs of innate immune activation. J Neuroinflammation. 2012;9:156.
87. Hametner S, Wimmer I, Haider L, Pfeifenbring S, Bruck W, Lassmann H. Iron and neurodegeneration in the multiple sclerosis brain. Ann Neurol. 2013;74(6):848–61.
88. Bagnato F, Hametner S, Yao B, van Gelderen P, Merkle H, Cantor FK, et al. Tracking iron in multiple sclerosis: a combined imaging and histopathological study at 7 Tesla. Brain. 2011;134(Pt 12):3602–15.
89. Bagnato F, Hametner S, Welch EB. Visualizing iron in multiple sclerosis. Magn Reson Imaging. 2013;31(3):376–84.
90. LeVine SM. The role of reactive oxygen species in the pathogenesis of multiple sclerosis. Med Hypotheses. 1992;39(3):271–4.
91. Mahad D, Lassmann H, Turnbull D. Review: mitochondria and disease progression in multiple sclerosis. Neuropathol Appl Neurobiol. 2008;34(6):577–89.
92. Mahad D, Ziabreva I, Lassmann H, Turnbull D. Mitochondrial defects in acute multiple sclerosis lesions. Brain. 2008;131(Pt 7):1722–35.
93. Mahad DJ, Ziabreva I, Campbell G, Lax N, White K, Hanson PS, et al. Mitochondrial changes within axons in multiple sclerosis. Brain. 2009;132(Pt 5):1161–74.
94. Zambonin JL, Zhao C, Ohno N, Campbell GR, Engeham S, Ziabreva I, et al. Increased mitochondrial content in remyelinated axons: implications for multiple sclerosis. Brain. 2011;134(Pt 7):1901–13.
95. Dimauro S, Rustin P. A critical approach to the therapy of mitochondrial respiratory chain and oxidative phosphorylation diseases. Biochim Biophys Acta. 2009;1792(12):1159–67.
96. Trapp BD, Stys PK. Virtual hypoxia and chronic necrosis of demyelinated axons in multiple sclerosis. Lancet Neurol. 2009;8(3):280–91.
97. Lublin FD, Reingold SC, Cohen JA, Cutter GR, Sorensen PS, Thompson AJ, et al. Defining the clinical course of multiple sclerosis: the 2013 revisions. Neurology. 2014;83(3):278–86.
98. Luchetti S, Fransen NL, van Eden CG, Ramaglia V, Mason M, Huitinga I. Progressive multiple sclerosis patients show substantial lesion activity that correlates with clinical disease severity and sex: a retrospective autopsy cohort analysis. Acta Neuropathol. 2018;135(4):511–28.

# Immunology of Multiple Sclerosis

# 9

Gregory F. Wu and Anne Haney Cross

**Key Points**

1. Multiple sclerosis (MS) is thought to be an autoimmune disease, but this is unproven and no single autoimmune target has been identified in MS.
2. Central nervous system (CNS) injury in MS results from an amalgamation of innate and adaptive immune responses—B cells, CD4 and CD8 T cells, and innate immune cells (such as monocytes, macrophages, and dendritic cells [DCs]), each contributes to the damage.
3. Resident central nervous system cells, such as microglia and astrocytes, as well as cells infiltrating from the periphery, contribute to the pathogenesis of MS.
4. Lymphoid and innate immune system responses can contribute to regulation of the immune response during MS.

G. F. Wu · A. H. Cross (✉)
Neuroimmunology Section, Department of Neurology, Washington University School of Medicine, Saint Louis, MO, USA
e-mail: crossa@wustl.edu

## Introduction

Multiple sclerosis (MS) is strongly linked to the immune system via multiple lines of evidence, which we will describe in this chapter. The neuropathology, genetics, cerebrospinal fluid (CSF) abnormalities, and beneficial responses to treatments that dampen the immune system, all support a link to the immune system, especially the cellular immune system. The neuropathology of most active MS lesions includes mononuclear inflammatory cell infiltration, antibodies, complement, and soluble immune mediators such as cytokines and chemokines (see Chap. 8 for pathology of MS) [1, 2]. International studies of MS genetics have identified associations of risk of MS with variants of multiple genes that encode immune system components (see Chap. 11 for genetics of MS) [3]. Although MS clearly involves the immune system, whether MS is an *autoimmune* disease has not been definitively proven.

Lessons regarding the immunology of MS have been learned, sometimes by accident, from clinical trials. An ill-fated clinical trial of interferon gamma (IFNγ[gamma]) led to clinical relapses in people with relapsing-remitting MS (RRMS). This study implicated interferon gamma in relapse induction. Additionally, a trial testing varying doses of altered peptide ligands of myelin basic protein (MBP) led to relapses with enhanced T cell responses to MBP peptide 83–99 [4]. The apparent induction of relapses with



A. L. Piquet, E. Alvarez (eds.), *Neuroimmunology*, https://doi.org/10.1007/978-3-030-61883-4_9

higher doses of altered MBP peptide added support to the notion that MS is truly autoimmune.

## Part 1. The Adaptive Immune System in Multiple Sclerosis: T and B Cells

The human immune system has two major divisions: the innate and the adaptive components. The adaptive immune system responds in a highly specific manner to specific antigenic targets (such as infectious agents) it encounters, developing "memory" for subsequent exposures and thereby protecting against future exposures to the same agent. The adaptive immune system consists primarily of T lymphocytes and B lymphocytes and their products (see Chap. 1). Despite the stringent mechanisms to eliminate and control autoreactive T and B cells, evidence indicates that healthy humans harbor such cells [5]. CD4 and CD8 T lymphocytes reactive to "self" myelin antigens (MBP, proteolipid protein [PLP], and myelin oligodendrocyte glycoprotein [MOG]), or neuronal antigens have been detected in similar numbers in healthy controls and in MS patients. However, many studies have pointed toward altered and more activated phenotypes of such self-reactive cells in people with MS versus healthy controls (reviewed in [6]). MS patients harbor more previously activated T cells that recognize several different myelin components, including myelin basic protein [7, 8], proteolipid protein [9], and myelin oligodendroglial cell protein, than healthy controls.

Current evidence strongly points toward T cells as playing a central role in MS pathogenesis. Among the lines of evidence implicating T cells in MS pathogenesis is that they are prominently present in active MS lesions in the central nervous system (CNS). The main T cell subsets are CD4 T lymphocytes, also called helper T cells (Th), and CD8 T lymphocytes, also known as cytotoxic T cells. Both types of T cells have been implicated in MS pathogenesis. CD4+ and CD8+ T cells are each present in active MS lesions, but are distributed differently [10]. In most studies CD4+ T cells are present at the leading edges of lesions, as well as in the center and with a few CD4+ T cells in the parenchyma of the CNS, whereas CD8+ T cells are mainly at the border of lesions [11]. The concentrations of CD4+ T cells and the number of CD8+ T cells present in MS lesions are each correlated with the degree of axon loss [12]. CD8+ T cells dominate in numbers within active lesions, as shown in several studies [13]. These CD8+ T cells are clonally expanded, which supports the notion that they are undergoing an antigen-driven response [14].

T cell lineages and their development are extremely complex and still being elucidated. CD4+ and CD8+ T cells are the two main types of T cells, most of which express a T cell receptor (TcR) comprised of an alpha and beta subunit, and the CD3 signaling protein is also part of the TcR complex. Both CD4+ and CD8+ T cells recognize their antigenic targets only as processed peptide antigens that are presented in context of major histocompatibility complex (MHC) class II or MHC class I molecules on antigen presenting cells (APCs), respectively. Subsets of CD4+ and CD8+ T cells can have many different functions, such as effector functions, antigen-specific memory, and immune system regulation.

### CD4+ T Cells

The main human gene associated with increased risk of MS is MHCII. Having one copy of HLA (human leukocyte antigen) DRB1*1501 or 1503 confers a two to threefold risk of MS development. As MHCII is required for antigen presentation to CD4+ T cells, this implicates CD4+ T cells in MS. Another key finding implicating CD4+ T lymphocytes in MS is that CD4+ T cells are sufficient to induce the primary animal model for MS, experimental autoimmune encephalomyelitis (EAE). EAE can be transferred to naïve syngeneic recipient animals (often mice) of susceptible strains by CD4+ T cells that target several different myelin antigens, including MBP, PLP, and MOG.

In MS, CD4+ T cells of differing subtypes likely play both effector roles and regulatory roles. CD4+ T cells subtypes can be defined by their expression of transcription factors, cytokines, and their distinct functions. T helper (Th) 1, Th2, Th17, and T-regulatory (T-reg) cells are the main CD4+ T cell subtypes. Th1 and Th17 cells are each independently capable of inducing EAE [15]. CD4+ Th2 cells do not induce EAE. Notably, the identity of Th cells as Th1 versus Th17 is not fixed—cells can shift from one subtype to the other depending upon cytokine milieu.

Despite the unequivocal role of CD4 T cells in the development of EAE, proving their role in human MS has been more difficult. Anti-CD4 depleting antibody therapy did not produce any clinical benefits in MS patients, although it is possible that studies thus far have been underpowered to detect an effect [16–18]. However, more recent larger studies with monoclonal antibodies targeting Th1 and Th17 cells are in accord. Recent appreciation of the contribution of non-T lymphocytes, in particular B cells, in MS pathogenesis has focused attention away from T cells.

## Th1 Cells

Th1 cells are a subtype of CD4 + T cells that, upon activation, secrete interleukin-2 (IL-2), IFNγ(gamma), and tumor necrosis factor-alpha (TNFα[alpha]). Naïve T cells become polarized toward Th1 cells in the presence of IL-12, which is mainly produced by myeloid lineage cells. T-bet is an essential and specific transcription factor for Th1 cell development in mice. As with all CD4+ T cells, Th1+ cells recognize processed antigen that is presented by another cell (the APC) in context of MHC class II. Several groups of investigators have shown that myelin-specific CD4 Th1 cells secreting IFNγ(gamma) are sufficient to transfer EAE into naïve mice [20, 21]. Moreover, mice lacking T-bet are resistant to EAE, confirming the key importance of Th1 type CD4+ T lymphocytes in EAE induction [19].

The role of Th1 cells in MS is less clear than it is in murine EAE, although a strong association of Th1 cells with MS has been observed. Th1 cells are present in increased numbers in MS lesions and in MS CSF compared with control tissues/CSF. Activated Th1 cells secrete copious amounts of IFNγ(gamma), which was shown to be deleterious in relapsing MS in an ill-fated study treating RRMS patients with exogenous IFNγ(gamma) [21]. On the other hand, TNFα(alpha), which is produced chiefly by activated macrophages but is also produced by CD4 Th1+ lymphocytes (and other cells such as natural killer [NK] cells, neutrophils, mast cells, and eosinophils), may be protective in MS based on data demonstrating neutralization of TNFα(alpha) in people with MS leading to clinical relapses [22].

## Th17 Cells

Th17 cells represent a more recently described CD4+ T cell lineage that is distinct from the Th1 and Th2 phenotypes [23, 24]. It appears that Th17 and Th1 cells may each play critical but different roles in EAE pathogenesis and, by extension, in MS. Recent evidence indicates that the cells responsible for inducing inflammation in many EAE mouse models are Th17 cells. In EAE, it has been proposed that a first wave of Th17 cells enter the CNS through the choroid plexus and trigger the entry of a second wave of T cells that migrate in large numbers into CNS by crossing activated parenchymal vessels [25].

Despite IL-17 mRNA first being reported in MS lesions in 2002, the roles of Th17 cells and their soluble products (IL-17, IL-21, and IL-22) in the pathogenesis of MS are yet to be fully elucidated [26]. During MS relapses, T cells characterized by Th17 expression are seen at increased frequency. Th17 cells may be more critical than Th1 cells for MS lesion development, although Th1 cells are tenfold more numerous than Th17 cells in the CSF and peripheral blood of MS patients. Evidence also exists that the two T cell subtypes are not strictly separate, with some CD4+ T cells present in MS lesions expressing both IL-17 and IFNγ(gamma) [27]. One interesting study showed expression of IL-17 receptors on human blood-brain barrier (BBB) endothelial cells in MS lesions, which would be expected to promote Th17 cell migration across the BBB [28].

The key soluble modulator IL-23, together with IL-6, transforming growth factor beta (TGF-β[beta]), and IL-21, leads to the differentiation, activation, and expansion of Th17 cells [29–31]. The transcription factor that specifically defines the Th17 subtype is retinoid-related orphan receptor-γ(gamma)t (ROR-γ[gamma]t) [31]. Elevated concentrations of sodium may be another trigger of development of pathogenic Th17 cells [32]. In an important study, it was noted that a high-salt diet in mice induced to develop EAE-enhanced Th17 responses and worsened clinical and histological EAE severity compared with mice on a regular diet [33]. Perhaps the modern world's typical high-salt diet is an environmental factor leading to increased risk of developing the disease via enhancing Th17 cells [34, 35].

Notably, a clinical trial has shed doubt on the importance of Th1 cells and Th17 cells in MS pathogenesis. Treatment with ustekinumab, a monoclonal antibody targeting the IL-12/IL-23 p40 chain, which is common to both IL-12 and IL-23 (cytokines required for differentiation of Th1 and Th17 cells, respectively), failed to stop magnetic resonance imaging (MRI) activity in RRMS patients in a multi-center phase 2 study [36].

## Th2 Cells

CD4+ Th2 cells are particularly important in humoral immunity in humans. Th2 cells secrete IL-4, IL-5, IL-10, and IL-13, and they do not alone induce EAE [37]. In fact, several different studies indicate a regulatory role of Th2 cells in MS and in the EAE model. This may be via production by Th2 cells of immunoregulatory cytokines such as IL-4 and IL-10, and/or neurotrophic factors [38, 39]. Treatments that increase Th2 over Th1 cells tend to ameliorate EAE, and by inference possibly MS [40]. A relative enhancement in Th2 cells over Th1 cells is thought to be a mechanism of action of glatiramer acetate in RRMS patients [41].

## T Follicular Helper (Tfh) Cells

T follicular helper (Tfh) cells constitute a subset of CD4 T cells that is essential for generation of memory B cells [42]. Tfh cells produce IL-21, a cytokine that supports the development of high affinity B cells in germinal centers. Bcl-6 is the transcription factor essential for Tfh cell generation in vivo [43]. Tfh cells express CXCR5, a chemokine receptor for CXCL13, which also is a chemoattract of B cells.

Tfh cells are found in ectopic lymphoid follicles (ELFs) that are frequently found at sites of chronic inflammation, including in the CNS of MS patients [44]. These collections are defined by the co-localization of Tfh cells, proliferating B cells, and CXCL13. Since Tfh cells (and other cell types such as B cells) express CXCR5, the presence of high concentrations of CXCL13 promotes Tfh cell migration into lymphoid follicles. Excessive concentrations of CXCL13 have been consistently found in CSF of MS patients, which would perhaps encourage ELF formation and maintenance [45]. ELFs are most commonly seen in deep meningeal recesses in progressive MS patients; their importance is underscored by their association with worse MS prognosis [46].

## CD8+ T Cells

CD8+ T cells are found in larger number within active lesions than CD4+ T cells, as shown in several studies. CD8+ T cells in MS lesions are clonally expanded, which supports an antigen-driven response [14]. The number of CD8+ T cells present in MS lesions is associated with the degree of axon loss [47]. Together, these findings suggest that CD8+ T cells play an important role in the human disease. In contrast to CD4+ T cells, CD8+ T cells recognize their cognate antigen in context of MHC class I. Compared with MHCII, MHCI can be expressed on a far larger range of cell types, including oligodendroglial cells and even neurons [48].

Normal mice harbor CD8+ MHC class I–restricted cytotoxic T cells that recognize MBP. These cells can also transfer an autoimmune encephalomyelitis, which is somewhat different from the CD4+ T cell–mediated EAE because the cerebellum, rather than the spinal cord, is a main target [49]. Thus far, no studies specifically depleting CD8+ T cells have been done in MS patients to our knowledge.

## Regulatory T Cells (T-regs)

Self-tolerance toward myelin antigens is clearly maintained in normal healthy people. Active suppression of T cells reactive with self-antigens by T-regs plays a critical role in the maintenance of this self-tolerance in vivo [50]. CD4+ T-regs are defined by high expression of the IL-2 receptor (CD25) and expression of the FoxP3 transcription factor. CD4+ T-regs secrete the immunoregulatory cytokine IL-10 and are protective against EAE [51]. However, despite numerous studies, it remains controversial as to whether T-regulatory cells are reduced in number or function in MS patients compared with healthy controls. Overall, studies to date do not consistently show altered numbers of T-regs in people with MS. Some studies have indicated that T-regs are dysfunctional in MS, but this will require confirmation [52].

## B Cells and Their Products

B cells have long been implicated in MS pathogenesis. Excessive B cells are commonly noted in MS CSF [52–54]. Yet, B cells are far less common in MS brains than T cells, and they are mostly localized around vessels in active lesions and rarely in CNS parenchyma. Nonetheless, the number of B cells in MS lesions has been correlated to the degree of axon loss [55]. Also, several groups of investigators have identified clonal overlap between B cell receptor sequences in peripheral blood, MS brain parenchyma, meningeal ELFs, and CSF, which indicates trafficking between these compartments [55–57].

Immunoglobulins (Ig), the product of B cells and of plasma cells (which differentiate from B cells), have been identified in excess in MS CSF and in MS CNS tissues for many decades [58]. Excessive oligoclonal immunoglobulins in CSF are used in MS diagnosis [59]. Plasma cells are found both in perivascular regions and in tissue parenchyma, often in large numbers, and even in inactive disease [60].

It was the advent of monoclonal antibodies that specifically deplete circulating B cells that led to the decisive establishment of a critical role of B cells in MS pathogenesis. Studies of four different monoclonal antibodies that lyse CD20+ B cells, rituximab, ocrelizumab, ofatumumab, and ublituximab have all shown a rapid and profound reduction in clinical relapses and in gadolinium-enhancing lesions on MRI. Interestingly, almost all other approved disease-modifying treatments for relapsing MS reduce B cells, especially memory B cells [61]. It is not known if the major role of B cells occurs in the CNS, the peripheral immune system, or both.

Exactly how B cells exert a pathogenic role in MS is not fully understood. Although B cells make antibody and differentiate into plasma cells, which make the majority of antibodies, reduction in antibody production is unlikely to be related. This is because the rapid reduction in MS activity seen after B cell depletion in the peripheral blood occurs prior to any reduction in IgG, IgM, or IgA levels in blood or CSF [62]. B cells are a source, directly or indirectly, of several key cytokines and chemokines that appear key to MS activity, and reduction in these factors may be part of why B cell depletion is so effective in relapsing MS [62–64]. Moreover, B cells are highly effective APCs for autoimmune CD4+ T cells, including T cells that recognize MBP and MOG—two CNS myelin proteins that are believed to be targets in MS [65]. A likely role that B cells play in inciting MS disease activity is through presentation of self-antigen to CD4+ T cells, and enhancing inflammation and recruiting other inflammatory cells via production of cytokines and chemokines. Also, as noted before, proliferating B cells, as well as TfH cells and CXCL13, are associated with ELFs in the meninges in MS patients. ELFs are associated with more aggressive clinical MS and earlier death from MS [66].

## Part 2: The Innate Immune System in Multiple Sclerosis

Several innate cell subtypes participate in the inflammatory changes that occur during MS. Historically, the characterization of MS plaques focused on phagocytic cells in active and chronic active lesions or "plaques" [67–69]. Macrophages have been observed to participate in the destruction of intact myelin within MS plaques. In fact, CD68+ macrophages are found in great abundance within active plaques [70], often engorged with lipids after stripping myelin from axons [71]. This effector process is thought to occur in concert with infiltration of monocytes from the periphery [71, 72] but has also been ascribed to endogenous microglial cells [74]. Microglia, rather than macrophages, preferentially express triggering receptor expressed on myeloid cells 2 (TREM2) within the CNS [75], enabling clearance of debris by microglia [76] and potentially contributing to their activation status [77]. Other functional distinctions between microglia and macrophages include differential production of distinct chemokines and cytokines, antigen processing and presentation, and response to aging [77, 78]. These differences may contribute to the relative timing and impact on myelin injury caused by macrophages and microglia during the genesis of MS lesions. Overall, phagocytic innate cells belonging to several cellular subtypes contribute to the end effector stages of myelin destruction in MS.

Macrophages and microglia also mediate axonal damage concomitant with, and ensuing, myelin injury [79, 80]. During formation of MS lesions, many microglia transition from a role in maintaining axonal integrity at steady state [81, 82] to exerting direct inflammatory damage to axons [83, 84]. Macrophages are also often associated with axonal injury in MS lesions [86]. Correlations between axonal damage in MS plaques and the density of human leukocyte antigen-DR+ (HLA-DR+) macrophages/microglia suggest a causal relationship [55]. Mechanisms of axonal damage by innate cells in MS lesions include production of reactive oxygen and nitrogen species, proteases, and complement [87]. Whether these same mechanisms contribute to the neuronal changes in cortical and deep grey matter in MS has yet to be definitively demonstrated.

### Macrophages

Similar to distinct polarization states of T cells, a dichotomy of macrophage polarization has been described. Macrophages have been categorized into M1 and M2 phenotypes [88]. Although this dichotomy represents an oversimplification, it can be used to broadly distinguish between macrophages (M1) which exhibit pro-inflammatory properties, including expression of IL-1β(beta), TNFα(alpha), and inducible nitric oxide synthase (iNOS), as well as up-regulation of MHCII and co-stimulatory molecules [89]. M1 markers have been detected in MS lesions, including iNOS and CD40 [89, 90, 91]. Using animal models of MS, blockade of macrophage co-stimulation [92, 93] or use of iNOS inhibitors [94–96] has reduced the severity of disease, implicating M1 macrophages in the pathogenesis of neuro-inflammation.

In contrast, M2 macrophages (also termed "alternatively activated" macrophages) are defined by the production of suppressive cytokines and the tendency to propagate anti-inflammatory adaptive immune responses [89]. M2 macrophages emerge in greater numbers after the peak of EAE, gradually replacing M1 macrophages. In the later phases of EAE development, M2 macrophages presumably quell the inflammatory milieu but also promote healing [96, 97]. M2 macrophages are identified by the production of cytokines such as IL-4, IL-13, and IL-10 and the expression of arginine (Arg) [98]. Segal and colleagues describe the accumulation of Arg1+ myeloid cells in the CNS that are significantly impaired in antigen presentation to CD4 T cells [99]. Myelin-containing macrophages within MS lesions have been shown to express M2 traits, suggesting that, once damage is done, macrophages can revert to a suppressive state [78]. The strict delineation between M1 and M2 macrophages is probably an oversimplification in vivo. For example, a majority of foamy macrophages present in active demyelinating lesions express both M1 and M2 markers [100]. Further, a true division of macrophage polarity and function has been questioned in part because a singular set of gene expression representing one or the other state (M1 vs. M2) has never been clearly demonstrated in vivo [101]. Nevertheless, subgroups of macrophages are critical for the crescendo as well as resolution of inflammation targeting the CNS during MS, and macrophages are recognized as participating in both myelin injury and removal throughout the genesis and repair of an MS lesion. Overall, it should be recognized that macrophages likely exhibit a broad and possibly fluid spectrum of function in MS.

## Microglia

Traditionally, microglial contributions to CNS injury, including during MS lesion genesis, occur after activation. "Resting" microglia are characterized by ramified processes with dynamic probing activity [102]. In contrast, activated microglia are rounded/amoeboid in shape and exhibit inflammatory features, including up-regulation of MHCII, production of nitric oxide (NO), and enhanced phagocytosis [103]. The cues responsible for activation of microglia are wide-ranging and remain undefined in MS [73, 103]. Like macrophages, microglia have been classified into M1 and M2 polar states [87, 104, 105]. This distinction is controversial for several reasons [106], including the idea that microglia are adapted specifically to their microenvironment and that individual microglia may exhibit a multitude of different functional phenotypes. Nevertheless, microglia can produce a variety of pro-inflammatory cytokines and chemokines, such as TNF-α(alpha), IL-6, IL-23, and IL-1β(beta), along with reactive oxygen species [106, 107]. In part via these soluble molecules, microglia can orchestrate pro-inflammatory T cell function in vivo, resulting in amplification of adaptive responses within the CNS [108, 109]. In contrast, microglia can also exhibit reparative qualities. For example, soluble factors produced by microglia promote proliferation and differentiation of oligodendrocytes [110] and promote remyelination of lesions in EAE [111]. Additionally, microglia may possibly induce the development of T-reg, leading to the suppression of inflammatory responses targeting the CNS [112]. Hence, microglia are capable of a wide range of immune responses, including phagocytic activity, production of soluble immune factors, and coordination of T cell responses that can both damage and repair CNS tissue in MS.

Consideration for how each subgroup of innate cells contributes to MS lesion formation and resolution must account for the barriers between the periphery and CNS. Macrophages of hematopoietic origin must traffic into the CNS compartment to exert myelin damage in MS. The dynamic interplay between the many different immune cell subsets, including peripheral innate and adaptive cells, is depicted in Fig. 9.1 with particular attention to the blood-brain barrier. Microglia share surface expression of markers with several myeloid-derived innate cell populations, including CD45, CD11b, CX3CR1, and F4/80 [113]. One significant pursuit in recent years has been to accurately distinguish microglial cells from peripherally derived monocytes and macrophages. The genetic and ontological basis for a division between microglia and monocytes/macrophages is undergoing refinement via high throughput gene expression analysis [75, 78, 112–114]. In murine systems, several gene pathways specifically define microglia, including TREM2 and TGFβ(beta) expression [107]. Additionally, using microglial transcriptome datasets, TMEM119 was identified as a useful surface marker specific to *human* microglia [115]. In more intensive analyses, a multitude of cells with phagocytic characteristics were distinguished in the *murine* CNS using a combination of single-cell proteomics, high parametric mass cytometry, and fate mapping [79]. Ultimately, the diversity of various innate cells within the CNS will assist in defining commonalities as well as unique aspects of inflammation both regionally as well as temporally during MS. It should be noted that the different subsets of innate cells within the naïve and inflamed CNS as identified

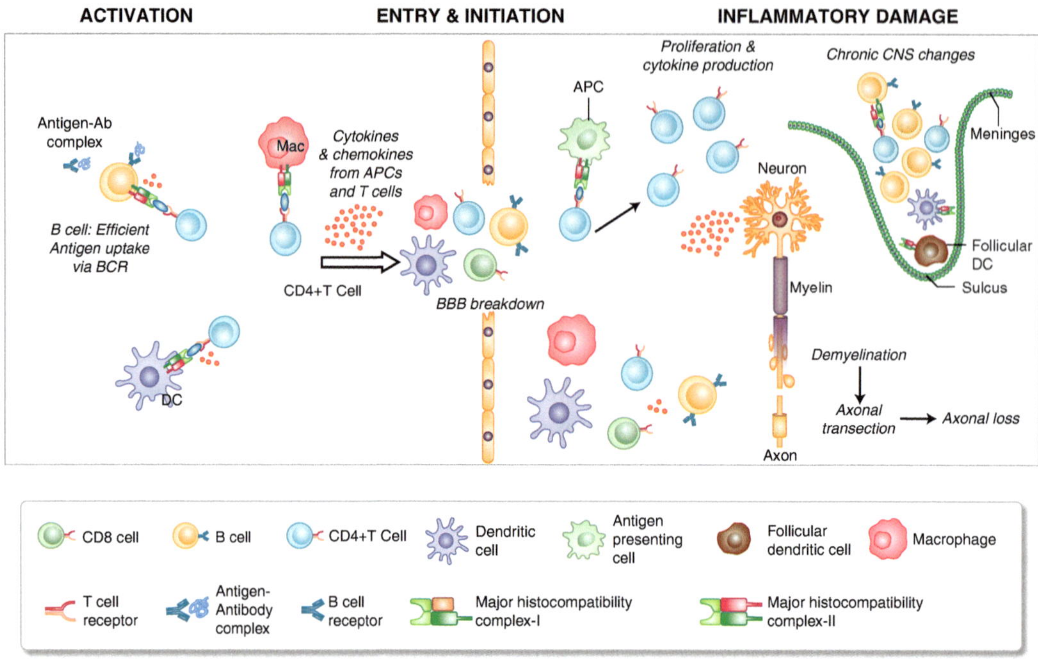


**Fig. 9.1** The pathogenesis of MS occurs in multiple phases and locations. In this simplified schematic of the immunopathogenesis of MS, CD4 T cells are central to each phase of disease. The coordinated immune response targeting CNS antigens in MS is initiated with activation of CD4 T cells by APCs, presumably outside of the CNS. Any number of different APCs can present target antigen via MHCII to CD4 T cells, including DCs, macrophages, and B cells. Migration by T cells through the blood-brain barrier (BBB) is essential. Cognate interactions between CD4 T cells and APCs are thought to be required again prior to fulminant inflammatory changes involving breakdown of the BBB, infiltration of other leukocytes, and production of cytokines and antibodies. Ultimately, myelin and nerve injury occur during the effector phase. Infiltration of monocytes through the BBB can result in the expansion of available innate cells during the propagation and resolution of lesions. Chronically, accumulation of immune cells in structures resembling lymphoid tissue and largely composed of lymphocytes including B cells can be observed in the meninges of patients with progressive MS

in mice may not always apply to the human. The degree of distinction between microglia and macrophages in the pathology of MS remains less clear. Ultimately, a better understanding of the diversity of various innate cells within the CNS will assist in defining commonalities as well as unique aspects of inflammation both regionally as well as temporally during MS.

## Antigen Processing and Presentation by Innate Immune Cells

Innate immune cells can perform another function that is essential to the pathogenesis of MS: the processing and presentation of antigens to T cells (Fig. 9.1). To participate in antigen-specific immune responses, naïve CD4 T cells must be activated by professional APCs that express processed antigen-MHCII complexes (Fig. 9.2). Dendritic cells (DCs), macrophages, and B cells are considered professional APCs and, along with microglia, have the capacity for activating naïve CD4 T cells due to expression of MHCII and co-stimulatory molecules. Antigen presentation by APCs coordinates cognate interactions that are required for neuro-inflammation to be propagated in MS [116]. Experimentally, there is considerable evidence that antigen-specific interactions between APCs and CD4 T cells are required once the CD4

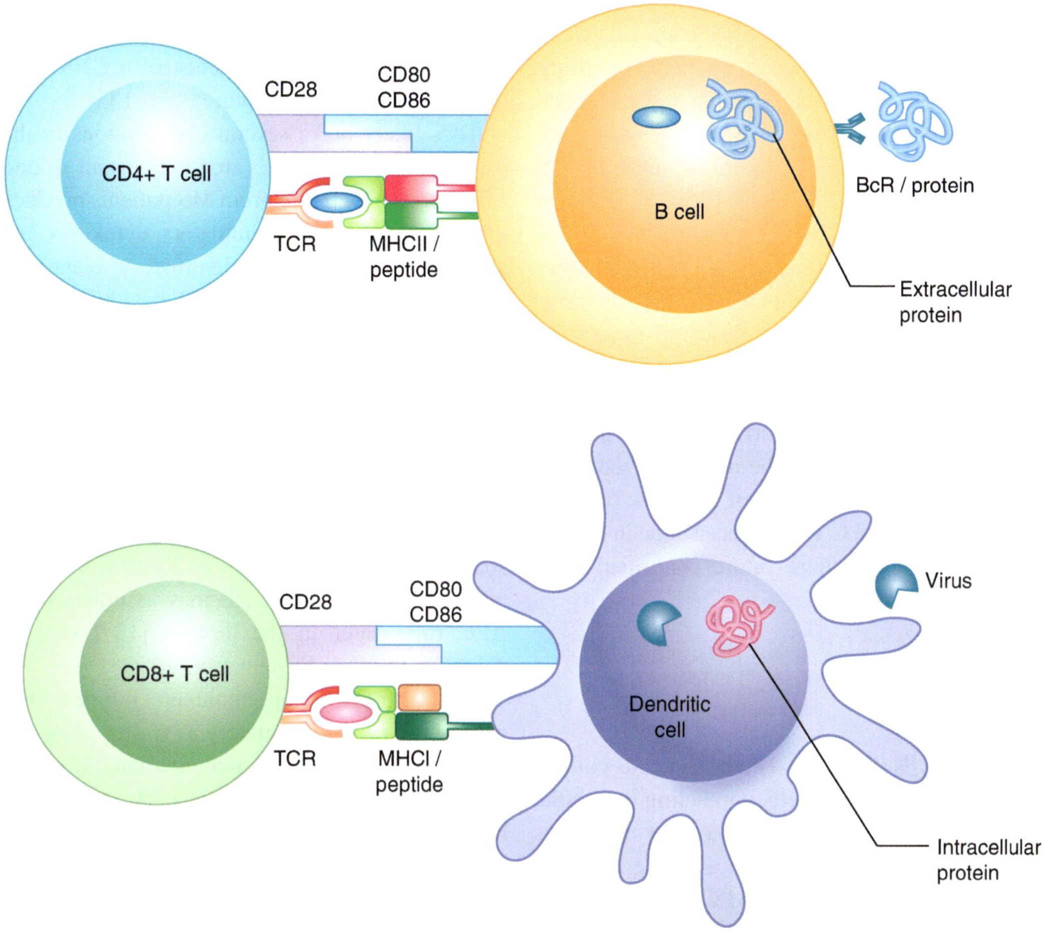


**Fig. 9.2** Antigen-specific interactions between the innate and adaptive immune system in MS. Target antigens, such as peptides from myelin proteins, are presented via MHCII to T cells and recognized by TcRs. For CD8 T cells antigen is recognized when presented by MHCI, while for CD4 T cells antigen targets are recognized when presented in the context of MHCII. Co-stimulatory interactions between APCs and T cells (e.g., via CD80/CD28) are required in order for inflammatory responses to be generated. Different APCs are capable of presenting antigen based on expression of MHCI and/or MHCII, location, and timing. B cells constitutively express MHCII and are optimal APCs for antigens in low abundance, such as myelin antigens

T cell has migrated to the CNS [116, 117], although in MS this has not been established. The identity and characteristics of CNS APCs remain unknown, as are the anatomic locations where this cognate interaction takes place. Different APC populations inhabit various sites within the CNS that can serve as entry points for CD4 T cells during MS, including the post-capillary venules, meninges, and choroid plexus [118, 119]. In EAE, microglia are dispensable as APCs, signifying the importance of hematopoietically derived APCs in propagating adaptive inflammatory activity in MS [120, 121]. Whether the bulk of MHCII expression by macrophages found in MS lesions indicates that they perform all APC functions within the CNS or whether other specialized cells such as DCs [122] or antigen-specific B cells [122, 123] make critical contributions is unclear. One cell subset poised to contribute to regulation of antigen-specific CD4 T cells within the CNS is border-associated macrophages (BAMs). BAMs include choroid plexus macrophages and perivascular macrophages [79]. Again, due to limited surface markers to allow specific distinctions between APCs such as DCs and BAMs, particularly during inflammation, identifying specific roles remains challenging. For example, BAMs and DCs share expression of CD11c under various inflammatory circumstances [78, 120, 124, 125]. A thorough review of APCs in MS concluded that the functionally relevant APC or APC subtype responsible for T cell reactivation within the CNS has not been definitively identified [126]. Which of these cells in fact act as APCs for T cell reactivation and to what extent they can exert this function has been studied intensively, but unfortunately with no firm conclusion. It should also be noted that the work identifying BAMs and most of the work on APC function within the CNS have been done in animal models and will require future confirmation in humans.

## Dendritic Cells

Dendritic cells (DCs) are a heterogeneous collection of innate immune cells populating lymphoid and non-lymphoid tissues throughout the body. DCs are critical regulators of adaptive immune responses [127]. Among all APCs, DCs—almost all of which express the integrin CD11c—are able to function as APCs during all phases of CD4 T cell activation due to their high proficiency at antigen processing, presentation of target antigens, and migratory properties [128]. In animal models of MS, DCs are responsible for initiation of myelin-targeted CNS autoimmunity [119, 122, 128]. In EAE, DCs not only initiate disease but also propagate autoimmunity via epitope spreading [129]. As part of their sentinel nature, DCs capture extravascular antigens from tissue and then migrate via lymphatic vessels to secondary lymphoid organs to coordinate adaptive immune responses. With respect to the immune-specialized characteristics of CNS tissue, the possibility that the CNS somehow restricts routine surveillance by DCs that occurs for other tissues has been debated for years [130]. DCs may be able to access CNS antigens and present them in draining lymph nodes [131]. Trafficking of APCs, including DCs, to and from the CNS is now a concept under intense scrutiny in part due to the re-emergence of a CNS lymphatic system in the literature [131–133].

DCs likely make several contributions to the pathogenesis of MS. DCs are found in and around MS lesions [121]. DCs from MS patients produce greater amounts of inflammatory cytokines and generate more robust inflammatory responses from CD4 T cells compared with healthy controls [134]. In steady state within the CNS, DCs are thought to populate barriers between the CNS and the periphery such as the meninges and microvasculature [134–136]. As such, they are positioned to initiate antigen-specific responses from entering T cells. In fact, DCs could serve as the first APC to locally initiate autoreactive adaptive immune responses as T cell traffic into the CNS [119, 122]. B cells may assume critical APC roles later in the disease [137] (see in Adaptive Immune System section).

As with other innate immune cell types, DCs can exhibit pro-inflammatory as well as suppressive behavior during immune-mediated diseases such as MS [122]. For example, early work in the EAE model demonstrated that the expression of co-stimulatory molecules was reduced by MHCII+ CD11c+ mononuclear cells isolated from the brains of mice with disease, leading to decreased capacity to induce T cell proliferation [138]. In MS patients, immuno-suppressive traits were identified in MHCII+ cells containing myelin antigens in cervical lymph nodes, suggesting a regulatory function of some migratory DCs [139]. Complicating matters is the tremendous rate of discovery of functional and developmental traits of DCs, which has led to a vast refinement of DC nomenclature [140]. Using contemporary analysis of human DC subsets will allow for greater resolution in determining the role of bona fide DCs in the pathogenesis of MS.

## Other Hematopoietically Derived Innate Cells

A special category of myeloid cells termed myeloid-derived suppressor cells (MDSCs) has the unique ability to suppress inflammatory immune responses. MDSCs emerge during states of chronic inflammation or injury [141] and have been found in greater abundance in MS patients during relapses [141–143]. A loss of negative regulation by MDSCs may play a role in the overall inflammatory nature of MS, as fewer MDSCs are found in MS patients compared to healthy controls [144]. This is further supported by the observation that MDSCs from MS patients exhibited reduced suppressive qualities. Further, in murine EAE, expansion of MDSCs reduced the severity of disease [144]. More recently, MDSCs were shown to influence the function of other immune cells in EAE. Korn and colleagues found that MDSCs serve to limit B cell infiltration into the CNS and dampen pro-inflammatory traits of B cells during EAE [145]. Notably, some characteristics are shared between MDSCs and neutrophils (PMNs) [146]. A role for PMNs in the pathogenesis of MS has been suggested by observations not only in animal models of MS [147] but also from CSF examination of MS patients [148]. However, MS plaques are almost completely devoid of neutrophils and are not observed in significant numbers within the CSF of MS patients [149], making it difficult to imagine how PMNs would be instrumental in the development of MS lesions. Hence, the influence of polymorphonuclear cells (PMNs) as well as MDSCs may all be exerted at a distance from the actual CNS lesions in MS.

## Astrocytes

The contribution by astrocytes has often been overlooked in terms of CNS immune surveillance and autoimmune demyelination in MS. Reactive astrocytes are a common feature of active lesion formation [150]. Astrocytes are capable of producing cytokines and chemokines that participate in inflammatory responses within the CNS [150]. In addition to signaling from astrocytes to various immune cells, inflammatory mediators produced by astrocytes such as NO and glutamate have the potential to directly injure myelin and axons [151]. In addition to employing soluble factors, astrocytes modulate immune reactivity within the CNS by influencing the BBB. Physical connections by astrocyte end-foot processes with microvasculature of the CNS impart regulatory properties on the BBB that influence immune cell entry [152]. As is becoming more the convention, in vivo and in vitro analyses have led to the application of pro- and anti-inflammatory subsets to reactive astrocytes, termed A1 and A2 [153]. Inflammatory features such as the production of IL-1α(alpha) and TNF define A1 astrocytes, whereas A2 astrocytes express several neurotrophic factors and are more able to promote neuronal and oligodendrocyte survival and growth. Interestingly, changes in polarization state are in part dependent upon microglia, demonstrating the entwined nature of glial responses during CNS injury [153]. How distinct each astrocyte polarization state remains to be seen, and how this relates to the inflammatory changes in MS remains undefined.

## Conclusion

Both the innate and adaptive branches of the immune response contribute to the pathological changes in MS, supporting the notion that MS is an autoimmune disease. Herein, we have attempted to summarize current knowledge regarding how each arm of the immune system contributes to MS.

## References


1. Popescu BFG, Lucchinetti CF. Neuropathology of multiple sclerosis. In: Multiple sclerosis: a mechanistic view. London: Academic Press; 2015. p. 181–200.
2. Esiri MM. Immunoglobulin-containing cells in multiple-sclerosis plaques. Lancet. 1977;310(8036):478–80.
3. Sawcer S, Hellenthal G, Pirinen M, Spencer CCA, Patsopoulos NA, Moutsianas L, et al. Genetic risk and a primary role for cell-mediated immune mechanisms in multiple sclerosis. Nature. 2011;476:214–9.
4. Bielekova B, Goodwin B, Richert N, Cortese I, Kondo T, Afshar G, et al. Encephalitogenic potential of the myelin basic protein peptide (amino acids 83–99) in multiple sclerosis: results of a phase II

clinical trial with an altered peptide ligand. Nat Med. 2000;6:1167.

5. Burns J, Rosenzweig A, Zweiman B, Lisak RP. Isolation of myelin basic protein-reactive T-cell lines from normal human blood. Cell Immunol. 1983;81(2):435–40.
6. Legroux L, Pittet CL, Beauseigle D, Deblois G, Prat A, Arbour N. An optimized method to process mouse CNS to simultaneously analyze neural cells and leukocytes by flow cytometry. J Neurosci Methods. 2015;247:23–31.
7. Lovett-Racke AE, Trotter JL, Lauber J, Perrin PJ, June CH, Racke MK. Decreased dependence of myelin basic protein-reactive T cells on CD28-mediated costimulation in multiple sclerosis patients: a marker of activated/memory T cells. J Clin Invest. 1998;101(4):725–30.
8. Allegretta M, Nicklas JA, Sriram S, Albertini RJ. T cells responsive to myelin basic protein in patients with multiple sclerosis. Science (80). 1990;247(4943):718–21.
9. Trotter JL, Damico CA, Cross AH, Pelfrey CM, Karr RW, Fu XT, et al. HPRT mutant T-cell lines from multiple sclerosis patients recognize myelin proteolipid protein peptides. J Neuroimmunol. 1997;75(1–2):95–103.
10. Traugott U, Reinherz EL, Raine CS. Multiple sclerosis: distribution of T cell subsets within active chronic lesions. Science (80). 1983;219(4582):308–10.
11. Hauser SL, Bhan AK, Gilles F, Kemp M, Kerr C, Weiner HL. Immunohistochemical analysis of the cellular infiltrate in multiple sclerosis lesions. Ann Neurol. 1986;19(6):578–87.
12. Frischer JM, Bramow S, Dal-Bianco A, Lucchinetti CF, Rauschka H, Schmidbauer M, et al. The relation between inflammation and neurodegeneration in multiple sclerosis brains. Brain. 2009;132(5):1175–89.
13. Booss J, Esiri MM, Tourtellotte WW, Mason DY. Immunohistological analysis of T lymphocyte subsets in the central nervous system in chronic progressive multiple sclerosis. J Neurol Sci. 1983;62(1–3):219–32.
14. Babbe H, Roers A, Waisman A, Lassmann H, Goebels N, Hohlfeld R, et al. Clonal expansions of CD8 + T cells dominate the t cell infiltrate in active multiple sclerosis lesions as shown by micromanipulation and single cell polymerase chain reaction. J Exp Med [Internet]. 2000;192(3):393–404. Available from: http://www.jem.org/lookup/doi/10.1084/jem.192.3.393.
15. Lovett-Racke AE, Yang Y, Racke MK. Th1 versus Th17: are T cell cytokines relevant in multiple sclerosis? Biochim Biophys Acta – Mol Basis Dis. 2011;1812(2):246–51.
16. Lindsey RJW, Hodgkinson S, Mehta R, Mitchell D, Enzmann D, Steinman L. Repeated treatment with chimeric anti-CD4 antibody in multiple sclerosis. Ann Neurol. 1994;36(2):183–9.
17. Rep MHG, Van Oosten BW, Roos MTL, Adèr HJ, Polman CH, Van Lier RAW. Treatment with depleting CD4 monoclonal antibody results in a preferential loss of circulating naive T cells but does not affect IFN-γ secreting TH1 cells in humans. J Clin Invest. 1997;99(9):2225–31.
18. Van Oosten BW, Lai M, Hodgkinson S, Barkhof F, Miller DH, Moseley IF, et al. Treatment of multiple sclerosis with the monoclonal anti-CD4 antibody cM-T412: results of a randomized, double-blind, placebo-controlled, MR- monitored phase II trial. Neurology. 1997;49(2):351–7.
19. Lovett-Racke AE, Rocchini AE, Choy J, Northrop SC, Hussain RZ, Ratts RB, et al. Silencing T-bet defines a critical role in the differentiation of autoreactive T lymphocytes. Immunity. 2004;21(5):719–31.
20. Fletcher JM, Lalor SJ, Sweeney CM, Tubridy N, Mills KHG. T cells in multiple sclerosis and experimental autoimmune encephalomyelitis. Clin Exp Immunol. 2010;162:1–11.
21. Panitch HS, Hirsch RL, Schindler J, Johnson KP. Treatment of multiple sclerosis with gamma interferon: exacerbations associated with activation of the immune system. Neurology [Internet]. 1987;37(7):1097–1097. Available from: http://www.neurology.org/cgi/doi/10.1212/WNL.37.7.1097.
22. Lenercept. TNF neutralization in MS: results of a randomized, placebo-controlled multicenter study. Neurology [Internet]. 1999;53(3):457. Available from: http://www.neurology.org/cgi/doi/10.1212/WNL.53.3.457.
23. Beecham AH, Patsopoulos NA, Xifara DK, Davis MF, Kemppinen A, Cotsapas C, et al. Analysis of immune-related loci identifies 48 new susceptibility variants for multiple sclerosis. Nat Genet. 2013;45(11):1353–62.
24. Lieberich N. Netzwerke in medizinischen bereichen. Teil 1: Grundlagen der vernetzung. Medizintechnik. 2000;120:64–9.
25. Reboldi A, Coisne C, Baumjohann D, Benvenuto F, Bottinelli D, Lira S, et al. C-C chemokine receptor 6-regulated entry of TH-17 cells into the CNS through the choroid plexus is required for the initiation of EAE. Nat Immunol. 2009;10(5):514–23.
26. Lock C, Hermans G, Pedotti R, Brendolan A, Schadt E, Garren H, et al. Gene-microarray analysis of multiple sclerosis lesions yields new targets validated in autoimmune encephalomyelitis. Nat Med. 2002;8(5):500–8.
27. Kebir H, Ifergan I, Alvarez JI, Bernard M, Poirier J, Arbour N, et al. Preferential recruitment of interferon-γ-expressing TH17 cells in multiple sclerosis. Ann Neurol. 2009;66(3):390–402.
28. Kebir H, Kreymborg K, Ifergan I, Dodelet-Devillers A, Cayrol R, Bernard M, et al. Human TH17 lymphocytes promote blood-brain barrier disruption and central nervous system inflammation. Nat Med. 2007;13(10):1173–5.

29. Cua DJ, Sherlock J, Chen Y, Murphy CA, Joyce B, Seymour B, et al. Interleukin-23 rather than interleukin-12 is the critical cytokine for autoimmune inflammation of the brain. Nature. 2003;421:744.
30. Tesmer LA, Lundy SK, Sarkar S, Fox DA. Th17 cells in human disease. Immunol Rev. 2008;223:87–113.
31. Ivanov II, McKenzie BS, Zhou L, Tadokoro CE, Lepelley A, Lafaille JJ, et al. The orphan nuclear receptor RORγt directs the differentiation program of proinflammatory IL-17+T helper cells. Cell. 2006;126(6):1121–33.
32. Wu C, Yosef N, Thalhamer T, Zhu C, Xiao S, Kishi Y, et al. Induction of pathogenic TH 17 cells by inducible salt-sensing kinase SGK1. Nature. 2013;496(7446):513–7.
33. Kleinewietfeld M, Manzel A, Titze J, Kvakan H, Yosef N, Linker RA, et al. Sodium chloride drives autoimmune disease by the induction of pathogenic TH 17 cells. Nature. 2013;496(7446):518–22.
34. Manzel A, Kleinewietfeld M, Titze J, Hafler D, Müller D, Linker R. Poster II. Mult Scler J [Internet]. 2012;18(4_suppl):279–508. Available from: http://journals.sagepub.com/doi/10.1177/1352458512459021.
35. Manzel A, Muller DN, Hafler DA, Erdman SE, Linker RA, Kleinewietfeld M. Role of "western diet" in inflammatory autoimmune diseases. Curr Allergy Asthma Rep. 2014;14(1):404.
36. Segal BM, Constantinescu CS, Raychaudhuri A, Kim L, Fidelus-Gort R, Kasper LH. Repeated subcutaneous injections of IL12/23 p40 neutralising antibody, ustekinumab, in patients with relapsing-remitting multiple sclerosis: a phase II, double-blind, placebo-controlled, randomised, dose-ranging study. Lancet Neurol. 2008;7(9):796–804.
37. Steinman L. A brief history of TH17, the first major revision in the TH1/TH2 hypothesis of T cell-mediated tissue damage. Nat Med. 2007;13:139–45.
38. Roy A, Liu X, Pahan K. Myelin basic protein-primed T cells induce neurotrophins in glial cells via α5β3 integrin. J Biol Chem. 2007;282(44):32222–32.
39. Zhang Y, Jalili F, Ouamara N, Zameer A, Cosentino G, Mayne M, et al. Glatiramer acetate-reactive T lymphocytes regulate oligodendrocyte progenitor cell number in vitro: role of IGF-2. J Neuroimmunol. 2010;227(1–2):71–9.
40. Kornberg MD, Smith MD, Shirazi HA, Calabresi PA, Snyder SH, Kim PM. Bryostatin-1 alleviates experimental multiple sclerosis. Proc Natl Acad Sci [Internet]. 2018;115(9):2186–91. Available from: http://www.pnas.org/lookup/doi/10.1073/pnas.1719902115.
41. Qin Y, Zhang DQ, Prat A, Pouly S, Antel J. Characterization of T cell lines derived from glatiramer-acetate-treated multiple sclerosis patients. J Neuroimmunol. 2000;108(1–2):201–6.
42. Crotty S. T follicular helper cell differentiation, function, and roles in disease. Immunity. 2014;41:529–42.
43. Kroenke MA, Eto D, Locci M, Cho M, Davidson T, Haddad EK, et al. Bcl6 and Maf cooperate to instruct human follicular helper CD4 T cell differentiation. J Immunol [Internet]. 2012;188(8):3734–44. Available from: http://www.jimmunol.org/cgi/doi/10.4049/jimmunol.1103246.
44. Serafini B, Rosicarelli B, Magliozzi R, Stigliano E, Aloisi F. Detection of ectopic B-cell follicles with germinal centers in the meninges of patients with secondary progressive multiple sclerosis. Brain Pathol. 2004;14(2):164–74.
45. Alvarez E, Piccio L, Mikesell RJ, Klawiter EC, Parks BJ, Naismith RT, et al. CXCL13 is a biomarker of inflammation in multiple sclerosis, neuromyelitis optica, and other neurological conditions. Mult Scler J. 2013;19(9):1204–8.
46. Magliozzi R, Howell O, Vora A, Serafini B, Nicholas R, Puopolo M, et al. Meningeal B-cell follicles in secondary progressive multiple sclerosis associate with early onset of disease and severe cortical pathology. Brain. 2007;130(4):1089–104.
47. Frischer JM, Bramow S, Dal-Bianco A, Lucchinetti CF, Rauschka H, Schmidbauer M, et al. The relation between inflammation and neurodegeneration in multiple sclerosis brains. Brain. 2009;132:1175.
48. Massa PT, Ozato K, McFarlin DE. Cell type-specific regulation of major histocompatibility complex (MHC) Class I gene expression in astrocytes, oligodendrocytes, and neurons. Glia. 1993;8(3):201–7.
49. Huseby ES, Liggitt D, Brabb T, Schnabel B, Öhlén C, Goverman J. A pathogenic role for myelin-specific Cd8 + T cells in a model for multiple sclerosis. J Exp Med [Internet]. 2001;194(5):669–76. Available from: http://www.jem.org/lookup/doi/10.1084/jem.194.5.669.
50. Viglietta V, Baecher-Allan C, Weiner HL, Hafler DA. Loss of functional suppression by CD4 + CD25 + regulatory T cells in patients with multiple sclerosis. J Exp Med. 2004;199(7):971–9.
51. Kohm AP, Carpentier PA, Anger HA, Miller SD. Cutting edge: CD4+CD25+ regulatory T cells suppress antigen-specific autoreactive immune responses and central nervous system inflammation during active experimental autoimmune encephalomyelitis. J Immunol [Internet]. 2002;169(9):4712–6. Available from: http://www.jimmunol.org/cgi/doi/10.4049/jimmunol.169.9.4712.
52. Viglietta V, Baecher-Allan C, Weiner HL, Hafler DA. Loss of functional suppression by CD4 + CD25 + regulatory T cells in patients with multiple sclerosis. J Exp Med [Internet]. 2004;199(7):971–9. Available from: http://www.jem.org/lookup/doi/10.1084/jem.20031579.
53. Wurth S, Kuenz B, Bsteh G, Ehling R, Di Pauli F, Hegen H, et al. Cerebrospinal fluid B cells and disease progression in multiple sclerosis – a longitudinal prospective study. PLoS One [Internet]. 2017;12(8):e0182462. Available from: http://www.ncbi.nlm.nih.gov/pubmed/28777826.

54. Cepok S. Patterns of cerebrospinal fluid pathology correlate with disease progression in multiple sclerosis. Brain [Internet]. 2001;124(11):2169–76. Available from: https://academic.oup.com/brain/article-lookup/doi/10.1093/brain/124.11.2169.
55. Frischer JM, Bramow S, Dal-Bianco A, Lucchinetti CF, Rauschka H, Schmidbauer M, et al. The relation between inflammation and neurodegeneration in multiple sclerosis brains. Brain. 2009;132(5):1175–89.
56. Obermeier B, Lovato L, Mentele R, Brück W, Forne I, Imhof A, et al. Related B cell clones that populate the CSF and CNS of patients with multiple sclerosis produce CSF immunoglobulin. J Neuroimmunol. 2011;233(1–2):245–8.
57. von Büdingen HC, Bar-Or A, Zamvil SS. B cells in multiple sclerosis: connecting the dots. Curr Opin Immunol. 2011;23:713–20.
58. Link H. Oligoclonal immunoglobulin G in multiple sclerosis brains. J Neurol Sci. 1972;16(1):103–14.
59. Link H, Huang YM. Oligoclonal bands in multiple sclerosis cerebrospinal fluid: an update on methodology and clinical usefulness. J Neuroimmunol. 2006;180:17–28.
60. Prineas JW, Wright RG. Macrophages, lymphocytes, and plasma cells in the perivascular compartment in chronic multiple sclerosis. Lab Invest. 1978;38:409.
61. Longbrake EE, Cross AH. Effect of multiple sclerosis disease-modifying therapies on b cells and humoral immunity. JAMA Neurol. 2016;73:219–25.
62. Cross A, Stark J, Lauber J, Ramsbottom MJ, Lyons JA. Rituximab reduces B cells and T cells in cerebrospinal fluid of multiple sclerosis patients. J Neuroimmunol [Internet]. 2006;180(1–2):63–70. Available from: http://www.sciencedirect.com/science/article/pii/S0165572806002700.
63. Alvarez E, Piccio L, Mikesell RJ, Trinkaus K, Parks BJ, Naismith RT, et al. Predicting optimal response to B-cell depletion with rituximab in multiple sclerosis using CXCL13 index, magnetic resonance imaging and clinical measures. Mult Scler J – Exp Transl Clin. 2015;1:2055217315623800.
64. Piccio L, Naismith RT, Trinkaus K, Klein RS, Parks BJ, Lyons JA, et al. Changes in B- and T-lymphocyte and chemokine levels with rituximab treatment in multiple sclerosis. Arch Neurol. 2010;67(6):707.
65. Fraussen J, Claes N, Van Wijmeersch B, van Horssen J, Stinissen P, Hupperts R, et al. B cells of multiple sclerosis patients induce autoreactive proinflammatory T cell responses. Clin Immunol. 2016;173:124.
66. Howell OW, Reeves CA, Nicholas R, Carassiti D, Radotra B, Gentleman SM, et al. Meningeal inflammation is widespread and linked to cortical pathology in multiple sclerosis. Brain. 2011;134(9):2755–71.
67. Lassmann H, Brück W, Lucchinetti CF. The immunopathology of multiple sclerosis: an overview. In: Brain pathology. 2007. p. 210–8.
68. Compston A, Coles A. Multiple sclerosis [Internet]. 2008/10/31. Lancet. Department of Clinical Neurosciences, University of Cambridge Clinical School, Addenbrooke's Hospital, Cambridge, UK. 2008;372:1502–17. Available from: http://www.ncbi.nlm.nih.gov/entrez/query.fcgi?cmd=Retrieve&db=PubMed&dopt=Citation&list_uids=18970977.
69. Frohman EM, Costello F, Stuve O, Calabresi P, Miller DH, Hickman SJ, et al. Modeling axonal degeneration within the anterior visual system: implications for demonstrating neuroprotection in multiple sclerosis [Internet]. Arch Neurol. 2008;65:26–35. Available from: http://archneur.ama-assn.org/cgi/content/abstract/65/1/26.
70. Lucchinetti C, Brück W, Parisi J, Scheithauer B, Rodriguez M, Lassmann H. Heterogeneity of multiple sclerosis lesions: Implications for the pathogenesis of demyelination. Ann Neurol. 2000;47:707.
71. Barnett MH, Henderson AP, Prineas JW. The macrophage in MS: just a scavenger after all? Pathology and pathogenesis of the acute MS lesion [Internet]. 2006/04/25. Vol. 12, Mult Scler. Department of Medicine, University of Sydney, NSW, Australia. mbarnett@mail.usyd.edu.au; 2006. p. 121–32. Available from: http://www.ncbi.nlm.nih.gov/entrez/query.fcgi?cmd=Retrieve&db=PubMed&dopt=Citation&list_uids=16629415.
72. Kouwenhoven M, Teleshova N, Ozenci V, Press R, Link H. Monocytes in multiple sclerosis: phenotype and cytokine profile [Internet]. 2000/12/08. Vol. 112, J Neuroimmunol. Neuroimmunology Unit, Division of Neurology, Karolinska Institutet, Huddinge University Hospital, Stockholm, Sweden. mathilde.kouwenhoven@neurotec.ki.se; 2001. p. 197–205. Available from: http://www.ncbi.nlm.nih.gov/entrez/query.fcgi?cmd=Retrieve&db=PubMed&dopt=Citation&list_uids=11108949.
73. Ajami B, Bennett JL, Krieger C, McNagny KM, Rossi FM. Infiltrating monocytes trigger EAE progression, but do not contribute to the resident microglia pool [Internet]. 2011/08/02. Vol. 14, Nat Neurosci. University of British Columbia, Biomedical Research Centre, Vancouver, British Columbia, Canada; 2011. p. 1142–9. Available from: http://www.ncbi.nlm.nih.gov/entrez/query.fcgi?cmd=Retrieve&db=PubMed&dopt=Citation&list_uids=21804537.
74. Zrzavy T, Hametner S, Wimmer I, Butovsky O, Weiner HL, Lassmann H. Loss of 'homeostatic' microglia and patterns of their activation in active multiple sclerosis [Internet]. Vol. 140, Brain. 2017. p. 1900–13. Available from: https://doi.org/10.1093/brain/awx113.
75. Piccio L, Buonsanti C, Mariani M, Cella M, Gilfillan S, Cross AH, et al. Blockade of TREM-2 exacerbates experimental autoimmune encephalomyelitis. Eur J Immunol. 2007;37(5):1290–301.
76. Zhao TS, Han YZ, Sun YH, Yang CH, Li WB. Synthesis of propylene carbonate catalyzed by metal oxide supported KI. Petrochemical Technol [Internet]. 2000;29(2):101–5. Available from: internal-pdf://244.176.220.78/art00004.
77. Cantoni C, Bollman B, Licastro D, Xie M, Mikesell R, Schmidt R, et al. TREM2 regulates microglial cell

activation in response to demyelination in vivo. Acta Neuropathol. 2015;129(3):429–47.
78. Amici SA, Dong J, Guerau-de-Arellano M. Molecular mechanisms modulating the phenotype of macrophages and microglia [Internet]. Vol. 8, Front Immunol. Dr Mireia Guerau-de-Arellano, School of Health and Rehabilitation Sciences, Division of Medical Laboratory Science, College of Medicine, Wexner Medical Center, The Ohio State University, United States, mireia.guerau@osumc.edu Dr Mireia Guerau-de-Arellano, In; 2017. Available from: https://www.frontiersin.org/article/10.3389/fimmu.2017.01520.
79. Mrdjen D, Pavlovic A, Hartmann FJ, Schreiner B, Utz SG, Leung BP, et al. High-dimensional single-cell mapping of central nervous system immune cells reveals distinct myeloid subsets in health, aging, and disease [Internet]. Vol. 48, Immunity. 2018. p. 599. Available from: http://www.sciencedirect.com/science/article/pii/S1074761318300736.
80. Trapp BD, Ransohoff R, Rudick R. Axonal pathology in multiple sclerosis: relationship to neurologic disability [Internet]. 1999/09/28. Vol. 12, Curr Opin Neurol. Department of Neurosciences, Cleveland Clinic Foundation, Cleveland, OH 44195, USA. trappb@ccf.org; 1999. p. 295–302. Available from: http://www.ncbi.nlm.nih.gov/entrez/query.fcgi?cmd=Retrieve&db=PubMed&dopt=Citation&list_uids=10499174.
81. Dutta R, Trapp BD. Pathogenesis of axonal and neuronal damage in multiple sclerosis [Internet]. 2007/07/20. Vol. 68, Neurology. Department of Neuroscience, Lerner Research Institute, Cleveland Clinic, Cleveland, OH 44195, USA; 2007. p. S22–31; discussion S43–54. Available from: http://www.ncbi.nlm.nih.gov/entrez/query.fcgi?cmd=Retrieve&db=PubMed&dopt=Citation&list_uids=17548565.
82. Schafer DP, Lehrman EK, Kautzman AG, Koyama R, Mardinly AR, Yamasaki R, et al. Microglia sculpt postnatal neural circuits in an activity and complement-dependent manner [Internet]. Vol. 74, Neuron. 2012. p. 691–705. Available from: http://www.sciencedirect.com/science/article/pii/S0896627312003340.
83. Eichhoff G, Brawek B, Garaschuk O. Microglial calcium signal acts as a rapid sensor of single neuron damage in vivo [Internet]. Vol. 1813, Biochim Biophys Acta. Institute of Physiology II, University of Tubingen, Keplerstr. 15, 72074 Tubingen, Germany; 2011. p. 1014–24. Available from: http://www.ncbi.nlm.nih.gov/pubmed/21056596.
84. Howell OW, Rundle JL, Garg A, Komada M, Brophy PJ, Reynolds R. Activated microglia mediate axoglial disruption that contributes to axonal injury in multiple sclerosis [Internet]. Vol. 69, J Neuropathol Exp Neurol. 2010. p. 1017–33. Available from: https://doi.org/10.1097/NEN.0b013e3181f3a5b1.
85. Ransohoff RM, El Khoury J. Microglia in health and disease [Internet]. Vol. 8, Cold Spring Harb Perspect Biol. Neuroinflammation Research Center, Department of Neurosciences, Cleveland Clinic Lerner Research Institute, Cleveland, Ohio 44195. Division of Infectious Diseases and Center for Immunology and Inflammatory Diseases, Massachusetts General Hospital, Harvard; 2015. Available from: http://www.ncbi.nlm.nih.gov/pubmed/26354893.
86. Bitsch A, Schuchardt J, Bunkowski S, Kuhlmann T, Bruck W. Acute axonal injury in multiple sclerosis. Correlation with demyelination and inflammation [Internet]. 2000/05/29. Vol. 123 (Pt 6, Brain. Department of Neurology, Georg-August-Universitat, Gottingen, Germany; 2000. p. 1174–83. Available from: http://www.ncbi.nlm.nih.gov/entrez/query.fcgi?cmd=Retrieve&db=PubMed&dopt=Citation&list_uids=10825356.
87. Luo C, Jian C, Liao Y, Huang Q, Wu Y, Liu X, et al. The role of microglia in multiple sclerosis [Internet]. Vol. 13, Neuropsychiatr Dis Treat. Department of Neurology, The First Affiliated Hospital of Guangxi Medical University. Department of Neurology, The Fifth Affiliated Hospital of Guangxi Medical University and The First People's Hospital of Nanning, Nanning, People's Republic of China; 2017. p. 1661–7. Available from: http://www.ncbi.nlm.nih.gov/pubmed/28721047.
88. Jha AK, Huang SC-C, Sergushichev A, Lampropoulou V, Ivanova Y, Loginicheva E, et al. Network integration of parallel metabolic and transcriptional data reveals metabolic modules that regulate macrophage polarization [Internet]. Vol. 42, Immunity. 2015. p. 419–30. Available from: http://www.sciencedirect.com/science/article/pii/S1074761315000801.
89. Murray PJ. Macrophage polarization [Internet]. Vol. 79, Ann Rev Physiol. 2017. p. 541–66. Available from: https://www.annualreviews.org/doi/abs/10.1146/annurev-physiol-022516-034339.
90. De Groot CJ, Ruuls SR, Theeuwes JW, Dijkstra CD, Van der Valk P. Immunocytochemical characterization of the expression of inducible and constitutive isoforms of nitric oxide synthase in demyelinating multiple sclerosis lesions [Internet]. Vol. 56, J Neuropathol Exp Neurol. Graduate School Neurosciences Amsterdam, Department of Pathology, Free University Academic Hospital, The Netherlands; 1997. p. 10–20. Available from: http://www.ncbi.nlm.nih.gov/pubmed/8990125.
91. Cross AH, Manning PT, Keeling RM, Schmidt RE, Misko TP. Peroxynitrite formation within the central nervous system in active multiple sclerosis. J Neuroimmunol. 1998;88:45.
92. Gerritse K, Laman JD, Noelle RJ, Aruffo A, Ledbetter JA, Boersma WJ, et al. CD40-CD40 ligand interactions in experimental allergic encephalomyelitis and multiple sclerosis [Internet]. Vol. 93, Proc Natl Acad Sci. 1996. p. 2499. Available from: http://www.pnas.org/content/93/6/2499.abstract.
93. Jin Y-H, Hou W, Kang HS, Koh C-S, Kim BS. The role of interleukin-6 in the expression of PD-1 and

PDL-1 on central nervous system cells following infection with Theiler's murine encephalomyelitis virus [Internet]. Vol. 87, Journal of Virology. 2013. p. 11538–51. Available from: https://jvi.asm.org/content/jvi/87/21/11538.full.pdf.
94. Zhao W, Tilton RG, Corbett JA, McDaniel ML, Misko TP, Williamson JR, et al. Experimental allergic encephalomyelitis in the rat is inhibited by aminoguanidine, an inhibitor of nitric oxide synthase. J Neuroimmunol. 1996;64(2):123–33.
95. Rose JW, Hill KE, Wada Y, CIB K, Tsunoda I, Fujinami RS, et al. Nitric oxide synthase inhibitor, aminoguanidine, reduces inflammation and demyelination produced by Theiler's virus infection. J Neuroimmunol. 1998;81(1–2):82–9.
96. Cross AH, Misko TP, Lin RF, Hickey WF, Trotter JL, Tilton RG. Aminoguanidine, an inhibitor of inducible nitric oxide synthase, ameliorates experimental autoimmune encephalomyelitis in SJL mice. J Clin Invest. 1994;93(6):2684–90.
97. Porcheray F, Viaud S, Rimaniol AC, Léone C, Samah B, Dereuddre-Bosquet N, et al. Macrophage activation switching: an asset for the resolution of inflammation [Internet]. Vol. 142, Clin Exp Immunol. 2005. p. 481–9. Available from: https://doi.org/10.1111/j.1365-2249.2005.02934.x.
98. Shin T, Ahn M, Matsumoto Y. Mechanism of experimental autoimmune encephalomyelitis in Lewis rats: recent insights from macrophages. Anat Cell Biol [Internet]. 2012;45(3):141–8. Available from: https://doi.org/10.5115/acb.2012.45.3.141.
99. Giles DA, Washnock-Schmid JM, Duncker PC, Dahlawi S, Ponath G, Pitt D, et al. Myeloid cell plasticity in the evolution of central nervous system autoimmunity [Internet]. Vol. 83, Ann Neurol. 2018. p. 131–41. Available from: https://doi.org/10.1002/ana.25128.
100. Vogel DYS, Vereyken EJF, Glim JE, Heijnen PDAM, Moeton M, van der Valk P, et al. Macrophages in inflammatory multiple sclerosis lesions have an intermediate activation status [Internet]. Vol. 10, J Neuroinflammation. 2013. p. 809. Available from: https://doi.org/10.1186/1742-2094-10-35.
101. Martinez FO, Gordon S. The M1 and M2 paradigm of macrophage activation: time for reassessment [Internet]. Vol. 6, F1000prime reports. 2014. p. 13. Available from: https://www.ncbi.nlm.nih.gov/pubmed/24669294 https://www.ncbi.nlm.nih.gov/pmc/PMC3944738/.
102. Panatier A, Robitaille R. The soothing touch: microglial contact influences neuronal excitability [Internet]. Vol. 23, Dev Cell. INSERM U862, Neurocentre Magendie, 33077 Bordeaux, France; Universite de Bordeaux, 33077 Bordeaux, France. aude.panatier@inserm.fr; 2012. p. 1125–6. Available from: http://www.ncbi.nlm.nih.gov/pubmed/23237948.
103. Town T, Nikolic V, Tan J. The microglial "activation" continuum: from innate to adaptive responses [Internet]. Vol. 2, J Neuroinflammation. 2005. p. 24. Available from: https://doi.org/10.1186/1742-2094-2-24.
104. Voet S, Prinz M, van Loo G. Microglia in central nervous system inflammation and multiple sclerosis pathology [Internet]. Trends Mol Med. 2019. Available from: https://doi.org/10.1016/j.molmed.2018.11.005.
105. Walker DG, Lue LF. Immune phenotypes of microglia in human neurodegenerative disease: challenges to detecting microglial polarization in human brains [Internet]. Vol. 7, Alzheimer's Res Ther. Laboratory of Neuroinflammation, Banner Sun Health Research Institute, 10515 West Santa Fe Drive, Sun City, AZ, 85351, USA. douglas.walker@bannerhealth.com. Laboratory of Neuroinflammation, Banner Sun Health Research Institute, 10515 West Santa Fe Drive; 2015. p. 56. Available from: http://www.ncbi.nlm.nih.gov/pubmed/26286145.
106. Ransohoff RM. A polarizing question: do M1 and M2 microglia exist [Internet]. Vol. 19, Nat Neurosci. Neuroimmunology, Biogen, Cambridge, Massachusetts USA; 2016. p. 987–91. Available from: http://www.ncbi.nlm.nih.gov/pubmed/27459405.
107. Butovsky O, Jedrychowski MP, Moore CS, Cialic R, Lanser AJ, Gabriely G, et al. Identification of a unique TGF-β-dependent molecular and functional signature in microglia. Nat Neurosci [Internet]. 2014;17(1):131–43. Available from: http://www.ncbi.nlm.nih.gov/pubmed/24316888.
108. Aloisi F, Ria F, Penna G, Adorini L. Microglia are more efficient than astrocytes in antigen processing and in Th1 but Not Th2 cell activation [Internet]. Vol. 160, J Immunol. 1998. p. 4671. Available from: http://www.jimmunol.org/content/160/10/4671.abstract.
109. Strachan-Whaley M, Rivest S, Yong VW. Interactions between microglia and T cells in multiple sclerosis pathobiology [Internet]. Vol. 34, J Interfer Cytokine Res. 2014. p. 615–22. Available from: https://doi.org/10.1089/jir.2014.0019
110. Miron VE, Boyd A, Zhao J-W, Yuen TJ, Ruckh JM, Shadrach JL, et al. M2 microglia and macrophages drive oligodendrocyte differentiation during CNS remyelination [Internet]. Vol. 16, Nat Neurosci. 2013. p. 1211. Available from: https://doi.org/10.1038/nn.3469.
111. Rice RA, Pham J, Lee RJ, Najafi AR, West BL, Green KN. Microglial repopulation resolves inflammation and promotes brain recovery after injury. Glia [Internet]. 2017 Jan 2;65(6):931–44. Available from: https://doi.org/10.1002/glia.23135.
112. Ebner F, Brandt C, Thiele P, Richter D, Schliesser U, Siffrin V, et al. Microglial Activation Milieu Controls Regulatory T Cell Responses [Internet]. Vol. 191, J Immunol. 2013. p. 5594. Available from: http://www.jimmunol.org/content/191/11/5594.abstract
113. Gautier EL, Shay T, Miller J, Greter M, Jakubzick C, Ivanov S, et al. Gene-expression profiles and transcriptional regulatory pathways that underlie the

identity and diversity of mouse tissue macrophages [Internet]. Vol. 13, Nat Immunol. 2012. p. 1118. Available from: https://doi.org/10.1038/ni.2419
114. Geirsdottir L, David E, Keren-Shaul H, Weiner A, Bohlen SC, Neuber J, et al. Cross-Species Single-Cell Analysis Reveals Divergence of the Primate Microglia Program. Cell. 2019 12;179(7):1609–22.e16. https://doi.org/10.1016/j.cell.2019.11.010.
115. Satoh J, Kino Y, Asahina N, Takitani M, Miyoshi J, Ishida T, et al. TMEM119 marks a subset of microglia in the human brain [Internet]. Vol. 36, Neuropathology. 2016. p. 39–49. Available from: https://doi.org/10.1111/neup.12235.
116. Becher B, Bechmann I, Greter M. Antigen presentation in autoimmunity and CNS inflammation: how T lymphocytes recognize the brain [Internet]. Vol. 84, J Mol Med. Neurology Department, Division for Neuroimmunology, University of Zurich, Winterthurerstrasse 190, CH-8057 Zurich, Switzerland. burkhard.becher@neuroimm.unizh.ch; 2006. p. 532–43. Available from: http://www.ncbi.nlm.nih.gov/entrez/query.fcgi?cmd=Retrieve&db=PubMed&dopt=Citation&list_uids=16773356.
117. Slavin AJ, Soos JM, Stuve O, Patarroyo JC, Weiner HL, Fontana A, et al. Requirement for endocytic antigen processing and influence of invariant chain and H-2M deficiencies in CNS autoimmunity. J Clin Invest [Internet]. 2001;108(8):1133–9. Available from: http://www.ncbi.nlm.nih.gov/entrez/query.fcgi?cmd=Retrieve&db=PubMed&dopt=Citation&list_uids=11602620.
118. Bartholomaus I, Kawakami N, Odoardi F, Schlager C, Miljkovic D, Ellwart JW, et al. Effector T cell interactions with meningeal vascular structures in nascent autoimmune CNS lesions [Internet]. 2009/10/16. Vol. 462, Nature. Max Planck Institute for Neurobiology, 82152 Martinsried, Germany; 2009. p. 94–8. Available from: http://www.ncbi.nlm.nih.gov/entrez/query.fcgi?cmd=Retrieve&db=PubMed&dopt=Citation&list_uids=19829296.
119. Ransohoff RM, Kivisakk P, Kidd G. Three or more routes for leukocyte migration into the central nervous system [Internet]. Vol. 3, Nat Rev Immunol. The Mellen Center for Multiple Sclerosis Treatment and Research, Department of Neurology, 9500 Euclid Avenue, The Cleveland Clinic Foundation, Cleveland, Ohio 44195, USA. ransohr@ccf.org; 2003. p. 569–81. Available from: http://www.ncbi.nlm.nih.gov/entrez/query.fcgi?cmd=Retrieve&db=PubMed&dopt=Citation&list_uids=12876559.
120. Wu GF, Shindler KS, Allenspach EJ, Stephen TL, Thomas HL, Mikesell RJ, et al. Limited sufficiency of antigen presentation by dendritic cells in models of central nervous system autoimmunity [Internet]. 2010/11/26. Vol. 36, J Autoimmun. Department of Neurology, University of Pennsylvania School of Medicine, Philadelphia, PA 19004, USA. wug@neuro.wustl.edu; 2011. p. 56–64. Available from: http://www.ncbi.nlm.nih.gov/entrez/query.fcgi?cmd=Retrieve&db=PubMed&dopt=Citation&list_uids=21095100.
121. Greter M, Heppner FL, Lemos MP, Odermatt BM, Goebels N, Laufer T, et al. Dendritic cells permit immune invasion of the CNS in an animal model of multiple sclerosis [Internet]. Vol. 11, Nat Med. Neuroimmunology Unit, Department of Neurology, University Hospital Zurich, Frauenklinkstrasse 10, CH-8091 Switzerland; 2005. p. 328–34. Available from: http://www.ncbi.nlm.nih.gov/entrez/query.fcgi?cmd=Retrieve&db=PubMed&dopt=Citation&list_uids=15735653
122. Wu GF, Laufer TM. The role of dendritic cells in multiple sclerosis [Internet]. Vol. 7, Curr Neurol Neurosci Rep. Department of Neurology, Hospital of the University of Pennsylvania, 745 BRB II/III, 421 Curie Boulevard, Philadelphia, PA 19104, USA. gregory.wu@uphs.upenn.edu; 2007. p. 245–52. Available from: http://www.ncbi.nlm.nih.gov/entrez/query.fcgi?cmd=Retrieve&db=PubMed&dopt=Citation&list_uids=17488591.
123. Parker Harp CR, Archambault AS, Sim J, Ferris ST, Mikesell RJ, Koni PA, et al. B cell antigen presentation is sufficient to drive neuroinflammation in an animal model of multiple sclerosis [Internet]. Vol. 194, J Immunol. Department of Neurology, Washington University School of Medicine, St. Louis, MO 63110; Department of Developmental Biology, Washington University School of Medicine, St. Louis, MO 63110; Department of Pathology and Immunology, Washington University Sch; 2015. p. 5077–84. Available from: http://www.ncbi.nlm.nih.gov/pubmed/25895531.
124. Molnarfi N, Schulze-Topphoff U, Weber MS, Patarroyo JC, Prod'homme T, Varrin-Doyer M, et al. MHC class II-dependent B cell APC function is required for induction of CNS autoimmunity independent of myelin-specific antibodies [Internet]. Vol. 210, J Exp Med. Department of Neurology and 2 Program in Immunology, University of California, San Francisco, San Francisco, CA 94158; 2013. p. 2921–37. Available from: http://www.ncbi.nlm.nih.gov/pubmed/24323356.
125. Anandasabapathy N, Victora GD, Meredith M, Feder R, Dong B, Kluger C, et al. Flt3L controls the development of radiosensitive dendritic cells in the meninges and choroid plexus of the steady-state mouse brain [Internet]. 2011/07/27. Vol. 208, J Exp Med. Laboratory of Cellular Physiology and Immunology, The Rockefeller University, New York, NY 10065, USA; 2011. p. 1695–705. Available from: http://www.ncbi.nlm.nih.gov/entrez/query.fcgi?cmd=Retrieve&db=PubMed&dopt=Citation&list_uids=21788405
126. Waisman A, Johann L. Antigen-presenting cell diversity for T cell reactivation in central nervous system autoimmunity [Internet]. Vol. 96, J Mol Med. 2018. p. 1279–92. Available from: https://doi.org/10.1007/s00109-018-1709-7.

127. Steinman RM. Decisions about dendritic cells: past, present, and future [Internet]. Vol. 30, Annu Rev Immunol. Laboratory of Cell Physiology and Immunology, The Rockefeller University, New York, NY 10021, USA; 2012. p. 1–22. Available from: http://www.ncbi.nlm.nih.gov/pubmed/22136168.
128. Shortman K, Naik SH. Steady-state and inflammatory dendritic-cell development [Internet]. Vol. 7, Nat Rev Immunol. The Walter and Eliza Hall Institute of Medical Research, 1G Royal Parade, Parkville, Victoria 3050, Australia. shortman@wehi.edu.au s.naik@nki.nl; 2007. p. 19–30. Available from: http://www.ncbi.nlm.nih.gov/entrez/query.fcgi?cmd=Retrieve&db=PubMed&dopt=Citation&list_uids=17170756.
129. McMahon EJ, Bailey SL, Castenada CV, Waldner H, Miller SD. Epitope spreading initiates in the CNS in two mouse models of multiple sclerosis [Internet]. Vol. 11, Nat Med. Department of Microbiology-Immunology and the Interdepartmental Immunobiology Center, Northwestern University Medical School, 303 East Chicago Avenue, Chicago, Illinois 60611, USA; 2005. p. 335–9. Available from: http://www.ncbi.nlm.nih.gov/entrez/query.fcgi?cmd=Retrieve&db=PubMed&dopt=Citation&list_uids=15735651.
130. Ousman SS, Kubes P. Immune surveillance in the central nervous system [Internet]. Vol. 15, Nat Neurosci. 2012. p. 1096–101. Available from: https://doi.org/10.1038/nn.3161.
131. Mohammad MG, Tsai VWW, Ruitenberg MJ, Hassanpour M, Li H, Hart PH, et al. Immune cell trafficking from the brain maintains CNS immune tolerance [Internet]. Vol. 124, J Clin Invest. 2014. p. 1228–41. Available from: https://doi.org/10.1172/JCI71544.
132. Louveau A, Smirnov I, Keyes TJ, Eccles JD, Rouhani SJ, Peske JD, et al. Structural and functional features of central nervous system lymphatic vessels [Internet]. Vol. 523, Nature. 1] Center for Brain Immunology and Glia, School of Medicine, University of Virginia, Charlottesville, Virginia 22908, USA [2] Department of Neuroscience, School of Medicine, University of Virginia, Charlottesville, Virginia 22908, USA. 1] Medical Scienti; 2015. p. 337–41. Available from: http://www.ncbi.nlm.nih.gov/pubmed/26030524.
133. Aspelund A, Antila S, Proulx ST, Karlsen TV, Karaman S, Detmar M, et al. A dural lymphatic vascular system that drains brain interstitial fluid and macromolecules [Internet]. Vol. 212, J Exp Med. 2015. p. 991. Available from: http://jem.rupress.org/content/212/7/991.abstract
134. Karni A, Abraham M, Monsonego A, Cai G, Freeman GJ, Hafler D, et al. Innate immunity in multiple sclerosis: myeloid dendritic cells in secondary progressive multiple sclerosis are activated and drive a proinflammatory immune response [Internet]. 2006/09/05. Vol. 177, J Immunol. Center for Neurologic Diseases, Brigham and Women's Hospital, Harvard Medical School, Harvard Medical School, Boston, MA 02115–5817, USA; 2006. p. 4196–202. Available from: http://www.ncbi.nlm.nih.gov/entrez/query.fcgi?cmd=Retrieve&db=PubMed&dopt=Citation&list_uids=16951385.
135. Matyszak MK, Perry VH. Dendritic cells in inflammatory responses in the CNS [Internet]. Vol. 417, Adv Exp Med Biol. Department of Pharmacology, Oxford University. Matyszak@vax.ox.ac.uk; 1997. p. 295–9. Available from: http://www.ncbi.nlm.nih.gov/pubmed/9286376.
136. McMenamin PG. Distribution and phenotype of dendritic cells and resident tissue macrophages in the dura mater, leptomeninges, and choroid plexus of the rat brain as demonstrated in wholemount preparations [Internet]. Vol. 405, J Comp Neurol. Department of Anatomy and Human Biology, University of Western Australia, Nedlands. mcmenamin@anhb.uwa.edu.au; 1999. p. 553–62. Available from: http://www.ncbi.nlm.nih.gov/pubmed/10098945.
137. Lehmann-Horn K, Schleich E, Hertzenberg D, Hapfelmeier A, Kumpfel T, von Bubnoff N, et al. Anti-CD20 B-cell depletion enhances monocyte reactivity in neuroimmunological disorders [Internet]. 2011/10/27. Vol. 8, J Neuroinflammation. Department of Neurology, Technische Universitat Munchen, Munich, Germany; 2011. p. 146. Available from: http://www.ncbi.nlm.nih.gov/entrez/query.fcgi?cmd=Retrieve&db=PubMed&dopt=Citation&list_uids=22027448.
138. Suter T, Biollaz G, Gatto D, Bernasconi L, Herren T, Reith W, et al. The brain as an immune privileged site: dendritic cells of the central nervous system inhibit T cell activation [Internet]. 2003/10/28. Vol. 33, Eur J Immunol. Section of Clinical Immunology, University Hospital, Zurich, Switzerland; 2003. p. 2998–3006. Available from: http://www.ncbi.nlm.nih.gov/entrez/query.fcgi?cmd=Retrieve&db=PubMed&dopt=Citation&list_uids=14579268.
139. van Zwam M, Huizinga R, Melief M-J, Wierenga-Wolf AF, van Meurs M, Voerman JS, et al. Brain antigens in functionally distinct antigen-presenting cell populations in cervical lymph nodes in MS and EAE [Internet]. Vol. 87, J Mol Med. 2009. p. 273–86. Available from: https://doi.org/10.1007/s00109-008-0421-4.
140. Murphy TL, Grajales-Reyes GE, Wu X, Tussiwand R, Briseño CG, Iwata A, et al. Transcriptional control of dendritic cell development [Internet]. Vol. 34, Ann Rev Immunol. 2016. p. 93–119. Available from: https://www.annualreviews.org/doi/abs/10.1146/annurev-immunol-032713-120204.
141. Gabrilovich DI. Myeloid-derived suppressor cells [Internet]. Vol. 5, Cancer Immunol Res. 2017. p. 3. Available from: http://cancerimmunolres.aacrjournals.org/content/5/1/3.abstract.
142. Iacobaeus E, Douagi I, Jitschin R, Marcusson-Ståhl M, Andrén AT, Gavin C, et al. Phenotypic and functional alterations of myeloid-derived suppressor

cells during the disease course of multiple sclerosis [Internet]. Vol. 96, Immunol Cell Biol. 2018. p. 820–30. Available from: https://doi.org/10.1111/imcb.12042.

143. Ioannou M, Alissafi T, Lazaridis I, Deraos G, Matsoukas J, Gravanis A, et al. Crucial role of granulocytic myeloid-derived suppressor cells in the regulation of central nervous system autoimmune disease [Internet]. Vol. 188, J Immunol. 2012. p. 1136. Available from: http://www.jimmunol.org/content/188/3/1136.abstract.
144. Cantoni C, Cignarella F, Ghezzi L, Mikesell B, Bollman B, Berrien-Elliott MM, et al. Mir-223 regulates the number and function of myeloid-derived suppressor cells in multiple sclerosis and experimental autoimmune encephalomyelitis. Acta Neuropathol. 2017;133(1):61–77.
145. Knier B, Hiltensperger M, Sie C, Aly L, Lepennetier G, Engleitner T, et al. Myeloid-derived suppressor cells control B cell accumulation in the central nervous system during autoimmunity [Internet]. Vol. 19, Nat Immunol. Department of Experimental Neuroimmunology, Klinikum rechts der Isar, Technical University of Munich, Munich, Germany. Department of Neurology, Klinikum rechts der Isar, Technical University of Munich, Munich, Germany. Institute of Molecular Oncology an; 2018. p. 1341–51. Available from: http://www.ncbi.nlm.nih.gov/pubmed/30374128.
146. Zhou J, Nefedova Y, Lei A, Gabrilovich D. Neutrophils and PMN-MDSC: Their biological role and interaction with stromal cells [Internet]. Vol. 35, Seminars in Immunology. 2018. p. 19–28. Available from: http://www.sciencedirect.com/science/article/pii/S1044532317300076.
147. Steinbach K, Piedavent M, Bauer S, Neumann JT, Friese MA. Neutrophils Amplify Autoimmune Central Nervous System Infiltrates by Maturing Local APCs [Internet]. Vol. 191, The Journal of Immunology. 2013. p. 4531. Available from: http://www.jimmunol.org/content/191/9/4531.abstract.
148. Rumble JM, Huber AK, Krishnamoorthy G, Srinivasan A, Giles DA, Zhang X, et al. Neutrophil-related factors as biomarkers in EAE and MS [Internet]. Vol. 212, The Journal of Experimental Medicine. 2015. p. 23–35. Available from: http://jem.rupress.org/content/jem/212/1/23.full.pdf.
149. Kuhlmann T, Lassmann H, Brück W. Diagnosis of inflammatory demyelination in biopsy specimens: a practical approach [Internet]. 01/04. Acta Neuropathol. 2008;115:275–87. Available from: https://www.ncbi.nlm.nih.gov/pubmed/18175128 https://www.ncbi.nlm.nih.gov/pmc/PMC2668559/.
150. Brosnan CF, Raine CS. The astrocyte in multiple sclerosis revisited [Internet]. Glia. 2013;61:453–65. Available from: https://doi.org/10.1002/glia.22443.
151. Hewett SJ, Misko TP, Keeling RM, Margarita Behrens M, Choi DW, Cross AH. Murine encephalitogenic lymphoid cells induce nitric oxide synthase in primary astrocytes. J Neuroimmunol. 1996;64(2):201.
152. Colombo E, Farina C. Astrocytes: key regulators of neuroinflammation [Internet]. Trends Immunol. 2016;37:608–20. Available from: http://www.sciencedirect.com/science/article/pii/S1471490616300722.
153. Liddelow SA, Guttenplan KA, Clarke LE, Bennett FC, Bohlen CJ, Schirmer L, et al. Neurotoxic reactive astrocytes are induced by activated microglia [Internet]. Nature. 2017;541:481. Available from: https://doi.org/10.1038/nature21029.

# Epidemiology of Multiple Sclerosis and Environmental Risk Factors

# 10

Kyla A. McKay and Helen Tremlett

**Key Points**

1. Between 2.0 and 2.3 million people are living with multiple sclerosis (MS) worldwide, with increased rates further from the equator.
2. The typical onset age is around 30 years, and more women are affected than men.
3. Lifestyle and environmental factors, in combination with genetic predisposition, are likely chief contributors to MS risk.
4. Strongest evidence suggests a role for markers of Epstein–Barr virus infections, low sunlight exposure and/or low vitamin D intake or serum levels, cigarette smoking, and adolescent obesity as factors that increase the risk of developing MS.
5. There appear to be critical risk periods for exposure, and these risk factors can also interact, sometimes in an additive way, to increase the risk of MS.

## Introduction

Worldwide, multiple sclerosis (MS) is the most common progressive neurological disease of young adults [1]. Three-quarters of those diagnosed are women, and MS can affect a wide range of individuals—from young children to older adults [2]. Despite substantial efforts to uncover its etiology, the exact cause(s) of MS at the individual level remain elusive, but there is a growing literature pointing toward a range of population-level risk factors [3]. As with many complex diseases, it is thought that MS arises from a combination of a genetic predisposition and exposure to one or more candidate environmental factors. These environmental factors may be required in combination (at the same time or in a specific sequence) and at critical time periods. In this chapter, we will summarize the current understanding of MS epidemiology, with a focus on the environmental risk factors for its development.

A central concept of epidemiology is *causality*, which is assessed through the careful analysis of the relationship between exposures (e.g., cigarette smoking) and outcomes (e.g., lung disease and cancer). A widely accepted definition of causality has not been achieved, as there are limitations to all current theories of causal inference [4]. However, most agree that in order for an exposure to be deemed a causal factor for disease, it must occur before the onset of disease (*temporality*) and be replicable across studies

K. A. McKay
Department of Clinical Neuroscience, Karolinska Institutet, Stockholm, Sweden

H. Tremlett (✉)
Department of Medicine (Neurology), University of British Columbia, Vancouver, BC, Canada
e-mail: helen.tremlett@ubc.ca



A. L. Piquet, E. Alvarez (eds.), *Neuroimmunology*, https://doi.org/10.1007/978-3-030-61883-4_10

(*consistent*). Observational research carries a set of unique challenges, and in interpreting findings one must consider several important points. By definition, epidemiological studies capture naturally occurring events, making them vulnerable to bias and imprecision, unless appropriate considerations are made to the study design, analyses, and interpretation [5]. For this chapter, we will focus on the best evidence to date, published in peer-reviewed journals. We will include a range of observational study designs, including case–control or cohort studies and systematic reviews that have collated relevant bodies of work related to the epidemiology and environmental risk factors for MS.

## Multiple Sclerosis Epidemiology

### Incidence and Prevalence

Patterns of MS prevalence have been reasonably well studied within certain regions, but lacking in others, and reliable estimates for MS incidence remain scarce [6]. Further, a lack of standardization of estimates to the age and sex distributions in the underlying population (which will influence risk estimates) and differences in ascertainment across studies can make comparing estimates or tracking changes over time difficult [7, 8]. Global estimates of MS in 2013 and 2015 suggest that between 2.0 and 2.3 million people are living with MS worldwide, equating to a global median prevalence of up to 33 per 100,000 [1, 2]. However, in many regions of the world, either formal studies have never been conducted or data remain suboptimal, including much of Africa and South America. Nonetheless, reports point to a rising prevalence over time, estimated at a 19% increase in people living with MS over a 10-year period (2005–2015) [1]. This increase is likely due to enhanced surveillance and recognition of MS combined with increased life expectancy [9–11]. Much less is known of MS incidence; two systematic reviews spanning Europe and North America concluded that inconsistencies in methodologies prevented any definitive conclusions [7, 8]. The incidence of MS may have increased in specific jurisdictions or over certain time periods [12, 13], although stable incidence rates have also been reported in Canada and the United Kingdom in recent decades [9, 10, 14–17].

### Global Distribution

There is a marked latitudinal gradient in the prevalence of MS across the globe, in which rates increase with increasing distance from the equator [2]. Access to health services (including magnetic resonance imaging [MRI] scanners and neurologists) and the comprehensive health data needed to accurately estimate population prevalence also tends to mirror this gradient [2]. Medium- to high-prevalence areas include Australasia, Europe, and North America, while low-prevalence areas include South Asia, Latin America, Africa, and the Caribbean [18]. Clearly, differences in case ascertainment and ability to recognize and diagnose MS exist between countries. One way of assessing the presence of a latitude gradient is to examine MS incidence/prevalence within a specific jurisdiction, the premise being that access to health systems is consistent throughout the area under study. For example, a latitudinal gradient for MS incidence (studied in women only) and prevalence for both sexes has been found within the USA [19, 20]. However, a 2018 study published by the US Burden of Disease Collaborators demonstrated that health and life expectancy are region-specific, even within the USA [21]. Nonetheless, a latitude gradient for MS prevalence has also been observed in Australia and New Zealand [18, 22]. Whether the global latitude gradient for MS incidence and/or prevalence has attenuated over time has been debated [18, 23]; it is feasible that it has (or will), as previously underserved regions develop more comprehensive health systems [12].

Some of the earliest epidemiological studies in MS focused on the potential relationship between migration to or from low- and high-risk regions and subsequent MS risk [24–28]. These early studies and some subsequent work have suggested that persons who were past adolescence at the time of migration maintained the MS risk of their country of origin, while younger individuals adopted the MS risk of the country to which they migrated [27, 28]. However, many methodological challenges face migration studies, as people who migrate are typically not the same as those who remain in their country of origin, and both immigration status and the age of migration/language ability may also influence an individual's ability to access their new country's health system [29]. Nonetheless, these migration studies have been interpreted by some as evidence of a population-level, but region-specific, exposure to environmental factors in the development of MS [24, 26].

## Demographic Characteristics

The majority of individuals with MS first develop symptoms in early adulthood, with the typical onset age being around 30 years. However, age is no barrier to developing MS; between 2% and 10% of MS patients develop the disease in childhood [30] and 4–10% after the age of 50 [31–33]. MS was historically considered a disease of white persons or persons of Northern European descent. However, emerging evidence suggests that the risk may be higher in black persons (living in the USA) than previously thought [34]. More women are affected by MS than men, and the female:male sex ratio has increased in some areas over certain time periods, with a systematic review estimating a mean sex ratio of 1.4 in 1955 and 2.3 in 2000 based on studies from Europe, North America, and Australia [12], although the sex ratio has been relatively steady in other places and more recent time periods, for example, across Canada in the last 2 decades [9, 10, 14, 16].

## Mortality

Although MS is not considered a fatal disease, it is the most commonly reported cause of death in persons with MS [35]. Population-based studies have shown that survival rates have improved over time, in line with the general population [36–38]. Still, on average persons with MS live 5–10 years less than the general population [35, 37, 38].

## Risk Factors for Multiple Sclerosis

In this section of the chapter, we will summarize our current understanding of the environmental factors associated with MS disease susceptibility. The evidence has been organized into the following subsections: "Infections"; "Vaccines"; "Sunlight Exposure and Vitamin D"; "Smoking"; "Dietary Macro- and Micronutrients, Alcohol, and Caffeine Consumption"; "Obesity"; "Socioeconomic Status"; "Stress"; and "Occupational Exposures." A schematic outlining the most consistent and emerging evidence for the environmental factors that may alter MS susceptibility is provided in Fig. 10.1. Last, we briefly describe the role of timing of exposures and gene–environment interactions in relation to MS onset.

## Infections

Infectious agents have long been considered a putative causal agent in MS given the immunological nature of the disease. Patterns of migration led researchers to hypothesize that childhood exposure to pathogens may alter MS risk. As such, the following pathogens have been studied the most extensively: Epstein–Barr virus (EBV), measles, mumps, rubella, varicella zoster virus (VZV), human herpes virus-6 (HHV-6), and *Chlamydia pneumoniae* [39].

### Epstein–Barr Virus

Epstein–Barr virus (EBV) is ubiquitous in the global population, with more than 90% of people exposed [40]. EBV exposure has also been consistently linked to MS risk [39]. Most persons are exposed to EBV in early childhood, typically

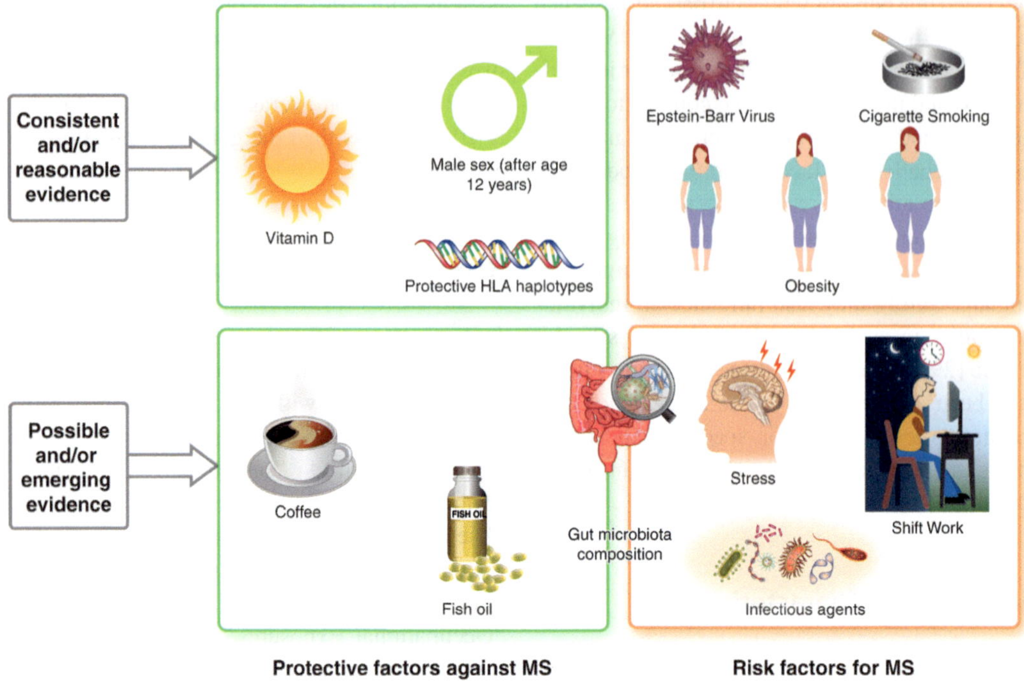


**Fig. 10.1** Illustration of factors considered protective or risk factors for multiple sclerosis (MS), categorized by level of evidence

resulting in an asymptomatic or mild flu-like illness. If a person is first exposed to EBV in adolescence or adulthood, it can manifest as infectious mononucleosis. Both high antibody titers against EBV and history of infectious mononucleosis have repeatedly been associated with an increased MS risk. The most striking examples of this include data from the USA's Department of Defense Serum Repository [41, 42]. The repository houses more than 40 million serum samples from military personnel collected approximately every 2 years [43]. This allows for an estimation of the possible timing of EBV infection (measured as the presence of antibodies) prior to the onset of MS. MS risk increased with increasing titers of antibodies against EBV, such that persons in the highest sextile of antibody levels had 36 times the risk of MS relative to persons in the lowest sextile, although the confidence intervals (CI) were wide (95% CI: 9.6–136) [41]. The same researchers followed all persons who were negative for EBV antibodies at their first serum testing and found that the 10 persons who eventually developed MS all developed EBV antibodies prior to the symptomatic onset of their disease [42].

A 2010 meta-analysis summarized the evidence for a potential causal association between prior infectious mononucleosis and MS risk, which included 18 studies, comprising 19,390 MS patients and 16,007 controls. Taken altogether, authors reported over a twofold increased risk of MS among persons with a history of infectious mononucleosis relative to no history of infectious mononucleosis (95% CI: 1.97–2.39) [44]. A meta-analysis of the published literature (prior to December 2011) concluded that EBV antibodies were present in 98.3% of adult MS patients tested, relative to 93.7% of controls not known to have MS. In pediatric-onset MS, lower EBV seropositivity has been reported relative to adults with MS; 8.3% of pediatric MS patients and 34.7% of control children were seronegative for EBV [45]. The authors interpreted findings to infer that EBV infection is necessary for the development of MS [45], although not all agree and it remains possible that EBV is an epiphenomenon rather than a causative agent [46, 47].

### Other Infectious Agents

Whether other childhood exposures to pathogens can alter the risk of MS is unclear, in part related to the quality of evidence [48]. Many studies conducted in adults have relied upon recall of prior infection, which may be susceptible to recall bias, or antibody levels from prevalent cases of MS, in which temporality cannot be established [48]. Studies of viral pathogens such as measles, mumps, rubella, and varicella zoster (chickenpox) have shown mixed results, although some of the best evidence suggests no association with MS [39]. HHV-6 is another common virus, present in virtually all humans by the age of 2 years [49]. Given the scarcity of HHV-6-negative individuals, and the very young age of exposure, it has been effectively impossible to perform a case–control study. Instead, the evidence for a role of HHV-6 has come largely from pathological data, which found the virus in MS lesions on autopsy [50]. Due to a lack of evidence surrounding temporality, it cannot be concluded that HHV-6 plays a causal role in MS. A single study found an increased immune response to HHV-6 among individuals in the early phase of MS (relapsing-remitting MS [$n$ = 42] and clinically isolated syndrome [$n$ = 53]) relative to individuals without MS ($n$ = 50) [51].

Interest in the bacterial species, *Chlamydia pneumoniae*, in relation to MS risk (and/or possible subsequent progression) piqued in the late 1990s/early 2000s after one team found evidence of *Chlamydia pneumoniae* antibodies and DNA in the cerebrospinal fluid of an MS patient [39, 52]. *Chlamydia pneumoniae* is thought to be relatively common in the general population, with seroconversion often occurring between the ages of 5 and 14 [53]. It is also estimated to cause 10% of community-acquired pneumonia and 5% of bronchitis and sinusitis cases in US adults [53]. However, while some studies reported an increased risk of MS in relation to markers of *Chlamydia pneumoniae* infection, others did not [54–56]. Replication of some of the earlier assays to detect the presence of the bacteria was problematic, and interest in *Chlamydia pneumoniae* in relation to MS risk appears to have waned [39].

Parasitic gut infection with helminths, such as *Trichuris trichiura* (human whipworm), has been considered as a potential protective factor against MS, in part based on the high levels of human parasite infection in areas of low MS prevalence [57]. However, studies to date have been rather limited, often involving small numbers of participants, so while an area of interest, no conclusions can be drawn.

In summary, demonstrating temporality between a prior infection or infestation and subsequent risk of MS is very challenging. Differences in the biomarkers or assays used to measure the presence of a prior infection also create challenges when comparing between studies. The best evidence is for a prior EBV infection, assessed as either presence of antibodies or infectious mononucleosis and increased risk of MS. Although even if causation was definitively demonstrated, and an effective vaccine was available, it is unclear as to the value in preventing MS as the primary goal; large numbers of EBV-negative individuals would have to be vaccinated to prevent a single case of MS [58].

## Vaccines

Following recommendations from the World Health Organization (WHO), France initiated a vaccination program against hepatitis B in 1994, targeting all students in their first year of secondary school [59]. By 1998, a series of case reports of MS had been described to the French Health Authorities that were thought to be related to the vaccine, leading to a suspension of the program [59]. Subsequently, a series of well-designed studies were conducted, resulting in a 2011 meta-analysis of hepatitis B and other vaccines in relation to MS risk. The authors of the meta-analysis concluded that there was no evidence for an altered risk of MS following vaccination against hepatitis B, tuberculosis (bacille Calmette–Guérin [BCG] vaccine), influenza, measles–mumps–rubella (MMR), polio, or typhoid fever [60]. A subsequent California, USA-based study compared 780 incident cases of CNS demyelinating syndromes (including MS) with 3885 matched controls and also found no relationship between hepatitis B vaccination and MS risk [61]. No relationships have been found between other vaccines targeting the human papilloma virus (HPV) or the H1N1 influenza A virus and MS risk, despite concerns raised regarding the adjuvants used in some vaccines and/or individual case reports of MS occurring after vaccination [62, 63]. The potential for some vaccines, such as diphtheria or tetanus, to have a protective effect against MS is intriguing but requires replication [60].

## Sunlight Exposure and Vitamin D

Ecological studies of latitude and MS prevalence prompted interest in climatic factors, with the suggestion that sunlight exposure may influence MS risk (incidence). Specific wave lengths within the ultraviolet (UV) spectrum stimulate human skin to initiate the synthesis of 25-hydroxyvitamin D (25(OH)D). Both 25(OH)D and direct sunlight exposure have systemic immunomodulatory effects, which provides a biological rationale [64]. Low sunlight exposure, low serum 25-OH-D levels, and/or low vitamin D intake have all been associated with a higher MS risk [65]. The literature is vast and growing, as additional important aspects are studied in further detail and across different populations. Here we include a selection of key studies examining these relationships in MS.

### Sunlight Exposure

A population-based case–control study based in Tasmania, Australia, provided crucial insights into the potential role of sunlight exposure in modulating MS risk. In total, 136 prevalent MS cases and 272 community controls were asked about their past sunlight exposure. In addition, actinic (skin) damage was measured, providing an objective measure of cumulative lifetime sun exposure [66]. Higher sun exposure between the ages of 6 and 15 years was associated with a reduced risk of MS (adjusted odds ratio = 0.31; 95% CI: 0.16–0.59), as was greater actinic damage [66]. Several subsequent studies have both replicated and extended these observations. For example, others have now demonstrated that the exposure window extends from birth to early adulthood [67]. In addition, while consistency in the relationship between higher sunlight and lower MS risk has been observed across racial/ethnic groups (living in Southern California, USA [68]; see also Sect. "Serum Vitamin D"), others have explored seasonal effects and shown that the region of residence can influence whether summer and/or winter sun is of greater relevance [67].

### Serum Vitamin D

One team accessed a valuable biobank of serum samples from the USA's Department of Defense Serum Repository, which included personnel entering the military, some of whom later developed MS. Overall, 257 MS cases had samples available prior to their disease onset. Higher serum 25(OH)D levels were associated with lower MS risk among the white participants (148 cases, 296 controls; odds ratio = 0.59; 95% CI: 0.36–0.97 for a 50 nmol/L increase in 25(OH)D). Authors reported a slightly stronger relationship when analyses were restricted to serum levels measured before age 20 years. Similar associations were not observed for blacks or Hispanics (109 cases, 218 controls) [69]. A later study from Southern California observed a similar racial disparity [68]. Although they found higher sunlight exposure, measured as lifetime ultraviolet radiation, associated with lower MS risk for all racial groups (whites 247 cases/267 controls, blacks 116/131, and Hispanics 183/197), higher serum 25(OH)D levels were only associated with a lower MS risk in whites [68].

Serum 25(OH)D levels during pregnancy and risk of MS in the offspring were assessed using the Finnish Maternity Cohort [70]. In this nested case–control study, 176 MS cases were matched to 326 controls on birth region, date of maternal serum sample collection, and date of the mother's and child's birth. Gestational 25(OH)D deficiency (<12.02 ng/mL vs. 12.02–<20.03 ng/mL) was associated with an almost twofold increased risk of MS in the offspring (relative risk = 1.90; 95% CI: 1.20–3.01).

One meta-analysis attempted to address the issue of serum 25(OH)D levels and MS risk. However, of the 11 studies included (comprising of 1007 MS cases and 829 controls), the majority examined serum levels some years after onset of disease [71]. Therefore, the observation that the MS cases had lower serum vitamin D levels than controls could have been a consequence of MS and not related to MS onset [71].

## Diet and Dietary Supplements as Sources of Vitamin D

Diet can contribute to serum vitamin D levels, although food and beverage intake is particularly challenging to study, especially retrospectively. The USA Nurses' Health Study comprises of two prospectively followed cohorts of female nurses, totalling more than 238,000 participants, with a diet-related questionnaire completed every 4 years [72]. By 2000, 173 women had developed MS, and when compared to controls [73], those in the highest (versus lowest) quintile of total vitamin D intake (sourced from food or supplements) were at a reduced risk of MS (relative risk = 0.67; 95% CI: 0.40–1.12; p for trend = 0.03). This relationship remained significant when supplementary intake alone was assessed but not for vitamin D sourced from food [73].

A subsequent investigation using the USA Nurses' Health Study cohorts reported a reduced risk of MS in the offspring of women with high dietary vitamin D intake during pregnancy, the primary source being from fortified milk [74]. For the women whose mothers consumed two to three glasses of milk per day (versus <3 glasses per month), while pregnant, the risk of MS was lower (adjusted rate ratio = 0.62, 95% CI; 0.40–0.95; p trend 0.001). The use of dietary supplements was not captured in sufficient detail to assess (no dose-related information were available).

## Mendelian Randomization

Mendelian randomization is a technique applied to observational studies that, if designed well, can eliminate reverse causation and confounding when assessing causality [75]. Careful selection of gene markers, single nucleotide polymorphisms (SNPs), which are independently associated with the exposure of interest, are studied in relation to disease risk. As genes are determined at fertilization/conception, the premise is that they cannot be affected by lifetime exposures and must occur prior to the onset of disease. Three SNPs known to be associated with higher 25(OH)D levels were found to be protective against MS based on a case–control study design involving non-Hispanic whites in California, USA, and participants in Sweden (pooled meta-analysis of 7391 MS cases and 14,777 controls, OR = 0.85; 95% CI: 0.76–0.94) [76].

## Summary Statement and Future Directions

Higher sunlight exposure and/or higher serum vitamin D levels appear to be consistently associated with a reduced risk of MS. While there has been much discussion surrounding how to use these observations to pursue preventive strategies against the development of MS, much still remains unknown, including fundamental aspects such as which intervention to use (sunlight and/or vitamin D supplements), dose, duration, and the optimal timing (gestation, childhood, and/or adulthood) [67]. Further, most work to date has been conducted in whites/people of Northern European ancestry, with women predominating in virtually all studies. Findings may not apply to other racial groups and/or men [34, 68, 69, 77]. Nonetheless, teams in Australia and New Zealand are already targeting specific high-risk populations to address some of these fundamental issues. Specifically, a phase II-b randomized placebo-controlled trial has been underway (2013–2019) to assess three different doses of oral vitamin D supplements in people with a first demyelinating event (clinically isolated syndrome). The overarching goal is to prevent or delay a second demyelinating event and hence delay or prevent a diagnosis of MS [78].

## Smoking

Cigarette smoking is the leading cause of preventable death in the world [79]. While the number of smokers is declining globally, cigarette use is increasing in parts of the developing world [79]. Cigarette smoking has been implicated in heightened autoimmunity and an increased risk of rheumatoid arthritis, systemic lupus erythematous, and MS [80]. Smoking is believed to initiate an immune response primarily through its inflammatory properties [80].

A meta-analysis collated published studies from 1960 to May 2010 and concluded that the risk of developing MS was 50% higher among smokers based on 3052 MS cases and 457,619 controls (risk ratio = 1.52, 95% CI: 1.39–1.66) [81]. Subsequent studies have found a greater risk of MS in persons exposed to passive smoking [82]. Parental smoking at home was associated with an increased incidence of childhood-onset MS in France [83], the USA [84], and Canada [85]. A Swedish record-linkage study found no association between maternal smoking during pregnancy (collected around the time of pregnancy) and risk of MS in the offspring [86], suggesting that the "at-risk" period for smoke exposure occurs in the postnatal period.

In summary, the results of the meta-analysis and subsequently published original observational studies, coupled with the biological plausibility of a relationship, provide strong evidence for an effect of smoking on MS risk.

## Dietary Macro- and Micronutrients, Alcohol, and Caffeine Consumption

While diet may play a role in the risk of many chronic diseases, it is particularly challenging to measure objectively. Further, it is worth remembering that the ability to assess whether a specific dietary component affects disease risk will depend on the frequency of that dietary component in the wider (general) population under study. If a population consumes a food or beverage item in low to modest amounts (e.g., fatty fish or alcohol or coffee/caffeine), then it is very unlikely that a relationship with the disease (MS) of interest will be found, even if such a relationship exists.

Some of the best available evidence related to diet and MS risk stems from the USA's Nurses' Health Study [87, 88]. One of the earliest studies conducted looked at dietary fat and concluded that there was no evidence to support a relationship between total fat intake nor specific types of fat (animal, vegetable, saturated, monounsaturated, polyunsaturated, *trans*unsaturated, cholesterol) and MS risk. Almost 2 decades later, the relationship with fat intake was assessed in more detail, with a focus on polyunsaturated fatty acids (PUFAs). Higher intake of PUFAs was associated with a lower MS risk (highest versus lowest quintile, hazard ratio = 0.67; 95% CI: 0.49–0.90), with the effect most prominent for the plant-based PUFA α(alpha)-linolenic acid. No significant relationship with MS risk was found for the primarily marine-derived PUFAs, the long-chain fatty acids eicosapentaenoic acid and docosahexaenoic acid [89].

Additional works also accessing the Nurses' Health Study cohorts have concluded that dietary intake of salt, carotenoids, vitamin C, and vitamin E did not alter MS risk [88, 90]. While these insights from women living in the USA are invaluable, there remains much scope to assess the potential role of diet on MS risk in other settings and populations. Also, as individual components of the diet might not act in isolation, there may be value in assessing the role of overall food intake and dietary patterns on MS risk.

Whether alcohol intake alters MS risk is unclear [91–93]; no association was found based on the prospective Nurses' Health Study cohorts [91]. A later retrospective Swedish study reported that high alcohol intake in the 1 year pre-MS onset reduced MS risk [92]. While reverse causation could explain the latter study findings, MS can cause both bladder and gait (walking) issues, which are evident up to 5 years prior to MS onset [94], and both may limit ability or desire for excessive alcohol intake. Nonetheless, the Swedish study authors also noted that most Nurses' Health Study participants were low-to-moderate drinkers, such that if a relationship existed with higher intakes, this could have been missed [92]. Authors of a United Kingdom-based study used a different approach by accessing hospital records to identify persons with MS and persons who had been admitted for an alcohol misuse disorder. They found that alcohol misuse disorders were associated with an increased MS risk, particularly in men [93]. However, hospitalized individuals can differ substantially from the wider population, making the generalizability of findings unclear. Overall, long-term prospective cohort studies are required to clarify the potential relationship between alcohol intake and MS risk.

Like alcohol, the study of coffee/caffeine in relation to MS risk has produced mixed evidence in the different populations studied [91, 95]. A single prospective study using the USA's Nurses' Health Study reported no altered risk of MS among caffeine consumers (coffee, tea, or cola). Subsequently, a USA- and Sweden-based case–control study garnered much interest after suggesting that high consumers of coffee were at a reduced risk of MS [95]. Among high coffee consumers (>900 mL [~6 cups] daily versus none), a lower MS risk was observed (OR = 0.70; 95% CI: 0.49–0.99 in Sweden; OR = 0.69; 95% CI: 0.50–0.96 in Northern California, USA). The authors note that their study was based on retrospective self-reported data, which is susceptible to both recall and misclassification bias [95]. Unmeasured confounding is another important consideration; coffee drinking may be associated with other health- and lifestyle-related factors that may also be associated with the presence of MS, such as migraine [96].

## Obesity

Globally, the prevalence of obesity has tripled since 1975, and the health and social consequences of this epidemic are myriad [97]. Obesity contributes to both the onset and progression of a number of immune-mediated diseases including type 1 diabetes, rheumatoid arthritis, and possibly also MS [98]. Adipose tissue is thought to be involved in the chemical signaling that can modulate immune responses [98].

One of the first studies providing a convincing link between obesity and MS risk was published in 2009 using the USA Nurses' Health Study data [99]. Women who reported higher body size in early adulthood (at age 20; relative to a "normal"/average body size) were at a significantly increased risk of later developing MS (relative risk = 1.96; 95% CI: 1.33–2.89). Body mass index (BMI) was available for these women, measured at age 18 years, and the presence of obesity ($\geq 30$ kg/m$^2$) was similarly associated with an increased risk of MS [99]. These findings were later corroborated by studies in Sweden [100], Norway, and Italy [101]. A study of pediatric MS (or CIS) found an increased risk among obese girls but not boys [102]. There may be a period in childhood, extending through to early adulthood, in which obesity increases MS risk.

## Socioeconomic Status

Socioeconomic status (SES) refers to a person's position in society due to social or economic factors [103]. Disease, disability, and mortality are all increased among persons in lower socioeconomic strata [103]. As such, high SES is typically associated with better health outcomes. However, at least historically, it has been suggested that a high SES is associated with a greater risk of MS. This led authors to perform a systematic review of the evidence. Of 21 studies (published until August 2013), 5 reported an association between high SES and increased risk of MS, 13 reported no evidence of an association, and 3 found that low SES was associated with increased MS risk [104]. Challenges in comparing across studies investigating SES included differences in SES metrics used, calendar time periods covered, and impact of SES within each country on access to healthcare or exposure to environmental factors (smoking, pollutants). The authors noted that there was evidence of a stronger effect of SES in countries with higher inequality [104]. Level of education may also contribute; some of the same authors conducted two subsequent studies, accessing two different cohorts of people with and without MS in Norway. Both studies showed that a higher level of education was associated with a lower risk of MS [105, 106]. The most recent, and largest, study accessed the Norwegian MS Registry and the National Education Registry, combined with population registries. A higher level of educational attainment was associated with a lower risk of MS (odds ratio = 0.73; 95% CI: 0.59–0.90). This observation appeared consistent across a 50-year period in Norway; whether these same effects are observed in other jurisdictions overtime might provide insights into what could be driving this relationship (such as lifestyle factors) [106].

## Stress

Psychological stress has long been considered a putative agent for the onset of chronic illness, perhaps related to the enhanced inflammatory response to chronic stress [107]. A systematic review summarized the published evidence linking stress to MS risk (1980 to November 2010) [108]. While the five studies included pointed to an increased risk of MS following stressful life events, the measures of stress were so heterogeneous that the authors could draw no firm conclusions [108]. A subsequent study found no relationship between stress and MS risk among women in the USA Nurses' Health Study. Stress was measured as work- and home-related (with questions asked prospectively, before onset of MS), and as physical and sexual abuse during childhood and adolescence (with questions asked retrospectively, after MS was diagnosed) [109].

## Occupational Exposures

Occupational exposure to organic solvents such as in paints, varnishes, adhesives, and cleaning agents is common. These organic solvents are also used in the production of dyes, polymers, plastics, textiles, printing inks, agricultural products, and pharmaceuticals [110]. Two meta-analyses have explored the relationship with MS risk; both concluded that exposure to organic solvents was associated with an increased risk of MS [111, 112]. The most recent study pooled data from 15 publications (1994–2012) and reported that the odds of developing MS were 1.5 times higher (95% CI 1.0–2.3) among workers exposed to organic solvents relative to unexposed persons [112].

There is a growing body of evidence to suggest that shift work may increase the risk of chronic disease [113]. The potential mechanism of action is not understood, but immune dysregulation [113] and the indirect effect of higher rates of adverse health behaviors associated with night shift work have been hypothesized [114]. Shift work could also just be a marker of an adverse health profile, not a direct cause of MS; individuals who become shift workers are more likely to have a lower socioeconomic status and engage in adverse behaviors, such as smoking [115]. Nonetheless, a population-based case–control study from Sweden asked participants to recall their working hours and subsequently found an association between prior shift work and an increased risk of MS (odds ratio = 1.6; 95% CI, 1.2–2.1). The odds of MS were slightly higher if the shift work had occurred before the age of 20 (odds ratio = 2.0; 95% CI, 1.2–3.6) [116, 117]. A later Danish study reached a similar conclusion, reporting that for every additional 100 night shifts, the odds ratio for developing MS was 1.20 (95% CI: 1.08–1.34) [118].

## Timing of Exposure and Gene–Environment Interactions for MS Onset

The timing of environmental exposures may play a role in the subsequent triggering of MS onset (Fig. 10.2) [119, 120]. While early childhood and adolescence appear to represent critical "at-risk" periods, there is evidence to suggest that risk factors may act in utero and into adulthood as well [67, 119]. Suggestions of a prenatal risk period include the reported associations between maternal prepregnancy obesity [84], low vitamin D (or low milk) intake during pregnancy [74], and possibly an older paternal age [121]. All have been associated with an increased risk of MS. While birth month or season has been reported as associated with MS risk (spring births associated with a higher MS risk and autumn births with lower risk) [122], challenges with these analyses, including factoring in the seasonality effects of births in the underlying general population, make these observations uncertain [123].

The strongest evidence has suggested that exposures during childhood and adolescence are important for the eventual development of MS [119]. Lower serum 25(OH)D levels [69], lower sunlight exposure [66], and higher BMI [99, 102, 124] have been shown to exert their effect during this time. The "at-risk" period may exist through until early adulthood, as reported with smoking [81], stress [108], shift work, low serum vitamin D levels [73], and low sunlight exposure while living in areas of low UV radiation [67].

The ability to combine both environmental exposure information and genetic risk factors for MS has led to the identification of key factors that might act synergistically to increase the risk of MS [125]. The interplay between human leukocyte antigen (HLA) genes, EBV, smoking, and/or organic solvents has been investigated. Individuals in Sweden who were HLA-DRB1*15 positive and HLA-A*02 negative and had been exposed to EBV were at a 16-fold higher odds for MS than those who did not carry any of these factors (odds ratio = 16.0; 95% CI: 9.4–27.3) [126]. Another Swedish study reported a 13-fold increased risk of MS among smokers who were HLADRB1*15 positive and HLA-A*02 negative relative to persons with none of these risk factors (odds ratio = 13.5; 95% CI: 8.1–22.6) [127]. In a third study, researchers pooled resources from three case–control studies in Australia, the USA, and Sweden to explore the interplay between

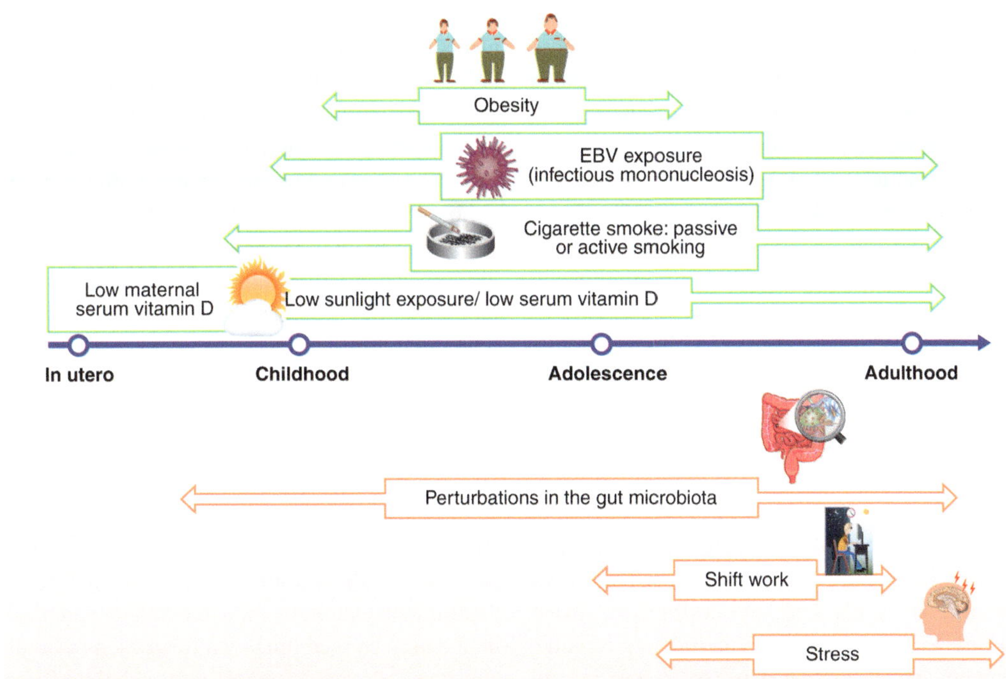


**Fig. 10.2** Illustration of the potential modifiable/environmental risk factors for adult-onset multiple sclerosis (MS) across the life span

smoking, EBV, and HLA-DRB1*15 status on MS risk [128]. They found that antibody levels against EBV were significantly higher among smokers, but no modification by HLA status was observed, suggesting that the HLA-DRB1*15 MS risk is independent of smoking [128].

A final case–control study from Sweden (2042 incident MS cases and 2947 controls) found that in individuals carrying HLA-DRB1*15 and lacking HLA-A*02, exposure to smoking and organic solvents was associated with a 30-fold increased risk of MS (OR = 30.3; 95% CI:11.7–78.3); 40 cases and 5 controls contributed to these risk estimates, meaning 40 MS cases were exposed to all 3 elements and 5 controls were exposed to none [129]. Overall, the multiplicative effect of some risk factors suggests that interactions between factors play an important role in MS development and may point to shared casual pathways [125].

## Conclusion

There is compelling evidence to suggest that MS risk is shaped, in part, by environmental exposures. The strongest evidence to date suggests a role for markers of EBV infections, reduced sunlight exposure/serum vitamin D levels, cigarette smoking, and adolescent obesity as factor(s) that may increase the risk of developing MS. There appear to be critical risk periods for exposure, and these factors can also interact, sometimes in an additive way, to increase the risk of MS [125].

There are many challenges in studying MS epidemiology and risk factors for MS [3]. Recognition of MS by both the general public and clinicians has increased over time, and the diagnostic criteria for MS have also been refined. Further, access to health services and neurologists specialized in MS can differ, such that summarizing information across decades or between regions can be challenging. Further, the exact time of MS onset for any individual remains unknown. Indeed, recent studies are suggesting that the disease processes and even changes in physical, cognitive, and other health behaviors may begin many years before current clinical recognition of MS [94, 130, 131], such that establishing temporality in MS can be challenging. Thus, it can be sometimes difficult to differentiate between a risk factor for MS and a change in behavior/lifestyle because the disease processes have already begun (in the yet to be diagnosed individual).

## Future Perspectives

Recent insights into the MS prodrome are helping narrow the etiologically relevant period when searching for factors triggering MS. A better understanding of the MS prodrome might also facilitate the more timely recognition of MS in the future [94, 130, 131]. Prospectively following very large cohorts of otherwise healthy people prior to onset of disease can represent an optimal study design, akin to the USA Nurses' Health Study [132]. However, these studies are also very costly and challenging to conduct well, with minimal loss to follow-up.

The identification of causal factors for MS can provide valuable insights into disease mechanisms and potentially be used to develop preventative strategies. Based on the current evidence, however, the total prevention of MS is not possible. For instance, some have estimated that 90% of MS cases might be preventable with the elimination of EBV [133]. However, with no suitable vaccine available [58] and uncertainties in possible unwanted sequelae with eliminating a common virus that humans have coevolved with for millennia, this approach is currently not feasible. Gene–environment studies have reported reasonably high odds ratios for MS risk; however, from a public health perspective, genetic testing is not useful or indicated. Around one-third of North America or European populations may carry the HLA "risk genes" for MS. Promoting the reduction of smoking and adolescent obesity and safely "optimizing" sunlight exposure and/or vitamin D intake may be recommended for persons who are considered "at risk" for MS (although the optimal doses for the latter have yet to be elucidated).

Despite best efforts, much remains unknown about the cause(s) of MS. Some new avenues of inquiry include interrogation of the gut microbiome and the role of the gut–brain axis in relation to MS risk [134]. Other potentially modifiable factors, such as diet and exercise, may also play a role and are currently not well understood. Specific interactions between genes and environmental factors have been identified, and the field of epigenetics may also shed further light on the pathways to MS onset [125]. Given the large populations required for the study of MS, international collaborative efforts combined with substantial and strategic investment in research will be necessary if the causes of MS are to be uncovered.

## References


1. Vos T, Allen C, Arora M, Barber RM, Brown A, Carter A, et al. Global, regional, and national incidence, prevalence, and years lived with disability for 310 diseases and injuries, 1990–2015: a systematic analysis for the Global Burden of Disease Study 2015. Lancet. 2016;388:1545–602.
2. Multiple Sclerosis International Federation. Atlas of MS 2013: mapping multiple sclerosis around the world. Mult Scler Int Fed. 2013;2013:1–28.
3. Belbasis L, Bellou V, Evangelou E, Ioannidis JPA, Tzoulaki I. Environmental risk factors and multiple sclerosis: an umbrella review of systematic reviews and meta-analyses. Lancet Neurol. 2015;14:263–73.
4. Rothman KJ, Greenland S. Causation and causal inference in epidemiology. Am J Public Health. 2005;95:144–50.
5. Rothman KJ, Greenland S, Lash TL. Validity in epidemiologic studies. In: Rothman KJ, Greenland S, Lash TL, editors. Mod. Epidemiol. Third, Philadelphia, PA US: 2008, p. 128–147.
6. Tremlett H, Lucas R. An 'epidemic' of multiple sclerosis and falling infection rates? Reflecting on comparisons made and the rising multiple sclerosis incidence in Bach's 2002 New England Journal of Medicine figure. Eur J Neurol. 2018;25:196–9.
7. Kingwell E, Marriott JJ, Jetté N, Pringsheim T, Makhani N, Morrow SA, et al. Incidence and prevalence of multiple sclerosis in Europe: a systematic review. BMC Neurol. 2013;13:128.
8. Evans C, Beland SG, Kulaga S, Wolfson C, Kingwell E, Marriott J. Incidence and prevalence of multiple sclerosis in the Americas: a systematic review. Neuroepidemiology. 2013;40:195–210.
9. Marrie RA, Yu N, Blanchard J, Leung S, Elliott L. The rising prevalence and changing age distribution of multiple sclerosis in Manitoba. Neurology. 2010;74:465–71.
10. Kingwell E, Zhu F, Marrie RA, Fisk JD, Wolfson C, Warren S, et al. High incidence and increasing prevalence of multiple sclerosis in British Columbia, Canada: findings from over two decades (1991–2010). J Neurol. 2015;262:2352–63.
11. Ahlgren C, Oden A, Lycke J. High nationwide prevalence of multiple sclerosis in Sweden. Mult Scler. 2011;17:901–8.
12. Alonso A, Hernán MA. Temporal trends in the incidence of multiple sclerosis. Neurology. 2008;71:129–35.
13. Koch-Henriksen N, Thygesen LC, Stenager E, Laursen B, Magyari M. Incidence of MS has increased markedly over six decades in Denmark particularly with late onset and in women. Neurology. 2018;90:e1954–63.
14. Marrie RA, Fisk JD, Stadnyk KJ, Yu BN, Tremlett H, Wolfson C, et al. The incidence and prevalence of multiple sclerosis in Nova Scotia, Canada. Can J Neurol Sci. 2013;40:824–31.
15. Rotstein DL, Chen H, Wilton AS, Kwong JC, Marrie RA, Gozdyra P, et al. Temporal trends in multiple sclerosis prevalence and incidence in a large population. Neurology. 2018;90:e1435–41.
16. Al-Sakran LH, Marrie RA, Blackburn DF, Knox KB, Evans CD. Establishing the incidence and prevalence of multiple sclerosis in Saskatchewan. Can J Neurol Sci. 2018;45:295–303.
17. Mackenzie IS, Morant SV, Bloomfield GA, MacDonald TM, O'Riordan J. Incidence and prevalence of multiple sclerosis in the UK 1990-2010: a descriptive study in the General Practice Research Database. J Neurol Neurosurg Psychiatry. 2014;85:76–84.
18. Simpson S, Blizzard L, Otahal P, Van der Mei I, Taylor B. Latitude is significantly associated with the prevalence of multiple sclerosis: a meta-analysis. J Neurol Neurosurg Psychiatry. 2011;82:1132–41.
19. Hernán MA, Olek MJ, Ascherio A. Geographic variation of MS incidence in two prospective studies of US women. Neurology. 1999;53:1711–8.
20. Wallin M, Culpepper W, Campbell J, Nelson L, Langer-Gould A, Marrie R, et al. The prevalence of multiple sclerosis in the United States: a population-based healthcare database approach. Mult Scler. 2017;23:125.
21. Mokdad AH, Ballestros K, Echko M, Glenn S, Olsen HE, Mullany E, et al. The state of US health, 1990–2016. JAMA. 2018;319:1444–72.
22. Taylor BV, Pearson JF, Clarke G, Mason DF, Abernethy DA, Willoughby E, et al. MS prevalence in New Zealand, an ethnically and latitudinally diverse country. Mult Scler. 2010;16:1422–31.
23. Koch-Henriksen N, Sørensen PS. The changing demographic pattern of multiple sclerosis epidemiology. Lancet Neurol. 2010;9:520–32.
24. Gale CR, Martyn CN. Migrant studies in multiple sclerosis. Prog Neurobiol. 1995;47:425–48.

25. Visscher BR, Detels R, Coulson AH, Malmgren RM, Dudley JP. Latitude, migration, and the prevalence of multiple sclerosis. Am J Epidemiol. 1977;6:470–5.
26. Kurtzke JF. Epidemiology in multiple sclerosis: a pilgrim's progress. Brain. 2013;136:2904–17.
27. Alter M, Leibowitz U, Speer J. Risk of multiple sclerosis related to age at immigration to Israel. Arch Neurol. 1966;15:234–7.
28. McLeod JG, Hammond SR, Kurtzke JF. Migration and multiple sclerosis in immigrants to Australia from United Kingdom and Ireland: a reassessment. I. Risk of MS by age at immigration. J Neurol. 2011;258:1140–9.
29. Singh Setia M, Quesnel-Vallee A, Abrahamowicz M, Tousignant P, Lynch J. Access to health-care in Canadian immigrants: a longitudinal study of the National Population Health Survey. Health Soc Care Community. 2011;19:70–9.
30. Chabas D, Green AJ, Waubant E. Pediatric multiple sclerosis. NeuroRx. 2006;3:264–75.
31. Martinelli V, Rodegher M, Moiola L, Comi G. Late onset multiple sclerosis: clinical characteristics, prognostic factors and differential diagnosis. Neurol Sci. 2004;25:s350–5.
32. Tremlett H, Devonshire V. Is late-onset multiple sclerosis associated with a worse outcome? Neurology. 2006;67:954–9.
33. Shirani A, Zhao Y, Petkau J, Gustafson P, Karim ME, Evans C, et al. Multiple sclerosis in older adults: the clinical profile and impact of interferon beta treatment. Biomed Res Int. 2015;2015:1–11.
34. Langer-Gould A, Brara SM, Beaber BE, Zhang JL. Incidence of multiple sclerosis in multiple racial and ethnic groups. Neurology. 2013;80:1734–9.
35. Manouchehrinia A, Tanasescu R, Tench CR, Constantinescu CS. Mortality in multiple sclerosis: meta-analysis of standardised mortality ratios. J Neurol Neurosurg Psychiatry. 2016;87:324–31.
36. Ragonese P, Aridon P, Salemi G, D'Amelio M, Savettieri G. Mortality in multiple sclerosis: a review. Eur J Neurol. 2008;15:123–7.
37. Kingwell E, Van Der Kop M, Zhao Y, Shirani A, Zhu F, Oger J, et al. Relative mortality and survival in multiple sclerosis: findings from British Columbia, Canada. J Neurol Neurosurg Psychiatry. 2012;83:61–6.
38. Lunde HMB, Assmus J, Myhr KM, Bø L, Grytten N. Survival and cause of death in multiple sclerosis: a 60-year longitudinal population study. J Neurol Neurosurg Psychiatry. 2017;88:621–5.
39. Ascherio A, Munger KL. Environmental risk factors for multiple sclerosis. Part I: the role of infection. Ann Neurol. 2007;61:288–99.
40. Cohen JI. Epstein–Barr virus infection. N Engl J Med. 2000;343:481–92.
41. Munger KL, Levin LI, O'Reilly EJ, Falk KI, Ascherio A. Anti-Epstein-Barr virus antibodies as serological markers of multiple sclerosis: a prospective study among United States military personnel. Mult Scler. 2011;17:1185–93.
42. Levin LI, Munger KL, Reilly EJO, Falk KI, Ascherio A. Primary infection with the Epstein-Barr Virus and risk of multiple sclerosis. Ann Neurol. 2010;67:824–30.
43. Rubertone MV, Brundage JF. The defense medical surveillance system and the department of defense serum repository: glimpses of the future of public health surveillance. Am J Public Health. 2002;92:1900–4.
44. Handel AE, Williamson AJ, Disanto G, Handunnetthi L, Giovannoni G, Ramagopalan SV. An updated meta-analysis of risk of multiple sclerosis following infectious mononucleosis. PLoS One. 2010;5:1–5. https://doi.org/10.1371/journal.pone.0012496.
45. Pakpoor J, Disanto G, Gerber JE, Dobson R, Meier UC, Giovannoni G, et al. The risk of developing multiple sclerosis in individuals seronegative for Epstein-Barr virus: a meta-analysis. Mult Scler. 2013;19:162–6.
46. Pakpoor J, Ramagopalan SV. Epstein-Barr virus is a necessary causative agent in the pathogenesis of multiple sclerosis: yes. Mult Scler. 2013;19:1690–1.
47. Salzer J, Myhr K-M. Epstein–Barr virus is a necessary causative agent in the pathogenesis of multiple sclerosis: no. Mult Scler. 2013;19:1692–3.
48. McKay KA, Jahanfar S, Duggan T, Tkachuk S, Tremlett H. Factors associated with onset, relapses or progression in multiple sclerosis: a systematic review. Neurotoxicology. 2017;61:189–212.
49. Braun DK, Dominguez G, Pellett PE. Human herpesvirus 6. Clin Microbiol Rev. 1997;10:521–67.
50. Opsahl ML, Kennedy PGE. Early and late HHV-6 gene transcripts in multiple sclerosis lesions and normal appearing white matter. Brain. 2005;128:516–27.
51. Villoslada P, Juste C, Tintore M, Llorenç V, Codina G, Pozo-Rosich P, et al. The immune response against herpesvirus is more prominent in the early stages of MS. Neurology. 2003;60:1944–8.
52. Sriram S, Mitchell W, Stratton C. Multiple sclerosis associated with chlamydia pneumoniae infection of the CNS. Neurology. 1998;50:571–2.
53. Kuo C, Jackson LA, Campbell LA, Grayston JT. Chlamydia pneumoniae (TWAR). Clin Microbiol Rev. 1995;8:451–61.
54. Munger KL, DeLorenze GN, Levin LI, Rubertone MV, Vogelman JH, Peck C a, et al. A prospective study of Chlamydia pneumoniae infection and risk of MS in two US cohorts. Neurology. 2004;62:1799–803.
55. Munger KL, Peeling RW, Hernán MA, Chasan-taber L, Olek MJ, Hankinson SE, et al. Infection with Chlamydia pneumoniae and risk of multiple sclerosis. Epidemiology. 2003;14:141–7.

56. Krone B, Pohl D, Rostasy K, Kahler E, Brunner E, Oeffner F, et al. Common infectious agents in multiple sclerosis: a case-control study in children. Mult Scler J. 2008;14:136–9.
57. Fleming JO, Cook TD. Multiple sclerosis and the hygiene hypothesis. Neurology. 2006;67:2085–6.
58. Cohen JI, Mocarski ES, Raab-Traub N, Corey L, Nabel GJ. The need and challenges for development of an Epstein-Barr virus vaccine. Vaccine. 2013;31S:B194–6.
59. Le Houézec D. Evolution of multiple sclerosis in France since the beginning of hepatitis B vaccination. Immunol Res. 2014;60:219–25.
60. Farez MF, Correale J. Immunizations and risk of multiple sclerosis: systematic review and meta-analysis. J Neurol. 2011;258:1197–206.
61. Langer-Gould A, Qian L, Tartof SY, Brara SM, Jacobsen SJ, Beaber BE, et al. Vaccines and the risk of multiple sclerosis and other central nervous system demyelinating diseases. JAMA Neurol. 2014;71:1506–13.
62. Meggiolaro A, Migliara G, La Torre G. Association between Human Papilloma Virus (HPV) vaccination and risk of multiple sclerosis: a systematic review. Hum Vaccin Immunother. 2018;14:1266–74.
63. Mahmud SM, Bozat-emre S, Mostaço-guidolin LC, Marrie RA. Registry cohort study to determine risk for multiple sclerosis after vaccination for pandemic influenza A (H1N1) with Arepanrix, Manitoba, Canada. Emerg Infect Dis. 2018;24:1267–74.
64. Hayes CE, Cantorna MT, DeLuca HF. Vitamin D and multiple sclerosis. Proc Soc Exp Biol Med. 1997;216:21–7.
65. Ascherio A, Munger KL, Simon KC. Vitamin D and multiple sclerosis. Lancet Neurol. 2010;9:599–612.
66. Van Der Mei IAF, Dwyer T, Blizzard L, Ponson AL, Simmons R, Taylor BV, et al. Past exposure to sun, skin phenotype, and risk of multiple sclerosis: case-control study. Br Med J. 2003;327:316–22. https://doi.org/10.1136/bmj.327.7410.316.
67. Tremlett H, Zhu F, Ascherio A, Munger KL. Sun exposure over the life course and associations with multiple sclerosis. Neurology. 2018;90:e1191–9.
68. Langer-Gould A, Lucas R, Xiang AH, Chen LH, Wu J, Gonzalez E, et al. MS Sunshine Study: sun exposure but not vitamin D is associated with multiple sclerosis risk in blacks and hispanics. Nutrients. 2018;10:268.
69. Munger KL, Levin LI, Hollis BW, Howard NS, Ascherio A. Serum 25-Hydroxyvitamin D levels and risk of multiple sclerosis. JAMA Neurol. 2006;296:2832–8.
70. Munger KL, Åivo J, Hongell K, Soilu-Hänninen M, Surcel HM, Ascherio A. Vitamin D status during pregnancy and risk of multiple sclerosis in offspring of women in the Finnish Maternity Cohort. JAMA Neurol. 2016;73:515–9. https://doi.org/10.1001/jamaneurol.2015.4800.
71. Duan S, Lv Z, Fan X, Wang L, Han F, Wang H, et al. Vitamin D status and the risk of multiple sclerosis: a systematic review and meta-analysis. Neurosci Lett. 2014;570:108–13. https://doi.org/10.1016/j.neulet.2014.04.021.
72. Colditz GA, Manson JE, Hankinson SE. The nurses' health study: 20-year contribution to the understanding of health among women. J Womens Health. 1997;6:49–62. https://doi.org/10.1089/jwh.1997.6.49.
73. Munger KL, Zhang SM, O'Reilly E, Hernan MA, Olek MJ, Willett WC, et al. Vitamin D intake and incidence of multiple sclerosis. Neurology. 2004;62:60–5. https://doi.org/10.1212/01.WNL.0000101723.79681.38.
74. Mirzaei F, Michels KB, Munger K, Reilly EO, Chitnis T, Forman MR, et al. Gestational vitamin D and the risk of multiple sclerosis in the offspring. Ann Neurol. 2011;70:30–40. https://doi.org/10.1002/ana.22456.Gestational.
75. Lawlor DA, Harbord RM, Sterne JAC, Timpson N, Smith GD. Mendelian randomization: using genes as instruments for making causal inferences in epidemiology. Stat Med. 2008;27:1133–63.
76. Rhead B, Bäärnhielm M, Gianfrancesco M, Mok A, Quach H, Shen L, et al. Mendelian randomization shows a causal effect of low vitamin D on multiple sclerosis risk. Neurol Genet. 2016;2:e97.
77. Wallin MT, Culpepper WJ, Coffman P, Pulaski S, Maloni H, Mahan CM, et al. The Gulf War era multiple sclerosis cohort: age and incidence rates by race, sex and service. Brain. 2012;135:1778–85. https://doi.org/10.1093/brain/aws099.
78. MS Research Australia. Vitamin D MS prevention trial – PrevANZ. MSRA 2018. https://msra.org.au/vitamin-d-ms-prevention-trial-prevanz/ (accessed 15 Aug 2018).
79. World Health Organization. Tobacco. Fact sheets 2018. http://www.who.int/en/news-room/fact-sheets/detail/tobacco (accessed 10 July 2018).
80. Costenbader KH, Karlson EW. Cigarette smoking and autoimmune disease: what can we learn from epidemiology? Lupus. 2006;15:737–45.
81. Handel AE, Williamson AJ, Disanto G, Dobson R, Giovannoni G, Ramagopalan SV. Smoking and multiple sclerosis: an updated meta-analysis. PLoS One. 2011;6:e16149. https://doi.org/10.1371/journal.pone.0016149.
82. Hedström AK, Bäärnhielm M, Olsson T, Alfredsson L. Exposure to environmental tobacco smoke is associated with increased risk for multiple sclerosis. Mult Scler. 2011;17:788–93. https://doi.org/10.1177/1352458511399610.
83. Mikaeloff Y, Caridade G, Tardieu M, Suissa S. Parental smoking at home and the risk of childhood-onset multiple sclerosis in children. Brain. 2007;130:2589–95. https://doi.org/10.1093/brain/awm198.
84. Gardener H, Munger KL, Chitnis T, Michels KB, Spiegelman D, Ascherio A. Prenatal and perinatal

factors and risk of multiple sclerosis. Epidemiology. 2009;20:611–8. https://doi.org/10.1097/EDE.0b013e31819ed4b9.Prenatal.
85. Lavery AM, Collins BN, Waldman AT, Hart CN, Bar-Or A, Marrie RA, et al. The contribution of secondhand tobacco smoke exposure to pediatric multiple sclerosis risk. Mult Scler. 2018;25:515–22.
86. Montgomery SM, Bahmanyar S, Hillert J, Ekbom A, Olsson T. Maternal smoking during pregnancy and multiple sclerosis amongst offspring. Eur J Neurol. 2008;15:1395–9. https://doi.org/10.1111/j.1468-1331.2008.02331.x.
87. Zhang SM, Willett WC, Hernán MA, Olek MJ, Ascherio A. Dietary fat in relation to risk of multiple sclerosis among two large cohorts of women. Am J Epidemiol. 2000;152:1056–64.
88. Zhang SM, Hernán MA, Olek MJ, Spiegelman D, Willett WC, Ascherio A. Intakes of carotenoids, vitamin C, and vitamin E and MS risk among two large cohorts of women. Neurology. 2001;57:75–80. https://doi.org/10.1212/WNL.57.1.75.
89. Bjørnevik K, Chitnis T, Ascherio A, Munger KL. Polyunsaturated fatty acids and the risk of multiple sclerosis. Mult Scler. 2017;23:1830–8. https://doi.org/10.1177/1352458517691150.
90. Cortese M, Yuan C, Chitnis T, Ascherio A, Munger KL. No association between dietary sodium intake and the risk of multiple sclerosis. Neurology. 2017;89:1322–9. https://doi.org/10.1212/WNL.0000000000004417.
91. Massa J, O'Reilly EJ, Munger KL, Ascherio A. Caffeine and alcohol intakes have no association with risk of multiple sclerosis in a prospective study of women. Mult Scler. 2013;19:53–8. https://doi.org/10.1126/scisignal.2001449.Engineering.
92. Hedström AK, Hillert J, Olsson T, Alfredsson L. Alcohol as a modifiable lifestyle factor affecting multiple sclerosis risk. JAMA Neurol. 2014;71:300–5. https://doi.org/10.1001/jamaneurol.2013.5858.
93. Pakpoor J, Goldacre R, Disanto G, Giovannoni G, Goldacre MJ. Alcohol misuse disorders and multiple sclerosis risk. JAMA Neurol. 2014;71:1088–189. https://doi.org/10.1001/jamaneurol.2014.1795.
94. Wijnands JM, Zhu F, Kingwell E, Zhao Y, Ekuma O, Lu X, et al. Five years before multiple sclerosis onset: phenotyping the prodrome. Mult Scler. 2018;5:1092–101.
95. Hedström AK, Mowry EM, Gianfrancesco MA, Shao X, Schaefer CA, Shen L, et al. High consumption of coffee is associated with decreased multiple sclerosis risk; results from two independent studies. J Neurol Neurosurg Psychiatry. 2016;87:454–60. https://doi.org/10.1136/jnnp-2015-312176.
96. Wijnands JMA, Kingwell E. Time to wake up and smell the coffee? Coffee consumption and multiple sclerosis. J Neurol Neurosurg Psychiatry. 2016;87:453. https://doi.org/10.1136/jnnp-2015-312431.
97. World Health Organization. Obesity and overweight. WHO fact sheets 2018. http://www.who.int/news-room/fact-sheets/detail/obesity-and-overweight (accessed 10 July 2018).
98. Versini M, Jeandel PY, Rosenthal E, Shoenfeld Y. Obesity in autoimmune diseases: not a passive bystander. Autoimmun Rev. 2014;13:981–1000. https://doi.org/10.1016/j.autrev.2014.07.001.
99. Munger KL, Chitnis T, Ascherio A. Body size and risk of MS in two cohorts of US women. Neurology. 2009;73:1543–50. https://doi.org/10.1212/WNL.0b013e3181c0d6e0.
100. Hedström AK, Olsson T, Alfredsson L. High body mass index before age 20 is associated with increased risk for multiple sclerosis in both men and women. Mult Scler. 2012;18:1334–6. https://doi.org/10.1177/1352458512436596.
101. Wesnes K, Riise T, Casetta I, Drulovic J, Granieri E, Holmøy T, et al. Body size and the risk of multiple sclerosis in Norway and Italy: the EnvIMS study. Mult Scler. 2015;21:388–95. https://doi.org/10.1177/1352458514546785.
102. Langer-Gould A, Brara SM, Beaber BE, Koebnick C. Childhood obesity and risk of pediatric multiple sclerosis and clinically isolated syndrome. Neurology. 2013;80:548–52.
103. Galobardes B, Shaw M, Lawlor DA, Lynch JW, Smith GD. Indicators of socioeconomic position (part 1). J Epidemiol Community Health. 2006;60:7–12. https://doi.org/10.1136/jech.2004.023531.
104. Goulden R, Ibrahim T, Wolfson C. Is high socioeconomic status a risk factor for multiple sclerosis? A systematic review. Eur J Neurol. 2015;22:899–911. https://doi.org/10.1111/ene.12586.
105. Bjørnevik K, Riise T, Cortese M, Holmøy T, Kampman MT, Magalhaes S, et al. Level of education and multiple sclerosis risk after adjustment for known risk factors: the EnvIMS study. Mult Scler. 2016;22:104–11. https://doi.org/10.1177/1352458515579444.
106. Bjørnevik K, Riise T, Benjaminsen E, Celius EG, Dahl OP, Kampman MT, et al. Level of education and multiple sclerosis risk over a 50-year period: registry-based sibling study. Mult Scler. 2017;23:213–9. https://doi.org/10.1177/1352458516646863.
107. Cohen S, Janicki-Deverts D, Doyle WJ, Miller GE, Frank E, Rabin BS, et al. Chronic stress, glucocorticoid receptor resistance, inflammation, and disease risk. Proc Natl Acad Sci. 2012;109:5995–9. https://doi.org/10.1073/pnas.1118355109.
108. Artemiadis AK, Anagnostouli MC, Alexopoulos EC. Stress as a risk factor for multiple sclerosis onset or relapse: a systematic review. Neuroepidemiology. 2011;36:109–20. https://doi.org/10.1159/000323953.
109. Riise T, Mohr DC, Munger KL, Rich-Edwards JW, Kawachi I, Ascherio A. Stress and the risk of multiple sclerosis. Neurology. 2011;76:1866–71.

110. The National Institute for Occupational Safety and Health. Organic solvents. Centers Dis Control Prev 2018. https://www.cdc.gov/niosh/topics/organsolv/default.html (accessed 24 July 2018).
111. Landtblom A-M, Flodin U, Söderfeldt B, Wolfson C, Axelson O. Organic solvents and multiple sclerosis: a synthesis of the current evidence. Epidemiology. 1996;7:429–33.
112. Barragán-Martínez C, Speck-Hernández CA, Montoya-Ortiz G, Mantilla RD, Anaya JM, Rojas-Villarraga A, et al. Organic solvents as risk factor for autoimmune diseases: a systematic review and meta-analysis. PLoS One. 2012;7:e51506. https://doi.org/10.1371/journal.pone.0051506.
113. Wang XS, Armstrong MEG, Cairns BJ, Key TJ, Travis RC. Shift work and chronic disease: the epidemiological evidence. Occup Med (Lond). 2011;61:78–89. https://doi.org/10.1093/occmed/kqr001.
114. Ramin C, Devore EE, Wang W, Pierre-Paul J, Wegrzyn LR, Schernhammer ES. Night shift work at specific age ranges and chronic disease risk factors. Occup Environ Med. 2015;72:100–7. https://doi.org/10.1136/oemed-2014-102292.
115. Ponsonby AL, Lucas RM. Shift work and multiple sclerosis. Ann Neurol. 2011;70:680–3. https://doi.org/10.1002/ana.22641.
116. Hedström AK, Åkerstedt T, Hillert J, Olsson T, Alfredsson L. Shift work at young age is associated with increased risk for multiple sclerosis. Ann Neurol. 2011;70:733–41. https://doi.org/10.1002/ana.22597.
117. Hedström AK, Åkerstedt T, Olsson T, Alfredsson L. Shift work influences multiple sclerosis risk. Mult Scler. 2015;21:1195–9. https://doi.org/10.1177/1352458514563592.
118. Gustavsen S, Søndergaard HB, Oturai DB, Laursen B, Laursen JH, Magyari M, et al. Shift work at young age is associated with increased risk of multiple sclerosis in a Danish population. Mult Scler Relat Disord. 2016;9:104–9. https://doi.org/10.1016/j.msard.2016.06.010.
119. Handel AE, Giovannoni G, Ebers GC, Ramagopalan SV. Environmental factors and their timing in adult-onset multiple sclerosis. Nat Rev Neurol. 2010;6:156–66. https://doi.org/10.1038/nrneurol.2010.1.
120. van der Mei IAF, Simpson S, Stankovich J, Taylor BV. Individual and joint action of environmental factors and risk of MS. Neurol Clin. 2011;29:233–55. https://doi.org/10.1016/j.ncl.2010.12.007.
121. Montgomery SM, Lambe M, Olsson T, Ekbom A. Parental age, family size, and risk of multiple sclerosis. Epidemiology. 2004;15:717–23. https://doi.org/10.1097/01.ede.0000142138.46167.69.
122. Torkildsen O, Grytten N, Aarseth J, Myhr K-M, Kampman MT. Month of birth as a risk factor for multiple sclerosis: an update. Acta Neurol Scand Suppl. 2012;126:58–62. https://doi.org/10.1111/ane.12040.
123. Fiddes B, Wason J, Kemppinen A, Ban M, Compston A, Sawcer S. Confounding underlies the apparent month of birth effect in multiple sclerosis. Ann Neurol. 2013;73:714–20. https://doi.org/10.1002/ana.23925.
124. Munger KL, Bentzen J, Laursen B, Stenager E, Koch-Henriksen N, Sørensen TI a, et al. Childhood body mass index and multiple sclerosis risk: a long-term cohort study. Mult Scler J. 2013;19:1323–9. https://doi.org/10.1177/1352458513483889.
125. Olsson T, Barcellos LF, Alfredsson L. Interactions between genetic, lifestyle and environmental risk factors for multiple sclerosis. Nat Rev Neurol. 2016;13:26–36. https://doi.org/10.1038/nrneurol.2016.187.
126. Sundqvist E, Sundström P, Lindén M, Hedström a K, Aloisi F, Hillert J, et al. Epstein-Barr virus and multiple sclerosis: interaction with HLA. Genes Immun. 2012;13:14–20. https://doi.org/10.1038/gene.2011.42.
127. Hedström AK, Sundqvist E, Bäärnhielm M, Nordin N, Hillert J, Kockum I, et al. Smoking and two human leukocyte antigen genes interact to increase the risk for multiple sclerosis. Brain. 2011;134:653–64. https://doi.org/10.1093/brain/awq371.
128. Simon KC, van der Mei IAF, Munger KL, Ponsonby AL, Dickinson J, Dwyer T, et al. Combined effects of smoking, anti-EBNA antibodies, and HLA-DRB1*1501 on multiple sclerosis risk. Neurology. 2010;74:1365–71.
129. Hedström AK, Hössjer O, Katsoulis M, Kockum I, Olsson T, Alfredsson L. Organic solvents and MS susceptibility: interaction with MS risk HLA genes. Neurology. 2018;91:e455–62.
130. Wijnands JMA, Kingwell E, Zhu F, Zhao Y, Högg T, Stadnyk K, et al. Health-care use before a first demyelinating event suggestive of a multiple sclerosis prodrome: a matched cohort study. Lancet Neurol. 2017;16:445–51.
131. Cortese M, Riise T, Bjørnevik K, Bhan A, Farbu E, Grytten N, et al. Preclinical disease activity in multiple sclerosis: a prospective study of cognitive performance prior to first symptom. Ann Neurol. 2016;80:616–24.
132. Morabia A. 120 000 nurses who shook public health. Am J Public Health. 2016;106:1528–9. https://doi.org/10.2105/AJPH.2016.303345.
133. Ascherio A, Munch M. Epstein-Barr virus and multiple sclerosis. Epidemiology. 2000;11:220–4.
134. Tremlett H, Bauer KC, Appel-Cresswell S, Finlay BB, Waubant E. The gut microbiome in human neurological disease: a review. Ann Neurol. 2017;81:369–82. https://doi.org/10.1002/ana.24901.

# The Genetics of Multiple Sclerosis

# 11

Annalise E. Miner, Neda Dastgheyb,
Miryam Palomino, and Jennifer S. Graves

**Key Points**

1. Of multiple sclerosis (MS) genetic heritability, 48% has now been accounted for.
2. The HLA allele *DRB1*15:01* accounts for up to 10.5% of genetic variance underlying MS susceptibility, making it the single largest genetic risk factor; however, this allele has also been found to have epistatic effects on other genetic haplotypes, demonstrating a nuanced and complex impact of gene-gene interactions in MS.
3. Genome-wide association studies (GWAS) have identified more than 200 non-HLA genetic risk factors of MS and have helped to demonstrate that these genetic loci associated with MS are generally related to T-cell activation and proliferation in the immune system.
4. MS has considerable genetic overlap with other autoimmune diseases.
5. Low levels of vitamin D and an increased body mass index (BMI) were found to significantly increase the risk of MS development in Mendelian randomization studies.

## Introduction

Multiple sclerosis (MS) is an autoimmune disease of the central nervous system (CNS) characterized by injury to the myelin sheath and subsequent neurodegeneration. The maladaptive autoimmune response in MS results in a myriad of neurological symptoms affecting motor, sensory, coordination, and cognitive functions. MS is a leading cause of disability among young adults and affects females more than males [1, 2]. The identified MS risk factors are varied and likely at interplay with one another. Among them are genetic factors (about 25% of the risk) and environmental factors including living at a high geographic latitude, smoking, obesity, low vitamin D levels, and Epstein-Barr virus (EBV) exposure [3–5].

There has been debate about whether MS is an "outside in" or an "inside out" disease, arising from an abnormal peripheral immune system or from abnormalities within the CNS that trigger an immune response, respectively. As outlined below, the study of the genetics of MS strongly

A. E. Miner · N. Dastgheyb · M. Palomino
Department of Neurosciences, University of California, San Diego, La Jolla, CA, USA

J. S. Graves (✉)
Department of Neurosciences, Rady's Pediatric MS Center, UCSD Neuroimmunology Research Program, University of California, San Diego, ACTRI, La Jolla, CA, USA
e-mail: jgraves@health.ucsd.edu



A. L. Piquet, E. Alvarez (eds.), *Neuroimmunology*, https://doi.org/10.1007/978-3-030-61883-4_11

implicates that it is primarily a disease of the immune system. MS genetic risk factors are more closely overlapping with those of other autoimmune diseases, including rheumatoid arthritis and type 1 diabetes than with diseases of the CNS [6].

With now more than 230 genetic risk factors identified for MS, leveraging tens of thousands of samples, questions related to gene-gene or gene-environment interactions, genetic influence on phenotype, and pharmacogenetic factors for treatment response have been raised [7]. However, these areas of inquiry have proven to be very challenging, requiring large sample sizes, unified approaches to phenotype, and better understanding of some medications' mechanisms of action. Nonetheless, some gene-environment interactions have been reported, and some genetic factors have been associated with relapse rate and regions of lesion burden. Mendelian randomization experiments have been successfully employed to use genetic drivers of environmental risk factors to support causal associations between these factors and MS risk, eliminating concerns for reverse causation.

The pursuit of genetic risk factors for MS has led to critical insights on the pathophysiology of the disease, and future work will help determine what shapes a person's experience of the disease.

## Heritability

Heritability in MS can be defined as the proportion of total variance in risk of disease development that could be explained by genetic factors [8]. Estimation of heritability has largely been based on studies investigating MS risk within families. First-degree relatives (siblings, children, parents) have an increased risk of developing MS (3–5%) [9]. Risk is notably increased to 25–30% in monozygotic twins [9–11]. One of the early twin studies in MS was a longitudinal, population-based study of twins in Canada, which established a 25% risk in monozygotic twins. These twin results have led to estimation that MS genetic susceptibility explains a quarter of the overall risk of the disease. Notably, the Canadian twin study also found that the risk of MS in dizygotic twins is doubled compared to non-twin siblings [11]. These latter results suggest that there may be an influence of the shared intrauterine environment since dizygotic twins and non-twin siblings share the same amount of genetic material, but the former also share very early environmental exposures. Nonetheless, additional studies investigating the MS risk in half-sibling, adoptee, and spouse studies have demonstrated that the primary responsible factor of MS occurrence within families is genetics, more so than the environment of early childhood and adulthood [12, 13]. In sum, there is clustering of MS in genetically related familial groups, and as expected the strongest risk is within monozygotic twins.

In concordant sibling studies, age of onset was determined to be a more important factor than year of onset of MS; this means that siblings who were both diagnosed with MS were more likely to be diagnosed at the same age, rather than in the same calendar year [14, 15]. Genetics was a more powerful predictor than cohabitation [14, 15]. If two siblings had both been diagnosed in the same calendar year at different ages, the environment would be suspected to have played the more important role in MS development. However, being diagnosed at the same chronological age points toward a genetic factor influencing MS predisposition.

As an update to estimations of heritability, a later meta-analysis evaluated eight twin studies from France, United Kingdom (UK), Denmark, North America, Italy, Finland, and Sweden [16]. A biometric multigroup analysis was conducted under a liability threshold model and found meta-analytic estimates for genetic heritability as close to 50% (0.50, 95% CI: 0.39–0.6) [16]. The model also found an estimate of 0.21 (95% CI: 0.11–0.30) for a shared environmental factor of MS and a 0.29 (95% CI: 0.26–0.33) for a unique environmental factor. These results suggest a higher heritability estimate of MS and also a shared very early environmental risk factor. In pediatric MS, perinatal environmental factors have been associated with risk of disease onset [17].

Sex dimorphism is a predominant feature of MS, with more females affected than males [10, 18, 19]. Sex chromosome contribution to the imbalance in risk has been challenging to study, as X and Y chromosomes are not typically included in genome-wide association studies (GWAS). In a recent large analysis by the International MS Genetics Consortium (IMSGC) of 47,000 cases and 68,000 controls, a locus on the X chromosome was identified with MS risk that reaches genome-wide statistical significance (rs2807267) [7]. The functional consequence of this polymorphism and relation to pathophysiology of MS is not yet fully understood.

Sex effects in MS also include higher transmission of risk through mothers as opposed to fathers, suggesting a genetic or epigenetic factor influencing this gender disparity [20]. Several studies have implicated mitochondrial DNA (mtDNA), which is transmitted to a child strictly from the mother, in MS susceptibility [21, 22]. Mitochondrial dysfunction has been linked to the pathophysiology of chronic neurodegenerative disorders [21, 22]. In regard to demyelinating disorders, a mouse model was established using mtDNA with double-strand breaks that were introduced into myelinating oligodendrocytes [21]. This introduction led to impairment of locomotor function, demyelination, glial activation, and axonal degradation in both male and female mice [21]. These findings suggest that mtDNA damage can cause chronic demyelination and axonal loss in mice [21]. A human haplotype analysis in which mtDNA was sequenced showed a trend toward an overrepresentation of super-haplogroup U in MS participants [23]. Tranah et al. observed in a pooled analysis from seven clinical sites a 1.15-fold higher risk of MS in JT haplotype carriers [24]. These results support the hypothesis that mtDNA variation contributes to MS susceptibility, with potential to explain maternal inheritance of risk.

With the efforts of the IMSGC and generous participation of more than 100,000 individuals, 48% of heritability of MS has now been explained [7]. The specific susceptibility factors are described below and include more than 30 alleles within the human leukocyte antigen region (HLA) and more than 200 non-HLA alleles.

## HLA Risk Factors

The human major histocompatibility complex (MHC) is located on chromosome 6 (6021.3) and contains a cluster of more than 200 genes, many of which encode proteins essential to a variety of immune responses [25, 26]. These include the human leukocyte antigen (HLA) class I and class II genes, which encode heterodimeric cell surface glycoproteins that bind and present peptide antigens to T-cell receptors (TCR) [26]. This process of antigen presentation is central to initiation of an adaptive immune response against pathogens and plays a major role in autoimmune disease pathogenesis [25, 26]. Humans have three class I genes, *HLA-A, HLA-B,* and *HLA-C,* which are highly polymorphic and encode the α(alpha) chain of the HLA-A, HLA-B, and HLA-C molecules, respectively [26]. There are three class II gene sets, HLA-DR, HLA-DP, and HLA-DQ, which encode the α(alpha) and β(beta) chains of their respective molecules and are also highly polymorphic [26]. The combination of HLA alleles inherited from each parent is termed the HLA haplotype. The expression of each HLA allele at a locus is codominant [26]. Due to the extensive diversity of HLA alleles, the World Health Organization (WHO) Nomenclature Committee for Factors of the HLA System has established a system of nomenclature for HLA genes based on locus, type, subtype, synonymous DNA substitutions within the coding region, differences in the noncoding region, and expression status [27]. Specific HLA alleles have been associated as risk factors for many diseases, including MS [25]. The association between HLA allelic variants and MS susceptibility and progression was first established in 1972 [28, 29]. Since then, many studies have been conducted to better understand how HLA genetic risk factors contribute to MS (Table 11.1) [23, 25–28, 30–44].

**Table 11.1** HLA class I and II alleles implicated in MS, referenced in this chapter

| | Gene sets | Allele | Associations | Population studied/effected | References |
|---|---|---|---|---|---|
| HLA Class II | HLA-DR | DRB1*15:01 | Increased risk for MS development; accounts for up to 10.5% of genetic variance underlying MS risk; associated with earlier age of onset in females, oligoclonal band presence, and high IgG levels in CSF | European; African, Asian | [23, 25–28] |
| | | DRB1*13:03 | Increased risk of MS development; associated with disease frequency in a population | Ashkenazi Jews | [30, 31] |
| | | DRB1*03:01 | Increased risk of MS development | European | [32–34] |
| | | DRB1*15:03 | Susceptibility to MS development | African | [35] |
| | | DRB5*01:01 | Risk of MS development; attenuates MS severity (DRB5*null patients at an increased risk of developing secondary progressive MS) | African | [35] |
| | | DRB1*14:01 | Associated with protective effects against MS development; dominant allele that outweighs the risk of carrying DRB1*15:01 | European | [36, 37] |
| | | DRB1*10:01 | Negatively correlated with MS occurrence | Middle Eastern | [38] |
| | | DRB1*04:05 | Associated with risk of opticospinal MS; severity of MS progression | Asian | [39–41] |
| | HLA-DQ | DQB1*03 | Carriers of homozygous genotype for DRB1*15 and DQB1*03 reached a higher rate of disability significantly faster | European | [42] |
| | | DQB1*15:15 | Less frequently found in patients with slowly progressing MS | European | [42] |
| | | DQB1*03:03 | Associated with slower MS disease progression | European | [42] |
| HLA Class I | HLA-A | A*02 | Protective effects against MS development, as long as not carrying DRB1*15:01; independently reduces susceptibility to MS | European | [43, 44] |
| | HLA-B | B*44 | Protective factors against MS; also associated with better radiologic outcomes when the disease was present | European | [44] |

*Abbreviations: HLA* human leukocyte antigen, *MS* multiple sclerosis, *IgG* immunoglobulin G, *CSF* cerebrospinal fluid

## HLA Factors of Increased Risk

Investigations into the potential genetic heritability of MS began in the 1970s by multiple research groups that used small-sized candidate gene studies to begin to analyze genetic factors related to MS [29, 45]. Among these early studies was investigation of HLA risk factors. The strongest genetic risk factor found to correlate with MS is the *HLA-DRB1*15:01* allele [32]. Carrying a single copy of this allele confers a threefold increase in MS risk in European populations [33]. *HLA-DRB1*15:01* is generally inherited with other HLA alleles as part of an extended *DRB1*1501-DQB1*0602* haplotype [32]. A Human Genome Epidemiology (HuGE) review examined 72 papers investigating the impact of these specific genetic loci published between 1993 and 2004; the review found that in all but a limited few articles, the frequency of *DRB1*15:01* was significantly increased in cases compared to control subjects [30]. The only studies that did not find this association were carried out in non-European populations, suggesting a differential genetic susceptibility among ancestral groups [30]. The large impact of *DRB1*15:01* was also demonstrated in a GWAS, which found that this allele in the class II region of the MHC accounts for up to 10.5% of the genetic variance underlying the risk of developing MS [31]. This is the largest effect of any single factor, which we know of, that contributes to MS risk.

Patients with MS who carry the *DRB1*15:01* haplotype are more likely to be female and have an earlier age of onset; additionally, this haplotype has also been associated with oligoclonal band presence and high immunoglobulin G (IgG) levels in the cerebrospinal fluid (CSF), both of which are markers of increased CNS autoimmune activity [36, 37]. A recent next-generation sequencing study investigated the *HLA-DQB1*, *HLA-DQA1*, and *HLA-DRB1* alleles in 1403 unrelated European-American MS patients and 1425 healthy controls. This study additionally confirmed that the *HLA-DRB1*15:01* allele occurred at a significantly higher frequency in patients compared to controls [37]. The *HLA-DRB1*15:01*, *HLA-DRB5*01:01:01*, and *HLA-DRB5*01:01:01vl_STR1* alleles were all present in significantly higher frequencies in MS patients compared with controls [37]. Finally, the study also determined that when the *HLA-DRB1*15:01* haplotypes were excluded, the *HLA-DQB1*03:02:01~HLA-DQA1*03:01:01~HHLA-DRB1*04:01:01SG~HLA-DRB4*01:03:01:01* haplotypes occurred at a higher frequency in MS patients compared to controls and were all significantly associated with MS susceptibility [37]. These results illustrate that while the *HLA-DRB1*15:01* allele has a strong impact on MS susceptibility, there are other important genetic loci in the HLA region that influence MS development.

The risk of MS is further modified by other HLA factors. The *DRB1*03:01* and **13:03* alleles have been associated with an increased risk of MS. The relationship between the occurrence of **03:01* and MS was first observed in a study using patients from Sardinia; patients who carried the **13:03* allele were identified to be at a higher risk of MS development [38, 46]. A 2012 Spanish study genotyped *HLA-B, HLA-DRB1*, and *HLA-DQA1* haplotypes in a population of 1069 MS patients and 624 ethnically matched healthy control patients; the results demonstrated that *DRB1*03:01* presence within a haplotype contributed to MS susceptibility [47]. In populations with an increased frequency of MS, the *DRB1*13:03* allele is also considered to be a risk factor for MS development. The **13:03* allele is most frequently seen in Ashkenazi Jewish patients; a study focusing on this population was the first to identify the association of *DRB1*13:03* with MS with a significant effect size [48]. Building from these findings, a study investigating Israeli families impacted by MS found that there was an association of the **13:03* allele with disease frequency [49]. While the effect sizes of the **03:01* and the **13:03* HLA alleles are not as large as the **15:01* genetic loci, they demonstrate important effects on disease susceptibility in certain populations and ethnicities.

The *HLA-DRB1*15:03* allele has also been implicated in MS susceptibility and especially in those of African descent [50]. African-Americans have greater HLA haplotype diversity, as well as unique patterns of linkage disequilibrium that separates their genetic predisposition to MS from Caucasian populations [51]. A 2008 study found that while *HLA-DRB1*15:01* was still found to be an important factor in MS development in African-American populations, *HLA-DRB1*15:03* was also significantly associated with susceptibility [50]. The *HLA-DRB5*01:01* haplotype is often linked to MS risk in patients of African descent. The *HLA-DRB5* allele also has been found to attenuate MS severity; this is supported by the evidence that *HLA-DRB5*null* subjects are at an increased risk of developing secondary progressive MS [50]. This finding is supported by evidence that *HLA-DRB5* acts as a modifier in experimental autoimmune encephalomyelitis [35]. Additionally, patients who present with *HLA-DRB5*null* carrier allele status are also always in the *DRB1*15:03* haplotypes; *the HLA-DRB1*15:03* haplotype is not observed in Caucasian populations [50].

The strong impact of the *HLA-DRB1*15:01* allele has been shown to interact with other genetic risk alleles within the HLA. The *DRB1*15:01* homozygous genotype was observed to have an epistatic effect on the *DRB1*08:01* allele; the presence of the *DRB1*15:01/*08:01* heterozygous genotype had an increased risk of MS development when compared to the other heterozygous **15:01* genotypes investigated [52]. An important finding from this study was that the *DRB1*08:01* genotype was not predisposing on its own; it only contributed significantly to risk of disease onset when the **15:01* allele was present in the same genotype [52]. These findings add nuance to the understanding of the **15:01* risk alleles and susceptible genotypes for MS and demonstrate the complexity behind genetic predisposition.

## HLA Protective Factors

While the aforementioned HLA alleles are associated with an increased risk for MS development, other HLA alleles have been associated with protective effects. These associations tend to be more challenging to identify than risk factors and seem to vary more widely between different ethnic groups. In European populations, an allele commonly associated with protective effects against MS development is *DRB1*14:01* [52, 53]. A UK study observed that protection stemming from this allele is dominant, and carrying this allele outweighed the increased risk from carrying *DRB1*15:01* [53]. In a study of 4347 MS patients, investigating the genetic complexity of MHC haplotypes, the *DRB1*14:01* allele was confirmed as a protective factor, along with *DRB1*01*, and suggested a common mechanism underlying the function of both [52].

Alleles that have been further linked to protective effects against MS development are found in the class I HLA loci, including HLA A and HLA B. The *HLA-A*02* allele was linked to protection against MS in a study in Portuguese MS patients [54]. The results of this study found that the *HLA-A*02* allele decreased the risk of developing MS and that this effect was independent from the impact of *HLA-DRB1*15:01* [54]. A later study found that the *HLA-A*02* allele independently reduced susceptibility to developing MS [39]. In a Scandinavian cohort study of more than 3000 patients and healthy controls, it was observed that all *HLA-A*02*-bearing haplotypes were protective against MS development, as long as they did not carry the *HLA-DRB1*15:01* haplotype as well [40]. This study also identified a single class I haplotype, which carried *A*02-C*05-B*12*, which negated the risk of carrying *HLA-DRB1*15:01*, demonstrating a complex role of class I alleles in MS risk and development [40].

The same study that demonstrated supporting evidence for the protective role of *HLA-A*02* also identified the *HLA-B*44* allele as a protective factor and found that it was associated with better radiologic outcomes when the disease was present [39]. These brain-related outcomes include better brain parenchymal function and decreased T2 hyperintense lesion volume [39]. Both the *HLA-A*02* and *HLA-B*44* alleles were assessed after accounting for the effect of *HLA-DRB1*15:01*.

A study that investigated *HLA-DRB1* and *HLA-DQB1* alleles in 120 Iranian MS patients found that the *DRB1*10:01* allele was negatively correlated with MS occurrence, suggesting a potential protective effect against the disease, but the sample size was very modest [55]. While a proportion of the healthy controls carried this gene, none of the MS patients did [55]. A separate case-control study in the Slovak population identified the *DRB1*07, DRB1*13*, and *DRB1*03* alleles to be protective against MS development, along with the *DRB1*13-DQB1*06* and *DRB1*11-DQB1*03* haplotypes [56]. Part of these results was replicated in a study using a Finnish cohort, which found the *DRB1*13~DQB1*06* haplotype to have protective properties against MS [57]. Two independent investigations from Brazil and Canada found protective effects of the *DRB1*11* allele, signifying an important multicultural correlation for this specific loci [58, 59]. The results of multiple studies investigating protective HLA factors against MS development have demonstrated that increased, region-specific research is needed in order to fully identify different factors.

In total, 32 loci within the MHC region have been confirmed to have association with increased or decreased risk to have MS [7]. While these loci represent a substantial proportion of MS heritability, the rest of the currently identified factors reside within the autosomal non-MHC genome.

## Genome-Wide Association Studies: Non-HLA Risk Factors

Although early studies of non-HLA genetic risk factors identified a few alleles of interest, the results were not initially reliably replicated, largely due to modest sample sizes. In 2007, a study by Fisher et al. demonstrated that MS is likely characterized by a diverse spectrum of risk allele frequencies underlying susceptibility—a theory termed the polygenic model of MS [60]. The risk in evidential support for the polygenic model of MS genetic heritability gave the theoretical support necessary to begin intensive genome-wide studies that utilized large datasets to effectively screen thousands of single nucleotide polymorphisms (SNPs) [61]. These GWAS were made possible through technological advancements and the development of novel methodological approaches such as the SNP array and have provided the foundation for later advances in our understanding of the genetic heritability of MS. GWAS are conducted by comparing the allele frequency at each position in the genome between case and control subjects, assuming an additive model of genetic effects, and a significant difference between the two highlights a genetic association to MS risk [62]. GWAS research investigations are agnostic, or hypothesis-free, and target the entire genome by tagging linkage disequilibrium (LD) blocks; this means that they require large sample sizes of participants, but the sample is not restricted to family-based relationships. Case-control samples can be used. Over time, statistical thresholds have been adopted in order to account for multiple comparisons and to ensure reproducibility in independent studies [63]. The current standard for genome-wide significance level is $P < 5*10^{-8}$, which is equivalent to $P < 0.05$ after the Bonferroni correction for the number of independent tests in the entire genome, given LD between known variants [63].

The first GWAS targeted non-HLA elements of MS and was completed by the International Multiple Sclerosis Genetics Consortium (IMSGC) in 2007 [64]. Using 1540 parent-affected offspring trios, this study was able to identify two genetic loci outside of the HLA genes that had previously not been implicated in MS: the interleukin-7 receptor (IL-7R) and the interleukin-2 receptor (Il-2R) [64]. Both of these genetic loci are linked to cytokine receptors, and replication studies have provided evidence for a causal link between these two receptors and MS pathogenesis [65–68]. The gene product of IL-7R, the interleukin-y receptor alpha chain, forms a complex with a gamma chain cytokine receptor that is critical for the proliferation and survival of T and B lymphocytes in the immune system [66, 69]. Genetic aberrations in this complex have been demonstrated to lead to immune deficiency syndromes [70]. An mRNA assay study using peripheral blood mononuclear cells (PBMCs) in MS patients found that there was an increased expression of the IL7 signaling pathway in the CSF of patients compared to controls [66]. This suggested that the disruption of this signaling pathway in the CNS could partially account for the onset of MS. Polymorphism in the IL-7R gene was classified as a significant risk factor for MS, as identified by four independent case-control and family-based datasets [65]. Additionally, the IL-2R gene, which was also implicated in MS through the preliminary GWAS, encodes a subunit of a receptor for the pro-inflammatory cytokine interleukin 2 (IL2); this cytokine has been associated with multiple autoimmune diseases in addition to MS, including rheumatoid arthritis and type 1 diabetes [64, 71, 72]. IL-2R helps to maintain suppressive functions of a T-cell subtype, which in turn facilitates effector and memory T-cell differentiation and plays an important role in immune system function [73]. Furthermore, the IL-2R gene is regulated by vitamin D in CD4+ T-cells, implicating a known environmental factor in MS development in genetics [71].

After the first GWAS was completed and the roles of IL-7R and IL-2R in MS were identified, the results were quickly replicated by independent studies; the buildup of investigation surrounding GWAS and heritability in MS led to a marked increase in genetic studies and sample sizes [65, 66]. In order to create a larger subject pool and achieve greater statistical power, the IMSGC expanded to include approximately 23 research groups from 15 countries. By 2010, a large series of GWAS and meta-analyses made possible by this expansion increased the number of confirmed genetic loci associated with MS to 26 [74]. While the study was able to maximize its subject pool by using case-controls instead of family-based, this method created confounding such as population stratification that limits interpretation of results [74]. To control for this, a novel approach called the variance component method was used to adjust for this genomic inflation bias [74]. Another GWAS was conducted by the IMSGC a few years later, which identified 52 loci that were definitely associated with the risk of developing MS, including 29 completely novel loci [75]. Of these identified genes, 23 were previously known to be involved in other autoimmune diseases; this suggested that a common mechanism may underlie several autoimmune disorders [75]. A GWAS in 2013 brought the number of established MS-associated genetic loci up to 110, with 103 of these loci existing outside of the HLA [76]. The most recent GWAS, conducted in 2017 by the IMSGC, included a sample size of more than 115,000 case and control subjects [77]. This study found that the total number of MS risk-associated genes is approximately 233, with 200 of these loci non-HLA related and 32 HLA-related [77]. The speed with which the multiple consecutive GWAS were able to identify increasingly large numbers of genetic loci outside of the HLA that are involved in this multifaceted disease is a testament to this study design in our understanding of MS susceptibility.

Pediatric-onset MS may represent an extreme example of genetic susceptibility given the young age of disease development. To date, many of the adult susceptibility alleles have been confirmed in this population, although some with potentially higher effect sizes than seen in adults [78]. MHC class III variants were also associated with MS risk in children [78]. A caveat of these studies is that sample sizes are limited in this more rare MS population.

A strength of the GWAS was their incredibly large datasets that generated important connections between genetic loci and disease prevalence, but they are not without limitations. LD is the idea that the regions where SNPs exist as indicated by GWAS are not overly specific; these regions can be expansive sections of the genome that include several genes that could potentially be implicated in MS [79]. Additionally, incredibly large sample sizes were required to detect even marginal genetic associations with minor effects, under any inheritance model; this means that the era of GWAS has mostly come to an end, due to the increasing statistical need for datasets of larger sizes. Follow-up experiments have helped to refine our understanding of the conclusions reached by GWAS in the past decade and look to new methodologies.

The conclusion from sustained GWAS investigations into MS has been that the loci associated with MS are generally related to T-cell activation and proliferation in the immune system [80, 81] and that MS susceptibility genes are enriched in all immune cell types [7]. Transcriptomic and epigenetic enrichment analyses demonstrate that T-cell biology is a major feature of MS but stops short of naming it the key characteristic of the disease; alternatively, it is noted that there are many active components in MS outside of T-cell activity, including adaptive and innate immunity in pathogenesis [10, 80–82]. However, this picture of the genetics of MS is not complete, and statistical analyses have demonstrated that there are potential remaining genetic factors in addition to the environmental factors that remain to be elucidated. Susceptibility studies have demonstrated that while MS has important genetic components, and these implicate the immune system as the origin of the disease, it is not an entirely heritable disease.

## Non-Caucasian Populations

The majority of research conducted on the genetics of MS has included participants with European ancestry; ethnic minority populations have been underrepresented in heritability research [83]. While historically MS has been reported to have the highest prevalence in Caucasians with Northern European ancestry, it has also been consistently reported in most ethnic groups [84]. In order to untangle the complex genetic interactions that lead to MS, it is imperative to investigate the genetic differences between populations that lead to the same disease.

When compared to Caucasian populations, people with African ancestry have a statistically smaller risk of MS development; however, if they do develop MS, they may be at risk for faster disease progression and increased disability [85]. These observations implicate genetic factors as potential modifiers of the differences in MS risk and phenotype in different populations, but environmental factors may also notably contribute.

The MHC class II *HLA-DRB1*15* and *HLA-DRB1*15:03* alleles are susceptibility factors in populations of African descent. As mentioned in a previous section, African-Americans have greater HLA haplotype diversity, in addition to patterns of LD that changes their genetic predisposition to MS as compared to Caucasian populations [51]. The *HLA-DRB1*15:03* allele has been significantly associated with MS susceptibility in cohorts with African descent [50]. The class II *HLA-DRB*15* alleles have been observed to indicate severity of disease progression in this cohort [50, 86]. A study investigating differences in MS clinical outcomes in 673 African-American and 717 European American participants demonstrated that African ancestry of HLA correlated with an earlier age of onset and increased disability, as measured by the Multiple Sclerosis Severity Score (MSSS) and cane dependency [87].

In order to investigate the MS susceptibility genetic profile in African-Americans, a replication study investigated allele frequencies of 19 SNPs in 12 MHC genetic loci in 918 patients and 656 controls [88]. The results showed both HLA and non-HLA factors were associated with MS risk; it also demonstrated consistent findings of SNP associations compared to the literature in Caucasian populations, showing common immunological mechanisms for MS development [88]. A follow-up study was conducted a few years later, which used an expanded population size of 1162 cases and 2092 controls to assess the association of MS risk variants in a population with African ancestry [89]. This study, using a significance threshold of $p < 0.01$, found none of the *HLA-DQB1* alleles were significantly associated with MS; conversely, researchers identified eight SNPS associated with MS outside of the MHC [89]. While research has suggested there are not large differences in HLA risk factors among African and Caucasian populations, more modest sample sizes in these studies limit the identification of factors that may be rare in the population or have small effect sizes.

In addition to those with European or African ancestry, there are distinct patterns of MS disease diagnosis and course in those with Asian genetic ancestry. A Japanese study found an association of *DRB1*15* alleless with a more typical course of MS; if a patient had no presence of this allele, they were more likely to have the clinically distinct opticospinal MS [90]. Opticospinal MS in Asian patients was associated with the risk variant *DRB1*04:05*; this allele was linked to a disease course characterized by an earlier age of onset, reduced severity, and a statistically smaller number of brain lesions compared to similar studies in Caucasian populations [90, 91]. However, these studies predate research surrounding neuromyelitis optica (NMO) and myelin oligodendrocyte (MOG) antibodies. A meta-analysis investigated the association between HLA genes and MS in Chinese populations and found that the *HLA-DR2.DRB1*15* haplotypes were associated with risk of MS in Chinese patients but to a lesser extent than in Western MS population [92]. Additionally, this study identifies *HLA-DR9* alleles as conferring resistance to MS in this population [92]. Wang et al. conducted a whole-exome sequencing study on MS patients from Southern China, in an attempt to identify genetic variants in a Chinese population compared to Caucasian populations [92]. They found 17 variants that were previously unreported in the literature as being related to MS that were shown to have significantly different frequencies between the MS patients and healthy controls, specifically a rare variant located on exon 7 of TRIOBP [92]. While this research has helped to fill in our understanding of the genetics of MS in patients with Asian ancestry, the small sample size ($n$ = 8) means that the study will have to be further supported by continuing investigation.

## Multiple Sclerosis Genetic Overlap with Other Autoimmune Diseases

Multiple sclerosis has been characterized as an autoimmune disorder, a definition that has been supported by the implication of HLA genetic loci in its etiopathogenesis. Autoimmune diseases, or AIDs, are known to have some amount of similar implicated biological mechanisms that lead to an overactive immune response (Fig. 11.1). In MS, the largest single risk factor is the presence of the *HLA DRB1*15:01* allele. As described previously, this allele follows what is called the additive model; the more copies of the allele present in the genome, the more genetically susceptible an individual is to developing MS [53, 93]. This phenomenon is also seen in other autoimmune diseases, such as celiac disease, narcolepsy, and type 1 diabetes; in all of these cases, homozygosity for a gene increases risk in comparison to heterozygous genetic groups [41, 42, 94]. The similarity between dosage effects in HLA genes of known autoimmune diseases and MS strongly suggests that MS arises primarily from an abnormal peripheral immune system and provides the theoretical foundation for further parallels between MS and other AIDs.

GWAS have provided researchers with the ability to use large sample sizes in order to unravel some of the genetic components of complex diseases, like MS [46]. The results from these studies demonstrate that MS-related genes overlap notably with genetic loci that have previously been associated with many other AIDs [95]. GWAS demonstrate that just over one-third of the known MS risk loci overlapped with regions that have previously been associated with other AIDs, including celiac disease, type 1 diabetes, and rheumatoid arthritis [72].

Autoimmune diseases, such as MS, are incredibly complex; the polygenic mechanisms underlying each individual pathology remain challenging to fully elucidate as the field is only beginning to reach sample sizes that allow for studies of gene-gene interactions. In addition to genes underlying risk factors present in multiple AIDs, there are also genes that act as protective factors for some AIDs but risk factors for others. Sirota et al. compared genetic profiles of six autoimmune diseases, including MS, and five non-autoimmune diseases, in order to identify SNPs that are both protective and susceptibility factors [96]. The study identified two broad classes of autoimmune diseases, where specific MHC SNPs that make an individual susceptible to one protect against the other. Rheumatoid arthritis and ankylosing spondylitis belong to one class, while MS and autoimmune thyroid diseases belong to the other, and they are differentiated by certain MHC polymorphisms [96]. The

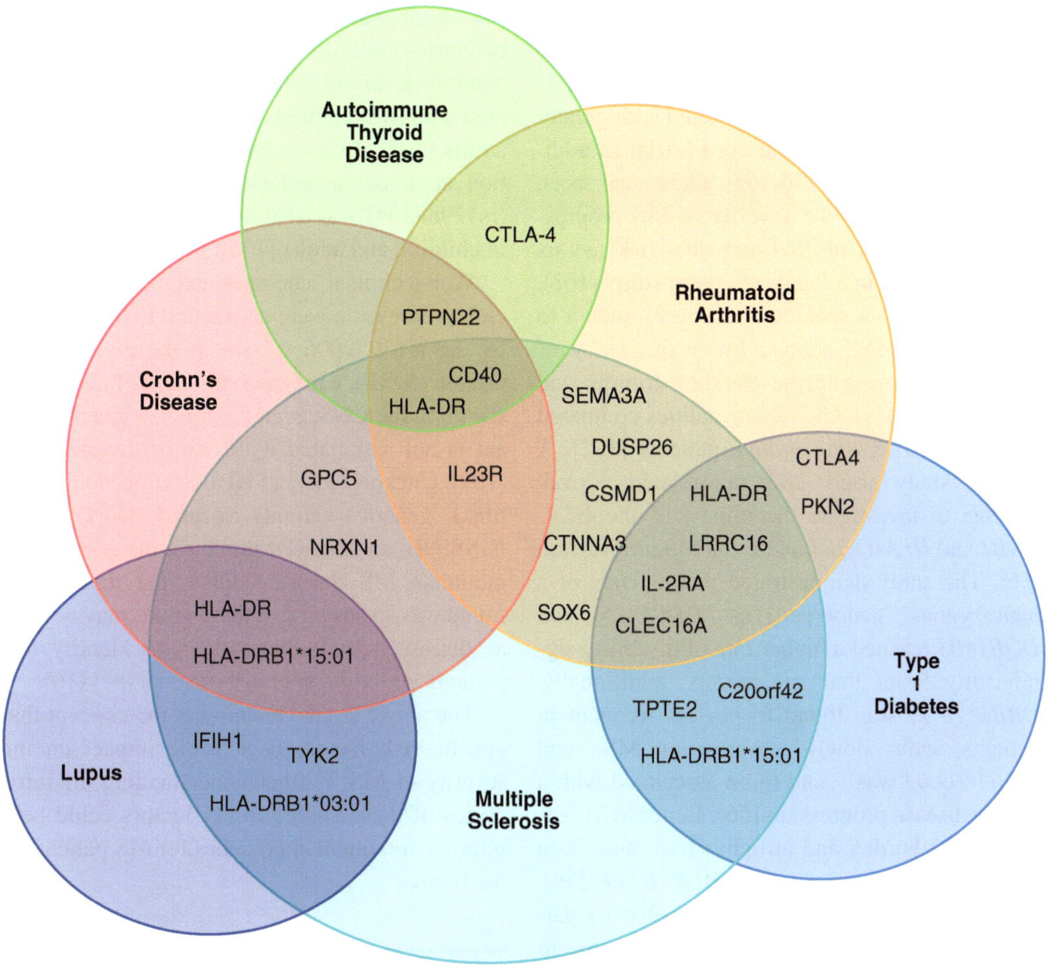


**Fig. 11.1** Examples of multiple sclerosis genetic overlap with other autoimmune diseases

study hypothesized that the SNP alleles associated with different autoimmune phenotypes may interact differently with both environmental and other genetic factors, thereby changing the biological context of the SNP in different people [96]. Further investigation is necessary to understand the relationships between these AIDs more clearly.

## Susceptibility Factors and MS Disease Course

In a multifaceted disease such as MS, there are a multitude of both environmental and genetic risk factors (as well as interactions among these) that may impact not only disease onset but also phenotype. Preserving function and day-to-day quality of life is an especially important driver of medical standard of care. In order to slow the progression of MS as best as possible, it is important to continue investigating the impact of different susceptibility factors on disease course.

The HLA is one of the most important sites of genetic heritability for MS diagnosis and may have importance also for phenotype. Carrying *HLA-DRB1*15:01* has been associated with earlier age of MS [97, 98]. The HLA alleles *HLA-DRB1*15:01*, *HLA-DQB1*03:01*, *HLA-DQB1*03:02*, *HLA-DQB1*06:02*, and *HLA-DQB1*06:03* have been associated with more severe MS progression, as indicated by patient magnetic resonance imaging (MRI) scans and evidence of increased inflammation and neurodegeneration [99, 100]. The presence of *HLA-DRB1*15:01* may also be associated with greater lesion burden in the spinal cord [101]. In addition, the *HLA-DRB1*04:05* allele has been shown to influence the severity of MS progression, independent of environmental risk factors such as latitude, in a Japanese cohort study [102]. Patients with this specific allele were shown to have lower MSSS scores, lower frequency of brain lesions meeting the Barkhof criteria, and decreased levels of CSF abnormalities compared to *HLA-DRB1*0405-negative* patients [102]. A similar study used 282 non-related Slovak patients to investigate the impact of the *HLA-DRB1* and *HLA-DQB1* alleles on MS progression [56]. The study demonstrated that carriers of a homozygous genotype for *DRB1*15* and *DQB1*03* reached a higher rate of disability significantly faster than noncarriers; additionally, *DRB1*15:15* was found to be less frequent in patients with slowly progressing MS, and *DQB1*03:03* was found to be associated with a slower disease progression [56]. Better MRI outcomes (T2 burden and atrophy) have also been observed in those carrying *HLA B*44* [39]. Genetic loci within the HLA may also be partially responsible for an individual's response to several disease-modifying therapies (DMTs). Fifteen specific HLA alleles were found to have modifying effects on DMTs, all of which led to a reduction in the MS severity score (MSSS), a type of disability status calculation for the severity of MS [103].

In addition to HLA-related genetic susceptibility factors impacting MS disease progression, non-HLA factors have been associated with disease course [104, 105]. In a set of approximately 800 participants, modest associations were found between some allelic variants outside the HLA and MS phenotype (age at onset, MSSS, brain parenchymal volume, and T2 lesion load), but many of these did not reach genome-wide significance [104]. A genetic risk or burden score of non-HLA susceptibility alleles has been associated with age at onset and the presence of oligoclonal bands [106]. A similar unweighted genetic risk score was not associated with relapses in pediatric-onset MS [107]. The non-HLA MS susceptibility variant within the *AH1* gene is associated with MS relapse hazard in both children and adults [108]. In a genome-wide analysis, variation at a loci associated with the gene *LRP2* (rs12988804) was also associated with relapses in children and adults [105].

Worse clinical outcomes and more severe disease progression were also found to be predicted by the rs12959006 variant in the myelin basic protein (MBP), a major component of the myelin sheath in the CNS, even though this genetic variant is not associated with risk of disease onset [109]. Unexpectedly, a SNP detection study identified genetic variants located in CPXM2, IGSSF9B, and NLRP9 that have the potential to modulate MS disease course; the relationship found was so strong that these genes may be used as disease activity biomarkers to identify MS patients with divergent disease courses [110].

The above research supports the concept that genetic risk factors have some impact on the severity of MS. Further understanding the intricacies of these susceptibility factors could help improve treatment approaches for MS patients in the future.

## Mendelian Randomization: Support for Environmental Risk Factors

Mendelian randomization studies offer the ability to study environmental risk factors with minimal concern for reverse causation. This kind of study assesses the impact of genetically determined variation in a biological pathway marker on a disease, which helps scientists to support or reject the pathway's hypothesized causal role in disease etiology [5]. Mendelian randomization studies are ecologically valid and natural studies that attempt to emulate randomized control trials, the gold standard for assessing causality in a relationship. The genetic alleles driving an environmental risk factor (cards dealt at birth before disease onset) can be used as a proxy for the risk factor in the analysis. Being upstream of the final risk factor level and non-mutable by the disease itself or other confounding environmental factors, these alleles greatly reduce concerns of reverse causation and other confounding.

## Vitamin D

In research cohorts across the world, people with MS exhibit lower vitamin D levels than healthy controls [111, 112]. In those with MS, individuals with more disability or higher relapse rates have lower vitamin D levels than those with less severe disease [113]. These observations may in part explain the historical observation that MS prevalence is greater in higher latitude areas with less sunlight [114]. Vitamin D has been demonstrated to have direct effects on the immune system, making it a plausible risk factor, but concerning among many of these studies is the possibility that the observation is explained by reverse causation—that those with MS-related disability venture outside less, causing low vitamin D levels.

In order to understand vitamin D mechanisms more thoroughly and how they relate to MS, Mendelian randomization studies have been undertaken. The study by Mokry et al. used almost 34,000 subjects to demonstrate that alleles of vitamin D metabolism genes caused genetically determined variation in vitamin D status; this means that vitamin D level is not just linked to sun exposure but also to baseline genetics [103]. The study found that individuals with genetically determined vitamin D deficiency had more than a 10x risk of developing MS compared to those with the highest quartile of genetically determined vitamin D levels [103]. A GRS of three variants associated with vitamin D level is associated with increased risk of pediatric-onset MS [115]. These studies demonstrate a causal link between the genetically determined vitamin D levels and the pathology of MS.

Mendelian randomization can also be used to assess associations with phenotype. In a cohort of approximately 200 pediatric participants with MS, a vitamin D genetic risk score was associated with relapse rate, demonstrating that those with highest risk for low vitamin D levels had 2.6-fold higher relapse rates [116]. These results were replicated in an independent cohort of children with MS [116].

## Obesity

In addition to vitamin D, the process of Mendelian randomization has also been used in order to determine the causality of the relationship of obesity with MS risk and progression. The justification for using this type of investigation is similar to the reasoning behind vitamin D studies: in order to establish causality and understand the nature of the relationship between obesity and MS in a naturalistic setting with unmeasured confounders.

A growing body of evidence has demonstrated that obesity has some level of impact on MS onset and progression. A body mass index (BMI) of more than 30 $kg/m^2$ during late adolescence and early adulthood is associated with a twofold increase in MS risk in women compared to those with a healthy weight [117]. Obesity in childhood has also been correlated with the diagnosis of MS in pediatric- and later-onset patients [118, 119].

In order to better understand the relationship of obesity and MS, Mendelian randomization studies were conducted. A 2016 investigation used a two-sample MR approach with summary statistics from both the Genetic Investigation of Anthropometric Traits (GIANT) consortium and the International MS Genetics Consortium (IMSGC); the study weighted the effect of SNPs on MS by their effect on BMI. The large collaborative effort found that genetically elevated BMI was causally associated with risk of MS development [120]. A separate-sample MR study using an international dataset found a causal effect of increased BMI on susceptibility of MS using 19 established variants that predict BMI [43]. This study also identified the fat mass and obesity-associated gene, FTO, to be implicated in MS; specifically, FTO alone increased the risk of MS. FTO has been previously associated with cancers, Alzheimer's disease, dementia, and cognitive decline in healthy adults [44, 121–123].

These results suggest that not only does BMI have an indirect impact on MS through inflammatory pathways, but there are characteristics of a genetically elevated BMI that directly increase disease risk. BMI-related genes are also associated with increased risk of pediatric-onset MS [115].

Obesity has also been shown to play a role in vitamin D levels in the body, thereby demonstrating an indirect impact that an increased BMI may have on MS [124]. A MR study investigated a proposed theory that lower vitamin D bioavailability in obese patients constitutes the biological mechanism through which obesity impacts MS [124, 125]. The study found that for every unit increase in BMI, there was a reduced 25OHD (vitamin D) concentration in the body by 1.15% [124]. A separate study importantly found that both vitamin D and obesity genetic variants have independent impacts on disease risk [115]. The results of these Mendelian randomization studies demonstrate important variables in an individual's risk for developing MS and have shown that BMI has both indirect and direct impacts on MS diagnosis and progression.

## Conclusion

The aforementioned research has clearly demonstrated that MS is a partially heritable disease. Genetic factors both within and outside the HLA region are associated with MS risk, the greatest influence of which is from carrying the *HLA-DRB1*15:01* allele. GWAS studies over the last decade have captured the strong genetic similarities of MS with other autoimmune diseases, suggesting MS is an "outside in" disease. More work is needed to understand the relationship of genetic factors with disease course, requiring more consistency across research cohorts in how investigators phenotype the disease. Mendelian randomization studies have grown in prevalence as a methodology with which to analyze the impact of environmental factors leveraging genetic drivers of those factors to minimize bias and reverse causation forms of confounding. With the larger sample sizes now available, it will also become possible to better study epistatic effects among susceptibility factors as well as begin to study critical gene-environment interactions that likely lead to onset of disease.

## References

1. Compston A, Coles A. Multiple Sclerosis. Lancet. 2008;372(9648):1502–17.
2. Hauser SL, Goodwin DS. Multiple sclerosis and other demyelinating diseases. In: Kasper D, Fauci A, Hauser S, Longo D, Jameson LJ, Loscalzo J, editors. Harrison's principles of internal medicine. 19th ed. New York: McGaw-Hill Education/Medical; 2015. p. 3395–409.
3. Hedstrom AK, Alfredsson L, Olsson T. Environmental factors and their interactions with risk genotypes in MS susceptibility. Curr Opin Neurol. 2016;29(3):293–8.
4. Hedstrom AK, Olsson T, Alfredsson L. Smoking is a major preventable risk factor for multiple sclerosis. Mult Scler. 2016;22(8):1021–6.
5. Lawlor DA, Harbord RM, Sterne JA, Timpson N, Davey Smith G. Mendelian randomization: using genes as instruments for making causal inferences in epidemiology. Stat Med. 2008;27:1133–63.
6. International Multiple Sclerosis Genetics Consortium; Welcome Trust Case Control Consortium 2, Sawcer S, Hellenthal G, Pirinen M, Spencer CCA, Patsopoulos NA, Moutsianas L, et al. Genetic risk and a primary role for cell-mediated immune mechanisms in multiple sclerosis. Nature. 2011;476:214–9.
7. International Multiple Sclerosis Genetics Consortium. Multiple sclerosis genomic map implicates peripheral immune cells and microglia in susceptibility. Science. 2019;365(6460):eaav7188.
8. Boomsma D, Busjahn A, Peltonen L. Classical twin studies and beyond. Nat Rev Genet. 2002;3(11):872–82. https://doi.org/10.1038/nrg932.
9. Sadovnick AD, Baird PA. The familial nature of multiple sclerosis: age-corrected empiric recurrence risks for children and siblings of patients. Neurology. 1988;38(6):990–1.
10. Willer CJ, Dyment DA, Risch NJ, Sadovnick AD, Ebers GC, Canadian Collaborative Study Group. Twin concordance and sibling recurrence rates in multiple sclerosis. Proc Natl Acad Sci U S A. 2003;100(22):12877–82.
11. Heltberg A, Holm NV. Concordance in twins and recurrence in siblings in multiple sclerosis. Lancet. 1982;1:1068.
12. Ebers GC, Sadovnick AD, Risch NJ. A genetic basis for familial aggregation in multiple sclerosis. Canadian Collaborative Study Group. Nature. 1995;377(6545):150–1.
13. Ebers GC, Yee IM, Sadovnick AD, Duquette P. Conjugal multiple sclerosis: population-based

prevalence and recurrence risks in offspring. Canadian Collaborative Study Group. Ann Neurol. 2000;48(6):927–31.
14. Bulman DE, Sadovnick AD, Ebers GC. Age of onset in siblings concordant for multiple sclerosis. Brain. 1991;114(pt 2):937–50.
15. Doolittle TH, Myers RH, Lehrich JR, Birnbaum G, Sheremata W, Franklin GM, et al. Multiple sclerosis sibling pairs: clustered onset and familial predisposition. Neurology. 1990;40(10):1546–52.
16. Fagnani C, Neale MC, Nistico L, Stazi MA, Ricigliano VA, Buscarinu MC, et al. Twin studies in multiple sclerosis: a meta-estimation of heritability and environmentality. Mult Scler. 2015;21(11):1404–13.
17. Graves JS, Chitnis T, Weinstock-Guttman B, Rubin J, Zelikovitch AS, Nourbakhsh B, Network of Pediatric Multiple Sclerosis Centers, et al. Maternal and perinatal exposures are associated with risk for pediatric-onset multiple sclerosis. Pediatrics. 2017;139:e20162838.
18. Sadovnick AD, Armstrong H, Rice GP, Bulman D, Hashimoto L, Paty DW, et al. A population-based study of multiple sclerosis in twins: update. Ann Neurol. 1993;33(3):281–5.
19. Mumford CJ, Wood NW, Kellar-Wood H, Thorpe JW, Miller DH, Compston DA. The British Isles survey of multiple sclerosis in twins. Neurology. 1994;44(1):11–5.
20. Küçükali Cİ, Kürtüncü M, Çoban A, Çebi M, Tüzün E. Epigenetics of multiple sclerosis: an updated review. NeuroMolecular Med. 2015;17:83–96.
21. Madsen PM, Pinto M, Patel S, McCarthy S, Gao H, Taherian M, et al. Mitochondrial DNA double-strand breaks in oligodendrocytes cause demyelination, axonal injury, and CNS inflammation. J Neurosci. 2017;37(42):10185–99.
22. Mahad D, Lassmann H, Turnbull D. Review: mitochondria and disease progression in multiple sclerosis. Neuropathol Appl Neurobiol. 2008;34:577–89.
23. Ban M, Elson J, Walton A, Turnbull D, Compston A, Chinnery P, Sawcer S. Investigation of the role of mitochondrial DNA in multiple sclerosis susceptibility. PLoS One. 2008;3(8):e2891.
24. Tranah GJ, Santaniello A, Caillier SJ, D'Alfonso S, Martinelli Boneschi F, Hauser SL, Oksenberg JR. Mitochondrial DNA sequence variation in multiple sclerosis. Neurology. 2015;85:325–30.
25. Nepom GT. The major histocompatibility complex. In: Jameson J, Fauci AS, Kasper DL, Hauser SL, Longo DL, Loscalzo J, editors. Harrison's principles of internal medicine, vol. 20e. New York: McGraw-Hill; 2018.
26. Murphy K, Travers P, Walport M, Janeway C. Janeway's immunobiology. New York: Garland Science; 2012.
27. Marsh SG, Albert ED, Bodmer WF, Bontrop RE, Dupont B, Erlich HA, et al. Nomenclature for factors of the HLA system, 2010. Tissue Antigens. 2010;75(4):291–455.
28. Bertrams J, Kuwert E, Liedtke U. HL-A antigens and multiple sclerosis. Tissue Antigens. 1972;2:405–8.
29. Naito S, Namerow N, Mickey MR, Terasaki PI. Multiple sclerosis: association with HL-A3. Tissue Antigens. 1972;2:1–4.
30. Schmidt H, Williamson D, Ashley-Koch A. HLA-DR15 haplotype and multiple sclerosis: a HuGE review. Am J Epidemiol. 2007;165:1097–109.
31. The International Multiple Sclerosis Genetics Consortium & The Wellcome Trust Case Control Consortium 2. Genetic risk and a primary role for cell-mediate immune mechanisms in multiple sclerosis. Nature. 2011;476:214–9.
32. Hollenbach JA, Oksenberg JR. The immunogenetics of multiple sclerosis: a comprehensive review. J Autoimmun. 2015;64:13–25.
33. Canto E, Oksenberg JR. Multiple sclerosis genetics. Mult Scler. 2018;24(1):75–9.
34. Hensiek AE, Sawcer SJ, Feakes R, Deans J, Mander A, Akesson E, et al. HLA-DR 15 is associated with female sex and younger age at diagnosis in multiple sclerosis. J Neurol Neurosurg Psychiatry. 2002;72:184–7.
35. Khare M, Mangalam A, Rodriguez M, David CS. HLA DR and DQ interaction in myelin oligodendrocyte glycoprotein-induced experimental autoimmune encephalomyelitis in HLA class II transgenic mice. J Neuroimmunol. 2005;169:1–12.
36. Celius EG, Harbo HF, Egeland T, Vartdal F, Vandivik B, Spurkland A. Sex and age at diagnosis are correlated with the HLA-DR2, DQ6 haplotype in multiple sclerosis. J Neurol Sci. 2000;178:132–5.
37. Creary LE, Mallempati KC, Gangavarapu S, Caillier SJ, Oksenberg JR, Fernandez-Vina MA. Deconstruction of HLA-DRB1*04:01:01 and HLA-DRB1*15:01:01 class II haplotypes using next-generation sequencing in European-Americans with multiple sclerosis. Mult Scler. 2018;25(6):772–82.
38. Marrosu MG, Murru MR, Costa G, Cucca F, Sotgiu S, Rosati G, et al. Multiple sclerosis in Sardinia is associated and in linkage disequilibrium with HLA-DR3 and –DR4 alleles. Am J Hum Genet. 1997;61:454–7.
39. Healy BC, Liguori M, Tran D, Chitnis T, Glanz B, Wolfish C, et al. HLA B*44: protective effects in MS susceptibility and MRI outcome measures. Neurology. 2010;75(7):634–40.
40. Link J, Kockum I, Lorentzen ÅR, Lie BA, Celius EG, Westerlind H, et al. Importance of human leukocyte antigen (HLA) class I and II alleles on the risk of multiple sclerosis. PLoS One. 2012;7(5):e36779.
41. Van der Heide A, Verduijn W, Haasnoot GW, Drabbels JJ, Lammers GJ, Claas FH. HLA dosage effect in narcolepsy with cataplexy. Immunogenetics. 2015;67:1–6.
42. Noble JA, Erlich HA. Genetics of type 1 diabetes. Cold Spring Harb Perspect Med. 2012;2:a007732.
43. Gianfrancesco MA, Glymour MM, Walter S, Rhead B, Shao X, Shen L, et al. Causal effect of genetic variants associated with body mass index on mul-

tiple sclerosis susceptibility. Am J Epidemiol. 2017;185:162–71.
44. Kaklamani V, Yi N, Sadim M, Siziopikou K, Zhang K, Xu Y, et al. The role of the fat mass and obesity associated gene (FTO) in breast cancer risk. BMC Med Genet. 2011;12:52.
45. Jersild C, Svejgaard A, Fog T. HL-A antigens and multiple sclerosis. Lancet. 1972;1:1240–1.
46. Cocco E, Sardu C, Pieroni E, Valentini M, Murru R, Costa C, et al. HLA-DRB1-DQB1 haplotypes confer susceptibility and resistance to multiple sclerosis in Sardinia. PLoS One. 2012;7:e33972.
47. de la Concha EG, Cavanillas ML, Cenit MC, Urcelay E, Arroyo R, Fernandez O, et al. DRB1*03:01 haplotypes: differential contribution to multiple sclerosis risk and specific association with the presence of intrathecal IgM bands. PLoS One. 2012;7:e31018.
48. Kwon OJ, Karni A, Israel S, Brautbar C, Amar A, Meiner Z, et al. HLA class II susceptibility to multiple sclerosis among Ashkenazi and non-Ashkenazi Jews. Arch Neurol. 1999;56:555–60.
49. Karni A, Kohn Y, Safirman C, Abramsky O, Barcellos L, Oksenberg JR, et al. Evidence for the genetic role of human leukocyte antigens in low frequency DRB1*1501 multiple sclerosis patients in Israel. Mult Scler. 1000;5:410–5.
50. Caillier SJ, Briggs F, Cree BA, Baranzini SE, Fernandez-Vina M, Ramsay PP, et al. Uncoupling the roles of HLA-DRB1 and HLA-DRB5 genes in multiple sclerosis. J Immunol. 2008;181(8):5473–80.
51. Howson JM, Roy MS, Zeitels L, Stevens H, Todd JA. HLA class II gene associations in African American type 1 diabetes reveal a protective HLA-DRB1*03 haplotype. Diabet Med. 2013;30(6):710–6.
52. Dyment DA, Herrera BM, Cader MZ, Willer CJ, Lincoln MR, Sadovnick AD, et al. Complex interactions among MHC haplotypes in multiple sclerosis: susceptibility and resistance. Hum Mol Genet. 2005;14:2019–26.
53. Barcellos LF, Sawcer S, Ramsay PP, Baranzini SE, Thomson G, Briggs F, et al. Heterogeneity at the HLA-DRB1 locus and risk for multiple sclerosis. Hum Mol Genet. 2006;15:2813–24.
54. Silva AM, Bettencourt A, Pereira C, Santos E, Carvalho C, Mendonça D, Costa PP, Monteiro L, Martins B. Protective role of the HLA-A*02 allele in Portuguese patients with multiple sclerosis. Mult Scler. 2009;15(6):771–4.
55. Kollaee A, Ghaffarpor M, Ghlichnia HA, Ghaffari SH, Zamani M. The influence of the HLA-DRB1 and HLA-DQB1 allele heterogeneity on disease risk and severity in Iranian patients with multiple sclerosis. Int J Immunogenet. 2012;39(5):414–22.
56. Cierny D, Lehotsky J, Kantorova E, Sivak S, Javor J, Kurca E, et al. The HLA-DRB1 and HLA-DQB1 alleles are associated with multiple sclerosis disability progression in Slovak population. Neurol Res. 2018;40(7):607–14.
57. Laaksonen M, Pastinen T, Sjoroos M, Kuokkanen S, Ruutiainen J, Sumelahti ML, et al. HLA class II associated risk and protection against multiple sclerosis – a Finnish family study. J Neuroimmunol. 2002;122:140–5.
58. Kaimen-Maciel DR, Reiche EM, Borelli SD, Morimoto HK, Melo FC, Lopes J, et al. HLA-DRB1* allele-associated genetic susceptibility and protection against multiple sclerosis in Brazilian patients. Mol Med Rep. 2009;2:993–8.
59. Ramagopalan SV, Morris AP, Dyment DA, Herrera BM, DeLuca GC, Lincoln MR, et al. The inheritance of resistance alleles in multiple sclerosis. PLoS Genet. 2007;3:1607–13.
60. Fisher RA. The correlations between relatives on the supposition of Mendelian inheritance. Trans R Soc Edinburgh. 2009;52:399–433.
61. Sawcer S. The complex genetics of multiple sclerosis: pitfalls and prospects. Brain. 2008;131(Pt 12):3118–31.
62. Cotsapas C, Mitrovic M. Genome-wide association studies of multiple sclerosis. Clin Transl Immunol. 2018;7:e1018. https://doi.org/10.1002/cti2.1018.
63. Pe'er I, Yelensky R, Altshuler D, Daly MJ. Estimation of the multiple testing burden for genomewide association studies of nearly all common variants. Genet Epidemiol. 2008;32:381–5.
64. International Multiple Sclerosis Genetics Consortium, Hafler DA, Compston A, Sawcer S, Lander ES, Daly MJ, et al. Risk alleles for multiple sclerosis identified by a genome wide study. N Engl J Med. 2007;357(9):851–62.
65. Gregory SG, Schmidt S, Seth P, Oksenberg JR, Hart J, Prokop A, et al. Interleukin 7 receptor alpha chain (IL7R) shows allelic and functional association with multiple sclerosis. Nat Genet. 2007;39:1083–91.
66. Lundmark F, Duvefelt K, Iacobaeus E, Kockum I, Wallstrom E, Khademi M, et al. Variation in interleukin 7 receptor alpha chain (IL7R) influences risk of multiple sclerosis. Nat Genet. 2007;39:1108–13.
67. Weber F, Fontaine B, Cournu-Rebeix I, Kroner A, Knop M, Lutz S, et al. IL2RA and IL7RA genes confer susceptibility for multiple sclerosis in two independent European populations. Genes Immun. 2008;9:259–63.
68. Matesanz F, Caro-Maldonado A, Fedetz M, Fernandez O, Milne RL, Guerrero M, et al. IL2RA/CD25 polymorphisms contribute to multiple sclerosis susceptibility. J Neurol. 2007;254:682–4.
69. Noguchi M, Nakamura Y, Russell SM, Ziegler SF, Tsang M, Cao X, et al. Interleukin-2 receptor gamma chain: a functional component of the interleukin-7 receptor. Science. 1993;262(5141):1877–80.
70. Ye SK, Maki K, Kitamura T, Sunaga S, Akashi K, Domen J, et al. Induction of germline transcription in the TCRgamma locus by Stat5: implications for accessibility control by the IL-7 receptor. Immunity. 1999;11(2):213–23.

71. Stahl EA, Raychaudhuri S, Remmers EF, Xie G, Eyre S, Thomson BP, et al. Genome-wide association study meta-analysis identifies seven new rheumatoid arthritis risk loci. Nat Genet. 2010;42:508–14.
72. International Multiple Sclerosis Genetics Consortium; Wellcome Trust Case Control Consortium 2, Sawcer S, Hellenthal G, Pirinen M, Spencer CC, Patsopoulos NA, Moutsianas L, et al. Genetic risk and a primary role for cell-mediated immune mechanisms in multiple sclerosis. Nature. 2011;476:214–9.
73. Boyman O, Sprent J. The role of interleukin-2 during homeostasis and activation of the immune system. Nat Rev Immunol. 2012;12:180–90.
74. The International Multiple Sclerosis Genetics Consortium (IMSGC). Evidence for polygenic susceptibility to multiple sclerosis—the shape of things to come. Am J Hum Genet. 2010;86:621–5.
75. International Multiple Sclerosis Genetics Consortium. Genome-wide association study of severity in multiple sclerosis. Genes Immun. 2011;12(8):615–25.
76. International Multiple Sclerosis Genetics Consortium (IMSGC), Beecham AH, Patsopoulos NA, Xifara DK, Davis MF, Kemppinen A, Cotsapas C, et al. Analysis of immune-related loci identifies 48 new susceptibility variants for multiple sclerosis. Nat Genet. 2013;45(11):1353–60.
77. International Multiple Sclerosis Genetics Consortium, Patsopoulos NA, Baranzini SE, Santaniello A, Shoostari P, Cotsapas C, Wong G, et al. The multiple sclerosis genomic map: role of peripheral immune cells and resident microglia in susceptibility. Science. 2019;365(6460):eaav7188. https://doi.org/10.1126/science.aav7188.
78. Gianfrancesco MA, Stridh P, Shao X, Rhead B, Graves JS, Chitnis T, et al. Genetic risk factors for pediatric-onset multiple sclerosis. Mult Scler. 2018;24:1825–34.
79. Altshuler D, Daly MJ, Lander ES. Genetic mapping in human disease. Science. 2008;322:881–8.
80. Farh KK, Marson A, Zhu J, Kleinewietfeld M, Housley WJ, Beik S, Shoresh N, Whitton H, Ryan RJ, Shishkin AA, Hatan M, Carrasco-Alfonso MJ, Mayer D, Luckey CJ, Patsopoulos NA, De Jager PL, Kuchroo VK, Epstein CB, Daly MJ, Hafler DA, Bernstein BE. Genetic and epigenetic fine mapping of causal autoimmune disease variants. Nature. 2015;518(7539):337–43.
81. Dendrou CA, Fugger L, Friese MA. Immunopathology of multiple sclerosis. Nat Rev Immunol. 2015;15:545–58.
82. Raj T, Rothamel K, Mostafavi S, Ye C, Lee MN, Replogle JM, et al. Polarization of the effects of autoimmune and neurodegenerative risk alleles in leukocytes. Science. 2014;344(6183):519–23.
83. Diaz V. Encouraging participation of minorities in research studies. Ann Fam Med. 2012;10:372–3.
84. Kingwell E, Marriott JJ, Jetté N, Pringsheim T, Makhani N, Morrow SA, et al. Incidence and prevalence of multiple sclerosis in Europe: a systematic review. BMC Neurol. 2013;13:128.
85. Cree BA, Khan O, Bourdette D, Goodin DS, Cohen JA, Marrie RA, et al. Clinical characteristics of African Americans vs Caucasian Americans with multiple sclerosis. Neurology. 2004;63:2039–45.
86. Oksenberg JR, Barcellos LF, Cree BA, Baranzini SE, Bugawan TL, Khan O, et al. Mapping multiple sclerosis susceptibility to the HLA-DR locus in African Americans. Am J Hum Genet. 2004;74(1):160–7.
87. Cree BA, Reich DE, Khan O, De Jager PL, Nakashima I, Takahashi T, et al. Modification of multiple sclerosis phenotypes by African ancestry at HLA. Arch Neurol. 2009;66(2):226–33.
88. Johnson BA, Wang J, Taylor EM, Caillier SJ, Herbert J, Khan OA, et al. Multiple sclerosis susceptibility alleles in African Americans. Genes Immun. 2010;11:343–50. https://doi.org/10.1038/gene.2009.81.
89. Isobe N, Gourraud PA, Harbo HF, Caillier SJ, Santaniello A, Khankhanian P, et al. Genetic risk variants in African Americans with multiple sclerosis. Neurology. 2013;81(3):219–27.
90. Yoshimura S, Isobe N, Yonekawa T, Matsushita T, Masaki K, Sato S, et al. Genetic and infectious profiles of Japanese multiple sclerosis patients. PLoS One. 2012;7:e48592.
91. Matsuoka T, Matsushita T, Osoegawa M, Kawano Y, Minohara M, Mihara F, et al. Association of the HLA-DRB1 alleles with characteristic MRi features of Asian multiple sclerosis. Mult Scler. 2008;14:1181–90.
92. Wang H, Pardeshi LA, Rong X, Li E, Wong KH, Peng Y, et al. Novel variants identified in multiple sclerosis patients from southern China. Front Neurol. 2018;9:582.
93. Barcellos LF, Oksenberg JR, Begovich AB, Martin ER, Schmidt S, Vittinghoff E, et al. HLA-DR2 dose effect on susceptibility to multiple sclerosis and influence on disease course. Am J Hum Genet. 2003;72:710–6.
94. Vader W, Stepniak D, Kooy Y, Mearin L, Thompson A, van Rood JJ, et al. The HLA-DQ2 gene dose effect in celiac disease is directly related to the magnitude and breadth of gluten-specific T cell responses. Proc Natl Acad Sci U S A. 2003;100:12390–5.
95. Baranzini SE. Revealing the genetic basis of multiple sclerosis: are we there yet? Curr Opin Genet Dev. 2011;21:317–24.
96. Sirota M, Schaub MA, Batzoglou S, Robinson WH, Butte AJ. Autoimmune disease classification by inverse association with SNP alleles. PLoS Genet. 2009;5:e1000792.
97. Masterman T, Ligers A, Olsson T, Andersson M, Olerup O, Hillert J. HLA-DR15 is associated with lower age at onset in multiple sclerosis. Ann Neurol. 2000;48:211–9.
98. Smestad C, Brynedal B, Jonasdottir G, Lorentzen AR, Masterman T, Akesson E, et al. The impact of HLA-A and -DRB1 on age at onset, disease course

and severity in Scandinavian multiple sclerosis patients. Eur J Neurol. 2007;14:835–40.
99. Werneck LC, Lorenzoni PJ, Kay CSK, Scola RH. Multiple sclerosis: disease modifying therapy and the human leukocyte antigen. Arq Neuropsiquiatr. 2018;76(10):697–704.
100. Okuda DT, Srinivasan R, Oksenberg JR, Goodin DS, Baranzini SE, Beheshtian A, et al. Genotype-phenotype correlations in multiple sclerosis: HLA genes influence disease severity inferred by 1HMR spectroscopy and MRI measures. Brain. 2009;132(Pt 1):250–9.
101. Qiu W, Raven S, James I, Luo Y, Wu J, Castley A, et al. Spinal cord involvement in multiple sclerosis: a correlative MRI and high-resolution HLA-DRB1 genotyping study. J Neurol Sci. 2011;300:114–9.
102. Nakamura Y, Matsushita T, Sato S, Niino M, Fukazawa T, Yoshimura S, et al. Japan Multiple Sclerosis Genetics Consortium. Latitude and HLA-DRB1*04:05 independently influence disease severity in Japanese multiple sclerosis: a cross-sectional study. J Neuroinflammation. 2016;13(1):239.
103. Mokry LE, Ross S, Ahmad OS, Forgetta V, Smith GD, Goltzman D, Leong A, Greenwood CM, Thanassoulis G, Richards JB. Vitamin D and risk of multiple sclerosis: a Mendelian randomization study. PLoS Med. 2015;12:e1001866.
104. Baranzini SE, Wang J, Gibson RA, Galwey N, Naegelin Y, Barkhof F, et al. Genome-wide association analysis of susceptibility and clinical phenotype in multiple sclerosis. Hum Mol Genet. 2009;18(4):767–78.
105. Zou ZY, Zhou ZR, Che CH, Liu CY, He RL, Huang HP. Genetic epidemiology of amyotrophic lateral sclerosis: a systematic review and meta-analysis. J Neurol Neurosurg Psychiatry. 2017;88:540–9.
106. Harbo HF, Isobe N, Berg-Hansen P, Bos SD, Caillier SJ, Gustavsen MW, et al. Oligoclonal bands and age at onset correlate with genetic risk score in multiple sclerosis. Mult Scler. 2014;20(6):660–8.
107. Graves JS, Barcellos LF, Shao X, Noble J, Mowry EM, Quach H, et al. Genetic predictors of relapse rate in pediatric MS. Mult Scler. 2016;22:1528–35.
108. Graves JS, Barcellos LF, Simpson S, Belman A, Lin R, Taylor BV, et al. The multiple sclerosis risk allele within the AHI1 gene is associated with relapses in children and adults. Mult Scler Relat Disord. 2018;19:161–5.
109. Zhou Y, Simpson S, Charlesworth JC, Van Der Mei I, Lucas RM, Ponsonby AL, et al. Variation within MBP gene predicts disease course in multiple sclerosis. Brain Behav. 2017;7:e00670. https://doi.org/10.1002/brb3.670.
110. Gil-Varea E, Urcelay E, Vilarino-Guell C, Costa C, Midaglia L, Matesanz F, et al. Exome sequencing study in patients with multiple sclerosis reveals variants associated with disease course. J Neuroinflammation. 2018;15(1):265.
111. Munger KL, Levin LI, Hollis BW, Howard NS, Ascherio A. Serum 25-hydroxyvitamin D levels and risk of multiple sclerosis. JAMA. 2006;296:2832–8.
112. Kragt JJ, van Amerongen BM, Killestein J, Dijkstra C, Uitdehaag B, Polman C, et al. Higher levels of 25-hydroxyvitamin D are associated with a lower incidence of multiple sclerosis only in women. Mult Scler. 2009;15(1):9–15.
113. Smolders J, Menheere P, Kessels A, Damoiseaux J, Hupperts R. Association of vitamin D metabolite levels with relapse rate and disability in multiple sclerosis. Mult Scler. 2008;14:1220–4.
114. Hart PH, Lucas RM, Booth DR, Carroll WM, Nolan D, Cole JM, et al. Narrowband UVB phototherapy for clinically isolated syndrome: a trial to deliver the benefits of vitamin D and other UVB-induced molecules. Front Immunol. 2017;8:3.
115. Gianfrancesco MA, Stridh P, Rhead B. Evidence for a causal relationship between low vitamin D, high BMI, and pediatric-onset MS. Neurology. 2017;88:1623–9.
116. Graves JS, Barcellos LF, Krupp L, Belman A, Shao X, Quach H, et al. Vitamin D genes influence MS relapses in children. Mult Scler. 2019;26(8):894–901.
117. Munger KL, Chitnis T, Ascherio A. Body size and risk of MS in two cohorts of US women. Neurology. 2009;73:1543–50.
118. Langer-Gould A, Brara SM, Beaber BE. Childhood obesity and risk of pediatric multiple sclerosis and clinically isolated syndrome. Neurology. 2013;80:548–52.
119. Munger KL, Bentzen J, Laursen B. Childhood body mass index and multiple sclerosis risk: a long-term cohort study. Mult Scler. 2013;19:1323–9.
120. Mokry LE, Ross S, Timpson NJ, Sawcer S, Davey Smith G, Richards JB. Obesity and multiple sclerosis: a mendelian randomization study. PLoS Med. 2016;13:e1002053.
121. Lewis SJ, Murad A, Chen L, Davey Smith G, Donovan J, Palmer T, et al. Associations between an obesity related genetic variant (*FTO* rs9939609) and prostate cancer risk. PLoS One. 2010;5(10):e13485.
122. Keller L, Xu W, Wang HX, Winblad B, Fratiglioni L, Graff C. The obesity related gene, *FTO*, interacts with APOE, and is associated with Alzheimer's disease risk: a prospective cohort study. J Alzheimers Dis. 2011;23(3):461–9.
123. Bressler J, Fornage M, Demerath EW, Knopman DS, Monda KL, North KE, et al. Fat mass and obesity gene and cognitive decline: the Atherosclerosis Risk in Communities Study. Neurology. 2013;80(1):92–9.
124. Vimaleswaran KS, Berry DJ. Lu C. Causal relationship between obesity and vitamin D status: Bi-directional Mendelian randomization analysis of multiple cohorts. PLoS Med. 2013;10:e1001383.
125. Pakpoor J, Pakpoor J. Childhood obesity and risk of pediatric multiple sclerosis and clinically isolated syndrome. Neurology. 2013;81:1366.

# Part III

# Autoimmune Neurology

# Autoimmune and Paraneoplastic Encephalitis

12

Anastasia Zekeridou

**Key Points**

1. Autoimmune encephalitis is a treatable and potentially reversible cause of cognitive dysfunction with incidence and prevalence similar to infectious causes.
2. Autoimmune encephalitis should be suspected in patients with subacute onset of cognitive difficulties or rapidly progressive dementia especially if associated with new-onset drug-resistant seizures, movement disorder, evidence of cancer, or personal history of autoimmunity.
3. Clinical phenotypes include limbic and brainstem encephalitis, prominent psychiatric manifestations, progressive encephalopathy with rigidity, and myoclonus or more restricted phenotypes, such as autoimmune epilepsy.
4. Brain magnetic resonance imaging (MRI), electroencephalography (EEG), and cerebrospinal fluid (CSF) can be useful in diagnosing autoimmune encephalitis but can also be normal.
5. Testing both serum and CSF for neural antibodies aids clinical diagnosis of both autoimmunity and cancer (if applicable) and improves diagnostic sensitivity. The absence of neural autoantibodies does not exclude autoimmune encephalitis.
6. Early recognition, appropriate immunotherapy treatment, and cancer screening are crucial for patients' outcomes.

## Introduction

Encephalitis of all etiologies causes more than 20,000 hospitalizations per year in the United States [1] with 35–50% of cases being of unknown etiology [1, 2]. Autoimmune encephalitis has been increasingly recognized as a treatable cause of encephalitis in patients with subacute onset of cognitive difficulties in the presence or absence of a tumor. Recent studies have suggested that incidence rates of autoimmune and infectious encephalitis in Minnesota were comparable (0.8/100,000 and 1.0/100,000 person-years, respectively) [3]. The frequency of autoimmune N-methyl-d-aspartate receptor (NMDAR) encephalitis among young adults in the California encephalitis project surpassed that of any individual viral etiology [4]. Recognition of autoimmune encephalitis (including in the

A. Zekeridou (✉)
Departments of Laboratory Medicine and Pathology and Neurology, Mayo Clinic College of Medicine, Rochester, MN, USA
e-mail: Zekeridou.anastasia@mayo.edu



A. L. Piquet, E. Alvarez (eds.), *Neuroimmunology*, https://doi.org/10.1007/978-3-030-61883-4_12

context of a paraneoplastic neurological syndrome and cancer) is crucial as it is a potentially treatable and reversible disorder, especially if recognized early [5]. Often the patients' symptoms are misdiagnosed as non-treatable neurodegenerative disorders or are thought to correspond to cancer progression in paraneoplastic cases. The discovery of neural autoantibody biomarkers has revolutionized how we diagnose and approach the neural autoantibody-positive encephalitis cases and has broadened the spectrum of recognizable syndromes now attributable to autoimmune or paraneoplastic encephalitides. Suggested criteria for neural autoantibody-negative cases are based on disease presentation and ancillary testing and are valuable aids to guide diagnosis and treatment [6] (Table 12.1).

The presence of cancer, central nervous system (CNS) infections, or systemic infections have been implicated as triggers in the pathogenesis of paraneoplastic, autoimmune, and postinfectious encephalitides [7–12]. In paraneoplastic cases, systemic tumors expressing onconeural antigens, often in mutated forms to bypass self-tolerance, may trigger autoimmunity [12–15]. A Paraneoplastic encephalitis is evidence of a potent antitumor immune response targeting either intracellular onconeural antigens (e.g., anti-neuronal nuclear antibody 1 [ANNA1] or anti-Hu encephalitis in patients with small-cell lung cancer [SCLC]) or neural synapses (e.g., NMDAR encephalitis in patients with ovarian teratomas) [15]. Autoimmune encephalitis, especially cases of NMDAR encephalitis, has also been described both in children and in adults after a herpes simplex virus (HSV) encephalitis (discussed further in Chap. 25) [7, 10]. Different pathophysiological mechanisms have been suggested in these cases, including viral-induced release of NMDAR and other neural proteins, or mechanisms of molecular mimicry [10]. Molecular mimicry has also been suggested as hypothesis in postinfectious acute disseminated encephalomyelitis (ADEM) [16].

In patients with neural autoantibody-positive autoimmune and paraneoplastic encephalitides, the cellular location of the respective antigen target (cell surface versus intracellular) hints to the pathophysiological mechanisms of the disease. In patients with autoantibodies targeting accessible proteins in neuronal, glial, or neuromuscular synapses (synaptic autoantibodies), the antibodies have pathogenic potential and thus antibody-depleting or B-cell-depleting therapies can be very efficient [17, 18]. On the other hand, in cases where the respective antigen is located intracellularly, the neural autoantibodies are just a biomarker of the cytotoxic T-cells of the same specificity targeting cells that present the respective antigenic peptides on their MHC-1 molecules [13]. Even though the latter cases have a higher association with malignancies, neural synaptic antibodies are also found in patients with cancer, e.g., α(alpha)-amino-3-hydroxy-5-methyl-4-isoxazolepropionic acid receptor (AMPAR) encephalitis in patients with small-cell lung cancer [8]. Neural autoantibodies, when present, guide diagnosis, cancer screening, and treatment modalities and may inform prognosis.

**Table 12.1** When to suspect an autoimmune/paraneoplastic encephalitis

| |
|---|
| *Clinical clues* |
| Subacute onset of neurological symptoms consistent with encephalitis<br>+/– fluctuations<br>+/– rapid progression |
| New-onset seizures, resistant to antiepileptic drugs (incl. faciobrachial dystonic seizures) |
| Coexisting psychiatric/behavioral manifestations |
| Coexisting autonomic dysfunction |
| Coexisting movement disorders (orofacial dyskinesias, chorea, etc.) |
| Viral prodrome |
| Personal history of autoimmunity |
| Personal history of current or past malignancy |
| Symptoms developing in the setting of immune checkpoint blockade treatment for cancer |
| *Ancillary testing clues* |
| MRI imaging of mesiotemporal lobe T2/FLAIR hyperintensities |
| Specific MRI patterns:<br>Linear perivascular radial enhancement around the ventricles (GFAP-IgG)<br>Multifocal non-enhancing cortical-subcortical T2/FLAIR hyperintensities ($GABA_AR$-IgG) |
| Inflammatory CSF (lymphocytic pleocytosis, oligoclonal bands, elevated IgG index; elevated protein alone is not specific) |
| Serological evidence of coexisting systemic or organ-specific autoimmunity |
| Presence of well-established neural autoantibodies detected with contemporary assays |
| EEG evidence of epileptiform activity or slow waves in temporal lobe(s); extreme delta brush |

## Clinical Manifestations

Neurological autoimmunity can affect any level of the neuraxis. CNS autoimmunity can present with psychiatric manifestations, cognitive impairment with or without altered consciousness, seizures, movement disorders, cerebellar and brainstem dysfunction, or myelitis. An autoimmune origin should be suspected in the setting of subacute onset of neurological manifestations especially in patients with evidence of systemic autoimmunity or cancer or after immune checkpoint inhibitor treatment for systemic malignancies (neurological complications of immune checkpoint inhibitors are discussed further in Chap. 25). Specific syndromes such as limbic encephalitis and brainstem encephalitis should raise suspicion for an autoimmune origin. Ancillary testing findings, such as an inflammatory CSF in the absence of infection, can also reinforce the suspicion of autoimmune encephalitis as can distinctive clinical manifestations that point toward a specific diagnosis (Tables 12.1 and 12.2).

This chapter focuses on the clinical and autoantibody manifestations of autoimmune and paraneoplastic encephalitis based on the pathophysiological mechanisms underlying these diseases. In addition, we will discuss some other well-defined autoimmune neurological syndromes and will provide a diagnostic and management algorithm for these challenging cases. Cerebellar (autoimmune ataxia in the context of paraneoplastic neurological syndromes) and basal ganglia syndromes (autoimmune movement disorders) are discussed in Chaps. 16 and 14, respectively.

## Neural Autoantibody-Positive Autoimmune and Paraneoplastic Encephalitides

A list of neural autoantibodies implicated in autoimmune encephalitis along with their distinctive manifestations and cancer associations is provided in Tables 12.3 and 12.4.

**Table 12.2** Clinical clues suggesting a specific diagnosis in patients with autoimmune/paraneoplastic encephalitis

| Clinical phenotype | Neural autoantibody specificities |
|---|---|
| Faciobrachial dystonic seizures | LGI-1 |
| "Dizziness spells" | LGI-1 |
| Status epilepticus, predominant seizures | $GABA_AR$, $GABA_BR$, GAD65 (and others) |
| Myoclonus | DPPX, glycine R |
| Opsoclonus myoclonus | ANNA2 (Ri) |
| Psychosis | NMDAR (and others) |
| Coexisting ataxia | GAD65, ANNA1 (Hu), Caspr2 (episodic ataxia) |
| Progressive encephalopathy with rigidity and myoclonus; exaggerated startle | Glycine R, DPPX |
| Stiff-person syndrome | Amphiphysin, GAD65, glycine R |
| Peripheral nerve hyperexcitability | Caspr2, LGI1 |
| Coexisting gastrointestinal dysmotility | ANNA1 (Hu) |
| Coexisting gastrointestinal hypermotility | DPPX |
| Morvan's syndrome (central, peripheral, and autonomic nervous system hyperexcitability) | Caspr2 |
| Coexisting sensory neuronopathy | ANNA1 (Hu) |
| Prominent meningeal involvement | GFAP |
| Coexisting myelopathy | GFAP, amphiphysin, ANNA1, CRMP5 |
| Children | Aquaporin 4, MOG, NMDAR, $GABA_AR$ |

**Table 12.3** Neural autoantibodies associated with encephalitis of synaptic specificities

| Name/ antigen | Common neurological presentations | Common oncological associations |
|---|---|---|
| VGKC (Kv1) | | Rare (thymoma) |
| LGI-1 | Limbic encephalitis, faciobrachial dystonic seizures, hyponatremia | Thymoma |
| Caspr2 | Peripheral nerve hyperexcitability, limbic encephalitis, dysautonomia, neuropathy | |

(continued)

**Table 12.3** (continued)

| Name/ antigen | Common neurological presentations | Common oncological associations |
|---|---|---|
| DPPX (Kv4.2, Kv4.3) | Dysautonomia with gastrointestinal hypermotility, psychiatric manifestations, limbic encephalitis, ataxia, oculomotor dysfunction, myoclonus, rigidity, exaggerated "startle" | B-cell neoplasia |
| NMDAR | Psychosis, catatonia, seizures, encephalitis, central dysautonomia | Ovarian teratoma |
| AMPAR | Limbic encephalitis, seizures | Thymoma, SCLC, breast carcinoma |
| mGluR5 | Limbic encephalitis, psychiatric manifestations | Hodgkin lymphoma |
| $GABA_AR$ | Encephalitis with prominent seizures | Thymoma, lymphoma |
| $GABA_BR$ | Limbic encephalitis with prominent seizures | SCLC |
| Glycine receptor | Progressive encephalomyelitis with rigidity and myoclonus, optic neuritis | Thymoma, lymphoma, breast cancer |
| Neurexin 3a | Encephalitis, seizures | No cancer association |
| IgLON5 | Sleep disorder (NREM and REM) and brainstem dysfunction | No cancer association |
| AQP4 | Neuromyelitis optica spectrum disorder (optic neuritis, transverse myelitis, circumventricular organ involvement), encephalitis in children (ADEM-like phenotype) | Breast carcinoma, thymoma, carcinoid, B-cell neoplasia |
| MOG | Optic neuritis, myelitis, acute disseminated encephalomyelitis | No cancer association |

*AMPAR* α(alpha)-amino-3-hydroxy-5-methyl-4-isoxazolepropionic acid receptor, *AQP4* aquaporin-4, *Caspr2* contactin-associated protein-2, *DPPX* dipeptidyl-peptidase–like protein-6, *GABA* γ(gamma) aminobutyric acid, *LGI1* leucine-rich, glioma inactivated 1 protein, *mGluR* metabotropic glutamate receptor, *MOG* myelin oligodendrocyte glycoprotein, *NMDAR* N-methyl-d-aspartate receptor, *NREM* non-rapid eye movement, *REM* rapid eye movement, *SCLC* small-cell lung carcinoma, *VGKC* voltage-gated potassium channel

## Antibodies Targeting Neural Synapses

One of the most common and well-described autoimmune encephalitides that has a distinctive clinical syndrome is *NMDAR encephalitis*. Even though it was first described in 2007 in young women presenting with psychiatric manifestations in the context of an ovarian teratoma, we now know that patients harboring autoantibodies targeting the NMDAR (NR1 subunit) include men and children [19–22]. It can be seen following an infectious HSV encephalitis [10]. Accurate diagnosis with testing of the autoantibodies in the CSF of patients who are suspected to have NMDAR encephalitis with assays specific for the NR1 subunit of the NMDAR is critical as there have been reports of false positivity. Median age at disease onset is 21 years with a vast majority (>80%) being women [20]. NMDAR encephalitis can be paraneoplastic; tumors found (38% of patients) are most often encountered in women between the age of 18 and 45 years, and the vast majority are ovarian teratomas [20]. Neural elements are found in ovarian teratomas of seropositive patients that are morphologically atypical but not in control ovarian teratomas [15]. The patients are often diagnosed based on the characteristic clinical presentation and confirmed by autoantibody testing. Patients present with psychiatric manifestations (new-onset psychosis, hallucinations, stereotypical behaviors, agitation, or catatonia) that can be misdiagnosed as a primary psychiatric disorder. They also develop, in the course of the disease, short-term memory loss, followed by seizures, movement disorders (orofacial dyskinesias are a hallmark of the disease), and autonomic instability with central hypoventilation [23]. Children present with seizures as their first symptom more frequently than adults and have significantly more abnormal movements [21]. Male patients present more often with seizures as initial manifestation of their disease [22]. Intensive care unit (ICU) stays have

**Table 12.4** Neural autoantibodies associated with encephalitis of nuclear and cytoplasmic specificities

| Name/antigen | Common neurological presentations | Common oncological associations |
|---|---|---|
| ANNA-1 (Hu) | Limbic encephalitis, diffuse encephalomyelitis, polyneuropathy and sensory neuronopathy, dysautonomia (with gastrointestinal dysmotility) | SCLC, neuroblastoma in children (rare), thymoma (rare) |
| ANNA-2 (Ri, Nova 1/ Nova2 | Opsoclonus myoclonus, jaw dystonia, laryngospasm, brainstem encephalitis | Carcinoma (lung [SCLC] and breast) |
| ANNA-3 | Limbic encephalitis, diffuse encephalomyelitis, polyneuropathy | SCLC, other lung and pharyngeal carcinomas |
| Ma1/Ma2 | Limbic encephalitis, brainstem encephalitis, cerebellar ataxia | Lung carcinoma, renal carcinoma |
| Ma2 | Limbic and brainstem encephalitis | Seminoma (testicular germ-cell tumor) |
| PCA-2 (MAP1B) | Encephalitis (limbic and/or other), cerebellar ataxia, peripheral neuropathy | SCLC |
| CRMP5 | Cranial neuropathies, uveoretinitis, encephalitis (limbic and/or other), chorea, myelitis, radiculoplexopathies, polyneuropathies | SCLC, thymoma |
| Amphiphysin | Encephalomyelopathy (limbic and/or other), stiff-man phenomena, neuropathy | Breast carcinoma, SCLC |
| GAD65 | Stiff-person syndrome, cerebellar ataxia, encephalitis (limbic and/or other); predominant seizures, myelopathy | Rare: thymoma, SCLC |
| GFAP | Meningoencephalomyelitis (predominant meningeal involvement) | Ovarian teratoma, diverse adenocarcinomas |
| AK5 | Limbic encephalitis with prominent short-term memory loss, anxiety | No associated cancer |
| NIF, light chain | Cerebellar ataxia, encephalopathy, myelopathy | Neuroendocrine (SCLC, Merkel cell carcinoma) |
| Septin 5 | Cerebellar ataxia with brainstem encephalitis and predominant oculomotor deficits | No associated cancer |

*AK5* adenylate kinase 5, *ANNA* anti-neuronal nuclear antibody, *CRMP* collapsin response-mediator protein, *GAD* glutamic acid decarboxylase, *GFAP* glial fibrillary acidic protein, *MAP1B* microtubule-associated protein 1B, *NIF* neuronal intermediate filaments, *PCA* Purkinje cell cytoplasmic antibody, *SCLC* small-cell lung carcinoma

been reported to be necessary in up to 75% of cases [20]. The majority of patients have a good outcome (80%), but the recovery can take multiple months; predictors of good outcome identified are early treatment and no admission to an intensive care unit [20]. MRI findings are not specific and can include T2/FLAIR hyperintensities in the hippocampi, cerebellar or cerebral cortex, basal ganglia, or brainstem [23]. Electroencephalogram (EEG) is often abnormal with findings of generalized slowing or epileptiform discharges. A subset of patients with NMDAR encephalitis have "extreme delta brush" [24].

Except for NMDAR encephalitis that has a specific disease phenotype, limbic encephalitis has been associated with multiple neuronal autoantibodies of both synaptic and intracellular specificities. Limbic encephalitis refers to the rapid development of short-term memory loss, psychiatric manifestations, seizures, and confusion and is often associated with EEG abnormalities, an inflammatory CSF, and T2/FLAIR mesial-temporal lobe hyperintensities (Fig. 12.1) [25]. Even in the absence of a neural-specific autoantibody, this clinical phenotype is evocative of an autoimmune encephalitis but can also be seen in infectious causes (e.g., HSV encephalitis).

*Voltage-gated potassium channel (VGKC) autoimmunity* can be seen in patients with limbic encephalitis. Autoantibodies targeting the leucine-rich, glioma inactivated-1 protein (LGI-1) and contactin-associated protein-2

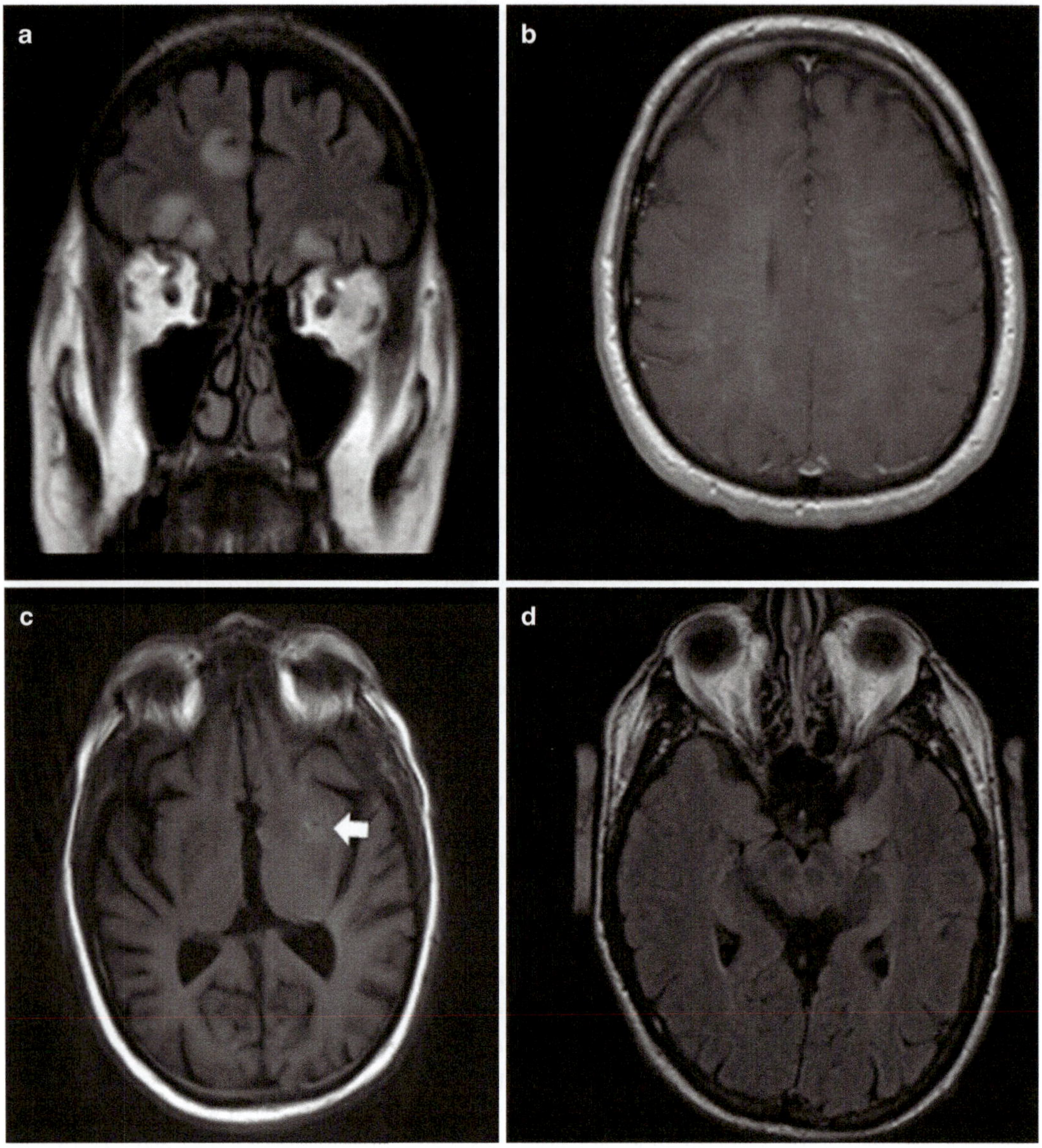


**Fig. 12.1** Magnetic resonance imaging (MRI) in patients with encephalitis. (**a**) $GABA_AR$ encephalitis with cortical and subcortical, non-enhancing (not shown) FLAIR hyperintensities. (**b**) Radial gadolinium enhancement in a patient with GFAP meningoencephalitis. (**c**) T1 basal ganglia hyperintensity (left) in a patient with LGI-1 encephalitis and faciobrachial dystonic seizures. (**d**) Mesiotemporal lobe FLAIR hyperintensity (left) in a patient with limbic encephalitis

(Caspr2) are more specific in the setting of VGKC autoimmunity; the significance of VGKC-immunoglobulin G (IgG) positivity in the absence of LGI1 and Caspr2 subspecificities is unclear [26]. Patients with LGI1-IgG can present early with faciobrachial dystonic seizures (FBDS), which is a hallmark of this disease. FBDS respond poorly to antiepileptic medication but do respond to immunotherapy [5, 27]. Cognitive impairment also occurs, and early immunotherapy is essential in order to avoid longer-term cognitive impairment [5].

Peripheral manifestations often accompany LGI1 encephalitis (or can be seen alone) including peripheral nerve hyperexcitability [27]. Hyponatremia is often seen [27]. Paroxysmal dizziness spells have been described in 14% of patients [27]. Patients with LGI1 encephalitis are older (medial age of 65 years), are more often male, and rarely have cancer (13%) [27]. The CSF in these patients is rarely inflammatory. MRI changes include FLAIR/T2 mesiotemporal lobe hyperintensities and in 13% of cases T1 basal ganglia hyperintensities (Fig. 12.1). A minority of patients with VGKC autoimmunity have autoantibodies targeting Caspr2. Even though in the past it was considered that Caspr2 autoantibodies associate with a peripheral phenotype (including peripheral nerve hyperexcitability), we now know that more than 60% of cases have CNS manifestations, especially in older adults [27]. Morvan's syndrome (central, peripheral, and autonomic hyperexcitability), even though rare, is a syndromic manifestation of Caspr2 autoimmunity.

Autoantibodies targeting *gamma-aminobutyric acid (GABA)$_B$R* are seen in patients with limbic encephalitis with prominent seizures [28]. A cancer is diagnosed in half of the patients, and it is most often small-cell lung cancer (SCLC). This disorder can respond favorably to immunotherapy, but the overall survival might be dictated by the presence of cancer [28]. *GABA$_A$R autoantibodies* are also seen in patients with encephalitis and prominent seizures with refractory status epilepticus and FLAIR/T2 cortical and subcortical hyperintensities (Fig. 12.1a) [29]. Less than one-quarter of the patients have a malignancy (mostly thymomas). It can coexist with other autoantibodies, such as glutamic acid decarboxylase-65 (GAD65)-IgG and LGI1-IgG [28, 29]. Even though the median age of the patients is 40 years, this is a disorder that can be seen in children [29].

Another example of paraneoplastic limbic encephalitis is *AMPAR encephalitis*. More than half of the patients present a cancer including thymoma, SCLC, and breast and ovarian tumors [30–32]. Even though this disorder can be paraneoplastic, the patients can respond favorably to immunotherapy. When AMPAR-IgG is accompanied by antibodies to intracellular neural antigens, then the prognosis is worse [30]. Examples of complete recovery after severe encephalopathy necessitating ICU stays and intubation reinforce the necessity of early immunotherapy and highlight the importance of giving these patients time to recover [30].

Metabotropic glutamate receptor-5 (mGluR5) antibodies were first described in patients with limbic encephalitis and Hodgkin's lymphoma [33]. The syndrome was described as "Ophelia's syndrome" (character from Shakespeare's play *Hamlet*) by Dr. Carr who described the first case of limbic encephalitis in his 15-year-old daughter who was found to have Hodgkin's lymphoma followed by improvement of her neurological syndrome after cancer treatment [34]. A series of mGluR5-IgG-positive patients confirmed the clinical phenotype and also described a patient with SCLC; patients in their majority respond favorably to immunotherapy and cancer treatment [35].

Encephalitis with *dipeptidyl-peptidase–like protein-6 (DPPX)-IgG* (Kv4.2 and Kv4.3 subunits of the potassium channel) can manifest with limbic or brainstem features and is associated with central hyperexcitability (myoclonus, exaggerated startle, rigidity, and hyperreflexia) as well as dysautonomia (mostly gastrointestinal hypermotility) [36]. A PERM phenotype (progressive encephalopathy with rigidity and myoclonus) that has been described with glycine receptor-IgG (see below) has also been seen in patients with DPPX-IgG [37]. B-cell neoplasms have been described, and in general the patients can respond favorably to immunotherapy (60%) [36].

Five patients were described with antibodies targeting *neurexin-3$\alpha$*(alpha) with a severe form of encephalitis that can be potentially treatable (three patients had improvement with immunotherapy) [38]. No cancer association has yet been reported.

*Glycine receptor-IgG* has been associated with PERM but also limited forms of stiff-person syndrome (SPS) that is immunotherapy responsive [39, 40]. Tumor associations are rare, but lymphoma and thymoma have been described [39, 40]. This is discussed further in Chap. 30.

While most of the aforementioned neurological syndromes respond favorably to immunotherapy as often seen in autoimmune neurological diseases mediated by potentially pathogenic antibodies (see above), *IgLON5-IgG* has been seen in patients with chronic progression of their neurological symptoms and often unfavorable response to immunotherapy [41]. Rapid eye movement (REM) and non-REM parasomnias were firstly described with sleep breathing disorders and are the hallmark of this disease. Brainstem dysfunction, findings compatible with hyperexcitability (myoclonus, cramps, exaggerated startle), neuropsychiatric disorders, and dysautonomia have been seen [42]. Even though some patients might have an improvement with immunotherapy, the clinical phenotype mostly resembles a neurodegenerative disorder, and the brain pathology reveals neuronal loss and deposits of hyperphosphorylated tau [41, 42].

Aquaporin-4 (AQP4) autoimmune astrocytopathy and myelin oligodendrocyte glycoprotein (MOG) autoimmune oligodendrogliopathy can be associated with encephalitis, especially in the pediatric population [11, 43]. ADEM is a syndromal manifestation of MOG autoimmunity and can also be a presenting manifestation of AQP4 autoimmunity in children [11, 43]. These diseases will be discussed in Chap. 15.

## Autoantibodies Specific for Intracellular Neural Antigens

Autoantibodies specific for intracellular antigens were the first described in cases of paraneoplastic CNS disorders. *ANNA1 autoimmunity (anti-Hu)* is a classic example of T-cell-mediated neurological autoimmunity. It is strongly associated with SCLC, but it can also rarely be seen in patients with thymoma or children with opsoclonus myoclonus and neuroblastoma (this syndrome is rarely associated with a neural autoantibody) [13]. Except for the limbic encephalitis phenotype, ANNA1 autoimmunity can present with other multiple central manifestations, including seizures, abnormal movements, brainstem and cerebellar dysfunction, and myelopathy [13]. Even though a mixed sensorimotor polyneuropathy is more common, a classic syndromic manifestation of ANNA1 autoimmunity is a sensory neuronopathy or ganglionopathy as well as dysautonomia with prominent gastrointestinal dysmotility [13]. As these manifestations can coexist in the same patient, they can give us clues to the diagnosis (Table 12.2).

Other examples of limbic encephalitis with autoantibodies targeting intracellular antigens include *adenylate kinase 5 (AK5)-IgG*; no malignancy has been described [44]. Encephalitic manifestations have also been described in patients with autoantibodies targeting the *Purkinje cell cytoplasmic antibody 2 (PCA2),* which is specific for the microtubule-associated protein 1B (MAP1B) that is most often seen in patients with diffuse neurological manifestations that include the peripheral nervous system and ataxia [45]. It is also seen in patients with SCLC. Other markers of SCLC that can be seen in patients with encephalopathy are *CRMP5-IgG* (also seen in thymoma), *amphiphysin-IgG* (also seen in women with breast cancer and other gynecological malignancies), and *neuronal intermediate filament (NIF), light chain-IgG* (also seen in other neuroendocrine tumors such as Merkel cell carcinomas) [46–48]. Amphiphysin-IgG autoimmunity is accompanied by stiff-person syndrome manifestations, myelopathy and peripheral neuropathy; NIF light chain-IgG by ataxia and myelopathy; and CRMP5-IgG by retinitis, optic neuritis, basal ganglionitis, myelopathy, and radiculoneuropathy [46–48].

Young male patients with seminoma can present with a severe encephalitis (limbic, diencephalic, and brainstem) associated with ataxia. The autoantibodies described to date are specific for *Ma2*. The patients often respond to cancer treatment [49]. Dual seropositivity for the homologous *Ma1* protein is seen mostly in women with a diverse clinical phenotype and cancer associations and predicts a worse neurological outcome [50].

A brainstem encephalitis is also seen in patients (both men and women) with *ANNA2 (or anti-Ri) antibodies*. The syndromic manifestations of ANNA2 autoimmunity are opsoclonus myoclonus associated with jaw dystonia and laryngospasm [51]. The cancers associated are breast (and gynecological tract more rarely) in women and lung (mostly SCLC) in both sexes [51]. Some patients improve with cancer treatment and immunotherapy [51]. *Septin-5-IgG* has also been described recently in five patients with cerebellar ataxia and encephalitis with brainstem symptoms without any cancer associations [52].

*GAD65 autoimmunity* is rarely associated with malignancies and can be seen in patients with autoimmune endocrinopathies such as diabetes type 1. Neurological manifestations are most often seen in patients with higher serum values of GAD65-IgG (>20 nmol/L). GAD65 autoimmunity syndromic CNS manifestations include autoimmune epilepsy, cerebellar ataxia, anxiety, and brainstem symptoms that can be associated with other encephalitic manifestations [53]. In addition, GAD65-IgG can be seen in patients with stiff-person syndrome, which is discussed in detail in Chap. 30 [53].

A recently described autoimmune astrocytopathy unified by the presence of *glial fibrillary acidic protein (GFAP)-IgG* in the patients' CSF manifests as a relapsing meningoencephalomyelitis with optic disk edema [54]. There is prominent meningeal involvement, and the patients can also present with myelitis [55]. They are generally responsive to prolonged steroid treatment. A characteristic MRI imaging with periventricular linear radial perivascular enhancement is seen in 50% of cases (Fig. 12.1) [55]. This disease can be paraneoplastic in more than 30% of cases [55].

## Other Encephalitides

Often patients will have a neural antibody-negative autoimmune encephalitis that can be suspected on the basis of clinical presentation and ancillary testing (Table 12.1). Other well-described entities are discussed below. ADEM will be discussed in Chap. 29.

### Steroid Responsive Encephalopathy Associated with Autoimmune Thyroiditis (SREAT, Also Known as Hashimoto's Encephalopathy)

Hashimoto's encephalopathy presents mostly in women with seizures, myoclonus, hallucinations, or stroke-like episodes in the presence of subclinical or mild thyroid disease (usually hypothyroidism) and serum thyroid (thyroid peroxidase [TPO], thyroglobulin) antibodies [6]. TPO antibodies are seen in 10% of normal patients and are not specific for Hashimoto's encephalitis [6]. Caution is advised in the diagnosis of patients without the classic clinical phenotype that have elevated TPO antibodies. Response to steroids is seen in more than 90% of cases [56].

### Bickerstaff Encephalitis

Bickerstaff encephalitis is characterized by progressive ophthalmoplegia and ataxia with disturbance of consciousness and is seen often in a postinfectious setting [6, 57]. Other features that can accompany this entity are pupillary abnormalities and facial diplegia, with other brainstem manifestations and generalized weakness with Guillain–Barré features of demyelinating radiculoneuropathy on electromyography (EMG) [6, 57]. CSF pleocytosis occurs in almost half of cases, and ganglioside Q1b antibodies are seen in more than 60%. This disease is most often monophasic and responsive to immunotherapy [6, 57].

## Ancillary Testing

*MRI* patterns that are consistent with limbic encephalitis include mesiotemporal lobe T2/FLAIR hyperintensities, with or without gadolinium enhancement (Fig. 12.1d). Distinctive patterns of MRI CNS abnormalities can be suggestive of a specific diagnosis. Cortical and subcortical multifocal, most often non-enhancing, T2/FLAIR hyperintensities can be seen in cases with $GABA_AR$ autoimmunity (Fig. 12.1a) [29]. Basal ganglia T2/FLAIR hyperintensities can be seen in patients with CRMP5 autoimmunity, while T1 hyperintensities can be seen in patients with LGI1 autoimmunity and FBDS (Fig. 12.1c) [58, 59]. In patients with autoimmune GFAP astrocytopathy, about half of the cases have associated periventricular linear radial perivascular enhancement (Fig. 12.1b) [55].

*2-Deoxy-2-[fluorine-18]fluoro-D-glucose integrated with computed tomography ($^{18}$F-FDG/PET-CT)* abnormalities are not specific but can be useful in cases of autoimmune encephalitis, especially if the MRI is negative [60]. Hypometabolism is the most common finding, followed by the presence of both hypo- and hypermetabolism in different brain regions [60]. Isolated hypermetabolism can also be seen [60]. FDG/PET-CT abnormalities can also be helpful in monitoring the response to immunotherapy in the absence of MRI abnormalities.

*Electroencephalogram (EEG)* is not diagnostic for autoimmune encephalitis, and the changes are not specific. EEG is very useful for detecting subclinical seizures [61]. Extreme delta brush (rhythmic delta activity with bursts of beta superimposed on the delta waves) has been described in 33% of cases with NMDAR encephalitis [62].

*CSF studies* should always include cell count and subtype of cells and total protein but also oligoclonal bands and IgG index. White blood cell count and IgG index can be elevated, and CSF-restricted oligoclonal bands can be present in cases of autoimmune encephalitis. An elevated protein alone is not specific in the diagnosis of autoimmune encephalitis. An elevated IgG index specific for the antibody in question can be useful to identify severe cases as seen in patients with LGI1 autoimmunity [63]. There are some cases of autoimmune encephalitis with a normal CSF profile, as reported in LGI1 encephalitis [27]. In these cases, establishing an autoimmune diagnosis can be difficult, especially in the absence of neural-specific autoantibodies. Clinical presentation and other ancillary testing can suggest an autoimmune origin (Table 12.1).

*Neural autoantibody testing* is discussed in Chap. 2. As in all cases of suspected neural autoimmunity, the testing for neural autoantibodies should be done in both serum and CSF.

## Autoimmune Encephalitis Management Approach

When an autoimmune or paraneoplastic encephalitis is suspected, an extensive workup including CNS imaging, CSF studies, and neural autoantibody testing is suggested to establish a baseline (Fig. 12.2). Cancer screening is essential, especially in cases where a cancer-predictive neural autoantibody is detected, such as ANNA-1. The detection of a specific neural autoantibody can guide cancer screening; for example, detection of Ma2 antibodies in a young male should suggest a testicular seminoma. FDG/PET-CT of the body increases the yield of malignancy detection in patients with negative screening by whole body computed tomography (CT) [64]. In women, screening should include a gynecological exam with ultrasound and a mammogram, while in men screening should include a testicular ultrasound to look for seminoma. Dermatological evaluations and gastrointestinal endoscopies are suggested in cases of high suspicion of a paraneoplastic neurological disorder. Depending on the clinical and the serological manifestations of autoimmunity, in cases of negative initial cancer screening, this can be repeated in regular intervals (e.g., in cases of ANNA1 autoimmunity in smokers).

Once an autoimmune encephalitis is suspected, then an immunotherapy trial can help define if this is an immunotherapy-responsive condition (Fig. 12.2). In addition, there are some encephalitides that are monophasic or others that are relapsing requiring long-term immunosuppression. If a malignancy is diagnosed, specific treatment is required in combination with immunotherapy, if needed. Treatment approaches to autoimmune encephalitis are discussed in Chap. 17.

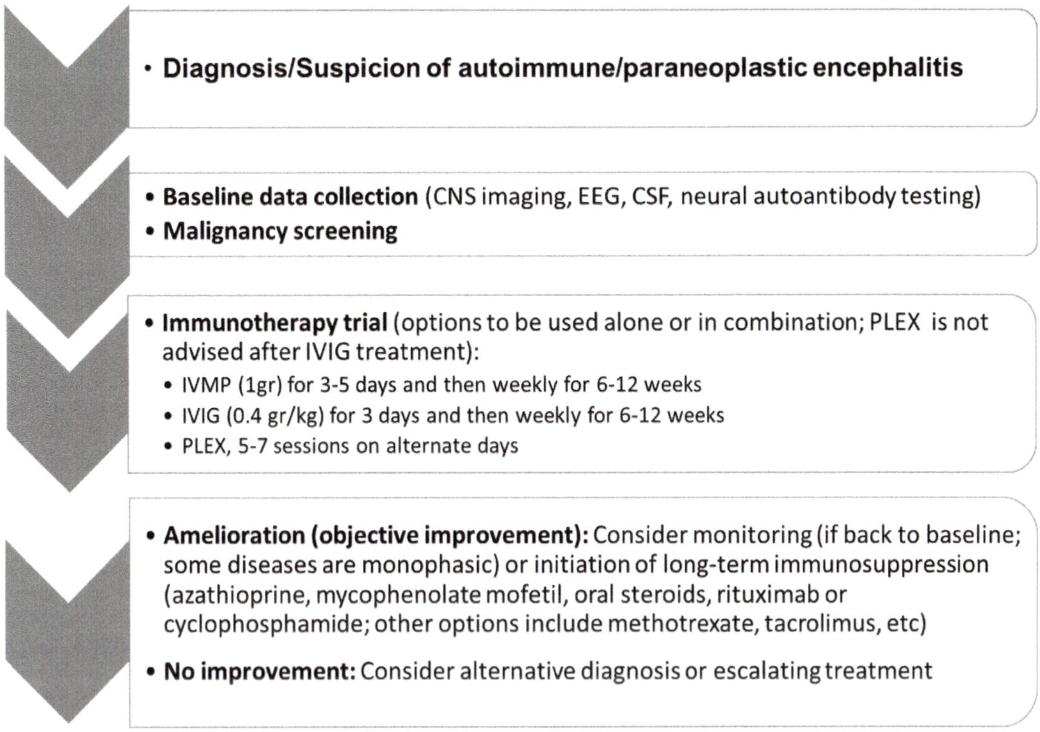


**Fig. 12.2** Management algorithm for suspected autoimmune/paraneoplastic encephalitis

## Conclusion

Autoimmune encephalitis is potentially treatable, especially if diagnosed early. Neural autoantibody testing and recognition of new autoantibodies have expanded the spectrum of manifestations related to autoimmune encephalitis. Cancer screening is essential, as is early immunotherapy treatment. Cognizance of the potential clinical and serological manifestations, cancer associations, and treatment modalities in autoimmune encephalitis can significantly improve patient outcomes.

## References

1. Vora NM, Holman RC, Mehal JM, Steiner CA, Blanton J, Sejvar J. Burden of encephalitis-associated hospitalizations in the United States, 1988–1997. Neurology. 2014;82:443–51.
2. Granerod J, Ambrose HE, Davies NWS, Clewley JP, Walsh AL, Morgan D, et al. Causes of encephalitis and differences in their clinical presentations in England: a multicentre, population-based prospective study. Lancet Infect Dis. 2010;10(12):835–44.
3. Dubey D, Pittock SJ, Kelly CR, McKeon A, Lopez-Chiriboga AS, Lennon VA, et al. Autoimmune encephalitis epidemiology and a comparison to infectious encephalitis. Ann Neurol. 2018;83(1):166–77.
4. Gable MS, Sheriff H, Dalmau J, Tilley DH, Glaser CA. The frequency of autoimmune N-methyl-D-aspartate receptor encephalitis surpasses that of individual viral etiologies in young individuals enrolled in the California encephalitis project. Clin Infect Dis. 2012;54(7):899–904.
5. Thompson J, Bi M, Murchison AG, Makuch M, Bien CG, Chu K, et al. The importance of early immunotherapy in patients with faciobrachial dystonic seizures. Brain. 2018;141(2):348–56.
6. Graus F, Titulaer MJ, Balu R, Benseler S, Bien CG, Cellucci T, et al. A clinical approach to diagnosis of autoimmune encephalitis. Lancet Neurol. 2016;15(4):391–404.
7. Linnoila JJ, Binnicker MJ, Majed M, Klein CJ, McKeon A. CSF herpes virus and autoantibody profiles in the evaluation of encephalitis. Neurol Neuroimmunol Neuroinflamm. 2016;3(4):e245.
8. McKeon A, Pittock SJ. Paraneoplastic encephalomyelopathies: pathology and mechanisms. Acta Neuropathol. 2011;122(4):381–400.

9. Dalmau J, Geis C, Graus F. Autoantibodies to synaptic receptors and neuronal cell surface proteins in autoimmune diseases of the central nervous system. Physiol Rev. 2017;97(2):839–87.
10. Armangue T, Spatola M, Vlagea A, Mattozzi S, Cárceles-Cordon M, Martinez-Heras E, et al. Frequency, symptoms, risk factors, and outcomes of autoimmune encephalitis after herpes simplex encephalitis: a prospective observational study and retrospective analysis. Lancet Neurol. 2018;17(9):760–72.
11. López-Chiriboga AS, Majed M, Fryer J, Dubey D, McKeon A, Flanagan EP, et al. Association of MOG-IgG serostatus with relapse after acute disseminated encephalomyelitis and proposed diagnostic criteria for MOG-IgG–associated disorders. JAMA Neurol. 2018;75(11):1355–63.
12. Zekeridou A, McKeon A, Lennon VA. Frequency of synaptic autoantibody accompaniments and neurological manifestations of thymoma. JAMA Neurol. 2016;73:853–9.
13. Iorio R, Lennon VA. Neural antigen-specific autoimmune disorders. Immunol Rev. 2012;248:104–21.
14. Zekeridou A, Griesmann G, Lennon VA. Mutated cancer autoantigen implicated cause of paraneoplastic myasthenia gravis. Muscle Nerve. 2018;58(4):600–4. https://doi.org/10.1002/mus.26166.
15. Day GS, Laiq S, Tang-Wai DF, Munoz DG. Abnormal neurons in teratomas in NMDAR encephalitis. JAMA Neurol. 2014;71(6):717–24.
16. Koelman DLH, Mateen FJ. Acute disseminated encephalomyelitis: current controversies in diagnosis and outcome. J Neurol. 2015;262(9):2013–24.
17. Dalmau JO, Graus F. Antibody-mediated encephalitis. N Engl J Med. 2018;378:840–51.
18. Crisp SJ, Kullmann DM, Vincent A. Autoimmune synaptopathies. Nat Rev Neurosci. 2016;17(2):103–17.
19. Dalmau J, Tüzün E, Wu H, Masjuan J, Rossi JE, Voloschin A, et al. Paraneoplastic anti-N-methyl-D-aspartate receptor encephalitis associated with ovarian teratoma. Ann Neurol. 2007;61(1):25–36.
20. Titulaer MJ, McCracken L, Gabilondo I, Armangué T, Glaser C, Iizuka T, et al. Treatment and prognostic factors for long-term outcome in patients with anti-NMDA receptor encephalitis: an observational cohort study. Lancet Neurol. 2013;12(2):157–65.
21. Zekeridou A, Karantoni E, Viaccoz A, Ducray F, Gitiaux C, Villega F, et al. Treatment and outcome of children and adolescents with N-methyl-d-aspartate receptor encephalitis. J Neurol. 2015;262:1859.
22. Viaccoz A, Desestret V, Ducray F, Picard G, Cavillon G, Rogemond V, et al. Clinical specificities of adult male patients with NMDA receptor antibodies encephalitis. Neurology. 2014;82(7):556–63.
23. Dalmau J, Lancaster E, Martinez-Hernandez E, Rosenfeld MR, Balice-Gordon R. Clinical experience and laboratory investigations in patients with anti-NMDAR encephalitis. Lancet Neurol. 2011;10(1):63–74.
24. Schmitt SE, Pargeon K, Frechette ES, Hirsch LJ, Dalmau J, Friedman D. Extreme delta brush: a unique EEG pattern in adults with anti-NMDA receptor encephalitis. Neurology. 2013;79:1094–100.
25. Tüzün E, Dalmau J. Limbic encephalitis and variants: classification, diagnosis and treatment. Neurologist. 2007;13(5):261–71.
26. Lang B, Makuch M, Moloney T, Dettmann I, Mindorf S, Probst C, et al. Intracellular and non-neuronal targets of voltage-gated potassium channel complex antibodies. J Neurol Neurosurg Psychiatry. 2017;88:353.
27. Gadoth A, Pittock SJ, Dubey D, Mckeon A, Jeff W, Schmeling JE, et al. Expanded phenotypes and outcomes among 256 LGI1/CASPR2-IgG positive patients. Ann Neurol. 2017;82:79–82.
28. Petit-pedrol M, Armangue T, Peng X, Bataller L, Cellucci T, Davis R, et al. Encephalitis with refractory seizures, status epilepticus, and antibodies to the GABA A receptor: a case series, characterisation of the antigen, and analysis of the eff ects of antibodies. Lancet Neurol. 2014;13(March):14–6.
29. Spatola M, Petit-Pedrol M, Mateus B, Simabukuro M, Armangue T, Castro FJ, et al. Investigations in GABA A receptor antibody-associated encephalitis. Neurology. 2017;88:1012–20.
30. Hoftberger R, Van Sonderen A, Leypoldt F, Houghton D, Geschwind MD, Gelfand J, et al. Encephalitis and AMPA receptor antibodies. Novel findings in a case series of 22 patients. Neurology. 2015;84:2403–12.
31. Joubert B, Kerschen P, Zekeridou A, Desestret V, Rogemond V, Chaffois M-O, et al. Clinical spectrum of encephalitis associated with antibodies against the α-amino-3-Hydroxy-5-Methyl-4-isoxazolepropionic acid receptor. JAMA Neurol. 2015;72(10):1163–9.
32. Lai M, Hughes EG, Peng X, Zhou L, Gleichman AJ, Shu H, et al. AMPA receptor antibodies in limbic encephalitis alter synaptic receptor location. Ann Neurol. 2009;65(4):424–34.
33. Lancaster E, Martinez-Hernandez E, Titulaer MJ, Boulos M, Weaver S, Antoine J-C, et al. Antibodies to metabotropic glutamate receptor 5 in the Ophelia syndrome. Neurology. 2011;77(18):1698–701.
34. Carr I. The ophelia syndrome: memory loss in Hodgkin's disease. Lancet. 1982;319:844–5.
35. Spatola M, Sabater L, Planagumà J, Martínez-Hernandez E, Armangué T, Prüss H, et al. Encephalitis with mGluR5 antibodies. Neurology. 2018;90(22):e1964–72.
36. Tobin WO, Lennon VA, Komorowski L, Probst C, Clardy SL, Aksamit AJ, et al. DPPX potassium channel antibody: frequency, clinical accompaniments, and outcomes in 20 patients. Neurology. 2014;83:1–7.
37. Balint B, Jarius S, Nagel S, Haberkorn U, Probst C, Blöcker IM, et al. Progressive encephalomyelitis with rigidity and myoclonus: a new variant with DPPX antibodies. Neurology. 2014;82(17):1521–8.
38. Gresa-Arribas N, Planagumà J, Petit-Pedrol M, Kawachi I, Katada S, Glaser CA, et al. Human neurexin-3α antibodies associate with encephalitis and alter synapse development. Neurology. 2016;86(24):2235–42.

39. Hinson SR, Lopez-Chiriboga AS, Bower JH, Matsumoto JY, Hassan A, Basal E, et al. Glycine receptor modulating antibody predicting treatable stiff-person spectrum disorders. Neurol Neuroimmunol Neuroinflamm. 2018;5(2):1–7.
40. Carvajal-González A, Leite MI, Waters P, Woodhall M, Coutinho E, Balint B, et al. Glycine receptor antibodies in PERM and related syndromes: characteristics, clinical features and outcomes. Brain. 2014;137(Pt 8):2178–92.
41. Sabater L, Gaig C, Gelpi E, Bataller L, Lewerenz J, Torres-vega E, et al. A novel non-rapid-eye movement and rapid-eye-movement parasomnia with sleep breathing disorder associated with antibodies to IgLON5: a case series, characterisation of the antigen, and post-mortem study. Lancet Neurol. 2014;1(14):1–12.
42. Honorat JA, Komorowski L, Josephs KA, Fechner K, Louis EKS, Hinson SR, et al. IgLON5 antibody Neurological accompaniments and outcomes in 20 patients. Neurol Neuroimmunol Neuroinflamm. 2017;4(5):e385.
43. Zekeridou A, Lennon VA. Aquaporin-4 autoimmunity. Neurol Neuroimmunol Neuroinflamm. 2015;2(4):e110.
44. Do LD, Chanson E, Desestret V, Joubert B, Ducray F, Brugière S, et al. Characteristics in limbic encephalitis with anti-adenylate kinase 5 autoantibodies. Neurology. 2017;88(6):514–24.
45. Gadoth A, Kryzer TJ, Fryer J, Mckeon A, Lennon VA, Ph D, et al. Microtubule associated protein (MAP) 1B : novel paraneoplastic biomarker. Ann Neurol. 2017;2(81):266–77.
46. Yu Z, Kryzer TJ, Griesmann GE, Kim KK, Benarroch EE, Lennon VA. CRMP-5 neuronal autoantibody: marker of lung cancer and thymoma-related autoimmunity. Ann Neurol. 2001;49(2):146–54.
47. Pittock SJ, Lucchinetti CF, Parisi JE, Benarroch EE, Mokri B, Stephan CL, et al. Amphiphysin autoimmunity: paraneoplastic accompaniments. Ann Neurol. 2005;58(1):96–107.
48. Basal E, Zalewski N, Kryzer TJ, Hinson SR, Guo Y, Dubey D, et al. Paraneoplastic neuronal intermediate filament autoimmunity. Neurology. 2018;91(18):e1677–89. https://doi.org/10.1212/WNL.0000000000006435.
49. Dalmau J, Graus F, Villarejo A, Posner JB, Blumenthal D, Thiessen B, et al. Clinical analysis of anti-Ma2-associated encephalitis. Brain. 2004;127(8):1831–44.
50. Rosenfeld MR, Eichen JG, Wade DF, Posner JB, Dalmau J. Molecular and clinical diversity in paraneoplastic immunity to Ma proteins. Ann Neurol. 2001;50(3):339–48.
51. Pittock SJ, Lucchinetti CF, Lennon VA. Anti-neuronal nuclear autoantibody type 2: Paraneoplastic accompaniments. Ann Neurol. 2003;53(5):580–7.
52. Honorat JA, Lopez-Chiriboga AS, Kryzer TJ, Fryer JP, Devine M, Flores A, et al. Autoimmune septin-5 cerebellar ataxia. Neurol Neuroimmunol Neuroinflamm. 2018;5(5):e474.
53. Pittock SJ, Yoshikawa H, Ahlskog JE, Tisch SH, Benarroch EE, Kryzer TJ, et al. Glutamic acid decarboxylase autoimmunity with brainstem, extrapyramidal, and spinal cord dysfunction. Mayo Clin Proc. 2006;81(9):1207–14.
54. Fang B, McKeon A, Hinson SR, Kryzer TJ, Pittock SJ, Aksamit AJ, et al. Autoimmune glial fibrillary acidic protein astrocytopathy: a novel meningoencephalomyelitis. JAMA Neurol. 2016;73:1297.
55. Flanagan EP, Hinson SR, Lennon VA, Fang B, Aksamit AJ, Morris PP, et al. Glial fibrillary acidic protein immunoglobulin G as biomarker of autoimmune astrocytopathy: analysis of 102 patients. Ann Neurol. 2017;81:298–309.
56. Chong JY, Rowland LP, Utiger RD. Hashimoto encephalopathy: syndrome or myth? Arch Neurol. 2003;60(2):164–71.
57. Odaka M, Yuki N, Yamada M, Koga M, Takemi T, Hirata K, et al. Bickerstaff's brainstem encephalitis: clinical features of 62 cases and a subgroup associated with Guillain-Barré syndrome. Brain. 2003;126(10):2279–90.
58. Vernino S, Tuite P, Adler C, Meschia J, Boeve B, Boasberg P, et al. Paraneoplastic chorea associated with CRMP-5 neuronal antibody and lung carcinoma. Ann Neurol. 2002;51(5):625–30.
59. Flanagan EP, Kotsenas AL, Britton JW, McKeon A, Watson RE, Klein CJ, et al. Basal ganglia T1 hyperintensity in LGI1-autoantibody faciobrachial dystonic Seizures. Neurol Neuroimmunol Neuroinflammation. 2015;2(6):e161.
60. Probasco JC, Solnes L, Nalluri A, Cohen J, Jones KM, Zan E, et al. Abnormal brain metabolism on FDG-PET/CT is a common early finding in autoimmune encephalitis. Neurol Neuroimmunol Neuroinflamm. 2017;4(4):1–10.
61. Aurangzeb S, Symmonds M, Knight RK, Kennett R, Wehner T, Irani SR. LGI1-antibody encephalitis is characterised by frequent, multifocal clinical and subclinical seizures. Seizure. 2017;50:14–7.
62. Schmidt LS, Kjær TW, Schmiegelow K, Born AP. EEG with extreme delta brush in young female with methotrexate neurotoxicity supports NMDA receptor involvement. Eur J Paediatr Neurol. 2017;21:795.
63. Gadoth A, Zekeridou A, Klein CJ, Thoreson CJ, Majed M, Dubey D, et al. Elevated LGI1-IgG CSF index predicts worse neurological outcome. Ann Clin Transl Neurol. 2018;5:646.
64. McKeon A, Apiwattanakul M, Lachance DH, Lennon VA, Mandrekar JN, Boeve BF, et al. Positron emission tomography-computed tomography in paraneoplastic neurologic disorders systematic analysis and review. Arch Neurol. 2010;67(3):322–9.

# Autoimmune Epilepsy

# 13

Khalil Husari and Divyanshu Dubey

**Key Points**

1. A considerable proportion of patients (10–16%) with epilepsy of unknown etiology may have an autoimmune or paraneoplastic cause.
2. Among patients with epilepsy, coexisting clinical features such as subacute progressive cognitive decline, psychiatric symptoms, viral prodrome, autonomic dysfunction, inflammatory cerebrospinal fluid (CSF), medial temporal lobes hyperintensities on magnetic resonance imaging (MRI), or presence of underlying malignancy are suggestive of an autoimmune etiology.
3. A predictive scoring system (antibody prevalence in epilepsy and encephalopathy [APE2] score) can be utilized in identifying those patients with autoimmune seizures or epilepsy.
4. Early diagnosis and initiation of immunotherapy is critical for favorable clinical outcomes.
5. The field of autoimmune epilepsy is likely to expand further with discovery of several novel autoantibodies and improved mechanistic understanding.

## Introduction

Epilepsy is a chronic debilitating disease affecting 0.5–1.0% of the world's population [1]. Although epilepsy can arise from different structural, metabolic, infectious, or genetic etiologies, the cause of a significant proportion of cases remains unknown [2]. The link between subset of epilepsies and neuroinflammation has been recognized for decades. This includes the suspected inflammatory pathogenesis of epilepsy syndromes such as Rasmussen's encephalitis [3] and favorable response to immunotherapy of others, for example, Landau-Kleffner syndrome [4]. Furthermore, in the 1960s, initial cases of paraneoplastic limbic encephalitis associated with epilepsy were described [5]. Over the past 2 decades, a plethora of neural autoantibodies targeting cell surface or intracellular antigens associated with encephalopathy and/or epilepsy have been discovered [6]. Many more biomarkers with specific clinical and/or oncological associations are likely to be discovered over the coming

K. Husari
Department of Neurology, Johns Hopkins University, Johns Hopkins Hospital, Baltimore, MD, USA

D. Dubey (✉)
Department of Laboratory Medicine & Pathology, Department of Neurology, Mayo Clinic, Rochester, MN, USA
e-mail: dubey.divyanshu@mayo.edu



A. L. Piquet, E. Alvarez (eds.), *Neuroimmunology*, https://doi.org/10.1007/978-3-030-61883-4_13

years. The rate of discovery may be fueled by development and validation of phage immunoprecipitation sequencing and immunoprecipitation–mass spectrometry techniques [7, 8].

Diagnosis of autoimmune epilepsy, in a majority of the cases, is based on their clinical characteristics, magnetic resonance imaging (MRI) results, cerebrospinal fluid (CSF) analysis, and/or response to immunotherapy trials [9]. The International League Against Epilepsy (ILAE) has recognized autoimmune epilepsy as a distinct entity in the 2017 epilepsy classification [10]. However, diagnostic criteria for autoimmune epilepsy are lacking. A subset of these patients who have coexisting encephalopathy may be characterized using the autoimmune encephalitis diagnostic criteria proposed by Graus et al. in 2016 [11]. Diagnostic criteria for autoimmune encephalitis is further discussed in Chap. 12.

## Epidemiology

The true incidence of autoimmune epilepsy remains unknown. Some information can be deduced from a population-based epidemiological study evaluating incidence and prevalence of autoimmune encephalitis. In Olmsted County, Minnesota, the incidence of autoimmune encephalitis was found to be 0.8/100,000 with a prevalence of 13.7/100,000 [12]. This study also showed significant increase in incidence of autoimmune encephalitis in the last decade, with increased recognition of neural-specific antibodies associated with autoimmune encephalitis [13]. However, in this study, proposed autoimmune encephalitis diagnostic criteria were utilized for selection of their cases. Among the selected cases, only a subset had epilepsy as a part of their syndrome. Additionally, autoimmune epilepsy cases without cognitive impairment were excluded.

A hospital-based prospective study reported that 20% of adult patients with epilepsy of unknown etiology were seropositive for neural-specific antibodies associated with autoimmune epilepsy or encephalopathy [14]. In another UK-based retrospective study, estimated frequency of neural-specific antibodies was 15% among patients without a genetic, structural, or metabolic etiology for epilepsy [14–16]. Epilepsies of unknown etiology are estimated to constitute one-third of all epilepsies among adults [17]. Therefore, the rate of autoimmune epilepsies based on these studies can be inferred to be around 5–7% of all epilepsies, at least in adults. The frequency of autoantibodies in pediatric epilepsy is more unclear. Wright et al. showed the presence of autoantibodies in about 10% of pediatric patients with new-onset epilepsy [18].

## Clinical Presentation

Clinical presentations of autoimmune epilepsy are variable and evolving as new antibodies are being discovered. However, a majority of these cases have coexisting features of autoimmune encephalitis including subacute progressive cognitive decline, psychiatric symptoms, viral prodrome, autonomic dysfunction, inflammatory CSF, oncological association, or brain MRI changes consistent with autoimmune encephalitis [11]. In this regard, a predictive model based on clinical features and initial neurological assessment (antibody prevalence in epilepsy and encephalopathy [APE2] score) may aid in identification of these patients with autoimmune epilepsy [14, 15, 19]. Furthermore, a scoring system for response to immunotherapy (response to immunotherapy in epilepsy and encephalopathy [RITE2] score) may also be utilized for immunotherapy trials (Tables 13.1 and 13.2). The APE2 score of greater or equal to 4 was 99% sensitive and 93% specific for neural-specific-antibodies, while a RITE2 score of greater than or equal to 7 had 96% sensitivity and 86% specificity for favorable initial immunotherapy response [19].

## Neural-Specific Antibodies Associated with Autoimmune Epilepsy

### Cell Surface Epitopes

N-methyl-D-aspartate receptor (NMDA-R) encephalitis typically affects young women with a reported median age of 22 years (range:

**Table 13.1** Components of the APE2 score. The assigned APE2 score is the sum of values for all components

| 1A: Antibody prevalence in epilepsy and encephalopathy (APE2score) | Value: |
|---|---|
| New onset, rapidly progressive mental status changes that developed over 1–6 weeks, or new onset seizure activity (within 1 year of evaluation) | (+1) |
| Neuropsychiatric changes; agitation, aggressiveness, emotional lability | (+1) |
| Autonomic dysfunction[a] | (+1) |
| Viral prodrome (rhinorrhea, sore throat, low-grade fever) to be scored in the absence of underlying systemic malignancy within 5 years of neurological symptom onset | (+2) |
| Faciobrachial dystonic seizures | (+3) |
| Facial dyskinesias, to be scored in the absence of faciobrachial dystonic seizures | (+2) |
| Seizure refractory to at least to two anti-seizure medications | (+2) |
| CSF findings consistent with inflammation[b] (elevated CSF protein >50 mg/dL and/or lymphocytic pleocytosis >5 cells/mcL, if the total number of CSF RBC is <1000 cells/mcL) | (+2) |
| Brain MRI suggesting encephalitis[b] (T2/FLAIR hyperintensity restricted to one or both medial temporal lobes, or multifocal in grey matter, white matter, or both compatible with demyelination or inflammation) | (+2) |
| Neoplasm diagnosed within 5 years of neurological symptom onset (excluding cutaneous squamous cell carcinoma, basal cell carcinoma, brain tumor, cancer with brain metastasis) | (+2) |
| | Total (max: 18) |

*Abbreviations: CSF* cerebrospinal fluid, *RBC* red blood cells, *MRI* magnetic resonance imaging, *FLAIR* fluid attenuated inversion recovery

Key: [a]Sustained atrial tachycardia or bradycardia, orthostatic hypotension (≥20 mm Hg fall in systolic pressure or ≥10 mm Hg fall in diastolic pressure within 3 minutes of quiet standing), hyperhidrosis, persistently labile blood pressure, ventricular tachycardia, cardiac asystole, or gastrointestinal dysmotility. Scored only if no history of autonomic dysfunction prior to onset of suspected autoimmune syndrome and the autonomic dysfunction not attributable to medications, hypovolemia, plasmapheresis, or infection. [b]Patients scored zero if MRI brain or CSF analysis not performed

2 months to 85 years) [20]. Clinical presentation usually begins with a prodrome of a headache or fever, followed by psychiatric manifestations including delusions, hallucinations, mania-like episodes, alternating episodes of extreme agitation, and catatonia. Patients then progress to develop

**Table 13.2** Components of the RITE2 score. RITE2 score included all the components of APE2 score and two additional variables: initiation of immunotherapy within 6 months of symptom onset and plasma membrane-specific autoantibody detected. The assigned RITE2 scores are the sum of values for all components

| 1B: Response to immunotherapy in epilepsy and encephalopathy score (RITE2 score) | Value: |
|---|---|
| New-onset, rapidly progressive mental status changes that developed over 1–6 weeks or new onset seizure activity (within 1 year of evaluation) | (+1) |
| Neuropsychiatric changes; agitation, aggressiveness, emotional labiality | (+1) |
| Autonomic dysfunction[a] | (+1) |
| Viral prodrome (rhinorrhea, sore throat, low-grade fever) only to be scored in the absence of underlying malignancy within 5 years of neurological symptom onset | (+2) |
| Faciobrachial dystonic movements | (+3) |
| Facial dyskinesias, to be scored in the absence of faciobrachial dystonic seizures | (+2) |
| Seizure refractory to at least two anti-seizure medications | (+2) |
| CSF findings consistent with inflammation[b] (elevated CSF protein >50 mg/dL and/or lymphocytic pleocytosis >5 cells/mcL, if the total number of CSF RBC is <1000 cells/mcL) | (+2) |
| Brain MRI suggesting encephalitis[b] (T2/FLAIR hyperintensity restricted to one or both medial temporal lobes, or multifocal in grey matter, white matter, or both compatible with demyelination or inflammation) | (+2) |
| Neoplasm diagnosed within 5 years of neurological symptom onset (excluding cutaneous squamous cell carcinoma, basal cell carcinoma, brain tumor, cancer with brain metastasis) | (+2) |
| Immunotherapy initiated within 6 months of symptom onset | (+2) |

(continued)

**Table 13.2** (continued)

| 1B: Response to immunotherapy in epilepsy and encephalopathy score (RITE2 score) | Value: |
|---|---|
| Neural plasma membrane autoantibody detected (NMDAR, $GABA_AR$, $GABA_BR$, AMPAR, DPPX, mGluR1, mGluR2, mGluR5, LGI1, CASPR2 or MOG) | (+2) |
| | Total (max: 22) |

*Abbreviations: AMPAR* amino-3-hydroxy-5-methyl-4-isoxazolepropionic, *ANNA-1* Antineuronal nuclear antibody-1, *ANNA-2* antineuronal nuclear antibody-2, *ANNA-3* antineuronal nuclear antibody-3, *CASPR-2* contactin-associated protein-2, *CRMP5* collapsin response-mediator protein-5, *CSF* cerebrospinal fluid, *DPPX* dipeptidyl-peptidase-like protein 6, *FLAIR* fluid attenuated inversion recovery, *GAD65* glutamic acid decarboxylase-65, *GABABR* γ(gamma)-aminobutyric acid-B receptor, *GFAP α(alpha)* glial fibrillary acidic protein, *LGI1* leucine-rich glioma-inactivated protein-1, *MOG* myelin oligodendrocyte glycoprotein, *MRI* magnetic resonance imaging, *NMDAR* N-methyl D-aspartate receptor, *PCA-1* Purkinje cell cytoplasmic antibody type 1, *PCA-2* Purkinje cell cytoplasmic antibody type 2, *RBC* red blood cells

Key: [a]Sustained atrial tachycardia or bradycardia, orthostatic hypotension (≥20 mmHg fall in systolic pressure or ≥10 mmHg fall in diastolic pressure within three minutes of quiet standing), hyperhidrosis, persistently labile blood pressure, ventricular tachycardia, cardiac asystole or gastrointestinal dysmotility. Scored only if no history of autonomic dysfunction prior to onset of suspected autoimmune syndrome and the autonomic dysfunction not attributable to medications, hypovolemia, plasmapheresis, or infection. [b]Patients scored zero if MRI brain or CSF analysis not performed

seizures, encephalopathy, oral dyskinesia, choreoathetosis, and autonomic dysfunction [21]. Seizures in NMDA-R encephalitis are usually focal non-motor seizures that might progress to refractory status epilepticus [22]. If untreated, patients will progress to a comatose state [21].

In about half of the patients, a trigger can be identified. The two main triggers are the presence of ovarian teratoma [23] and a history of herpes simplex virus (HSV) encephalitis [24]. Approximately two-thirds of adult women between the ages of 18 to 45 years with NMDA-R encephalitis have been reported to have ovarian teratoma [21]. However, the presence of this tumor is extremely rare in children younger than 12 years or older adults (≥45 years) [6, 21, 25]. Furthermore, prospective evaluation of HSV encephalitis patients showed that 17% of these cases developed NMDA-R encephalitis during follow-up. Three additional patients in this cohort were positive for NMDA-R immunoglobulin G (IgG) without any clinical features of autoimmune encephalitis on follow-up evaluation [24].

Leucine-rich, glioma-inactivated 1 (LGI1) immunoglobulin G is typically associated with seizures and memory deficits usually among older patients (>40 years). However, a few pediatric cases have also been described [26]. One characteristic phenotype described among the adult patients is faciobrachial dystonic seizures (FBDS). These are brief focal dystonic motor seizures and occur multiple times a day. They have a characteristic stereotypic contraction of the face, arm, and leg [27]. Another characteristic seizure semiology is unilateral piloerections episodes. More recently, paroxysmal dizzy spells have also been described in a subset of patients [28]. These "dizzy spells" or "out of body experiences" may precede encephalopathy by 2–12 months.

A minority (~2%) of patients with voltage-gated potassium channel-complex (VGKC) antibodies have coexisting contactin-associated protein-like 2 (CASPR-2) IgG. Peripheral nervous system involvement is more common (neuromyotonia, myokymia, or dysautonomia) among these patients. However, a considerable proportion of patients, especially older patients, may have coexisting epilepsy or encephalitis. Recent studies have highlighted that VGKC IgG in the absence of LGI1 and/or CASPR-2 IgG seropositivity is not a specific biomarker of autoimmunity [29].

α(alpha)-amino-3-hydroxy-5-methyl-4-isoxazolepropionic acid (AMPA) receptor antibody-associated encephalitis typically presents with classic limbic encephalitis symptoms (anterograde memory deficits, retrograde amnesia, mood changes, and temporal lobe seizures). Recent studies have supported direct antibody-mediated pathogenicity [30, 31]. Median age of onset is around 60 years old (range: 23–81 years), and it occurs more commonly in females (64%) [21]. Two-thirds of the patients have underlying malignancy, mainly small cell lung cancer and thymoma [32]. A considerable proportion of patients have a refractory course and go on to develop diffuse cortical atrophy [30, 31, 33].

γ(gamma)-aminobutyric acid type B (GABA-B) receptor encephalitis usually presents as refractory non-convulsive status epilepticus [21]. Median age of onset is 61 years (range: 16–77 years) and tends to occur more commonly in males [6]. A subset of these cases have underlying malignancy, most commonly small cell lung carcinoma.

Fulminant encephalitis and refractory seizures or status epilepticus have been associated with GABA-A receptor encephalitis. These patients have characteristic multifocal cortical and subcortical T2/FLAIR hyperintensities [22]. Age of symptom onset tends to be younger (median age: 40 years) than cases with GABA-B encephalitis [6].

Patients with dipeptidyl-peptidase-like-protein 6 (DPPX) antibody-associated encephalitis can also have seizures as part of the syndrome. The usual clinical manifestations include gastrointestinal dysfunction, weight loss followed by cognitive dysfunction, hyperekplexia, myoclonus, parasomnias, and occasionally progressive encephalomyelitis with rigidity and myoclonus (PERM) [21, 34].

Metabotropic glutamate receptor 5 (mGluR5) IgG encephalitis patients usually present with subacute onset of encephalopathy, mood changes, movement disorder, and seizures [35]. Status epilepticus has been reported to be a common presenting feature especially among pediatric cases [36].

## Intracellular Epitopes

Glutamic acid decarboxylase (65 kd, GAD65) antibodies (serum titers >20 nmol/L or detection in CSF) are associated with various autoimmune neurological diseases including autoimmune epilepsy, stiff person syndrome, cerebellar ataxia, limbic encephalitis, and PERM [37, 38]. Women are more frequently affected than men, and the median age of symptom onset is 30 years (range: 5–80 years) [22]. In a study of 112 patients with unexplained adult-onset focal epilepsy, 5.4% were found to have high titers of GAD65 antibodies (>1000 U/mL) [39]. Usually patients with GAD65 antibodies (serum titers >20 nmol/L or detection in CSF) are associated with a treatment-refractory course [38, 39]. Refractory nature of the disease is postulated to be secondary to cell-mediated cytotoxicity rather than a direct antibody-mediated pathogenesis.

Patients with antineuronal nuclear antibody type-1(ANNA-1, a.k.a. anti-Hu) IgG antibodies present with various central and peripheral nervous system manifestations. ANNA-1 IgG seropositivity has a strong association with small cell cancer (81%). Sensory neuronopathy and autonomic dysfunction, especially gastroparesis, are hallmarks of ANNA-1 autoimmunity [13]. However, a considerable proportion (10–17%) of cases present with limbic encephalitis or refractory seizures. Both temporal and extratemporal localization of the seizures have been reported [40].

Seizures and limbic encephalitis are less common among patients with ANNA-2 IgG (a.k.a anti-Ri). These patients usually present with a brainstem or cerebellar syndrome. Initial manifestations include opsoclonus-myoclonus, jaw opening dystonia, and laryngospasms [41].

Ma2 IgG patients usually have a limbic encephalitis or brainstem encephalitis phenotype. In a retrospective study, bilateral tonic-clonic or focal unaware seizures occurred in 12 out of 27 (44%) patients [42]. A majority of Ma2 IgG seropositive have testicular germ cell tumor [43].

Collapsin response-mediator protein-5 (CRMP5) IgG is a paraneoplastic biomarker of small cell lung cancer or thymoma [44]. Patients with CRMP-5 IgG usually manifest with various neurologic signs including chorea, cranial neuropathy, dementia, cerebellar ataxia, myelopathy, and peripheral neuropathy [22, 45, 46]. Focal aware and unaware seizures have also been rarely reported in patients with CRMP-5. Management of underlying malignancy and early initiation of immunotherapy may be associated with favorable outcomes [47].

## Rasmussen's Encephalitis

Rasmussen's encephalitis is a rare chronic neurological disorder characterized by drug-resistant focal motor epilepsy (epilepsia partialis continua), cognitive decline, hemiplegia, and unilateral hemispheric brain atrophy [5]. Clinical onset is usually during childhood, although it has been reported during adulthood. The disease progresses over three stages, with the first being a "prodromal stage" with a relatively low seizure frequency and rarely a mild hemiparesis. Following that, the patient will enter the "acute stage," which is characterized by frequent intractable seizures along with progressive neurological decline (hemiparesis, hemianopia, cognitive deterioration, and aphasia if dominant hemisphere). During the final "residual" stage, the patient develops permanent and stable neurological deficits and intractable focal motor seizures [48]. CSF examination may be normal or may show inflammatory changes (lymphocytic pleocytosis and elevated CSF protein). Electroencephalogram (EEG) shows unilateral inter-ictal epileptiform discharges, slowing, and ictal rhythms with occasional spread to the contralateral side due to bilateral synchrony. No specific electrographic signatures have been associated with Rasmussen's encephalitis. MRI brain scan shows FLAIR/T2 hyperintensity and atrophy involving unilateral cortical and/or subcortical regions with a predilection for perisylvian area. Fluorodeoxyglucose positron emission tomography (FDG-PET) scan may show increased uptake in the affected hemisphere [5, 48]. Multiple case reports have demonstrated favorable response to immunotherapy. In 2013, a randomized trial of tacrolimus and intravenous immunoglobulin showed slowing down of tissue and function loss with either therapies, but without improvement in seizures [49]. At present, hemispherectomy remains the best option of seizure control and arresting the neurological decline [50].

## New-Onset Refractory Status Epilepticus

In new-onset refractory status epilepticus (NORSE), a previously healthy individual develops refractory de novo seizures and status epilepticus with no readily identifiable etiology. A retrospective study exploring NORSE in the adult population has shown a significant percentage of these patients to have immune-mediated etiologies—primarily antibody-mediated encephalitis with anti-NMDA receptor being the most common etiology [51]. Various treatments have been tried, including anti-seizure medications, achieving burst suppression with anesthetics, and dietary therapy with modest and variable effects [52]. Use of immunotherapy has been associated with favorable outcomes in a few cases (5–33%) [52].

## Steroid Responsive Encephalopathy with Autoimmune Thyroiditis (SREAT) or Hashimoto's Encephalopathy

Clinical characteristics of Hashimoto's encephalopathy include encephalopathy, seizures, stroke-like episodes, and myoclonus [13]. These patients typically have thyroid peroxidase (TPO) antibodies but may or may not have a history of thyroiditis. Seizure presentations are variable including new-onset refractory status epilepticus or progressive myoclonic epilepsy [5, 53]. A triad of encephalopathy, evidence of thyroid autoimmunity (clinically or serologically), and a favorable response to steroids have been traditionally utilized for identification of these cases [11].

# Electroencephalogram

EEG plays a vital role in diagnosis and management of autoimmune epilepsy and encephalitis. It is essential to look for epileptiform or seizure activity. Long-term monitoring is utilized among patients with subclinical or clinical status epilepticus [54]. Additionally, EEG can also be utilized to evaluate response to immunotherapy and antiepileptic drugs in some instances.

EEG findings in autoimmune encephalitis are variable and may be nonspecific. Extreme delta brush (EDB) was first described in NMDA encephalitis patients by Schmitt et al. [55]. This EEG pattern consists of rhythmic delta activity at 1–3 Hz with superimposed burst of rhythmic beta activity at 20–30 Hz riding on each delta wave [55]. EDB has been reported in about 30% of acutely ill or hospitalized NMDA encephalitis patients and is associated with a more prolonged illness. Although it was thought to be fairly specific initially, recent studies have described the presence of EDB with other metabolic and structural causes of encephalopathy [56].

FBDS, a pathognomonic feature of LGI1 autoimmune epilepsy, usually has no ictal EEG correlate. Occasionally preceding electrodecrement or sharply contoured rhythmic delta activity has been reported over the contralateral frontotemporal region [57]. In a small series of LGI1 patients, frequent subclinical temporal lobe seizures associated with hyperventilation were reported. This ictal activity had similar morphology to subclinical rhythmic electrographic discharges of adults (SREDA) [58]. A more recent study of EEGs with 16 LGI1 encephalitis patients reported multiple frequent seizure semiologies or subclinical seizures associated with temporal and frontal discharges [57]. Additionally, multifocal interictal epileptiform discharges and interictal slow-wave activity were observed in 25% and 69%, respectively.

## Imaging

Brain MRI is usually included in initial evaluation of autoimmune epilepsy and encephalitis. Features considered suggestive of autoimmune encephalitis include T2/FLAIR hyperintensity restricted to one or both medial temporal lobes (Fig. 13.1) or multifocal T2/FLAIR hyperintensities in the gray matter, white matter, or both compatible with demyelination or inflammation [11]. However, MRI may be normal, especially early in the course of the disease [59, 60]. Brain MRI also provides valuable information regarding differential diagnosis of new-onset epilepsy and/or subacute onset cognitive dysfunction such as tumors, brain abscess, neuro-sarcoidosis, and other inflammatory and infectious diseases. Beside brain MRI, abnormalities in functional MRI [61], diffusion tensor imaging (DTI) [61], FDG-PET/CT [62], and single photon emission tomography (SPECT) [63] have been described in patients with autoimmune epilepsy and can provide valuable diagnostic and at times prognostic values.

In the initial phase of NMDA-R encephalitis, the brain MRI abnormalities are typically discrete and nonspecific [20, 64]. However, resting state functional MRI shows disrupted hippocampal functional connectivity. Moreover, DTI has detected a more widespread white matter damage that correlated with disease severity [61]. Decreased occipital lobe metabolism on FDG-PET/CT has been described as a unique finding in these patients [65]. Resolution of lateral and medial occipital hypometabolism may correlate with better outcomes.

Brain MRI findings in patients with LGI1 encephalitis varies depending on the stage of the disease and the progression [65]. In the early phase of the disease, brain MRI is typically normal, although basal ganglia abnormalities including increased FLAIR signal, restricted diffusion, and contrast enhancement have been reported [66]. Patients with FBDS may develop T1 hyperintensity in region of basal ganglia [67]. As the disease progresses, unilateral or bilateral T2/FLAIR hyperintensities of the medial temporal lobes and basal ganglia are observed. On follow-up imaging, hippocampal atrophy is frequently seen [65]. FDG-PET/CT reveals basal ganglia hypermetabolism, a specific finding, and frequently the earliest finding on imaging studies [68].

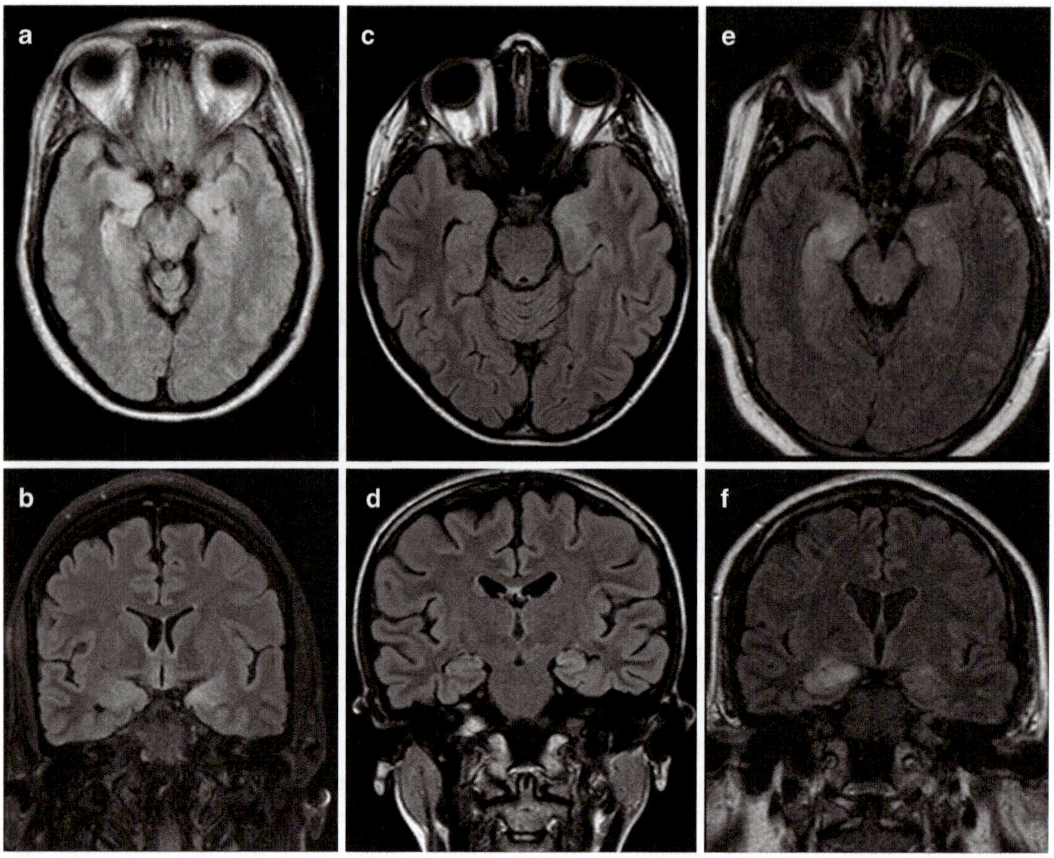


**Fig. 13.1** Patient 1 with LGI1 IgG limbic encephalitis. MRI brain (FLAIR sequence) showing bilateral medial temporal hyperintensities on axial (**a**) and sagittal sections (**b**). Patient 2 with ANNA1 IgG limbic encephalitis. MRI brain (FLAIR sequence) showing bilateral medial temporal hyperintensities on axial (**c**) and sagittal sections (**d**). Patient 3 with Ma2 IgG limbic encephalitis. MRI brain (FLAIR sequence) showing bilateral medial temporal (right greater than left) hyperintensities on axial (**e**) and sagittal sections (**f**). *Abbreviations:* ANNA-1 antineuronal nuclear antibody-1, MRI magnetic resonance imaging, FLAIR fluid attenuated inversion recovery, LGI1 leucine-rich, glioma inactivated 1

Among patients with GAD65 antibody-associated autoimmune epilepsy, brain MRI demonstrates disproportionate parenchymal atrophy for age and abnormal cortical/subcortical T2 hyperintensities. Hippocampal abnormalities are seen only in a minority (26%) of patients [69].

Patients with mGluR5 antibodies have abnormal brain MRI in 45% of the cases, with involvement of both limbic and extralimbic (thalamus, pons, cerebral, and cerebellar cortices) regions. Additionally, FDG-PET in some of these cases demonstrates hypometabolism of the temporoparietal cortex or cerebellum [36]. Whereas GABA-A receptor encephalitis has a unique pattern of widespread and extensive cortical and subcortical FLAIR hyperintensity [70].

Medial temporal lobe involvement has been reported in multiple antibody specificities including AMPA-R [71], GABA-B receptor IgG [72], ANNA-1 IgG [40], Ma2-IgG [42], adenylate kinase 5 [73], etc. A majority of these cases do not have associated gadolinium enhancement except for Ma2 IgG-associated limbic encephalitis [42].

## Cancer Screening

CT of the chest, abdomen, and pelvis with contrast is recommended as initial evaluation for cancer association. Scrotal ultrasounds should be performed in all males, especially those presenting with brainstem, diencephalic or limbic encephalitis. In women, mammograms should be performed for evaluation of breast cancer. Pelvic sonography and pelvic MRI are recommended for ovarian teratoma or adenocarcinoma screening. If initial radiological evaluations did not reveal any malignancies and clinical suspicion for paraneoplastic neurological syndrome is high or the patient has neural specific antibody with strong oncological association (Table 13.3), PET-CT should be pursued [74, 75]. If the patient's evaluation reveals a neoplasm other than that predicted by the antibody present, further cancer evaluation should be performed as more than one cancer can coexist [74]. Paraneoplastic syndromes and associated malignancies are further discussed in Chap. 16.

## Treatment

Management of autoimmune epilepsy is focused on immunotherapies. Multiple studies have demonstrated favorable effects of early immunotherapy on seizure frequency and cognition [9, 15, 76, 77]. However, randomized control trials evaluating efficacy immunotherapy in autoimmune epilepsy are limited. Therefore, immunotherapy recommendations are largely based on case series and clinical experience [20, 78]. Recently, a placebo-controlled trial of IVIg for LGI1 and CASPR2 IgG-associated autoimmune epilepsy was published [79]. In this study of 17 patients, 75% of the patients receiving IVIg achieved more than 50% seizure reduction by 5 weeks compared to only 22% of patients in the placebo arm [79]. In some instances, a positive response to a treatment trial of immunotherapy can aid in the diagnosis of seronegative autoimmune epilepsy [9]. In this regard, the RITE2 score may be a useful scoring system for managing clinician prior to immunotherapy initiation [15].

Immunotherapeutic agents are classically divided into first-line and second-line therapies (Table 13.4). First-line therapies include high-dose intravenous methylprednisolone (IVMP), intravenous immunoglobulin, or plasmapheresis. Second-line agents such as rituximab, cyclophosphamide, mycophenolate, azathioprine, or bortezomib are used in refractory cases or as a maintenance therapy to prevent relapses.

Treatment of autoimmune epilepsy should be based on the severity of the clinical course (Fig. 13.2). In patients with rapid progression and refractory course, more aggressive immunotherapy is needed including both first- and second-line therapies. Conversely, some patients with autoimmune epilepsy have a more benign course, and their epilepsy can be controlled with antiepileptic drugs and a short course of immunotherapy. In all cases, cancer surveillance, as discussed previously should be pursued as treatment of the underlying cancer, is pivotal for the successful treatment of autoimmune epilepsy.

Proposed immunotherapy trials include 6 or 12 weeks of high-dose IV methylprednisolone (IVMP). Intravenous methylprednisolone 1000 mg per day for 3 days followed by once weekly for 5 weeks (6 IVMP week trial), followed by once every 2 weeks for 6 weeks (12 IVMP week trial). If the patient has contraindications for IVMP (active infections, poorly controlled diabetes, chronic hepatitis, or tuberculosis), a 6- or 12-week course of intravenous immunoglobulin[1] may be considered. These include 0.4 g/kg IVIg daily for 3 days followed by 0.4 g/kg every week for 6 weeks (6 IVIg week trial) and then every 2 weeks for 6 weeks (12 IVIg week trial). A treatment response can be ascertained using a seizure diary to assess seizure frequency and/or change in semiology and neurological examination including screening mental status examination after completion of immunotherapy trial. The quality of life in epilepsy (QOLIE-31) can be utilized as well. EEG,

**Table 13.3** Clinical features of specific neural autoantibody-associated syndromes

| Antibody | Neurological presentations | Epilepsy association (3+, 2+, 1+) | Epilepsy presentations | Brain MRI | Cancer association (3+, 2+, 1+) | Specific cancer type |
|---|---|---|---|---|---|---|
| LGI1 | FBDS, piloerection seizures, peripheral nervous system disease symptoms, limbic encephalitis | 3+ | FBDS, unilateral piloerection, paroxysmal dizzy spells | medial temporal FLAIR hyperintensity, T1 basal ganglia hyperintensity (FBDS cases) | 1+, 2+[a] | Thymoma |
| GABA-B | SE, limbic encephalitis | 3+ | Crescendo seizures, SE is frequent. | medial temporal FLAIR hyperintensity | 2+ | Small cell lung cancer |
| GABA-A | SE, autoimmune encephalitis | 3+ | Crescendo seizures and status epilepticus | Multifocal cortical and subcortical hyperintensity | 1+ | Thymoma |
| NMDA-R | Oral dyskinesia, catatonia, neuropsychiatric dysfunction, autonomic dysfunction, refractory epilepsy | 2+ | Crescendo seizures, status epilepticus and encephalopathy (EEG: extreme delta-brush) | normal or nonspecific cortical and/or subcortical changes | 2+ | Ovarian teratoma |
| GAD-65 | SPS, hyperekplexia, brainstem encephalitis (especially African Americans) | 2+ | Multifocal epilepsy, drug-resistant epilepsy, rarely SE | multifocal cortical and subcortical hyperintensity or brainstem hyperintensity | 1+ | Thymoma |
| AMPA-R | Limbic encephalitis | 2+ | SE has been reported | Cortical atrophy, deep gray nuclei FLAIR hyperintensity | 2+ | Thymoma, small cell lung cancer, breast adenocarcinoma |
| mGluR5 | Encephalopathy, mood changes, movement disorder and seizures | 2+ | Seizures are common, SE in children. | Normal in 50%, limbic/ cortical FLAIR changes | 2–3+ | Hodgkin lymphoma |
| ANNA-1/Hu | Limbic encephalitis, Sensory neuronopathy, autonomic dysfunction | 2+ | Temporal and/or extra-temporal seizures. Rarely SE | normal or medial temporal FLAIR hyperintensity | 3+ | Small cell lung cancer, neuroendocrine tumors |
| ANNA-2/Ri | Stridor, laryngospasm, jaw dystonia, opsoclonus myoclonus | 1+ | Seizures are rare | brainstem FLAIR hyperintensity and/or atrophy | 3+ | Small cell lung cancer, breast cancer |
| Ma-1/Ma-2 | Limbic encephalitis, Brainstem encephalitis | 2+ | Focal unaware seizure or bilateral tonic-clonic | brainstem FLAIR hyperintensity or medial temporal FLAIR hyperintensity | 3+ | Testicular germ cell tumor[b], small cell lung cancer[c] |
| Amphiphysin | SPS, PERM, transverse myelitis | 2+ | Limbic encephalitis and seizures can occur in up to 30% of patients. | normal or nonspecific cortical and/or subcortical changes | 2+ | Small cell lung cancer, breast cancer |
| CASPR2 | Neuromyotonia, Morvan's syndrome, limbic encephalitis, refractory epilepsy, sleep disorder | 1+ | Focal unware seizures or focal motor/sensory seizures. | normal or medial temporal FLAIR hyperintensity | 1, 2+[a] | Thymoma |

| | | | | | | |
|---|---|---|---|---|---|---|
| Glycine | SPS, PERM | 1+ | Rarely associated with seizures. | normal or nonspecific cortical and/or subcortical changes | Rare (<5%) | Thymoma |
| DPPX | Diarrhea, hyperekplexia, Ambiguous sleep, parasomnias, PERM, | 1+ | Rarely associated with seizures. | normal or nonspecific cortical and/or subcortical changes | Rare (<10%) | Lymphoma |
| GFAPα(alpha) | Meningo-encephalomyelitis, tremor, ataxia, autonomic dysfunction | 1+ | Rarely associated with seizures | Peri-radial/patchy enhancement or diffuse subcortical hyperintensity | 1+ | Ovarian teratoma |
| CRMP5 | Choreo-athetosis, optic neuritis, retinitis, limbic encephalits, ataxia, transverse myelitis, polyradiculoneuropathy | 1+ | Rarely associated with seizures (Focal aware and focal unaware) | normal or medial temporal FLAIR hyperintensity | 3+ | Small cell lung cancer, thymoma |
| IgLON5 | Parasomnias, REM and NREM dysfunction, brainstem dysfunction, hyper-excitability disorder | Rare | Rarely associated with nocturnal frontal lobe epilepsy | normal or nonspecific cortical and/or subcortical changes | Unknown | – |
| Neurexin-3a: | prodrome of fever, headache and gastrointestinal symptoms followed by development of encephalopathy and seizures | 1+ | Limited data available. | Normal | None | None |
| MOG | ADEM, ON, TM | 1+ | Rarely associated with refractory seizures and status epilepticus | Multifocal demyelination, involvement of corpus callosum, deep gray nuclei | – | – |
| Adenylate kinase 5 | Limbic encephalitis | Rare | Limited data available. One patient reported to have seizure 6 months after disease onset. | Bilateral medial temporal FLAIR hyperintensity | None | None |

*Abbreviations: ADEM* acute disseminated encephalomyelitis, *AMPA-R* amino-3-hydroxy-5-methyl-4-isoxazolepropionic, *ANNA-1* antineuronal nuclear antibody-1, *ANNA-2* antineuronal nuclear antibody-2, *CBA* cell based assay, *CASPR2* contactin-associated protein-like 2, *CRMP-5* collapsin response-mediator protein 5, *DPPX* dipeptidyl-peptidase-like protein-6, *EEG* electroencephalogram, *EMG* electromyography, *FBDS* faciobrachial dystonic seizures, *FLAIR* fluid attenuated inversion recovery, *GABA-A* gamma-aminobutyric acid type A, *GABA-B* gamma-aminobutyric acid type B, *GAD-65* glutamic acid decarboxylase 65, *GFAP* glial fibrillary acidic protein, *LGI1* leucine-rich: glioma inactivated-1, *MOG* myelin oligodendrocyte glycoprotein, *MRI* magnetic resonance imaging, *NMDA-R* N-methyl-D-aspartate receptor, *ON* optic neuritis, *PERM* progressive encephalomyelitis with rigidity and myoclonus, *REM* rapid eye movement, *SE* status epilepticus, *SPS* stiff person syndrome, *TM* transverse myelitis, *WB* western blot

Key: 1+, 10–30%; 2+, 30–60%; 3+, >60%

[a]coexisting LGI1 and CASPR-2 antibodies

[b]Ma2 antibodies

[c]Ma1 antibodies with or without Ma2 antibodies

**Table 13.4** Review of dosing, mechanism of action, adverse effects, and monitoring of various immunomodulatory agents

| Immunomodulatory Agent | Route of Administration | Dosing/Regimen | Mechanism of action | Adverse effects | Monitoring and Prophylaxis |
|---|---|---|---|---|---|
| Corticosteroids | IV or PO | Initial dose: IV methylprednisolone (1 g per day for 3–5 days)<br>6 week trial: 1 g per day for 3 days followed by once weekly for 5 weeks<br>12 week trial: 1 g per day for 3 days followed by once weekly for 5 weeks, followed by once every 2 weeks for 6 weeks.<br>Oral Maintenance: 60–80 mg prednisone daily, duration of taper variable (usually over 4–8 weeks) | Acts on nuclear glucocorticoids receptors to reduce cytokine and chemokine production<br>Reduces migration of leukocytes to the target tissue | Insomnia, psychiatric dysfunction, hyperglycemia, electrolyte imbalances, fluid retention, hypertension, peptic ulcer, Cushing syndrome, cataracts, infections, osteoporosis and avascular necrosis.<br>Addisonian crisis on rapid withdrawal of corticosteroids | Blood pressure, serum electrolytes and glucose monitoring<br>PJP ppx: TMP/SMX<br>osteoporosis ppx: Vitamin D + calcium<br>GI ppx: PPI or H2 blocker |
| Plasmapheresis | IV | 1 exchange every other day for 10–14 days | Extracorporeal blood filtration designed to remove large molecular weight molecules, including immunoglobulins, immune complex, and complements | Hypotension, electrolyte imbalance, perioral paresthesia (hypocalcemia), coagulopathy, central line infection, hemorrhage, thrombosis, and pneumothorax | PT, INR, PTT, Fibrinogen |
| IVIg | IV | Initial dose: 0.4 g/kg daily for 5 days<br>6 week trial: 0.4 g/kg daily for 3 days followed by 0.4 g/kg every week for 5 weeks.<br>12 week trial: 0.4 g/kg daily for 3 days followed by 0.4 g/kg every week for 6 weeks and then every 2 weeks for 6 weeks. | Remains unclear<br>Interaction with antigen-binding fragment on the antibodies and/or crystalizable fragment on the antibodies or the antigen presenting cells | Headache, aseptic meningitis, acute renal failure, thrombotic/thromboembolic events anaphylaxis due to IgA deficiency | Electrolytes, renal function |

| | | | | | |
|---|---|---|---|---|---|
| Rituximab | IV | 1000 mg followed by 2nd dose in 2 weeks, 375 mg/m² weekly for 4 weeks | B-cell depletion by antibody dependent cellular cytotoxicity, complement dependent cytotoxicity and apoptosis | Allergic reaction, opportunistic infection, reactivation of tuberculosis infection or hepatitis B infection, PML | Hepatitis B antibodies, Quatiferon test (for latent tuberculosis), pregnancy test, Liver function test |
| Cyclophosphamide | IV or PO | IV: 600–1000 mg/m².<br>PO dosage based on renal function:<br>GFR >100: 2 mg/Kg<br>GFR 50–99: 1.5 mg/Kg<br>GFR 25–49: 1.2 mg/Kg<br>GFR 15–24: 1.0 mg/Kg<br>GFR less than 15 or on dialysis 0.8 mg/Kg | Alkylating agent with interferes with DNA synthesis | Gastrointestinal (nausea, vomiting), hair loss, mucositis, hemorrhagic cystitis, infertility and myelosuppression | CBC, liver function test, creatinine<br>CBC at 8–14 days post infusion. CBC and urinalysis (for red blood cells) should be checked weekly for a month, every 2 weeks for 2 months and then once a month. Increased hydration recommended |
| Mycophenolate | PO | Initially 500 mg twice daily, target 1000–1500 mg twice daily. | Inhibition of inosine monophosphate dehydrogenase-mediated guanosine nucleotide synthesis | Gastrointestinal (nausea, vomiting, diarrhea), hypertension, peripheral edema, infections, myelosuppression, lymphoma | CBC, creatinine, urine pregnancy test<br>CBC once per week for 1 month, then once every 2 weeks for 2 months then once every 1–3 months for the duration of therapy |
| Azathioprine | PO | Initially 1.5 mg/kg/day, target 2–3 mg/kg/ day (guided by 5-point MCV increase from baseline) | Converted to cytotoxic 6-thioguanine nucleotides, which leads to incorporation as a false base into DNA inducing lymphocyte apoptosis | Gastrointestinal symptoms (nausea, vomiting, diarrhea), hypersensitivity reactions, hair loss, cytopenia, hepatotoxicity, lymphoma and infections | CBC, liver function test, creatinine, TPMT and urine pregnancy.<br>CBC should be checked once per week for 1 month, then once every 2 weeks for 2 months, then once every 1–3 months for the duration of therapy |
| Bortezomib | SC | 1–6 cycles, each cycle consists of 4 injections (1.3 mg/m²) | Proteasome inhibitor, decreasing levels of plasma cells. | Bone marrow suppression, GI side effects, peripheral neuropathy, opportunistic infections. | CBC, liver function and electrolytes |

*Abbreviations:* GFR glomerular filtration rate, *GI* gastrointestinal, *IV* intravenous, *PO* per oral, *PLEX* plasmapheresis, *IVIg* intravenous immunoglobulin, *PT* prothrombin time, *INR* international normalized ratio, *PTT* partial thromboplastin time, *CBC* complete blood count, *TPMT* thiopurine S-methyltransferase, *TMP* Trimethoprim, *SMX* Sulfamethoxazole, *PPI* proton pump inhibitor, *PJP Pneomocystis jiroveci* pneumonia, *ppx* prophylaxis, *SC* subcutaneous

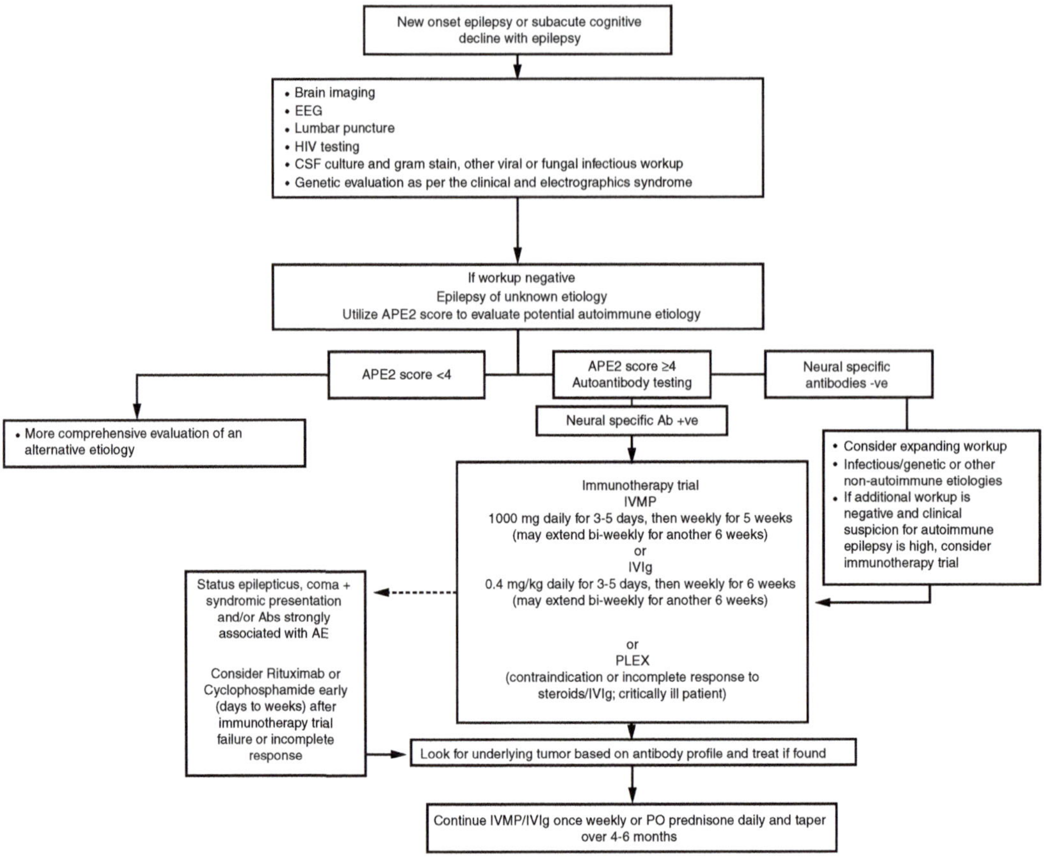


**Fig. 13.2** Management algorithm for autoimmune epilepsy. *Abbreviations:* Ab antibody, CSF cerebrospinal fluid, EEG electroencephalogram, HIV human immunodeficiency virus, IVIg intravenous immunoglobulin, IVMP intravenous methylprednisone, PLEX plasmapheresis, PO per oral

brain MRI with gadolinium, PET brain, and formal cognitive tests are additional parameters that can be monitored. Seizures in autoimmune epilepsy may show early improvement within 4–6 weeks of initiating immunotherapy. Conversely, cognitive impairment and amnesia, if present, recover more slowly. For patients who have incomplete or lack of response to IVMP or IV, or those who have contraindication to IVMP and/or IVIg, a course of plasmapheresis (5–7 cycles) should be considered followed by a second-line immunotherapy (rituximab or cyclophosphamide). A gradual taper of prednisone if feasible is advised following the initial treatment, as abrupt discontinuation might lead to relapses. Chronic maintenance immunotherapy should be initiated to decrease the likelihood of a relapse. Azathioprine, mycophenolate mofetil, rituximab, and cyclophosphamide are agents typically utilized. The exact duration required for maintenance immunotherapy is not known, although a trial of immunotherapy withdrawal may be considered after 2 years of treatment if the patient has not had any relapses.

## Antiepileptic Drugs (AEDS)

Even though seizures in autoimmune epilepsy are characteristically resistant to antiepileptic drugs (AEDs) alone, they continue to play an important role in symptomatic management. In all autoimmune epilepsy patients, AEDs should be used along with immunotherapy treatment. There are no randomized trial data to support one AED over another. Levetiracetam is commonly employed for management of seizures given the favorable side effect profile and minimal drug-to-drug interaction. In some instances, it can be difficult to ascertain if the psychiatric manifestations are due to levetiracetam adverse effects [80] or due to disease pathology. Anti-inflammatory effect for levetiracetam has been hypothesized to play a beneficial role [81]. However, a recent retrospective study evaluating AED in autoimmune epilepsy found none of the patients on levetiracetam achieved seizure freedom [82]. Whereas seizure freedom rates were considerably higher with the use of sodium channel blocking AEDs (carbamazepine, phenytoin, oxcarbazepine, and lacosamide). [82]. The reason for better efficacy of sodium channel-blocking AEDs remains unclear. Interestingly, both carbamazepine and oxcarbazepine have been shown to reduce levels of interleukin-1 (IL)-1 and IL-2 in healthy subjects [81]. Medications such as carbamazepine and phenytoin have enzyme induction properties, which can alter the pharmacokinetics of immunosuppressive therapies. Therefore, newer sodium channel blocking AEDs with more favorable pharmacokinetic profiles (such as oxcarbazepine and lacosamide) could be preferred in management of autoimmune epilepsy.

## Follow-Up

Patients with autoimmune epilepsy should be followed up regularly, preferably by an epileptologist in conjunction with a neuroimmunologist. On long-term follow-up, some patients continue to have drug-resistant epilepsy, despite initial treatment with immunotherapy and continued treatment with AEDs. Whether the disease is now at a "burn out" stage or an acute inflammatory process continues is hard to discern. Ancillary data including repeating brain MRI with gadolinium contrast, CSF analysis, and PET brain can be helpful in guiding treatment decisions. Serum and/or CSF antibody titers have poor relationship to clinical course. Epilepsy surgery has been tried in select cases of autoimmune epilepsy [83]. However, outcomes seem to be worse than that expected in other etiologies of drug-resistant epilepsy.

## Conclusion

### Future Directions

Many challenges continue to face us as we move forward in the field of autoimmune epilepsy. Several novel biomarkers are likely to be discovered over the next few years. The use of phage immunocepitation sequencing and protein microarrays might help us to identify novel antibodies that were not discovered using traditional immunoprecipitation mass spectrometry techniques [84–86].

Further insights into the mechanisms of antibody-mediated epilepsy syndromes will allow us to better utilize the various AEDs and immunotherapies. Additionally, investigations into the role of cytokines and chemokines in diagnosis and prognostication of autoimmune epilepsy will also aid in seizure management and relapse prevention [87, 88]. Finally, randomized controlled trials are needed to determine optimal therapeutic agents, doses, and treatment duration for autoimmune epilepsy management.

## References

1. Ong MS, Kohane IS, Cai T, Gorman MP, Mandl KD. Population-level evidence for an autoimmune etiology of epilepsy. JAMA Neurol. 2014;71(5):569–74.
2. Brodie MJ, Zuberi SM, Scheffer IE, Fisher RS. The 2017 ILAE classification of seizure types and the epilepsies: what do people with epilepsy and their caregivers need to know? Epileptic Disord. 2018;20(2):77–87.
3. Granata T, Andermann F. Rasmussen encephalitis. Handb Clin Neurol. 2013;111:511–9.
4. Sinclair DB, Snyder TJ. Corticosteroids for the treatment of Landau-kleffner syndrome and continuous spike-wave discharge during sleep. Pediatr Neurol. 2005;32(5):300–6.
5. Britton J. Autoimmune epilepsy. Handb Clin Neurol. 2016;133:219–45.
6. Dalmau J, Graus F. Antibody-mediated encephalitis. N Engl J Med. 2018;378(9):840–51.
7. Larman HB, Zhao Z, Laserson U, Li MZ, Ciccia A, Gakidis MA, et al. Autoantigen discovery with

a synthetic human peptidome. Nat Biotechnol. 2011;29(6):535–41.
8. Scharf M, Miske R, Kade S, Hahn S, Denno Y, Begemann N, et al. A spectrum of neural autoantigens, newly identified by histo-immunoprecipitation, mass spectrometry, and recombinant cell-based indirect immunofluorescence. Front Immunol. 2018;9:1447.
9. Toledano M, Britton JW, McKeon A, Shin C, Lennon VA, Quek AM, et al. Utility of an immunotherapy trial in evaluating patients with presumed autoimmune epilepsy. Neurology. 2014;82(18):1578–86.
10. Scheffer IE, Berkovic S, Capovilla G, Connolly MB, French J, Guilhoto L, et al. ILAE classification of the epilepsies: position paper of the ILAE Commission for Classification and Terminology. Epilepsia. 2017;58(4):512–21.
11. Graus F, Titulaer MJ, Balu R, Benseler S, Bien CG, Cellucci T, et al. A clinical approach to diagnosis of autoimmune encephalitis. Lancet Neurol. 2016;15(4):391–404.
12. Dubey D, Pittock SJ, Kelly CR, McKeon A, Lopez-Chiriboga AS, Lennon VA, et al. Autoimmune encephalitis epidemiology and a comparison to infectious encephalitis. Ann Neurol. 2018;83(1):166–77.
13. Dubey D, Toledano M, McKeon A. Clinical presentation of autoimmune and viral encephalitides. Curr Opin Crit Care. 2018;24(2):80–90.
14. Dubey D, Alqallaf A, Hays R, Freeman M, Chen K, Ding K, et al. Neurological autoantibody prevalence in epilepsy of unknown etiology. JAMA Neurol. 2017;74(4):397–402.
15. Dubey D, Singh J, Britton JW, Pittock SJ, Flanagan EP, Lennon VA, et al. Predictive models in the diagnosis and treatment of autoimmune epilepsy. Epilepsia. 2017;58(7):1181–9.
16. Brenner T, Sills GJ, Hart Y, Howell S, Waters P, Brodie MJ, et al. Prevalence of neurologic autoantibodies in cohorts of patients with new and established epilepsy. Epilepsia. 2013;54(6):1028–35.
17. Abramovici S, Bagic A. Epidemiology of epilepsy. Handb Clin Neurol. 2016;138:159–71.
18. Wright S, Geerts AT, Jol-van der Zijde CM, Jacobson L, Leang B, Waters P, van Tol MJ, Stroink H, Neuteboom RF, Brouwer OF, Vincent A. Neuronal antibodies in pediatric epilepsy: clinical features and long-term outcomes of a historical cohort not treated with immunotherapy. Epilepsia. 2016;57(5):823–31.
19. Dubey D, Kothapalli N, McKeon A, Flanagan EP, Lennon VA, Klein CJ, et al. Predictors of neural-specific autoantibodies and immunotherapy response in patients with cognitive dysfunction. J Neuroimmunol. 2018;323:62–72.
20. Titulaer MJ, McCracken L, Gabilondo I, Armangue T, Glaser C, Iizuka T, et al. Treatment and prognostic factors for long-term outcome in patients with anti-NMDA receptor encephalitis: an observational cohort study. Lancet Neurol. 2013;12(2):157–65.
21. Dalmau J, Geis C, Graus F. Autoantibodies to synaptic receptors and neuronal cell surface proteins in autoimmune diseases of the central nervous system. Physiol Rev. 2017;97(2):839–87.
22. Quek AML, O'Toole O. Autoimmune epilepsy: the evolving science of neural autoimmunity and its impact on epilepsy management. Semin Neurol. 2018;38(3):290–302.
23. Dalmau J, Tuzun E, Wu HY, Masjuan J, Rossi JE, Voloschin A, et al. Paraneoplastic anti-N-methyl-D-aspartate receptor encephalitis associated with ovarian teratoma. Ann Neurol. 2007;61(1):25–36.
24. Armangue T, Spatola M, Vlagea A, Mattozzi S, Carceles-Cordon M, Martinez-Heras E, et al. Frequency, symptoms, risk factors, and outcomes of autoimmune encephalitis after herpes simplex encephalitis: a prospective observational study and retrospective analysis. Lancet Neurol. 2018;17:760.
25. Titulaer MJ, McCracken L, Gabilondo I, Iizuka T, Kawachi I, Bataller L, et al. Late-onset anti-NMDA receptor encephalitis. Neurology. 2013;81(12):1058–63.
26. Sebastian Lopez-Chiriboga A, Klein C, Zekeridou A, McKeon A, Dubey D, Flanagan EP, et al. LGI1 and CASPR2 neurological autoimmunity in children. Ann Neurol. 2018;84(3):473–80.
27. Irani SR, Michell AW, Lang B, Pettingill P, Waters P, Johnson MR, et al. Faciobrachial dystonic seizures precede Lgi1 antibody limbic encephalitis. Ann Neurol. 2011;69(5):892–900.
28. Gadoth A, Pittock SJ, Dubey D, McKeon A, Britton JW, Schmeling JE, et al. Expanded phenotypes and outcomes among 256 LGI1/CASPR2-IgG-positive patients. Ann Neurol. 2017;82(1):79–92.
29. van Sonderen A, Schreurs MW, de Bruijn MA, Boukhrissi S, Nagtzaam MM, Hulsenboom ES, et al. The relevance of VGKC positivity in the absence of LGI1 and Caspr2 antibodies. Neurology. 2016;86(18):1692–9.
30. Lai M, Hughes EG, Peng X, Zhou L, Gleichman AJ, Shu H, et al. AMPA receptor antibodies in limbic encephalitis alter synaptic receptor location. Ann Neurol. 2009;65(4):424–34.
31. Haselmann H, Mannara F, Werner C, Planaguma J, Miguez-Cabello F, Schmidl L, et al. Human autoantibodies against the AMPA receptor subunit GluA2 induce receptor reorganization and memory dysfunction. Neuron. 2018;100:91.
32. Hoftberger R, van Sonderen A, Leypoldt F, Houghton D, Geschwind M, Gelfand J, et al. Encephalitis and AMPA receptor antibodies: novel findings in a case series of 22 patients. Neurology. 2015;84(24):2403–12.
33. Joubert B, Kerschen P, Zekeridou A, Desestret V, Rogemond V, Chaffois MO, et al. Clinical spectrum of encephalitis associated with antibodies against the alpha-amino-3-hydroxy-5-methyl-4-isoxazolepropionic acid receptor: case series and review of the literature. JAMA Neurol. 2015;72(10):1163–9.
34. Boronat A, Gelfand JM, Gresa-Arribas N, Jeong HY, Walsh M, Roberts K, et al. Encephalitis and antibodies to

dipeptidyl-peptidase-like protein-6, a subunit of Kv4.2 potassium channels. Ann Neurol. 2013;73(1):120–8.
35. Lancaster E, Martinez-Hernandez E, Titulaer MJ, Boulos M, Weaver S, Antoine JC, et al. Antibodies to metabotropic glutamate receptor 5 in the Ophelia syndrome. Neurology. 2011;77(18):1698–701.
36. Spatola M, Sabater L, Planaguma J, Martinez-Hernandez E, Armangue T, Pruss H, et al. Encephalitis with mGluR5 antibodies: symptoms and antibody effects. Neurology. 2018;90(22):e1964–e72.
37. Pittock SJ, Yoshikawa H, Ahlskog JE, Tisch SH, Benarroch EE, Kryzer TJ, et al. Glutamic acid decarboxylase autoimmunity with brainstem, extrapyramidal, and spinal cord dysfunction. Mayo Clin Proc. 2006;81(9):1207–14.
38. Peltola J, Kulmala P, Isojarvi J, Saiz A, Latvala K, Palmio J, et al. Autoantibodies to glutamic acid decarboxylase in patients with therapy-resistant epilepsy. Neurology. 2000;55(1):46–50.
39. Lilleker JB, Biswas V, Mohanraj R. Glutamic acid decarboxylase (GAD) antibodies in epilepsy: diagnostic yield and therapeutic implications. Seizure. 2014;23(8):598–602.
40. Rudzinski LA, Pittock SJ, McKeon A, Lennon VA, Britton JW. Extratemporal EEG and MRI findings in ANNA-1 (anti-Hu) encephalitis. Epilepsy Res. 2011;95(3):255–62.
41. Pittock SJ, Lucchinetti CF, Lennon VA. Anti-neuronal nuclear autoantibody type 2: paraneoplastic accompaniments. Ann Neurol. 2003;53(5):580–7.
42. Dalmau J, Graus F, Villarejo A, Posner JB, Blumenthal D, Thiessen B, et al. Clinical analysis of anti-Ma2-associated encephalitis. Brain. 2004;127(Pt 8):1831–44.
43. Voltz R, Gultekin SH, Rosenfeld MR, Gerstner E, Eichen J, Posner JB, et al. A serologic marker of paraneoplastic limbic and brain-stem encephalitis in patients with testicular cancer. N Engl J Med. 1999;340(23):1788–95.
44. Yu Z, Kryzer TJ, Griesmann GE, Kim K, Benarroch EE, Lennon VA. CRMP-5 neuronal autoantibody: marker of lung cancer and thymoma-related autoimmunity. Ann Neurol. 2001;49(2):146–54.
45. Dubey D, Lennon VA, Gadoth A, Pittock SJ, Flanagan EP, Schmeling JE, et al. Autoimmune CRMP5 neuropathy phenotype and outcome defined from 105 cases. Neurology. 2018;90(2):e103–e10.
46. Vernino S, Tuite P, Adler CH, Meschia JF, Boeve BF, Boasberg P, et al. Paraneoplastic chorea associated with CRMP-5 neuronal antibody and lung carcinoma. Ann Neurol. 2002;51(5):625–30.
47. Quek AM, Britton JW, McKeon A, So E, Lennon VA, Shin C, et al. Autoimmune epilepsy: clinical characteristics and response to immunotherapy. Arch Neurol. 2012;69(5):582–93.
48. Bien CG, Granata T, Antozzi C, Cross JH, Dulac O, Kurthen M, et al. Pathogenesis, diagnosis and treatment of Rasmussen encephalitis: a European consensus statement. Brain. 2005;128(Pt 3):454–71.
49. Bien CG, Tiemeier H, Sassen R, Kuczaty S, Urbach H, von Lehe M, et al. Rasmussen encephalitis: incidence and course under randomized therapy with tacrolimus or intravenous immunoglobulins. Epilepsia. 2013;54(3):543–50.
50. Vining EP, Freeman JM, Pillas DJ, Uematsu S, Carson BS, Brandt J, et al. Why would you remove half a brain? The outcome of 58 children after hemispherectomy-the Johns Hopkins experience: 1968 to 1996. Pediatrics. 1997;100(2 Pt 1):163–71.
51. Gaspard N, Foreman BP, Alvarez V, Cabrera Kang C, Probasco JC, Jongeling AC, et al. New-onset refractory status epilepticus: etiology, clinical features, and outcome. Neurology. 2015;85(18):1604–13.
52. Gaspard N, Hirsch LJ, Sculier C, Loddenkemper T, van Baalen A, Lancrenon J, et al. New-onset refractory status epilepticus[2] and febrile infection-related epilepsy syndrome (FIRES): state of the art and perspectives. Epilepsia. 2018;59(4):745–52.
53. Arya R, Anand V, Chansoria M. Hashimoto encephalopathy presenting as progressive myoclonus epilepsy syndrome. Eur J Paediatr Neurol. 2013;17(1):102–4.
54. Kaplan PW, Sutter R. Electroencephalography of autoimmune limbic encephalopathy. J Clin Neurophysiol. 2013;30(5):490–504.
55. Schmitt SE, Pargeon K, Frechette ES, Hirsch LJ, Dalmau J, Friedman D. Extreme delta brush: a unique EEG pattern in adults with anti-NMDA receptor encephalitis. Neurology. 2012;79(11):1094–100.
56. Baykan B, Gungor Tuncer O, Vanli-Yavuz EN, Baysal Kirac L, Gundogdu G, Bebek N, et al. Delta brush pattern is not unique to NMDAR encephalitis: evaluation of two independent long-term EEG cohorts. Clin EEG Neurosci. 2018;49(4):278–84.
57. Aurangzeb S, Symmonds M, Knight RK, Kennett R, Wehner T, Irani SR. LGI1-antibody encephalitis is characterised by frequent, multifocal clinical and subclinical seizures. Seizure. 2017;50:14–7.
58. Steriade C, Mirsattari SM, Murray BJ, Wennberg R. Subclinical temporal EEG seizure pattern in LGI1-antibody-mediated encephalitis. Epilepsia. 2016;57(8):e155–60.
59. Escudero D, Guasp M, Arino H, Gaig C, Martinez-Hernandez E, Dalmau J, et al. Antibody-associated CNS syndromes without signs of inflammation in the elderly. Neurology. 2017;89(14):1471–5.
60. Dubey D, Sawhney A, Greenberg B, Lowden A, Warnack W, Khemani P, et al. The spectrum of autoimmune encephalopathies. J Neuroimmunol. 2015;287:93–7.
61. Finke C, Kopp UA, Scheel M, Pech LM, Soemmer C, Schlichting J, et al. Functional and structural brain changes in anti-N-methyl-D-aspartate receptor encephalitis. Ann Neurol. 2013;74(2):284–96.
62. Solnes LB, Jones KM, Rowe SP, Pattanayak P, Nalluri A, Venkatesan A, et al. Diagnostic value of (18)F-FDG PET/CT versus MRI in the setting of antibody-specific autoimmune encephalitis. J Nucl Med. 2017;58(8):1307–13.

63. Ohta K, Seki M, Dalmau J, Shinohara Y. Perfusion IMP-SPECT shows reversible abnormalities in GABA(B) receptor antibody associated encephalitis with normal MRI. Brain Behav. 2011;1(2):70–2.
64. Heine J, Pruss H, Bartsch T, Ploner CJ, Paul F, Finke C. Imaging of autoimmune encephalitis–relevance for clinical practice and hippocampal function. Neuroscience. 2015;309:68–83.
65. Probasco JC, Solnes L, Nalluri A, Cohen J, Jones KM, Zan E, et al. Decreased occipital lobe metabolism by FDG-PET/CT: an anti-NMDA receptor encephalitis biomarker. Neurol Neuroimmunol Neuroinflamm. 2018;5(1):e413.
66. Irani SR, Vincent A. Voltage-gated potassium channel-complex autoimmunity and associated clinical syndromes. Handb Clin Neurol. 2016;133:185–97.
67. Flanagan EP, Kotsenas AL, Britton JW, McKeon A, Watson RE, Klein CJ, et al. Basal ganglia T1 hyperintensity in LGI1-autoantibody faciobrachial dystonic seizures. Neurol Neuroimmunol Neuroinflamm. 2015;2(6):e161.
68. Shin YW, Lee ST, Shin JW, Moon J, Lim JA, Byun JI, et al. VGKC-complex/LGI1-antibody encephalitis: clinical manifestations and response to immunotherapy. J Neuroimmunol. 2013;265(1–2):75–81.
69. Fredriksen JR, Carr CM, Koeller KK, Verdoorn JT, Gadoth A, Pittock SJ, et al. MRI findings in glutamic acid decarboxylase associated autoimmune epilepsy. Neuroradiology. 2018;60(3):239–45.
70. Spatola M, Petit-Pedrol M, Simabukuro MM, Armangue T, Castro FJ, Barcelo Artigues MI, et al. Investigations in GABAA receptor antibody-associated encephalitis. Neurology. 2017;88(11):1012–20.
71. Dogan Onugoren M, Deuretzbacher D, Haensch CA, Hagedorn HJ, Halve S, Isenmann S, et al. Limbic encephalitis due to GABAB and AMPA receptor antibodies: a case series. J Neurol Neurosurg Psychiatry. 2015;86(9):965–72.
72. Hoftberger R, Titulaer MJ, Sabater L, Dome B, Rozsas A, Hegedus B, et al. Encephalitis and GABAB receptor antibodies: novel findings in a new case series of 20 patients. Neurology. 2013;81(17):1500–6.
73. Do LD, Chanson E, Desestret V, Joubert B, Ducray F, Brugiere S, et al. Characteristics in limbic encephalitis with anti-adenylate kinase 5 autoantibodies. Neurology. 2017;88(6):514–24.
74. Pittock SJ, Palace J. Paraneoplastic and idiopathic autoimmune neurologic disorders: approach to diagnosis and treatment. Handb Clin Neurol. 2016;133:165–83.
75. Graus F, Delattre JY, Antoine JC, Dalmau J, Giometto B, Grisold W, et al. Recommended diagnostic criteria for paraneoplastic neurological syndromes. J Neurol Neurosurg Psychiatry. 2004;75(8):1135–40.
76. Irani SR, Stagg CJ, Schott JM, Rosenthal CR, Schneider SA, Pettingill P, et al. Faciobrachial dystonic seizures: the influence of immunotherapy on seizure control and prevention of cognitive impairment in a broadening phenotype. Brain. 2013;136(Pt 10):3151–62.
77. Byun JI, Lee ST, Jung KH, Sunwoo JS, Moon J, Lim JA, et al. Effect of immunotherapy on seizure outcome in patients with autoimmune encephalitis: a prospective observational registry study. PLoS One. 2016;11(1):e0146455.
78. Thompson J, Bi M, Murchison AG, Makuch M, Bien CG, Chu K, et al. The importance of early immunotherapy in patients with faciobrachial dystonic seizures. Brain. 2018;141(2):348–56.
79. Dubey D, Britton J, McKeon A, Gadoth A, Zekeridou A, Lopez Chiriboga SA, et al. Randomized placebo-controlled trial of intravenous immunoglobulin in autoimmune LGI1/CASPR2 epilepsy. Ann Neurol. 2020;87:313–23.
80. Abou-Khalil BW. Antiepileptic drugs. Continuum (Minneap Minn). 2016;22(1 Epilepsy):132–56.
81. Beghi E, Shorvon S. Antiepileptic drugs and the immune system. Epilepsia. 2011;52(Suppl 3):40–4.
82. Feyissa AM, Lopez Chiriboga AS, Britton JW. Antiepileptic drug therapy in patients with autoimmune epilepsy. Neurol Neuroimmunol Neuroinflamm. 2017;4(4):e353.
83. Carreno M, Bien CG, Asadi-Pooya AA, Sperling M, Marusic P, Elisak M, et al. Epilepsy surgery in drug resistant temporal lobe epilepsy associated with neuronal antibodies. Epilepsy Res. 2017;129:101–5.
84. Schubert RD, Wilson MR. A tale of two approaches: how metagenomics and proteomics are shaping the future of encephalitis diagnostics. Curr Opin Neurol. 2015;28(3):283–7.
85. Mohan D, Wansley DL, Sie BM, Noon MS, Baer AN, Laserson U, et al. PhIP-Seq characterization of serum antibodies using oligonucleotide-encoded peptidomes. Nat Protoc. 2018;13(9):1958–78.
86. Larman HB, Salajegheh M, Nazareno R, Lam T, Sauld J, Steen H, et al. Cytosolic 5′-nucleotidase 1A autoimmunity in sporadic inclusion body myositis. Ann Neurol. 2013;73(3):408–18.
87. Vezzani A, Fujinami RS, White HS, Preux PM, Blumcke I, Sander JW, et al. Infections, inflammation and epilepsy. Acta Neuropathol. 2016;131(2):211–34.
88. Vezzani A. Epilepsy and inflammation in the brain: overview and pathophysiology. Epilepsy Curr. 2014;14(1 Suppl):3–7.

# 14 Autoimmune and Paraneoplastic Movement Disorders

A. Sebastian López-Chiriboga
and Andrew McKeon

**Key Points**

1. Movement disorders are a common manifestation of many antibody-mediated central system diseases.
2. Autoimmune movement disorders should be considered in the differential diagnosis for all movement disorder phenotypes.
3. Autoantibody testing in serum and cerebrospinal fluid (CSF) helps confirm the diagnosis and assist in the prediction of immunotherapy response, cancer association, and prognosis.
4. Early recognition and appropriate cancer screening is important.
5. Immunotherapy and symptomatic therapy can be effective.

## Introduction

Autoimmune movement disorders are a common feature of many antibody-associated neurological disorders [1, 2]. These disorders are secondary to an aberrant immune response against neural antigens [1]. Multiple autoantibodies have been associated with autoimmune movement disorders. The autoantibody targets are diverse and include neuronal surface proteins such as N-methyl-D-aspartate receptor (NMDAR) [3]; leucine-rich, glioma-inactivated protein type 1 (LGI1) [4] and glycine receptors [5]; and also intracellular antigens such as glutamic acid decarboxylase 65-kilodalton isoform (GAD65) enzyme [6] and Purkinje cell cytoplasmic antibody (PCA-1), also known as anti-Yo [6]. The onset is typically subacute, and they can be the first manifestation of a more broad or multifocal neurologic disorder, usually accompanied by encephalopathy, neuropsychiatric symptoms, brainstem dysfunction, dysautonomia, or myelopathy.

Adults and children can be affected. The gender and age of onset can be helpful in the diagnosis of specific syndromes as some autoimmune movement disorders occur more frequently in females with malignancy (such as ataxia associated to anti-Yo antibodies and gynecologic neoplasms) [7] or children and adolescents (postinfectious chorea or orofacial dyskinesias in NMDA receptor autoimmunity) [3, 8]. These disorders can occur as a postinfectious or paraneoplastic phenomenon or sometimes they can be

A. S. López-Chiriboga (✉)
Department of Neurology, Division of Multiple Sclerosis and Autoimmune Neurology, Mayo Clinic, Florida, FL, USA
e-mail: lopez.alfonso@mayo.edu

A. McKeon
Department of Neurology, Department of Laboratory Medicine and Pathology, Division of Multiple Sclerosis and Autoimmune Neurology, Mayo Clinic, Florida, FL, USA



A. L. Piquet, E. Alvarez (eds.), *Neuroimmunology*, https://doi.org/10.1007/978-3-030-61883-4_14

idiopathic [1, 2, 9]. Autoantibody testing in serum and CSF is crucial as it helps confirm the diagnosis and assist in the prediction of immunotherapy response, cancer association, and prognosis [2, 9–11].

When diagnosed early, many patients respond to immunotherapy; however, some patients have a refractory, progressive course despite treatment [9, 12]. In this chapter, we review the clinical features, antibody, cancer association, and the treatment approach to autoimmune movement disorders.

## Epidemiology and Demographics

Epidemiologic studies that comprehensively assess autoimmune movement disorders are limited. However, a recent population-based study in Olmsted County, Minnesota, found that GAD65 antibodies are one of the most common antibodies in patients with autoimmune encephalitis [13]. Stiff person syndrome (SPS) is a rare disorder, with an estimated prevalence of approximately 1 in 1,250,000 [14]. Autoimmune chorea was estimated to affect 1.5 per million persons-years in one study [15]. In a prospective study that evaluated 1500 patients, GAD65 autoimmunity was the second most common cause of ataxia, following "celiac disease-gluten related ataxia" [16].

## Clinical Syndromes

Autoimmunity should be considered in the differential diagnosis of all movement disorder phenotypes. The clinical presentation is typically broad as multiple neuroanatomical structures can be affected such as the cortex, basal ganglia, brainstem, cerebellum, and spinal cord. Broadly speaking, movement disorders can be classified into two main categories: hyperkinetic (fast movements) and hypokinetic disorders (slow movements) [17]. Incoordination or ataxia or disorders of movement execution (apraxia) and mixed disorders can also occur [1, 17]. Finally, disorders characterized by stiffness and spasms as manifestations of central nervous system (CNS) hyperexcitability are known as the stiff-person spectrum (SPS) disorders [18].

### Hyperkinetic Disorders

#### Chorea

Chorea is characterized by brief, semi-directed, irregular movements that are not repetitive or rhythmic, but appear to flow from one muscle to another; the involuntary movements can also have a writhing quality referred to as choreoathetosis [19, 20].

Autoimmune chorea in children is usually a post infectious phenomenon. It occurs as a complication of group A beta-hemolytic *Streptococcus* (GABHS) infection; it is also known as Sydenham's chorea (SC), and it is the most common cause of autoimmune chorea worldwide [19, 21]. It affects approximately 30% of patients with acute rheumatic fever, and it can occur several months after the initial infection. The diagnosis is supported by demonstration of a recent streptococcal infection using throat cultures, or positive antistreptolysin O or anti-DNAse antibodies [10]. Adult onset of Sydenham's chorea is rare, and most adult cases are associated with exacerbation of chorea following prior childhood Sydenham's chorea.

Molecular mimicry to streptococcal antigens leading to autoantibody targeting the basal ganglia has been suggested as the possible underlying mechanism; antibodies in serum targeting the dopamine 2 receptor have been identified in some patients with Sydenham's chorea [8]. Treatment includes antibiotic therapy with penicillin, immunotherapy with steroids, or intravenous immunoglobulin (IVIg) has also been reported to be effective [21]. Autoimmune chorea can also be seen in patients with antiphospholipid antibodies (aPL); it commonly affects children and women in a subacute and often monophasic course [22]. In patients with idiopathic suspected autoimmune chorea, lupus and antiphospholipid syndrome are common accompaniments [15].

It is now well recognized that chorea, ballismus, and athetosis can occur after herpes simplex virus (HSV) encephalitis, typically 2–6 weeks following the initial infection. These abnormal movements are now recognized as occurring due to post-infectious

NMDAR encephalitis and other unclassified neural autoantibodies in up to 30% of patients [23, 24].

In adults, several neural autoantibodies have been associated with chorea (Table 14.1). Of these, the classic one is collapsin response mediator protein 5 (CRMP-5) IgG [15, 25]. CRMP5 autoimmunity is usually a paraneoplastic disorder that occurs in association with small cell lung cancer (SCLC) and thymoma [26] (Table 14.1). Most patients have a broad multifocal neurologic disorder that can include optic neuropathy, vitritis, encephalitis, and myelopathy [25]; hyperintensities in the striatum can be seen in some patients (Fig. 14.1). Antineuronal nuclear anti-

**Table 14.1** Antibodies association, movement disorder phenotype, and other clinical features

| *Clinical Syndromes and Antibody Associations* | | |
|---|---|---|
| Nuclear and cytoplasmic targets | | |
| *Target antigen* | *Movement disorder* | *Other clinical features* |
| ANNA-1 (anti-Hu) | Chorea | Limbic encephalitis, sensory neuronopathy, gastrointestinal dysmotility |
| ANNA-2 (anti-Ri) | Opsoclonus-myoclonus, jaw dystonia | Brainstem encephalitis, laryngospasm |
| ANNA-3 | Ataxia | Sensory/sensorimotor neuropathies, myelopathy, brain stem and limbic encephalopathy |
| Amphiphysin | SPS | Encephalopathy, myeloradiculopathy |
| AGNA | Ataxia | LES |
| CRMP-5 | Chorea, ataxia | Optic neuropathy, peripheral and autonomic neuropathy, retinitis, myelopathy |
| GAD65 | Ataxia, SPS | Encephalitis, autoimmune epilepsy, myelopathy, neuropathy |
| GFAP | Tremor | Meningoencephalitis, myelitis, bilateral optic disc edema |
| GRAF1 | Ataxia | Neuropathy, encephalopathy |
| Kelch like protein 11 | Ataxia | Hearing loss, brainstem encephalitis |
| ITPR1 | Ataxia | Neuropathy |
| Ma1 Ma2 | Ataxia, parkinsonism | Limbic encephalitis, diencephalic syndrome, brainstem encephalitis |
| PCA-1 | Ataxia | Brainstem encephalitis, neuropathy |
| PCA-2 (MAP1B) | Ataxia | Encephalopathy, neuropathy |
| Synaptic targets | | |
| *Target antigen* | *Movement disorder* | *Other clinical features* |
| CASPR2 | Episodic ataxia | Peripheral nerve hyperexcitability (Isaac's syndrome), encephalitis, dysautonomia, neuropathy, Morvan's syndrome |
| DPPX | Ataxia, myoclonus | Severe weight loss, diarrhea, dysautonomia, cardiac arrhythmia, encephalomyelitis |
| Dopamine 2 | Chorea | Psychiatric symptoms |
| $GABA_BR$ | Chorea | Limbic encephalitis, frequent seizures |
| mGluR1 | Ataxia | Encephalopathy, dysgeusia |
| Glycine | SPS, myoclonus | Encephalopathy, Seizures |
| IgLON5 | Chorea, Ataxia, Stiff limb (SPS) | Sleep disorders, encephalopathy |
| LGI1 | Faciobrachial dystonic seizures | Limbic encephalitis, hyponatremia, Peripheral nerve hyperexcitability |
| Neurexin-3 | Dyskinesias | Encephalitis |
| NMDAR | Dyskinesias, dystonia | Catatonia, seizures, psychosis, dysautonomia |
| VGCC | Ataxia | LES, encephalitis, neuropathy |

*Abbreviations: ANNA* Anti-neuronal nuclear antibody, *SPS* stiff person syndrome, *LES* Lambert-Eaton syndrome, *CRMP5* collapsin response mediator protein, *GAD* glutamic acid decarboxylase, *GFAP* glial fibrillary acidic protein, *GRAF1* GTPase regulator associated with focal adhesion kinase 1, *ITPR1* inositol 1,4,5-trisphosphate receptor type 1, *PCA* Purkinje cell cytoplasmic antibody, *Caspr2* contactin-associated protein-2, *DNER* δ(delta) Notch-like epidermal growth factor–related receptor, *DPPX* Dipeptidyl-peptidase–like protein-6, *GABA* γ(gamma) aminobutyric acid, *LGI1* leucine-rich, glioma inactivated 1 protein, *mGluR* metabotropic glutamate receptor, *NMDAR* N-methyl-d-aspartate receptor, *VGCC* voltage-gated calcium channel

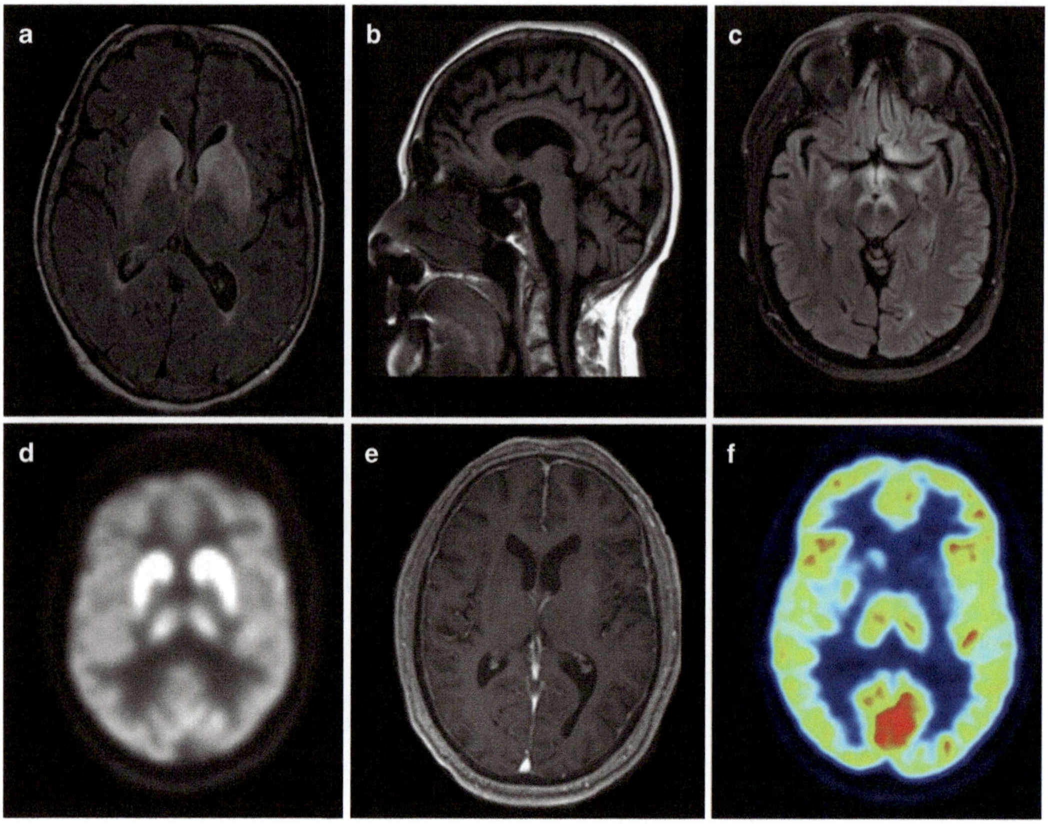


**Fig. 14.1** Neuroimaging findings in paraneoplastic and autoimmune movement disorders. (**a**) Axial fluid-attenuated inversion recovery image showing increased T2 signal involving the striatum bilaterally in a 66-year-old female patient with chorea and encephalopathy seropositive for PCA-2, CRMP-5, and amphiphysin antibodies and small cell lung cancer. (**b**) Sagittal T1- weighted images showing prominent cerebellar atrophy in a 74-year-old woman with cerebellar ataxia and positive PCA-1 antibodies and fallopian cancer. (**c**) Axial fluid-attenuated inversion recovery image showing increased T2 signal involving the limbic system, midbrain, and hypothalamus in a 42-year-old male patient presenting with hypersomnia, ataxia, and encephalopathy and seropositivity for Ma2 antibodies and a testicular seminoma. (**d**) Bilateral striatal hypermetabolism in a 67-year-old male with LGI1 antibodies and FBDS. (**e**) Axial T1-weighted image showing T1 hyperintensities in the left basal ganglia in a 72-year-old male with LGI1 antibodies and FBDS. (**f**) FDG/PET showing left > right hypometabolism in the striatum in a 23-year-old patient with NMDAR encephalitis

body type 1 (ANNA-1, or anti-Hu) is also associated with a multifocal neurologic disorder typically characterized by sensory neuronopathy and limbic encephalitis, although chorea can also be a presenting feature [15, 27]. Similar to CRMP-5 autoimmunity, patients harboring ANNA-1 antibodies typically have a paraneoplastic syndrome in the setting of an underlying neuroendocrine neoplasm [15].

Other autoantibodies associated with chorea/hemichorea include contactin-associated protein 2 (CASPR2) [28], GAD65 [6], leucine-rich glioma inactivated 1(LGI1) [29], and IgLON5 [22]. These antibodies very rarely occur as in a paraneoplastic context.

Table 14.1 summarizes the clinical syndromes, antibody association, and accompanying neurologic features.

## Dyskinesia

Dyskinesias are involuntary repetitive muscle movements, resembling chorea. They can affect one body part, such as an arm, leg, or the head, or can involve the entire body. Orofacial dyskinesias are a hallmark clinical feature of patients with NMDA-R encephalitis, and involvement of the tongue and the limbs can be seen in some patients [30]. They very rarely occur in isolation and often occur as prodrome of or in association with neuropsychiatric symptoms, autonomic disturbances, and decreased responsiveness [30, 31]. A very similar syndrome has been reported in patients harboring antibodies targeting a cell adhesion molecule involved in the development and function of synapses known as neurexin-3α(alpha) [32].

## Dystonia

Dystonia is characterized by sustained or repetitive muscle contractions resulting in twisting and repetitive movements or abnormal fixed postures [17].

Prominent limb dystonia can be a feature of NMDA-R encephalitis [3, 30, 31]. In adults, brainstem encephalitis associated with antineuronal nuclear antibody, type 2 (ANNA-2), also known as anti-Ri, can be associated with prominent jaw closing dystonia. This disorder is usually encountered in women with breast cancer. The symptoms can be severe and sometimes associated with laryngeal muscle compromise requiring prophylactic tracheostomy [33]. Craniocervical dystonia has been reported in some patients with IgLON5 antibodies [22]. Axial or generalized dystonia can be encountered in stiff person syndrome patients with elevated GAD65 antibodies [6].

Faciobrachial dystonic seizures (FBDS) are a very common manifestation of LGI1 autoimmunity. These are brief (less than 3 seconds) stereotyped movements affecting the face, arm, and sometimes the ipsilateral lower extremity. They often occur multiple times per day and precede the onset of encephalopathy or accompany a neuropsychiatric presentation in some patients [4, 34]. It has been suggested that early cessation of FBDS by treatment with immunotherapy can help prevent long-term cognitive dysfunction, as early immunotherapy is associated with better cognitive outcomes [35]. It continues to be unclear whether these movements are secondary to seizures or if they represent a movement disorder, as surface electroencephalogram (EEG) is often normal during the events [34]. Bilateral striatal hypermetabolism on positron emission tomography/computed tomography (PET/CT) and T1 hyperintensities in the basal ganglia have been described as imaging features in patients with FBDS (Fig. 14.1) [36, 37].

## Myoclonus

Myoclonus is characterized by brief, irregular, shock-like, or lightning-like jerks [9]. Opsoclonus-myoclonus syndrome (OMS) is a classic autoimmune CNS disorder. In children, OMS may be associated with ANNA-1 antibodies and neuroblastoma, and most patients are five or younger at the time of the diagnosis [38]. In adults, OMS is sometimes paraneoplastic in etiology, but it can also be a parainfectious or postinfectious phenomenon, or idiopathic autoimmune phenomenon [39]. Only 30% of patients have detectable neural autoantibodies. Responses to immunotherapy are better in idiopathic autoimmune etiologies compared to paraneoplastic cases; older age and encephalopathy are predictors of paraneoplastic OMS [40]. OMS has been recently reported in association with antibodies targeting the glutamate receptor δ(delta)2 [41].

Myoclonus has been reported as a clinical finding, without opsoclonus, in patients with antibodies targeting dipeptidyl-peptidase-like protein-6 (DPPX)—a regulatory subunit of neuronal Kv4.2 potassium channels—commonly with prominent diarrhea as a prodrome and accompanying encephalopathy [41]. Myoclonus is also a characteristic feature of the syndrome of progressive encephalopathy with limb rigidity and myoclonus (PERM), classically associated with glycine receptor antibodies [5, 42, 43]. Orthostatic myoclonus has been reported in patients with CASPR2 antibodies [44].

## Tremor

Tremor can be a clinical feature of many antibody-mediated neurological disorders; however, it very rarely occurs in isolation. It can be a predominant clinical finding in patients with autoimmune glial fibrillary acidic protein (GFAP) astrocytopathy, with accompanying encephalitis or myelitis [45, 46]. Other antibodies associated prominently with tremor include DPPX and LGI1/CASPR2 [4, 47]. Corticosteroid treatment for autoimmune neurological disorders can cause tremor.

## Hypokinetic Disorders

### Parkinsonism

Parkinsonism is characterized by bradykinesia, rigidity, tremor, and postural imbalance [17]. Atypical parkinsonian features such as poor response to levodopa or additional clinical features such as rapid onset, accompanying narcolepsy, hypothalamic, and other brainstem symptoms should prompt evaluation for a paraneoplastic or immune-mediated cause. Patients fulfilling criteria for probable and possible clinical diagnostic criteria for multiple system atrophy (MSA) and progressive supranuclear palsy (PSP) have been reported in association with LGI1 antibodies and other neural autoantibodies and good response to immunotherapy [17].

Patients with Ma1 and Ma2 antibodies can present with prominent parkinsonian features in association with hypersomnia, narcolepsy–cataplexy, hyperthermia, hyperphagia, hypothalamic dysfunction, and classic limbic encephalitis [27]. Patients with Ma2 antibodies typically have a severe hypokinetic syndrome with a tendency to eye closure; dramatic reduction of verbal output has also been described [48]. Young males affected by this disorder almost always have a testicular seminoma [48]. Imaging is typically abnormal with prominent involvement of limbic structures, brainstem, and hypothalamus (Fig. 14.1). Other autoantibodies associated with parkinsonian features include CRMP5, ANNA-2 [33], DPPX [47], IgLON5 [49, 50], and GAD 65 [48].

### Stiff Person Spectrum Disorders

Classic stiff person syndrome (SPS) is characterized by fluctuating truncal and limb muscle rigidity and spasms [14]. Other symptoms include falls and hyperekplexia. In fact, many patients have fixed spinal deformities (exaggerated lordosis) from long-standing rigidity. Neurophysiologic findings often reflect motor neuron hyperexcitability [18], and 80% of SPS patients are positive for GAD65 antibodies. Limited forms of SPS include "stiff limb" and "stiff trunk." A more widespread form is known as progressive encephalomyelitis with rigidity and myoclonus (PERM), because as well as stiffness and spasms, which are widespread, myoclonus and encephalopathy are also present. GAD65 antibody positivity and negativity can be encountered in patients with classical and limited forms and also PERM [18]. GAD65 antibody positive patients are more likely to be female and have systemic autoimmunity—usually one or more of thyroid disease, vitiligo, pernicious anemia, and type I diabetes [51].

Unlike most patients with GAD65-associated systemic autoimmune disorders, patients with SPS have values in the hundreds of nanomoles per liter (nmol/L) [2, 18, 52]. A paraneoplastic cause is rarely encountered; adenocarcinomas (lung and breast) and thymomas have been reported, typically in older SPS patients with poor immunotherapy response [18, 52].

Antibodies targeting the glycine receptor alpha subunit (GlyRα[alpha]1) are the second most common antibodies found in patients with SPS spectrum [47, 48]. These patients are more likely to present with a PERM phenotype or limited forms of SPS. GlyRα(alpha)1antibodies have also been reported in other neurologic disorders such as optic neuritis and patients with functional neurologic disorders [43, 53], where the clinical relevance is less clear. A recent study showed that GlyRα(alpha)1-modulating antibodies can be helpful to improve the diagnostic specificity and treatment response in patients with SPS [43].

Patients with antibodies targeting amphiphysin—a 128 kDa synaptic vesicle-bound protein that works with dynamin to retrieve membrane constituents after neurotransmitter exocytosis—can present with SPS variant, usually in the setting of breast cancer or small cell lung cancer and myeloradiculopathy [54, 55].

Other autoantibodies described in SPS spectrum patients include DPPX (PERM phenotype only) [47] and IgLON5 (a solitary case) [22].

The management of SPS is detailed in Chap. 30.

### Ataxia

Ataxia is characterized by a lack of muscle control and incoordination of voluntary movements [17]. It is one of the most commonly encountered disorders in autoimmune neurology clinical practice. More than 20 autoantibodies have been reported in association with autoimmune ataxia [56]. Although the differential diagnosis of ataxia is broad, paraneoplastic cerebellar degeneration (PCD) should be considered in most cases, particularly in adults [57]. Patients with PCD typically present subacutely (weeks to months) with a pancerebellar syndrome, often with severe disability [12, 16, 27, 57].

PCA-1 antibody is the most commonly found antibody in patients with PCD [57]. These patients are usually women with breast cancer or adenocarcinomas of Müllerian origin (uterus, ovaries, or fallopian tubes) [57]. Although rarely seen in men, when identified, evaluation for a possible gastrointestinal carcinoma should be undertaken as this is the most common oncological accompaniment [58]. Immunotherapy response is often limited, but cancer treatment and cyclophosphamide can sometimes stabilize disease progression [12]. Imaging typically shows pancerebellar atrophy (Fig. 14.1).

Microtubule-associated protein (MAP) 1B, the antigen of the previously described Purkinje cell cytoplasmic antibody type 2 (PCA-2) antibody, is a paraneoplastic biomarker usually seen in patients with SCLC; neurological accompaniments include neuropathy, ataxia, and encephalopathy [59].

GAD 65 antibody is the most common antibody associated with non-paraneoplastic autoimmune cerebellar ataxia. Patients may present with ataxia overlapping with other GAD65 neurological syndromes, such as SPS, encephalitis, and intractable epilepsy [6, 52].

In contrast to the poor response to immunotherapy observed in most patients with PCD, cerebellar ataxia in association with some neural surface antibodies may improve upon receiving immunotherapy. Other patients have autoimmune ataxia as a component of a broader encephalitic disorder. Examples include episodic ataxia in CASPR2-positive patients [60], ataxia as a component of a meningoencephalitic syndrome in autoimmune GFAP astrocytopathy [45], or accompanying seizures in GABA-B receptor autoimmunity [36].

Patients harboring antibodies targeting the metabotropic glutamate type-1 receptor (mGluR1) can present with ataxia and dysgeusia [61]. mGluR1 is part of the glutamate/calcium signaling pathway of Purkinje cells. In addition to mGluR1, autoimmunity targets multiple other antigenic targets in this pathway. Individual disorders are rare, and the antibodies have a similar and characteristic staining pattern of murine cerebellum using indirect immunofluorescence and are often known as "Medusa-head" antibodies because of distinctive staining of molecular layer dendrites [56, 62, 63]. Examples of antigens pertinent to "Medusa-head" ataxia include Homer-3, ITPR 1, CARP VIII, PKCγ(gamma), ARHGAP26, GluRδ(delta)2, and DNER [2, 9, 10, 62]. Immunotherapy response is variable, and cancer associations are broad (Table 14.2) [56, 62, 63].

Neuronal calcium channel antibodies (VGCC--P/Q and VGCC-N types) are associated with cerebellar ataxia with or without other clinical features such as myasthenic syndromes or encephalopathy; small cell carcinoma is a well-known association, but the cancer types are diverse; many of these patients have coexistent neural autoantibodies [64, 65].

Disease-specific immunoglobulin G (IgG) biomarkers of neurological autoimmunity continue to be discovered, increasing the spectrum of diagnosable immune-mediated ataxia. Recently, antibodies targeting Septin-5, a cytoskeletal GTP-binding protein that participates in neurotransmitter exocytosis was described in six patients with cerebellar ataxia [50].

## Clinical Evaluation

A detailed description of the time course, identification of accompanying features such as neuropsychiatric features, dysautonomia, brainstem dysfunction, or sensory neuronopathy should raise the suspicion for an immune-mediated cause. Most patients present subacutely (weeks or months), however, with more chronic and indolent presentations [2, 10]. Patient sex and age are helpful to increase the suspicion for specific disorders. Medical history such as the coexistence of type I diabetes mellitus, hypothyroidism, and pernicious anemia in patients with GAD65 antibodies, or recent streptococcal or HSV infec-

**Table 14.2** Antibodies that target nuclear or antibodies that target plasma membrane proteins cytoplasmic proteins and their oncologic association

| Antibody | Oncologic association | Antibody | Oncologic association |
|---|---|---|---|
| Amphiphysin | Small-cell carcinoma, breast adenocarcinoma | DPPX | Lymphoproliferative disorders |
| ANNA-1 (anti-Hu) | Small-cell carcinoma, Neuroblastoma (pediatric) | $GABA_BR$ | Small-cell lung |
| ANNA-2 (anti-Ri) | Small-cell carcinoma, breast adenocarcinoma | mGluR1 | Hodgkin's lymphoma |
| CRMP-5 (anti-CV2) | Small-cell carcinoma, thymoma | Glycine receptor | Thymoma, lymphoma, ovarian |
| GAD65 | Rare-thymoma, breast adenocarcinoma | NMDAR | Ovarian teratoma, others* |
| GFAP | Ovarian teratoma, adenocarcinomas | (LGI1/ CASPR2) | Small-cell lung carcinoma, thymoma or adenocarcinomas |
| ITPR | Lung adenocarcinoma | P/Q- and N-type VGCC | Small-cell carcinoma |
| Anti Ma1-Anti-Ma2 (anti Ta) | Testicular germ cell tumors, lung cancer, other solid neoplasms | PCA-Tr (DNER) | Hodgkin lymphoma |
| PCA-1 (anti-Yo) | Mullerian, adenocarcinomas, breast | | |
| PCA-2 (MAP1B) | Small-cell carcinoma | | |
| ARHGAP26/GRAF | Ovarian, squamous cell carcinoma | | |

*ANNA* antineuronal nuclear antibody; 4; *CASPR* contactin-associated protein; *CRMP-5* collapsin response mediator protein 5; *DPPX* dipeptidyl-peptidase-like protein-6 antibodies, *GABA* gamma amino butyric acid, *GAD65* 65 kDa isoform of glutamic acid decarboxylase, *GFAP* glial fibrillary acidic protein, *GRAF1* GTPase regulator associated with focal adhesion kinase 1, *ITPR* inositol 1,4,5-trisphosphate receptor type 1, *LGI1* leucine rich glioma, *mGluR* metabotropic glutamate receptor, *MAP1B* Microtubule-associated protein 1B, *NMDA* N-methyl-D-aspartate, *PCA* Purkinje cytoplasmic antibody, *VGCC* voltage-gated calcium channel

tion in NMDA autoimmunity, can also be important clues. Strong family history of autoimmunity is also an important clue.

History of cancer or strong family history of cancer or a personal smoking history is an important clinical clue to a paraneoplastic disorder. Identification of an autoantibody can help tailor the specific cancer evaluation, but in cases where no autoantibody is identified or initial cancer search is unrevealing, a broad search is recommended. This could include whole body $^{18}$F-fluorodeoxyglucose ($^{18}$F-FDG) PET/CT, upper and lower gastrointestinal (GI) endoscopies, breast and gynecological-directed examinations in women, and prostate examinations in men [66]. Testicular seminoma, small cell cancers, ovarian and breast neoplasms, neuroblastoma, and thymomas are common oncological associations in paraneoplastic movement disorders.

Table 14.2 summarizes the autoantibodies and their common oncological associations.

## Neurological Examination

A well-documented detailed neurological examination is important. Movement disorders can be difficult to objectively quantify, and thus video recordings are very helpful to document abnormalities and can be used for comparison to evaluate the response to immunotherapy [2].

## Imaging

Different neuroimaging abnormalities have been reported in patients with antibody-mediated movement disorders [11]. The initial test of choice in the evaluation of patients with autoimmune or paraneoplastic movement disorders is magnetic resonance imaging (MRI) with and without contrast. A normal MRI does not exclude the possibility of an immune-mediated disorder, and functional brain imaging with PET/CT can be helpful to identify hyper- or hypometabolism in the basal ganglia or cerebellum in some cases [66]. Figure 14.1 shows examples of brain MRI and PET/CT findings of patients with autoimmune movement disorders.

## Electrophysiological Evaluation

Surface electromyography (EMG) helps characterize certain movement disorders and objectively quantify abnormalities, particularly tremor and myoclonus, and to differentiate these from functional movement disorders. Measurement of acoustic startle and exteroceptive responses can assist in diagnosing SPS spectrum disorders [67].

## Laboratory Testing

Serum testing for non-neurological markers of autoimmunity such as nonorgan-specific autoantibodies (SS-A, SS-B, ds-DNA, etc.) can be helpful clues in suspected autoimmune movement disorders. CSF should be tested for markers of inflammation including white cell count, protein, oligoclonal bands, IgG index, and synthesis rate. Elevation of these markers, although nonspecific, can help support an autoimmune diagnosis. A normal CSF does not exclude an autoimmune etiology for the neurologic presentation [2, 9, 11].

### Autoantibody Detection

Antibody testing should be performed in serum and CSF to improve the yield of detection. Immunotherapy can affect the results, thus it is important to obtain serum or CSF prior to the initiation of immunotherapy, and repeat testing should be considered if negative results are seen in the setting of recent immunotherapy administration. Seronegative patients with a highly suspected autoimmune etiology may harbor unclassified (yet to be characterized) antibodies, thus sending samples to research centers that assess for such antibodies is recommended if initial screening for well-characterized antibodies is negative. Indirect tissue immunofluorescence and immunohistochemistry serve as screening tools for antibody detection, typically confirmed with Western blot for antibodies that bind cytosolic or nuclear antigens. Cell-based assays are utilized for detection of autoantibodies that recognize cell surface proteins [2, 11].

A detailed description of antibody identification methods can be found in Chap. 2.

## Treatment

Class 1 treatment recommendations based on randomized clinical trials in autoimmune movement disorders and autoimmune neurology in general are lacking, and treatment strategies are based on expert opinion or case series. One crossover, randomized, double-blind controlled clinical trial showed improvement in stiffness and heightened-sensitivity scores in patients with SPS with the use of intravenous immunoglobulin [67]. Symptomatic therapy with benzodiazepines or neuroleptics can help ameliorate the severity of some movement disorders, but it is rarely sufficient for complete cessation, and most patients require the use of immunotherapy.

Not all autoimmune movement disorders are accompanied by an antibody, and therefore after excluding competing infectious, metabolic, or genetic causes, a trial of immunotherapy could be considered in carefully selected patients. Oncologic therapy for patients with paraneoplastic syndromes is the first stage of management. Cancer treatment (surgery, chemotherapy, radiation therapy) should be instituted as early as possible as this will stabilize or improve the neurologic symptoms; the concomitant use of immunotherapy can help improve the neurologic symptoms [2, 9–12, 68].

The reader is referred to Chap. 17 "Treatment Approaches in Autoimmune Neurology" for more information regarding immunotherapy in antibody-mediated neurological diseases and Chap. 30 "Stiff Person Syndrome" for treatment and management approaches in SPS.

## Symptomatic Therapy

Symptomatic therapy should be recommended for most patients; the agent utilized depending on the underlying movement disorder. For hyperkinetic movement disorders such as chorea, dopamine-depleting agents include tetrabenazine 12.5–50 mg 3 times per day and deutetrabenazine 6–12 mg twice per day. Alternatively, dopamine antagonists can be used, such as risperidone 2–8 mg/day (by mouth). Monitoring for sedation, insomnia, depression, akathisia, and drug-induced parkinsonism is also important [10].

Benzodiazepines can be effective for treatment of myoclonus and opsoclonus. Typical starting doses are (in divided doses) clonazepam 1–2 mg/day or diazepam 10–15 mg/day. Drowsiness, fatigue, ataxia, and hypotonia are common side effects. Levetiracetam can be used as an alternative for treating myoclonus, with less sedation compared to benzodiazepines.

Antiepileptic drugs (AEDs) such as valproate and carbamazepine can be used to ameliorate chorea [69].

Muscle relaxants such as baclofen, orally or intrathecally, can be used in SPS; benzodiazepines can be quite effective, and patients often tolerate high doses (up to 100 mg/day in divided doses) [2, 9, 10]. The management of SPS is detailed in Chap. 30.

## Immunotherapy

### Acute Treatment

The mainstay of acute treatment for immune-mediated neurologic disorders is high-dose corticosteroids—typically intravenous methylprednisolone 1000 mg for 5 days. Corticosteroids have pleiotropic effects including reducing blood-brain barrier leakage and thereby trafficking of leukocytes, decreasing peripheral lymphocyte numbers, and reducing the production of inflammatory substances and inflammatory cells [70]. IVIg is an alternative option. Recently, a prospective study showed that the early use of IVIg in patients with paraneoplastic neurological syndromes with onconeural antibodies may improve neurologic outcome [70].

Plasma exchange (PLEX) is effective for acute, severely affected patients who fail to respond to corticosteroids. A typical regimen is apheresis of 1.5 plasma volumes every other day for 5–7 treatments. The exact mechanism is unknown and difficult to study given the non-selectivity of serum replacement with plasma exchange, but removal of the pathogenic antibody is believed to contribute, though removal of inflammatory mediators such as cytokines may also help. Steroids, IVIg, and PLEX are considered first-line therapy; medications such as rituximab or cyclophosphamide can be considered if a patient fails to respond to first-line therapy [71].

### Maintenance Treatment

A maintenance immunotherapy medication is recommended for patients who have high risk of relapse. Retrospective studies have shown that azathioprine and mycophenolate reduce the annualized relapse rate in neuromyelitis optica (NMO) and are used for the treatment of other autoimmune neurologic disorders [12, 27, 71]. Both do not take full effect until as long as 6 months after initiation; therefore, patients should also receive moderate doses of oral corticosteroids (prednisone, 20–60 mg per day) for 6 months to prevent attacks.

Rituximab, a chimeric anti-CD20 monoclonal antibody, is commonly used as first-line therapy for relapse prevention. Rituximab has several advantages over other immunosuppressive treatments including easily monitored adherence and rapid onset of action. It results in B-cell depletion within 2 weeks of completing the initial course of 2 infusions of 1 g intravenously separated by 2 weeks. In addition to intravenous methylprednisolone just prior to administration of rituximab to avert potential allergic reactions, patients should receive some steroid treatment for 1 month following the initial administration of rituximab to avert exacerbations resulting from lysis of B cells and accompanying transient increases in antibody titers [72].

Table 14.3 summarizes commonly used maintenance treatments, side effects, and recommended monitoring.

Other drugs less commonly used but potentially effective immunosuppressant treatments include methotrexate, tacrolimus, and cyclosporine [71].

**Table 14.3** Commonly used maintenance treatments, side effects, and monitoring

| Drug | Dose | Route | Frequency | Side effects | Screening and monitoring laboratory parameters |
|---|---|---|---|---|---|
| Azathioprine | 2–3 mg/kg/d | PO | QD/BID | Fever, malaise, myalgias, nausea, vomit, diarrhea.<br>Leukopenia, anemia, thrombocytopenia, liver toxicity, hypersensitivity reaction, rash, lymphoma | TMPT/pregnancy testing prior to first dose. CBC w/diff, LFTs weekly first month; decrease testing, interval thereafter<br>Limit sun and ultraviolet light exposure<br>Discontinue for neutrophil count below 1000/uL. |
| Glucocorticoids (prednisone) | 20–60 mg | PO | Daily, may use 10–20 mg as an adjunct for individuals with breakthrough attacks on other maintenance treatments; or in tapering dosage after an acute attack after initiating azathioprine or mycophenolate | Insomnia, increased appetite, psychiatric disturbance-psychosis, diabetes, cataracts, osteoporosis, hip avascular necrosis, delayed wound healing | CBC, electrolytes, Blood pressure<br>Glucose<br>*Pneumocystis jiroveci* pneumonia<br>Prophylaxis Patients with GERD may need proton pump inhibitors to prevent gastric ulcers in patients at risk for gastrointestinal bleeding<br>Osteoporosis prevention/monitoring |
| Mycophenolate mofetil (MMF) | 500 mg/d to 3000 mg/d<br>Dosage is adjusted based on kidney function | PO | BID | Vomiting, diarrhea, hypertension, creatinine elevation, edema, lymphoma, diarrhea, myelotoxicity, teratogenicity | CBC w/dif, renal function, pregnancy test prior. CBC weekly fist month, increase interval thereafter<br>Limit sun and ultraviolet light exposure<br>Dermatology evaluation at least yearly |
| Rituximab | 1000 mg<br>Repeat in 2 weeks or 375 mg/ $m^2 \times 4$ doses | IV | Without monitoring: Administer every 6 months.<br>With monitoring: Check CD19 count monthly beginning 5–6 months after last dose and redoes when CD19 >1% | Infusion reactions, rash, pruritus, edema, hypertension, fever, fatigue, chills, headache, diarrhea, cytopenias, neutropenic fever, liver toxicity, hepatitis B reactivation<br>Consider institution of IVIg if hypogammaglobulinemia occurs that leads to recurrent sinopulmonary infection | Hepatitis B and tuberculosis screening<br>Periodic CBC w/diff<br>Immunoglobulin G baseline and periodically |

*Abbreviations: BID* twice per day, *CBC w/diff* complete blood count with differential, *GERD* gastroesophageal reflux disease, *IV* intravenous, *IVIg* intravenous immunoglobulin, *LFTs* liver function tests, *MCV* mean corpuscular volume, *PO* per oral, *QD* once per day, *TPMT* thiopurine S-methyltransferase

## Conclusion and Future Directions

Improved diagnosis of autoimmune movement disorders has been greatly facilitated by the discovery of antibodies that are highly specific diagnostic biomarkers. Autoimmune movement disorders should be considered in the differential diagnosis of diverse movement disorder phenotypes, particularly where features are atypical for the classical degenerative forms. Early recognition, cancer screening, and treatment along with early initiation of immunotherapy are important.

Developing clinical trials is key, as the understanding of the pathophysiology of these disorders continues to improve.

## References


1. Mckeon A. Autoimmune movement disorders. Handb Clin Neurol. 2016;133:301–15.
2. Honorat JA, Mckeon A. Autoimmune movement disorders: a clinical and laboratory approach. 1910;
3. Varley JA, Webb AJS, Balint B, Fung VSC, Sethi KD, Tijssen MAJ, et al. The Movement disorder associated with NMDAR antibody-encephalitis is complex and characteristic: an expert video-rating study. J Neurol Neurosurg Psychiatry. 2018;90(6):724–6.
4. Gadoth A, Pittock SJ, Dubey D, McKeon A, Britton JW, Schmeling JE, et al. Expanded phenotypes and outcomes among 256 LGI1/CASPR2-IgG–positive patients. Ann Neurol. 2017;82(1):79–92.
5. Carvajal-González A, Leite MI, Waters P, Woodhall M, Coutinho E, Balint B, et al. Glycine receptor antibodies in PERM and related syndromes: characteristics, clinical features and outcomes. Brain. 2014;137(Pt 8):2178–92.
6. McKeon A, Tracy JA. GAD65 neurological autoimmunity. Muscle Nerve. 2017;56(1):15–27.
7. Peterson K, Rosenblum MK, Kotanides H, Posner JB. Paraneoplastic cerebellar degeneration. I. a clinical analysis of 55 anti-Yo antibody-positive patients. Neurology. 1992;42(10):1931–7.
8. Dale RC, Merheb V, Pillai S, Wang D, Cantrill L, Murphy TK, et al. Antibodies to surface dopamine-2 receptor in autoimmune movement and psychiatric disorders. Brain. 2012;135(11):3453–68.
9. Balint B, Vincent A, Meinck H-M, Irani SR, Bhatia KP. Movement disorders with neuronal antibodies: syndromic approach, genetic parallels and pathophysiology. Brain. 2018;141(1):13–36.
10. Fidel Baizabal-Carvallo J, Jankovic J. Autoimmune and paraneoplastic movement disorders: an update. J Neurol Sci. 2018;385:175–84.
11. López-Chiriboga A, Flanagan E. Diagnostic and therapeutic approach to autoimmune neurologic disorders. Semin Neurol. 2018;38(03):392–402.
12. Jones AL, Flanagan EP, Pittock SJ, Mandrekar JN, Eggers SD, Ahlskog JE, et al. Responses to and outcomes of treatment of autoimmune cerebellar ataxia in adults. JAMA Neurol. 2015;72(11):1304.
13. Dubey D, Pittock SJ, Kelly CR, McKeon A, Lopez-Chiriboga AS, Lennon VA, et al. Autoimmune encephalitis epidemiology and a comparison to infectious encephalitis. Ann Neurol. 2018;83(1):166–77.
14. Espay AJ, Chen R. Rigidity and spasms from autoimmune encephalomyelopathies: stiff-person syndrome. Muscle Nerve. 2006;34(6):677–90.
15. O'Toole O, Lennon VA, Ahlskog JE, Matsumoto JY, Pittock SJ, Bower J, et al. Autoimmune chorea in adults. Neurology. 2013;80(12):1133–44.
16. Hadjivassiliou M, Martindale J, Shanmugarajah P, Grünewald RA, Sarrigiannis PG, Beauchamp N, et al. Causes of progressive cerebellar ataxia: prospective evaluation of 1500 patients. J Neurol Neurosurg Psychiatry. 2017;88(4):301–9.
17. Abdo WF, Van De Warrenburg BP, Burn DJ, Quinn NP, Bloem BR. The clinical approach to movement disorders. Nat Rev Neurol. 2010;6:29–37.
18. McKeon A, Robinson MT, McEvoy KM, Matsumoto JY, Lennon VA, Ahlskog JE, et al. Stiff-man syndrome and variants. Arch Neurol. 2012;69(2):230.
19. Bhidayasiri R, Truong DD. Chorea and related disorders. Postgrad Med J. 2004;80(947):527–34.
20. Walker RH. The non-Huntington disease choreas: five new things. Neurol Clin Pract. 2016;6(2):150–6.
21. Dean SL, Singer HS. Treatment of Sydenham's chorea: a review of the current evidence. Tremor Other Hyperkinet Mov (N Y). 2017;7:456.
22. Honorat JA, Komorowski L, Josephs KA, Fechner K, St Louis EK, Hinson SR, et al. IgLON5 antibody: neurological accompaniments and outcomes in 20 patients. Neurol Neuroimmunol Neuroinflamm. 2017;4(5):e385.
23. Armangue T, Spatola M, Vlagea A, Mattozzi S, Cárceles-Cordon M, Martinez-Heras E, et al. Frequency, symptoms, risk factors, and outcomes of autoimmune encephalitis after herpes simplex encephalitis: a prospective observational study and retrospective analysis. Lancet Neurol. 2018;17(9):760–72.
24. Linnoila JJ, Binnicker MJ, Majed M, Klein CJ, McKeon A. CSF herpes virus and autoantibody profiles in the evaluation of encephalitis. Neurol Neuroimmunol Neuroinflamm. 2016;3(4):e245.
25. Yu Z, Kryzer TJ, Griesmann GE, Kim K, Benarroch EE, Lennon VA. CRMP-5 neuronal autoantibody: marker of lung cancer and thymoma-related autoimmunity. Ann Neurol. 2001;49(2):146–54.
26. Vernino S, Tuite P, Adler CH, Meschia JF, Boeve BF, Boasberg P, et al. Paraneoplastic chorea associated with CRMP-5 neuronal antibody and lung carcinoma. Ann Neurol. 2002;51(5):625–30.

27. Panzer J, Dalmau J. Movement disorders in paraneoplastic and autoimmune disease. Curr Opin Neurol. 2011;24(4):346–53.
28. Vynogradova I, Savitski V, Heckmann JG. Hemichorea associated with CASPR2 antibody. Tremor Other Hyperkinet Mov (N Y). 2014;4:239.
29. Tofaris GK, Irani SR, Cheeran BJ, Baker IWS, Cader ZM, Vincent A. Immunotherapy-responsive chorea as the presenting feature of LGI1-antibody encephalitis. Neurology. 2012;79(2):195–6.
30. Titulaer MJ, McCracken L, Gabilondo I, Armangué T, Glaser C, Iizuka T, et al. Treatment and prognostic factors for long-term outcome in patients with anti-NMDA receptor encephalitis: an observational cohort study. Lancet Neurol. 2013;12(2):157–65.
31. Baizabal-Carvallo JF, Stocco A, Muscal E, Jankovic J. The spectrum of movement disorders in children with anti-NMDA receptor encephalitis. Mov Disord. 2013;28(4):543–7.
32. Gresa-Arribas N, Planagumà J, Petit-Pedrol M, Kawachi I, Katada S, Glaser CA, et al. Human neurexin-3α antibodies associate with encephalitis and alter synapse development. Neurology. 2016;86(24):2235–42.
33. Pittock SJ, Lucchinetti CF, Lennon VA. Anti-neuronal nuclear autoantibody type 2: paraneoplastic accompaniments. Ann Neurol. 2003;53(5):580–7.
34. Irani SR, Michell AW, Lang B, Pettingill P, Waters P, Johnson MR, et al. Faciobrachial dystonic seizures precede Lgi1 antibody limbic encephalitis. Ann Neurol. 2011;69(5):892–900.
35. Thompson J, Bi M, Murchison AG, Makuch M, Bien CG, Chu K, et al. The importance of early immunotherapy in patients with faciobrachial dystonic seizures. Brain. 2018;141(2):348–56.
36. López Chiriboga AS, Siegel JL, Tatum WO, Shih JJ, Flanagan EP. Striking basal ganglia imaging abnormalities in LGI1 ab faciobrachial dystonic seizures. Neurol Neuroimmunol Neuroinflamm. 2017;4(3):e336.
37. Flanagan EP, Kotsenas AL, Britton JW, McKeon A, Watson RE, Klein CJ, et al. Basal ganglia T1 hyperintensity in LGI1-autoantibody faciobrachial dystonic seizures. Neurol Neuroimmunol Neuroinflamm. 2015;2(6):e161.
38. Antunes NL, Khakoo Y, Matthay KK, Seeger RC, Stram DO, Gerstner E, et al. Antineuronal antibodies in patients with neuroblastoma and paraneoplastic opsoclonus-myoclonus. J Pediatr Hematol Oncol. 2000;22(4):315–20.
39. Klaas JP, Ahlskog JE, Pittock SJ, Matsumoto JY, Aksamit AJ, Bartleson JD, et al. Adult-onset opsoclonus-myoclonus syndrome. Arch Neurol. 2012;69(12):1598.
40. Armangué T, Sabater L, Torres-Vega E, Martínez-Hernández E, Ariño H, Petit-Pedrol M, et al. Clinical and immunological features of opsoclonus-myoclonus syndrome in the era of neuronal cell surface antibodies. JAMA Neurol. 2016;73(4):417–24.
41. Berridge G, Menassa DA, Moloney T, Waters PJ, Welding I, Thomsen S, et al. Glutamate receptor δ2 serum antibodies in pediatric opsoclonus myoclonus ataxia syndrome. Neurology. 2018;91(8):e714–23.
42. McKeon A, Martinez-Hernandez E, Lancaster E, Matsumoto JY, Harvey RJ, McEvoy KM, et al. Glycine receptor autoimmune spectrum with stiff-man syndrome phenotype. JAMA Neurol. 2013;70(1):44–50.
43. Hinson SR, Lopez-Chiriboga AS, Bower JH, Matsumoto JY, Hassan A, Basal E, et al. Glycine receptor modulating antibody predicting treatable stiff-person spectrum disorders. Neurol Neuroimmunol Neuroinflamm. 2018;5(2):e438.
44. Gövert F, Witt K, Erro R, Hellriegel H, Paschen S, Martinez-Hernandez E, et al. Orthostatic myoclonus associated with Caspr2 antibodies. Neurology. 2016;86(14):1353–5.
45. Flanagan EP, Hinson SR, Lennon VA, Fang B, Aksamit AJ, Morris PP, et al. Glial fibrillary acidic protein immunoglobulin G as biomarker of autoimmune astrocytopathy: analysis of 102 patients. Ann Neurol. 2017;81(2):298–309.
46. Dubey D, Hinson SR, Jolliffe EA, Zekeridou A, Flanagan EP, Pittock SJ, et al. Autoimmune GFAP astrocytopathy: prospective evaluation of 90 patients in 1 year. J Neuroimmunol. 2018;321:157–63.
47. Tobin WO, Lennon VA, Komorowski L, Probst C, Clardy SL, Aksamit AJ, et al. DPPX potassium channel antibody: frequency, clinical accompaniments, and outcomes in 20 patients. Neurology. 2014;83(20):1797–803.
48. Dalmau J, Graus F, Villarejo A, Posner JB, Blumenthal D, Thiessen B, et al. Clinical analysis of anti-Ma2-associated encephalitis. Brain. 2004;127(8):1831–44.
49. Sabater L, Gaig C, Gelpi E, Bataller L, Lewerenz J, Torres-Vega E, et al. A novel non-rapid-eye movement and rapid-eye-movement parasomnia with sleep breathing disorder associated with antibodies to IgLON5: a case series, characterisation of the antigen, and post-mortem study. Lancet Neurol. 2014;13(6):575–86.
50. Honorat JA, Lopez-Chiriboga AS, Kryzer TJ, Fryer JP, Devine M, Flores A, et al. Autoimmune septin-5 cerebellar ataxia. Neurol Neuroimmunol Neuroinflamm. 2018;5(5):e474.
51. Martinez-Hernandez E, Ariño H, McKeon A, Iizuka T, Titulaer MJ, Simabukuro MM, et al. Clinical and immunologic investigations in patients with stiff-person spectrum disorder. JAMA Neurol. 2016;73(6):714.
52. Pittock SJ, Yoshikawa H, Ahlskog JE, Tisch SH, Benarroch EE, Kryzer TJ, et al. Glutamic acid decarboxylase autoimmunity with brainstem, extrapyramidal, and spinal cord dysfunction. Mayo Clin Proc. 2006;81(9):1207–14.
53. Martinez-Hernandez E, Sepulveda M, Rostásy K, Höftberger R, Graus F, Harvey RJ, et al. Antibodies to aquaporin 4, myelin-oligodendrocyte glycoprotein, and the glycine receptor α1 subunit in patients with isolated optic neuritis. JAMA Neurol. 2015;72(2):187.

54. Pittock SJ, Lucchinetti CF, Parisi JE, Benarroch EE, Mokri B, Stephan CL, et al. Amphiphysin autoimmunity: paraneoplastic accompaniments. Ann Neurol. 2005;58(1):96–107.
55. Murinson BB, Guarnaccia JB. Stiff-person syndrome with amphiphysin antibodies: distinctive features of a rare disease. Neurology. 2008;71(24):1955–8.
56. Jarius S, Wildemann B. 'Medusa head ataxia': the expanding spectrum of Purkinje cell antibodies in autoimmune cerebellar ataxia. Part 1: anti-mGluR1, anti-Homer-3, anti-Sj/ITPR1 and anti-CARP VIII. J Neuroinflammation. 2015;12(1):166.
57. Shams'ili S, Grefkens J, de Leeuw B, van den Bent M, Hooijkaas H, van der Holt B, et al. Paraneoplastic cerebellar degeneration associated with antineuronal antibodies: analysis of 50 patients. Brain. 2003;126(Pt 6):1409–18. https://pubmed.ncbi.nlm.nih.gov/32744608/.
58. Linnoila J, Guo Y, Gadoth A, Raghunathan A, Parks B, McKeon A, et al. Purkinje cell cytoplasmic antibody type I (anti-Yo): predictive of gastrointestinal adenocarcinomas in men. J Neurol Neurosurg Psychiatry. 2018;89(10):1116–7.
59. Gadoth A, Kryzer TJ, Fryer J, McKeon A, Lennon VA, Pittock SJ. Microtubule-associated protein 1B: novel paraneoplastic biomarker. Ann Neurol. 2017;81(2):266–77.
60. Joubert B, Gobert F, Thomas L, Saint-Martin M, Desestret V, Convers P, et al. Autoimmune episodic ataxia in patients with anti-CASPR2 antibody-associated encephalitis. Neurol Neuroimmunol Neuroinflamm. 2017;4(4):e371.
61. Lopez-Chiriboga AS, Komorowski L, Kümpfel T, Probst C, Hinson SR, Pittock SJ, et al. Metabotropic glutamate receptor type 1 autoimmunity. Neurology. 2016;86(11):1009–13.
62. Jarius S, Wildemann B. 'Medusa head ataxia': the expanding spectrum of Purkinje cell antibodies in autoimmune cerebellar ataxia. Part 3: anti-Yo/CDR2, anti-Nb/AP3B2, PCA-2, anti-Tr/DNER, other antibodies, diagnostic pitfalls, summary and outlook. J Neuroinflammation. 2015;12(1):168.
63. Jarius S, Wildemann B. 'Medusa head ataxia': the expanding spectrum of Purkinje cell antibodies in autoimmune cerebellar ataxia. Part 2: anti-PKC-gamma, anti-GluR-delta2, anti-Ca/ARHGAP26 and anti-VGCC. J Neuroinflammation. 2015;12(1):167.
64. Zalewski N, Lennon VA, Pittock SJ, Mckeon A. Calcium channel autoimmunity: cerebellar ataxia and lambert-eaton syndrome coexisting. Muscle Nerve. 2018;58(1):29–35.
65. Zalewski NL, Lennon VA, Lachance DH, Klein CJ, Pittock SJ, Mckeon A. P/Q- and N-type calcium-channel antibodies: oncological, neurological, and serological accompaniments. Muscle Nerve. 2016;54(2):220–7.
66. Probasco JC, Solnes L, Nalluri A, Cohen J, Jones KM, Zan E, et al. Abnormal brain metabolism on FDG-PET/CT is a common early finding in autoimmune encephalitis. Neurol Neuroimmunol Neuroinflamm. 2017;4(4):e352.
67. Matsumoto JY, Caviness JN, McEvoy KM. The acoustic startle reflex in stiff-man syndrome. Neurology. 1994;44(10):1952–5.
68. Damato V, Balint B, Kienzler A-K, Irani SR. The clinical features, underlying immunology, and treatment of autoantibody-mediated movement disorders. Mov Disord. 2018;33(9):1376–89.
69. Siniscalchi A, Gallelli L, De Sarro G. Use of antiepileptic drugs for hyperkinetic movement disorders. Curr Neuropharmacol. 2010;8(4):359–66.
70. Berzero G, Karantoni E, Dehais C, Ducray F, Thomas L, Picard G, et al. Early intravenous immunoglobulin treatment in paraneoplastic neurological syndromes with onconeural antibodies. J Neurol Neurosurg Psychiatry. 2018;89(7):789–92.
71. McKeon A. Immunotherapeutics for autoimmune encephalopathies and dementias. Curr Treat Options Neurol. 2013;15(6):723–37.
72. Kosmidis ML, Dalakas MC. Practical considerations on the use of rituximab in autoimmune neurological disorders. Ther Adv Neurol Disord. 2010;3(2):93–105.

# Autoimmune Demyelinating Syndromes: Aquaporin-4-IgG-positive NMOSD and MOG - IgG Associated Disorder

15

Elia Sechi and Eoin P. Flanagan

**Key Points**

1. Aquaporin-4-immunoglobulin G (AQP4-IgG) and myelin oligodendrocyte glycoprotein (MOG)-IgG autoantibodies are associated with central nervous system (CNS) demyelinating disease entities that are distinct from multiple sclerosis (MS).
2. Each disorder has characteristics clinical, laboratory, and magnetic resonance imaging (MRI) features, although some features overlap.
3. In the correct clinical context, testing for these autoantibodies is critical to avoid misdiagnosis.
4. Distinction between the differing types of inflammatory CNS demyelinating diseases is important as it has treatment and prognostic implications.

E. Sechi
Department of Neurology, Mayo Clinic, Rochester, MN, USA

E. P. Flanagan (✉)
Department of Neurology, and Laboratory Medicine and Pathology, Mayo Clinic, Rochester, MN, USA
e-mail: flanagan.eoin@mayo.edu

## Part I: Introduction to AQP4-IgG and MOG-IgG as Autoantibody Biomarkers

### History and Definitions

The simultaneous occurrence of optic neuritis (ON) and longitudinally extensive transverse myelitis (LETM), without apparent brain involvement, was first reported in 1865 by the British neurologist Jacob Augustus Lockhart Clarke, in a 17-year-old girl who eventually died for respiratory failure [1]. About 30 years later (1894), Eugene Devic and his student Fernand Gault reviewed the existing literature for similar cases and described the disease with the name of neuromyelitis optica (NMO) [2]. For a century, Devic's disease was considered by many as a more aggressive variant of multiple sclerosis (MS) confined to optic nerves and spinal cord, and in Asia a similar disease was termed optic-spinal MS [3].

In 2004, the identification of aquaporin-4-immunoglobulin G (AQP4-IgG), an antibody directed against the main central nervous system (CNS) water-channel with high specificity for NMO, gave the disease a definite identity distinct from MS and traced a new route in the field of neuroimmunology [4, 5]. It soon became clear that the spectrum of clinical and radiological manifestations related to AQP4-IgG was much broader than previously thought, ranging from partial forms (e.g., recurrent optic neuritis or recurrent myelitis) to predominant brain



A. L. Piquet, E. Alvarez (eds.), *Neuroimmunology*, https://doi.org/10.1007/978-3-030-61883-4_15

involvement (e.g., area postrema syndrome, encephalopathy), and the concept of neuromyelitis optica spectrum disorders (NMOSD) was introduced [6].

In 2015, diagnostic criteria for NMOSD were published by an international panel of experts [7]. The criteria allowed NMOSD diagnosis by AQP4-IgG seropositivity and involvement of at least one CNS region among optic nerve, cerebral hemispheres, diencephalic region, brainstem, or spinal cord, and introduced the concept of AQP4-IgG seronegative NMOSD, defined by more stringent clinical and radiological requirements than AQP4-IgG seropositive NMOSD [7]. The absence of AQP4-IgG in these patients raised the possibility that other antibody biomarkers of demyelination could account for some of this subgroup, which was later shown to be the case with myelin oligodendrocyte glycoprotein (MOG)-IgG.

MOG-IgG is an autoantibody directed against a minor component of CNS myelin located on oligodendrocytes that was historically studied as potential target of autoimmunity in MS with inconsistent results and detection in both MS patients and healthy controls [8]. However, in the last decade with the use of cell-based assays (CBA) expressing MOG in its conformational status rather than assays employing denatured MOG proteins (e.g., western blot, enzyme-linked immunosorbent assay [ELISA]), we have been able to define a CNS demyelinating disease associated with MOG-IgG that is distinct from MS [8]. While MOG-IgG is rarely detected in sera of MS patients and generally with a low titer, its spectrum of clinical and radiological manifestations has some overlap with AQP4-IgG, with MOG-IgG being detected in up to 42% of seronegative NMOSD cases [9]. To avoid diagnostic ambiguity, an alternative molecular target-defined nomenclature (i.e., AQP4 autoimmunity and MOG autoimmunity) has also been proposed [10]. In this chapter, the term NMOSD will be used to indicate the AQP4-IgG seropositive forms, unless specified otherwise.

## Epidemiology of NMOSD- and MOG-IgG-Associated Disorders

When the 2015 diagnostic criteria are applied, the overall prevalence of NMOSD (both seropositive and seronegative cases) is very similar in Western countries and Asia, accounting for 3.9/100,000 (USA) and 4.1/100,000 (Japan), respectively. These numbers decrease to 3.3/100,000 (USA) and 3.2/100,000 (Japan) if AQP4-IgG-positive cases only are considered [11, 12]. By contrast, the disease is more common among Afro-Caribbeans with an estimated prevalence of 10/100,000 (7.9/100,000 AQP4-IgG seropositive cases) [11]. The relative frequency of NMOSD as a proportion of CNS demyelinating disorders seems to be inversely related to that of MS, being higher in countries where MS is less represented (e.g., some Asian countries and regions close to the equator) [13]. The epidemiology of MOG-IgG has been less well elucidated, but the frequency of the antibody seems higher among Caucasians [14]. Both NMOSD- and MOG-IgG-associated disorders may potentially affect any age with a sharp female predominance for the former (up to 9 times more prevalent than in males) [11].

## Part II: AQP4-IgG Neuromyelitis Optica Spectrum Disorders

### Pathophysiology

There is a strong body of evidence suggesting AQP4-IgG as a pathogenic effector rather than only a specific disease biomarker in NMOSD [15]. AQP4 is a tetrameric transmembrane protein mainly expressed in the CNS but also present in the collecting ducts of the kidney, parietal cells of the stomach, lungs, and skeletal muscles [16]. In the CNS, AQP4 is mainly located at the surface of astrocytic end-feet that abut capillaries and pia and take part in the constitution of the blood-brain and blood-cerebrospinal fluid (CSF) barriers playing a crucial role in water homeostasis [16]. The subpial and subependymal regions around the ventricles are particularly rich in AQP4 and represent sites of characteristic NMOSD lesions on brain MRI, although brain abnormalities may also occur in areas where AQP4 expression is lower (see later section on "MRI") [17, 18].

AQP4-IgG is an IgG1, and the exact mechanisms by which it exerts its pathogenic effect are diverse and still largely unclear, but complement activation seems to play a major role, mainly via the classical pathway [19, 20]. AQP4 tetramers are constituted by two main AQP4 isoforms, M1 and M23, which have identical extracellular domain residues but differ by 22 amino acids that are exclusively present in the M1 isoform at the cytoplasmic N terminus. A low M1/M23 ratio seems to favor AQP4 aggregation into supramolecular complexes called orthogonal arrays of particles (OAPs), with a conformational status that would facilitate binding of AQP4-IgG and C1q, and subsequent complement activation, inflammation, blood-brain barrier dysfunction, and necrosis [21–23]. The higher proportion of these supramolecular complexes in the optic nerves and spinal cord might partly account for their frequent involvement in NMOSD and the characteristic necrotic changes that are observed in these sites pathologically [24]. Monoclonal antibodies targeting key proteins of the complement cascade have been proven to be highly effective either acutely and for relapse prevention in NMOSD (see "Treatment" section, subheadings "Acute Treatment" and "Long-Term Treatment."). Conversely, a predominance of the M1 isoform would favor AQP4 internalization by antibody binding without complement activation, resulting in vasogenic edema and less-destructive lesions that are often reversible on MRI (see later sections on "Pathology" and "MRI") [24, 25]. Demyelination and neuronal loss in NMOSD would therefore represent a secondary phenomenon occurring as a consequence of astrocytic damage, differing from MOG-IgG-associated disorders where oligodendrocytes would represent the primary disease target [26]. Glutamate excitotoxicity by dysfunction of the excitatory amino acid transporter 2 (EAAT2), the main glutamate transporter predominantly expressed in astroglial cells and physically linked to AQP4 on astrocytic membrane, is likely to play a major role in tissue damage secondary to astrocytes loss [25–28].

AQP4-IgG is produced by B cells outside the CNS, and its concentration is therefore higher in serum than in CSF [16]. Antibody detection in serum of patients years before they become symptomatic suggests the presence of AQP4-IgG is not sufficient to induce NMOSD and that additional factors are likely to be involved in NMOSD pathogenesis [29]. To corroborate this, antibodies that bind to the brain microvascular endothelial cell's glucose-regulated protein 78 ("GRP78") have been recently proposed as potential mediators of blood-brain barrier disruption, thus allowing AQP4-IgG to enter the CNS and cause harm ("double-hit theory") [30]. The reason why AQP4-expressing extra-CNS organs are generally not affected by AQP4-IgG autoimmunity is not completely understood but might rely on the presence of complement-inhibiting proteins (e.g., CD59) in peripheral tissues that would not be expressed at astrocytic foot processes in the CNS. Dysfunction of these proteins may account for the rarely reported cases of peripheral manifestations related to AQP4-IgG [31, 32].

The trigger for AQP4-IgG production is still not clear, although up to 5–15% of cases (generally elderly patients) are thought to be paraneoplastic [33, 34]. Familial forms have rarely been described as well as associations with specific human leukocyte antigen alleles (e.g., HLA-DRB1*03) [15, 35]. However, like other autoimmune diseases that frequently coexist with NMOSD (see later section on "Coexisting Autoimmunity"), the interaction between still unclear genetic and environmental factors likely plays a major role in the generation of AQP4-IgG autoimmunity [36–38]. The contribution of CD4+ cells of the Th17 subtype and interleukin-6 (Il-6) seems to be fundamental to autoantibody production in NMOSD, representing an optimal target for immunotherapies (see later section on "Treatment") [39, 40].

## Pathology

The pathology of NMOSD is characterized by extensive loss of immunoreactivity for AQP4 and glial fibrillary acidic protein—GFAP (the main cytoskeletal protein of astrocytes)—in lesional tissue, and the absence of cortical demyelinating lesions, a frequent pathological finding in MS [41, 42]. Unlike MS, high GFAP levels indicating astrocytic damage can be detected in the CSF of NMOSD patients during disease flares [43].

Two main subtypes of NMOSD lesions have been described [44]. The typical NMOSD lesions are characterized by a unique vasculocentric pattern of demyelination, extensive complement activation, immunoglobulin deposition (both IgG and IgM subtypes), and inflammatory infiltrates variably including eosinophils/neutrophils, but also T cells, B cells, and macrophages. Severe axonal and astrocyte losses are also typically encountered. The perivascular presence (rim and rosette pattern) of C9neo antigen (indicating activation of the membrane attack complex), immunoglobulin deposition, and activated macrophages observed in active NMOSD lesions correspond to the AQP4-enriched astrocytic end-feet that surrounds the blood vessels and represents the initial site of complement activation in the brain [19]. The spinal cord is typically affected over multiple contiguous vertebral segments and mainly in its central part, although both gray and white matters show extensive necrotic changes in the advanced stages [19, 24]. The second lesion subtype is less destructive with the absence of demyelination, vacuolated myelin, and limited axonal injury and inflammation [24]. Clinical and MRI manifestations of these less-destructive lesions are likely due to functional impairment rather than structural damage (see previous section "Pathophysiology").

## Clinical Features

A comparison between the general characteristics of NMOSD, MOG-IgG-associated disorders, and MS is shown in Table 15.1.

The most typical manifestations of NMOSD are ON, transverse myelitis, and, less frequently, area postrema syndrome [7]. Involvement of other brain regions is less common but more heterogeneous and may rarely occur in isolation at onset [45]. The clinical course is generally relapsing-remitting, but, in contrast to MS, attacks are typically more severe, and long-term sequelae are frequent [46, 47].

*Optic neuritis* in NMOSD is generally recurrent and/or bilateral, with AQP4-IgG being very rarely detected in patients with isolated unilateral ON [48, 49]. Patients typically present acutely with more severe visual impairment and less complete recovery compared to MS patients with ON; unilateral or bilateral blindness is not uncommon after years [50, 51]. Since AQP4-IgG preferentially targets the posterior optic pathway including the chiasm and optic tracts, NMOSD may sometimes manifest with bitemporal or homonymous hemianopia with normal fundoscopy [51].

The *transverse myelitis* in NMOSD is characterized by acute/subacute onset (from hours to a few weeks) and severe functional impairment, with tetra- or para-plegia being not infrequent at the nadir of attack [52]. Signs and symptoms of spinal cord dysfunction include weakness and numbness, which are frequently bilateral and symmetric, but a partial, MS-like myelitis may occur [53]; bowel and bladder dysfunction and Lhermitte's phenomenon are additional features often encountered. Tonic spasms are paroxysmal, unilateral, stereotyped painful tonic postures of the limbs usually lasting 1–3 minutes. These episodes are triggered by movement or hyperventilation and responsive to low-dose carbamazepine. Studies have found these are more common in NMOSD than MS [54]. Paroxysmal attacks of pruritus with intractable localized segmental itchiness have also been described after NMOSD myelitis [55]. When infectious, rheumatologic, vascular, and neoplastic causes are excluded, AQP4-IgG autoimmunity represents the most common cause of both monophasic (50%) and recurrent (93%) LETM [48].

**Table 15.1** Comparison of demographics, clinical, laboratory, and MRI characteristics in MS-, NMOSD-, and MOG-IgG-associated disorders

| | MS | AQP4-IgG | MOG-IgG |
|---|---|---|---|
| Demographics | | | |
| Most affected age | 20–40 | 30–50 | 0–40 |
| Sex (F:M) | 2:1 | 9:1 | 1:1 |
| Ethnicity | Any, mostly Caucasian | Any, African-American/ Afro-Caribbean at higher risk | Any |
| Clinical features | | | |
| Antecedent infection/ immunization | Rare | Rare | Common |
| Disease course | Relapsing 85%; progressive15% at onset; relapsing patients may develop secondary prorgession. | Generally relapsing, progressive disability is rare. | Monophasic in about 40% (mainly children); progressive disability not reported. |
| Optic neuritis | ++ | +++ | +++ |
| Myelitis | +++ | +++ | ++ |
| Area postrema syndrome | Rare | ++ | Rare |
| Encephalopathy | Rare | + | ++ |
| Seizures | Rare | Rare | + |
| CSF | | | |
| Oligoclonal bands | 90% (persistent) | 10–30% (transient) | <15% (transient) |
| White cell count >50/μ(mu)l | Rare | 13–35% | 13–35% |
| MRI[a] | | | |
| Optic nerve | Generally unilateral, short lesions, mainly along the intraorbital tract | Uni-/bilateral, long lesions (> 50% of nerve), mainly posterior segments including chiasm | Uni-/bilateral, long lesions, mainly anterior segments, enhancement of the optic nerve sheath and perineural fat |
| Spinal cord | Multiple short lesions; periphery of cord (dorsal, lateral column). Ring or homogeneous enhancement | Longitudinally extensive, generally single (85%). Central/diffuse on axial images. Ring or patchy enhancement | Longitudinally extensive (60–80%), generally >1 lesion, conus involved, H-sign on axial, enhancement in ~50% of cases |
| Brain | Ovoid periventricular, infratentorial, juxtacortical, inferior temporal pole, T1 hypointense | Nonspecific, area postrema, peri- 3rd/4th ventricle, corticospinal tracts, sometimes extensive white matter lesions | Deep gray matter, ADEM-like, 4th ventricle and middle cerebellar peduncles |
| Treatment | | | |
| Acute | IV methylprednisolone; PLEX generally not necessary | IV methylprednisolone; low threshold to follow with PLEX | IV methylprednisolone; PLEX if severe episode; IVIg possible alternative to PLEX, especially for children |
| Recovery | Generally complete | Often incomplete | Generally good despite severe attacks |
| Maintenance | Large variety of MS medications FDA approved medications in USA (see Chapter 32) | **FDA approved in USA**: Eculizumab, satralizumab, inebilizumab. **Others**: azathioprine, methotrexate, mycophenolate, prednisone, rituximab, tocilizumab | For relapsing disease; No FDA approved medications; Attack-prevention treatment may be considered if severe attack with poor recovery or persistently elevated antibody titer: Azathioprine, mycophenolate, methotrexate, rituximab, IVIg |

*ADEM* acute disseminated encephalomyelitis, *AQP4* aquaporin-4, *CSF* cerebrospinal fluid, *F* female, *IV* intravenous, *Ig* immunoglobulins, *M* male, *MS* multiple sclerosis, *MOG* myelin oligodendrocyte glycoprotein, *MRI* magnetic resonance imaging, *PLEX* plasma exchange

[a]Considering adult patients only, MRI abnormalities in children tend to be more extensive and heterogeneous in all the three subgroups

*Area postrema syndrome* is defined by episodes of intractable nausea, vomiting and hiccups [56–58]. Like other circumventricular organs, the area postrema is a site of high AQP4 expression where the blood-brain barrier is particularly permeable and represents a frequent target for AQP4-IgG autoimmunity [59]. Area postrema syndrome can be the initial manifestation of NMOSD in approximately 12% of patients (often leading to inconclusive gastroenterological evaluations) and ultimately will occur in up to 40% of patients [60]. The three cardinal symptoms of nausea, vomiting, and hiccups can be concomitantly observed in 43% of cases, and they typically last 48 hours or more [58]. Response to immunotherapy is generally excellent [58]. Area postrema lesions are pathologically characterized by nondestructive changes that are consistent with the reversibility of both MRI findings and symptoms related to involvement of this region (see previous section "Pathology") [61].

*Less common NMOSD manifestations* include brainstem syndromes (e.g., diplopia, ophthalmoplegia), diencephalic syndromes with thalamic/hypothalamic dysfunction (e.g., narcolepsy, eating disorders, inappropriate antidiuretic hormone secretion [SIADH], hypothermia, hypotension, and endocrinopathies), and cerebral syndromes (e.g., acute disseminated encephalomyelitis [ADEM], posterior reversible encephalopathy syndrome [PRES]) [45, 62–64]. Similar to MS, rare patients may develop epileptic seizures (typically partial and well responsive to antiepileptic drugs) secondary to brain lesions [65]. Other rare NMOSD manifestations include myeloradiculitis with cauda equina involvement [66], obstructive hydrocephalus (from ependymal inflammation) [67], and myositis with hyperCKemia (from skeletal muscle involvement, which is known to contain AQP4; see earlier section "Pathophysiology") [31].

## MRI

Characteristic MRI abnormalities of NMOSD are shown and compared with those of MOG-IgG associated disorders and MS in Figs. 15.1, 15.2, and 15.3.

***Optic nerve MRI*** When compared to MS, ON in NMOSD is more likely to show bilateral T2-hyperintensity of the optic nerves during the

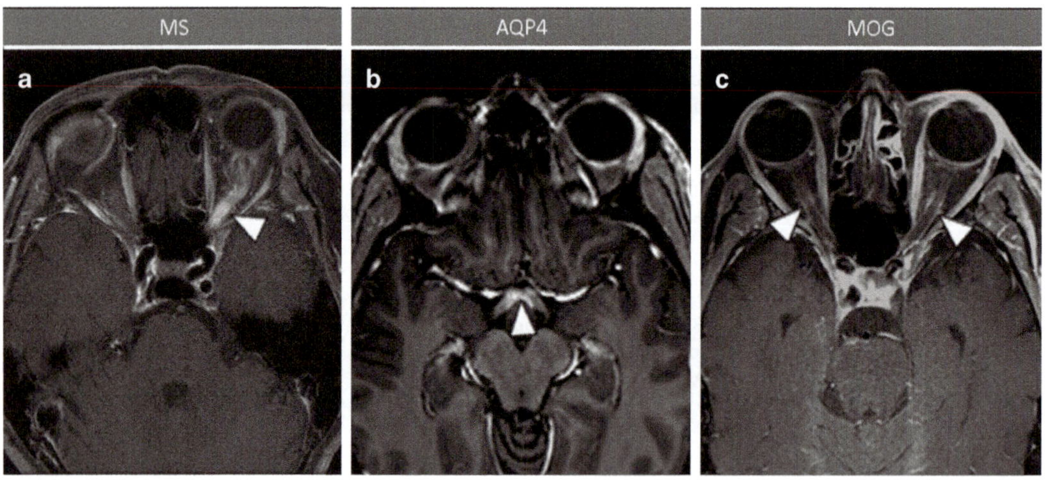


**Fig. 15.1** Characteristic optic nerve MRI findings in MS, AQP4-IgG seropositive NMOSD, and MOG-IgG-associated disorders. Axial fat-desaturated post-gadolinium T1-weighted images showing focal enhancement of the left optic nerve in an MS patient (**a**) and of the posterior optic chiasm in a patient with AQP4-IgG seropositive NMOSD (**b**). Contrast enhancement is bilaterally appreciable for nearly the entire length of the optic nerve sheaths in a patient with MOG-IgG associated disorders (**c**)

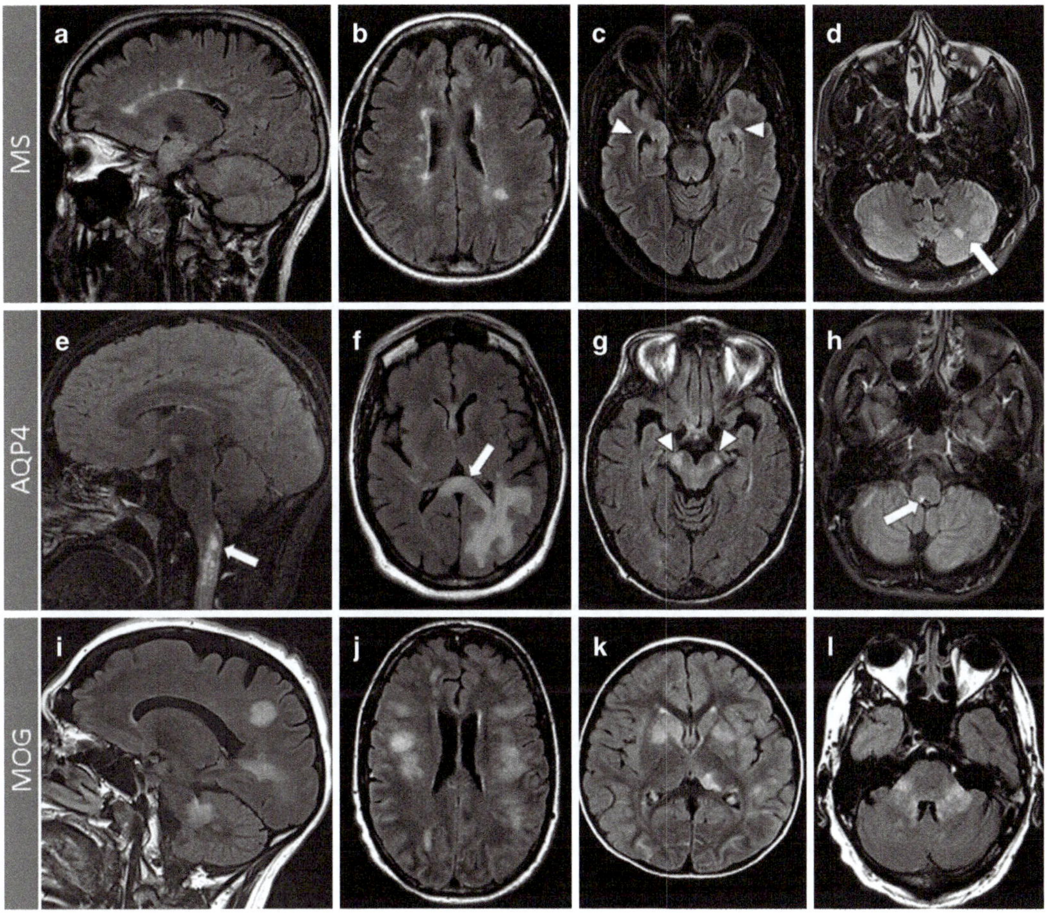


**Fig. 15.2** Characteristic brain MRI findings in patients with MS (top row, **a–d**), AQP4-IgG seropositive NMOSD (middle row, **e–h**), and MOG-IgG-associated disorders (bottom row, **i–l**). Sagittal FLAIR images showing small ovoid lesions on the callosal-septal interface in an MS patient (**a**); a cervical spinal cord lesion extending cranially to the dorsal medulla and area postrema in an NMOSD patient (**e**); and three large demyelinating lesions in the cerebral hemisphere and cerebellar peduncle of a patient with MOG-IgG-associated disorders (**i**). Axial FLAIR images of the hemispheric periventricular white matter showing small, ovoid lesions perpendicularly oriented to the surface of the lateral ventricles (Dawson's fingers) typically found in MS (**b**); an extensive hemispheric white matter lesion involving the splenium of the corpus callosum for its entire thickness and resembling an arch-bridge in NMOSD (**f**, arrow); "fluffy" multifocal lesions in a patient with MOG-IgG-positive ADEM (**j**). Axial FLAIR images also showing characteristics temporal pole (**c**, arrowheads) and cerebellar (**d**, arrow) lesions in MS; bilateral corticospinal tract cerebral peduncles (**g**, arrowheads) and area postrema (**h**, arrow) involvement in NMOSD (the patient in **h** had concomitant area postrema syndrome); and involvement of the deep gray nuclei (**k**), periaqueductal region and bilateral cerebellar peduncles (**l**) in MOG-IgG-associated disorders

acute phase (with or without abnormal gadolinium enhancement) with predominant involvement of the posterior segments, including chiasm and optic tracts (Fig. 15.1). However, ON lesions can sometimes be indistinguishable between MS and NMOSD, and searching for the concomitant presence of specific MRI abnormalities in other CNS regions might help (Fig. 15.2) [68, 69].

***Spinal cord MRI*** A longitudinally extensive (≥3 contiguous vertebral body segments) hyperintense lesion on T2-weighted sequences is the typical accompaniment of NMOSD myelitis [52]. Similar to what can be observed in other inflammatory myelopathies, a cervical spinal cord lesion may occasionally extend cranially to the medulla and the fourth ventricle (Fig. 15.2e);

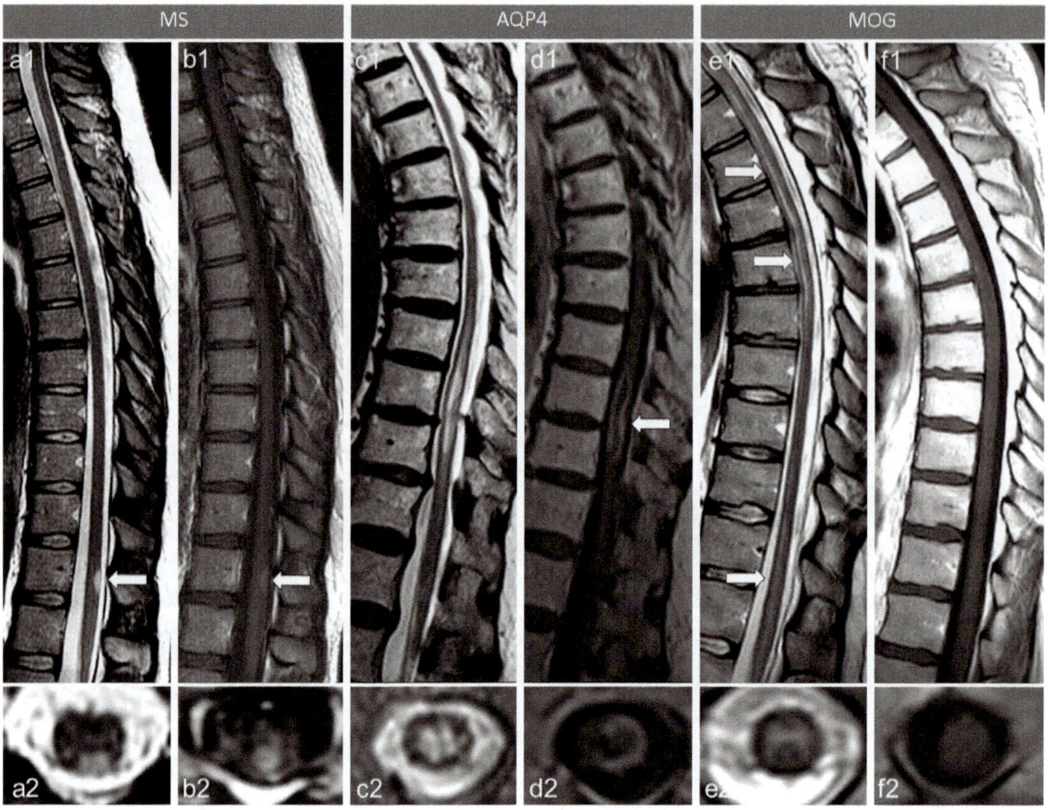


**Fig. 15.3** Representative myelitis lesions in MS, AQP4-IgG seropositive NMOSD, and MOG-IgG associated disorders. Sagittal (top row) and axial (bottom row) T2-weighted (**a**, **c**, **e**) and post-contrast T1-weighted (**b**, **d**, **f**) images from three different patients are shown: (**a**, **b**) short, enhancing demyelinating lesion involving the dorsal columns of the conus medullaris is typical of MS; (**c**, **d**) "bright" longitudinally extensive lesion in the lower thoracic spinal cord with ring enhancement visible on both sagittal (**d1**, arrow) and axial (**d2**) images in a patient with NMOSD; (**e**, **f**) longitudinally extensive lesion with predominant involvement of the thoracic spinal cord and conus medullaris (**e1**, arrows) in a MOG-IgG-positive patient that does not enhance after gadolinium (**f**). The concomitant presence of a ventral hyperintense line on sagittal images (**e1**) and of an H-shaped (H-sign) hyperintensity on axial images (**e2**) is typical of MOG-IgG autoimmunity

however, when this occurs in association with symptoms of area postrema syndrome, it is highly suggestive of NMOSD [70, 71]. Short myelitis lesions may be encountered in about 14% of cases according to the timing of MRI [53]. It is therefore important to be aware that imaging the myelitis too early, when the lesion formation is not complete, or too late (especially in case of steroid administration), when the lesion is resolving, may lead to the detection of short myelitis lesions or resolution of the lesion completely [72]. In patients with short transverse myelitis, certain factors help predict those likely to be AQP4-IgG positive including non-Caucasian ethnicity, coexisting autoimmunity, tonic spasms, centrally located lesions on T2-weighted axial images, absence of typical MS brain lesions, and lack of oligoclonal bands [53]. In those patients, the subsequent myelitis is longitudinally extensive in more than 90% of cases [53]. NMOSD myelitis lesions can be particularly bright on T2-weighted images with signal intensity similar to that of the surrounding CSF ("bright spotty lesions"; Fig. 15.3c) [52, 73]. T1-hypointensity and atrophic evolution (long atrophic spinal cord segments) are also common [74]. Contrast enhancement is generally patchy or ring-like (Fig. 15.2d) [75].

***Brain MRI*** Brain abnormalities are detected in more than 60% of NMOSD patients during the course of their disease, and they are generally nonspecific [76]. The typical morphology of MS lesions (e.g., ovoid periventricular, juxta-cortical, or inferior temporal pole; Fig. 15.2a–d) is generally not encountered in NMOSD, although 13% of patients fulfill Barkhof criteria for dissemination in space on initial brain MRI [77, 78]. Characteristic NMOSD brain abnormalities are commonly identified at sites on high AQP4 expression: (1) periependymal regions around the lateral ventricles, especially the splenium of the corpus callosum, which can be diffusely involved in its entire thickness ("arch-bridge sign"; Fig. 15.2f); (2) periependymal lesions surrounding the third ventricles and cerebral aqueduct (generally asymptomatic but may manifest with symptoms of thalamic/hypothalamic dysfunction); and (3) periependymal lesions adjacent to the fourth ventricle, with or without area postrema syndrome (Fig. 15.2h). For unclear reasons, the corticospinal tracts are also frequently affected in NMOSD (up to 44%) despite the lower AQP4 concentration; these lesions typically extend longitudinally from the internal capsule to the cerebral peduncle and are often bilateral (Fig. 15.2g) [17, 77]. Extensive hemispheric white matter lesions (>3 cm in diameter) can be also encountered (Fig. 15.2f). They can appear in isolation as tumefactive demyelinating lesions or as multiple confluent/bilateral lesions resembling acute disseminated encephalomyelitis (ADEM), posterior reversible encephalopathy syndrome (PRES), or other leukoencephalopathies/leukoedystrophies [63, 79–81]. Extensive white matter lesions in NMOSD typically show high diffusivity on diffusion-weighted images (DWI) and apparent diffusion coefficient (ADC)-map consistent with vasogenic edema by AQP4 impairment, although areas of restricted diffusion and necrosis may occur [63, 82]. Mass effect is generally absent [63, 80, 83]. Abnormal gadolinium enhancement is common during attacks and frequently displayed in a poorly marginated and multiple patchy pattern ("cloud-like"). Other patterns of enhancement include nodular, leptomeningeal, and linear enhancement along the ependymal surface of the lateral ventricles ("pencil-thin") [74, 84].

## Cerebrospinal Fluid Findings and Autoantibody Detection

During disease flares, CSF analysis typically shows elevated protein and high white cell count (up to 75% of untreated patients), altough a non-inflammatory CSF is common during attacks of isolated optic neuritis [85, 86]. The pleocytosis is typically lymphocytic predominant, although high proportions of monocytes, neutrophils, and eosinophils have been reported [85]. CSF-restricted oligoclonal bands are detected in 16% of cases, and they generally disappear after the acute phase of the disease, in contrast to MS where oligoclonal bands persist over time [85].

AQP4-IgG detection in either serum or CSF confirms NMOSD diagnosis [7]. The antibody was originally discovered in 2004 by indirect immunofluorescence on rodent brain tissue, a technique later superseded by ELISA essays, which had greater sensitivity [5, 87]. However, cell-based assay by either using fluorescence activated cell sorting (FACS) or direct immunofluorescence is now recognized as the preferred technique for AQP4-IgG detection (particularly using the M1 isoform), yielding a sensitivity of 75–80% and a specificity of >99% [88]. Care is needed when interpreting ELISA results as the frequency of false-positive results is higher (up to 5 times), especially in the presence of low antibody titer and atypical clinical/MRI findings for NMOSD [89]. Serum is the optimal and preferred specimen for AQP4-IgG testing as it has greater sensitivity than CSF. However, cases of isolated CSF positivity have rarely been reported [90].

## Coexisting Autoimmunity

The coexistence of NMOSD with other systemic or neurological autoimmune disorders is not uncommon [91]. Both nonorgan-specific (e.g., antinuclear, extractable nuclear antibodies) and organ-specific antibodies (e.g., anti-thyroid peroxidase) can be frequently detected in patients with NMOSD, accounting for up to 50% of cases [92]. Among neurologic autoimmunity, myasthenia gravis seems the most frequently associated with NMOSD, with the myasthenia gravis typically preceding NMOSD diagnosis by years to decades [93]. Anti-NMDA (N-Methyl-D-aspartate) receptor antibody encephalitis has also been found to coexist (either preceding, concurrent, or subsequent to NMOSD) more frequently than would be expected [94]. while the co-occurrence of AQP4-IgG and MOG-IgG is exceptionally rare [159]. In patients with systemic autoimmune disorders (e.g., systemic lupus erythematosus, Sjögren's syndrome), the occurrence of neurological manifestations that is typical of NMOSD (e.g., optic neuritis, LETM) should prompt testing for AQP4-IgG as a high proportion will be AQP4-IgG positive. In these cases, symptoms would be attributable to the coexisting neurologic autoimmune disorder (NMOSD), rather than represent an unusual manifestation of the systemic autoimmune disorder [95]. A recent study highlighted that half of patients historically diagnosed with antiphospholipid antibody-associated LETM were AQP4-IgG seropositive [96]. Coexisting antiphospholipid antibodies, which are found in almost 50% of NMOSD patients, may contribute to the increased risk of miscarriage and venous thromboembolism reported with NMOSD [97, 98]. Systemic autoimmunity is less commonly encountered in MOG-IgG-positive patients [99].

## Prognosis

Morbidity and mortality in NMOSD are greater than MS and are related to the severity of relapses and cumulative sequelae, while the development of a progressive phase of the disease is extremely rare, versus MS where it occurs in the majority [46, 100]. Permanent mono- or paraplegia and blindness of at least one eye are reported to occur in more than half of relapsing cases with a mean follow-up of about 8 years. Respiratory failure from severe cervical spinal cord involvement is a rare severe complication of a myelitis attack particularly those with preexisting quadriparesis [100]. However, it is likely that NMOSD patients have improved over time and that patients currently have a better prognosis due to greater awareness and faster recognition and treatment of the disease in the neurological community [101].

## Treatment

For many years, recommendations for acute and long-term treatment in NMOSD have largely derived from clinical practice and retrospective/prospective uncontrolled studies [102, 103]. Only recently, results from the first randomized clinical trials on long-term treatment have been published [104].

### Acute Treatment

Treatment of relapses generally begins with intravenous methylprednisolone (IVMP) 1 g/day for 5 consecutive days. However, steroids alone are often ineffective, and complete recovery is observed in only one-third of patients. Plasma exchange (PLEX), generally 5–7 exchanges every other day, seems more effective than steroids either in isolation or as an add-on therapy but carries a higher rate of complications (e.g., sepsis) [105]. The efficacy of PLEX for treatment of acute CNS demyelinating inflammatory disorders, including NMO, ADEM, monophasic, and recurrent transverse myelitis, has been proven before the discovery of AQP4-IgG [106]. Recent findings suggest that maximum improvement with PLEX is achieved within 5 days of symptoms onset [107]. Given that corticosteroids effect is often not immediate (3–5 days), concurrent PLEX administration should be considered in those with severe deficit (e.g., paraplegia, blindness) since the expected benefit in these cases largely outweigh the risk of PLEX-related complications (e.g., infections) [108]. Intravenous immunoglobulins (IVIg) 0.4 grams/kilogram/day for 5 days, and then weekly for 6–12 weeks, may be considered an alternative to PLEX [109]. Cyclophosphamide and the C1-esterase inhibitor Cinryze® have also been shown to be potentially effective acutely with IVMP [110, 111].

## Long-Term Treatment

Preventing relapses in NMOSD is fundamental to prevent disability, and maintenance treatment is strongly recommended after the acute phase [103]. Azathioprine, mycophenolate, or rituximab has been commonly used empirically as first-line therapy according to patients' characteristics and preferences [112–115]. While the first two options require up to 6 months to become fully effective and oral steroid tapering is necessary in the meanwhile (generally prednisone 20–60 mg/day), rituximab is frequently preferred for the shorter time-action profile and greater tolerability. An open-label, randomized comparative trial has shown rituximab is superior to azathioprine in terms of relapse prevention and long-term morbidity [116], while the superiority of rituximab vs placebo has been recently confirmed in a small randomized, double-blind trial in Japan [160]. Rituximab is commonly administered in one of the two of the following treatment regimens: (1) two 1000 mg intravenous infusions separated by 2 weeks, repeated every 6 months, or (2) 375 mg/m2 once per week for 4 weeks and then every 6 months. Periodic monitoring of serum CD19 B cell-count with a target of zero might be an alternative approach to guide the frequency of reinfusions [117]. Progressive multifocal leukoencephalopathy has been reported with rituximab in NMOSD [118]. Unfortunately, whether and when immunosuppression should be discontinued in relapse-free patients still needs to be determined. Other potentially effective long-term therapies include methotrexate, tacrolimus, cyclosporine, and mitoxantrone.

Three drugs have recently been FDA approved in the USA after being shown to be highly effective in preventing NMOSD relapses and disability with large randomized, placebo-controlled trials. The C5 complement inhibitor eculizumab (administered intravenous at a dosage of 900 mg/week for the first 4 weeks, then 1200 mg every 2 weeks, in adults with relapsing disease despite other prior or concomitant immunotherapies) showed a rate of relapses of 3% vs. 43% rate observed in the placebo group (relative risk reduction 94%) and was very well-tolerated [119]. Of note, all patients in the trial were AQP4-IgG positive (i.e., the results do not apply to patients with seronegative NMOSD or MOG-IgG associated disorders), and all received meningococcal vaccination (eculizumab increases the risk of meningococcal and encapsulated bacterial infections). Similarly, the anti-CD19 inebilizumab (MEDI-551) showed a 77% relapse risk reduction in NMOSD patients (including a minority of seronegative cases) and a similar adverse-event rate compared to placebo [120]. Lastly, the IL-6 inhibitor satralizumab (SA237), administered subcutaneous 120 mg every 2 weeks for the first 6 weeks and every 4 weeks thereafter, showed a 79% relapse risk reduction in NMOSD patients compared to placebo (34% if seronegative NMOSD cases only are considered) [121].

## Pregnancy

AQP4 is expressed in human placenta, and both preeclampsia and fetal loss are more common in NMOSD patients [98, 122]. Unlike MS, pregnancy does not seem to have any protective effect on relapses in NMOSD [122]. Although there are currently no strong recommendations on whether or when to maintain NMOSD treatment during pregnancy, mycophenolate, methotrexate, and cyclophosphamide must be avoided given their teratogenic effects while azathioprine and rituximab are probably safer options [122]. Further studies are needed to fully clarify the best therapeutic approach during pregnancy.

# Part III: MOG-IgG-Associated Disorders

## Pathophysiology

The pathophysiology of MOG-IgG-associated disorders is still poorly understood [8, 123]. In humans, MOG is exclusively found in the CNS, specifically on the outermost surface of the myelin sheath and the plasma membrane of oligodendrocytes, and thus its potential for being a specific target of CNS autoimmunity is high [8]. However, the peripheral MOG expression in rats and primates and reports of MOG-IgG-positive cases or radiculo-neuropathy may suggest a possible extra-CNS effect of MOG-IgG, which deserves further investigations [124]. Similar to AQP4-IgG, MOG-IgG is of the IgG1 isotype and thus able to activate complement via the direct pathway [125]. Some convincing evidences supporting MOG-IgG as a pathogenic effector include the following: (1) MOG-IgG mediates in vitro destruction of MOG expressing cells via natural killer cells activation [126]; (2) human MOG-IgG1 induces a reversible brain edema in mice that is consistent with the complete resolution of MRI lesions frequently observed in MOG-IgG-associated disorders (see later section "MRI") [127]; (3) transgenic mice with MOG-specific T and B cells spontaneously develop a severe form of experimental autoimmune encephalomyelitis characterized by predominant optic nerves and spinal cord involvement [128], and (4) oligodendrocytes incubated with MOG-IgG purified from seropositive patients show loss of cytoskeletal organization [167]. Notably, the so-called MRZ reaction (i.e., intrathecal antibody response to measles, rubella, and varicella zoster viruses), frequently detectable in MS patients, is absent in both patients with MOG-IgG and AQP4-IgG [166]; while antibodies against the human endogeneous retrovirus-W are similarly identified in patients with MS and MOG-IgG-associated disorder, but not in those with NMOSD [168].

## Pathology

Pathology reports in MOG-IgG-associated disorders show coexisting perivenous and confluent demyelination, complement deposition and cortical demyelination with intracortical lesions predominating. A CD4+ Tcell predominant inflammatory reaction with granulocytic inflammation is noted. Radially expanding smoldering white matter lesions of MS are not encountered [129–136, 161, 162]. In one pathology series, selective loss of MOG staining was reported while in another study that was not encountered [161, 162]. In contrast to AQP4-IgG seropositive NMOSD, astrocytes are typically preserved in MOG-IgG-associated disorders and AQP4 immunostaining is also preserved. Overall, demyelinating lesions associated with MOG-IgG might be less destructive than those associated with AQP4-IgG; as suggested but the higher tendency to resolve completely over time on MRI, and the lower serum levels of neurofilament light chains (i.e., a marker of neuronal injury) released during attacks [175].

## Clinical Features

The spectrum of possible clinical manifestations of MOG-IgG-associated disorders is broad (Table 15.1) [8]. Extension of CNS inflammation seems strongly related to age with an ADEM-like phenotype typically observed among children and more limited forms (e.g., ON, myelitis) more commonly encountered in adults, although any age can potentially be affected by any clinical phenotype [49, 137–143]. Despite the abundant overlap, the relative frequency of the possible clinical manifestations is different between MOG and AQP4 autoimmunity, with both monophasic and recurrent ON (either unilateral or bilateral) being more commonly encountered with MOG-IgG-associated disorders, while myelitis is more frequent in NMOSD [48, 49, 144]. A relapsing disease course is not uncommon, accounting for about 60% of cases. In contrast to AQP4-IgG seropositive NMOSD and MS, the clinical presentation of MOG-IgG-associated disorders is often preceded by prodromal viral-like symptoms or, less frequently, a vaccination [142].

*Optic neuritis* is the most common manifestation at onset and can occur in isolation or in association with other CNS manifestations (e.g., encephalopathy, myelitis) [49, 145]. Involvement of the optic nerves can be unilateral, bilateral (more than one-third of cases), monophasic, or recurrent, sometimes fulfilling diagnostic criteria for CRION (chronic relapsing inflammatory optic neuropathy) when accompanied by a steroid-dependent course [145, 146]. Although clinical attacks tend to be severe, visual outcome is typically better than AQP4-IgG seropositive NMOSD and permanent blindness is rare among treated patients [145]. Optic disc edema, due to frequent involvement of the orbital optic nerve portion (see next section "MRI"), can be detected in up to 86% and is infrequently encountered in MS patients [145].

*Spinal cord involvement* can occur with MOG-IgG either in isolation or concurrently with other manifestations (e.g., ADEM, optic neuritis) [147]. Importantly, isolated recurrent myelitis is rare in MOG-IgG-associated disorders and should prompt consideration for AQP4-IgG testing [48]. When taken together, AQP4-IgG and MOG-IgG account for 50% of cases of isolated idiopathic LETM in the population [148]. The severity of myelitis attacks is comparable to that of NMOSD myelitis, but recovery is better, and residual gait impairment is uncommon. Other factors that favor MOG-IgG-associated myelitis compared to AQP4-IgG seropositive NMOSD myelitis include male sex, Caucasian ethnicity, a preceding viral prodrome or vaccination, and sexual dysfunction (which is consistent with the more frequent involvement of conus medullaris) [137, 149, 150].

*Area postrema syndrome* may occur in MOG-IgG-associated disorders but seems less frequent than in NMOSD and less frequently accompanied by discrete area postrema lesions on MRI [140, 163].

*Other* possible manifestations include brainstem symptoms, diencephalic symptoms, and encephalopathy (especially in the context of ADEM) [137, 140, 141]. Rare cases of hemi- or bilateral cortical encephalitis, mainly manifesting with seizures or focal deficit from cortical inflammation, have been reported and may only show very slight abnormalities on brain MRI (see next section "MRI") [130, 131, 151].

## MRI

MRI abnormalities in MOG-IgG-associated disorders vary according to the clinical phenotypes (e.g., ADEM vs. CRION) although they may frequently overlap. Overall, MRI lesions are generally "fluffy" T2-hyperintense lesions with poorly defined margins (Fig. 15.2i–l), which less frequently enhance with gadolinium administration than NMOSD and frequently resolve after immunotherapy [78].

***Optic Nerve MRI*** Typical findings during an optic neuritis episode include contrast enhancement, perineural enhancement with extension to the surrounding orbital tissues (50% of cases), and involvement of more than half of the prechiasmic optic nerve length (80% of cases; Fig. 15.1c). In contrast to AQP4-IgG seropositive NMOSD, involvement of chiasm and optic tracts is uncommon (14% of cases) [145].

***Brain MRI*** Children with MOG-IgG autoimmunity are more likely to show ADEM-like abnormalities with large (>2 cm) T2-hyperintense lesions involving the periventricular/subcortical white matter, the cortex, and the deep gray nuclei (Fig. 15.2i–k) [78, 152]. With MOG-IgG, brainstem lesions typically surround the third ventricle and involve the middle cerebellar peduncles (Fig. 15.1l). Rare patients with cortical encephalitis may show focal cortical hyperintensity on fluid-attenuated inversion recovery (FLAIR) sequences, sometimes corresponding to regions of hyperperfusion on brain SPECT (single-photon emission computed tomography) [151]. Lesions that have been shown to be more suggestive of MS than MOG-IgG include inferior temporal pole periventricular lesions, ovoid periventricular, and T1-hypointense lesions [78, 153].

***Spinal Cord MRI*** Myelitis episodes are often accompanied by multiple lesions (compared to AQP4-IgG seropositive NMOSD, which typically has a solitary lesion), one of which is usually longitudinally extensive ($\geq$3 vertebral segments on sagittal T2-weighted sequence), and the presence of which helps distinguish from MS [164]. However, up to 29% of patients have one or more short lesions without a longitudinally extensive lesion and thus more difficult to distinguish from MS. The lesions frequently involve the conus medullaris. In about 30% of cases, the concomitant presence of a T2-hyperintense line on sagittal images and a central "H-shaped" T2-hyperintensity on axial images (restricted to the gray matter) can be detected (Fig. 15.3e) and helps distinguish from AQP4-IgG where this is much less frequently encountered (8%). However, there is some overlap between MOG-IgG and AQP4-IgG on axial images as both can have central lesions involving gray and white matter, and this differs from MS cord lesions that are almost always peripheral and located in the dorsal and lateral columns. Gadolinium enhancement is less frequent occurring in about half of patients with MOG-IgG acutely compared to more than three-quarters of MS and NMOSD patients (Fig. 15.3f) [149]. Of note, the initial spinal cord MRI can be normal in up to 10% of patients during a myelitis attack [165].

## Cerebrospinal Fluid Findings and Autoantibody Detection

CSF findings are frequently indistinguishable from those observed in NMOSD (see previous section "Cerebrospinal Fluid Findings and Autoantibody Detection" in Part II) [99]. An abnormal white cell count (>5 cells/mcl) is observed in 50–70% of cases, but is uncommon in patients with isolated optic neuritis [166, 174]. Oligoclonal bands are infrequently detected (<15%) in most studies and may be transient like in NMOSD [137, 139, 140, 143, 174]. Elevation of myelin basic protein (MBP) but not GFAP levels can be observed in the CSF during relapses, reflecting oligodendrocytes damage [154].

Serum testing for MOG-IgG by cell-based assay on transfected cells with direct visual immunofluorescence (fixed cells) or by fluorescence-activated cell sorting (live cells) is optimal [155]. In our experience, MOG conformational status might be rarely altered in fixed cells thus slightly increasing the risk of false-positive or false-negative results. MOG-IgG can be detected (generally with a borderline titer) in up to 2% of patients with typical MS, and the reason for this is not clear [155]. Consequently, MOG-IgG testing should be avoided in patients with typical clinical and MRI features of MS since it reduces the positive predictive value of the test, particularly as MS appears to be much more common than MOG-IgG-associated diseases [156]. Serum antibody titer can normalize with immunotherapy, and it frequently decreases between relapses [141]. Similar to NMOSD, isolated antibody positivity on CSF may rarely occur, and CSF testing should be considered despite negative result on serum when the diagnostic suspicion is high [157].

## Treatment and Prognosis

Data to guide treatment in MOG-IgG-associated disorders are still scarce, and randomized controlled trials are lacking. Similarly to AQP4-IgG seropositive NMOSD, relapses are generally treated with high-dose IVMP, PLEX, or IVIg (particularly in children) and with severe episodes (e.g., ADEM) may be followed by a slow oral taper. A steroid-dependent course with frequent relapses during tapering is not uncommon (e.g., CRION). Maintenance steroid-sparing therapy may be recommended in those with relapsing disease, and options include azathioprine, methotrexate, mycophenolate mofetil, monthly IVIg, and rituximab [169–171]. Retrospective data suggest maintenance IVIg to be more effective in reducing relapses [169], while the effect of rituximab seems less compared to that observed in patients with NMOSD [171]. The long-term outcome in MOG-IgG-associated disorders is generally favorable, with a final disability that may remain mild to moderate even in patients with a long-lasting relapsing course [172]. However, an unfavorable outcome is observed in a minority of patients, the early identification of whom would be important in the future to tailor optimal treatment strategies [173]. In those with a single attack, the decision of whether to initiate maintenance treatment is more difficult as some patients will be destined to have a monophasic course. There are a variety of factors that may play a role in the decision to initiate maintenance immunosuppression after the first attack including clinical judgment, patient preference, severity of the initial episode, and recovery from the first attack. While some have recommended treating all MOG-IgG positive patients from the first attack, others will often observe if the patient relapses off of treatment, prior to initiating maintenance immunotherapy [140]. Persistent MOG-IgG seropositivity beyond 3–6 months seems to portray an increased risk of relapse, although relapses can be also observed in some patients with transient seropositivity [141, 158].

## Conclusion

In conclusion, the discovery of AQP4-IgG and MOG-IgG has led to a new era in the field of inflammatory demyelinating diseases of the CNS. The antibodies help distinguish their respective diseases from MS and help with guiding treatment decisions and prognostication. Ongoing and future studies will likely lead to the use of more targeted immunotherapies in these diseases.

## References

1. Jarius S, Wildemann B. An early case of neuromyelitis optica: on a forgotten report by Jacob Lockhart Clarke, FRS. Mult Scler. 2011;17(11):1384–6.
2. Miyazawa I, Fujihara K, Itoyama Y. Eugene Devic (1858–1930). J Neurol. 2002;249(3):351–2.
3. Eikelenboom MJ, Killestein J, Uitdehaag BM, Polman CH. Opticospinal multiple sclerosis: a pathogenetically distinct form? Mult Scler. 2004;10(3):334–5.
4. Lennon VA, Kryzer TJ, Pittock SJ, Verkman AS, Hinson SR. IgG marker of optic-spinal multiple sclerosis binds to the aquaporin-4 water channel. J Exp Med. 2005;202(4):473–7.
5. Lennon VA, Wingerchuk DM, Kryzer TJ, Pittock SJ, Lucchinetti CF, Fujihara K, et al. A serum autoantibody marker of neuromyelitis optica: distinction from multiple sclerosis. Lancet. 2004;364(9451):2106–12.
6. Wingerchuk DM, Lennon VA, Lucchinetti CF, Pittock SJ, Weinshenker BG. The spectrum of neuromyelitis optica. Lancet Neurol. 2007;6(9):805–15.
7. Wingerchuk DM, Banwell B, Bennett JL, Cabre P, Carroll W, Chitnis T, et al. International consensus diagnostic criteria for neuromyelitis optica spectrum disorders. Neurology. 2015;85(2):177–89.
8. Reindl M, Di Pauli F, Rostasy K, Berger T. The spectrum of MOG autoantibody-associated demyelinating diseases. Nat Rev Neurol. 2013;9(8):455–61.
9. Hamid SHM, Whittam D, Mutch K, Linaker S, Solomon T, Das K, et al. What proportion of AQP4-IgG-negative NMO spectrum disorder patients are MOG-IgG positive? A cross sectional study of 132 patients. J Neurol. 2017;264(10):2088–94.
10. Pittock SJ, Lucchinetti CF. Neuromyelitis optica and the evolving spectrum of autoimmune aquaporin-4 channelopathies: a decade later. Ann N Y Acad Sci. 2016;1366(1):20–39.
11. Flanagan EP, Cabre P, Weinshenker BG, St Sauver J, Jacobson DJ, Majed M, et al. Epidemiology of aquaporin-4 autoimmunity and neuromyelitis optica spectrum. Ann Neurol. 2016;79(5):775–83.
12. Houzen H, Kondo K, Niino M, Horiuchi K, Takahashi T, Nakashima I, et al. Prevalence and clinical features of neuromyelitis optica spectrum disorders in northern Japan. Neurology. 2017;89(19):1995–2001.
13. Etemadifar M, Nasr Z, Khalili B, Taherioun M, Vosoughi R. Epidemiology of neuromyelitis optica in the world: a systematic review and meta-analysis. Mult Scler Int. 2015;2015:174720.
14. Papais-Alvarenga RM, Neri VC, de Araujo EAACR, da Silva EB, Alvarenga MP, Pereira A, et al. Lower frequency of antibodies to MOG in Brazilian patients with demyelinating diseases: an ethnicity influence? Mult Scler Relat Disord. 2018;25:87–94.
15. Ratelade J, Verkman AS. Neuromyelitis optica: aquaporin-4 based pathogenesis mechanisms and new therapies. Int J Biochem Cell Biol. 2012;44(9):1519–30.
16. Papadopoulos MC, Verkman AS. Aquaporin 4 and neuromyelitis optica. Lancet Neurol. 2012;11(6):535–44.
17. Kim W, Park MS, Lee SH, Kim SH, Jung IJ, Takahashi T, et al. Characteristic brain magnetic resonance imaging abnormalities in central nervous system aquaporin-4 autoimmunity. Mult Scler. 2010;16(10):1229–36.
18. Pittock SJ, Weinshenker BG, Lucchinetti CF, Wingerchuk DM, Corboy JR, Lennon VA. Neuromyelitis optica brain lesions localized at sites of high aquaporin 4 expression. Arch Neurol. 2006;63(7):964–8.
19. Lucchinetti CF, Mandler RN, McGavern D, Bruck W, Gleich G, Ransohoff RM, et al. A role for humoral mechanisms in the pathogenesis of Devic's neuromyelitis optica. Brain. 2002;125(Pt 7):1450–61.
20. Nytrova P, Potlukova E, Kemlink D, Woodhall M, Horakova D, Waters P, et al. Complement activation in patients with neuromyelitis optica. J Neuroimmunol. 2014;274(1–2):185–91.
21. Crane JM, Lam C, Rossi A, Gupta T, Bennett JL, Verkman AS. Binding affinity and specificity of neuromyelitis optica autoantibodies to aquaporin-4 M1/M23 isoforms and orthogonal arrays. J Biol Chem. 2011;286(18):16516–24.
22. Mader S, Lutterotti A, Di Pauli F, Kuenz B, Schanda K, Aboul-Enein F, et al. Patterns of antibody binding to aquaporin-4 isoforms in neuromyelitis optica. PLoS One. 2010;5(5):e10455.

23. Phuan PW, Ratelade J, Rossi A, Tradtrantip L, Verkman AS. Complement-dependent cytotoxicity in neuromyelitis optica requires aquaporin-4 protein assembly in orthogonal arrays. J Biol Chem. 2012;287(17):13829–39.
24. Lucchinetti CF, Guo Y, Popescu BF, Fujihara K, Itoyama Y, Misu T. The pathology of an autoimmune astrocytopathy: lessons learned from neuromyelitis optica. Brain Pathol. 2014;24(1):83–97.
25. Hinson SR, Romero MF, Popescu BF, Lucchinetti CF, Fryer JP, Wolburg H, et al. Molecular outcomes of neuromyelitis optica (NMO)-IgG binding to aquaporin-4 in astrocytes. Proc Natl Acad Sci U S A. 2012;109(4):1245–50.
26. Marignier R, Nicolle A, Watrin C, Touret M, Cavagna S, Varrin-Doyer M, et al. Oligodendrocytes are damaged by neuromyelitis optica immunoglobulin G via astrocyte injury. Brain. 2010;133(9):2578–91.
27. Hinson SR, Clift IC, Luo N, Kryzer TJ, Lennon VA. Autoantibody-induced internalization of CNS AQP4 water channel and EAAT2 glutamate transporter requires astrocytic Fc receptor. Proc Natl Acad Sci U S A. 2017;114(21):5491–6.
28. Hinson SR, Roemer SF, Lucchinetti CF, Fryer JP, Kryzer TJ, Chamberlain JL, et al. Aquaporin-4-binding autoantibodies in patients with neuromyelitis optica impair glutamate transport by down-regulating EAAT2. J Exp Med. 2008;205(11):2473–81.
29. Shimizu F, Sano Y, Takahashi T, Haruki H, Saito K, Koga M, et al. Sera from neuromyelitis optica patients disrupt the blood-brain barrier. J Neurol Neurosurg Psychiatry. 2012;83(3):288–97.
30. Shimizu F, Schaller KL, Owens GP, Cotleur AC, Kellner D, Takeshita Y, et al. Glucose-regulated protein 78 autoantibody associates with blood-brain barrier disruption in neuromyelitis optica. Sci Transl Med. 2017;9(397):eaai9111.
31. Guo Y, Lennon VA, Popescu BF, Grouse CK, Topel J, Milone M, et al. Autoimmune aquaporin-4 myopathy in neuromyelitis optica spectrum. JAMA Neurol. 2014;71(8):1025–9.
32. Yao X, Verkman AS. Complement regulator CD59 prevents peripheral organ injury in rats made seropositive for neuromyelitis optica immunoglobulin G. Acta Neuropathol Commun. 2017;5(1):57.
33. Annus A, Bencsik K, Obal I, Kincses ZT, Tiszlavicz L, Hoftberger R, et al. Paraneoplastic neuromyelitis optica spectrum disorder: a case report and review of the literature. J Clin Neurosci. 2018;48:7–10.
34. Pittock SJ, Lennon VA. Aquaporin-4 autoantibodies in a paraneoplastic context. Arch Neurol. 2008;65(5):629–32.
35. Matiello M, Kim HJ, Kim W, Brum DG, Barreira AA, Kingsbury DJ, et al. Familial neuromyelitis optica. Neurology. 2010;75(4):310–5.
36. Bo M, Niegowska M, Arru G, Sechi E, Mariotto S, Mancinelli C, et al. Mycobacterium avium subspecies paratuberculosis and myelin basic protein specific epitopes are highly recognized by sera from patients with Neuromyelitis optica spectrum disorder. J Neuroimmunol. 2018;318:97–102.
37. Brum DG, Barreira AA, dos Santos AC, Kaimen-Maciel DR, Matiello M, Costa RM, et al. HLA-DRB association in neuromyelitis optica is different from that observed in multiple sclerosis. Mult Scler. 2010;16(1):21–9.
38. Varrin-Doyer M, Spencer CM, Schulze-Topphoff U, Nelson PA, Stroud RM, Cree BA, et al. Aquaporin 4-specific T cells in neuromyelitis optica exhibit a Th17 bias and recognize Clostridium ABC transporter. Ann Neurol. 2012;72(1):53–64.
39. Barros PO, Dias ASO, Kasahara TM, Ornelas AMM, Aguiar RS, Leon SA, et al. Expansion of IL-6(+) Th17-like cells expressing TLRs correlates with microbial translocation and neurological disabilities in NMOSD patients. J Neuroimmunol. 2017;307:82–90.
40. Chihara N, Aranami T, Sato W, Miyazaki Y, Miyake S, Okamoto T, et al. Interleukin 6 signaling promotes anti-aquaporin 4 autoantibody production from plasmablasts in neuromyelitis optica. Proc Natl Acad Sci U S A. 2011;108(9):3701–6.
41. Popescu BF, Guo Y, Jentoft ME, Parisi JE, Lennon VA, Pittock SJ, et al. Diagnostic utility of aquaporin-4 in the analysis of active demyelinating lesions. Neurology. 2015;84(2):148–58.
42. Popescu BF, Parisi JE, Cabrera-Gomez JA, Newell K, Mandler RN, Pittock SJ, et al. Absence of cortical demyelination in neuromyelitis optica. Neurology. 2010;75(23):2103–9.
43. Misu T, Takano R, Fujihara K, Takahashi T, Sato S, Itoyama Y. Marked increase in cerebrospinal fluid glial fibrillar acidic protein in neuromyelitis optica: an astrocytic damage marker. J Neurol Neurosurg Psychiatry. 2009;80(5):575–7.
44. Misu T, Hoftberger R, Fujihara K, Wimmer I, Takai Y, Nishiyama S, et al. Presence of six different lesion types suggests diverse mechanisms of tissue injury in neuromyelitis optica. Acta Neuropathol. 2013;125(6):815–27.
45. Kim W, Kim SH, Lee SH, Li XF, Kim HJ. Brain abnormalities as an initial manifestation of neuromyelitis optica spectrum disorder. Mult Scler. 2011;17(9):1107–12.
46. Aungsumart S, Apiwattanakul M. Clinical outcomes and predictive factors related to good outcomes in plasma exchange in severe attack of NMOSD and long extensive transverse myelitis: case series and review of the literature. Mult Scler Relat Disord. 2017;13:93–7.
47. Wingerchuk DM, Pittock SJ, Lucchinetti CF, Lennon VA, Weinshenker BG. A secondary progressive clinical course is uncommon in neuromyelitis optica. Neurology. 2007;68(8):603–5.
48. Jitprapaikulsan J, Lopez Chiriboga AS, Flanagan EP, Fryer JP, McKeon A, Weinshenker BG, et al. Novel glial targets and recurrent longitudinally extensive transverse myelitis. JAMA Neurol. 2018;75(7):892–5.

49. Kim SM, Woodhall MR, Kim JS, Kim SJ, Park KS, Vincent A, et al. Antibodies to MOG in adults with inflammatory demyelinating disease of the CNS. Neurol Neuroimmunol Neuroinflamm. 2015;2(6):e163.
50. Fernandes DB, Ramos Rde I, Falcochio C, Apostolos-Pereira S, Callegaro D, Monteiro ML. Comparison of visual acuity and automated perimetry findings in patients with neuromyelitis optica or multiple sclerosis after single or multiple attacks of optic neuritis. J Neuroophthalmol. 2012;32(2):102–6.
51. Levin MH, Bennett JL, Verkman AS. Optic neuritis in neuromyelitis optica. Prog Retin Eye Res. 2013;36:159–71.
52. Flanagan EP, Kaufmann TJ, Krecke KN, Aksamit AJ, Pittock SJ, Keegan BM, et al. Discriminating long myelitis of neuromyelitis optica from sarcoidosis. Ann Neurol. 2016;79(3):437–47.
53. Flanagan EP, Weinshenker BG, Krecke KN, Lennon VA, Lucchinetti CF, McKeon A, et al. Short myelitis lesions in aquaporin-4-IgG-positive neuromyelitis optica spectrum disorders. JAMA Neurol. 2015;72(1):81–7.
54. Kim SM, Go MJ, Sung JJ, Park KS, Lee KW. Painful tonic spasm in neuromyelitis optica: incidence, diagnostic utility, and clinical characteristics. Arch Neurol. 2012;69(8):1026–31.
55. Elsone L, Townsend T, Mutch K, Das K, Boggild M, Nurmikko T, et al. Neuropathic pruritus (itch) in neuromyelitis optica. Mult Scler. 2013;19(4):475–9.
56. Apiwattanakul M, Popescu BF, Matiello M, Weinshenker BG, Lucchinetti CF, Lennon VA, et al. Intractable vomiting as the initial presentation of neuromyelitis optica. Ann Neurol. 2010;68(5):757–61.
57. Misu T, Fujihara K, Nakashima I, Sato S, Itoyama Y. Intractable hiccup and nausea with periaqueductal lesions in neuromyelitis optica. Neurology. 2005;65(9):1479–82.
58. Shosha E, Dubey D, Palace J, Nakashima I, Jacob A, Fujihara K, et al. Area postrema syndrome: frequency, criteria, and severity in AQP4-IgG-positive NMOSD. Neurology. 2018;91(17):e1642–51.
59. Kaur C, Ling EA. The circumventricular organs. Histol Histopathol. 2017;32(9):879–92.
60. Iorio R, Lucchinetti CF, Lennon VA, Farrugia G, Pasricha PJ, Weinshenker BG, et al. Intractable nausea and vomiting from autoantibodies against a brain water channel. Clin Gastroenterol Hepatol. 2013;11(3):240–5.
61. Popescu BF, Lennon VA, Parisi JE, Howe CL, Weigand SD, Cabrera-Gomez JA, et al. Neuromyelitis optica unique area postrema lesions: nausea, vomiting, and pathogenic implications. Neurology. 2011;76(14):1229–37.
62. Kremer L, Mealy M, Jacob A, Nakashima I, Cabre P, Bigi S, et al. Brainstem manifestations in neuromyelitis optica: a multicenter study of 258 patients. Mult Scler. 2014;20(7):843–7.
63. Sechi E, Addis A, Batzu L, Mariotto S, Ferrari S, Conti M, et al. Late presentation of NMOSD as rapidly progressive leukoencephalopathy with atypical clinical and radiological findings. Mult Scler. 2018;24(5):685–8.
64. Viegas S, Weir A, Esiri M, Kuker W, Waters P, Leite MI, et al. Symptomatic, radiological and pathological involvement of the hypothalamus in neuromyelitis optica. J Neurol Neurosurg Psychiatry. 2009;80(6):679–82.
65. Nakano H, Tanaka M, Kinoshita M, Tahara M, Matsui M, Tanaka K, et al. Epileptic seizures in Japanese patients with multiple sclerosis and neuromyelitis optica. Epilepsy Res. 2013;104(1–2):175–80.
66. Takai Y, Misu T, Nakashima I, Takahashi T, Itoyama Y, Fujihara K, et al. Two cases of lumbosacral myeloradiculitis with anti-aquaporin-4 antibody. Neurology. 2012;79(17):1826–8.
67. Clardy SL, Lucchinetti CF, Krecke KN, Lennon VA, O'Toole O, Weinshenker BG, et al. Hydrocephalus in neuromyelitis optica. Neurology. 2014;82(20):1841–3.
68. Khanna S, Sharma A, Huecker J, Gordon M, Naismith RT, Van Stavern GP. Magnetic resonance imaging of optic neuritis in patients with neuromyelitis optica versus multiple sclerosis. J Neuroophthalmol. 2012;32(3):216–20.
69. Storoni M, Davagnanam I, Radon M, Siddiqui A, Plant GT. Distinguishing optic neuritis in neuromyelitis optica spectrum disease from multiple sclerosis: a novel magnetic resonance imaging scoring system. J Neuroophthalmol. 2013;33(2):123–7.
70. Dubey D, Pittock SJ, Krecke KN, Flanagan EP. Association of extension of cervical cord lesion and area postrema syndrome with neuromyelitis optica spectrum disorder. JAMA Neurol. 2017;74(3):359–61.
71. Iorio R, Damato V, Mirabella M, Evoli A, Marti A, Plantone D, et al. Distinctive clinical and neuroimaging characteristics of longitudinally extensive transverse myelitis associated with aquaporin-4 autoantibodies. J Neurol. 2013;260(9):2396–402.
72. Asgari N, Skejoe HP, Lennon VA. Evolution of longitudinally extensive transverse myelitis in an aquaporin-4 IgG-positive patient. Neurology. 2013;81(1):95–6.
73. Yonezu T, Ito S, Mori M, Ogawa Y, Makino T, Uzawa A, et al. "Bright spotty lesions" on spinal magnetic resonance imaging differentiate neuromyelitis optica from multiple sclerosis. Mult Scler. 2014;20(3):331–7.
74. Kim HJ, Paul F, Lana-Peixoto MA, Tenembaum S, Asgari N, Palace J, et al. MRI characteristics of neuromyelitis optica spectrum disorder: an international update. Neurology. 2015;84(11):1165–73.
75. Zalewski NL, Morris PP, Weinshenker BG, Lucchinetti CF, Guo Y, Pittock SJ, et al. Ring-enhancing spinal cord lesions in neuromyelitis optica spectrum disorders. J Neurol Neurosurg Psychiatry. 2017;88(3):218–25.
76. Pittock SJ, Lennon VA, Krecke K, Wingerchuk DM, Lucchinetti CF, Weinshenker BG. Brain abnor-

malities in neuromyelitis optica. Arch Neurol. 2006;63(3):390–6.
77. Huh SY, Min JH, Kim W, Kim SH, Kim HJ, Kim BJ, et al. The usefulness of brain MRI at onset in the differentiation of multiple sclerosis and seropositive neuromyelitis optica spectrum disorders. Mult Scler. 2014;20(6):695–704.
78. Jurynczyk M, Geraldes R, Probert F, Woodhall MR, Waters P, Tackley G, et al. Distinct brain imaging characteristics of autoantibody-mediated CNS conditions and multiple sclerosis. Brain. 2017;140(3):617–27.
79. Ayrignac X, Daliere CC, Nerrant E, Vincent T, De Seze J, Labauge P. Extensive cerebral white matter involvement in a patient with NMO spectrum disorder. Mult Scler. 2014;20(10):1401–3.
80. Eichel R, Meiner Z, Abramsky O, Gotkine M. Acute disseminating encephalomyelitis in neuromyelitis optica: closing the floodgates. Arch Neurol. 2008;65(2):267–71.
81. Magana SM, Matiello M, Pittock SJ, McKeon A, Lennon VA, Rabinstein AA, et al. Posterior reversible encephalopathy syndrome in neuromyelitis optica spectrum disorders. Neurology. 2009;72(8):712–7.
82. Ishida C, Ikeda T, Komai K, Takahashi K, Noguchi-Shinohara M, Yamada M. Contiguous and symmetrical disease expansion and massive necrosis of the cerebral white matter in a patient with neuromyelitis optica. Mult Scler. 2018;24(9):1258–61.
83. Matsushita T, Isobe N, Matsuoka T, Ishizu T, Kawano Y, Yoshiura T, et al. Extensive vasogenic edema of anti-aquaporin-4 antibody-related brain lesions. Mult Scler. 2009;15(9):1113–7.
84. Ito S, Mori M, Makino T, Hayakawa S, Kuwabara S. "Cloud-like enhancement" is a magnetic resonance imaging abnormality specific to neuromyelitis optica. Ann Neurol. 2009;66(3):425–8.
85. Jarius S, Paul F, Franciotta D, Ruprecht K, Ringelstein M, Bergamaschi R, et al. Cerebrospinal fluid findings in aquaporin-4 antibody positive neuromyelitis optica: results from 211 lumbar punctures. J Neurol Sci. 2011;306(1–2):82–90.
86. Sellner J, Boggild M, Clanet M, Hintzen RQ, Illes Z, Montalban X, et al. EFNS guidelines on diagnosis and management of neuromyelitis optica. Eur J Neurol. 2010;17(8):1019–32.
87. Waters PJ, McKeon A, Leite MI, Rajasekharan S, Lennon VA, Villalobos A, et al. Serologic diagnosis of NMO: a multicenter comparison of aquaporin-4-IgG assays. Neurology. 2012;78(9):665–71; discussion 9.
88. Fryer JP, Lennon VA, Pittock SJ, Jenkins SM, Fallier-Becker P, Clardy SL, et al. AQP4 autoantibody assay performance in clinical laboratory service. Neurol Neuroimmunol Neuroinflamm. 2014;1(1):e11.
89. Jolliffe EA, Keegan BM, Flanagan EP. Trident sign trumps Aquaporin-4-IgG ELISA in diagnostic value in a case of longitudinally extensive transverse myelitis. Mult Scler Relat Disord. 2018;23:7–8.
90. Majed M, Fryer JP, McKeon A, Lennon VA, Pittock SJ. Clinical utility of testing AQP4-IgG in CSF: guidance for physicians. Neurol Neuroimmunol Neuroinflamm. 2016;3(3):e231.
91. Pittock SJ, Lennon VA, de Seze J, Vermersch P, Homburger HA, Wingerchuk DM, et al. Neuromyelitis optica and non organ-specific autoimmunity. Arch Neurol. 2008;65(1):78–83.
92. Zekeridou A, Lennon VA. Aquaporin-4 autoimmunity. Neurol Neuroimmunol Neuroinflamm. 2015;2(4):e110.
93. McKeon A, Lennon VA, Jacob A, Matiello M, Lucchinetti CF, Kale N, et al. Coexistence of myasthenia gravis and serological markers of neurological autoimmunity in neuromyelitis optica. Muscle Nerve. 2009;39(1):87–90.
94. Titulaer MJ, Hoftberger R, Iizuka T, Leypoldt F, McCracken L, Cellucci T, et al. Overlapping demyelinating syndromes and anti-N-methyl-D-aspartate receptor encephalitis. Ann Neurol. 2014;75(3):411–28.
95. Graus F, Titulaer MJ, Balu R, Benseler S, Bien CG, Cellucci T, et al. A clinical approach to diagnosis of autoimmune encephalitis. Lancet Neurol. 2016;15(4):391–404.
96. Guerra H, Pittock SJ, Moder KG, Fryer JP, Gadoth A, Flanagan EP. Frequency of aquaporin-4 immunoglobulin G in longitudinally extensive transverse myelitis with antiphospholipid antibodies. Mayo Clin Proc. 2018;93(9):1299–304.
97. Farber RS, Gross R, Zakin E, Fabian M. Risk of venous thromboembolism in neuromyelitis optica patients hospitalized for acute relapse. Mult Scler. 2017;23(7):988–94.
98. Nour MM, Nakashima I, Coutinho E, Woodhall M, Sousa F, Revis J, et al. Pregnancy outcomes in aquaporin-4-positive neuromyelitis optica spectrum disorder. Neurology. 2016;86(1):79–87.
99. Sato DK, Callegaro D, Lana-Peixoto MA, Waters PJ, de Haidar Jorge FM, Takahashi T, et al. Distinction between MOG antibody-positive and AQP4 antibody-positive NMO spectrum disorders. Neurology. 2014;82(6):474–81.
100. Wingerchuk DM, Hogancamp WF, O'Brien PC, Weinshenker BG. The clinical course of neuromyelitis optica (Devic's syndrome). Neurology. 1999;53(5):1107–14.
101. Jiao Y, Fryer JP, Lennon VA, McKeon A, Jenkins SM, Smith CY, et al. Aquaporin 4 IgG serostatus and outcome in recurrent longitudinally extensive transverse myelitis. JAMA Neurol. 2014;71(1):48–54.
102. Kessler RA, Mealy MA, Levy M. Treatment of neuromyelitis optica spectrum disorder: acute, preventive, and symptomatic. Curr Treat Options Neurol. 2016;18(1):2.
103. Trebst C, Jarius S, Berthele A, Paul F, Schippling S, Wildemann B, et al. Update on the diagnosis and treatment of neuromyelitis optica: recommendations of the Neuromyelitis Optica Study Group (NEMOS). J Neurol. 2014;261(1):1–16.

104. Fyfe I. Breakthrough neuroimmunology trials impress at AAN meeting. Nat Rev Neurol. 2019;15(7):368.
105. Abboud H, Petrak A, Mealy M, Sasidharan S, Siddique L, Levy M. Treatment of acute relapses in neuromyelitis optica: steroids alone versus steroids plus plasma exchange. Mult Scler. 2016;22(2):185–92.
106. Weinshenker BG, O'Brien PC, Petterson TM, Noseworthy JH, Lucchinetti CF, Dodick DW, et al. A randomized trial of plasma exchange in acute central nervous system inflammatory demyelinating disease. Ann Neurol. 1999;46(6):878–86.
107. Bonnan M, Valentino R, Debeugny S, Merle H, Ferge JL, Mehdaoui H, et al. Short delay to initiate plasma exchange is the strongest predictor of outcome in severe attacks of NMO spectrum disorders. J Neurol Neurosurg Psychiatry. 2018;89(4):346–51.
108. Shemin D, Briggs D, Greenan M. Complications of therapeutic plasma exchange: a prospective study of 1,727 procedures. J Clin Apher. 2007;22(5):270–6.
109. Elsone L, Panicker J, Mutch K, Boggild M, Appleton R, Jacob A. Role of intravenous immunoglobulin in the treatment of acute relapses of neuromyelitis optica: experience in 10 patients. Mult Scler. 2014;20(4):501–4.
110. Levy M, Mealy MA. Purified human C1-esterase inhibitor is safe in acute relapses of neuromyelitis optica. Neurol Neuroimmunol Neuroinflamm. 2014;1(1):e5.
111. Yaguchi H, Sakushima K, Takahashi I, Nishimura H, Yashima-Yamada M, Nakamura M, et al. Efficacy of intravenous cyclophosphamide therapy for neuromyelitis optica spectrum disorder. Intern Med. 2013;52(9):969–72.
112. Costanzi C, Matiello M, Lucchinetti CF, Weinshenker BG, Pittock SJ, Mandrekar J, et al. Azathioprine: tolerability, efficacy, and predictors of benefit in neuromyelitis optica. Neurology. 2011;77(7):659–66.
113. Cree BA, Lamb S, Morgan K, Chen A, Waubant E, Genain C. An open label study of the effects of rituximab in neuromyelitis optica. Neurology. 2005;64(7):1270–2.
114. Damato V, Evoli A, Iorio R. Efficacy and safety of rituximab therapy in neuromyelitis optica spectrum disorders: a systematic review and meta-analysis. JAMA Neurol. 2016;73(11):1342–8.
115. Jacob A, Matiello M, Weinshenker BG, Wingerchuk DM, Lucchinetti C, Shuster E, et al. Treatment of neuromyelitis optica with mycophenolate mofetil: retrospective analysis of 24 patients. Arch Neurol. 2009;66(9):1128–33.
116. Nikoo Z, Badihian S, Shaygannejad V, Asgari N, Ashtari F. Comparison of the efficacy of azathioprine and rituximab in neuromyelitis optica spectrum disorder: a randomized clinical trial. J Neurol. 2017;264(9):2003–9.
117. Evangelopoulos ME, Andreadou E, Koutsis G, Koutoulidis V, Anagnostouli M, Katsika P, et al. Treatment of neuromyelitis optica and neuromyelitis optica spectrum disorders with rituximab using a maintenance treatment regimen and close CD19 B cell monitoring. A six-year follow-up. J Neurol Sci. 2017;372:92–6.
118. Flanagan EP, Aksamit AJ, Kumar N, Morparia NP, Keegan BM, Weinshenker BG. Simultaneous PML-IRIS and myelitis in a patient with neuromyelitis optica spectrum disorder. Neurol Clin Pract. 2013;3(5):448–51.
119. Pittock SJ, Berthele A, Fujihara K, Kim HJ, Levy M, Palace J, et al. Eculizumab in aquaporin-4-positive neuromyelitis optica spectrum disorder. N Engl J Med. 2019;381(7):614–25.
120. Cree B, Bennett J, Kim HJ, Weinshenker B, Pittock S, Wingerchuk D, Fujihara K, Paul F, Cutter G, Marignier R, Green A, Aktas O, Hartung H, Drappa J, Barron G, Madhani S, Ratchford J, She D, Katz E. A double-masked, placebo-controlled study with open-label period to evaluate the efficacy and safety of inebilizumab in adult subjects with neuromyelitis optica spectrum disorders–top line efficacy and safety results (Plen02.001). Neurology. 2019;92(15 Supplement)
121. Yamamura T, Kleiter I, Fujihara K, Palace J, Greenberg B, Zakrzewska-Pniewska B, Patti F, Tsai CP, Saiz A, Haramura M, Terada Y, Kawata Y, De Seze J. Efficacy of satralizumab (SA237) in subgroups of patients in SAkuraSky: a phase III double-blind, placebo-controlled, add-on study in patients with neuromyelitis optica spectrum disorder (NMOSD) (S43.008). Neurology. 2019;92(15 Supplement)
122. Shosha E, Pittock SJ, Flanagan E, Weinshenker BG. Neuromyelitis optica spectrum disorders and pregnancy: interactions and management. Mult Scler. 2017;23(14):1808–17.
123. Peschl P, Bradl M, Hoftberger R, Berger T, Reindl M. Myelin oligodendrocyte glycoprotein: deciphering a target in inflammatory demyelinating diseases. Front Immunol. 2017;8:529.
124. Vazquez Do Campo R, Stephens A, Marin Collazo IV, Rubin DI. MOG antibodies in combined central and peripheral demyelination syndromes. Neurol Neuroimmunol Neuroinflamm. 2018;5(6):e503.
125. Peschl P, Schanda K, Zeka B, Given K, Bohm D, Ruprecht K, et al. Human antibodies against the myelin oligodendrocyte glycoprotein can cause complement-dependent demyelination. J Neuroinflammation. 2017;14(1):208.
126. Brilot F, Dale RC, Selter RC, Grummel V, Kalluri SR, Aslam M, et al. Antibodies to native myelin oligodendrocyte glycoprotein in children with inflammatory demyelinating central nervous system disease. Ann Neurol. 2009;66(6):833–42.
127. Saadoun S, Waters P, Owens GP, Bennett JL, Vincent A, Papadopoulos MC. Neuromyelitis optica MOG-IgG causes reversible lesions in mouse brain. Acta Neuropathol Commun. 2014;2:35.

128. Bettelli E, Baeten D, Jager A, Sobel RA, Kuchroo VK. Myelin oligodendrocyte glycoprotein-specific T and B cells cooperate to induce a Devic-like disease in mice. J Clin Invest. 2006;116(9):2393–402.
129. Di Pauli F, Hoftberger R, Reindl M, Beer R, Rhomberg P, Schanda K, et al. Fulminant demyelinating encephalomyelitis: insights from antibody studies and neuropathology. Neurol Neuroimmunol Neuroinflamm. 2015;2(6):e175.
130. Fujimori J, Takai Y, Nakashima I, Sato DK, Takahashi T, Kaneko K, et al. Bilateral frontal cortex encephalitis and paraparesis in a patient with anti-MOG antibodies. J Neurol Neurosurg Psychiatry. 2017;88(6):534–6.
131. Ikeda T, Yamada K, Ogawa R, Takai Y, Kaneko K, Misu T, et al. The pathological features of MOG antibody-positive cerebral cortical encephalitis as a new spectrum associated with MOG antibodies: a case report. J Neurol Sci. 2018;392:113–5.
132. Jarius S, Metz I, Konig FB, Ruprecht K, Reindl M, Paul F, et al. Screening for MOG-IgG and 27 other anti-glial and anti-neuronal autoantibodies in 'pattern II multiple sclerosis' and brain biopsy findings in a MOG-IgG-positive case. Mult Scler. 2016;22(12):1541–9.
133. Kortvelyessy P, Breu M, Pawlitzki M, Metz I, Heinze HJ, Matzke M, et al. ADEM-like presentation, anti-MOG antibodies, and MS pathology: TWO case reports. Neurol Neuroimmunol Neuroinflamm. 2017;4(3):e335.
134. Spadaro M, Gerdes LA, Mayer MC, Ertl-Wagner B, Laurent S, Krumbholz M, et al. Histopathology and clinical course of MOG-antibody-associated encephalomyelitis. Ann Clin Transl Neurol. 2015;2(3):295–301.
135. Wang JJ, Jaunmuktane Z, Mummery C, Brandner S, Leary S, Trip SA. Inflammatory demyelination without astrocyte loss in MOG antibody-positive NMOSD. Neurology. 2016;87(2):229–31.
136. Lucchinetti CF, Bruck W, Rodriguez M, Lassmann H. Distinct patterns of multiple sclerosis pathology indicates heterogeneity on pathogenesis. Brain Pathol. 1996;6(3):259–74.
137. Cobo-Calvo A, Ruiz A, Maillart E, Audoin B, Zephir H, Bourre B, et al. Clinical spectrum and prognostic value of CNS MOG autoimmunity in adults: the MOGADOR study. Neurology. 2018;90(21):e1858–e69.
138. Hennes EM, Baumann M, Schanda K, Anlar B, Bajer-Kornek B, Blaschek A, et al. Prognostic relevance of MOG antibodies in children with an acquired demyelinating syndrome. Neurology. 2017;89(9):900–8.
139. Jarius S, Ruprecht K, Kleiter I, Borisow N, Asgari N, Pitarokoili K, et al. MOG-IgG in NMO and related disorders: a multicenter study of 50 patients. Part 2: epidemiology, clinical presentation, radiological and laboratory features, treatment responses, and long-term outcome. J Neuroinflammation. 2016;13(1):280.
140. Jurynczyk M, Messina S, Woodhall MR, Raza N, Everett R, Roca-Fernandez A, et al. Clinical presentation and prognosis in MOG-antibody disease: a UK study. Brain. 2017;140(12):3128–38.
141. Lopez-Chiriboga AS, Majed M, Fryer J, Dubey D, McKeon A, Flanagan EP, et al. Association of MOG-IgG serostatus with relapse after acute disseminated encephalomyelitis and proposed diagnostic criteria for MOG-IgG-associated disorders. JAMA Neurol. 2018;75(11):1355–63.
142. Mariotto S, Ferrari S, Monaco S, Benedetti MD, Schanda K, Alberti D, et al. Clinical spectrum and IgG subclass analysis of anti-myelin oligodendrocyte glycoprotein antibody-associated syndromes: a multicenter study. J Neurol. 2017;264(12):2420–30.
143. Ramanathan S, Mohammad S, Tantsis E, Nguyen TK, Merheb V, Fung VSC, et al. Clinical course, therapeutic responses and outcomes in relapsing MOG antibody-associated demyelination. J Neurol Neurosurg Psychiatry. 2018;89(2):127–37.
144. Chen JJ, Tobin WO, Majed M, Jitprapaikulsan J, Fryer JP, Leavitt JA, et al. Prevalence of myelin oligodendrocyte glycoprotein and aquaporin-4-IgG in patients in the optic neuritis treatment trial. JAMA Ophthalmol. 2018;136(4):419–22.
145. Chen JJ, Flanagan EP, Jitprapaikulsan J, Lopez-Chiriboga ASS, Fryer JP, Leavitt JA, et al. Myelin oligodendrocyte glycoprotein antibody (MOG-IgG)-positive optic neuritis: clinical characteristics, radiologic clues and outcome. Am J Ophthalmol. 2018;195:8–15.
146. Petzold A, Plant GT. Chronic relapsing inflammatory optic neuropathy: a systematic review of 122 cases reported. J Neurol. 2014;261(1):17–26.
147. Dubey D. Myelin oligodendrocyte glycoprotein autoantibody myelitis; clinical characteristics, MRI clues and discriminators from other demyelinating myelitis etiologies. JAMA Neurol. 2019;92(15 Supplement)
148. Sechi ES, Williams JP, Pittock SJ, Weinshenker BG, Keegan BM, Zalewski NL, Lopez-Chiriboga AS, Jitprapaikulsan J, Flanagan EP. Aquaporin-4 and MOG autoantibody discovery and idiopathic transverse myelitis epidemiology. Neurology. 2019;93(4):e414–20.
149. Dubey D, Pittock SJ, Krecke KN, Morris PP, Sechi E, Zalewski NL, et al. Clinical, radiologic, and prognostic features of myelitis associated with myelin Oligodendrocyte glycoprotein autoantibody. JAMA Neurol. 2018;76(3):301–9.
150. Kitley J, Leite MI, Kuker W, Quaghebeur G, George J, Waters P, et al. Longitudinally extensive transverse myelitis with and without aquaporin 4 antibodies. JAMA Neurol. 2013;70(11):1375–81.
151. Ogawa R, Nakashima I, Takahashi T, Kaneko K, Akaishi T, Takai Y, et al. MOG antibody-positive, benign, unilateral, cerebral cortical encephalitis with epilepsy. Neurol Neuroimmunol Neuroinflamm. 2017;4(2):e322.

152. Kitley J, Waters P, Woodhall M, Leite MI, Murchison A, George J, et al. Neuromyelitis optica spectrum disorders with aquaporin-4 and myelin-oligodendrocyte glycoprotein antibodies: a comparative study. JAMA Neurol. 2014;71(3):276–83.
153. Jurynczyk M, Tackley G, Kong Y, Geraldes R, Matthews L, Woodhall M, et al. Brain lesion distribution criteria distinguish MS from AQP4-antibody NMOSD and MOG-antibody disease. J Neurol Neurosurg Psychiatry. 2017;88(2):132–6.
154. Ikeda K, Kiyota N, Kuroda H, Sato DK, Nishiyama S, Takahashi T, et al. Severe demyelination but no astrocytopathy in clinically definite neuromyelitis optica with anti-myelin-oligodendrocyte glycoprotein antibody. Mult Scler. 2015;21(5):656–9.
155. Waters PJ, Komorowski L, Woodhall M, Lederer S, Majed M, Fryer J, et al. A multicenter comparison of MOG-IgG cell-based assays. Neurology. 2019;92(11):e1250–e5.
156. Jarius S, Paul F, Aktas O, Asgari N, Dale RC, de Seze J, et al. MOG encephalomyelitis: international recommendations on diagnosis and antibody testing. J Neuroinflammation. 2018;15(1):134.
157. Mariotto S, Gajofatto A, Batzu L, Delogu R, Sechi GP, Leoni S, et al. Relevance of antibodies to myelin oligodendrocyte glycoprotein in CSF of seronegative cases. Neurology 2020;93(20):e1867–e1872.
158. Waters P, Fadda G, Woodhall M, O'Mahony J, Brown RA, Castro DA, et al. Serial Anti-Myelin Oligodendrocyte Glycoprotein Antibody Analysis and Outcomes in Children with Demyelinating Syndromes. JAMA Neurol 2020;77(1):82–93.
159. Kunchok A, Chen JJ, McKeon A, et al. Coexistence of Myelin Oligodendrocyte Glycoprotein and Aquaporin-4 Antibodies in Adult and Pediatric Patients. JAMA Neurol 2020;77(2):257–9.
160. Tahara M, Oeda T, Okada K, et al. Safety and efficacy of rituximab in neuromyelitis optica spectrum disorders (RIN-1 study): a multicentre, randomised, double-blind, placebocontrolled trial. Lancet Neurol 2020;19(4):298–306.
161. Hoftberger R, Guo Y, Flanagan EP, et al. The pathology of central nervous system inflammatory demyelinating disease accompanying myelin oligodendrocyte glycoprotein autoantibody. Acta Neuropathologica 2020;139:875–892.
162. Takai Y, Misu T, Kaneko K, et al. Myelin oligodendrocyte glycoprotein antibody-associated disease: an immunopathological study. Brain 2020;143:1431–1446.
163. Amy Kunchok, Karl N. Krecke, Eoin P. Flanagan, Jiraporn Jitprapaikulsan, A. Sebastian Lopez-Chiriboga, John J. Chen, Brian G. Weinshenker, Sean J. Pittock, (2020) Does area postrema syndrome occur in myelin oligodendrocyte glycoprotein-IgG–associated disorders (MOGAD)?. Neurology 94 (2):85–88.
164. Asnafi S, Morris PP, Sechi E, et al. The frequency of longitudinally extensive transverse myelitis in MS: A population-based study. Mult Scler Relat Disord 2020;37:101487.
165. Sechi E, Krecke KNK, Pittock SJ, et al. Frequency and characteristics of MRI-negative myelitis associated with MOG autoantibodies. Mult Scler 2020. Online ahead of print. DOI:10.1177/1352458520907900.
166. Jarius S, Pellkofer H, Siebert N, et al. Cerebrospinal fluid findings in patients with myelin oligodendrocyte glycoprotein (MOG) antibodies. Part 1: Results from 163 lumbar punctures in 100 adult patients. J Neuroimflammation 2020. Online ahead of print. DOI:10.1186/s12974-020-01824-2.
167. Dale RC, Tantsis EM, Merheb V, et al. Antibodies to MOG have a demyelination phenotype and affect oligodendrocyte cytoskeleton. Neurol Neuroimmunol Neuroinflamm 2014;1(1):e12.
168. Arru G, Sechi E, Mariotto S, et al. Antibody response against HERV-W in patients with MOG-IgG associated disorders, multiple sclerosis and NMOSD. J Neuroimmunol 2019;338:577110.
169. John J. Chen, Eoin P. Flanagan, M. Tariq Bhatti, Jiraporn Jitprapaikulsan, Divyanshu Dubey, Alfonso (Sebastian) S. Lopez Chiriboga, James P. Fryer, Brian G. Weinshenker, Andrew McKeon, Jan-Mendelt Tillema, Vanda A. Lennon, Claudia F. Lucchinetti, Amy Kunchok, Collin M. McClelland, Michael S. Lee, Jeffrey L. Bennett, Victoria S. Pelak, Gregory Van Stavern, Ore-Ofe O. Adesina, Eric R. Eggenberger, Marie D. Acierno, Dean M. Wingerchuk, Byron L. Lam, Heather Moss, Shannon Beres, Aubrey L. Gilbert, Veeral Shah, Grayson Armstrong, Gena Heidary, Dean M. Cestari, Hadas Stiebel-Kalish, Sean J. Pittock, (2020) Steroid-sparing maintenance immunotherapy for MOG-IgG associated disorder. Neurology 95 (2):e111–e120.
170. Li S, Ren H, Xu Y, et al. Long-term efficacy of mycophenolate mofetil in myelin oligodendrocyte glycoprotein antibody-associated disorders: a prospective study. Neurol Neuroimmunol Neuroinflamm 2020; 7(3).
171. Durozard P, Rico A, Boutiere C, et al. Comparison of the Response to Rituximab between Myelin Oligodendrocyte Glycoprotein and Aquaporin-4 Antibody Diseases. Ann Neurol 2020;87(2):256-266.
172. Lopez-Chiriboga S, Sechi E, Buciuc M, et al. Long-term Outcomes in Patients With Myelin Oligodendrocyte Glycoprotein Immunoglobulin G-Associated Disorder. JAMA Neurol 2020. Online ahead of print. DOI:10.1001/jamaneurol.2020.3115.
173. Buciuc M, Sechi E, Flanagan EP, et al. Unfavorable outcome in highly relapsing MOGAD encephalitis. J Neurol Sci 2020;418:117088.
174. Sechi E, Buciuc M, Flanagan EP, et al. Variability of cerebrospinal fluid findings by attack phenotype in myelin oligodendrocyte glycoprotein-IgG-associated disorder. Mult Scler Relat Disord 2020
175. Mariotto S, Ferrari S, Gastaldi M, et al. Neurofilament light chain serum levels reflect disease severity in MOG-Ab associated disorders. J Neurol Neurosurg Psychiatry 2019;90(11):1293–1296.

# Paraneoplastic Neurological Syndromes

# 16

Marianna Spatola

**Key Points**

1. Paraneoplastic neurological syndromes (PNS) can affect any part of the nervous system. Classical PNS include limbic encephalitis, cerebellar degeneration, opsoclonus-myoclonus syndrome, encephalomyelitis, sensory neuropathy, and Lambert-Eaton myasthenic syndrome (LEMS).
2. PNS occur in association with cancer; however, they are not due to cancer invasion or complication of its treatment, but rather to the immune response directed against proteins shared between tumor cells and neurons.
3. PNS are characterized by the detection in the serum and cerebrospinal fluid (CSF) of antibodies reacting with neuronal intracellular antigens (e.g., Hu or Yo). Neuronal dysfunction is not caused by these antibodies, which are considered as markers of the underlying tumor, but is due to T-cell cytotoxicity.
4. The diagnosis of PNS relies on recognition of the clinical syndrome and detection of the associated cancer and autoantibodies in serum and CSF.
5. Response of PNS to tumor treatment and immunotherapy is generally poor.

## Introduction

Paraneoplastic neurological syndromes (PNS) are immune-mediated disorders that can affect any part of the neuraxis, from the central nervous system (CNS), including the retina, to peripheral nerves and neuromuscular junction. PNS can manifest with a variety of neurological symptoms, which can derive from dysfunction of one area of the nervous system (e.g., limbic encephalitis or brainstem encephalitis), single cell population (e.g., Purkinje cells in paraneoplastic cerebellar degeneration or ganglionic cells in paraneoplastic retinopathy), or multiple areas (e.g., encephalomyelitis).

These disorders occur in association with cancer; however, they are not due to cancer invasion or complication of its treatment, but rather to the immune response directed against proteins shared between tumor cells and neurons, which causes neuronal dysfunction or damage. Importantly, the onset of neurological symptoms often precedes (or leads to) the diagnosis of cancer by months or

M. Spatola (✉)
Massachusetts General Hospital Research Institute, Ragon Institute of Massachusetts General Hospital, Massachusetts Institute of Technology, Harvard Medical School, Boston-Cambridge, MA, USA
e-mail: mspatola@mgh.harvard.edu



A. L. Piquet, E. Alvarez (eds.), *Neuroimmunology*, https://doi.org/10.1007/978-3-030-61883-4_16

even years. Thus, considering also that in most cases the underlying cancer is of small size and limited stage, detection of PNS represents an opportunity for early cancer diagnosis and treatment.

PNS are rare disorders that occur in 1 out of 10,000 patients with cancer, with the exception of some PNS affecting the neuromuscular junction, such as Lambert-Eaton myasthenic syndrome (LEMS), which can be much more common [1]. Not all cancer types have the same propensity to trigger PNS; some, such as small-cell lung cancer (SCLC), breast and gynecologic cancers, and non-solid tumors (in particular Hodgkin's lymphoma), are more frequently associated with PNS than others.

PNS are characterized by the detection in the serum and cerebrospinal fluid (CSF) of antibodies reacting with neuronal antigens, which are also expressed by tumor cells. These antibodies are also called onconeuronal because of their relation to PNS and cancer. Some onconeuronal antibodies are associated with a specific neurological syndrome, for instance, Yo antibodies with paraneoplastic cerebellar degeneration [2] and Ma2 antibodies with limbic or brainstem encephalitis [3], whereas others can be associated with a wider range of neurological manifestations, such as Hu antibodies, which can be found in patients with paraneoplastic limbic encephalitis, neuronopathy, encephalomyelitis, or others [4, 5].

This chapter focuses on the main PNS involving the CNS; those affecting the peripheral nervous system and neuromuscular junction will be discussed in separate sections (Part 4: Chaps. 18, 19, 20, and 21). This chapter aims to (1) facilitate recognition and diagnosis of these disorders by providing clinical features, CSF analysis, magnetic resonance imaging (MRI) findings, and tumor association, exemplified by two clinical cases; (2) describe the associated onconeuronal antibodies and discuss their role in the pathogenesis of neuronal damage; and (3) summarize the therapeutic approach (discussed in detail in Chap. 17) and response of neurological symptoms to tumor treatment and immunotherapy.

## Clinical Cases

### Clinical Case 1

A 59-year-old woman was referred for a 1-month history of diplopia and vertigo, followed by progressive development of gait instability, limb incoordination, and speech difficulties. Neurological symptoms progressed over the following weeks to severe ataxia with inability to stand, walk, or eat unassisted. Neurological examination showed a severe pancerebellar syndrome with trunk, limb, and gait ataxia, downbeat nystagmus, and dysarthria. Brain MRI was unremarkable. CSF analysis showed no pleocytosis but elevated immunoglobulin G (IgG) index and protein. Oncologic screening revealed breast cancer with axillary lymph node involvement. Antibodies against Yo were found at high titers in serum and CSF, confirming the diagnosis of Yo antibody-associated paraneoplastic cerebellar degeneration. Cancer treatment, including surgery, chemotherapy, and radiotherapy, followed by immunotherapy with corticoids and rituximab, resulted in neurological stabilization. At 15 months from onset, because of neurological worsening, rituximab was changed to mycophenolate mofetil without further progression of her cerebellar syndrome. Brain MRI at 26 months from onset showed diffuse cerebellar atrophy (Fig. 16.1). At the last follow-up, 3.5 years from onset, the patient was clinically stable, with no evidence of tumor relapse.

### Clinical Case 2

A 52-year-old woman, current smoker, presented with gradual onset of sensory disturbances in her hands and feet, including numbness, prickling, and tingling, which progressively spread to the arms and legs. Sensory symptoms were associated with intermittent sharp, throbbing pain, and extreme sensitivity to touch. She also developed limb ataxia and gait instability, especially in poor lighting conditions, and complained of excessive sweating and urinary retention. Neurological

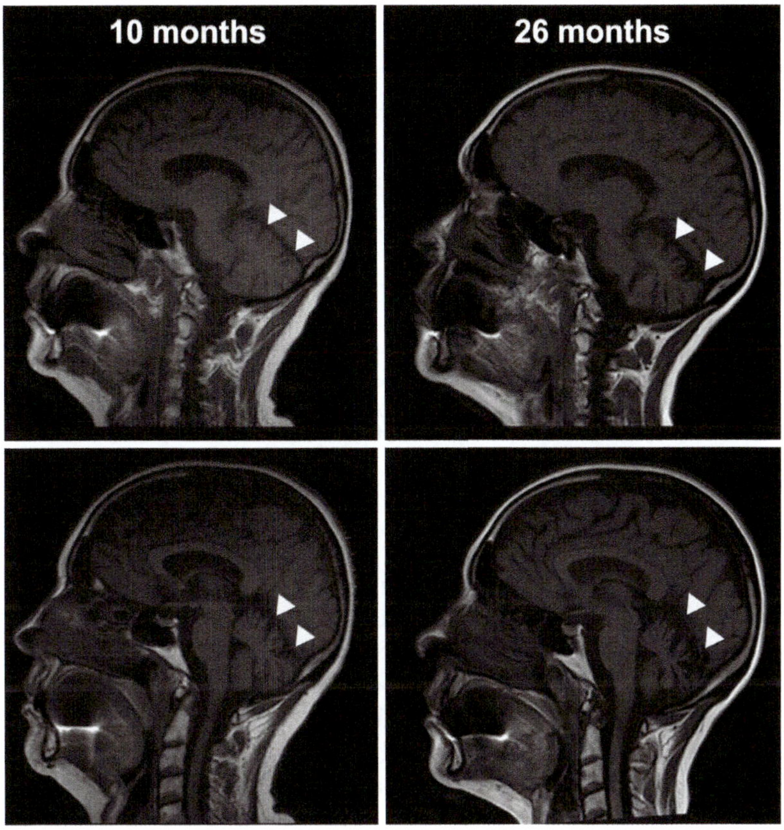


**Fig. 16.1** Evolution of cerebellar atrophy at 10 and 26 months from onset in a woman with anti-Yo paraneoplastic cerebellar degeneration (*Clinical Case 1*). Notice the progressive enlargement of the cerebellar interlobular fissures (upper panels, arrow heads) and CSF space under the tentorium cerebelli (lower panels, arrow heads), indicating cerebellar atrophy. (*Courtesy of Dr. Francesc Graus*)

examination revealed alteration of all sensory modalities in the lower limbs and to a lesser extent also the upper limbs, dysdiadochocinesia, severe sensory ataxia with abolished vibratory and position sense, and altered thermoalgesia of the right face. Osteotendinous reflexes were normal in the upper limbs and absent in the lower limbs, plantar reflex was bilaterally flexor. Muscle strength was normal. Nerve conduction studies revealed the absence of sensory nerve responses in the lower limbs and right trigeminal nerve and responses of decreased amplitude in the upper limbs, with normal motor studies. Oncologic screening found oat cell carcinoma of the lung and elevated Hu antibodies in the serum and CSF (Fig. 16.2), confirming the diagnosis of paraneoplastic neuronopathy associated with Hu antibodies. Tumor removal, chemotherapy, and radiotherapy resulted in partial improvement of neurological symptoms and disappearance of Hu antibodies from the serum. After 8 years, the patient progressively developed severe motor weakness and atrophy, which predominated in the lower limbs, obliging her to ambulate in a wheelchair. Hu antibodies reappeared in the serum at high titers, but there was no evidence of tumor relapse. The patient was treated with corticoids without improvement but stabilization of the neurological syndrome.

## General Concepts

### Diagnosis

The diagnosis of PNS relies on recognition of the clinical syndrome and detection of the associated cancer and onconeuronal antibodies in serum and CSF [6].

When evaluating a patient with suspected PNS, clinicians should carefully consider the patient's age and sex, clinical features (some neurological syndromes, such as limbic encephalitis, being more suggestive of a paraneoplastic origin

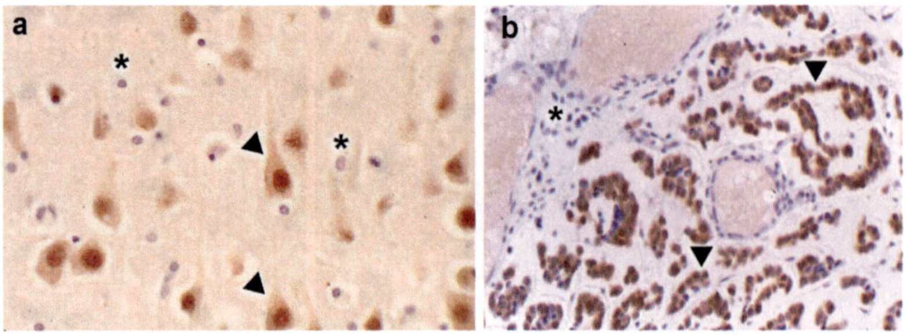


**Fig. 16.2** Immunolabeling of rat brain (**a**) by anti-Hu serum from a patient with paraneoplastic encephalomyelitis and breast cancer. Hu antibodies strongly react with the nucleus (and to a lesser extent also with the cytoplasm) of neurons (**a**, arrow heads) but do not label glial cells (**a**, star). These antibodies also react with breast cancer cells (**b**, arrow heads), which express Hu proteins, but do not label the surrounding normal breast tissue (**b**, star). Immunoperoxidase technique, slightly counterstained with hematoxylin; original magnification ×800 and ×200. (*Courtesy of Dr. Francesc Graus*)

than others), cancer risk factor (e.g.., smoking), and previous history of cancer. Findings that support diagnosis of PNS include CSF analysis showing low to moderate lymphocytic pleocytosis, increased IgG index, and oligoclonal bands. Although these findings are not specific to PNS, as they can be found in patients with other inflammatory disorders, CSF analysis is fundamental to rule out other pathologies and cancer complications, such as leptomeningeal carcinomatosis. Brain MRI abnormalities are also usually nonspecific, with the exception of unilateral or bilateral hyperintensities of mesial-temporal lobes in patients with limbic encephalitis. However, both CSF analysis and brain MRI can be completely normal in patients with PNS. In these cases, brain 18F-fluorodeoxyglucose position emission tomography (18F-FDG-PET) can be helpful as it can reveal metabolic abnormalities in brain areas that appear structurally normal by MRI [7]. In patients for whom the diagnosis remains uncertain, biopsy of lesions identified by brain MRI or 18F-FDG-PET might help to exclude other diagnoses or support an immune-mediated process.

Detection of onconeuronal antibodies in the CSF confirms the diagnosis of PNS. However, it is important to underline that the presence of these antibodies only in serum is not confirmatory, as they can be found, most often at low titers, in the serum of patients with cancer and no PNS. Identification of onconeuronal antibodies is also important because it can suggest the most likely underlying tumor. For instance, Ma2 antibodies are highly suggestive of an underlying testicular tumor, whereas identification of Hu antibodies should prompt search of SCLC.

Patients with suspected PNS should undergo careful tumor screening, taking into consideration that the associated tumor might be small in size and difficult to detect. Tumor search should be guided by the clinical syndrome and detected onconeuronal antibody. Some paraneoplastic syndromes (e.g., cerebellar degeneration with Yo antibodies) are so closely associated with a specific tumor type (breast or ovarian cancer) that if the tumor found does not correspond to the typically expected, a second neoplasm should be suspected [8]. Serum oncologic markers such as Ca-125, carcinoembryonic antigen, or prostate-specific antigen can be helpful. Whole-body computed tomography (CT) and 18F-FDG-PET have been suggested to be the best methods to identify occult cancers [9]. Pelvic and testicular tumors are best investigated by ultrasound, whereas mammography is the first choice to identify breast cancer [10]. If no tumor is found, and clinical suspicion of PNS remains high, it is recommended to repeat tumor screening every 6 months for up to 4 years [10], although in the vast majority of the patients the underlying tumor is identified within the first year after PNS onset [11].

## Clinical Syndromes

PNS can manifest with a variety of neurological syndromes, and their association with cancer varies according to the detected neuronal antibody. For example, stiff person syndrome (see later) typically occurs in patients without cancer who have antibodies to glutamic acid decarboxylase (GAD65); however, this syndrome can also occur in association with amphiphysin antibodies, which most likely occur in patients with cancer. Overall, neurological syndromes that are most frequently associated with cancer are called *classical PNS*, as opposed to *non-classical PNS*, which can frequently occur in the absence of cancer. It is important to consider that both *classical* and *non-classical PNS* can occur in patients without cancer, but what makes them paraneoplastic is their increased occurrence in patients with cancer. Table 16.1 provides an overview of the main *classical* and *non-classical* PNS and associated onconeuronal antibodies.

**Table 16.1** The main paraneoplastic syndromes of the central and peripheral nervous system

| Paraneoplastic syndrome | Onconeuronal antibodies |
|---|---|
| *Brain, retina, brainstem, and cerebellum* | |
| Limbic encephalitis[a] | Hu, Ma2, GAD65, AK5 |
| Cerebellar degeneration[a] | Yo, PCA-2, Ri, Tr, Hu |
| Retinopathy (CAR, MAR), optic neuritis[a] | Recoverin, bipolar ganglionic cells, CV2/CRMP5 |
| Basal ganglia and diencephalon | Hu, Ma2, CV2/CRMP5 |
| Brainstem encephalitis | Ma2, CV2/CRMP5 |
| Opsoclonus-myoclonus syndrome[a] | Ri, Hu, Ma2, amphiphysin, CV2/CRMP5 |
| *Spinal cord* | |
| Encephalomyelitis[a] | Hu, CV2/CRMP5, amphiphysin |
| Amyotrophic lateral sclerosis | CV2/CRMP5, Ma2, Yo, amphiphysin |
| Inflammatory myelitis | Hu, CV2/CRMP5, GAD65, Ri, PCA-2 |
| Stiff person syndrome | GAD65, amphiphysin, Ri, gephyrin |
| *Dorsal roots, peripheral nerve, and neuromuscular junction* | |
| Sensory neuronopathy[a] | Hu, CV2/CRMP5 |
| Subacute motor neuropathy[a] | Hu |
| Autonomic neuropathy | Hu, PCA-2 |
| Chronic gastrointestinal pseudo-obstruction[a] | – |
| Polyradiculopathy (acute or chronic) | CV2/CRMP5 |
| Lambert-Eaton myasthenic syndrome[a] | P/Q VGCC, SOX1 |

[a]Classical paraneoplastic syndromes

## Pathogenesis

The pathogenic mechanisms underlying PNS are not fully understood. It is thought that tumor cells, which express ectopic proteins, can trigger an anti-tumor immune response against these proteins, which are also expressed in neurons. Apoptotic tumor cells might indeed be phagocytosed by dendritic cells, which then present tumor antigens to B and T cells at local lymph nodes (Fig. 16.3) [12]. It is currently believed that PNS are mainly mediated by cytotoxic T-cell responses against intracellular antigens and that onconeuronal antibodies against the same antigens participate to, but are not directly causing, neuronal damage [13–15]. The idea that these antibodies might be implicated in the development of PNS has been suggested by the observation that Hu antibodies, for instance, accumulate inside neurons and they are found at higher concentration in injured brain areas, correlating with the main clinical symptoms [16]. However, several observations and experimental evidence support a minor role of these antibodies: (1) antibody titers do not correlate with neurological symptom severity and may persist despite improvement or resolution of PNS [17]; (2) target antigens, being localized inside the neuron, are not directly accessible to circulating antibodies; (3) patients' Hu antibodies can enter neurons but do not cause neuronal dysfunction/cytotoxicity [18, 19]; and (4) animal models obtained by intrathecal infusion of patients' antibodies or immunization with Hu or Yo recombinant proteins with production of neuronal antibodies have failed to reproduce the human disorder [20, 21]. Overall, all onconeuronal antibodies are thought to be not directly

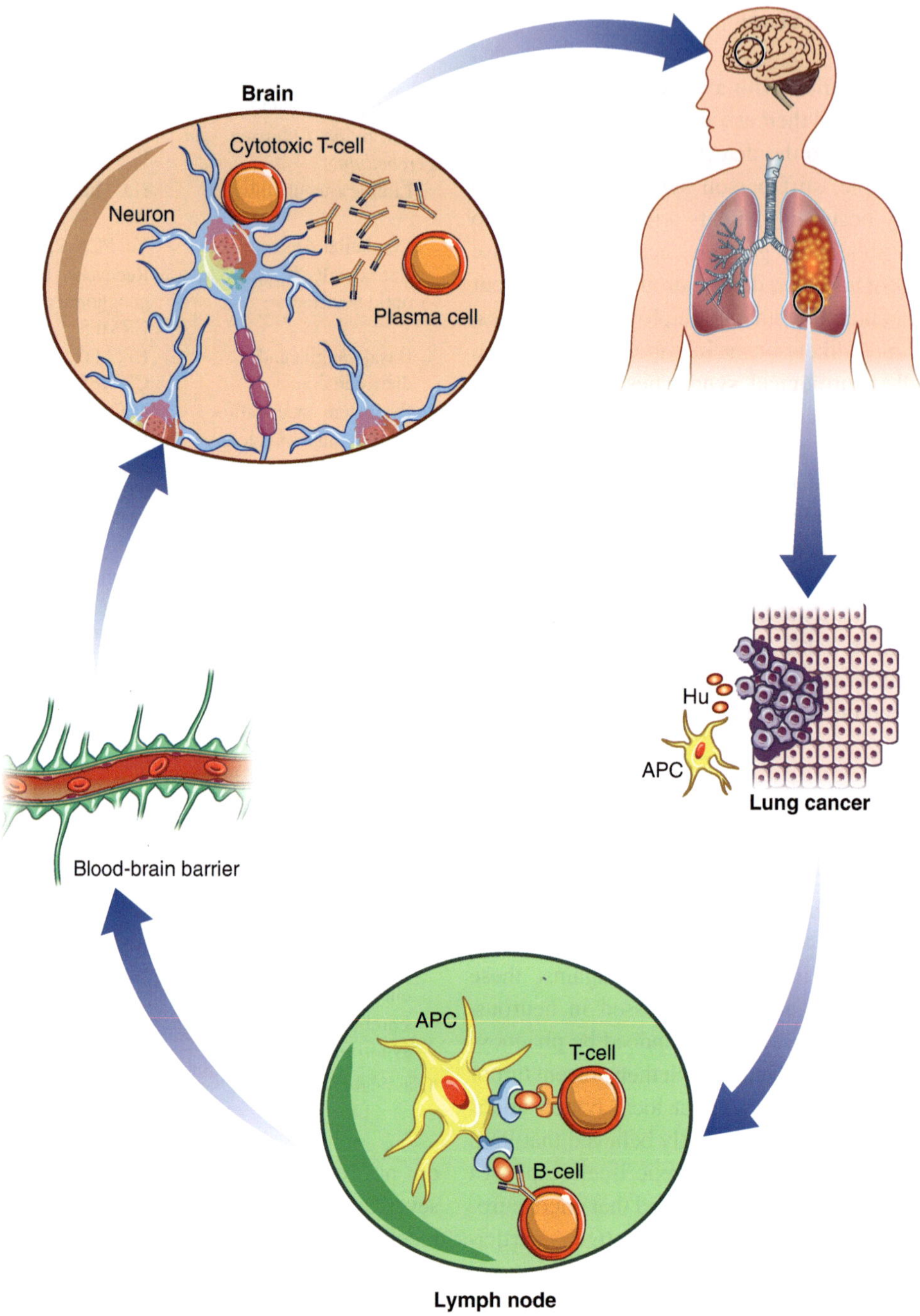


**Fig. 16.3** Possible pathogenesis of paraneoplastic neurological syndromes. Neuronal antigens (e.g., Hu), which are ectopically expressed by tumor cells (e.g., small-cell lung cancer) undergoing apoptosis, are uptaken by dendritic cells (antigen-presenting cell or APC). APC then migrates to local lymph nodes and presents these antigens to T cells and B cells. Activated T cells and B cells enter the blood stream, pass the blood-brain barrier, penetrate into the brain parenchyma, and react with neuronal antigens. Although antibody-producing plasma cells participate to the autoimmune response, T cells (in particular cytotoxic CD8+ lymphocytes) are the main responsible for neuronal damage and death

pathogenic, with the exception of those targeting intracellular synaptic proteins, such as amphiphysin. Indeed, there is evidence that these antibodies may have access to their target proteins during synaptic vesicle fusion and thus cause direct neuronal damage (see later).

## Therapeutic Approach and Prognosis

Treatment of PNS (discussed in detail in Chap. 17) is based on oncologic therapies, in cancer-associated cases, and can be complemented with immunotherapies, including corticosteroids, intravenous immunoglobulins, plasmapheresis, or immunosuppressive therapy (such as rituximab or cyclophosphamide). The response of neurological symptoms to these therapies is variable and depends on the specific PNS and associated onconeuronal antibody, as shown in Table 16.2 [1, 3, 5, 22–28]. Overall, patients with PNS show poor response to successful tumor treatment and immunotherapies (with some exception, such as GAD65 antibody-associated stiff person syndrome), likely in relation to early irreversible neuronal damage [5, 29]. However, it

**Table 16.2** The main onconeuronal antibodies, associated paraneoplastic neurological syndromes and tumor, and response to tumor treatment and immunotherapy

| Antibody | Cell-type target | Neurological syndrome | Tumor frequency (% patient) and type | Response to treatment |
|---|---|---|---|---|
| *Nuclear antibodies* | | | | |
| Hu (ANNA-1) | All neuron type of the central and peripheral nervous system | Limbic encephalitis<br>Encephalomyelitis<br>Cerebellar degeneration<br>Sensory neuropathy<br>Autonomic dysfunction | 90% [5]<br>Small-cell lung cancer, neuroblastoma, gynecologic, breast, prostate cancer | Poor, some patients show stabilization of neurological symptoms |
| Ri (ANNA-2) | All neuron type of the central nervous system | Opsoclonus-myoclonus syndrome<br>Brainstem encephalitis<br>Stiff person syndrome | 85% [22]<br>Breast, Small-cell and non-small-cell lung cancer, gynecologic, bladder, lymphoma | 60% of patients show moderate improvement, some have resolution |
| Ma1 and Ma2 | All neuron type of the central and peripheral nervous system | Limbic encephalitis<br>Diencephalic or brainstem encephalitis | 90% [3]<br>Testis (mostly germ cell cancer) | 30–50% of patients improve |
| ZIC4 | All neuron type, particularly in the cerebellum | Encephalomyelitis<br>Cerebellar degeneration | 95% [23] | Poor |
| SOX1 (AGNA-1) | Bergman glia, Golgi neurons | LEMS<br>Limbic encephalitis<br>Cerebellar ataxia | 90% [1]<br>Small-cell lung cancer | Poor |
| *Cytoplasmic antibodies* | | | | |
| Yo (PCA-1) | Purkinje neurons, deep cerebellar nuclei | Cerebellar degeneration | 90% [24]<br>Ovarian cancer, breast cancer, lung cancer | Poor |

(continued)

**Table 16.2** (continued)

| Antibody | Cell-type target | Neurological syndrome | Tumor frequency (% patient) and type | Response to treatment |
|---|---|---|---|---|
| PCA-2 (MAP1B) | Purkinje neurons | Limbic encephalitis<br>Brainstem encephalitis<br>Cerebellar degeneration<br>LEMS<br>Autonomic neuropathy | 80% [25]<br>Small-cell lung cancer | Poor |
| Tr (PCA-Tr or DNER) | Purkinje neurons | Cerebellar degeneration | 90% [26]<br>Hodgkin's lymphoma | 20% of patients respond |
| CV2 (CRMP5) | Oligodendrocytes ≫ neurons of the cortex, cerebellum, and optic nerve | Encephalomyelitis<br>Cerebellar degeneration<br>Chorea<br>Sensory neuropathy<br>Retinopathy, optic neuropathy | 90% [27]<br>Small-cell lung cancer (60%), malignant thymoma, uterine sarcoma | Some patients respond with substantial improvement or resolution |
| Recoverin, anti-retinal bipolar cells, and other retinal proteins | Photoreceptors, bipolar neurons (recoverin), ganglionic cells | Retinopathy (CAR, MAR) | Small-cell lung cancer, gynecologic cancer (CAR), and melanoma (MAR) [82, 83] | Few cases reported to have moderate response with stabilization |
| *Synaptic antibodies* | | | | |
| Amphiphysin | Presynaptic nerve terminal, central, and peripheral neurons | Stiff person syndrome<br>Encephalomyelitis | 90% [28]<br>Small-cell lung cancer, melanoma, breast cancer | Moderate, 60% of the patients show some neurological improvement |

has been suggested that prompt diagnosis and treatment while inflammatory process is still active and neurological symptoms are still progressing might be beneficial and even result in clinical improvement [13, 30, 31].

## Onconeuronal Antibodies and Associated PNS

### Antibodies, Nomenclature, and Target Antigens

The nomenclature of onconeuronal antibodies varies among different groups or authors. Traditionally, these antibodies have been named after the last name of the patient from whom the antibody was firstly identified (e.g., Hu, Yo, Ri, etc.). Some of these antibodies are also known by the name of the protein they target, for example, CV2 antibodies are also called collapsin response mediator protein (CRMP) 5. Some authors [32] proposed a generic nomenclature based on the target cell type (neuronal, glial) and location of the intracellular target antigen (nuclear, cytoplasmic), followed by the chronological order of discovery, for example, Hu antibody has been the first nuclear antibody described, and it is thus referred to as ANNA-1 (antinuclear neuronal antibody-1). Similarly, Yo antibodies are also referred to as PCA-1 (Purkinje cell cytoplasmic antibody type 1). Table 16.2 shows both the traditional and generic nomenclatures for each of these antibodies.

Onconeuronal antibodies are mainly of IgG class [33], although some patients can harbor antibodies of IgM and IgA classes. These antibodies react with a variety of intracellular antigens, located in the nucleus or cytoplasm of neurons, although some of these antigens (such as Tr or delta/notch-like epidermal growth factor-related receptor [DNER]) carry transmembrane or extracellular portions. Although the identity of these antigens (and the related gene) is known for most onconeuronal antibodies, the exact function of these proteins in neurons and in tumor cells is still unclear [34]. It is important to underline that these antibodies differ from neuronal cell surface antibodies discussed in Chap. 12, which target antigens located in the outer part of the neuronal cell membrane, such as synaptic neurotransmitter receptors (e.g., N-methyl-D-aspartate receptor or NMDAR). Indeed, these neuronal cell surface antibodies have also been identified in patients with neurological syndromes, but their association with cancer is less constant than for onconeuronal antibodies.

Most onconeuronal antibodies are detected in both serum and CSF, with higher titers in CSF (intrathecal synthesis) [35], and in patients with CNS involvement compared to those with peripheral neuronopathy [36]. Overall, antibody titers do not seem to correlate with the severity of neurological symptoms at onset or with outcome, and antibodies might persist despite successful treatment of the tumor and neurological improvement [17]. Thus, serial determination of antibody titers is not recommended [37].

Some of these antibodies are considered markers of the presence of a tumor, such as Hu antibodies, which can be identified in the serum of patients with small-cell lung cancer (SCLC), with or without PNS, although titers are much higher in patients with PNS [38]. On the other hand, other antibodies, such as Yo, are markers of the paraneoplastic neurological syndrome, as they are found in patients with gynecologic tumors only if associated with PNS.

In the following section, we will discuss onconeuronal antibodies associated with PNS involving the CNS, divided according to the localization of their targets (nuclear, cytoplasmic, or synaptic), then we will summarize those associated with PNS involving the visual system, and finally we will briefly describe those antibodies that are only rarely associated with cancer.

## Antinuclear Neuronal Antibodies

### Hu (ANNA-1) Antibodies

Patients with PNS associated with Hu antibodies can manifest symptoms involving any part of the central and peripheral nervous system. The most frequent anti-Hu PNS is sensory neuronopathy (see *Clinical Case 2*). It occurs in half of the patients and typically starts distally in the limbs, asymmetrically, and progressively extends to the trunk (and face). It generally involves all sensory modalities, affecting simultaneously both small and large nerve fibers. However, in some cases, one type of fiber can be predominantly involved, which results, for instance, in small fiber painful neuropathy or ataxic sensory ganglionopathy. Motor involvement is frequent but does not usually occur in isolation or as a prominent symptom. Other frequent clinical presentations are cerebellar degeneration, limbic encephalitis (confusion, memory loss, behavioral/personality changes, and MRI bilateral mesiotemporal lobe hyperintensities), encephalomyelitis (combining symptoms of brain and spinal cord involvement), brainstem encephalitis, and autonomic dysfunction, each of these syndromes occurring in 10–20% of the patients [5, 39, 40]. Less typical neurological findings include, among others, cranial nerve palsy, ophthalmoplegia, and orolingual tremor. Overall, neurological symptoms develop subacutely, with a median of 8 weeks from onset to peak of disease [40].

Cancer is found in more than 90% of the patients, and it is often diagnosed 4–6 months after onset of neurological symptoms [5, 40]. SCLC is the most frequently identified tumor, whereas extrathoracic tumors are much less common [40].

Response of PNS to tumor treatment is generally poor and at best prevents further impairment of neurological symptoms. Complete remission of the tumor has been identified as the only predictor of PNS stabilization, whereas adding immunotherapy does not appear to affect survival nor neurological outcome [41]. Patients older than 60 years, with a multifocal or more severe neurological involvement (Rankin scale >3) and who have not received tumor treatment, are at higher risk of mortality [5]. In a study analyzing autopsies from patients with anti-Hu PNS, the cause of death did not appear to be related to the SCLC, given that the tumor was small and most often limited to the chest at onset and during the disease course, but was rather related to brainstem dysfunction (central hypoventilation, aspiration pneumonia due to dysphagia) or severe dysautonomia [40]. This critical neurologic involvement might explain why patients with anti-Hu PNS and SCLC, who seem to have an efficient immune control of the tumor, have a mortality rate similar to patients with SCLC without PNS, who most die from tumor progression [40, 41].

Hu antibodies target a family of proteins that are highly expressed in the central and peripheral nervous system and that are also expressed by SCLC and other tumor cells (Fig. 16.2) [42]. The function of these proteins has only been partly elucidated, but it seems to be related to alternate splicing and regulation of mRNA stability [43, 44]. In neurons, Hu proteins are involved in cellular differentiation and plasticity and are thought to play a role in learning and memory processes [45–47].

### Ri (ANNA-2) Antibodies

The typical PNS associated with Ri antibodies, which are much less common than Hu antibodies, is opsoclonus-myoclonus syndrome (OMS). OMS is a rare syndrome characterized by multidirectional, conjugate, erratic, rapid movements of the eyes (opsoclonus), associated with myoclonic jerks and trunk ataxia. A study on 28 patients with anti-Ri PNS described additional symptoms, including brainstem dysfunction, apart from OMS, such as dysphagia, laryngospasm, ophthalmoplegia, and cranial nerves palsies; myelopathy; movement disorders (cervical and jaw-opening dystonia); or seizures [22]. Patients manifesting with stiff person syndrome have also been reported [48].

OMS has been typically associated with neuroblastoma in children, whose serum is most often negative or in some cases harbor Hu or Yo antibodies. By contrast, Ri antibodies are found almost exclusively in adults and are associated with other tumors such as breast cancer, lung cancer, and others. Ri antibodies frequently co-occur with other onconeuronal antibodies (such as Hu, ANNA-3, CV2/CRMP-5, and P/Q voltage-gated calcium channel [VGCC]) [22]. Unlike anti-Hu PNS, tumor treatment, with or without immunotherapy, often results in neurological improvement and decrease of antibodies [22, 49]. In few cases in whom no tumor is found, OMS may remit spontaneously. This different response to therapies of anti-Hu compared to anti-Ri PNS might be related to irreversible (versus reversible) effects on neurons. Indeed, in vitro studies have shown that, although both Hu and Ri antibodies are internalized by live slice-cultured neurons, only Hu antibodies are associated with neuronal death, whereas Ri antibodies seem to cause reversible neuronal dysfunction without affecting cell survival [50].

Ri antibodies react with the nuclei of neurons within the central, but not peripheral, nervous system [51]. Their target proteins (Nova-1 and Nova-2) are highly expressed in the brainstem, spinal cord, and cortex and also by several tumors (SCLC, breast, ovarian, and lymphoma) [52–55]. These proteins modulate inhibitory synaptic receptors (such as gamma-aminobutyric acid $[GABA]_A$ or glycine receptors) and seem involved in long-term potentiation and motor responses [56].

### Ma2 (and Ma1) Antibodies

PNS associated with Ma2 antibodies typically occur in young previously healthy men who develop limbic encephalitis, diencephalic encephalitis with hypothalamic dysfunction (excessive daytime sleepiness, narcolepsy-cataplexy episodes), basal ganglia (chorea, parkinsonism), or brainstem encephalitis, either isolated or in combination. Brainstem dysfunction includes prominent eye movement abnormalities (occurring in more than 90% of the patients) ranging from focal oculomotor palsies or vertical gaze palsy to complete external ophthalmoplegia. Brain MRI is abnormal in more than 70% of patients, showing T2/FLAIR (fluid-attenuated inversion recovery) hyperintensities in the hippocampus, hypothalamus, or brainstem [3].

Tumor is found in 90% of patients, half of them having a testicular germ cell tumor. If no testicular tumor (or other Ma2-expressing cancer) is found, elective orchiectomy should be considered in patients younger than 50 years, with clinical and MRI findings suggestive of Ma2 encephalitis, confirmed Ma2 antibodies, and who show progressive neurological symptoms, new testicular enlargement, or risk factors for testicular tumor, such as cryptorchidism or testicular microcalcifications [11].

Ma2 antibodies target proteins that are highly expressed in the CNS and in testicular germ cells. These antibodies are usually found in both serum and CSF of patients, especially those manifesting with limbic encephalitis. Some patients may harbor additional Ma1 antibodies and are more likely to develop cerebellar symptoms and to have an underlying tumor other than testicular, including lymphoma [3].

Tumor treatment may lead to neurological improvement in a proportion of patients. In a study of 38 patients with anti-Ma2 encephalitis, cancer treatment or immunotherapy resulted in symptom improvement or stabilization in 60% of them (few patients showing a complete recovery) [3]. This study found that age younger than 45, male gender, complete (testicular) tumor response, and absence of additional Ma1 antibodies were factors associated with favorable outcome.

### Other Neuronal (and Glial) Nuclear Antibodies

Antibodies targeting zinc-finger protein 4 (ZIC4) are found in patients with SCLC and PNS [57]. In most patients these antibodies coexist with Hu or CV2/CRMP5 antibodies and are associated with encephalomyelitis. However, when they occur in isolation, ZIC4 antibodies are associated with prominent cerebellar dysfunction [23].

SOX1 antibodies are onconeuronal antibodies that are highly predictive for PNS associated with SCLC. They are also known as AGNA-1 (anti-glial/neuronal nuclear antibody type 1) because they not only react with nuclei of neurons but also with glial cells. SOX1 antibodies most frequently identify patients with SCLC and LEMS but can also associate with other neurological syndromes including limbic encephalitis, cerebellar ataxia, sensory neuronopathy, and OMS. Also SOX2 antibodies have been identified in patients with similar neurological syndromes and SCLC [58]. Importantly, SOX1 (or SOX2) antibodies are not found in patients with LEMS or other PNS without cancer [59]. Thus, SOX1 seropositivity can be helpful in identifying patients at risk for SCLC, and, if no cancer is apparent at initial workup, it should prompt search for occult SCLC.

## Neuronal Anticytoplasmic Antibodies

### Yo (PCA-1) Antibodies

Yo antibodies are one of the most common and characteristic onconeuronal antibodies. The typical anti-Yo PNS is a rapidly progressive pancerebellar syndrome occurring in middle-aged previously healthy women (see *Clinical Case 1*). Cerebellar symptoms usually start as mild difficulties in walking on irregular floors or in high heels and evolve over days or weeks to inability to walk or stand unassisted, with many patients becoming severely disabled (mRS > 3) within the first 3 months of the disease [24]. Some patients may show additional neurological symptoms including dysphagia, bilateral facial palsy, and movement or motor disorders. Brain MRI is usually normal at diagnosis but shows cerebellar atrophy as disease progresses (Fig. 16.2), correlating with dramatic loss of Purkinje cells observed in autopsies.

Anti-Yo PNS usually precede the diagnosis of breast or gynecologic tumor, which is found in more than 90% of female cases. Occurrence of anti-Yo PNS is exceptional in men, in whom it has been associated with gastrointestinal tumors [60]. Unlike Hu antibodies, which are found in a significant proportion of patients with SCLC and no PNS, Yo antibodies are only rarely found (<2%) in patients with gynecologic tumors without neurological syndrome [61]. Yo antibodies usually occur in isolation and do not coexist with other onconeuronal antibodies [62].

Yo antibodies target cerebellar degeneration-related antigen 2 (CDR2) and its paralog CDR2L, which is involved in c-Myc-dependent regulation of cell cycle. These antibodies intensively react with the cytoplasm of Purkinje cells, visualized as a granular staining, and also with neurons of the deep cerebellar nuclei.

Treatment of the underlying tumor may result in stabilization of the neurological syndrome. Adding immunotherapy has been shown to decrease the levels of Yo antibodies in serum, but not CSF [35], and to improve neurological symptoms in some cases [62–64], especially if started early after disease onset, when patients are not yet severely disabled [29]. However, a series of 34 patients with long-term follow-up failed to demonstrate any beneficial effect of immunotherapy on survival or neurological outcome [24]. In this study, patients younger than 60 years and who had breast cancer had longer survival compared to older women with gynecologic cancer (8 years versus less than 2 years). The cause of death was related to disabling PNS in a quarter of the patients and to tumor progression in half of them.

### Tr (DNER) Antibodies

Tr antibodies, also known as DNER, have been associated with paraneoplastic cerebellar degeneration and Hodgkin's lymphoma and are not found in patients with PNS without Hodgkin's lymphoma or with Hodgkin's lymphoma without PNS [26]. These antibodies react with Purkinje cells of rat, but, unlike Yo antibodies, they show a characteristic punctate/dotted staining, which also involve the molecular layer of the cerebellum, but not the deep cerebellar nuclei. They are generally found in both serum and CSF, although few patients have been reported to harbor only CSF antibodies [65]. The target antigen of these antibodies is unknown.

Tumor treatment results in antibody titer decrease and stabilization of cerebellar symptoms, which most often remain disabling [66, 67]. Immunotherapy, and in particular plasmapheresis, may lead to neurological improvement in patients who did not respond to chemotherapy and corticosteroids or immunoglobulins [68].

### PCA-2 (MAP1B) Antibodies

Most patients with PCA-2 antibodies develop limbic encephalitis, cerebellar degeneration, sensorimotor neuronopathy, dysautonomia, or LEMS, generally associated with SCLC. PCA-2 antibodies react with Purkinje neurons, dentate cerebellar nucleus, and enteric neurons with a reticular pattern, which, unlike Yo antibodies, extends to dendrites. PCA-2 antibodies often co-occur with other neuronal antibodies, including P/Q-type voltage-gated calcium channels (VGCC), Hu, CV2/CRMP5, and others [25, 69]. PCA-2 antibodies have been recently recognized to target microtubule-associated protein-1-B (MAP1B) [25], which is thought to be involved in neuronal development and differentiation, including dendritic spine formation and synaptic maturation. Treatment of SCLC with or without immunotherapy has been reported to stabilize or, in some cases, improve the neurological symptoms [25].

### CV2 (CRMP5) Antibodies

CV2 antibodies, also known as CRMP5, have been associated with several neurological paraneoplastic manifestations involving the central and peripheral nervous system. These include limbic encephalitis, cerebellar degeneration, myelitis, peripheral neuropathy, dysautonomia, and chorea [70]. However, the most characteristic finding is ocular involvement, most frequently optic neuropathy, but also retinitis, uveitis, or vitreitis. These PNS occur in both women and men, and the most frequent associated tumors are SCLC and malignant thymoma, although other tumors have also been reported. In patients showing CNS involvement, CV2/CRMP5 antibodies are found in higher titer in CSF compared to serum [71]. At low titers they can be identified in the serum of patients with tumor but no PNS (5% of patients with SCLC and 12% of those with thymoma), although their presence does not seem to affect tumor outcome [72].

These antibodies react with glial cells and neurons of the neocortex, cerebellum, and optic nerve, and they target one of the CRMP proteins.

Anti-CV2/CRMP5 PNS can respond to tumor treatment and immunosuppression and in some cases result in complete resolution of the neurological syndrome [73].

### Antibodies to Protein Kinases

PNS associated with antibodies to protein kinases have also been described, including anti-protein kinase C gamma (PKCγ[gamma]) in two patients with lung or liver adenocarcinoma and cerebellar degeneration and anti-serine/threonine kinase (BRSK2) in a patient with limbic encephalitis and SCLC [74–76].

## Neuronal Antibodies to Synaptic Targets

### Antibodies Targeting Amphiphysin and Other Synaptic Proteins

Amphiphysin is a vesicular protein highly concentrated at the synaptic terminal. Antibodies targeting amphiphysin are typically found in patients with stiff person syndrome (SPS) associated with SCLC or breast cancer [28]. This syndrome is less common than the non-paraneoplastic form associated with GAD65 antibodies (see below). Patients with both paraneoplastic and non-paraneoplastic SPS (described in detail in Chap. 30) develop progressive symmetric muscle rigidity involving axial and proximal limbs, associated with painful spasms. Muscle stiffness is usually symmetric, although cases with prominent asymmetric or distal involvement or partial syndromes (e.g., "stiff limb syndrome") have also been reported. Compared to GAD65 antibody-related SPS, patients harboring amphiphysin antibodies are more likely to be female, to experience early severe pain, to have distal muscle or cervical involvement, and to be refractory to spasmolytic treatment [77].

Amphiphysin antibodies can coexist with other onconeuronal antibodies (CV2/CRMP5 and P/Q VGCC) and have also been reported in patients manifesting with paraneoplastic neuronopathy, myelitis, encephalomyelitis, and cerebellar syndrome [28, 78].

Symptomatic treatment of SPS includes high-dose benzodiazepines or other GABA-enhancing drugs (such as valproate, vigabatrin, gabapentin, levetiracetam, tiagabine) and spasmolytic agents (such as baclofen). Partial syndromes may benefit from local botulin toxin injection. Tumor treatment and immunotherapy often result in improvement and, in some cases, resolution of neurological deficits [28].

Antibodies to other synaptic proteins have been reported in few patients and include: gephyrin, associated with SPS and mediastinal carcinoma; synaptophysin, associated with sensory-motor and autonomic neuronopathy and SCLC; and synaptotagmin, associated with LEMS and SCLC [79–81].

## Antibodies Associated with Visual System PNS

PNS may affect all segments of the visual system, especially the retina and optic nerves. The most common PNS affecting the retina is cancer-associated retinopathy (CAR). CAR occurs with a variety of tumors (most frequently SCLC and gynecologic cancers) and is associated with antibodies targeting the photoreceptor protein recoverin and less frequently also several other targets (alpha-enolase, rhodopsin, etc.). Patients with CAR develop painless visual loss over days or weeks, affecting both eyes, usually asymmetrically, and are accompanied by photosensitivity, night blindness, photopsia, loss of color vision, and scotomas, resulting from damage to both rods and cones [82]. Recoverin antibody-associated CAR usually precedes the diagnosis of cancer, which is not the case for CAR associated with other retinal antibodies or for melanoma-associated retinopathy (MAR), which is more likely to occur in patients with known melanoma. Compared to patients with CAR, those with MAR show less severe visual loss and can develop exudative retinal detachment [83]. MAR is associated with antibodies against bipolar cells, which are more likely in patients with advanced stage of melanoma. Tumor treatment and immunotherapies have a little effect on both CAR and MAR.

## Antibodies Associated with Neurological Symptoms That Are Rarely Paraneoplastic

Although most neuronal antibodies targeting intracellular antigens are associated with paraneoplastic neurological syndromes, few of them are found in patients with neurological manifestations that are only rarely accompanied by cancer. This is the case of antibodies targeting GAD65 or adenylate kinase 5 (AK5).

### GAD65 Antibodies

GAD65 is an enzyme that coverts glutamate into GABA. It is abundantly expressed in neurons of the CNS and in pancreatic islet cells. GAD65 antibodies have been found in patients with autoimmune type I diabetes, as well as patients with autoimmune neurological syndromes [84], including stiff person syndrome, cerebellar ataxia, limbic encephalitis, and autoimmune epilepsy [85]. In patients with autoimmune neurological manifestations, GAD65 antibodies are found in both serum and CSF (with evidence of intrathecal synthesis) and at 100–1000 times higher titers than in patients with autoimmune diabetes, in whom these antibodies are usually found only in serum [86, 87].

Less than 5% of the patients with GAD65 antibody-associated neurological syndromes, and in particular stiff person syndrome, have an underlying cancer.

Unlike the aforementioned onconeuronal antigens, which are intracellular and thus inaccessible to circulating antibodies, synaptic targets such as GAD65 (or amphiphysin) might be accessible to circulating antibodies during synaptic vesicle fusion and endocytosis. This suggests that GAD65 antibodies and amphiphysin antibodies might play a pathogenic role, as supported by experimental evidence of both antibodies affecting neuronal function in vitro and amphiphysin antibodies causing neurological symptoms in animal models [88].

### AK5 Antibodies

Antibodies to AK5 were firstly identified in the serum and CSF of two patients without cancer who developed limbic encephalitis not responsive to immunotherapy [89]. A recent study confirmed these findings in ten patients [90], most of whom showed inflammatory CSF with elevated tau protein levels, indicating neuronal death. Immunotherapy was ineffective in all but one patient, and in most cases the disease evolved to severe cognitive dysfunction, associated in some patients with hippocampal atrophy.

## Conclusion

PNS manifest with several clinically defined neurological syndromes associated with cancer. Recognition of the associated neuronal antibody is important because it can orientate the search for an underlying tumor and has prognostic implications, some antibodies (e.g., Ri- or CV2/CRMP5-associated PNS) being associated with better outcomes than others (e.g., Hu or Yo antibodies). Although most PNS have a poor response to immunotherapy and cancer treatment, early diagnosis and treatment can allow stabilization of the symptoms and prevent progression of neurological disability.

**Acknowledgments** The author kindly thanks Dr. Francesc Graus, Hospital Clinic of Barcelona, for his critical revision of the text and his help for the figures.

## References

1. Titulaer MJ, Klooster R, Potman M, Sabater L, Graus F, Hegeman IM, et al. SOX antibodies in small-cell lung cancer and Lambert-Eaton myasthenic syndrome: frequency and relation with survival. J Clin Oncol. 2009;27:4260–7.
2. Rojas-Marcos I, Rousseau A, Keime-Guibert F, Reñé R, Cartalat-Carel S, Delattre JY, et al. Spectrum of paraneoplastic neurologic disorders in women with breast and gynecologic cancer. Medicine (Baltimore). 2003;82:216–23.

3. Dalmau J, Graus F, Villarejo A, Posner JB, Blumenthal D, Thiessen B, et al. Clinical analysis of anti-Ma2-associated encephalitis. Brain. 2004;127:1831–44.
4. Yu Z, Kryzer TJ, Griesmann GE, Kim K, Benarroch EE, Lennon VA. CRMP-5 neuronal autoantibody: marker of lung cancer and thymoma-related autoimmunity. Ann Neurol. 2001;49:146–54.
5. Graus F, Keime-Guibert F, Reñe R, Benyahia B, Ribalta T, Ascaso C, et al. Anti-Hu-associated paraneoplastic encephalomyelitis: analysis of 200 patients. Brain. 2001;124:1138–48.
6. Graus F, Delattre JY, Antoine JC, Dalmau J, Giometto B, Grisold W, et al. Recommended diagnostic criteria for paraneoplastic neurological syndromes. J Neurol Neurosurg Psychiatry. 2004;75:1135–40.
7. Ances BM, Vitaliani R, Taylor RA, Liebeskind DS, Voloschin A, Houghton DJ, et al. Treatment-responsive limbic encephalitis identified by neuropil antibodies: MRI and PET correlates. Brain. 2005;128:1764–77.
8. Miyamoto K, Kato T, Watanabe H, Miyamoto E, Suzuki S. A case of paraneoplastic syndrome accompanied by two types of cancer. J Neurol Neurosurg Psychiatry. 2002;72:408–9.
9. Antoine JC, Cinotti L, Tilikete C, Bouhour F, Camdessanché JP, Confavreux C, et al. [18F]fluorodeoxyglucose positron emission tomography in the diagnosis of cancer in patients with paraneoplastic neurological syndrome and anti-Hu antibodies. Ann Neurol. 2000;48:105–8.
10. Titulaer MJ, Soffietti R, Dalmau J, Gilhus NE, Giometto B, Graus F, et al. European Federation of Neurological Societies. Screening for tumours in paraneoplastic syndromes: report of an EFNS Task Force. Eur J Neurol. 2011;18:19–27.
11. Mathew RM, Vandenberghe R, Garcia-Merino A, Yamamoto T, Landolfi JC, Rosenfeld MR, et al. Orchiectomy for suspected microscopic tumor in patients with anti-Ma2-associated encephalitis. Neurology. 2007;68:900–5.
12. Albert ML, Jegathesan M, Darnell RB. Dendritic cell maturation is required for the cross-tolerization of CD8+ T cells. Nat Immunol. 2001;2:1010–7.
13. Blumenthal DT, Salzman KL, Digre KB, Jensen RL, Dunson WA, Dalmau J. Early pathologic findings and long-term improvement in anti-Ma2-associated encephalitis. Neurology. 2006;67:146–9.
14. Giometto B, Marchiori GC, Nicolao P, Scaravilli T, Lion A, Bardin PG, et al. Sub-acute cerebellar degeneration with anti-Yo autoantibodies: Immunohistochemical analysis of the immune reaction in the central nervous system. Neuropathol Appl Neurobiol. 1997;23:468–74.
15. Bernal F, Graus F, Pifarré A, Saiz A, Benyahia B, Ribalta T. Immunohistochemical analysis of anti-Hu-associated paraneoplastic encephalomyelitis. Acta Neuropathol. 2002;103:509–15.
16. Dalmau J, Furneaux HM, Rosenblum MK, Graus F, Posner JB. Detection of the anti-Hu antibody in specific regions of the nervous system and tumor from patients with paraneoplastic encephalomyelitis/sensory neuronopathy. Neurology. 1991;41:1757–64.
17. Manz RA, Löhning M, Cassese G, Thiel A, Radbruch A. Survival of long-lived plasma cells is independent of antigen. Int Immunol. 1998;10:1703–11.
18. Hormigo A, Ball C, Lieberman F. Internalization of anti-Hu IgG is not Fc gamma receptor mediated. Neurology. 1996;47:607–9.
19. Hormigo A, Lieberman F. Nuclear localization of anti-Hu antibody is not associated with in vitro cytotoxicity. J Neuroimmunol. 1994;55:205–12.
20. Tanaka M, Tanaka K, Onodera O, Tsuji S. Trial to establish an animal model of paraneoplastic cerebellar degeneration with anti-Yo antibody. 1. Mouse strains bearing different MHC molecules produce antibodies on immunization with recombinant Yo protein, but do not cause Purkinje cell loss. Clin Neurol Neurosurg. 1995;97:95–100.
21. Sillevis Smitt PA, Manley GT, Posner JB. Immunization with the paraneoplastic encephalomyelitis antigen HuD does not cause neurologic disease in mice. Neurology. 1995;45:1873–8.
22. Pittock SJ, Lucchinetti CF, Lennon VA. Anti-neuronal nuclear autoantibody type 2: Paraneoplastic accompaniments. Ann Neurol. 2003;53:580–7.
23. Bataller L, Wade DF, Graus F, Stacey HD, Rosenfeld MR, Dalmau J. Antibodies to Zic4 in paraneoplastic neurologic disorders and small-cell lung cancer. Neurology. 2004;62:778–82.
24. Rojas I, Graus F, Keime-Guibert F, Reñé R, Delattre JY, Ramón JM, et al. Long-term clinical outcome of paraneoplastic cerebellar degeneration and anti-Yo antibodies. Neurology. 2000;55:713–5.
25. Gadoth A, Kryzer TJ, Fryer J, McKeon A, Lennon VA, Pittock SJ. Microtubule-associated protein 1B: novel paraneoplastic biomarker. Ann Neurol. 2017;81:266–77.
26. Graus F, Dalmau J, Valldeoriola F, Ferrer I, Reñe R, Marin C, et al. Immunological characterization of a neuronal antibody (anti-Tr) associated with paraneoplastic cerebellar degeneration and Hodgkin's disease. J Neuroimmunol. 1997;74:55–61.
27. Rogemond V, Honnorat J. Anti-CV2 autoantibodies and paraneoplastic neurological syndromes. Clin Rev Allergy Immunol. 2000;19:51–9.
28. Pittock SJ, Lucchinetti CF, Parisi JE, Benarroch EE, Mokri B, Stephan CL, et al. Amphiphysin autoimmunity: paraneoplastic accompaniments. Ann Neurol. 2005;58:96–107.
29. Keime-Guibert F, Graus F, Fleury A, René R, Honnorat J, Broet P, et al. Treatment of paraneoplastic neurological syndromes with antineuronal antibodies (Anti-Hu, Anti-Yo) with a combination of immunoglobulins, cyclophosphamide, and methylprednisolone. J Neurol Neurosurg Psychiatry. 2000;68:479–82.
30. Vernino S, O'Neill BP, Marks RS, O'Fallon JR, Kimmel DW. Immunomodulatory treatment trial for paraneoplastic neurological disorders. Neuro-Oncology. 2004;6:55–62.

31. Fadul CE, Stommel EW, Dragnev KH, Eskey CJ, Dalmau JO. Focal paraneoplastic limbic encephalitis presenting as orgasmic epilepsy. J Neuro-Oncol. 2005;72:195–8.
32. Lennon VA. Paraneoplastic autoantibodies: the case for a descriptive generic nomenclature. Neurology. 1994;44:2236–40.
33. Jean WC, Dalmau J, Ho A, Posner JB. Analysis of the IgG subclass distribution and inflammatory infiltrates in patients with anti-Hu-associated paraneoplastic encephalomyelitis. Neurology. 1994;44:140–7.
34. Darnell RB. Onconeural antigens and the paraneoplastic neurologic disorders: at the intersection of cancer, immunity, and the brain. Proc Natl Acad Sci. 1996;93:4529–36.
35. Furneaux HF, Reich L, Posner JB. Autoantibody synthesis in the central nervous system of patients with paraneoplastic syndromes. Neurology. 1990;40:1085–91.
36. Vega F, Graus F, Chen QM, Poisson M, Schuller E, Delattre JY. Intrathecal synthesis of the anti-Hu antibody in patients with paraneoplastic encephalomyelitis or sensory neuronopathy: clinical-immunologic correlation. Neurology. 1994;44:2145–7.
37. Lladó A, Mannucci P, Carpentier AF, Paris S, Blanco Y, Saiz A, et al. Value of Hu antibody determinations in the follow-up of paraneoplastic neurologic syndromes. Neurology. 2004;63:1947–9.
38. Dalmau J, Furneaux HM, Gralla RJ, Kris MG, Posner JB. Detection of the anti-Hu antibody in the serum of patients with small cell lung cancer? A quantitative western blot analysis. Ann Neurol. 1990;27:544–52.
39. Lucchinetti CF, Kimmel DW, Lennon VA. Paraneoplastic and oncologic profiles of patients seropositive for type 1 antineuronal nuclear autoantibodies. Neurology. 1998;50:652–7.
40. Dalmau J, Graus F, Rosenblum MK, Posner JB. Anti-Hu--associated paraneoplastic encephalomyelitis/sensory neuronopathy. A clinical study of 71 patients. Medicine (Baltimore). 1992;71(2):59–72.
41. Keime-Guibert F, Graus F, Broët P, Reñé R, Molinuevo JL, Ascaso C, et al. Clinical outcome of patients with anti-Hu-associated encephalomyelitis after treatment of the tumor. Neurology. 1999;53:1719–23.
42. Torà M, Graus F, de Bolòs C, Real FX. Cell surface expression of paraneoplastic encephalomyelitis/sensory neuronopathy-associated Hu antigens in small-cell lung cancers and neuroblastomas. Neurology. 1997;48:735–41.
43. Zhu H, Hasman RA, Barron VA, Luo G, Lou H. A nuclear function of Hu proteins as neuron-specific alternative RNA processing regulators. Mol Biol Cell. 2006;17(12):5105–14.
44. Myer VE, Fan XC, Steitz JA. Identification of HuR as a protein implicated in AUUUA-mediated mRNA decay. EMBO J. 1997;16:2130–9.
45. Deschênes-Furry J, Perrone-Bizzozero N, Jasmin BJ. The RNA-binding protein HuD: a regulator of neuronal differentiation, maintenance and plasticity. BioEssays. 2006;28:822–33.
46. Bolognani F, Qiu S, Tanner DC, Paik J, Perrone-Bizzozero NI, Weeber EJ. Associative and spatial learning and memory deficits in transgenic mice overexpressing the RNA-binding protein HuD. Neurobiol Learn Mem. 2007;87:635–43.
47. Tanner DC, Qiu S, Bolognani F, Partridge LD, Weeber EJ, Perrone-Bizzozero NI. Alterations in mossy fiber physiology and GAP-43 expression and function in transgenic mice overexpressing HuD. Hippocampus. 2008;18:814–23.
48. de Albóniga-Chindurza A, Riva E, Jiménez-Huete A, Graus F, Franch O. Paraneoplastic stiff person syndrome with small cell carcinoma of the bladder and anti-Ri antibodies. Clin Neurol Neurosurg. 2018;173:194–5.
49. Leypoldt F, Eichhorn P, Saager C, Münchau A, Lewerenz J. Successful immunosuppressive treatment and long-term follow-up of anti-Ri-associated paraneoplastic myelitis. J Neurol Neurosurg Psychiatry. 2006;77:1199–200.
50. Greenlee JE, Clawson SA, Hill KE, Wood B, Clardy SL, Tsunoda I, et al. Neuronal uptake of anti-Hu antibody, but not anti-Ri antibody, leads to cell death in brain slice cultures. J Neuroinflammation. 2014;11:160.
51. Hormigo A, Dalmau J, Rosenblum MK, River ME, Posner JB. Immunological and pathological study of anti-Ri-associated encephalopathy. Ann Neurol. 1994;36:896–902.
52. Relógio A, Ben-Dov C, Baum M, Ruggiu M, Gemund C, Benes V, et al. Alternative splicing microarrays reveal functional expression of neuron-specific regulators in Hodgkin lymphoma cells. J Biol Chem. 2005;280:4779–84.
53. Luque FA, Furneaux HM, Ferziger R, Rosenblum MK, Wray SH, Schold SC Jr, et al. Anti-Ri: an antibody associated with paraneoplastic opsoclonus and breast cancer. Ann Neurol. 1991;29:241–51.
54. Buckanovich RJ, Posner JB, Darnell RB. Nova, the paraneoplastic Ri antigen, is homologous to an RNA-binding protein and is specifically expressed in the developing motor system. Neuron. 1993;11:657–72.
55. Knudsen A, Monstad SE, Dørum A, Lønning PE, Salvesen HB, Drivsholm L, et al. Ri antibodies in patients with breast, ovarian or small cell lung cancer determined by a sensitive immunoprecipitation technique. Cancer Immunol Immunother. 2006;55:1280–4.
56. Dredge BK, Stefani G, Engelhard CC, Darnell RB. Nova autoregulation reveals dual functions in neuronal splicing. EMBO J. 2005;24:1608–20.
57. Bataller L, Wade DF, Fuller GN, Rosenfeld MR, Dalmau J. Cerebellar degeneration and autoimmunity to zinc-finger proteins of the cerebellum. Neurology. 2002;59:1985–7.
58. Maddison P, Titulaer MJ, Verschuuren JJ, Gozzard P, Lang B, Irani SR, et al. The utility of anti-SOX2 antibodies for cancer prediction in patients with paraneoplastic neurological disorders. J Neuroimmunol. 2018;326:14–8.

59. Graus F, Vincent A, Pozo-Rosich P, Sabater L, Saiz A, Lang B, et al. Anti-glial nuclear antibody: marker of lung cancer-related paraneoplastic neurological syndromes. J Neuroimmunol. 2005;165:166–71.
60. Linnoila J, Guo Y, Gadoth A, Raghunathan A, Parks B, McKeon A, et al. Purkinje cell cytoplasmic antibody type I (anti-Yo): predictive of gastrointestinal adenocarcinomas in men. J Neurol Neurosurg Psychiatry. 2018;89:1116–7.
61. Drlicek M, Bianchi G, Bogliun G, Casati B, Grisold W, Kolig C, et al. Antibodies of the anti-Yo and anti-Ri type in the absence of paraneoplastic neurological syndromes: a long-term survey of ovarian cancer patients. J Neurol. 1997;244:85–9.
62. Pittock SJ, Kryzer TJ, Lennon VA. Paraneoplastic antibodies coexist and predict cancer, not neurological syndrome. Ann Neurol. 2004;56:715–9.
63. Phuphanich S, Brock C. Neurologic improvement after high-dose intravenous immunoglobulin therapy in patients with paraneoplastic cerebellar degeneration associated with anti-Purkinje cell antibody. J Neuro-Oncol. 2007;81:67–9.
64. Thöne J, Hohaus A, Lamprecht S, Bickel A, Erbguth F. Effective immunosuppressant therapy with cyclophosphamide and corticosteroids in paraneoplastic cerebellar degeneration. J Neurol Sci. 2008;272:171–3.
65. Bernal F, Shams'ili S, Rojas I, Sanchez-Valle R, Saiz A, Dalmau J, et al. Anti-Tr antibodies as markers of paraneoplastic cerebellar degeneration and Hodgkin's disease. Neurology. 2003;60:230–4.
66. Ypma PF, Wijermans PW, Koppen H, Sillevis Smitt PA. Paraneoplastic cerebellar degeneration preceding the diagnosis of Hodgkin's lymphoma. Neth J Med. 2006;64:243–7.
67. Geromin A, Candoni A, Marcon G, Ferrari S, Sperotto A, De Luca S, et al. Paraneoplastic cerebellar degeneration associated with anti-neuronal anti-Tr antibodies in a patient with Hodgkin's disease. Leuk Lymphoma. 2006;47(9):1960–3.
68. Taniguchi Y, Tanji C, Kawai T, Saito H, Marubayashi S, Yorioka N. A case report of plasmapheresis in paraneoplastic cerebellar ataxia associated with anti-Tr antibody. Ther Apher Dial. 2006;10:90–3.
69. Vernino S, Lennon VA. New Purkinje cell antibody (PCA-2): marker of lung cancer-related neurological autoimmunity. Ann Neurol. 2000;47:297–305.
70. Antoine JC, Honnorat J, Vocanson C, Koenig F, Aguera M, Belin MF, et al. Posterior uveitis, paraneoplastic encephalomyelitis and auto-antibodies reacting with developmental protein of brain and retina. J Neurol Sci. 1993;117:215–23.
71. Stich O, Jarius S, Kleer B, Rasiah C, Voltz R, Rauer S. Specific antibody index in cerebrospinal fluid from patients with central and peripheral paraneoplastic neurological syndromes. J Neuroimmunol. 2007;183:220–4.
72. Monstad SE, Drivsholm L, Skeie GO, Aarseth JH, Vedeler CA. CRMP5 antibodies in patients with small-cell lung cancer or thymoma. Cancer Immunol Immunother. 2008;57:227–32.
73. de la Sayette V, Bertran F, Honnorat J, Schaeffer S, Iglesias S, Defer G. Paraneoplastic cerebellar syndrome and optic neuritis with anti-CV2 antibodies: clinical response to excision of the primary tumor. Arch Neurol. 1998;55:405–8.
74. Sabater L, Gómez-Choco M, Saiz A, Graus F. BR serine/threonine kinase 2: a new autoantigen in paraneoplastic limbic encephalitis. J Neuroimmunol. 2005;170:186–90.
75. Sabater L, Bataller L, Carpentier AF, Aguirre-Cruz ML, Saiz A, Benyahia B, et al. Protein kinase C autoimmunity in paraneoplastic cerebellar degeneration and non-small-cell lung cancer. J Neurol Neurosurg Psychiatry. 2006;77:1359–62.
76. Höftberger R, Kovacs GG, Sabater L, Nagy P, Racz G, Miquel R, et al. Protein kinase Cγ antibodies and paraneoplastic cerebellar degeneration. J Neuroimmunol. 2013;256:91–3.
77. Murinson BB, Guarnaccia JB. Stiff-person syndrome with amphiphysin antibodies: distinctive features of a rare disease. Neurology. 2008;71:1955–8.
78. Moon J, Lee ST, Shin JW, Byun JI, Lim JA, Shin YW, et al. Non-stiff anti-amphiphysin syndrome: clinical manifestations and outcome after immunotherapy. J Neuroimmunol. 2014;274:209–14.
79. Takamori M, Komai K, Iwasa K. Antibodies to calcium channel and synaptotagmin in Lambert-Eaton myasthenic syndrome. Am J Med Sci. 2000;319:204–8.
80. Tschernatsch M, Klotz M, Probst C, Hosch J, Valtorta F, Diener M, et al. Synaptophysin is an autoantigen in paraneoplastic neuropathy. J Neuroimmunol. 2008;197:81–6.
81. Harada H, Hayashi A, Watanabe M, Tamaoka A, Shoji S. A case of stiff-man syndrome with head retraction like reflex myoclonus and jerky myoclonus of bilateral lower extremities which responded well to removal of mediastinal carcinoma. Rinsho Shinkeigaku. 1999;39:1025–8.
82. Thirkill CE, Keltner JL, Tyler NK, Roth AM. Antibody reactions with retina and cancer-associated antigens in 10 patients with cancer-associated retinopathy. Arch Ophthalmol. 1993;111:931–7.
83. Keltner JL, Thirkill CE, Yip PT. Clinical and immunologic characteristics of melanoma-associated retinopathy syndrome: eleven new cases and a review of 51 previously published cases. J Neuroophthalmol. 2001;21:173–87.
84. Solimena M, Folli F, Denis-Donini S, Comi GC, Pozza G, De Camilli P, et al. Autoantibodies to glutamic acid decarboxylase in a patient with stiff-man syndrome, epilepsy, and type I diabetes mellitus. N Engl J Med. 1988;318:1012–20.
85. Saiz A, Blanco Y, Sabater L, González F, Bataller L, Casamitjana R, et al. Spectrum of neurological syndromes associated with glutamic acid decarboxylase antibodies: diagnostic clues for this association. Brain. 2008;131:2553–63.

86. Honnorat J, Saiz A, Giometto B, Vincent A, Brieva L, de Andres C, et al. Cerebellar ataxia with anti-glutamic acid decarboxylase antibodies: study of 14 patients. Arch Neurol. 2001;58:225–30.
87. Taplin CE, Barker JM. Autoantibodies in type 1 diabetes. Autoimmunity. 2008;41:11–8.
88. Sommer C, Weishaupt A, Brinkhoff J, Biko L, Wessig C, Gold R, et al. Paraneoplastic stiff-person syndrome: passive transfer to rats by means of IgG antibodies to amphiphysin. Lancet. 2005;365:1406–11.
89. Tüzün E, Rossi JE, Karner SF, Centurion AF, Dalmau J. Adenylate kinase 5 autoimmunity in treatment refractory limbic encephalitis. J Neuroimmunol. 2007;186:177–80.
90. Do LD, Chanson E, Desestret V, Joubert B, Ducray F, Brugière S, et al. Characteristics in limbic encephalitis with anti-adenylate kinase 5 autoantibodies. Neurology. 2017;88:514–24.

# Treatment Approaches in Autoimmune Neurology: Focus on Autoimmune Encephalitis with Neuronal Cell Surface Antibodies

17

Juna M. de Vries and Maarten J. Titulaer

**Key Points**

1. In autoimmune encephalitis (AE), the neuronal antibodies targeted at extracellular antigens have a direct pathogenic effect, which can be successfully reversed with immunotherapy in most patients.
2. A timely diagnosis and subsequent start of treatment in patients with autoimmune encephalitis are important for good clinical outcome, as the rule is "time saves brain."
3. The frequency of tumors is in general lower in autoimmune encephalitis as compared to the classical paraneoplastic syndromes.
4. The decision to start second-line immunotherapy or maintenance immunotherapy should be based upon antibody type, the severity of disease, and the risk for relapse.
5. Multicenter clinical trials are needed to investigate whether the available alternative treatments have an additional treatment effect as compared to standard care alone.
6. In the future, it is desirable that individualized treatment strategies will be developed based on prognostic models, which take into account both the variations in severity of disease between AE patients and the individualized risk for adverse events.

J. M. de Vries · M. J. Titulaer (✉)
Academic Center for Neuroinflammatory Disorders, Department of Neurology, Erasmus MC University Medical Center, Rotterdam, The Netherlands
e-mail: m.titulaer@erasmusmc.nl

## Introduction

Since the discovery of the anti-N-methyl-D-aspartate receptor (NMDAR) antibodies in 2007 [1], multiple other cell surface or synaptic antigens have been detected in patients previously not identified as suffering from a neurological autoimmune disease. Anti-NMDAR encephalitis is still recognized as the most prevalent subtype of autoimmune encephalitis (AE), but over the last years, more and more autoantibodies directed at neuronal cell surface antigens have been identified and recognized as being responsible for specific neurological phenotypes, for instance, the anti-leucine-rich glioma-inactivated protein 1 (anti-LGI1), anti-contactin-associated protein-


A. L. Piquet, E. Alvarez (eds.), *Neuroimmunology*, https://doi.org/10.1007/978-3-030-61883-4_17

like 2 (Caspr2), or anti-alpha-amino-3-hydroxy--5-methyl-4-isoxazolepropionic acid receptor (AMPAR) antibodies [2–4]. This new entity of antibody-mediated neurological diseases has been one of the most exciting recent evolutions in clinical neurology. Within this group of AE, in contrast to the classical paraneoplastic syndromes (PNS; further discussed in Chap. 16), the circulating neuronal antibodies are aimed at extracellular proteins. These novel antibodies can reach their target protein in the absence of cell destruction and influence the antigen function or cause antigen internalization, thereby having a directly pathogenic effect [5–7]. By counteracting the autoimmune response with immunosuppressive drugs, most patients, even those who are severely affected by the disease, make a substantial improvement [1, 8–10]. This is in contrast with earlier unsatisfactory experience with treatments of patients with PNS related to onconeural antibodies directed to intracellular proteins such as Hu, Ri, and Yo [11, 12]. The antibody responses in these patients are regarded as an epiphenomena. The occurrence of irreversible neuronal damage in these patients is presumed to be caused by a cytotoxic T-cell-mediated immune response, explaining the overall poor response to immunotherapy. Furthermore, the strong association to malignant tumors adds to the eventual poor prognosis in these patients.

In AE caused by antibodies targeted at cell surface antigens, the incidence of tumors is lower as observed in PNS and varies per type of antibody. Some are more commonly associated with tumors, whereas in others, tumor rates are comparable to the general population [13]. All patients should be screened at least once, also given that tumor removal or chemotherapy will ameliorate the neurological outcome [9]. In this chapter, it is specified for each antibody how to screen for malignancies, including frequency and time frame (see next section "Tumor Association and Screening in Autoimmune Encephalitis" and Table 17.1 [14]).

There is no doubt whether immunosuppressive treatments have beneficial effects in AE, although (randomized) clinical trials studying the effects of immunosuppressive therapy are lacking. Available evidence is mainly based on retrospective data. This is both due to the relative newness of AE and the rarity of these diseases. Most evidence has been collected for anti-NMDAR encephalitis, containing one large, partially prospective cohort study [9]. For the other AE types, mainly smaller cohorts or case series have been published [10, 15–18].

Present-day treatment guidelines are therefore based on clinical experience and on data from clinical trials performed in other autoimmune neurological diseases, like Guillain-Barré syndrome and myasthenia gravis, or even on treatment regimens used in other specialties such as rheumatology and hematology-oncology. Within the current treatment guidelines, a distinction is made between first-line immunotherapy, second-line immunotherapy, and maintenance immunotherapy. In the last part of the chapter, a selection of promising emerging therapies is discussed.

With regard to the safety of immunotherapy, frequently encountered adverse effects are infections, viral reactivation, bone marrow suppression, and increased risk of malignancies. All intravenously administered immunosuppressive agents, some more than others, have a risk of infusion-related adverse events. In each patient, a "risk-benefit" analysis should be made to decide what type of therapy should be used in the best interest of the patient.

As the several types of AE give rise to a myriad of symptoms, rehabilitative strategies and symptomatic therapy should be tailored to the underlying phenomenology to meet the individual patient's needs. Favorably, treatment is carried out by a multidisciplinary team, including a neurologist, rehabilitation physician, psychiatrist, and if indicated an oncologist. In cases of pediatric-onset AE, a pediatrician should be involved too.

## Tumor Association and Screening in Autoimmune Encephalitis

The frequency of tumors is in general lower in AE with neuronal cell surface antibodies as compared to the PNS associated with antibodies

**Table 17.1** Tumor association in autoimmune encephalitis with neuronal cell surface antibodies

| Antigen | Tumor association | Type of tumor screening | Follow-up |
|---|---|---|---|
| NMDAR | Rarely in females <12 years old and males up to 45 years old<br>50% ovarian teratoma in females 12–45 years old<br>25% SCLC, thymoma, breast cancer, testicular cancer, pancreatic cancer in females and males >45 years old | Males <18 years old: testicular ultrasound<br>Females <12 years old: pelvic ultrasound<br>Males 18–45 years old: ultrasound testis and CT chest/abdomen<br>Females 12–45 years old: pelvic ultrasound, pelvic MRI<br>Females and males >45 years old: CT chest/abdomen and females also mammography; in negative cases also FDG-PET | One-time screening in females <12 years old and males 0–45 years old<br>Negative initial screening in females and males >45 years old: repeat screening once<br>Negative initial screening in females 12–45 years old with no improvement: repeat second screening after 3 months<br>Negative initial screening in females 12–45 years old with improvement: repeat screening after 6 months<br>subsequently repeat screening every 12 months up to 4 years in females 12–45 years old |
| LGI1 | 5–10% thymoma, SCLC, mesothelioma | CT chest/abdomen, if negative also FDG-PET | One-time screening[a] |
| Caspr2 | < 20% thymoma, SCLC | CT chest/abdomen, if negative also FDG-PET | One-time screening, repeat once after 3–6 months[a] |
| $GABA_BR$ | SCLC (50%)[b] | CT chest/abdomen, if negative also FDG-PET | Negative initial screening and no improvement: repeat screening after 3 months<br>Negative initial screening and improvement: repeat screening after 6 months<br>Subsequently every 6 months up to 4 years |
| DPPX | 10% B-cell lymphoma, mantle cell lymphoma (CLL and GI lymphoma) | Consultation of a hematologist | Depending on the advice of the hematologist, basically one-time screening |
| AMPAR | SCLC, thymoma, breast cancer | CT chest/abdomen and females also mammography; if negative perform FDG-PET | Negative initial screening and no improvement: repeat screening after 3 months<br>Negative initial screening and improvement: repeat screening after 6 months<br>Subsequently every 6 months up to 4 years |
| GlycR | <10% SCLC, thymoma | CT chest/abdomen, if negative also FDG-PET | One-time screening[a] |
| $GABA_AR$ | Unknown | Depending on clinical suspicion | |
| IgLON5 | Unknown | Depending on clinical suspicion | |
| mGluR1/ mGluR5 | Hodgkin lymphoma | Consultation of a hematologist | Depending on the advice of the hematologist, basically one-time screening |

*Abbreviations*: *NMDAR* anti-N-methyl-D-aspartate receptor encephalitis, *SCLC* small-cell lung carcinoma, *CT* computed tomography, *MRI* magnetic resonance imaging, *FDG-PET* fluorodeoxyglucose positron emission tomography, *LGI1* anti-leucine-rich glioma-inactivated protein 1 encephalitis, *Caspr2* anti-contactin-associated protein-like 2, *$GABA_BR$* anti-gamma-aminobutyric acid B receptor encephalitis, *DPPX* dipeptidyl-peptidase-like protein-6 encephalitis, *AMPAR* anti-alpha-amino-3-hydroxy-5-methyl-4-isoxazolepropionic acid receptor, *GlycR* anti-glycine receptor encephalitis, *$GABA_AR$* anti-gamma-aminobutyric acid A receptor encephalitis, *IgLON5* anti-IgLON5 encephalitis, *mGluR1/5* anti-metabotropic glutamate receptor 1/5 encephalitis, *CLL* chronic lymphocytic leukemia, *GI* gastrointestinal

[a]With clinical deterioration or lack of improvement consider to repeat tumor screening

[b]>90% if also KCTD16 antibodies are found [14]

against extracellular antigens [19]. This is illustrated by the rather low tumor rates in anti-LGI1 encephalitis, which are comparable to the rates observed in the general population [13, 16]. However, in specific anti-LGI1 cases, tumor removal might still be essential [20]. On the other hand, in anti-NMDAR receptor encephalitis occurring in female patients of childbearing age, the risk for an underlying tumor is fairly high (~50%) [9]. Early detection and treatment of the underlying tumor is of utmost importance for clinical outcome, both increasing the chance of curative cancer treatment and offering the patient a better chance to respond to immunotherapy. Patients' Karnofsky performance scores (KPS), a measure of a patient's functional impairments, are often low due to neurological deficits caused by the associated AE. For this reason, one should not refrain patients from oncological treatment, as most neurologic deficits can be reversible with immunotherapy.

Follow-up and type of tumor screening are dependent on the type of autoantibody found. Table 17.1 shows the tumor associations based on the available literature so far and a suggested tumor screening protocol [14, 21]. In 90% of patients who have an underlying malignancy, the tumor is not yet known and found due to the PNS or AE. Initial screening is abnormal in 80–90%. Advice for continued screening is stratified for the different neuronal autoantibodies in Table 17.1. In cases with a clear relapse not explained otherwise, or lack of improvement after adequate immunotherapy, one should consider to repeat or to broaden tumor screening. Tumors are often small and/or only pathological lymph nodes are identified by imaging. Also smaller abnormalities should be considered cautiously as the pre-scan likelihood for malignancy is increased in PNS or AE. For example, chalk spots on testicular ultrasound or small ovarian cysts can be the only evidence for a paraneoplastic disease, warranting removal. However, in the opinion of the writers, there is no indication to perform a laparoscopy in patients with anti-NMDAR encephalitis with a completely normal ultrasound and MRI of the pelvis.

## Immunotherapy in Autoimmune Encephalitis

In the acute phase, immunotherapy may consist of first-line and second-line immunotherapy. First-line immunotherapy involves high-dose intravenously or orally administered steroids (glucocorticoids), immunoglobulins (IVIg), and plasmapheresis (PLEX), all treatments with a broad mechanism of action. For second-line immunotherapy rituximab (RTX), a combination of RTX and cyclophosphamide (CPA) is used, while new drugs might be added in future years.

First-line and second-line immunotherapies used in AE have a broad immune suppressive or immune modulatory effect mostly via the influence on peripheral immune cells or proteins. The used approach is fairly similar to the first- and second-line treatment regiments of other autoimmune and inflammatory disorders. In the section "First-Line Immunotherapy in Anti-N-Methyl-D-Aspartate Receptor Encephalitis," we first describe more in depth our first- and second-line treatment strategy for anti-NMDAR encephalitis as summarized in the flowchart presented in Fig. 17.1a. Subsequently, the somewhat different treatment strategy for the other AEs caused by neuronal cell surface antibodies is discussed in the section "First-Line and Second-Line Immunotherapy in Encephalitis Caused by Other Neuronal Cell Surface Antibodies." In Fig. 17.1b our treatment approach for anti-LGI1 encephalitis is presented, one of the most prevalent types of AE after anti-NMDAR encephalitis, as an example for the treatment of milder subtypes of AE. Table 17.2 comprises the various used immunosuppressive agents and the way these are prescribed, listed for adults and children.

## First-Line Immunotherapy in Anti-N-Methyl-D-Aspartate Receptor Encephalitis

In patients suffering from anti-NMDAR encephalitis, first-line immunotherapy (with tumor removal when indicated) results in considerable clinical improvement within a month in approxi-

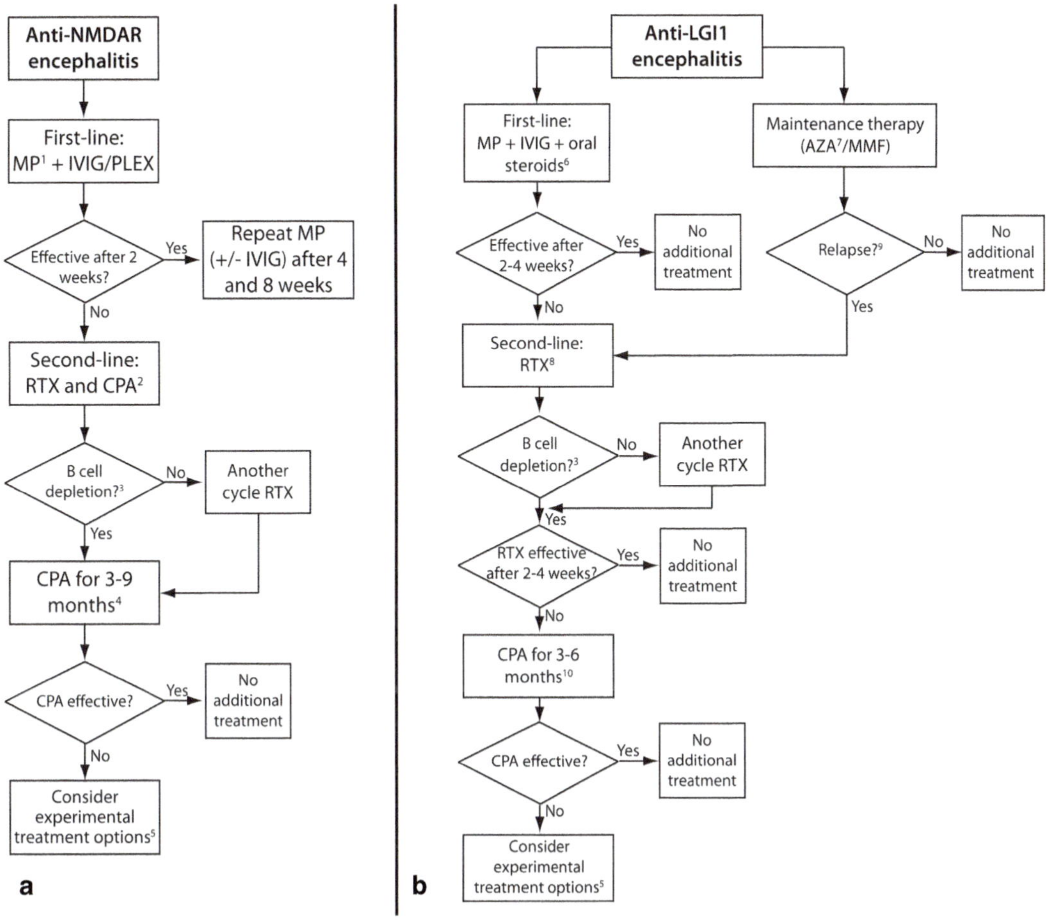


**Fig. 17.1** Flowchart with advised treatment strategies for patients' encephalitis caused by antibodies against extracellular antigens. (**a**) Treatment of anti-NMDA receptor encephalitis. (**b**) Treatment of anti-LGI1 encephalitis. In the case the clinical presentation of a patient with AE with other neuronal cell surface antibodies is severe and more like an anti-NMDAR encephalitis, it is advised to follow treatment guideline (**a**). For relatively milder presentations of AE, one can follow the treatment guideline as shown in (**b**). Depending on the relapse rate, one should or should not add maintenance therapy (discussed in chapter section "Maintenance Immunotherapy"). In patients with contraindications for chronic immunotherapy, one can decide on a watchful waiting policy. (1) Pulse therapy is preferred over oral steroids because of lower risk for neurocognitive and infectious adverse effects during the rehabilitation period and less interactions with the effect of second-line treatments. (2) In children the preference is for the use of RTX only, because of physician's experience, risk of premature gonadal failure, and long-term risk of malignancy due to CPA. (3) Two weeks after the last administration of RTX, order for B-cell analysis. When full B-cell depletion is not reached, extend the RTX treatment with another cycle. (4) CPA dosing regimens vary, but usually intravenous infusions are administrated for 4–6 months depending on clinical recovery. In extreme refractory cases of anti-NMDAR encephalitis, CPA can be given for a period of 12 months. (5) In severe refractory cases, one can consider adding one of the experimental treatments as described in chapter section "Alternative Treatment Strategies." (6) Start with prednisolone 30–40 mg/day, taper with 5 mg per 2 weeks to a daily dose of 20 mg, subsequently taper with 2.5 mg per 2 weeks to a daily dose of 10 mg/day, and subsequently taper with 2.5 mg per 4 weeks until stop. (7) When adverse effects due to AZA occur, check for a TPMT mutation. In case the patient is a poor metabolizer, one should lower the standard dose or switch to another type of maintenance therapy. (8) When RTX is administered, maintenance therapy can be stopped. (9) When a relapse occurs, one also repeats first-line treatment with IVMP and IVIg, previous to the treatment with RTX. (10) Treatment with CPA is seldom necessary in patients with anti-LGI1 encephalitis; only in cases with no to minimal clinical improvement, one should consider treatment with CPA. Cognitive symptoms take significant more time to recover as compared to the seizures; in patients with initial minimal cognitive recovery with adequate seizure control, one should not escalate treatment. Abbreviations: MP intravenous methylprednisolone, IVIg intravenous immunoglobulins, PLEX plasmapheresis, RTX rituximab, CPA cyclophosphamide, AZA azathioprine, MMF mycophenolate mofetil

**Table 17.2** Therapeutic agents used in autoimmune encephalitis

| Treatment | Adult regimen | Children <16 years of regimen |
|---|---|---|
| *First-line immunotherapy* | | |
| Methylprednisolone | 1000 mg/day intravenous (IV), for 3–5 days | 20 mg/kg/day IV, max. 1000 mg/day, for 3 days |
| Immunoglobulin | 0.4 g/kg/day IV, for 5 days | 0.4 g/kg/day IV, for 5 days |
| Plasmapheresis | 1 session every (other) day for 5–7 cycles | 1 session every (other) day for 5–7 cycles |
| *Second-line immunotherapy* | | |
| Rituximab | 1000 mg IV, two infusions with an interval of 14 days between them | 500 mg/m$^2$ two infusions with an interval of 14 days between them OR 375 mg/m$^2$ IV, weekly infusions for 4 weeks |
| Cyclophosphamide[a] | 750 mg/m$^2$ IV monthly for 3–6 months OR 15 mg/kg IV, max. 1200 mg/day, three infusions with an interval of 14 days, if necessary followed by six 500 mg/day infusions with an interval of 14 days between the infusions | 750 mg/m$^2$ IV monthly for 3–6 months |
| *Maintenance immunotherapy* | | |
| Azathioprine[b] | Initially 50 mg twice daily, followed by 75 mg twice daily, target 2–3 mg/kg/day in 1 or 2 doses. | 1–3 mg/kg/day in 1 or 2 doses |
| Mycophenolate mofetil | Initially 500 mg twice daily, target 1000 mg twice daily | Body surface area of <1.25 m$^2$: 600 mg/m$^2$ twice daily; body surface area of 1.25–1.5 m$^2$: 750 mg twice daily; body surface area of >1.5 m$^2$: 1000 mg twice daily |
| Mycophenolic acid | Initially 500 mg twice daily, target 750 mg twice daily | Not prescribed in children |

[a]Cyclophosphamide (CPA) dosing regimens vary, but usually intravenous infusions are administrated for 4 to 6 months depending on clinical recovery. In extreme refractory cases of anti-NMDAR encephalitis, CPA can be given for a period of 12 months

[b]When a patient develops adverse effects, such as hepatotoxicity or bone marrow depression, check whether the patient has a thiopurine S-methyltransferase (TPMT) mutation. In case the patient is a poor metabolizer, one should lower the standard dose or switch to another type of maintenance therapy

mately half of the patients, but improvement can be protracted [9, 22–24]. Of these patients, 97% had a good outcome (modified Rankin Scale (mRS) 0–2) after 2 years [9]. Without treatment, progressive neurologic deterioration and death can occur. However, spontaneous recovery has also been described in a few patients after several months or years of severe symptoms [9, 25]. Severely affected patients, including patients needing intensive care admission, have a poorer final outcome. It is important to start treatment early, as this leads to better outcomes and lowers the relapse risk [8, 9, 26]. In all patients, including the milder affected ones, the saying "time saves brain" is applicable, as residual neuropsychiatric deficits are more severe in those with delayed treatment [27]. This implies that the early part of the disease may be critical in terms of neuronal damage and long-term sequelae. Over recent years, the average treatment delay has become shorter over time due to better recognition [28].

When there is high suspicion of anti-NMDAR encephalitis, first-line therapy should be started once diagnostic samples have been obtained and alternative disorders, such as herpes simplex virus type 1 (HSV-1) encephalitis, have been excluded. In 2016 the Graus criteria formalized the diagnosis of "probable anti-NMDAR encephalitis" in patients presenting with a proper clinical picture and progressive course of disease in combination with compatible findings on electroencephalogram (EEG) and in cerebrospinal fluid (CSF) [18]. This way, treatment can be started before the antibody results become available.

Initial treatment is often started with *intravenous methylprednisolone pulse (IVMP)* in combination with IVIg or PLEX. IVMP belongs to the class of glucocorticoids; it acts on the glucocorticoid receptor and plays a central role in numerous physiological processes, including homeostasis, behavior, and bone mineral metabolism. With regard to immune function, it has anti-inflammatory, immunosuppressive, anti-proliferative, and vasoconstrictive properties [29]. *Intravenous immunoglobulins (IVIg)* is a preparation containing polyclonal human G isotype immunoglobulins (IgG) extracted from plasma of thousands of healthy blood donors [30]. Its mode of action is not fully understood, but is broad as illustrated by the many immune-mediated conditions that respond to IVIg. The mechanism of immunomodulatory actions is thought to be through antibody binding-mediated effects with neutralization of cytokines and complement via immune cell receptors (Fc receptor) and with increased clearance of pathogenic circulating antibodies [30]. During *plasmapheresis (PLEX)*, the blood of the patient is passed through a medical device separating plasma from other components of blood. The plasma containing the pathogenic antibodies is put aside, and the blood cells are returned to the patient in donor plasma or an artificial fluid similar to plasma [31]. PLEX offers the quickest short-term answer to removing pathological autoantibodies; however, the production of autoantibodies is not inhibited, and therefore it cannot be used in isolation for treatment. In the acute phase of the disease, patients are generally treated with five to seven plasma exchanges.

The choice to treat with IVIg or PLEX is dependent on patient-specific features, experience of the treating physician, and insurance coverage. In the management of Guillain-Barré syndrome and myasthenia gravis, there is no evidence for superiority in the efficacy of IVIg or PLEX, although PLEX might act somewhat quicker [32]. Other aspects can be pivotal for the decision to treat with IVIg instead of PLEX or vice versa. For instance, IVIg is associated with a lower risk of infections, and it is easier to administer in children and in patients with severe movement disorders and neuropsychiatric symptoms. During severe autonomic storms in patients admitted to the intensive care unit (ICU), PLEX might increase hemodynamically instability [22]. Plasma exchange is contraindicated in patients with hypocalcemia or allergies to heparin, frozen albumin, or frozen plasma. This process can also remove clotting factors requiring the monitoring of coagulation labs.

Considering the aforementioned arguments, most anti-NMDAR patients receive a combination of IVMP and IVIg. But there are exceptions, as immunoglobulins are expensive and subject to an increasing demand worldwide leading to shortages. PLEX is also best to use in patients with severe renal insufficiency, with a thrombotic tendency, or when treatment with IVIg leads to moderate to severe infusion-associated reactions or hyperviscosity syndrome.

When first-line treatment leads to clinical improvement, there is no need for second-line treatment. In these patients, we advise to repeat another IVMP pulse—if needed in combination with IVIg—after 4 and 8 weeks to maintain adequate suppression of antibody formation and hereby preventing treatment-related fluctuations. We prefer this treatment strategy above tapering with oral steroids, for the reason of less severe (e.g., neuropsychiatric and infectious) adverse effects and less interaction with the effects of second-line immunotherapy when using pulsed therapy.

## Second-Line Immunotherapy in Anti-N-Methyl-D-Aspartate Receptor Encephalitis

Second-line immunotherapy—RTX or a combination of RTX and CPA—is usually effective when first-line treatments fail. In patients without or with minimal clinical improvement after 2 weeks after the start of the initial therapies, we proceed with second-line therapies. The explanation for this deviation - as we advise to proceed after 2 weeks instead of the period of 4 weeks as described in the literature [9] - is that our experience is that patients do not show marked improvements between week 2 and 4 after the start of first-line therapy. Second-line treatment has beneficial effects on the final outcome, and in patients with multiple relapses, it reduces the likelihood of further relapses [9]. This has resulted in the current policy to treat patients with a relapse with a combination of first- and second-line treatment independent of the response to first-line therapy alone to prevent further relapses. Because most relapses are milder, additional treatment with RTX is often sufficient. Only, in more refractory cases, CPA should be added as well.

*Rituximab (RTX)* is a partially humanized monoclonal antibody directed against CD20, a glycoprotein primarily found on the surface of B cells, initially approved for the treatment of non-Hodgkin B-cell lymphomas. It reduces both naïve and memory B cells through antibody-mediated cellular toxicity, complement activation, and induction of apoptosis [33]. Besides B-cell depletion, there are findings suggestive of RTX affecting B-cell-T-cell interactions and direct inhibition of T cell activation [34]. Circulating B cells are usually beneath the detectable range for 6–8 months after treatment, which has also been shown in patients with AE, and subsequently short-lived plasmablasts are also depleted in anti-NMDAR encephalitis [35–37]. RTX treatment monitoring is via determination of B-cell count, which is determined 2 weeks after the last RTX infusion. In the case B cells are not fully depleted, another cycle of RTX is administered. See Table 17.2 for typical dosing schedule. In a retrospective study of 161 patients with AE (± proven antibody) treated additionally with RTX, adverse events of RTX were infusion-related reactions in 6.7% and infections, all pneumonia, in 11.3%, but no life-threatening or recurrent infectious occurred [37]. In another study in pediatric patients with anti-NMDAR encephalitis, 11 patients (7.6%) had infectious complications, including 2 life-threatening or disabling infections and 2 deaths [35]. Before the start of RTX, it is important to screen for chronic/latent infections, including hepatitis B screening and quantiferon gold in high-risk patients to prevent a flare-up during treatment. Long-lived plasma cells might cause a long-term lack of effect.

*Cyclophosphamide (CPA)* is an alkylating agent that impairs DNA replication or transcription, eventually leading to cell apoptosis. The effects of CPA are cell cycle independent and will affect rapidly proliferating cells the most. T cells, B cells, and NK cells are particularly sensitive to high-dose CPA because of their relatively low levels of aldehyde dehydrogenase, leading to direct suppression of lymphocyte proliferation (unlike first-line immunotherapy) [38].

Treatment in children is comparable to adults, as a similar escalation approach is being used; however, in children the dose of medications is less clear and often has been adapted from the use of the same medications in other autoimmune conditions. Results of immunotherapy in children are slightly better than in adults, probably because treating physicians are inclined to treat earlier and more aggressively [9, 39]. No difference in efficacy has been shown between RTX and CPA, although the available studies were not designed to identify any differences. However, the safety profile of RTX is considered more favorable. There is preference for treatment with RTX only over CPA as initial second-line immunotherapy in children younger than 16 years old, because of potential adverse effects, including the risk of premature gonadal failure, infertility, and long-term risk of occurrence of malignancies. Gonadotropin-releasing hormone agonist administration or egg/sperm collection may be employed to preserve fertility following CPA use [40]. Although physicians should be very careful about long-term risks, most studies about the risks of CPA originate from the 1950s and 1960s, using oral medication in far higher cumulative doses. If RTX treatment has no effect, escalation to CPA should be advised, despite its potential long-term risks. There is no convincing evidence for repeating RTX or chronic maintenance immunotherapy in anti-NMDAR encephalitis. An exception can be made in the rare patients with relapses despite second-line immunotherapy. In those, repetition of RTX could be considered to avoid new relapses. Of course, a still undetected paraneoplastic cause should be considered as well.

Favorable outcome is associated with an early start of treatment and a low severity of disease [8, 9, 26]. Probably due to earlier and more aggressive therapy with increased disease recognition over time, prognosis has become better over the years, bearing in mind that results are based on cohort studies with heterogeneous follow-up durations [8, 9, 23, 26, 28, 41].

After recovery, about one out of ten patients relapsed within the first 2 years after the initial episode. Patients without a tumor and those who did not receive second-line immunotherapy are at a greater risk for relapse [9]. Relapses are treated in the same way as newly diagnosed patients, with a very low threshold to initiate second-line immunotherapy early in the course of the relapse (to avoid a new relapse).

## First-Line and Second-Line Immunotherapy in Encephalitis Caused by Other Neuronal Cell Surface Antibodies

The proposed therapeutic schedules or other types of AE are based on mainly retrospective cohort studies and expert opinions. It deviates somewhat from the treatment of anti-NMDAR encephalitis, because the disease course is in most types less fulminant and relapse rates might be different. As anti-LGI1 encephalitis is one of the most common types of AE with a significant relapse rate, we choose to show in detail our treatment schedule for anti-LGI1 encephalitis in Fig. 17.1b. For relatively milder presentations of AE with other neuronal cell surface antibodies, one can follow the treatment schedule as shown in Fig. 17.1b. In case the clinical presentation of a patient with AE is severe/requires ICU admission, it is advised to follow the treatment schedule for anti/NMDAR encephalitis (Fig. 17.1a). Depending on the relapse rate, one should or should not add maintenance therapy (see section "Maintenance Immunotherapy").

As with anti-NMDAR encephalitis, first-line immunotherapy using IVMP in combination with IVIg or PLEX is initiated. This treatment is effective in 50–80% of the patients [10, 15]. In patients without an underlying malignancy or in patients with antibodies associated with a fairly high relapse rate, treatment with high-dose oral steroids is advised for several weeks, followed by a gradual tapering of the oral steroids. If it is anticipated that steroids will be used for a period longer than 3 months in a dose higher than 7.5 mg/day, one should consider osteoporosis prophylaxis. In patients treated for more than 3 weeks with oral steroids in doses higher than 20 mg/day, one should add *Pneumocystis jirovecii* prophylaxis. First-line treatment effects can be noticed within days but might take 3 to 4 weeks to reach an effect. In case of a temporary improvement, the first-line treatment cycle can be repeated. In case of no or inadequate treatment response, switching to another type of first-line treatment or proceeding to second-line treatment (RTX, CPA, or a combination) is advised. The choice to escalate and the timing of this escalation depend on the disease severity, symptoms to improve, and the antibody involved.

In anti-LGI1 encephalitis, the vast majority of patients develop subtle focal seizures or faciobrachial dystonic seizures (FBDS) fairly early in the disease course. It can also mimic dementia, in which the subtle seizures are easily overlooked. Figure 17.2 shows the disease course in anti-LGI1 encephalitis with the symptoms in most common order of appearance. As disease progresses, often tonic clonic seizures and additional irreversible cognitive impairment can arise [10, 42, 43]. Anti-epileptic drugs usually have little effect on these types of seizures, while FBDS and focal seizures tend to disappear within days to weeks after the start of first-line treatment [10, 16, 17, 44–46]. This superiority of immunotherapy over anti-epileptic drugs is also seen in patients with anti-NMDAR and anti-gamma-aminobutyric acid-A receptor (anti-$GABA_B$R) encephalitis. This was demonstrated in a cohort study including 153 patients with either anti-LGI1, anti-NMDAR, or anti-gamma-aminobutyric acid-B receptor (anti-$GABA_B$R) encephalitis, which showed that the chance to achieve seizure freedom was higher after the use

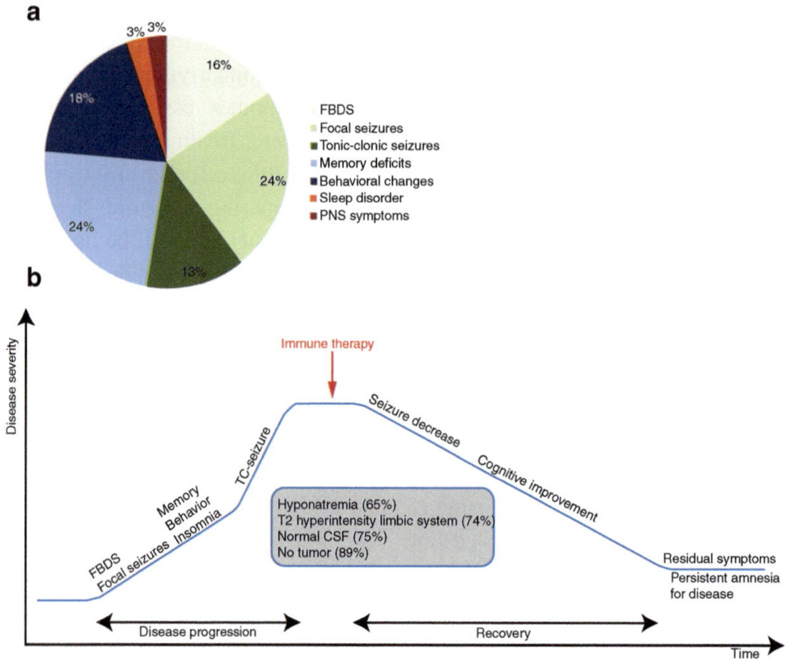


**Fig. 17.2** Disease course in anti-LGI1 encephalitis. Timeline: median disease progression 22 weeks, median treatment delay 25 weeks, median start of improvement 2 weeks after treatment, median time of recovery 33 weeks. FBDS faciobrachial dystonic seizures, TC-seizure tonic-clonic seizure. (This figure was derived from Ref. [10])

of immunotherapy than after the use of AEDs (53% of patients became seizure-free shortly after the start of immunotherapy versus 14% with the use of AEDs only; Fig. 17.3) [46]. After a 2-year follow-up, 98% of the patients alive had reached seizure freedom; 14% of these patients were still using AEDs.

Taken together, this underlines the importance of a timely diagnosis and timely start of immunotherapy, also with regard to achieving seizure control. The risk for developing chronic symptomatic seizures in the long run is fairly low [16].

## Maintenance Immunotherapy

In patients with a high relapse risk, there is an indication for maintenance therapy with immunosuppressants with the so-called steroid-sparing agents such as mycophenolate mofetil (MMF) or azathioprine (AZA). Due to their slow effect, these drugs are less useful in the acute phase of the disease. While steroids have been the mainstay of maintenance therapy in neuroimmunological disorders for a long time, the steroid-sparing agents are used to limit the side effects of long-term steroid use. Tapering of oral steroids can be initiated a few weeks after the start of steroid-sparing agents. In patients who relapse after an initial good response to RTX, one could consider intermittent administration of RTX at regular intervals (often every 6 months) or based on circulating B-cell numbers [47]. Six-month intervals are suggested for neuromyelitis optica spectrum disorder (NMOSD), although more recent literature might hint that longer intervals might be possible in selected patients. However, there is currently no evidence to support frequent repetition of administration in AE.

*Azathioprine (AZA)* and *mycophenolate mofetil (MMF)* are most commonly used as oral steroid-sparing agents for maintenance therapy in AE, as with autoimmune neurological disorders such as myasthenia gravis and NMOSD. Active

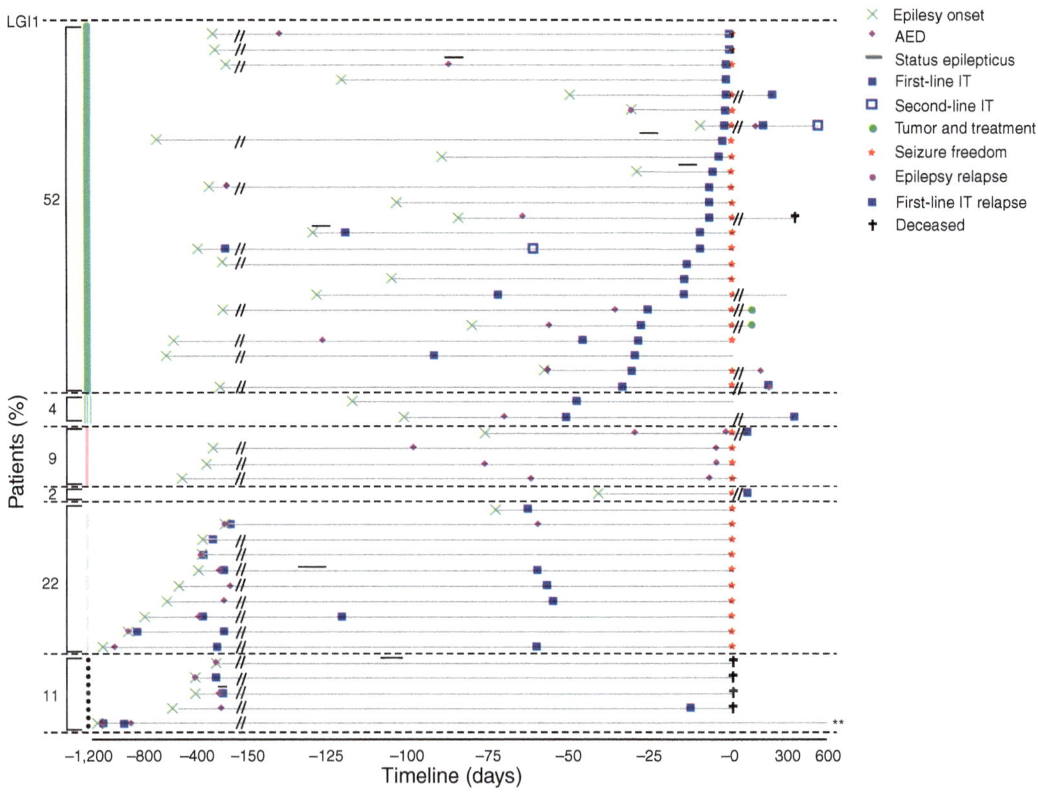


**Fig. 17.3** Timelines of anti-leucine-rich glioma-inactivated 1 encephalitis patients with epileptic seizures. The percentages shown on the left correspond to patients (1) reaching seizure freedom after the use of immunotherapy (green), (2) reaching seizure freedom probably after the use of immunotherapy (triple green), (3) reaching seizure freedom after the use of antiepileptic drugs (AEDs) (red), (4) reaching seizure freedom probably after the use of AEDs (double red), (5) who could not be categorized (gray stripes), and (6) who did not reach seizure freedom (black dots). If patients were treated with another immunomodulating treatment >1 month after the initial treatment (e.g., intravenous immunoglobulin after prednisolone), this is shown as a new blue square. Treatment with an additional AED or dosage increase after >1 month is shown as a second purple diamond. Relapses are only shown if patients had seizures. Median time of follow-up from the onset was 33 months (interquartile range [IQR], 19–52; range, 8–119). Median time of seizure freedom was 23 months (IQR, 14–40; range, 4–102). The median interval between the start of AEDs and the start of immunotherapy was 57 days (IQR, 27–152). **Timeline of the only patient who developed epilepsy after resolved encephalitis. The symbols in this timeline are not fitted to scale. The onset of seizures was in 2009, the patient was treated with prednisone (and AEDs), leading to reversibility of cognitive signs, but he still has temporal epilepsy. IT immunotherapy. (This figure was derived from Ref. [46])

metabolites of AZA—a synthetic purine analog derived from 6-mercaptopurine—are incorporated into the DNA, halting replication and disrupting function of endogenous purines [48]. AZA thus mostly affects proliferating cells, such as the T cells and B cells, clarifying its immunosuppressive effect. Mycophenolate mofetil (MMF) is a prodrug of mycophenolic acid (MPA). It acts as an inhibitor on de novo synthesis of guanosine nucleotides, as B and T lymphocytes are more dependent on the synthesis of guanosine nucleotides via this pathway than other cell types. Inosine monophosphate dehydrogenase (IMPDH) is required for lymphocytic clonal expansion, as MPA selectively inhibits IMPDH type II isoform, which is expressed in activated lymphocytes, rather than type I, which is expressed in most other cells. Therefore, MPA has a more potent cytostatic effect on lymphocytes than on other cells [49].

Maintenance therapy is not indicated in all patients with AE, as the disease course is monophasic in many AE patients. The decision who to start on maintenance therapy is based on the relapse risk, which is highly variable between the different types of AE [19]. Relapse rates might be underestimated, given the relatively recent discovery of the neuronal autoantibodies and scarcity of long-term follow-up data. Physicians should be aware that relapses can occur up to 5–10 years (or even longer) after the initial disease episode [10, 15, 50].

On the other hand, relapse rates might also be overestimated as relapsing cases are more easily identified as autoimmune, while monophasic cases might be considered of "viral" or "unknown" etiology. Furthermore, relapsing cases by definition have a longer follow-up, calculated from the initial episode (before discovery of the antibodies), than incidental cases, also inflating relapse risk. Therefore, relapse rates have dropped in subsequent case series after the initial publication. However, the use of first- and second-line immunotherapy in the initial episode of anti-NMDAR encephalitis already results in a lowering of the likelihood of further relapses [9]. This is exemplified by the reduction in relapse over time: relapses occurred in 15–24% of patients in the initial cohorts described in 2008 [41] and 2011 [50] and lowered to 9% in 2013 [9]. However, as anti-NMDAR encephalitis is usually monophasic, there is in the majority of patients no need for long-term immunosuppressive therapy. The risk for recurrence in anti-LGI1 encephalitis is higher. The initial studies with short-term follow-up indicated relapses in 0–18% of the patients [4, 51–53]. However, more recent studies with longer follow-up (>2 years) reported that 27–35% of the patients relapsed [10, 15]. In general, patients with anti-LGI1 encephalitis should be considered for maintenance therapy, especially those who suffer from cognitive symptoms or already proved to have a relapsing disease course. In patients with contraindications for chronic immunotherapy, one can decide on a watchful waiting policy. As relapses in anti-LGI1 almost invariably start identically to the initial period, patients and relatives can be instructed to monitor for symptoms, and often relapses are caught earlier than the initial presentation. The relapse rate in anti-Caspr2 encephalitis might be similar or perhaps somewhat lower (25–30%) [18]. It is best to inform the patient about the relapse risk as well as the risks and benefits of maintenance therapy and subsequently act on their preference. In case one chooses for watchful waiting, the patient and their caregivers need to be precisely informed about the various clinical syndromes caused by anti-Caspr2 antibodies, as relapses in anti-Caspr2 patients often present in a different way [18]. Overall relapse rates for anti-AMPAR, anti-glycine receptor (anti-GlyR), anti-$GABA_BR$, gamma-aminobutyric acid-A receptor (anti-$GABA_AR$), and anti-dipeptidyl-peptidase-like protein-6 (anti-DPPX) seem lower than for patients with anti-LGI1 and anti-Caspr2 encephalitis. However, limited data are available, and most are retrospective case series or small cohort studies [19]. For this reason, it seems valid to initiate maintenance therapy in these types of AE only in patients who have suffered a relapse. As relapse can occur after long intervals, it is difficult to determine the duration of maintenance immunotherapy. It has been suggested that medication withdrawal can be undertaken after at least 3 years without relapses or signs of active disease, but this rule of thumb might be appreciated differently based on the severity and reversibility of previous episodes, side effects of the maintenance therapy, and remaining high titers or absence of antibodies.

## Alternative Treatment Strategies

There is no doubt about the beneficial effectiveness of the current immunotherapeutic regimens for AE on final clinical outcome. Nevertheless, AE can be very disabling, and patients may need long-term hospitalization, including admission to the intensive care unit with a substantial mortality risk [14, 19]. In about one out of three AE patients, severe to moderate neurological sequelae were reported, despite adequate first- and second-line treatments [19]. Moreover, a considerable number of the patients with a perceived "good outcome" suffer from persisting neurocognitive symptoms and behavioral changes with a substantial impact on their quality of life [42, 54]. Clearly, there is a need for more effective therapies. Besides optimization of the currently used regimen and dosing schedules, immunotherapeutic regimens for AE could be more tailored to the type of associated antibody, as treatment response, length of treatment, relapse rate, and outcome vary depending on the type of associated antibody. The shortcomings of the current therapeutic protocol could be in part explained by the fact that most therapies affect peripheral immune cells or proteins, rather than the immune system in the central nervous system (CNS) itself. By developing treatments that are better at crossing the blood-brain barrier, the potency is increased to directly diminish the inflammatory process in the brain. In a study of three children diagnosed with anti-NMDAR encephalitis, who were considered "refractory" to treatment with RTX, the blood-brain barrier was bypassed by the intrathecal administration of methotrexate [55]. All three clinically improved, but it is difficult to assess the specific effects of MTX as RTX can act (very) slowly in patients, and in one patient CPA was added to the treatment regimen, which could be a potential confounder. Intrathecal treatment could be considered as an add-on therapy in AE that does not respond to first- and second-line treatment, but evidence is currently not sufficient to be easily recommended.

As in multiple sclerosis, new and more effective immune-modulating therapies are emerging rapidly, and it is expected that this evolution will pass over to the AE field as well, enabling to expand the treatment options in the near future.

## Monoclonal Antibodies Targeting the IL-6 Receptor

*Tocilizumab* is a monoclonal antibody that inhibits the binding of interleukin 6 (IL-6) to the IL-6 receptor, consequently repressing the B-cell proliferation and differentiation into antibody-producing cells. IL-6 facilitates also other inflammatory cascades involving cytotoxic T cells, T helper cells, and regulatory T cells that all contribute to autoimmunity [56]. The therapeutic effect of tocilizumab is proven in rheumatoid arthritis and demonstrated in therapy-resistant patients with neuromyelitis optica spectrum disorders (NMOSD) [57, 58]. Recently, *satralizumab* (a different anti-IL6 receptor monoclonal antibody) was shown to significantly reduce the relapse rate in NMOSD with 62% (hazard ratio = 0.38 95% CI 0.16–0.88) compared to placebo in a phase III randomized clinical trial (RCT) with a favorable safety profile during 2 years of follow-up (NCT02028884: the SAkuraSky trial) [59]. A reduction of 79% in the relapse rate was achieved in aquaporin-4 (AQP4)-positive patients.

With regard to AE, a case of anti-Caspr2 encephalitis was successfully treated with tocilizumab [60], and an observational study showed that tocilizumab potentially improves clinical outcome in patients, who were in poor condition 1 month after initiation of RTX [61]. Although the study shows promising results, the data suggest selection bias. In addition, another serious shortcoming is that two-thirds of the patients studied had seronegative AE. Confirmation by further studies in larger samples with more uniformity in diagnosis is necessary. Tocilizumab increases the risk of infection, and it hampers the recognition of an infection by diminishing the fever response and the levels of C-reactive protein [62]. For this reason, clinicians must be ultimately aware of systemic infection in treated patients, especially in those treated with multiple immunomodulating drugs.

## Bortezomib: A Proteasome Inhibitor

*Bortezomib*, a proteasome inhibitor particularly effective at depleting plasma cells, is approved for the treatment of multiple myeloma [63]. Given that plasma cells as mature, non-dividing antibody-secreting cells are unaffected by B-cell-depleting agents, steroids, and CPA [64], bortezomib may be an alternative option for refractory cases. Recent case reports and small case series have explored the treatment of severe anti-NMDAR encephalitis with bortezomib [65–69]. These cases showed that bortezomib was well tolerated in most patients. Hence, it remains unclear whether the clinical responses could be ascribed to bortezomib, rather than that the achieved remission could be part of the natural course of the disease or a (late) effect of preceding administered immunosuppressive drugs. To answer the additional value of bortezomib, a randomized clinical trial is planned with standard treatment regimens with and without bortezomib as an add-on treatment.

## Low-Dose IL-2 Therapy and Treg Modulation

In multiple sclerosis and other autoimmune conditions, the number and function of regulatory T cells are dysregulated [70]. IL-2 is a key regulator in the immune cascade of regulatory T cells and therefore has a key role in keeping tolerance over autoimmunity [71]. Low-dose IL-2 administration can selectively expand regulatory T cells without promoting effector T-cell responses because of the lower thresholds for activation of the regulatory T cells [72]. That is why low-dose IL-2 therapy is a promising new therapy for autoimmune and inflammatory disorders. Ten patients (four anti-NMDAR and six seronegative AE) with refractory symptoms were treated with low-dose IL-2 therapy in a pilot study. They showed modest responses after four to five treatment cycles, with the best response seen in the anti-NMDAR encephalitis patients [73]. However, randomized clinical trials are needed to establish whether there is a significant additional treatment effect of low-dose IL-2 therapies as compared to standard treatment only.

## Inebilizumab

*Inebilizumab* is a humanized IgG1 monoclonal antibody against CD19 that binds to and depletes CD19+ B cells including plasma cells and plasmablasts. Inebilizumab is believed to be effective for B-cell-related malignancies and autoimmune diseases. Inebilizumab has a broader efficiency to deplete circulating plasmablasts than other B-cell-targeted monoclonal antibodies, as more B cells harbor CD19 than CD20 [74]. A recently finished phase 2 and 3 study for treating NMOSD (NCT02200770 N-Momentum trial) showed to be very effective, providing a 77% relapse rate reduction by inebilizumab as compared to placebo in AQP4-positive patients [75].

## Eculizumab

*Eculizumab* is a recombinant humanized monoclonal antibody that specifically binds to the terminal complement component, hereby inhibiting the complement cascade. Eculizumab is already used to treat atypical hemolytic uremic syndrome and paroxysmal nocturnal hemoglobinuria. In AQP4-positive NMOSD, eculizumab significantly reduced the relapse rate as compared to placebo, with an annual relapse rate of 0.02 in the eculizumab group versus 0.35 in the placebo group [76]. Eculizumab may be an option for treatment of subtypes of AE, although the evidence for complement-mediated neuronal toxicity occurring in AE is not substantial. There is some evidence from a post-mortem study for complement-mediated neuronal toxicity in anti-Caspr2 and anti-LGI1 AE; however, this does not apply to anti-NMDAR encephalitis [77].

# Conclusion and Future Directions

The first step in improving the prognosis of patients confronted with AE is to increase disease awareness among medical specialists, as early recognition will lead to a prompt start of immunosuppressive therapy and may hereby prevent irreversible neurological damage.

The next step is to optimize the currently used acute treatment regimens and dosing schedules. It is important to tailor these regimens to the type of neuronal antibody causing the AE but also to patient-specific characteristics that can predict outcome. A first step has been the anti-NMDAR

Encephalitis One-Year Functional Status (NEOS) score, a prediction rule in anti-NMDAR encephalitis that provides information of outcome 1 month after the onset of the disease [78]. Ideally, the prediction of treatment response becomes available already at diagnosis and aids in the decision of whether to initiate immunotherapy. A third step would be the addition or development of more targeted and effective treatments with a favorable safety profile. Finally, the appropriate duration of maintenance immunotherapy for sustained remission should become known, preferably at the patient level, not only at the group level.

Given the rarity of AE, designing well-powered clinical trials is challenging. To succeed in obtaining valuable data to improve treatment strategies, international collaboration is valuable. In this setting, standardized, disease-specific outcome measures, which are sensitive in all stages of disease, are necessary. The currently used mRS score is not developed, nor sensitive, to adequately capture the complex deficits seen in AE patients. A step in the right direction is the recent development of the Clinical Assessment Scale in Autoimmune Encephalitis (CASE), although this score is heavily skewed toward the acute phase of anti-NMDAR encephalitis [79]. In this setting, one could test the efficacy and safety of the emerging immunosuppressive agents. It would be innovative to design individualized treatment strategies using prognostic models, which take both into account the variations in severity of disease between AE patients and the individualized risk for adverse events.

To conclude, exciting years are ahead of us in the rapidly evolving field of AE. The goal for the near future is that the upcoming clinical treatment trials and prognostic models will improve individualized treatment strategies with better clinical outcomes.

## References


1. Dalmau J, Tuzun E, Wu HY, Masjuan J, Rossi JE, Voloschin A, et al. Paraneoplastic anti-N-methyl-D-aspartate receptor encephalitis associated with ovarian teratoma. Ann Neurol. 2007;61(1):25–36.
2. Lai M, Hughes EG, Peng X, Zhou L, Gleichman AJ, Shu H, et al. AMPA receptor antibodies in limbic encephalitis alter synaptic receptor location. Ann Neurol. 2009;65(4):424–34.
3. Irani SR, Alexander S, Waters P, Kleopa KA, Pettingill P, Zuliani L, et al. Antibodies to Kv1 potassium channel-complex proteins leucine-rich, glioma inactivated 1 protein and contactin-associated protein-2 in limbic encephalitis, Morvan's syndrome and acquired neuromyotonia. Brain. 2010;133(9):2734–48.
4. Lai M, Huijbers MG, Lancaster E, Graus F, Bataller L, Balice-Gordon R, et al. Investigation of LGI1 as the antigen in limbic encephalitis previously attributed to potassium channels: a case series. Lancet Neurol. 2010;9(8):776–85.
5. Lancaster E, Dalmau J. Neuronal autoantigens--pathogenesis, associated disorders and antibody testing. Nat Rev Neurol. 2012;8(7):380–90.
6. van Coevorden-Hameete MH, de Graaff E, Titulaer MJ, Hoogenraad CC, Sillevis Smitt PA. Molecular and cellular mechanisms underlying anti-neuronal antibody mediated disorders of the central nervous system. Autoimmun Rev. 2014;13(3):299–312.
7. Vincent A, Bien CG, Irani SR, Waters P. Autoantibodies associated with diseases of the CNS: new developments and future challenges. Lancet Neurol. 2011;10(8):759–72.
8. Irani SR, Bera K, Waters P, Zuliani L, Maxwell S, Zandi MS, et al. N-methyl-D-aspartate antibody encephalitis: temporal progression of clinical and paraclinical observations in a predominantly non-paraneoplastic disorder of both sexes. Brain. 2010;133(Pt 6):1655–67.
9. Titulaer MJ, McCracken L, Gabilondo I, Armangue T, Glaser C, Iizuka T, et al. Treatment and prognostic factors for long-term outcome in patients with anti-NMDA receptor encephalitis: an observational cohort study. Lancet Neurol. 2013;12(2):157–65.
10. van Sonderen A, Thijs RD, Coenders EC, Jiskoot LC, Sanchez E, de Bruijn MA, et al. Anti-LGI1 encephalitis: clinical syndrome and long-term follow-up. Neurology. 2016;87(14):1449–56.
11. Graus F, Delattre JY, Antoine JC, Dalmau J, Giometto B, Grisold W, et al. Recommended diagnostic criteria for paraneoplastic neurological syndromes. J Neurol Neurosurg Psychiatry. 2004;75(8):1135–40.
12. Planaguma J, Leypoldt F, Mannara F, Gutierrez-Cuesta J, Martin-Garcia E, Aguilar E, et al. Human N-methyl D-aspartate receptor antibodies alter memory and behaviour in mice. Brain. 2015;138(Pt 1):94–109.
13. Leypoldt F, Armangue T, Dalmau J. Autoimmune encephalopathies. Ann N Y Acad Sci. 2015;1338:94–114.
14. van Coevorden-Hameete MH, de Bruijn M, de Graaff E, Bastiaansen D, Schreurs MWJ, Demmers JAA, et al. The expanded clinical spectrum of anti-GABABR encephalitis and added value of KCTD16 autoantibodies. Brain. 2019;142(6):1631–43.

15. Arino H, Armangue T, Petit-Pedrol M, Sabater L, Martinez-Hernandez E, Hara M, et al. Anti-LGI1-associated cognitive impairment: presentation and long-term outcome. Neurology. 2016;87(8):759–65.
16. Irani SR, Stagg CJ, Schott JM, Rosenthal CR, Schneider SA, Pettingill P, et al. Faciobrachial dystonic seizures: the influence of immunotherapy on seizure control and prevention of cognitive impairment in a broadening phenotype. Brain. 2013;136(Pt 10):3151–62.
17. Shin YW, Lee ST, Shin JW, Moon J, Lim JA, Byun JI, et al. VGKC-complex/LGI1-antibody encephalitis: clinical manifestations and response to immunotherapy. J Neuroimmunol. 2013;265(1–2):75–81.
18. van Sonderen A, Arino H, Petit-Pedrol M, Leypoldt F, Kortvelyessy P, Wandinger KP, et al. The clinical spectrum of Caspr2 antibody-associated disease. Neurology. 2016;87(5):521–8.
19. Nosadini M, Mohammad SS, Ramanathan S, Brilot F, Dale RC. Immune therapy in autoimmune encephalitis: a systematic review. Expert Rev Neurother. 2015;15(12):1391–419.
20. van Sonderen A, Roelen DL, Stoop JA, Verdijk RM, Haasnoot GW, Thijs RD, et al. Anti-LGI1 encephalitis is strongly associated with HLA-DR7 and HLA-DRB4. Ann Neurol. 2017;81(2):193–8.
21. Titulaer MJ, Soffietti R, Dalmau J, Gilhus NE, Giometto B, Graus F, et al. Screening for tumours in paraneoplastic syndromes: report of an EFNS task force. Eur J Neurol. 2011;18(1):19–e3.
22. Dalmau J, Lancaster E, Martinez-Hernandez E, Rosenfeld MR, Balice-Gordon R. Clinical experience and laboratory investigations in patients with anti-NMDAR encephalitis. Lancet Neurol. 2011;10(1):63–74.
23. Florance NR, Davis RL, Lam C, Szperka C, Zhou L, Ahmad S, et al. Anti-N-methyl-D-aspartate receptor (NMDAR) encephalitis in children and adolescents. Ann Neurol. 2009;66(1):11–8.
24. Gresa-Arribas N, Titulaer MJ, Torrents A, Aguilar E, McCracken L, Leypoldt F, et al. Antibody titres at diagnosis and during follow-up of anti-NMDA receptor encephalitis: a retrospective study. Lancet Neurol. 2014;13(2):167–77.
25. Iizuka T, Sakai F, Ide T, Monzen T, Yoshii S, Iigaya M, et al. Anti-NMDA receptor encephalitis in Japan: long-term outcome without tumor removal. Neurology. 2008;70(7):504–11.
26. Byrne S, Walsh C, Hacohen Y, Muscal E, Jankovic J, Stocco A, et al. Earlier treatment of NMDAR antibody encephalitis in children results in a better outcome. Neurol Neuroimmunol Neuroinflamm. 2015;2(4):e130.
27. McKeon GL, Robinson GA, Ryan AE, Blum S, Gillis D, Finke C, et al. Cognitive outcomes following anti-N-methyl-D-aspartate receptor encephalitis: a systematic review. J Clin Exp Neuropsychol. 2018;40(3):234–52.
28. Zekeridou A, Karantoni E, Viaccoz A, Ducray F, Gitiaux C, Villega F, et al. Treatment and outcome of children and adolescents with N-methyl-D-aspartate receptor encephalitis. J Neurol. 2015;262(8):1859–66.
29. Hapgood JP, Avenant C, Moliki JM. Glucocorticoid-independent modulation of GR activity: implications for immunotherapy. Pharmacol Ther. 2016;165:93–113.
30. Lunemann JD, Nimmerjahn F, Dalakas MC. Intravenous immunoglobulin in neurology--mode of action and clinical efficacy. Nat Rev Neurol. 2015;11(2):80–9.
31. Reeves HM, Winters JL. The mechanisms of action of plasma exchange. Br J Haematol. 2014;164(3):342–51.
32. Ortiz-Salas P, Velez-Van-Meerbeke A, Galvis-Gomez CA, Rodriguez QJ. Human immunoglobulin versus plasmapheresis in Guillain-Barre syndrome and myasthenia gravis: a meta-analysis. J Clin Neuromuscul Dis. 2016;18(1):1–11.
33. Abulayha A, Bredan A, El Enshasy H, Daniels I. Rituximab: modes of action, remaining dispute and future perspective. Future Oncol. 2014;10(15):2481–92.
34. Stroopinsky D, Katz T, Rowe JM, Melamed D, Avivi I. Rituximab-induced direct inhibition of T-cell activation. Cancer Immunol Immunother. 2012;61(8):1233–41.
35. Dale RC, Brilot F, Duffy LV, Twilt M, Waldman AT, Narula S, et al. Utility and safety of rituximab in pediatric autoimmune and inflammatory CNS disease. Neurology. 2014;83(2):142–50.
36. Hachiya Y, Uruha A, Kasai-Yoshida E, Shimoda K, Satoh-Shirai I, Kumada S, et al. Rituximab ameliorates anti-N-methyl-D-aspartate receptor encephalitis by removal of short-lived plasmablasts. J Neuroimmunol. 2013;265(1–2):128–30.
37. Lee WJ, Lee ST, Byun JI, Sunwoo JS, Kim TJ, Lim JA, et al. Rituximab treatment for autoimmune limbic encephalitis in an institutional cohort. Neurology. 2016;86(18):1683–91.
38. Brodsky RA. High dose cyclophosphamide treatment for autoimmune disorders. ScientificWorldJournal. 2002;2:1808–15.
39. Huang Q, Wu Y, Qin R, Wei X, Ma M. Clinical characteristics and outcomes between children and adults with anti-N-methyl-D-aspartate receptor encephalitis. J Neurol. 2016;263(12):2446–55.
40. Hickman LC, Valentine LN, Falcone T. Preservation of gonadal function in women undergoing chemotherapy: a review of the potential role for gonadotropin-releasing hormone agonists. Am J Obstet Gynecol. 2016;215(4):415–22.
41. Dalmau J, Gleichman AJ, Hughes EG, Rossi JE, Peng X, Lai M, et al. Anti-NMDA-receptor encephalitis: case series and analysis of the effects of antibodies. Lancet Neurol. 2008;7(12):1091–8.
42. Finke C, Pruss H, Heine J, Reuter S, Kopp UA, Wegner F, et al. Evaluation of cognitive deficits and structural hippocampal damage in encephalitis with leucine-rich, glioma-inactivated 1 antibodies. JAMA Neurol. 2017;74(1):50–9.

43. Andrade DM, Tai P, Dalmau J, Wennberg R. Tonic seizures: a diagnostic clue of anti-LGI1 encephalitis? Neurology. 2011;76(15):1355–7.
44. Flanagan EP, Kotsenas AL, Britton JW, McKeon A, Watson RE, Klein CJ, et al. Basal ganglia T1 hyperintensity in LGI1-autoantibody faciobrachial dystonic seizures. Neurol Neuroimmunol Neuroinflamm. 2015;2(6):e161.
45. Gao L, Liu A, Zhan S, Wang L, Li L, Guan L, et al. Clinical characterization of autoimmune LGI1 antibody limbic encephalitis. Epilepsy Behav. 2016;56:165–9.
46. de Bruijn M, van Sonderen A, van Coevorden-Hameete MH, Bastiaansen AEM, Schreurs MWJ, Rouhl RPW, et al. Evaluation of seizure treatment in anti-LGI1, anti-NMDAR, and anti-GABABR encephalitis. Neurology. 2019;92(19):e2185–96.
47. Kim SH, Huh SY, Lee SJ, Joung A, Kim HJ. A 5-year follow-up of rituximab treatment in patients with neuromyelitis optica spectrum disorder. JAMA Neurol. 2013;70(9):1110–7.
48. Hoffmann M, Rychlewski J, Chrzanowska M, Hermann T. Mechanism of activation of an immunosuppressive drug: azathioprine. Quantum chemical study on the reaction of azathioprine with cysteine. J Am Chem Soc. 2001;123(26):6404–9.
49. Allison AC, Eugui EM. Mycophenolate mofetil and its mechanisms of action. Immunopharmacology. 2000;47(2–3):85–118.
50. Gabilondo I, Saiz A, Galan L, Gonzalez V, Jadraque R, Sabater L, et al. Analysis of relapses in anti-NMDAR encephalitis. Neurology. 2011;77(10):996–9.
51. Irani SR, Gelfand JM, Bettcher BM, Singhal NS, Geschwind MD. Effect of rituximab in patients with leucine-rich, glioma-inactivated 1 antibody-associated encephalopathy. JAMA Neurol. 2014;71(7):896–900.
52. Malter MP, Frisch C, Schoene-Bake JC, Helmstaedter C, Wandinger KP, Stoecker W, et al. Outcome of limbic encephalitis with VGKC-complex antibodies: relation to antigenic specificity. J Neurol. 2014;261(9):1695–705.
53. Wegner F, Wilke F, Raab P, Tayeb SB, Boeck AL, Haense C, et al. Anti-leucine rich glioma inactivated 1 protein and anti-N-methyl-D-aspartate receptor encephalitis show distinct patterns of brain glucose metabolism in 18F-fluoro-2-deoxy-d-glucose positron emission tomography. BMC Neurol. 2014;14:136.
54. de Bruijn M, Aarsen FK, van Oosterhout MP, van der Knoop MM, Catsman-Berrevoets CE, Schreurs MWJ, et al. Long-term neuropsychological outcome following pediatric anti-NMDAR encephalitis. Neurology. 2018;90(22):e1997–2005.
55. Tatencloux S, Chretien P, Rogemond V, Honnorat J, Tardieu M, Deiva K. Intrathecal treatment of anti-N-methyl-D-aspartate receptor encephalitis in children. Dev Med Child Neurol. 2015;57(1):95–9.
56. Kang S, Tanaka T, Kishimoto T. Therapeutic uses of anti-interleukin-6 receptor antibody. Int Immunol. 2015;27(1):21–9.
57. Smolen JS, Beaulieu A, Rubbert-Roth A, Ramos-Remus C, Rovensky J, Alecock E, et al. Effect of interleukin-6 receptor inhibition with tocilizumab in patients with rheumatoid arthritis (OPTION study): a double-blind, placebo-controlled, randomised trial. Lancet. 2008;371(9617):987–97.
58. Ringelstein M, Ayzenberg I, Harmel J, Lauenstein AS, Lensch E, Stogbauer F, et al. Long-term therapy with interleukin 6 receptor blockade in highly active neuromyelitis optica spectrum disorder. JAMA Neurol. 2015;72(7):756–63.
59. Kleiter I, Fujihara K, Palace J, editors. Efficacy of satralizumab (SA237) in subgroups of patients in SAkuraSky: a Phase III double-blind, placebo-controlled, add-on study in patients with neuromyelitis optica spectrum disorder (NMOSD). Annual Meeting 2019 of the American Association of Neurology. Philidelphia; 2019.
60. Krogias C, Hoepner R, Muller A, Schneider-Gold C, Schroder A, Gold R. Successful treatment of anti-Caspr2 syndrome by interleukin 6 receptor blockade through tocilizumab. JAMA Neurol. 2013;70(8):1056–9.
61. Lee WJ, Lee ST, Moon J, Sunwoo JS, Byun JI, Lim JA, et al. Tocilizumab in autoimmune encephalitis refractory to rituximab: an institutional cohort study. Neurotherapeutics. 2016;13(4):824–32.
62. Rossi JF, Lu ZY, Jourdan M, Klein B. Interleukin-6 as a therapeutic target. Clin Cancer Res. 2015;21(6):1248–57.
63. Meister S, Schubert U, Neubert K, Herrmann K, Burger R, Gramatzki M, et al. Extensive immunoglobulin production sensitizes myeloma cells for proteasome inhibition. Cancer Res. 2007;67(4):1783–92.
64. Hiepe F, Radbruch A. Plasma cells as an innovative target in autoimmune disease with renal manifestations. Nat Rev Nephrol. 2016;12(4):232–40.
65. Behrendt V, Krogias C, Reinacher-Schick A, Gold R, Kleiter I. Bortezomib treatment for patients with anti-N-methyl-d-aspartate receptor encephalitis. JAMA Neurol. 2016;73(10):1251–3.
66. Scheibe F, Pruss H, Mengel AM, Kohler S, Numann A, Kohnlein M, et al. Bortezomib for treatment of therapy-refractory anti-NMDA receptor encephalitis. Neurology. 2017;88(4):366–70.
67. Schroeder C, Back C, Koc U, Strassburger-Krogias K, Reinacher-Schick A, Gold R, et al. Breakthrough treatment with bortezomib for a patient with anti-NMDAR encephalitis. Clin Neurol Neurosurg. 2018;172:24–6.
68. Sveinsson O, Granqvist M, Forslin Y, Blennow K, Zetterberg H, Piehl F. Successful combined targeting of B- and plasma cells in treatment refractory anti-NMDAR encephalitis. J Neuroimmunol. 2017;312:15–8.
69. Shin YW, Lee ST, Kim TJ, Jun JS, Chu K. Bortezomib treatment for severe refractory anti-NMDA receptor encephalitis. Ann Clin Transl Neurol. 2018;5(5):598–605.

70. Buckner JH. Mechanisms of impaired regulation by CD4(+)CD25(+)FOXP3(+) regulatory T cells in human autoimmune diseases. Nat Rev Immunol. 2010;10(12):849–59.
71. Liao W, Lin JX, Leonard WJ. Interleukin-2 at the crossroads of effector responses, tolerance, and immunotherapy. Immunity. 2013;38(1):13–25.
72. Yu A, Snowhite I, Vendrame F, Rosenzwajg M, Klatzmann D, Pugliese A, et al. Selective IL-2 responsiveness of regulatory T cells through multiple intrinsic mechanisms supports the use of low-dose IL-2 therapy in type 1 diabetes. Diabetes. 2015;64(6):2172–83.
73. Lim JA, Lee ST, Moon J, Jun JS, Park BS, Byun JI, et al. New feasible treatment for refractory autoimmune encephalitis: low-dose interleukin-2. J Neuroimmunol. 2016;299:107–11.
74. Agius MA, Klodowska-Duda G, Maciejowski M, Potemkowski A, Li J, Patra K, et al. Safety and tolerability of inebilizumab (MEDI-551), an anti-CD19 monoclonal antibody, in patients with relapsing forms of multiple sclerosis: results from a phase 1 randomised, placebo-controlled, escalating intravenous and subcutaneous dose study. Mult Scler. 2019;25(2):235–45.
75. Cree B, Bennett J, Kim HJ, Weinshenker B, Pittock S, Wingerchuk D, et al. Inebilizumab for the treatment of neuromyelitis optica spectrum disorder (NMOmentum): a double-blind, randomised placebo-controlled phase 2/3 trial. Lancet. 2019;12(394):1352–63.
76. Pittock S, Berthele A, Fujihara K, Kim HJ, Levy M, Nakashima I, et al., editors. Efficacy and safety of eculizumab in aquaporin-4 antibody positive (AQP4-IgG+) neuromyelitis optica spectrum disorder (NMOSD): a phase 3, randomized, double-blind, placebo-controlled, multicenter trial (PREVENT). Annual Meeting 2019 of Academy of Neurology. Philadelphia; 7 May 2019.
77. Bien CG, Vincent A, Barnett MH, Becker AJ, Blumcke I, Graus F, et al. Immunopathology of autoantibody-associated encephalitides: clues for pathogenesis. Brain. 2012;135(Pt 5):1622–38.
78. Balu R, McCracken L, Lancaster E, Graus F, Dalmau J, Titulaer MJ. A score that predicts 1-year functional status in patients with anti-NMDA receptor encephalitis. Neurology. 2019;92(3):e244–e52.
79. Lim JA, Lee ST, Moon J, Jun JS, Kim TJ, Shin YW, et al. Development of the clinical assessment scale in autoimmune encephalitis. Ann Neurol. 2019;85(3):352–8.

# Part IV

# Peripheral Nervous System Disorders

# 18 Acute and Chronic Immune Neuropathies and Radiculopathies

Anson W. Wilks and Robert C. Bucelli

**Key Points**

1. Guillain-Barré syndrome (GBS) is an immune-mediated monophasic disease often associated with antecedent infection and is characterized by weakness that is ascending and symmetric, areflexia, cytoalbuminologic dissociation, and typical electrodiagnostic features that vary by GBS subtype.
2. Antibody cross-reactivity between epitopes on infectious pathogens and peripheral nerve components has been convincingly demonstrated to be pathogenic in axonal variants of GBS, substantiating an immune-mediated mechanism of disease.
3. Intravenous immunoglobulin (IVIg) and plasma exchange (PLEX) are the mainstay of treatment of GBS; both hasten recovery and likely improve clinical outcomes, but there is no evidence to support combining these therapeutic interventions in GBS.
4. Chronic inflammatory/immune demyelinating polyradiculoneuropathy (CIDP) is a chronic condition with both overlapping and distinctive features relative to GBS (i.e., it is not simply a chronic form of GBS) and often follows a relapsing-remitting or progressive course.
5. Electrodiagnostics are requisite for diagnosis and important in differentiating CIDP from mimics of the disease such as hereditary demyelinating neuropathies and paraproteinemic neuropathies.
6. IVIg and corticosteroids are first-line therapies for CIDP.

## Introduction

Neuroimmunology, as a field, has traditionally focused on the central nervous system (CNS), but facility with various immune-mediated peripheral nervous system (PNS) disorders is essential to the astute neuroimmunologist. Therefore, the focus of this chapter is immune disorders of the nerve roots, plexus, and peripheral nerves. Disorders of the neuromuscular junction and muscle are addressed in Chaps. 19 and 20, respectively. Radiculopathy, plexopathy, and

A. W. Wilks
Department of Neurology, Washington University School of Medicine in Saint Louis, Saint Louis, MO, USA

R. C. Bucelli (✉)
Department of Neurology, Neuromuscular Section, Barnes Jewish Hospital, Washington University School of Medicine in Saint Louis, Saint Louis, MO, USA
e-mail: bucellir@wustl.edu



A. L. Piquet, E. Alvarez (eds.), *Neuroimmunology*, https://doi.org/10.1007/978-3-030-61883-4_18

polyneuropathy can be due to immune/inflammatory, vascular, infectious, nutritional, metabolic, neoplastic, toxic, or hereditary etiologies, and in many instances, the etiology is idiopathic. Appropriate diagnosis of immune-mediated neuropathy, however, is critical, as neuropathies due to an immunologic etiology are among the most likely to be treatment responsive.

Guillain-Barré syndrome (GBS) is the classic immune-mediated radiculopathy/polyneuropathy and is an umbrella term used to encompass many acute immune neuropathies. These include the two most common forms: acute inflammatory/immune demyelinating polyradiculoneuropathy (AIDP) and acute motor axonal neuropathy (AMAN). The chronic counterpart to GBS is chronic inflammatory/immune demyelinating polyradiculoneuropathy (CIDP), and it represents the most common chronic immune neuropathy. The focus of this chapter will be a comprehensive discussion of GBS, CIDP, and their variants, but a discussion of differential diagnosis will also be included.

## Guillain-Barré Syndrome

GBS is the most common cause of acute flaccid paralysis worldwide [1] and is classically characterized by a symmetric, ascending pattern of weakness frequently accompanied by sensory symptoms with minimal corresponding signs and cranial nerve deficits and as a delayed manifestation, autonomic dysfunction [1, 2]. Weakness progresses over the course of weeks and can result in neuromuscular respiratory failure [3], highlighting the importance of early diagnosis. Cardinal features of the disease include antecedent infection, hypo-/areflexia, elevated cerebrospinal fluid (CSF) protein in the absence of pleocytosis (cytoalbuminologic dissociation), a monophasic disease course, and characteristic electrodiagnostic (EDX) findings on electromyography and nerve conduction studies (EMG/NCS) [4]. Aside from AIDP and AMAN, there exist several additional GBS variants, including acute motor sensory axonal neuropathy (AMSAN), Miller Fisher syndrome (MFS), the pharyngeal-cervical-brachial (PCB) variant, the paraparetic GBS variant, idiopathic acute sensory neuronopathy/ganglionopathy, acute small fiber sensory neuropathy, and acute autonomic neuropathy/ganglionopathy.

### Epidemiology

The incidence of GBS is 1.1 to 1.8 per 100,000 per year. Incidence increases with age, and there is a slight male predominance [5]. It is less common in children but does occur [6]. There is a marked global variability in the incidence of specific variants [7]. For instance, the incidence of AMAN has been demonstrated in several studies to be much higher in China (65%) [8], Japan (38%) [9], and Mexico (38%) [10]. This is in contrast with North America and Europe where AIDP is the most common variant with only 4% of patients classified as axonal per one large patient cohort [11]. It is worth mentioning that AMAN may be underdiagnosed given the issues arising from current electrodiagnostic criteria [12] (see later). There has been some speculation that this regional variability is related to climate and its effect on the infectivity of *Campylobacter jejuni* (*C. jejuni*) [13], the most common antecedent infection for AMAN. Indeed, while seasonal variability in occurrence of GBS favors winter months, this is counterbalanced by outliers favoring the summer months in certain areas with a higher incidence of AMAN such as Northern China [14].

### Antecedent Events

GBS, a post-infectious phenomenon, is among the few neurologic disorders to fulfill all of Witebsky's postulates in that the pathophysiology is related to antibody cross-reactivity between epitopes on infectious agents with similar epitopes on peripheral nerve components, such as Schwann cells and the node of Ranvier [15]. This has been clearly demonstrated in AMAN, whereby cross-reactivity occurs with lipopolysaccharide on *C. jejuni* and the ganglioside GM1, an antigen heavily expressed at the node of Ranvier [16, 17] (see later). Further support of a post-infectious autoimmune mechanism is the fact that antecedent infection, most commonly upper respiratory or gastrointestinal, is reported in roughly two-thirds of patients with GBS [18]. The most commonly reported antecedent infections per one case-control study listed in descending order are *C. jejuni*, cytomegalovirus, Epstein-Barr virus, and *Mycoplasma pneumoniae* [19]. Influenza [20], hepatitis E [21, 22], and Zika virus [23] infections are reported antecedents as well. Vaccination is another antecedent, although there is some controversy surrounding this issue [24]. A large retrospective study found no association with vaccination (influenza, pneumococcal, and tetanus-diphtheria vaccination) [25], while some studies on H1N1 vaccination have demonstrated a slightly increased risk of GBS [26, 27]. Notably, the increased risk of GBS following influenza vaccination when reported is on the order of one case per million inoculated—several magnitudes less than that of antecedent influenza infection [28]. This relatively low risk of GBS following vaccination even applies to those with a history of GBS per one study that identified no cases of recurrent GBS attributable to vaccination [29].

**Table 18.1** Key diagnostic criteria and Brighton case definitions for Guillain-Barré syndrome

| | Level of diagnostic certainty | | | |
|---|---|---|---|---|
| Diagnostic criteria | 1 | 2 | 3 | 4 |
| Bilateral and flaccid weakness of limbs | + | + | + | +/− |
| Decreased or absent deep tendon reflexes in weak limbs | + | + | + | +/− |
| Monophasic course and time between onset-nadir 12 hours and 28 days | + | + | + | +/− |
| CSF cell count <50/μ(mu)l | + | +[a] | − | +/− |
| CSF protein concentration > normal value | + | +/−[a] | − | +/− |
| NCS findings consistent with one of the subtypes of GBS | + | +/− | − | +/− |
| Absence of alternative diagnosis for weakness | + | + | + | + |

Reprinted with permission from Fokke et al. [30]
+ present; − absent; +/− present or absent;
*CSF* cerebrospinal fluid, *NCS* nerve conduction studies, *GBS* Guillain-Barré syndrome
[a]If CSF is not collected or results not available, nerve electrophysiology results must be consistent with the diagnosis Guillain-Barré syndrome

## Clinical Features

There is a considerable overlap in the clinical features of AIDP and AMAN, and their differentiation is often reliant on EDX. Therefore, subtle distinguishing features will be highlighted as we discuss GBS as a class. The clinical features discussed below have been encapsulated in a clinically validated set of criteria called the Brighton criteria (Table 18.1) [30]. In AIDP, reflexes are depressed or absent. However, exceptions can occur in AMAN, in which reflexes can be preserved and hyperreflexia may develop, particularly in the recovery phase [31]. The weakness in GBS is typically symmetric. Marked asymmetry should raise suspicion for alternate etiologies of acute flaccid paresis such as infectious motor neuron disease (e.g., West Nile neuroinvasive disease or historically poliomyelitis) or vasculitic neuropathy (i.e., fulminant mononeuritis multiplex) [32]. While classically associated with an ascending pattern, weakness in some patients may be "descending" and manifests initially as craniobulbar palsy or arm weakness [30], which can masquerade as botulism. The clinical severity runs the gamut from mild paresthesia and hyporeflexia to quadriparesis and neuromuscular respiratory failure.

Progression occurs over days to weeks but reaches its nadir within 4 weeks of onset. AMAN is characterized by a more rapid progression and an earlier clinical nadir [33]. Eighty percent of patients will have reached their clinical nadir by 2 weeks and 97% by 4 weeks [30]. Continued progression beyond 4 weeks is atypical and should raise suspicion for acute-onset CIDP or a mimic of GBS. Progression of an acute-onset demyelinating neuropathy beyond 8 weeks, with supportive electrophysiologic data, is diagnostic of acute-onset CIDP [34]. This temporal criterion creates a diagnostic interface between AIDP and CIDP for the minority of patients who reach their nadir between 4 and 8 weeks. The term subacute inflammatory demyelinating polyneuropathy (SIDP) has been proposed to capture this group [35]. However, many do not consider this a distinct entity but rather a temporary diagnosis until patients longitudinally declare themselves to have AIDP or CIDP. Several weeks after reaching their nadir, patients will begin a slow recovery period [36].

Sensory symptoms are primarily "positive" in the form of dysesthesia (including back pain), and objective sensory deficits, if present at all, are typically mild. The majority of patients report pain usually localizing to the back and extremities in AIDP [37]. Sensory involvement or pain is minimal if present at all in AMAN, but subclinical sensory disturbances are often found on EDX [38]. Craniobulbar palsies do occur in AIDP. This is most commonly facial nerve palsy, which can be bilateral, but can also include ophthalmoparesis or dysphagia [1, 30]. Autonomic dysfunction occurs in approximately two-thirds of patients of AIDP, which can manifest as urinary retention, constipation, tachy-/bradyarrhythmia, and labile blood pressure [39–41]. Cranial nerve palsy is less common in AMAN [42], and if present at all, dysautonomia is mild [43]. It is important to be cognizant of cardiovascular complications frequently encountered in AIDP and to monitor hemodynamics closely as overzealous treatment of hypertension can cause profound hypotension. On the other hand, clinical judgment is mandated as cases have been reported of posterior reversible encephalopathy syndrome (PRES) associated with GBS, presumably due to acute hypertension [44, 45].

Close monitoring for neuromuscular respiratory failure is paramount. This is best monitored with serial measurements (ideally multiple times daily) of negative inspiratory force/forced vital capacity (NIF/FVC) and to a lesser extent neck flexion/extension strength, a surrogate measure of diaphragmatic strength. FVC can be approximated at the bedside by having the patient count out loud (one number per second) for the duration of a single breath (each number counted approximates 100 cc of vital capacity). Up to 30% of patients may require mechanical ventilation [46]. Short duration from symptom onset to hospitalization, FVC less than 60% predicted value, inability to lift the head, and presence of bulbar or facial weakness are all predictive of the need for intubation per several studies [46–48]. It has been suggested that demyelinating GBS (i.e., AIDP) has a higher risk for developing neuromuscular respiratory failure and that patients with AMAN without proximal weakness are at low risk for needing mechanical ventilation [49]. However, since patients with AMAN can require artificial ventilation and there is difficulty in stratifying patients upon initial presentation, respiratory monitoring of all patients with suspected GBS is warranted.

## Diagnostic Studies

CSF analysis classically shows cytoalbuminologic dissociation, practically defined by many clinicians as elevated CSF protein (albumin) with less than ten nucleated cells (NC)/$mm^3$. However, a more permissive cell count of less than 50 NC/$mm^3$ is utilized in diagnostic criteria for GBS [50]. Half of patients will not have this feature within 1 week of symptom onset, but after 3 weeks, roughly 75% of patients will have cytoalbuminologic dissociation [51, 52]. Thus, timing of CSF sampling is an important consideration, and the absence of cytoalbuminologic dissociation should not preclude diagnosis. Moreover, since this is not necessary for diagnosis, repeat lumbar puncture to confirm its presence is not recommended.

Electrophysiological findings can be supportive of the diagnosis of GBS. NCS will show early abnormalities in late responses (i.e., H-reflexes and F-waves), reflecting early nerve root involvement [53–55]. An absent H-reflex is the most sensitive electrodiagnostic finding for early GBS [55], and this is the electrophysiological correlate to an absent Achilles reflex. Prolonged distal motor latencies, conduction velocity slowing, conduction block, and increased temporal dispersion are all criteria for demyelination. Abnormal late responses, prolonged distal motor latencies, or slow conduction velocities need to be present in at least two nerves, at non-entrapment sites, to be considered supportive of the diagnosis. Sural sparing (i.e., normal sural sensory NCS with abnormal median/ulnar sensory NCSs) is a characteristic feature of AIDP and CIDP and can be useful to distinguish from mimics [56]. Low compound muscle action potential (CMAP) amplitudes or inexcitable CMAPs are characteristic of AMAN [11]. Repeat testing is reasonable both for prognostic purposes and to definitively distinguish between axonal and demyelinating features [57] (see later).

Lumbosacral nerve root enhancement may be visualized when gadolinium-enhanced magnetic resonance imaging (MRI) of the spine is obtained (Fig. 18.1) [58], but the primary role for imaging is to exclude alternative etiologies of acute paralysis.

## Atypical Features

Alarming features that are suggestive of an alternative diagnosis include early urinary incontinence or sphincter dysfunction, saddle anesthesia, a spinal sensory level, or abnormal abdominal reflexes [59, 60]. Any of these features should prompt spinal MRI, and clinicians should be wary of interpreting reflexes when acute myelopathy is suspected due to the possibility of spinal shock (i.e., early diminished reflexes prior to development of hyperreflexia). In general, regardless of reflexes, clinicians should have a

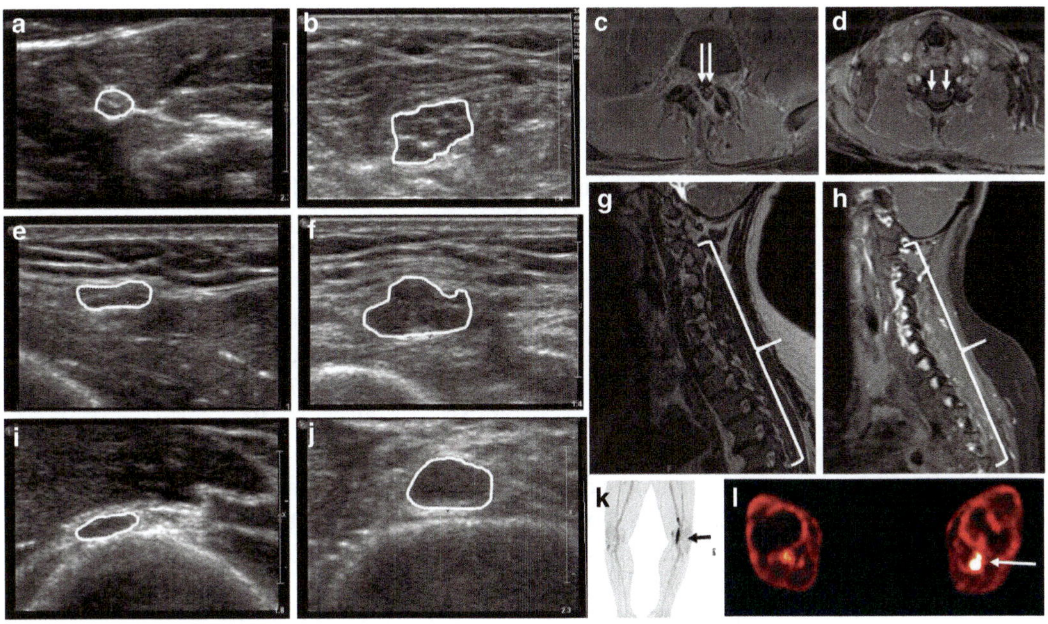


**Fig. 18.1** Imaging features of immune neuropathies. Nerve ultrasound images from a inflammatory demyelinating polyradiculoneuropathy (CIDP) patient demonstrating marked multifascicular enlargement of the median nerve in the forearm (**b**), ulnar nerve in the arm (**f**), and radial nerve at the spiral groove (**j**) as compared to images (**a**), (**e**), and (**i**), respectively, which were obtained at identical sites in a healthy control. Cross-sectional area (patient versus control) at the median nerve in the forearm (20.2 mm$^2$ versus 4.4 mm$^2$), ulnar nerve in the arm (24.6 mm$^2$ versus 7.8 mm$^2$), and radial nerve at spiral groove (18.4 mm$^2$ versus 5.5 mm$^2$). Nerve root enhancement (arrows) in a patient with IgG GM1-positive acute motor axonal neuropathy is evident on a contrast-enhanced T1 MRI of the lumbosacral (**c**) and cervical (**d**) spine. A patient with chronic immune sensory polyradiculopathy (CISP) also has nerve root enhancement (bracket) on T1 MRI post-contrast (**h**) but carries an additional feature , consistent with chronicity, of marked diffuse nerve root enlargement on T2 imaging (**g**, bracket). PET imaging (**k**, **l**) highlighting neurolymphomatosis, an entity commonly misdiagnosed as CIDP, involving the left tibial nerve (arrows) in the popliteal fossa (the right leg is displayed in both images for comparison)

low threshold for pursuing spine imaging in patients presenting with acute flaccid paralysis. Fever at onset is suspect [50]. Prominent encephalopathy is atypical for AMAN and AIDP and warrants consideration of Bickerstaff brainstem encephalitis (BBE), which shares a similar pathophysiological mechanism to MFS [61] (see later in GBS variants). Hearing loss would suggest an alternative diagnosis and has been described in neurosarcoidosis [62]. Pleocytosis in excess of 50 NC/mm$^3$ is atypical and warrants consideration of an infectious etiology, including human immunodeficiency virus (HIV) seroconversion as this can manifest with a GBS-like syndrome [63]. Finally, arsenic toxicity, tick paralysis, diphtheria, and porphyric neuropathy [32] can mimic GBS, but a careful history and exam along with clinically appropriate laboratory evaluation are often adequate in preventing misdiagnosis. Notably, porphyric neuropathy is axonal, thus more likely to be mistaken for axonal forms of GBS, and has prominent proximal weakness [64]. Diphtheric neuropathy may be mistaken for AIDP given the presence of demyelinating features on EDX, but it is more likely to have bulbar involvement at onset and a biphasic course with a second deterioration several weeks after onset [65].

## GBS Variants

Other variants of GBS deserve special consideration given that their presentation is distinct from the "classic" GBS picture. Acute motor sensory axonal neuropathy (AMSAN) along with AMAN has been termed axonal GBS. AMSAN is less common and thought to be a more extensive, severe manifestation of a similar disease that involves sensory as well as motor nerves [66]. An association between AMSAN and systemic lupus erythematosus (SLE) has been demonstrated [67, 68]. MFS is known by its clinical triad of ophthalmoparesis, ataxia, and areflexia, all occurring in the absence of prominent weakness in its classic form [69]. The anatomical localization of ataxia in MFS has been debated, and while cerebellar involvement has been proposed, it has also been posited that sensory ataxia results from proprioceptive loss due to selective involvement of muscle spindle afferents, which would also account for the additional feature of areflexia [70]. Ophthalmoplegia without ataxia and acute ataxic neuropathy represent limited forms of this disorder (i.e., incomplete Fisher syndrome) [71, 72]. Patients with MFS frequently have complete recovery without any residual symptoms, as compared to "classic" GBS patients [73]. The pharyngeal-cervical-brachial variant of GBS is a limited form of axonal GBS that shares some clinical and serologic overlap with MFS. As implied, the syndrome is characterized by oropharyngeal, neck, shoulder, and upper limb weakness [74]. The paraparetic variant is another limited form of GBS that has preferential lower limb involvement although upper limb reflexes and motor NCSs are often abnormal [75].

Idiopathic acute sensory neuronopathy presents with acute sensory ataxia and diminished vibration and joint position sense. As opposed to classic GBS, this syndrome is often asymmetric with upper extremity involvement at onset [76, 77]. Important differential diagnostic considerations for this disorder include pyridoxine toxicity, Sjögren's syndrome, and paraneoplastic disorders (particularly with anti-Hu/ANNA-1 or CRMP-5 antibodies, which are further discussed in Chap. 16) [78]. A more limited sensory presentation of GBS is acute small fiber sensory neuropathy (ASFSN) [79]. Small fiber neuropathy (SFN) as a class usually presents more insidiously with pain as a predominant feature and exam features of diminished pinprick sensation with preserved vibration and joint position sense. It can present with non-length dependent features (i.e., hand numbness concurrent to or preceding foot numbness). This has been attributed to small fiber ganglionopathy (i.e., selective involvement of ganglia subserving small fibers) [80]. More common etiologies of SFN are diabetes, Sjögren's syndrome, SLE, paraneoplastic (ANNA-1), toxins (including alcohol and iatrogenic etiologies), HIV, a number of hereditary conditions, and amyloidosis, but up to half of cases are idiopathic [81]. Acute onset of these clinical features warrants consideration of the relatively rare variant of GBS, ASFSN. Finally, acute autonomic neuropathy/ganglionopathy represents a variant of GBS in which the autonomic nerves and ganglia are selectively involved and have predominant autonomic features classically associated with AIDP. Orthostatic hypotension, gastrointestinal symptoms (e.g., constipation, nausea, diarrhea), heat intolerance, sicca symptoms, bladder dysfunction, and sexual dysfunction have all been reported [82].

## Pathophysiology

Multiple forms of GBS are unequivocally antibody mediated [83]. However, there are no defined antibodies for AIDP, and unlike many other forms of GBS, the pathogenesis of AIDP remains incompletely characterized. It is thought that antibody binding occurs at some as of yet unidentified epitope resulting in complement-mediated demyelination [84]. Some degree of secondary axonal degeneration, which is a major determinant of patient outcomes, may occur [85]. A cell-mediated immune process has been implicated as well [86]. This is supported in that experimental allergic neuritis (EAN) approximates the clinical and pathological characteristics of AIDP. EAN is achieved in animal models by transferring T cells sensitized to epitopes on myelin including P0 and P2 [87]. However, to date, convincing evidence is lacking that would suggest that any of these antigens are the targets of antibody-mediated demyelination in humans with AIDP.

Axonal GBS, on the other hand, has clearly defined antibodies. Anti-GM1, anti-GD1a, anti-GD1b, or anti-GalNac-GD1 IgG antibodies are found in the majority of patients (Table 18.2), particularly in cohorts outside of North America [9, 88–90]. Antibody positivity to complexes of the aforementioned gangliosides (e.g., GD1a/GM1) without reactivity to the isolated antigen has also been reported [91]. Some forms of AMAN are likely associated with antibody-mediated disruption of sodium channel function at the nodes of Ranvier, resulting in "conduction failure" without demyelination. This can result in conduction block characteristic of acute motor conduction block neuropathy (AMCBN), thought to be a limited form of AMAN [92].

Nodo-paranodopathy, in lieu of axonal or demyelinating classification, has been proposed for this disorder so as to focus on antibody-

**Table 18.2** Ganglioside autoantibodies and associated immune neuropathies

| Antibody | Associated immune neuropathy | Notes |
|---|---|---|
| IgG anti-GM1<br>IgG anti-GD1a<br>IgG anti-GalNac-GD1 | AMAN/AMSAN (GD1a rarely present in PBC) | Antigen expressed at node/paranode; autoantibodies instrumental in reversible conduction failure |
| IgG anti-GQ1b | MFS, BBE | Antigen enriched in ocular motor nerves; antibody has immunostained muscle spindle fibers in MFS |
| IgG anti-GT1b | PBC > MFS | Antigen enriched in ocular motor, glossopharyngeal, and vagus nerves |
| IgG anti-GD1b | Idiopathic acute sensory neuropathy | Antigen enriched in dorsal root ganglia |
| IgG anti-ganglionic AChR (alpha-3 subunit) | Acute autonomic neuropathy | Seropositivity predictive of orthostatic hypotension and cholinergic dysautonomia |
| IgM anti-GM1 | MMN | Specificity improved with assay technique and strict titer threshold |
| IgM anti-disialosyl gangliosides (GD1b) | CANOMAD/CANDA | GD3, GT1b, and GQ1b contain disialosyl epitopes; seropositivity likely represents cross-reactivity with GD1b |

*AChR* acetylcholine receptor, *AMAN* acute motor axonal neuropathy, *AMSAN* acute motor sensory axonal neuropathy, *BBE* Bickerstaff brainstem encephalitis, *CANDA* chronic ataxic neuropathies with disialosyl antibodies, *CANOMAD* chronic ataxic neuropathy ophthalmoplegia, *Ig* immunoglobulin, *IgM* paraprotein, cold agglutinins, and disialosyl antibodies, *MFS* Miller Fisher syndrome, *MMN* multifocal motor neuropathy, *PBC* pharyngeal-cervical-brachial

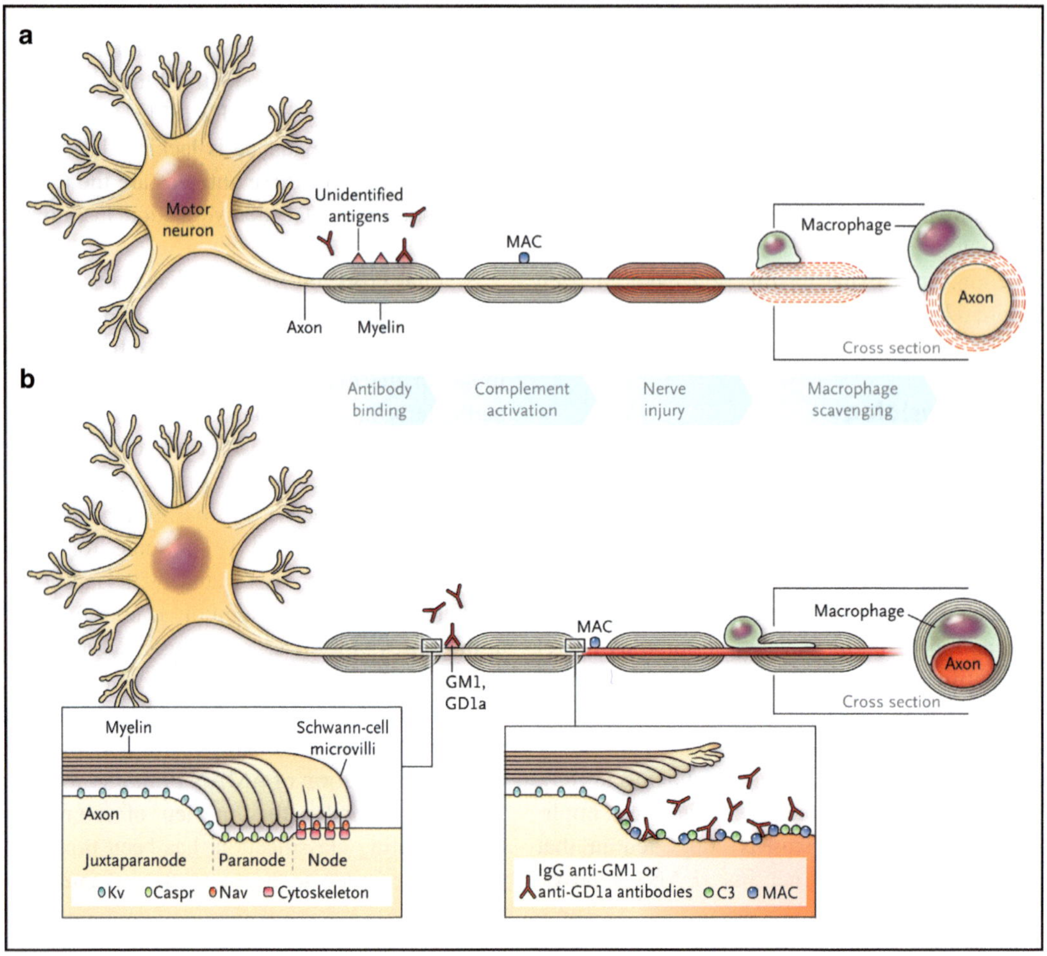


**Fig. 18.2** Possible immunopathogenesis of the Guillain-Barré syndrome. Panel (**a**) shows the immunopathogenesis of acute inflammatory demyelinating polyneuropathy. Although autoantigens have yet to be unequivocally identified, autoantibodies may bind to myelin antigens and activate complement. This is followed by the formation of membrane attack complex (MAC) on the outer surface of Schwann cells and the initiation of vesicular degeneration. Macrophages subsequently invade myelin and act as scavengers to remove myelin debris. Panel (**b**) shows the immunopathogenesis of acute motor axonal neuropathy. Myelinated axons are divided into four functional regions: the nodes of Ranvier, paranodes, juxtaparanodes, and internodes. Gangliosides GM1 and GD1a are strongly expressed at the nodes of Ranvier, where the voltage-gated sodium (Nav) channels are localized. Contactin-associated protein (Caspr) and voltage-gated potassium (Kv) channels are, respectively, present at the paranodes and juxtaparanodes. Immunoglobulin G (IgG) anti-GM1 or anti-GD1a autoantibodies bind to the nodal axolemma, leading to MAC formation. This results in the disappearance of Nav clusters and the detachment of paranodal myelin, which can lead to nerve conduction failure and muscle weakness. Axonal degeneration may follow at a later stage. Macrophages subsequently invade from the nodes into the periaxonal space, scavenging the injured axons. (Reprinted with permission from Yuki and Hartung [3])

mediated damage and dysfunction at the node and paranode (Fig. 18.2 [3]) [93]. Conduction block, in the context of axolemma antibody binding, has been termed reversible conduction failure. Some patients have true reversal of this conduction block upon repeat EDX. Conduction block is a hallmark of demyelinating disease, and thus, AMCBN can initially be mistaken for AIDP [94]. This highlights the importance of repeat EDX several weeks after the initial study to evaluate for more specific findings of AIDP such as increased temporal dispersion or conduction velocity slowing on NCS. The absence of these specific features of demyelination on a follow-up study several weeks after symptom onset in suspected GBS should raise suspicion for reversible conduction failure and axonal GBS [57]. In classic AMAN, repeat EDX performed weeks after onset will demonstrate interval, complement-mediated, axonal degeneration [95]. Evidence in support of this has been demonstrated in a murine model in which administration of eculizumab, a terminal complement pathway inhibitor (C5a and the membrane attack complex, C5b-9), conferred protective effects against anti-ganglioside antibodies [96]. Results of human trials evaluating the efficacy of eculizumab in GBS may alter the therapeutic landscape of the disorder [97]. Some consider axonal GBS (AMCBN, AMAN, AMSAN) to be along a continuum and representative of the same disorder, with AMSAN being a more severe manifestation that involves sensory as well as motor nerves [57]. No differences in respective levels of major gangliosides, including GM1 and GD1a, between motor and sensory nerves have been conclusively demonstrated [98], so the pathophysiological basis for motor nerve preference in limited forms of axonal GBS (AMCBN and AMAN) remains unknown.

Anti-GQ1b and anti-GT1b IgG (Table 18.2), with which it cross-reacts, are common in patients with MFS [3]. Given the ocular findings of MFS, it follows that GQ1b has been shown to be enriched in the ocular motor nerves [99]. A study demonstrated intrafusal fiber immunostaining (in a pattern consistent with muscle spindle involvement) with monoclonal antibody targeting GQ1b, which could be an anatomical explanation for the syndrome's ataxia and areflexia [100]. Anti-GQ1b antibodies likely target the CNS as well, given that these antibodies are also found in cases of Bickerstaff brainstem encephalitis (BBE), and many experts consider BBE and MFS to represent a PNS-CNS spectrum, resulting in the term "Fisher-Bickerstaff syndrome" for patients with combined CNS and PNS features [101]. The pharyngeal-cervical-brachial variant is thought to be an overlap syndrome anatomically between MFS and AMAN, given its clinical appearance. It is associated with antibodies against GT1a (most frequently) as well as GQ1b and GD1a (less commonly) (Table 18.2), the latter supporting an overlap with AMAN [74]. Notably, the GT1a ganglioside is expressed in ocular motor, glossopharyngeal, and vagus nerves, which when seropositive in the absence of cross-reactivity with GQ1b may lend specificity to this limited variant of GBS [102]. Anti-GD1b IgG antibodies (Table 18.2) have been associated with idiopathic acute sensory neuropathy [103]. The GD1b ganglioside is expressed in the dorsal root ganglia, and experimental sensory neuropathy has been induced by immunization of monoclonal antibody targeting GD1b in a rabbit model [104]. Acute autonomic neuropathy has been associated with IgG antibodies against the alpha-3 subunit of the ganglionic acetylcholine receptor, and seropositivity is more likely to be associated with orthostatic hypotension and cholinergic dysautonomia [105] (discussed in Chap. 21).

## Histopathology

Biopsies are not routinely recommended in the evaluation of GBS. However, the pathologic features of AIDP and axonal GBS variants have been well characterized in the literature [85, 106, 107]. AIDP nerve specimens show mononuclear cell infiltrates with segmental demyelination, which often occurs earliest at the paranode via macrophage-mediated myelin stripping. The degree of secondary axonal degeneration is a prognostic indicator for residual disability following the recovery period. Recovery is associated with remyelination, represented pathologically by a diminished internodal distance (i.e., an increase in the number of internodes). In contrast, AMAN is associated with deposition of immunoglobulin and complement on the axolemma, the latter of which is associated with macrophage-mediated axonal degeneration in more severe cases. Demyelination has been demonstrated in the limited histopathology available in MFS [108].

## Prognosis

The prognosis of GBS is generally but not universally favorable. One year after symptom onset, 60% of patients will have full recovery of motor function, while 14% of patients will have prominent residual deficits [109]. Poor prognostic factors are rapid progression with short interval between symptom onset to hospitalization, need for mechanical ventilation, preceding diarrhea, older age, and reduced CMAP amplitude on EDX [110, 111]. A convenient prognostication tool called the Erasmus GBS Outcome Score (EGOS) has been developed that accurately predicts outcome at 6 months based on age, preceding diarrhea, and GBS disability score at 2 weeks [112]. Of note, AMAN has a dichotomous improvement timeline with certain patients having rapid recovery and others going on to have a slow recovery period, the former of which likely represents AMCBN [113, 114].

## Treatment

The mainstay of treatment of GBS, aside from supportive care, is immunomodulatory therapy. Of note, many treatment trials have focused on the AIDP variant of GBS, and their primary endpoints have often focused on short-term outcomes. Plasma exchange (PLEX) was investigated first (i.e., before intravenous immunoglobulin [IVIg]), historically and became standard of care. Therefore, there exist no placebo-controlled trials in adults for IVIg, and most trials were designed as non-inferiority trials comparing IVIg with PLEX. A meta-analysis of PLEX in the treatment of GBS showed that it hastened recovery (primary outcomes) and full recovery without residual weakness at one year was more likely as compared to placebo (secondary outcomes) [115]. IVIg and PLEX subsequently were shown to be equivalent at hastening recovery by 4 weeks in another meta-analysis evaluating IVIg in the treatment of GBS [116]. The greatest benefit of both of these interventions is observed in patients who receive treatment within 2 weeks of symptom onset and in those who are non-ambulatory [112, 117–119]. Medical comorbidities, side effect profiles, and availability of resources should all be considered when deciding between these two treatment modalities.

IVIg is administered as 2 g/kg administered over 2 to 5 days, and a standard course of PLEX typically exchanges of 200–250 mL/kg over 7 to 10 days. Corticosteroids are of no benefit and may even be harmful in AIDP [120]. There is no evidence to suggest added benefit of IVIg following PLEX [121], and there is never a role for PLEX following IVIg given possible resultant diminished efficacy of IVIg. However, the issue of repeat treatment does arise when patients deteriorate after treatment, and typically a second trial of the same modality, whether PLEX or IVIg, is administered in this situation [122]. These deteriorations have been termed treatment-related fluctuations (TRFs). Of note, three or more TRFs should raise suspicion for acute-onset CIDP [123].

There have not been specific treatment trials for AMAN. However, based on a subgroup analysis of patients with "motor GBS" in the Dutch Guillain-Barré Study Group, IVIg was demonstrated to hasten recovery over PLEX [124], but no difference in clinical outcome between the two treatment modalities was demonstrated in another large study [11]. There is little data to guide treatment of AMSAN and other less common variants of GBS, and treatment is extrapolated from the more common AIDP and AMAN (i.e., PLEX or IVIg). The natural history of AMCBN is unclear in that it is unknown whether RCF improves as a result of therapy or as a natural course of the disease. It is reasonable to treat AMCBN as one would AIDP (i.e., IVIg or PLEX). Most cases of MFS improve as part of its natural course. IVIg hastens recovery without affecting outcome [125].

# Chronic Inflammatory/Immune Demyelinating Polyradiculoneuropathy

Chronic inflammatory/immune demyelinating polyradiculoneuropathy (CIDP) has several features that overlap with GBS, including its distribution of PNS involvement (i.e., nerve roots, plexus, and peripheral nerve) and a classification scheme that includes a classic form and several variants. Both conditions are motor-predominant with symmetric, proximal, and distal weakness and hypo-/areflexia. They both demonstrate features of acquired demyelination on electrodiagnostic testing (i.e., the AIDP variant of GBS), cytoalbuminologic dissociation on CSF analysis, and nerve root enhancement on MRI and are responsive to immunomodulatory therapy [126].

There are, however, a number of key distinguishing features between these two entities [127, 128]. For instance, CIDP is classically more insidious than GBS. Only a minority of patients with CIDP come to medical attention soon after symptom onset. The relapsing-remitting and/or progressive natural history of CIDP is distinct from the monophasic course characteristic of GBS (although a minority of patients with CIDP can present acutely). Cranial nerve involvement, respiratory failure, and autonomic features are all far less common in CIDP relative to AIDP. A history of antecedent events (e.g., infections, vaccinations) is also far less common in CIDP than GBS.

This section will focus on "classic CIDP" and its better-known variants: sensory predominant CIDP, distal acquired demyelinating symmetric (DADS) neuropathy, chronic immune sensory polyradiculopathy (CISP), and multifocal acquired demyelinating sensory and motor (MADSAM) neuropathy, also known as multifocal CIDP or Lewis-Sumner syndrome. We will also discuss multifocal motor neuropathy (MMN), a distinct entity that shares some features with CIDP, particularly the MADSAM variant. The differential diagnosis for CIDP, including acquired demyelinating polyneuropathies with and without monoclonal gammopathy, will also be discussed.

## Epidemiology

CIDP has a prevalence of 1–8.9 per 100,000 depending on the diagnostic criteria used and the population studied [129–132]. As with GBS, there is a male predominance, and prevalence increases with age [133]. There appears to be a relationship between age and clinical course. Much like that of multiple sclerosis (MS), the clinical course of CIDP may evolve in a "relapsing-remitting" or "progressive" fashion. Younger patients with CIDP tend to have relapses, while older patients more commonly have insidious progression [134]. While an association with antecedent events is less established in CIDP as compared to GBS, there are associations between exposure to some drugs (e.g., tumor necrosis factor alpha inhibitors) and development of CIDP, thereby bearing further resemblance to MS. While rare, there are patients with CIDP that may exhibit features of concurrent CNS demyelinating disease and even meet diagnostic criteria for MS. This syndrome is sometimes referred to as combined central and peripheral nervous system demyelinating disease or CCPD. Many of these patients have been found to have antibodies against nodal and paranodal antigens, like neurofascin (discussed in more detail later). There are also controversial associations between CIDP and other systemic diseases including HIV, melanoma, connective tissue diseases, organ transplantation, monoclonal gammopathy of undetermined significance (MGUS), hepatitis C, and diabetes mellitus [127]. However, the lattermost association has been called into question [130].

## Pathophysiology

It is clear that both cellular and humoral immune mechanisms play a role in CIDP, but the exact interplay between these mechanisms remains incompletely elucidated [135]. Cell-mediated immunity is supported by the fact that populations of CD4+ and CD8+ T cells are found in biopsy specimens of CIDP [136]. Among the first steps in the immunopathogenesis of CIDP are invasion of peripheral nerves and breakdown of the blood-nerve barrier (BNB) by activated T cells. BNB permeability in CIDP and resultant immune cell penetration of peripheral nerve have been evidenced by several findings. Firstly, downregulation or aberrant expression of tight junction proteins (e.g., claudin 5 and ZO-1) on sural nerve biopsy specimens of CIDP has been shown [137]. Increased levels of cytokines including chemokines have been demonstrated in CSF and sera of patients with CIDP [138, 139]. Finally, downregulation of integrins has been demonstrated in biopsy specimens in EAN [140], and increased expression of matrix metalloproteinases has been shown in biopsy specimens from CIDP [141]. Once T cells have invaded the nerve, the putative mechanism is clonal expansion by antigen presentation of some as of yet to be defined epitope. There is evidence that antigen-presenting cells are instrumental in development of CIDP since a spontaneous immune neuropathy with similar characteristics to CIDP has been shown to develop in non-obese diabetic mice deficient in the B7-2 costimulatory molecule [142]. This molecule is found on antigen-presenting cells and is integral to costimulation, which results in the terminal differentiation of T cells and regulation of local immune response. Notably, increased BNB permeability also allows access to molecules that are normally prevented from accessing the nerve roots and peripheral nerves, such as antibodies.

Involvement of the humoral arm of the immune system is supported by the fact that some CIDP variants are associated with antibodies directed against nodal/paranodal proteins such as contactin-1, contactin-associated protein 1 (CASPR1), and different isoforms of neurofascin (NF) (e.g., NF140, NF155, NF186) (Fig. 18.2) [143–145]. It is important to note that autoantibodies are rare in CIDP (i.e., the majority of patients are seronegative) [146]. Nonetheless, the presence of these autoantibodies suggests antibody-mediated disruption of non-compact myelin, and its interactions with the axon at the node/paranode may be instrumental for development of CIDP in a certain subpopulation. Indeed as discussed with AMAN, conduction failure, in this case due to compromised integrity of the Schwann cell-axon interaction, may be one potential mechanism by which weakness develops in CIDP and may explain the rapid recovery sometimes seen following treatment [147].

However, in most cases of CIDP, early weakness results from demyelinating conduction block (i.e., slow conduction velocities in isolation do not result in weakness). Much like AIDP, macrophages are believed to be the terminal antigen-presenting cells involved in the cell-mediated pathway discussed previously, and macrophage-mediated demyelination is a characteristic pathologic feature of classic CIDP [148, 149]. As the disease progresses, and particularly in patients with treatment refractory disease or for whom treatment is delayed, motor deficits may result from secondary axonal loss (similar to that seen in severe forms of AIDP) [150]. Perivascular macrophages and focal collections of lymphocytes, primarily T cells, are common [151]. In fact, impaired T-cell regulation is thought to underlie the persistent or recurrent immune response characteristic of CIDP distinguishing it from GBS [152].

## Clinical Features

The cardinal signs of CIDP are symmetric, non-length-dependent (i.e., proximal and distal) weakness and sensory loss preferentially affecting large fibers (loss of joint position and vibratory sensation) with frequent involvement of the hands soon after the feet [153, 154]. This distinguishes it from a classic length-dependent axonal polyneuropathy in which motor deficits are typically minimal and sensory symptoms in the fingertips would only be expected after lower extremity symptoms have ascended to near the level of the knees. These objective sensory deficits may be helpful in distinguishing acute-onset CIDP from AIDP [155]. Dysesthesia may occur but is less pronounced (consistent with relative sparing of unmyelinated C fibers). Postural tremor is fairly common in CIDP, reported in up to 80% of patients per one study [156]. A pure motor form exists in which sensory involvement is entirely spared [157]. A sensory predominant form exists as well, but EDX often show subclinical motor involvement, or this develops in a delayed fashion [158, 159]. Autonomic dysfunction, respiratory failure/insufficiency, and craniobulbar deficits are all much less common in CIDP, relative to AIDP [123, 160].

## Histopathology and Neuroimaging

Nerve biopsies are not routinely performed in the evaluation of CIDP. In the majority of cases, accurate diagnosis can be reached based on established clinical criteria, obviating the need for biopsy. Furthermore, the potentially low yield of a biopsy from the sural or superficial peroneal nerve (the two most commonly selected nerves for diagnostic biopsies) in an often motor-predominant condition may not outweigh the inherent invasiveness of the procedure. Further, the most prominent changes often occur in proximal nerve segments, which are not easily accessible to biopsy, and "sural sparing" is also a common finding in CIDP.

Nerve biopsy, nonetheless, is useful in atypical cases for which the diagnosis is not certain, and in these cases, pathology can distinguish between CIDP and mimics such as certain hereditary demyelinating neuropathies, sarcoidosis, amyloidosis, and vasculitis. Nerve pathology in CIDP includes segmental demyelination and remyelination with resultant onion bulb formation, a hallmark of chronic demyelinating disorders of the PNS (both acquired and hereditary) [161]. These findings can be visualized on routine histopathology or on teased nerve fibers or via electron microscopy. Mononuclear cell infiltrates can be visualized in the endoneurium. As with AIDP, macrophages may be visualized actively stripping away superficial myelin layers from the myelin sheath with eventual phagocytosis of myelin, a process that has been termed macrophage-mediated demyelination [107, 162, 163]. Secondary axonal degeneration may also be seen, which better correlates with poor prognosis than the presence of demyelination [164].

On MRI/magnetic resonance neurography, nerve root enhancement along with enlargement may be visualized [165] (Fig. 18.1). Nerve ultrasound, particularly of proximal nerve segments, may show nerve enlargement, increased echogenicity, increased vascularization, and fascicular enlargement [166] (Fig. 18.1). Moreover, nerve enlargement has been shown to correlate with disease activity with reduction in size occurring during remission and increase in size in refractory cases [167].

## Electrodiagnostics

In contrast to GBS, most diagnostic criteria for CIDP, each with its own sensitivity and specificity, mandate electrophysiologic evidence of acquired demyelination [153, 168, 169]. The European Federation of Neurological Societies and Peripheral Nerve Society (EFNS/PNS) criteria (Table 18.3) are considered best to capture patients with CIDP for both clinical and research purposes [170]. These criteria were validated in a

**Table 18.3** Electrodiagnostic criteria

| |
|---|
| (1) Definite: at least one of the following |
| (a) Motor distal latency prolongation ≥50% above ULN in two nerves (excluding median neuropathy at the wrist from carpal tunnel syndrome) |
| (b) Reduction of motor conduction velocity ≥30% below LLN in two nerves |
| (c) Prolongation of F-wave latency ≥30% above ULN in two nerves (≥50% if amplitude of distal negative peak CMAP <80% of LLN values) |
| (d) Absence of F-waves in two nerves if these nerves have distal negative peak CMAP amplitudes ≥20% of LLN + ≥ one other demyelinating parameter[a] in ≥ one other nerve |
| (e) Partial motor conduction block: ≥50% amplitude reduction of the proximal negative peak CMAP relative to distal, if distal negative peak CMAP ≥20% of LLN, in two nerves, or in one nerve + ≥ one other demyelinating parameter[a] in ≥ one other nerve |
| (f) Abnormal temporal dispersion (>30% duration increase between the proximal and distal negative peak CMAPs) in ≥ two nerves, or |
| (g) Distal CMAP duration (interval between onset of the first negative peak and return to baseline of the last negative peak) increase in ≥ one nerve (median ≥6.6 ms, ulnar ≥6.7 ms, peroneal ≥7.6 ms, tibial ≥8.8 ms)[b] + ≥ one other demyelinating parameter[a] in ≥ one other nerve |
| (2) Probable |
| ≥30% amplitude reduction of the proximal negative peak CMAP relative to distal, excluding the posterior tibial nerve, if distal negative peak CMAP ≥ 20% of LLN, in two nerves, or in one nerve + ≥ one other demyelinating parameter[a] in ≥ one other nerve |
| (3) Possible |
| As in (1) but in only one nerve |

Reprinted with permission from Van den Bergh et al. [170]

*ULN* upper limit of normal, *LLN* lower limit of normal, *CMAP* compound muscle action potential multicenter study, which demonstrated a sensitivity and specificity of 81% and 96%, respectively [171]. The hallmarks of acquired demyelination on EDX are nonuniform slowing of conduction velocities, conduction block, prolonged distal latencies, increased temporal dispersion, and abnormalities on late responses (i.e., F-wave and H-reflex studies) [172]. As noted previously, NCSs may show abnormalities that may be subclinical. The pattern of a small median sensory response with a normal sural sensory response, also known as "sural sparing," is a phenomenon common to both CIDP and AIDP. The reason for this observation is not well defined, but working hypotheses center around the location of where the studies are performed on the limb. The distal leg is the recording site for the sural antidromic sensory NCS, where the nerve is larger and has not yet tapered, and the recording site for the median and ulnar antidromic sensory NCSs is the distal digits, where the nerves have tapered and are thin. Whether this phenomenon is the result of early distal involvement or preferential involvement of the smaller myelinated fibers (both possibilities could account for selective involvement of median and ulnar nerves) is uncertain [173]. An exception to these typical electrodiagnostic findings exists for the variant referred to as chronic immune sensory polyradiculopathy (CISP), in which electrophysiologic abnormalities may be restricted to the late responses (i.e., absent or abnormal H-reflexes) [174].

## CIDP Variants

DADS neuropathy is a variant of CIDP manifesting with distal, symmetric sensory loss and sensory ataxia with minimal motor deficits, particularly early in the disease course (Table 18.4). It is a slowly progressive condition [175]. Distinct electrodiagnostic features include markedly prolonged distal latencies without conduction block and mild slowing of conduction velocities. Prolonged residual latency (RL) and a low terminal latency index (TLI), both electrophysiologic measures of selectively distal demyelination, are defining features of this condition [176]. A distinction is made based on the presence or absence of monoclonal protein (M-protein), which when present is most often IgM kappa. Monoclonal anti-myelin-associated glycoprotein (MAG) IgM is present in the majority of patients with DADS with M-protein, and these MAG autoantibodies are likely pathogenic [177]. This entity is often termed anti-MAG neuropathy in the literature and is considered by many experts to be distinct from CIDP. Notably, DADS patients without M-protein may respond to standard therapies utilized in classic CIDP [175], whereas DADS with M-protein is poorly responsive (or only transiently responsive) to standard CIDP therapies such as IVIg [178]. Rituximab is considered by many to be a better treatment option for DADS with M-protein. A randomized controlled trial (RCT) [179] that has been critiqued as providing low-quality evidence [178] showed a measurable benefit with rituximab. However, this was not replicated in a larger follow-up trial [180]. Nonetheless, experts have cited disparate outcome measures in the RCTs as compared to those of a previous uncontrolled study that demonstrated benefit [181], which may account for the discrepancy in results. It is also worth noting that the aforementioned uncontrolled trial administered maintenance dosing of rituximab, another inconstant variable across studies. As it happens, there is evidence to support the addition of interventions such as fludarabine or plasma exchange in patients with forms of IgM autoantibody-associated demyelinating polyneuropathies that are refractory to rituximab monotherapy [182, 183]. Thus, some patients with DADS with M-protein or anti-MAG neuropathy may benefit from rituximab, but given differences in dosing protocols across studies showing variable degrees of efficacy based on outcome measures tested, additional RCTs are warranted.

CISP is a pure sensory syndrome involving the dorsal roots and as such is the one CIDP variant with normal motor and sensory NCSs (Table 18.4). However, H-reflex studies and somatosensory evoked potentials (SSEPs) are often abnormal [174]. Evidence in favor of clas-

**Table 18.4** CIDP variants and mimics

| | Classic CIDP | DADS | CISP | MADSAM | *MMN*[a] | *CANOMAD/ CANDA* | *POEMS* |
|---|---|---|---|---|---|---|---|
| Distribution | Proximal/Distal | Distal | Proximal (nerve roots) | Multifocal | Multifocal | Distal | Legs > arms |
| Symmetry | Yes | Yes | Yes | No | No | Yes | Yes |
| Conduction block | Yes | No | No | Yes | Yes | No | No |
| Gammopathy | Absent or IgA/IgG | Usually IgM kappa | Usually absent | Usually absent | Usually absent | Usually IgM kappa | Usually lambda-restricted |
| Cytoalbuminologic dissociation | Present | Present | Present | Present | Absent | May be present | Usually present |
| Preferred treatment | IVIg or corticosteroids[b] | Possibly rituximab | IVIg or corticosteroids | IVIg > corticosteroids | IVIg | IVIg, rituximab | High-dose melphalan with auto-SCT |
| Autoantibody | Anti-NF, CASPR1, or contactin[c] | Anti-MAG | – | – | Anti-GM1 | – | – |

*CANDA* chronic ataxia neuropathies with disialosyl antibodies, *CANOMAD* chronic ataxic neuropathy, ophthalmoplegia, IgM paraprotein, cold agglutinins, and disialosyl antibodies, *CASPR1* contactin associated protein 1, *CIDP* chronic inflammatory demyelinating polyradiculoneuropathy, *CISP* chronic immune sensory polyradiculopathy, *DADS* distal acquired demyelinating symmetric, *Ig* immunoglobulin, *IVIg* intravenous immunoglobulin, *MADSAM* multifocal acquired demyelinating sensory and motor, *MAG* myelin-associated glycoprotein, *MMN* multifocal motor neuropathy, *NF* neurofascin, *POEMS* polyneuropathy organomegaly, endocrinopathy, M-protein, and skin changes, *SCT* stem cell transplantation

[a]Italics indicate CIDP mimic

[b]Possibly rituximab for CIDP with concurrent MGUS, systemic autoimmune disease, or nodal/paranodal autoantibody seropositivity

[c]Nodal/paranodal autoantibodies only positive in a minority of patients with CIDP

sifying this disorder as a CIDP variant (restricted to the dorsal roots) is the presence of cytoalbuminologic dissociation, nerve root enhancement and enlargement on MRI (Fig. 18.1), and evidence in the literature for pathologic features of acquired demyelination in cases in which dorsal nerve root biopsy was pursued. As with CIDP as a whole, biopsy is usually not necessary to confirm diagnosis.

MADSAM neuropathy is a CIDP variant that is unique in its relatively asymmetric, multifocal presentation (Table 18.4). The clinical syndrome is that of an upper limb predominant, mononeuropathy multiplex with demyelinating features [184], in contradistinction to the axonal features characteristic of peripheral nervous system vasculitis, which also often presents as mononeuropathy multiplex [185]. Both motor and sensory involvement occur in individual nerve distributions but can progress to the point of confluence, at which point the condition more closely resembles a severe form of classic CIDP. EDX abnormalities in affected nerves are similar to those typical of classic CIDP. However, unlike classic CIDP, there is literature to suggest that IVIg is preferred over corticosteroids in MADSAM [186, 187]. Some experts consider MADSAM and multifocal motor neuropathy (MMN, discussed in more detail in the next section) to be related disorders on a disease spectrum, given the preferential response to IVIg over corticosteroids [188].

## Multifocal Motor Neuropathy

MMN is an acquired, immune-mediated, demyelinating neuropathy that shares many features with CIDP and particularly MADSAM (as noted previously) but has notable disparate features including absent or minimal sensory involvement, frequent antiganglioside seropositivity, and lack of response to steroids. In fact, many consider MMN and CIDP to be on ends of a spectrum with MADSAM serving as an intermediate link [188–190]. MMN typically presents with upper extremity distal weakness in the distribution of individual nerves, rather than the myotomal distribution more typical of degenerative motor neuron disease, such as amyotrophic lateral sclerosis (ALS). It is often associated with anti-GM1 IgM [191], but seronegative forms are common. Indeed, the role of anti-GM1 IgM in the pathogenesis of MMN is controversial [192]. These antibodies may also be found in the sera of patients with motor neuron disease and non-neural autoimmune disorders [193]. Specificity for MMN may be improved by assay technique and titer threshold [194]. Electrodiagnostic testing plays a vital role in distinguishing MMN from a lower motor neuron predominant form of ALS in that a hallmark electrodiagnostic feature of MMN is the presence of conduction block [195]. Cytoalbuminologic dissociation is notably absent in MMN [196]. There is evidence for efficacy of IVIg in the treatment of MMN [121, 197, 198]. Both corticosteroids [199] and PLEX [200] can actually precipitate worsening in some patients. Patients refractory to IVIg, or for whom IVIg loses its efficacy, may respond to rituximab [201] or cyclophosphamide [202].

Another exceedingly rare entity to consider in the differential for MMN is what has been termed multifocal acquired motor axonopathy (MAMA), which is an immunotherapy-responsive disorder characterized by weakness in the distribution of individual nerves but without conduction block or other demyelinating features on EDX and without associated GM1 autoantibodies [203]. The absence of paraspinal abnormalities on EMG has been shown to be useful in distinguishing this from its primary mimic, motor neuron disease [204]. The recent advent of CSF and serum biomarkers, such as phosphorylated neurofilament heavy chain (pNFH), as a specific means of distinguishing ALS from disease mimics like MAMA will certainly aid in selecting out individuals more likely to respond to immunomodulatory therapies [205, 206]. Caution is warranted when MAMA is a diagnostic consideration as the absence of conduction block has been shown to be predictive of immune therapy non-responders and is found more commonly in patients with motor neuron disease, who were initially misdiagnosed [207].

## Differential Diagnosis

CIDP is an overdiagnosed entity [208]. One of the reasons for this is overreliance on supportive diagnostic findings, such as cytoalbuminologic dissociation, in the context of equivocal clinical features. Another common pitfall is misinterpretation of, or technically suboptimally performed, electrodiagnostic studies resulting in an inappropriate designation of a polyneuropathy as having acquired, primary demyelinating features [209]. Length-dependent axonal polyneuropathy can have conduction slowing due to dropout of the fastest-conducting nerve fibers—a finding often misinterpreted as evidence of primary demyelination [210]. The risk for misdiagnosis and inappropriate administration of immunomodulatory therapy is compounded by inconsistent adherence to diagnostic criteria for CIDP by nonspecialized neurologists [211]. The population that is perhaps most vulnerable to misdiagnosis is diabetics, given that cytoalbuminologic dissociation is common in this population, and features supportive of primary demyelination may be seen on electrodiagnostic testing, classically slow conduction velocities without concurrent increased temporal dispersion, or conduction block [212].

Once a primary demyelinating neuropathy is confirmed with confidence, a conclusion based upon clinical and electrophysiologic evidence, one must rule out hereditary neuropathies. Features suggestive of a hereditary etiology include a positive family history and specific electrodiagnostic features. Most forms of demyelinating hereditary motor sensory neuropathy (HMSN, also known as Charcot-Marie-Tooth [CMT]) show uniform slowing, lack conduction block, or increased temporal dispersion and show prolonged F-wave latencies but preserved persistence and normal chronodispersion [213]. Hereditary neuropathy with liability to pressure palsies (HNPP) will manifest with multiple painless entrapment neuropathies [214]. In rare circumstances, amyloidosis, particularly transthyretin familial amyloid polyneuropathies (TTR-FAPs), may have demyelinating features. Distinguishing features from CIDP include prominent pain, dysautonomia (associated with involvement of C and Aδ[delta] fibers), and concurrent axonal loss on EDX, which can be severe [215].

Gammopathy deserves special attention with respect to its role in immune-mediated neuropathies. Indeed, MGUS has an association with polyneuropathy in general [216], and it follows that serum screening for gammopathy is one of the three diagnostic tests (along with serum B12 with metabolites and a screening test for diabetes/impaired glucose tolerance) recommended by the American Academy of Neurology (AAN) in the evaluation of patients with a distal, symmetric polyneuropathy [217]. CIDP associated with IgA or IgG M-protein has a similar clinical course (including a similar response to immunomodulatory therapy) to that of classic CIDP [218]. An important caveat exists for CIDP with M-protein that has an associated lambda light chain. Thorough screening for POEMS syndrome is warranted in this case (see later). Demyelinating polyneuropathy with an IgM M-protein has distinct clinical features from those associated with IgA or IgG M-protein and warrants additional evaluation. As noted previously, an IgM kappa M-protein is common in patients with MAG autoantibodies and clinical features of a sensory predominant polyneuropathy with distal demyelinating features (i.e., DADS) [175]. A thorough evaluation for hematologic disorders is indicated in patients with the DADS phenotype, with or without MAG antibodies, as they may have comorbid mixed cryoglobulinemia (often associated with hepatitis C infection) or Waldenstrom's macroglobulinemia, neither of which will respond to the standard immunomodulatory therapies utilized for CIDP [219, 220].

An IgM M-protein is also a feature of the syndrome referred to as chronic ataxic neuropathy, ophthalmoplegia, cold agglutinins, and disialosyl antibodies (CANOMAD) syndrome (Table 18.4) [221, 222]. This syndrome has some clinical overlap with other immune neuropathies (e.g., the sensory ataxia of DADS and the ophthalmoparesis and cranial neuropathies of MFS/GBS) and as the name implies is associated with IgM against disialosyl gangliosides (i.e., primarily GD1b). Sensory ataxia resulting in a severe gait disorder and ophthalmoparesis (either fixed or episodic) dominate the clinical picture. Clinical phenotypes that do not have all features of CANOMAD (e.g., the IgM M-protein is a cold agglutinin in only about 50% of patients) have been called chronic ataxic neuropathy with disialosyl antibodies (CANDA). Nerve ultrasound shows nerve enlargement. IVIg and rituximab have been shown to have treatment efficacy.

As noted previously, a lambda-restricted monoclonal gammopathy (usually IgA lambda or IgG lambda M-protein) [223] in a patient presenting with clinical and electrodiagnostic features of CIDP should raise concern for polyneuropathy, organomegaly, endocrinopathy, monoclonal protein, and skin changes (POEMS) syndrome (Table 18.4). POEMS syndrome by diagnostic criteria is always associated with polyneuropathy and a monoclonal plasmaproliferative disorder and in the vast majority of cases is associated with osteosclerotic myeloma (evaluation for which is best accomplished by bone survey) or Castleman disease [223]. A monoclonal protein cannot be identified on serum or urine immunofixation in a minority of patients, in which case it must be identified by immunohistochemical analysis of a bone marrow biopsy. Therefore, a high index of suspicion is warranted. Nevertheless, in a large case series, bone lesions were identified in the vast majority of patients, and when rarely absent, patients had four out of five of the features enumerated by the acronym [223], implying that the absence of monoclonal protein and a negative bone survey without multiple minor criteria would likely suffice in excluding the diagnosis.

Unlike "classic CIDP," the polyneuropathy of POEMS is characterized by lower extremity predominance and earlier, often severe, concurrent axonal loss [224, 225]. Furthermore, there are certain electrodiagnostic features that can aid in distinguishing POEMS from CIDP. While both may have electrophysiological evidence of demyelination, POEMS is more likely to have uniform demyelination (i.e., less temporal dispersion or conduction block) and intermediate nerve segment conduction slowing resulting in a higher TLI, to have more axonal loss (smaller motor responses and a length-dependent pattern of active denervation on EMG), and to lack sural sparing [226, 227]. CSF protein is elevated. Papilledema is rare in CIDP and common in POEMS [223, 227, 228]. Markedly elevated vascular endothelial growth factor (VEGF) levels are also a useful diagnostic and therapeutic biomarker of POEMS syndrome, and VEGF is suggested to be pathogenic in the neuropathy [229]. In the absence of a solitary bone lesion amenable to targeted radiation therapy, high-dosage melphalan chemotherapy with autologous stem cell transplantation, for those who are able to tolerate it, results in significant improvement of neuropathy [230]. As a treatable disease with a distinct therapeutic approach from that of CIDP, POEMS should always remain in the differential diagnosis for atypical or refractory CIDP.

## Treatment

The two first-line therapies for CIDP are corticosteroids [231] and IVIg [232]. Corticosteroids, notably relatively contraindicated in GBS, can be administered in several different regimens. A common practice is to administer 1–1.5 mg/kg/day of oral prednisone (or prednisone equivalents) up to 100 mg [233]. Patients are then tapered based on the rate and the degree of response. An alternative to daily oral therapy is pulse corticosteroids. The most commonly utilized pulse corticosteroid regimens are either weekly IV methylprednisolone (e.g., an initial loading dose of 3–5 grams over 3–5 days followed by 1 gram weekly for 8–12 weeks before tapering the dose or frequency) [234] or monthly oral dexamethasone (e.g., 40 mg daily for 4 days every 28 days) [233]. While both daily and pulse corticosteroids have demonstrated similar efficacy, pulse therapy may be preferable as it is less likely to cause weight gain, Cushingoid features, and insomnia [233, 234].

IVIg has similar efficacy to corticosteroids and is typically administered as a "loading" dose of 2 g/kg administered over 2 to 5 days, after which maintenance dosing is 1 g/kg administered every 3 to 4 weeks. Adjustments in dose and frequency should be made based on the response and duration of benefit [235]. There is evidence that subcutaneous immunoglobulin (SCIG) is also effective as maintenance therapy in CIDP. The maintenance dosage is similar to that of IVIg but is divided weekly. Its ease of administration makes it an attractive alternative, and it was shown to have a similar rate of relapse as seen with IVIg and a good safety profile [236].

PLEX has been demonstrated to have short-term benefit in CIDP [237]. However, deterioration may occur after its transient effects have waned, and PLEX is associated with a notable risk of adverse events making maintenance therapy ill-advised (for neurological disorders in general). Thus, PLEX is typically reserved for patients who are refractory to all other forms of treatment.

The goal of therapy in CIDP, as in many other chronic immune-mediated disorders, is to achieve "remission." After initiating therapy, it is important to have relatively close monitoring of the patient to evaluate their response. A suggested frequency is 3 months. The disease state while on treatment can be classified in a cross-sectional fashion according to the following schema: cure (>5 years off treatment without disease activity), remission (<5 years off treatment without disease activity), stable active disease (>1 year on treatment without disease activity), improvement (>3 months and <1 year on treatment without disease activity), and unstable active disease [238]. Treatment should be continued for a minimum of 1 year after clinical stability is achieved, after which cessation of immunomodulatory therapy can be considered. Continued monitoring for clinical evidence of relapse is critical. If a patient fails to respond to corticosteroids or IVIg monotherapy, the remaining first-line agent or combination therapy should be attempted before consideration of second-line therapies. Forms of CIDP that fail to respond to the aforementioned agents are considered treatment refractory and warrant a re-evaluation of the diagnosis to ensure that a mimic of CIDP has not been missed.

There is little utility in trialing azathioprine [239] or mycophenolate mofetil [240] in patients who fail to respond to corticosteroids, as the main role for these drugs in CIDP is as steroid-sparing agents. Furthermore, the 6–12 months of therapy that is often necessary to achieve an observed effect with these agents is not practical for a treatment refractory population. Evidence supporting the use of cyclosporine A [241, 242], methotrexate [243], and cyclophosphamide [244] in refractory CIDP is limited to studies with relatively low levels of evidence.

Rituximab warrants specific consideration in patients with treatment refractory CIDP associated with contactin-1 or neurofascin autoantibodies (often IgG4 predominant). There have been promising preliminary data for rituximab in the treatment of this patient population [245], which is perhaps unsurprising given the known efficacy of rituximab in other IgG4-mediated disorders (e.g., MuSK myasthenia gravis) [246]. There is also recent evidence to suggest that rituximab may be of benefit in treatment refractory CIDP with comorbid systemic autoimmune disease [247] or hematologic disease [248].

## Conclusion

Immune-mediated demyelinating neuropathies constitute a heterogeneous class of treatable peripheral nervous system disorders. While the last century has witnessed great advances in our understanding of the pathophysiology of many of these disorders, many elements remain unknown and are the subject of ongoing investigations. The establishment of reliable diagnostic criteria and the addition of diagnostic modalities beyond electrodiagnostic testing, including autoantibody testing and peripheral nerve imaging, have aided in establishing earlier diagnosis and distinguishing rare variants that predict a response to specific forms of immunomodulatory therapy. Continued growth in our understanding of the pathophysiology of these conditions should translate, in the not-too-distant future, into more disease-specific, targeted immunomodulatory therapies for these disorders, akin to the rapid expansion of therapies for MS in recent years.

## References


1. Willison HJ, Jacobs BC, van Doorn PA. Guillain-Barré syndrome. Lancet. 2016;388(10045):717–27.
2. Hughes RA, Wijdicks EF, Benson E, Cornblath DR, Hahn AF, Meythaler JM, et al. Supportive care for patients with Guillain-Barré syndrome. Arch Neurol. 2005;62(8):1194–8.
3. Yuki N, Hartung HP. Guillain-Barré syndrome. N Engl J Med. 2012;366(24):2294–304.
4. van den Berg B, Walgaard C, Drenthen J, Fokke C, Jacobs BC, van Doorn PA. Guillain-Barré syndrome: pathogenesis, diagnosis, treatment and prognosis. Nat Rev Neurol. 2014;10(8):469–82.
5. Sejvar JJ, Baughman AL, Wise M, Morgan OW. Population incidence of Guillain-Barré syndrome: a systematic review and meta-analysis. Neuroepidemiology. 2011;36(2):123–33.
6. Morris AMS, Elliott EJ, D'Souza RM, Antony J, Kennett M, Longbottom H. Acute flaccid paralysis in Australian children. J Paediatr Child Health. 2003;39(1):22–6.
7. Doets AY, Verboon C, van den Berg B, Harbo T, Cornblath DR, Willison HJ, et al. Regional variation of Guillain-Barré syndrome. Brain. 2018;141(10):2866–77.
8. Ho TW, Mishu B, Li CY, Gao CY, Cornblath DR, Griffin JW, et al. Guillain-Barré syndrome in northern China relationship to Campylobacter jejuni infection and anti-glycolipid antibodies. Brain. 1995;118(3):597–605.
9. Ogawara K, Kuwabara S, Mori M, Hattori T, Koga M, Yuki N. Axonal Guillain-Barré syndrome: relation to anti-ganglioside antibodies and Campylobacter jejuni infection in Japan. Ann Neurol. 2000;48(4):624–31.
10. Nachamkin I, Arzarte Barbosa P, Ung H, Lobato C, Gonzalez Rivera A, Rodriguez P, et al. Patterns of Guillain-Barré syndrome in children: results from a Mexican population. Neurology. 2007;69(17):1665–71.
11. Hadden RD, Cornblath DR, Hughes RA, Zielasek J, Hartung HP, Toyka KV, et al. Electrophysiological classification of Guillain-Barré syndrome: clinical associations and outcome. Plasma exchange/Sandoglobulin Guillain-Barré Syndrome Trial Group. Ann Neurol. 1998;44(5):780–8.
12. Uncini A, Manzoli C, Notturno F, Capasso M. Pitfalls in electrodiagnosis of Guillain-Barré syndrome subtypes. J Neurol Neurosurg Psychiatry. 2010;81(10):1157–63.
13. Bae JS, Yuki N, Kuwabara S, Kim JK, Vucic S, Lin CS, et al. Guillain-Barré syndrome in Asia. J Neurol Neurosurg Psychiatry. 2014;85(8):907–13.
14. Webb AJ, Brain SA, Wood R, Rinaldi S, Turner MR. Seasonal variation in Guillain-Barré syndrome: a systematic review, meta-analysis and Oxfordshire cohort study. J Neurol Neurosurg Psychiatry. 2015;86(11):1196–201.
15. Goodfellow JA, Willison HJ. Guillain-Barré syndrome: a century of progress. Nat Rev Neurol. 2016;12(12):723–31.
16. Nachamkin I, Allos BM, Ho T. Campylobacter species and Guillain-Barré syndrome. Clin Microbiol Rev. 1998;11(3):555–67.
17. Yuki N, Susuki K, Koga M, Nishimoto Y, Odaka M, Hirata K, et al. Carbohydrate mimicry between human ganglioside GM1 and Campylobacter jejuni lipooligosaccharide causes Guillain-Barré syndrome. Proc Natl Acad Sci U S A. 2004;101(31):11404–9.
18. Hadden RDM, Karch H, Hartung HP, Zielasek J, Weissbrich B, Schubert J, et al. Preceding infections, immune factors, and outcome in Guillain-Barré syndrome. Neurology. 2001;56(6):758–65.
19. Jacobs BC, Rothbarth PH, van der Meche FGA, Herbrink P, Schmitz PIM, de Klerk MA, et al. The spectrum of antecedent infections in Guillain-Barré syndrome: a case-control study. Neurology. 1998;51(4):1110–5.
20. Sivadon-Tardy V, Orlikowski D, Porcher R, Sharshar T, Durand MC, Enouf V, et al. Guillain-Barré syndrome and influenza virus infection. Clin Infect Dis. 2009;48(1):48–56.
21. van den Berg B, van der Eijk AA, Pas SD, Hunter JG, Madden RG, Tio-Gillen AP, et al. Guillain-Barré syndrome associated with preceding hepatitis E virus infection. Neurology. 2014;82(6):491–7.
22. Geurtsvankessel CH, Islam Z, Mohammad QD, Jacobs BC, Endtz HP, Osterhaus AD. Hepatitis E and Guillain-Barré syndrome. Clin Infect Dis. 2013;57(9):1369–70.
23. Parra B, Lizarazo J, Jimenez-Arango JA, Zea-Vera AF, Gonzalez-Manrique G, Vargas J, et al. Guillain-Barré syndrome associated with Zika virus infection in Colombia. N Engl J Med. 2016;375(16):1513–23.
24. Hachinski V, R F. The swine influenza vaccination controversy. Arch Neurol. 1985;42(11):1092.
25. Baxter R, Bakshi N, Fireman B, Lewis E, Ray P, Vellozzi C, et al. Lack of association of Guillain-Barré syndrome with vaccinations. Clin Infect Dis. 2013;57(2):197–204.
26. Salmon DA, Proschan M, Forshee R, Gargiullo P, Bleser W, Burwen DR, et al. Association between Guillain-Barré syndrome and influenza A (H1N1) 2009 monovalent inactivated vaccines in the USA: a meta-analysis. Lancet. 2013;381(9876):1461–8.
27. De Wals P, Deceuninck G, Toth E, Boulianne N, Brunet D, Boucher RM, et al. Risk of Guillain-Barré syndrome following H1N1 influenza vaccination in Quebec. JAMA. 2012;308(2):175–81.
28. Vellozzi C, Iqbal S, Broder K. Guillain-Barré syndrome, influenza, and influenza vaccination: the epidemiologic evidence. Clin Infect Dis. 2014;58(8):1149–55.
29. Baxter R, Lewis N, Bakshi N, Vellozzi C, Klein NP, Network C. Recurrent Guillain-Barré syndrome following vaccination. Clin Infect Dis. 2012;54(6):800–4.

30. Fokke C, van den Berg B, Drenthen J, Walgaard C, van Doorn PA, Jacobs BC. Diagnosis of Guillain-Barré syndrome and validation of Brighton criteria. Brain. 2014;137(Pt 1):33–43.
31. Kuwabara S, Ogawara K, Koga M, Mori M, Hattori T, Yuki N. Hyperreflexia in Guillain-Barré syndrome: relation with acute motor axonal neuropathy and anti-GM1 antibody. J Neurol Neurosurg Psychiatry. 1999;67(2):180–4.
32. Wakerley BR, Yuki N. Mimics and chameleons in Guillain-Barré and Miller Fisher syndromes. Pract Neurol. 2015;15(2):90–9.
33. Hiraga A, Mori M, Ogawara K, Hattori T, Kuwabara S. Differences in patterns of progression in demyelinating and axonal Guillain-Barré syndromes. Neurology. 2003;61(4):471–4.
34. Sung JY, Tani J, Park SB, Kiernan MC, Lin CS. Early identification of 'acute-onset' chronic inflammatory demyelinating polyneuropathy. Brain. 2014;137(Pt 8):2155–63.
35. Oh SJ, Kurokawa K, de Almeida DF, Ryan HF, Claussen GC. Subacute inflammatory demyelinating polyneuropathy. Neurology. 2003;61(11):1507–12.
36. Ho TW, Li CY, Cornblath DR, Gao CY, Asbury AK, Griffin JW, et al. Patterns of recovery in the Guillain-Barré syndromes. Neurology. 1997;48(3):695–700.
37. Ruts L, Drenthen J, Jongen JL, Hop WC, Visser GH, Jacobs BC, et al. Pain in Guillain-Barré syndrome: a long-term follow-up study. Neurology. 2010;75(16):1439–47.
38. Capasso M, Notturno F, Manzoli C, Uncini A. Involvement of sensory fibres in axonal subtypes of Guillain-Barré syndrome. J Neurol Neurosurg Psychiatry. 2011;82(6):664–70.
39. van Doorn PA, Ruts L, Jacobs BC. Clinical features, pathogenesis, and treatment of Guillain-Barré syndrome. Lancet Neurol. 2008;7(10):939–50.
40. Anandan C, Khuder SA, Koffman BM. Prevalence of autonomic dysfunction in hospitalized patients with Guillain-Barré syndrome. Muscle Nerve. 2017;56(2):331–3.
41. Zochodne DW. Autonomic involvement in Guillain-Barré-syndrome - a review. Muscle Nerve. 1994;17(10):1145–55.
42. Kuwabara S, Yuki N. Axonal Guillain-Barré syndrome: concepts and controversies. Lancet Neurol. 2013;12(12):1180–8.
43. Asahina M, Kuwabara S, Suzuki A, Hattori T. Autonomic function in demyelinating and axonal subtypes of Guillain-Barré syndrome. Acta Neurol Scand. 2002;105(1):44–50.
44. Rigamonti A, Basso F, Scaccabarozzi C, Lauria G. Posterior reversible encephalopathy syndrome as the initial manifestation of Guillain-Barré syndrome: case report and review of the literature. J Peripher Nerv Syst. 2012;17(3):356–60.
45. Chen A, Kim J, Henderson G, Berkowitz A. Posterior reversible encephalopathy syndrome in Guillain-Barré syndrome. J Clin Neurosci. 2015;22(5):914–6.
46. Lawn ND, Fletcher DD, Henderson RD, Wolter TD, Wijdicks EF. Anticipating mechanical ventilation in Guillain-Barré syndrome. Arch Neurol. 2001;58(6):893–8.
47. Sharshar T, Chevret S, Bourdain F, Raphael JC. French Cooperative Group on plasma exchange in Guillain-Barré S. Early predictors of mechanical ventilation in Guillain-Barré syndrome. Crit Care Med. 2003;31(1):278–83.
48. Walgaard C, Lingsma HF, Ruts L, Drenthen J, van Koningsveld R, Garssen MJ, et al. Prediction of respiratory insufficiency in Guillain-Barré syndrome. Ann Neurol. 2010;67(6):781–7.
49. Durand M-C, Porcher R, Orlikowski D, Aboab J, Devaux C, Clair B, et al. Clinical and electrophysiological predictors of respiratory failure in Guillain-Barré syndrome: a prospective study. Lancet Neurol. 2006;5(12):1021–8.
50. Asbury AK, Cornblath DR. Assessment of current diagnostic criteria for Guillain-Barré syndrome. Ann Neurol. 1990;27(S1):S21–S4.
51. Nishimoto Y, Odaka M, Hirata K, Yuki N. Usefulness of anti-GQ1b IgG antibody testing in Fisher syndrome compared with cerebrospinal fluid examination. J Neuroimmunol. 2004;148(1–2):200–5.
52. Wong AH, Umapathi T, Nishimoto Y, Wang YZ, Chan YC, Yuki N. Cytoalbuminologic dissociation in Asian patients with Guillain-Barré and Miller Fisher syndromes. J Peripher Nerv Syst. 2015;20(1):47–51.
53. Mills KR, Murray NM. Proximal conduction block in early Guillain-Barré syndrome. Lancet. 1985;2(8456):659.
54. Lachman T, Shahani BT, Young RR. Late responses as aids to diagnosis in peripheral neuropathy. J Neurol Neurosurg Psychiatry. 1980;43(2):156–62.
55. Gordon PH, Wilbourn AJ. Early Electrodiagnostic findings in Guillain-Barré syndrome. Arch Neurol. 2001;58(6):913.
56. Derksen A, Ritter C, Athar P, Kieseier BC, Mancias P, Hartung HP, et al. Sural sparing pattern discriminates Guillain-Barré syndrome from its mimics. Muscle Nerve. 2014;50(5):780–4.
57. Uncini A, Kuwabara S. Electrodiagnostic criteria for Guillain-Barré syndrome: a critical revision and the need for an update. Clin Neurophysiol. 2012;123(8):1487–95.
58. Gorson KC, Ropper AH, Muriello MA, Blair R. Prospective evaluation of MRI lumbosacral nerve root enhancement in acute Guillain-Barré syndrome. Neurology. 1996;47(3):813–7.
59. Podnar S. Epidemiology of cauda equina and conus medullaris lesions. Muscle Nerve. 2007;35(4):529–31.
60. Gosavi TD, Lo YL. Images in clinical medicine. Superficial abdominal reflex. N Engl J Med. 2014;370(18):e29.
61. Shahrizaila N, Yuki N. Bickerstaff brainstem encephalitis and Fisher syndrome: anti-GQ1b antibody syndrome. J Neurol Neurosurg Psychiatry. 2013;84(5):576–83.

62. Souliere CR Jr, Kava CR, Barrs DM, Bell AF. Sudden hearing loss as the sole manifestation of neurosarcoidosis. Otolaryngol Head Neck Surg. 1991;105(3):376–81.
63. Brannagan TH, Zhou Y. HIV-associated Guillain–Barré syndrome. J Neurol Sci. 2003;208(1–2):39–42.
64. Albers JW, Fink JK. Porphyric neuropathy. Muscle Nerve. 2004;30(4):410–22.
65. Logina I, Donaghy M. Diphtheritic polyneuropathy: a clinical study and comparison with Guillain-Barré syndrome. J Neurol Neurosurg Psychiatry. 1999;67(4):433–8.
66. Griffin JW, Li CY, Ho TW, Tian M, Gao CY, Xue P, et al. Pathology of the motor-sensory axonal Guillain-Barré syndrome. Ann Neurol. 1996;39(1):17–28.
67. Ubogu EE, Zaidat OO, Suarez JI. Acute motor-sensory axonal neuropathy associated with active systemic lupus erythematosus and anticardiolipin antibodies. J Clin Rheumatol. 2001;7(5):326–31.
68. Santiago-Casas Y, Peredo RA, Vila LM. Efficacy of low-dose intravenous cyclophosphamide in systemic lupus erythematosus presenting with Guillain-Barré syndrome-like acute axonal neuropathies: report of two cases. Lupus. 2013;22(3):324–7.
69. Fisher M. An unusual variant of acute idiopathic polyneuritis (syndrome of ophthalmoplegia, ataxia and areflexia). N Engl J Med. 1956;255(2):57–65.
70. Kuwabara S, Asahina M, Nakajima M, Mori M, Fukutake T, Hattori T, et al. Special sensory ataxia in Miller Fisher syndrome detected by postural body sway analysis. Ann Neurol. 1999;45(4):533–6.
71. Chiba A, Kusunoki S, Obata H, Machinami R, Kanazawa I. Serum anti-GQ1b IgG antibody is associated with ophthalmoplegia in Miller Fisher syndrome and Guillain-Barré syndrome: clinical and immunohistochemical studies. Neurology. 1993;43(10):1911.
72. Ito M, Matsuno K, Sakumoto Y, Hirata K, Yuki N. Ataxic Guillain-Barré syndrome and acute sensory ataxic neuropathy form a continuous spectrum. J Neurol Neurosurg Psychiatry. 2011;82(3):294–9.
73. Mori M, Kuwabara S, Fukutake T, Yuki N, Hattori T. Clinical features and prognosis of Miller Fisher syndrome. Neurology. 2001;56(8):1104–6.
74. Nagashima T, Koga M, Odaka M, Hirata K, Yuki N. Continuous spectrum of pharyngeal-cervical-brachial variant of Guillain-Barré syndrome. Arch Neurol. 2007;64(10):1519–23.
75. van den Berg B, Fokke C, Drenthen J, van Doorn PA, Jacobs BC. Paraparetic Guillain-Barré syndrome. Neurology. 2014;82(22):1984–9.
76. Sterman AB, Schaumburg HH, Asbury AK. The acute sensory neuronopathy syndrome: a distinct clinical entity. Ann Neurol. 1980;7(4):354–8.
77. Knazan M, Bohlega S, Berry K, Eisen A. Acute sensory neuronopathy with preserved SEPs and long-latency reflexes. Muscle Nerve. 1990;13(5):381–4.
78. Kuntzer T, Antoine JC, Steck AJ. Clinical features and pathophysiological basis of sensory neuronopathies (ganglionopathies). Muscle Nerve. 2004;30(3):255–68.
79. Seneviratne U, Gunasekera S. Acute small fibre sensory neuropathy: another variant of Guillain-Barré syndrome? J Neurol Neurosurg Psychiatry. 2002;72(4):540–2.
80. Gorson KC, Herrmann DN, Thiagarajan R, Brannagan TH, Chin RL, Kinsella LJ, et al. Non-length dependent small fibre neuropathy/ganglionopathy. J Neurol Neurosurg Psychiatry. 2008;79(2):163–9.
81. Terkelsen AJ, Karlsson P, Lauria G, Freeman R, Finnerup NB, Jensen TS. The diagnostic challenge of small fibre neuropathy: clinical presentations, evaluations, and causes. Lancet Neurol. 2017;16(11):934–44.
82. Koike H, Atsuta N, Adachi H, Iijima M, Katsuno M, Yasuda T, et al. Clinicopathological features of acute autonomic and sensory neuropathy. Brain. 2010;133(10):2881–96.
83. Willison HJ, Yuki N. Peripheral neuropathies and anti-glycolipid antibodies. Brain. 2002;125(12):2591–625.
84. Hafer-Macko CE, Sheikh KA, Li CY, Ho TW, Cornblath DR, McKhann GM, et al. Immune attack on the Schwann cell surface in acute inflammatory demyelinating polyneuropathy. Ann Neurol. 1996;39(5):625–35.
85. Feasby TE, Hahn AF, Brown WF, Bolton CF, Gilbert JJ, Koopman WJ. Severe axonal degeneration in acute Guillain-Barré syndrome: evidence of two different mechanisms? J Neurol Sci. 1993;116(2):185–92.
86. Hughes RAC, Hadden RDM, Gregson NA, Smith KJ. Pathogenesis of Guillain–Barré syndrome. J Neuroimmunol. 1999;100(1–2):74–97.
87. Hartung HP, Pollard JD, Harvey GK, Toyka KV. Immunopathogenesis and treatment of the Guillain-Barré syndrome--part I. Muscle Nerve. 1995;18(2):137–53.
88. Ho TW, Mishu B, Li CY, Gao CY, Cornblath DR, Griffin JW, et al. Guillain-Barré syndrome in northern China. Relationship to Campylobacter jejuni infection and anti-glycolipid antibodies. Brain. 1995;118(Pt 3):597–605.
89. Islam Z, Jacobs BC, van Belkum A, Mohammad QD, Islam MB, Herbrink P, et al. Axonal variant of Guillain-Barré syndrome associated with Campylobacter infection in Bangladesh. Neurology. 2010;74(7):581–7.
90. Kuijf ML, Godschalk PC, Gilbert M, Endtz HP, Tio-Gillen AP, Ang CW, et al. Origin of ganglioside complex antibodies in Guillain-Barré syndrome. J Neuroimmunol. 2007;188(1–2):69–73.
91. Kaida K, Morita D, Kanzaki M, Kamakura K, Motoyoshi K, Hirakawa M, et al. Ganglioside complexes as new target antigens in Guillain-Barré syndrome. Ann Neurol. 2004;56(4):567–71.
92. Kokubun N, Nishibayashi M, Uncini A, Odaka M, Hirata K, Yuki N. Conduction block in acute motor axonal neuropathy. Brain. 2010;133(10):2897–908.

93. Uncini A, Susuki K, Yuki N. Nodo-paranodopathy: beyond the demyelinating and axonal classification in anti-ganglioside antibody-mediated neuropathies. Clin Neurophysiol. 2013;124(10):1928–34.
94. Kokubun N, Shahrizaila N, Hirata K, Yuki N. Reversible conduction failure is distinct from neurophysiological patterns of recovery in mild demyelinating Guillain-Barré syndrome. J Neurol Sci. 2013;326(1–2):111–4.
95. Hiraga A, Kuwabara S, Ogawara K, Misawa S, Kanesaka T, Koga M, et al. Patterns and serial changes in electrodiagnostic abnormalities of axonal Guillain-Barré syndrome. Neurology. 2005;64(5):856–60.
96. Halstead SK, Zitman FM, Humphreys PD, Greenshields K, Verschuuren JJ, Jacobs BC, et al. Eculizumab prevents anti-ganglioside antibody-mediated neuropathy in a murine model. Brain. 2008;131(Pt 5):1197–208.
97. Misawa S, Kuwabara S, Sato Y, Yamaguchi N, Nagashima K, Katayama K, et al. Safety and efficacy of eculizumab in Guillain-Barré syndrome: a multicentre, double-blind, randomised phase 2 trial. Lancet Neurol. 2018;17(6):519–29.
98. Ogawa-Goto K, Abe T. Gangliosides and glycosphingolipids of peripheral nervous system myelins - a minireview. Neurochem Res. 1998;23(3):305–10.
99. Chiba A, Kusunoki S, Obata H, Machinami R, Kanazawa I. Serum anti-GQ1b IgG antibody is associated with ophthalmoplegia in Miller Fisher syndrome and Guillain-Barré syndrome: clinical and immunohistochemical studies. Neurology. 2011;77(7):676.
100. Liu JX, Willison HJ, Pedrosa-Domellof F. Immunolocalization of GQ1b and related gangliosides in human extraocular neuromuscular junctions and muscle spindles. Invest Ophthalmol Vis Sci. 2009;50(7):3226–32.
101. Ito M, Kuwabara S, Odaka M, Misawa S, Koga M, Hirata K, et al. Bickerstaff's brainstem encephalitis and Fisher syndrome form a continuous spectrum: clinical analysis of 581 cases. J Neurol. 2008;255(5):674–82.
102. Koga M, Yoshino H, Morimatsu M, Yuki N. Anti-GT1a IgG in Guillain-Barré syndrome. J Neurol Neurosurg Psychiatry. 2002;72(6):767–71.
103. Yuki N, Hirata K. Postinfection sensory neuropathy associated with IgG anti-GD1b antibody. Ann Neurol. 1998;43(5):685–7.
104. Kusunoki S, Shimizu J, Chiba A, Ugawa Y, Hitoshi S, Kanazawa I. Experimental sensory neuropathy induced by sensitization with ganglioside GD1b. Ann Neurol. 1996;39(4):424–31.
105. Sandroni P, Vernino S, Klein CM, Lennon VA, Benrud-Larson L, Sletten D, et al. Idiopathic autonomic neuropathy: comparison of cases seropositive and seronegative for ganglionic acetylcholine receptor antibody. Arch Neurol. 2004;61(1):44–8.
106. Prineas JW. Pathology of the Guillain-Barré syndrome. Ann Neurol. 1981;9(S1):6–19.
107. Bilbao JM, Schmidt RE. Biopsy diagnosis of peripheral neuropathy. Cham: Springer; 2015.
108. Phillips MS, Stewart S, Anderson JR. Neuropathological findings in Miller Fisher syndrome. J Neurol Neurosurg Psychiatry. 1984;47(5):492–5.
109. Rajabally YA, Uncini A. Outcome and its predictors in Guillain-Barré syndrome. J Neurol Neurosurg Psychiatry. 2012;83(7):711–8.
110. The Italian Guillain-Barré Study Group. The prognosis and main prognostic indicators of Guillain-Barré syndrome: a multicentre prospective study of 297 patients. Brain. 1996;119(6):2053–61.
111. Walgaard C, Lingsma HF, Ruts L, van Doorn PA, Steyerberg EW, Jacobs BC. Early recognition of poor prognosis in Guillain-Barré syndrome. Neurology. 2011;76(11):968–75.
112. van Koningsveld R, Steyerberg EW, Hughes RAC, Swan AV, van Doorn PA, Jacobs BC. A clinical prognostic scoring system for Guillain-Barré syndrome. Lancet Neurol. 2007;6(7):589–94.
113. Tamura N, Kuwabara S, Misawa S, Kanai K, Nakata M, Sawai S, et al. Time course of axonal regeneration in acute motor axonal neuropathy. Muscle Nerve. 2007;35(6):793–5.
114. Hiraga A, Mori M, Ogawara K, Kojima S, Kanesaka T, Misawa S, et al. Recovery patterns and long term prognosis for axonal Guillain-Barré syndrome. J Neurol Neurosurg Psychiatry. 2005;76(5):719–22.
115. Raphaël JC, Chevret S, Hughes RAC, Annane D. Plasma exchange for Guillain-Barré syndrome. Cochrane Database Syst Rev. 2002;2:CD001978.
116. Hughes RA, Swan AV, van Doorn PA. Intravenous immunoglobulin for Guillain-Barré syndrome. Cochrane Database Syst Rev. 2012;7:CD002063.
117. van der Meche FG, Schmitz PI. A randomized trial comparing intravenous immune globulin and plasma exchange in Guillain-Barré syndrome. Dutch Guillain-Barré Study Group. N Engl J Med. 1992;326(17):1123–9.
118. Plasma Exchange/Sandoglobulin Guillain-Barré Syndrome Trial Group. Randomised trial of plasma exchange, intravenous immunoglobulin, and combined treatments in Guillain-Barré syndrome. Lancet. 1997;349(9047):225–30.
119. French Cooperative Group on Plasma Exchange in Guillain-Barré syndrome. Efficiency of plasma exchange in Guillain-Barré syndrome: role of replacement fluids. Ann Neurol. 1987;22(6):753–61.
120. Hughes RAC, van der Meché FGA, Hughes R. Corticosteroids for Guillain-Barré syndrome. Cochrane Database Syst Rev. 2000;3:CD001446.
121. Patwa HS, Chaudhry V, Katzberg H, Rae-Grant AD, So YT. Evidence-based guideline: intravenous immunoglobulin in the treatment of neuromuscular disorders: report of the therapeutics and technology assessment Subcommittee of the American Academy of neurology. Neurology. 2012;78(13):1009–15.
122. Kleyweg RP, van der Meche FG. Treatment related fluctuations in Guillain-Barré syndrome after high-

dose immunoglobulins or plasma-exchange. J Neurol Neurosurg Psychiatry. 1991;54(11):957–60.
123. Ruts L, Drenthen J, Jacobs BC, van Doorn PA, Dutch GBSSG. Distinguishing acute-onset CIDP from fluctuating Guillain-Barré syndrome: a prospective study. Neurology. 2010;74(21):1680–6.
124. Visser LH, Van Der Meché FGA, Van Doorn PA, Meulstee J, Jacobs BC, Oomes PG, et al. Guillain-Barré syndrome without sensory loss (acute motor neuropathy): a subgroup with specific clinical, electrodiagnostic and laboratory features. Brain. 1995;118(4):841–7.
125. Mori M, Kuwabara S, Fukutake T, Hattori T. Intravenous immunoglobulin therapy for Miller Fisher syndrome. Neurology. 2007;68(14):1144–6.
126. Koller H, Kieseier BC, Jander S, Hartung HP. Chronic inflammatory demyelinating polyneuropathy. N Engl J Med. 2005;352(13):1343–56.
127. Vallat J-M, Sommer C, Magy L. Chronic inflammatory demyelinating polyradiculoneuropathy: diagnostic and therapeutic challenges for a treatable condition. Lancet Neurol. 2010;9(4):402–12.
128. Dalakas MC. Medscape. Advances in the diagnosis, pathogenesis and treatment of CIDP. Nat Rev Neurol. 2011;7(9):507–17.
129. Rajabally YA, Simpson BS, Beri S, Bankart J, Gosalakkal JA. Epidemiologic variability of chronic inflammatory demyelinating polyneuropathy with different diagnostic criteria: study of a UK population. Muscle Nerve. 2009;39(4):432–8.
130. Laughlin RS, Dyck PJ, Melton LJ 3rd, Leibson C, Ransom J, Dyck PJ. Incidence and prevalence of CIDP and the association of diabetes mellitus. Neurology. 2009;73(1):39–45.
131. Lunn MPT, Manji H, Choudhary PP, Hughes RAC, Thomas PK. Chronic inflammatory demyelinating polyradiculoneuropathy: a prevalence study in south East England. J Neurol Neurosurg Psychiatry. 1999;66(5):677–80.
132. Iijima M, Koike H, Hattori N, Tamakoshi A, Katsuno M, Tanaka F, et al. Prevalence and incidence rates of chronic inflammatory demyelinating polyneuropathy in the Japanese population. J Neurol Neurosurg Psychiatry. 2008;79(9):1040–3.
133. McLeod JG, Pollard JD, Macaskill P, Mohamed A, Spring P, Khurana V. Prevalence of chronic inflammatory demyelinating polyneuropathy in New South Wales, Australia. Ann Neurol. 1999;46(6):910–3.
134. Hattori N, Misu K, Koike H, Ichimura M, Nagamatsu M, Hirayama M, et al. Age of onset influences clinical features of chronic inflammatory demyelinating polyneuropathy. J Neurol Sci. 2001;184(1):57–63.
135. Mathey EK, Park SB, Hughes RA, Pollard JD, Armati PJ, Barnett MH, et al. Chronic inflammatory demyelinating polyradiculoneuropathy: from pathology to phenotype. J Neurol Neurosurg Psychiatry. 2015;86(9):973–85.
136. Schmidt B, Toyka KV, Kiefer R, Full J, Hartung HP, Pollard J. Inflammatory infiltrates in sural nerve biopsies in Guillain-Barré syndrome and chronic inflammatory demyelinating neuropathy. Muscle Nerve. 1996;19(4):474–87.
137. Kanda T. Chronic inflammatory demyelinating polyneuropathy: decreased claudin-5 and relocated ZO-1. J Neurol Neurosurg Psychiatry. 2004;75(5):765–9.
138. Rentzos M, Angeli AV, Rombos A, Kyrozis A, Nikolaou C, Zouvelou V, et al. Proinflammatory cytokines in serum and cerebrospinal fluid of CIDP patients. Neurol Res. 2012;34(9):842–6.
139. Kieseier BC, Tani M, Mahad D, Oka N, Ho T, Woodroofe N, et al. Chemokines and chemokine receptors in inflammatory demyelinating neuropathies: a central role for IP-10. Brain. 2002;125(4):823–34.
140. Previtali SC, Archelos JJ, Hartung HP. Expression of integrins in experimental autoimmune neuritis and Guillain-Barré syndrome. Ann Neurol. 1998;44(4):611–21.
141. Leppert D, Hughes P, Huber S, Erne B, Grygar C, Said G, et al. Matrix metalloproteinase upregulation in chronic inflammatory demyelinating polyneuropathy and nonsystemic vasculitic neuropathy. Neurology. 1999;53(1):62–70.
142. Salomon B, Rhee L, Bour-Jordan H, Hsin H, Montag A, Soliven B, et al. Development of spontaneous autoimmune peripheral polyneuropathy in B7-2–deficient Nod mice. J Exp Med. 2001;194(5):677–84.
143. Querol L, Nogales-Gadea G, Rojas-Garcia R, Martinez-Hernandez E, Diaz-Manera J, Suarez-Calvet X, et al. Antibodies to contactin-1 in chronic inflammatory demyelinating polyneuropathy. Ann Neurol. 2013;73(3):370–80.
144. Querol L, Nogales-Gadea G, Rojas-Garcia R, Diaz-Manera J, Pardo J, Ortega-Moreno A, et al. Neurofascin IgG4 antibodies in CIDP associate with disabling tremor and poor response to IVIg. Neurology. 2014;82(10):879–86.
145. Delmont E, Manso C, Querol L, Cortese A, Berardinelli A, Lozza A, et al. Autoantibodies to nodal isoforms of neurofascin in chronic inflammatory demyelinating polyneuropathy. Brain. 2017;140(7):1851–8.
146. Mathey EK, Garg N, Park SB, Nguyen T, Baker S, Yuki N, et al. Autoantibody responses to nodal and paranodal antigens in chronic inflammatory neuropathies. J Neuroimmunol. 2017;309:41–6.
147. Pollard JD, Armati PJ. CIDP - the relevance of recent advances in Schwann cell/axonal neurobiology. J Peripher Nerv Syst. 2011;16(1):15–23.
148. Dyck PJ, Lais AC, Ohta M, Bastron JA, Okazaki H, Groover RV. Chronic inflammatory polyradiculoneuropathy. Mayo Clin Proc. 1975;50(11):621–37.
149. Vital C, Vital A, Lagueny A, Ferrer X, Fontan D, Barat M, et al. Chronic inflammatory demyelinating polyneuropathy: immunopathological and ultrastructural study of peripheral nerve biopsy in 42 cases. Ultrastruct Pathol. 2000;24(6):363–9.
150. Harbo T, Andersen H, Jakobsen J. Length-dependent weakness and electrophysiological signs of secondary axonal loss in chronic inflammatory demy-

elinating polyradiculoneuropathy. Muscle Nerve. 2008;38(2):1036–45.
151. Sommer C, Koch S, Lammens M, Gabreels-Festen A, Stoll G, Toyka KV. Macrophage clustering as a diagnostic marker in sural nerve biopsies of patients with CIDP. Neurology. 2005;65(12):1924–9.
152. Hughes RA, Allen D, Makowska A, Gregson NA. Pathogenesis of chronic inflammatory demyelinating polyradiculoneuropathy. J Peripher Nerv Syst. 2006;11(1):30–46.
153. Sander HW, Latov N. Research criteria for defining patients with CIDP. Neurology. 2003;60(8 Suppl 3):S8–15.
154. Dyck PJB, Tracy JA. History, diagnosis, and management of chronic inflammatory demyelinating polyradiculoneuropathy. Mayo Clin Proc. 2018;93(6):777–93.
155. Dionne A, Nicolle MW, Hahn AF. Clinical and electrophysiological parameters distinguishing acute-onset chronic inflammatory demyelinating polyneuropathy from acute inflammatory demyelinating polyneuropathy. Muscle Nerve. 2010;41(2):202–7.
156. Cao Y, Menon P, Ching-Fen Chang F, Mahant N, Geevasinga N, Fung VS, et al. Postural tremor and chronic inflammatory demyelinating polyneuropathy. Muscle Nerve. 2017;55(3):338–43.
157. Sabatelli M, Madia F, Mignogna T, Lippi G, Quaranta L, Tonali P. Pure motor chronic inflammatory demyelinating polyneuropathy. J Neurol. 2001;248(9):772–7.
158. Oh SJ, Joy JL, Kuruoglu R. "Chronic sensory demyelinating neuropathy": chronic inflammatory demyelinating polyneuropathy presenting as a pure sensory neuropathy. J Neurol Neurosurg Psychiatry. 1992;55(8):677–80.
159. Berger AR, Herskovitz S, Kaplan J. Late motor involvement in cases presenting as "chronic sensory demyelinating polyneuropathy". Muscle Nerve. 1995;18(4):440–4.
160. Stamboulis E, Katsaros N, Koutsis G, Iakovidou H, Giannakopoulou A, Simintzi I. Clinical and subclinical autonomic dysfunction in chronic inflammatory demyelinating polyradiculoneuropathy. Muscle Nerve. 2006;33(1):78–84.
161. Tracy JA, Dyck PJ, Klein CJ, Engelstad JK, Meyer JE, Dyck PJB. Onion-bulb patterns predict acquired or inherited demyelinating polyneuropathy. Muscle Nerve. 2019;59:665.
162. Prineas JW, McLeod JG. Chronic relapsing polyneuritis. J Neurol Sci. 1976;27(4):427–58.
163. Koike H, Nishi R, Ikeda S, Kawagashira Y, Iijima M, Katsuno M, et al. Ultrastructural mechanisms of macrophage-induced demyelination in CIDP. Neurology. 2018;91(23):1051–60.
164. Bouchard C, Lacroix C, Plante V, Adams D, Chedru F, Guglielmi JM, et al. Clinicopathologic findings and prognosis of chronic inflammatory demyelinating polyneuropathy. Neurology. 1999;52(3):498.
165. Kronlage M, Baumer P, Pitarokoili K, Schwarz D, Schwehr V, Godel T, et al. Large coverage MR neurography in CIDP: diagnostic accuracy and electrophysiological correlation. J Neurol. 2017;264(7):1434–43.
166. Telleman JA, Grimm A, Goedee S, Visser LH, Zaidman CM. Nerve ultrasound in polyneuropathies. Muscle Nerve. 2018;57(5):716–28.
167. Zaidman CM, Pestronk A. Nerve size in chronic inflammatory demyelinating neuropathy varies with disease activity and therapy response over time: a retrospective ultrasound study. Muscle Nerve. 2014;50(5):733–8.
168. Hughes R, Bensa SN, Willison H, Van Den Bergh P, Comi G, Illa I, et al. Randomized controlled trial of intravenous immunoglobulin versus oral prednisolone in chronic inflammatory demyelinating polyradiculoneuropathy. Ann Neurol. 2001;50(2):195–201.
169. Barohn RJ, Kissel TK, Warmolts JR, Mendell JR. Chronic inflammatory demyelinating polyradiculoneuropathy. Arch Neurol. 1989;46(8):878.
170. Van den Bergh PY, Hadden RD, Bouche P, Cornblath DR, Hahn A, Illa I, et al. European Federation of Neurological Societies/Peripheral Nerve Society guideline on management of chronic inflammatory demyelinating polyradiculoneuropathy: report of a joint task force of the European Federation of Neurological Societies and the Peripheral Nerve Society - first revision. Eur J Neurol. 2010;17(3):356–63.
171. Rajabally YA, Nicolas G, Pieret F, Bouche P, Van den Bergh PY. Validity of diagnostic criteria for chronic inflammatory demyelinating polyneuropathy: a multicentre European study. J Neurol Neurosurg Psychiatry. 2009;80(12):1364–8.
172. Bromberg MB. Comparison of electrodiagnostic criteria for primary demyelination in chronic polyneuropathy. Muscle Nerve. 1991;14(10):968–76.
173. Bromberg MB, Albers JW. Patterns of sensory nerve conduction abnormalities in demyelinating and axonal peripheral nerve disorders. Muscle Nerve. 1993;16(3):262–6.
174. Sinnreich M, Klein CJ, Daube JR, Engelstad J, Spinner RJ, Dyck PJB. Chronic immune sensory polyradiculopathy: a possibly treatable sensory ataxia. Neurology. 2004;63(9):1662–9.
175. Katz JS, Saperstein DS, Gronseth G, Amato AA, Barohn RJ. Distal acquired demyelinating symmetric neuropathy. Neurology. 2000;54(3):615.
176. Trojaborg W, Hays AP, van den Berg L, Younger DS, Latov N. Motor conduction parameters in neuropathies associated with anti-MAG antibodies and other types of demyelinating and axonal neuropathies. Muscle Nerve. 1995;18(7):730–5.
177. Dalakas MC. Pathogenesis and treatment of anti-MAG neuropathy. Curr Treat Options Neurol. 2010;12(2):71–83.
178. Lunn MP, Nobile-Orazio E. Immunotherapy for IgM anti-myelin-associated glycoprotein paraprotein-

associated peripheral neuropathies. Cochrane Database Syst Rev. 2016;10:CD002827.
179. Dalakas MC, Rakocevic G, Salajegheh M, Dambrosia JM, Hahn AF, Raju R, et al. Placebo-controlled trial of rituximab in IgM anti-myelin-associated glycoprotein antibody demyelinating neuropathy. Ann Neurol. 2009;65(3):286–93.
180. Leger JM, Viala K, Nicolas G, Creange A, Vallat JM, Pouget J, et al. Placebo-controlled trial of rituximab in IgM anti-myelin-associated glycoprotein neuropathy. Neurology. 2013;80(24):2217–25.
181. Pestronk A, Florence J, Miller T, Choksi R, Al-Lozi M, Levine T. Treatment of IgM antibody associated polyneuropathies using rituximab. J Neurol Neurosurg Psychiatry. 2003;74(4):485–9.
182. Koya J, Iwata A, Nakamura F, Nagashima Y, Ichikawa M, Tsuji S, et al. Fludarabine may overcome resistance to rituximab in IgM-related neuropathy. J Neurol Sci. 2012;315(1–2):150–2.
183. Baron M, Lozeron P, Harel S, Bengoufa D, Vignon M, Asli B, et al. Plasma exchanges for severe acute neurological deterioration in patients with IgM anti-myelin-associated glycoprotein (anti-MAG) neuropathy. J Neurol. 2017;264(6):1132–5.
184. Saperstein DS, Amato AA, Wolfe GI, Katz JS, Nations SP, Jackson CE, et al. Multifocal acquired demyelinating sensory and motor neuropathy: the Lewis-Sumner syndrome. Muscle Nerve. 1999;22(5):560–6.
185. Davies L, Spies JM, Pollard JD, McLeod JG. Vasculitis confined to peripheral nerves. Brain. 1996;119(Pt 5):1441–8.
186. Viala K, Renie L, Maisonobe T, Behin A, Neil J, Leger JM, et al. Follow-up study and response to treatment in 23 patients with Lewis-Sumner syndrome. Brain. 2004;127(Pt 9):2010–7.
187. Attarian S, Verschueren A, Franques J, Salort-Campana E, Jouve E, Pouget J. Response to treatment in patients with Lewis-Sumner syndrome. Muscle Nerve. 2011;44(2):179–84.
188. Martinez-Thompson JM, Snyder MR, Ettore M, McKeon A, Pittock SJ, Roforth MM, et al. Composite ganglioside autoantibodies and immune treatment response in MMN and MADSAM. Muscle Nerve. 2018;57(6):1000–5.
189. Verschueren A, Azulay JP, Attarian S, Boucraut J, Pellissier JF, Pouget J. Lewis-Sumner syndrome and multifocal motor neuropathy. Muscle Nerve. 2005;31(1):88–94.
190. Oh SJ, Claussen GC, Kim DS. Motor and sensory demyelinating mononeuropathy multiplex (multifocal motor and sensory demyelinating neuropathy): a separate entity or a variant of chronic inflammatory demyelinating polyneuropathy? J Peripher Nerv Syst. 1997;2(4):362–9.
191. Pestronk A, Chaudhry V, Feldman EL, Griffin JW, Cornblath DR, Denys EH, et al. Lower motor neuron syndromes defined by patterns of weakness, nerve conduction abnormalities, and high titers of antiglycolipid antibodies. Ann Neurol. 1990;27(3):316–26.
192. Parry GJ. Antiganglioside antibodies do not necessarily play a role in multifocal motor neuropathy. Muscle Nerve. 1994;17(1):97–9.
193. Pestronk A, Adams RN, Clawson L, Cornblath D, Kuncl RW, Griffin D, et al. Serum antibodies to GM1 ganglioside in amyotrophic lateral sclerosis. Neurology. 1988;38(9):1457.
194. Pestronk A, Choksi R. Multifocal motor neuropathy: serum IgM anti-GM1 ganglioside antibodies in most patients detected using covalent linkage of GM1 to ELISA plates. Neurology. 1997;49(5):1289–92.
195. Joint Task Force of the EFNS and the PNS. European Federation of Neurological Societies/Peripheral Nerve Society guideline on management of multifocal motor neuropathy. Report of a joint task force of the European Federation of Neurological Societies and the Peripheral Nerve Society--first revision. J Peripher Nerv Syst. 2010;15(4):295–301.
196. Vlam L, van der Pol WL, Cats EA, Straver DC, Piepers S, Franssen H, et al. Multifocal motor neuropathy: diagnosis, pathogenesis and treatment strategies. Nat Rev Neurol. 2011;8(1):48–58.
197. Federico P, Zochodne DW, Hahn AF, Brown WF, Feasby TE. Multifocal motor neuropathy improved by IVIg: randomized, double-blind, placebo-controlled study. Neurology. 2000;55(9):1256–62.
198. Leger JM, Chassande B, Musset L, Meininger V, Bouche P, Baumann N. Intravenous immunoglobulin therapy in multifocal motor neuropathy: a double-blind, placebo-controlled study. Brain. 2001;124(Pt 1):145–53.
199. Donaghy M, Mills KR, Boniface SJ, Simmons J, Wright I, Gregson N, et al. Pure motor demyelinating neuropathy: deterioration after steroid treatment and improvement with intravenous immunoglobulin. J Neurol Neurosurg Psychiatry. 1994;57(7):778–83.
200. Carpo M, Cappellari A, Mora G, Pedotti R, Barbieri S, Scarlato G, et al. Deterioration of multifocal motor neuropathy after plasma exchange. Neurology. 1998;50(5):1480–2.
201. Pestronk A, Florence J, Miller T, Choksi R, Al-Lozi MT, Levine TD. Treatment of IgM antibody associated polyneuropathies using rituximab. J Neurol Neurosurg Psychiatry. 2003;74(4):485–9.
202. Feldman EL, Bromberg MB, Albers JW, Pestronk A. Immunosuppressive treatment in multifocal motor neuropathy. Ann Neurol. 1991;30(3):397–401.
203. Katz JS, Barohn RJ, Kojan S, Wolfe GI, Nations SP, Saperstein DS, et al. Axonal multifocal motor neuropathy without conduction block or other features of demyelination. Neurology. 2002;58(4):615–20.
204. Fischer D, Grothe C, Schmidt S, Schroder R. On the early diagnosis of IVIG-responsive chronic multifocal acquired motor axonopathy. J Neurol. 2004;251(10):1204–7.
205. De Schaepdryver M, Jeromin A, Gille B, Claeys KG, Herbst V, Brix B, et al. Comparison of elevated phosphorylated neurofilament heavy chains in serum and cerebrospinal fluid of patients with amyotrophic

lateral sclerosis. J Neurol Neurosurg Psychiatry. 2018;89(4):367–73.
206. Steinacker P, Feneberg E, Weishaupt J, Brettschneider J, Tumani H, Andersen PM, et al. Neurofilaments in the diagnosis of motoneuron diseases: a prospective study on 455 patients. J Neurol Neurosurg Psychiatry. 2016;87(1):12–20.
207. Van den Berg-Vos RM, Franssen H, Wokke JHJ, Van Es HW, Van den Berg LH. Multifocal motor neuropathy: diagnostic criteria that predict the response to immunoglobulin treatment. Ann Neurol. 2000;48(6):919–26.
208. Allen JA, Lewis RA. CIDP diagnostic pitfalls and perception of treatment benefit. Neurology. 2015;85(6):498–504.
209. Allen JA, Ney J, Lewis RA. Electrodiagnostic errors contribute to chronic inflammatory demyelinating polyneuropathy misdiagnosis. Muscle Nerve. 2018;57(4):542–9.
210. Raynor EM, Ross MH, Shefner JM, Preston DC. Differentiation between axonal and demyelinating neuropathies: identical segments recorded from proximal and distal muscles. Muscle Nerve. 1995;18(4):402–8.
211. Allen JA, Gorson KC, Gelinas D. Challenges in the diagnosis of chronic inflammatory demyelinating polyneuropathy. Brain Behav. 2018;8(3):e00932.
212. Gorson KC, Ropper AH, Adelman LS, Weinberg DH. Influence of diabetes mellitus on chronic inflammatory demyelinating polyneuropathy. Muscle Nerve. 2000;23(1):37–43.
213. Pareyson D, Scaioli V, Laurà M. Clinical and electrophysiological aspects of Charcot-Marie-tooth disease. NeuroMolecular Med. 2006;8(1–2):3–22.
214. Andersson PB, Yuen E, Parko K, So YT. Electrodiagnostic features of hereditary neuropathy with liability to pressure palsies. Neurology. 2000;54(1):40.
215. Lozeron P, Mariani LL, Dodet P, Beaudonnet G, Theaudin M, Adam C, et al. Transthyretin amyloid polyneuropathies mimicking a demyelinating polyneuropathy. Neurology. 2018;91(2):e143–e52.
216. Kelly JJ Jr, Kyle RA, O'Brien PC, Dyck PJ. Prevalence of monoclonal protein in peripheral neuropathy. Neurology. 1981;31(11):1480–3.
217. England JD, Gronseth GS, Franklin G, Carter GT, Kinsella LJ, Cohen JA, et al. Practice parameter: evaluation of distal symmetric polyneuropathy: role of laboratory and genetic testing (an evidence-based review). Report of the American Academy of Neurology, American Association of Neuromuscular and Electrodiagnostic Medicine, and American Academy of Physical Medicine and Rehabilitation. Neurology. 2009;72(2):185–92.
218. Magy L, Chassande B, Maisonobe T, Bouche P, Vallat JM, Leger JM. Polyneuropathy associated with IgG/IgA monoclonal gammopathy: a clinical and electrophysiological study of 15 cases. Eur J Neurol. 2003;10(6):677–85.
219. Apartis E, Leger JM, Musset L, Gugenheim M, Cacoub P, Lyon-Caen O, et al. Peripheral neuropathy associated with essential mixed cryoglobulinaemia: a role for hepatitis C virus infection? J Neurol Neurosurg Psychiatry. 1996;60(6):661–6.
220. Svahn J, Petiot P, Antoine J-C, Vial C, Delmont E, Viala K, et al. Anti-MAG antibodies in 202 patients: clinicopathological and therapeutic features. J Neurol Neurosurg Psychiatry. 2018;89(5):499–505.
221. Willison HJ. The clinical and laboratory features of chronic sensory ataxic neuropathy with anti-disialosyl IgM antibodies. Brain. 2001;124(10):1968–77.
222. Garcia-Santibanez R, Zaidman CM, Sommerville RB, Lopate G, Weihl CC, Pestronk A, et al. CANOMAD and other chronic ataxic neuropathies with disialosyl antibodies (CANDA). J Neurol. 2018;265(6):1402–9.
223. Dispenzieri A, Kyle RA, Lacy MQ, Rajkumar SV, Therneau TM, Larson DR, et al. POEMS syndrome: definitions and long-term outcome. Blood. 2003;101(7):2496–506.
224. Sung JY, Kuwabara S, Ogawara K, Kanai K, Hattori T. Patterns of nerve conduction abnormalities in POEMS syndrome. Muscle Nerve. 2002;26(2):189–93.
225. Ohi T, Kyle RA, Dyck PJ. Axonal attenuation and secondary segmental demyelination in myeloma neuropathies. Ann Neurol. 1985;17(3):255–61.
226. Mauermann ML, Sorenson EJ, Dispenzieri A, Mandrekar J, Suarez GA, Dyck PJ, et al. Uniform demyelination and more severe axonal loss distinguish POEMS syndrome from CIDP. J Neurol Neurosurg Psychiatry. 2012;83(5):480–6.
227. Nasu S, Misawa S, Sekiguchi Y, Shibuya K, Kanai K, Fujimaki Y, et al. Different neurological and physiological profiles in POEMS syndrome and chronic inflammatory demyelinating polyneuropathy. J Neurol Neurosurg Psychiatry. 2012;83(5):476–9.
228. Nakanishi T, Sobue I, Toyokura Y, Nishitani H, Kuroiwa Y, Satoyoshi E, et al. The Crow-Fukase syndrome: a study of 102 cases in Japan. Neurology. 1984;34(6):712–20.
229. Watanabe O, Maruyama I, Arimura K, Kitajima I, Arimura H, Hanatani M, et al. Overproduction of vascular endothelial growth factor - vascular permeability factor is causative in Crow-Fukase (POEMS) syndrome. Muscle Nerve. 1998;21(11):1390–7.
230. Karam C, Klein CJ, Dispenzieri A, Dyck PJ, Mandrekar J, D'Souza A, et al. Polyneuropathy improvement following autologous stem cell transplantation for POEMS syndrome. Neurology. 2015;84(19):1981–7.
231. Hughes RA, Mehndiratta MM, Rajabally YA. Corticosteroids for chronic inflammatory demyelinating polyradiculoneuropathy. Cochrane Database Syst Rev. 2017;11:CD002062.
232. Eftimov F, Winer JB, Vermeulen M, de Haan R, van Schaik IN. Intravenous immunoglobulin for chronic inflammatory demyelinating polyradiculoneuropathy. Cochrane Database Syst Rev. 2013;12:CD001797.
233. van Schaik IN, Eftimov F, van Doorn PA, Brusse E, van den Berg LH, van der Pol WL, et al. Pulsed

high-dose dexamethasone versus standard prednisolone treatment for chronic inflammatory demyelinating polyradiculoneuropathy (PREDICT study): a double-blind, randomised, controlled trial. Lancet Neurol. 2010;9(3):245–53.
234. Lopate G, Pestronk A, Al-Lozi M. Treatment of chronic inflammatory demyelinating polyneuropathy with high-dose intermittent intravenous methylprednisolone. Arch Neurol. 2005;62(2):249–54.
235. Hughes RAC, Donofrio P, Bril V, Dalakas MC, Deng C, Hanna K, et al. Intravenous immune globulin (10% caprylate-chromatography purified) for the treatment of chronic inflammatory demyelinating polyradiculoneuropathy (ICE study): a randomised placebo-controlled trial. Lancet Neurol. 2008;7(2):136–44.
236. van Schaik IN, Bril V, van Geloven N, Hartung H-P, Lewis RA, Sobue G, et al. Subcutaneous immunoglobulin for maintenance treatment in chronic inflammatory demyelinating polyneuropathy (PATH): a randomised, double-blind, placebo-controlled, phase 3 trial. Lancet Neurol. 2018;17(1):35–46.
237. Mehndiratta MM, Hughes RA. Plasma exchange for chronic inflammatory demyelinating polyradiculoneuropathy. Cochrane Database Syst Rev. 2012;9:CD003906.
238. Gorson KC, van Schaik IN, Merkies IS, Lewis RA, Barohn RJ, Koski CL, et al. Chronic inflammatory demyelinating polyneuropathy disease activity status: recommendations for clinical research standards and use in clinical practice. J Peripher Nerv Syst. 2010;15(4):326–33.
239. Dyck PJ, O'Brien P, Swanson C, Low P, Daube J. Combined azathioprine and prednisone in chronic inflammatory-demyelinating polyneuropathy. Neurology. 1985;35(8):1173.
240. Gorson KC, Amato AA, Ropper AH. Efficacy of mycophenolate mofetil in patients with chronic immune demyelinating polyneuropathy. Neurology. 2004;63(4):715–7.
241. Barnett MH, Pollard JD, Davies L, McLeod JG. Cyclosporin A in resistant chronic inflammatory demyelinating polyradiculoneuropathy. Muscle Nerve. 1998;21(4):454–60.
242. Matsuda M, Hoshi K, Gono T, Morita H, Ikeda S. Cyclosporin A in treatment of refractory patients with chronic inflammatory demyelinating polyradiculoneuropathy. J Neurol Sci. 2004;224(1–2):29–35.
243. Group RT. Randomised controlled trial of methotrexate for chronic inflammatory demyelinating polyradiculoneuropathy (RMC trial): a pilot, multicentre study. Lancet Neurol. 2009;8(2):158–64.
244. Good JL, Chehrenama M, Mayer RF, Koski CL. Pulse cyclophosphamide therapy in chronic inflammatory demyelinating polyneuropathy. Neurology. 1998;51(6):1735–8.
245. Querol L, Rojas-Garcia R, Diaz-Manera J, Barcena J, Pardo J, Ortega-Moreno A, et al. Rituximab in treatment-resistant CIDP with antibodies against paranodal proteins. Neurol Neuroimmunol Neuroinflamm. 2015;2(5):e149.
246. Hehir MK, Hobson-Webb LD, Benatar M, Barnett C, Silvestri NJ, Howard JF Jr, et al. Rituximab as treatment for anti-MuSK myasthenia gravis: multicenter blinded prospective review. Neurology. 2017;89(10):1069–77.
247. Roux T, Debs R, Maisonobe T, Lenglet T, Delorme C, Louapre C, et al. Rituximab in chronic inflammatory demyelinating polyradiculoneuropathy with associated diseases. J Peripher Nerv Syst. 2018;23(4):235–40.
248. Benedetti L, Briani C, Franciotta D, Fazio R, Paolasso I, Comi C, et al. Rituximab in patients with chronic inflammatory demyelinating polyradiculoneuropathy: a report of 13 cases and review of the literature. J Neurol Neurosurg Psychiatry. 2011;82(3):306–8.

# 19 Autoimmune Diseases of the Neuromuscular Junction: Myasthenia Gravis and Lambert-Eaton Myasthenic Syndrome

Dery Miller, Jenny Joseph, and Rocio Garcia-Santibanez

**Key Points**

1. In autoimmune neuromuscular junction disorders, antibodies target presynaptic and postsynaptic proteins with disruption of neuromuscular transmission.
2. Myasthenia gravis is an autoimmune postsynaptic disorder that causes fatigable weakness and can be complicated with life-threatening crises.
3. Lambert-Eaton myasthenic syndrome is an autoimmune presynaptic disorder that causes fatigable weakness and autonomic dysfunction and is often associated with malignancy.
4. The treatment of both disorders is with immunomodulators, immunosuppression, and treatment of tumors associated with them.
5. Certain antineoplastic drugs can also trigger or worsen myasthenia gravis.

D. Miller · J. Joseph · R. Garcia-Santibanez (✉)
Department of Neurology, Emory University School of Medicine, Atlanta, GA, USA
e-mail: rcgarci@emory.edu

## Introduction

The neuromuscular junction is the target of two autoimmune disorders. In myasthenia gravis the target is in the postsynaptic membrane, and in Lambert-Eaton myasthenic syndrome, the target is in the presynaptic membrane. In this chapter we will discuss the epidemiology, pathophysiology, clinical manifestations, diagnosis, and treatment of these two disorders. We will also briefly review the neuromuscular junction disorders in the setting of immune checkpoint inhibitors.

## The Neuromuscular Junction

To understand the pathophysiology of the neuromuscular junction disorders, we first need to understand the anatomy and normal function of the neuromuscular junction.

The neuromuscular junction (Fig. 19.1) is composed of the nerve terminal of a motor neuron axon, the synaptic cleft, and a muscle fiber membrane. Acetylcholine, the neurotransmitter that facilitates depolarization and subsequent contraction of the postsynaptic muscle fiber, is contained in synaptic vesicles at the presynaptic motor neuron axon. Synaptic vesicles release acetylcholine at sites called active zones, which contain P/Q-type voltage-gated calcium channels (VGCC) and other proteins that work together to usher the synaptic vesicle to the presynaptic



A. L. Piquet, E. Alvarez (eds.), *Neuroimmunology*, https://doi.org/10.1007/978-3-030-61883-4_19

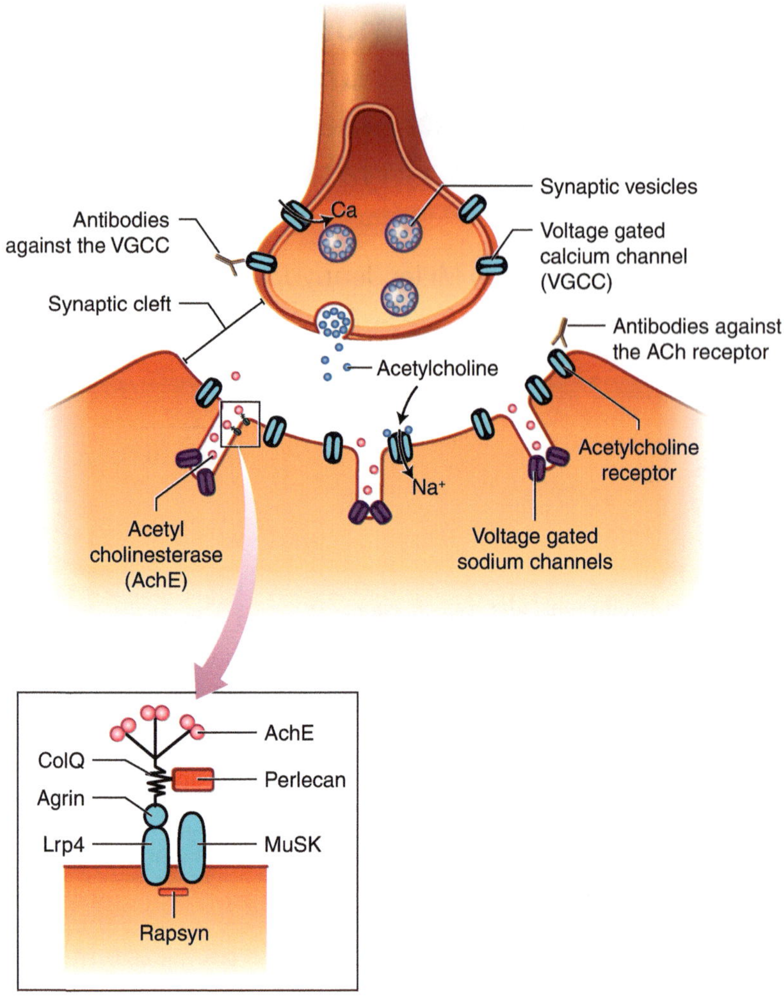


**Fig. 19.1** The neuromuscular junction. This illustration shows the different components of the neuromuscular junction, some of which are commonly targeted by antibodies

terminal. An action potential arriving at the presynaptic terminal triggers the opening of VGCC; the subsequent influx of calcium into the active zones facilitates acetylcholine release from their vesicles into the synaptic cleft [1]. The postsynaptic membrane has folds to increase the surface area. The nicotinic acetylcholine receptors (AchR) are more concentrated on the tips of the folds and the voltage-gated sodium channels deeper in the clefts. Released acetylcholine binds to the nicotinic AchR on the postsynaptic muscle membrane causing a graded depolarization, which opens the voltage-gated sodium channels. Once a threshold is reached and an action potential occurs, the muscle fiber contracts. Acetylcholinesterase then hydrolyzes the acetylcholine, which terminates the transmission [1].

Other important components of the postsynaptic membrane are proteins that together form a complex important in mediating the clustering of the AchR (zoomed-in section, Fig. 19.1) [2]. The proteins in this complex include muscle-specific receptor tyrosine kinase (MuSK), low-density lipoprotein receptor-related protein 4 (Lrp4), agrin, rapsyn, and docking protein 7 (Dok7). AchE is also closely related to this complex as it binds perlecan and MuSK via its collagen Q (ColQ) tail [2].

## Myasthenia Gravis

### Epidemiology

Myasthenia gravis (MG) is a rare autoimmune disorder that affects both sexes and all races. It has an incidence of approximately 7–23 new cases per million and a prevalence of 70–320 per million [3]. Myasthenia, although rare, can also affect children (under the age of 18); it is called juvenile myasthenia gravis and has an incidence one to five cases per million person-years [4]. As with most diseases, the prevalence of this disease has been increasing, likely due to increased recognition and longevity. Although myasthenia can affect any age group, there is a bimodal distribution with young women (20s–30s) and older men (60s–80s) being most affected [5].

### Pathophysiology

#### Antibodies in Myasthenia Gravis

Myasthenia is an autoimmune disorder of T-cell-mediated B-cell activation in which antibodies are produced against different components of the postsynaptic membrane of the neuromuscular junction. The antibodies cross-link the receptors and cause either endocytosis, complement-mediated destruction, or inactivation. Approximately 80% of patients with MG have antibodies against the nicotinic acetylcholine receptor (AchR), which are immunoglobulin G1 and G3 (IgG1 and IgG3) subtypes [6]. Their pathogenicity has been confirmed with animal studies of experimentally induced MG due to passive transfer of antibodies [7]. About 10% of myasthenic patients will have antibodies against muscle-specific kinase (MuSK), which are Ig4 subtype [6, 8]. Similarly to AchR antibodies, MuSK antibodies have been proven to be pathogenic on animal models [9]. Other antibodies such as antibodies against lipoprotein-related protein 4 (LRP4) and cortactin have also been found in previously seronegative patients [10, 11]. Cortactin regulates actin polymerization and is involved in acetylcholine receptor clustering downstream from the agrin/MuSK complex [11].

#### The Thymus in Myasthenia Gravis

The thymus plays a role in the pathophysiology of autoimmunity in AchR antibody myasthenia. AchR antibodies are thought to originate from the hypercellular germinal matrix of the thymus where myoid cells are expressed. Approximately 10–20% of patients with AchR antibody-positive generalized myasthenia gravis will have a thymoma, and around 50% of patients with thymoma have MG. Thymic hyperplasia can be seen in up to 70% of patients with AchR antibody-positive generalized MG. Thymus pathology is rare in MuSK myasthenia [12, 13]. There are no guidelines for the screening of thymomas, but it is recommended that all AchR antibody-positive patients undergo computed tomography (CT) or magnetic resonance imaging (MRI) of the chest for evaluation of thymoma. Both CT and MRI are highly sensitive to identify thymomas, but they are much less reliable in differentiating thymic hyperplasia from normal thymic tissue [14]. Reports suggest that CT chest with contrast and without contrast is equally sensitive in detecting thymoma [14]. Though thymomas are associated with seropositive patients, there are reports of thymomatous seronegative myasthenia gravis [15], and obtaining imaging in these patients (and in MuSK-positive patients) is done on a case-by-case basis.

### Clinical Manifestations

The most characteristic feature of MG is fatigable weakness due to failure of neuromuscular junction transmission with repetitive use. MG can be classified clinically in two groups depending on the pattern of weakness: ocular and generalized. Half of the patients with MG present with ocular weakness, usually with asymmetric ptosis or diplopia that worsens at the end of the day [16]. About 50–60% of the patients with ocular symptoms progress to have generalized myasthenia within 2–3 years of onset [17]. Overall, proximal muscles tend to be more affected than distal [16], but not uncommonly distal finger and wrist extensor muscles are affected, especially in advanced or untreated patients. Axial weakness, presenting as head drop, can be a feature in late-onset cases [18]. In about 66% of the patients with generalized myasthenia, symptoms progress to involve bulbar muscles, and patients will complain of dysarthria, dysphagia, dysphonia, or difficulty chewing [19]. Bulbar weakness is very common in MuSK antibody myasthenia gravis, and these patients tend to have less ocular symptoms. Respiratory muscle weakness occurs in approximately 40% of MG cases [13].

A feared complication of myasthenia gravis is the myasthenic crisis, which is defined as worsening weakness resulting in respiratory failure and need of respiratory support (noninvasive ventilation or intubation). About 15–20% of myasthenia patients will have a crisis at some point in their lives with a mortality rate of about 4% [20]. Common triggers for crisis are infection, surgery, pregnancy, electrolyte abnormalities, and certain medications (Table 19.1) [21]. Monitoring of pulmonary function tests is recommended. A vital capacity of less than 20 ml/kg and an inspiratory pressure of less than −40 cm $H_2 0$ are associated with the need for respiratory support [22]. A myasthenic crisis should be differentiated from a cholinergic crisis, which occurs from excess use of anticholinergic inhibitors and presents with both nicotinic (weakness) and muscarinic (increased secretions, diarrhea) toxic effects.

**Table 19.1** Medications that may worsen myasthenia gravis [21]

| |
|---|
| Fluoroquinolones |
| Aminoglycosides |
| Ketolides (telithromycin) |
| Magnesium (intravenous) |
| Macrolide antibiotics |
| Curare |
| Steroids |
| Botulinum toxin |
| Immune checkpoint inhibitors |
| Quinine |
| Procainamide[a] |
| Beta-blockers[a] |
| Calcium channel blockers[a] |

[a]Only anecdotal evidence and/or in vitro studies

## Diagnosis

MG is often diagnosed with a combination of history, clinical exam, and diagnostic tests. Diagnostic tests include the bedside use of edrophonium or an ice pack, serology (antibodies), and electrophysiologic testing.

Edrophonium was once commonly used for diagnosis of myasthenia but now its use is rare. Edrophonium is a fast-acting acetylcholinesterase inhibitor that when injected intravenously causes improvement of weakness (most commonly of ptosis) within minutes [23]; the patients need to be monitored for cholinergic side effects, which may need atropine for reversal. The ice pack test uses cold to demonstrate improvement of ptosis because acetylcholinesterase is inhibited at lower temperatures. Patients are asked to hold an ice pack over the affected eye for about 2 min after which an improvement in the ptosis is seen [24].

Antibody positivity percentage depends on the clinical subtype of myasthenia gravis. In ocular myasthenia gravis, approximately 50% of cases will have positive antibodies, most commonly AchR antibodies. In generalized myasthenia, approximately 80% of the patients will have positive AchR antibodies and 10% MuSK antibodies [25, 26]. The remainder may have Lrp4 antibodies, cortactin antibodies, or none. The latter is what is called seronegative myasthenia gravis.

Electrophysiologic testing includes repetitive nerve stimulation (RNS) and single-fiber electromyography (SFEMG). In repetitive nerve stimulation, a characteristic decrement in the amplitude of the action potential is seen with 2–3 Hz stimulation rates. Decrement is seen typically within the first four to five stimuli, then there is cessation of decrement or increment (increase in amplitude) with subsequent stimuli, representing the release of acetylcholine from secondary stores [27]. In SFEMG, pairs of single fibers are recorded and the presence of increased jitter and blocking is documented. Jitter is the measure of variability of time from when the end plate potential reaches threshold; there are established normal values of jitter depending on age and muscles tested [28]. SFEMG and RNS can be performed in facial and limb muscles. Both are more likely to be abnormal in generalized versus ocular myasthenia. SFEMG is the most sensitive test for the diagnosis of MG (abnormal in 92% of cases) with RNS being abnormal in 77% of the cases [25]. Though routine electromyography (EMG) is usually normal in myasthenia gravis, occasionally, muscle unit action potentials may be unstable; in rare cases they may have low amplitude, short duration, or be polyphasic as seen in myopathy [29]. A more detailed description of these electrophysiologic tests is beyond the scope of this chapter and can be found elsewhere [29].

## Treatment of Myasthenia Gravis

The approach to the treatment of MG depends on the clinical phenotype (ocular versus generalized), antibody status, type of antibody, and whether the patient is in a myasthenic crisis or not. Management includes symptomatic therapy and immunosuppressive or immunomodulatory therapy.

### Symptomatic Management

The main symptomatic therapy for myasthenia is pyridostigmine, an acetylcholinesterase inhibitor that reversibly binds acetylcholinesterase. The utility of acetylcholinesterase inhibitors in MG was first demonstrated by Walker in 1934 [30]. In 1954, pyridostigmine was shown to be effective in managing symptoms with less muscarinic side effects than the previously used formulations and since has been the mainstay of symptomatic control, obtaining the US Food and Drug Administration (FDA) approval in 1955 [31, 32]. The typical dosing is 60 mg every 6 hours but can be taken more frequently if tolerated. An extended-release formulation of 180 mg is often taken at bedtime.

### Immune Therapy

As with many autoimmune disorders, long-term immunosuppression is required. Corticosteroids are the immune therapy of choice in ocular MG. The nonsteroidal agents are often reserved for generalized MG.

#### Corticosteroids

Prednisone is the most common corticosteroid used. Dosing is usually done with a slow titration followed by a slow taper to the minimal effective dose that can be daily or on alternate days [33]. Slow titration is recommended to prevent worsening, which can occur in the first days of starting the medication. Ocular myasthenia usually requires lower doses than generalized myasthenia, and steroids can sometimes be tapered off [34]. The side effects of long-term steroid therapy often limit their use for chronic control of MG, and other long-term nonsteroidal immunosuppressive agents are preferred. Side effects include osteopenia/osteoporosis, hypertension, diabetes mellitus, cataracts, and weight gain, among others.

#### Nonsteroidal Immunosuppressive Agents

The most commonly used oral medications are azathioprine, mycophenolate mofetil, methotrexate, cyclosporine, and tacrolimus. The intravenous medications are rituximab and eculizumab. Side effects mostly include dose-dependent myelosuppression, but each medication has its own side effect profile.

Azathioprine given in combination with prednisolone has been associated with lower prednisolone maintenance dose, longer remissions, and fewer side effects [35]. An open-label trial and several case series have suggested that mycophenolate is useful in the control of MG, [36, 37], but two randomized controlled trials did not find it to be superior to placebo in controlling MG in combination with prednisone [38]. A single-blinded trial showed that methotrexate was similar in efficacy to azathioprine, [39] but a randomized clinical trial did not find it better than placebo in reducing the prednisone dose [40] though this second trial recruited milder cases requiring lower median doses of prednisone. Cyclosporine in combination with steroids has been found to cause improved strength when compared to placebo, but no significant reduction of steroid dose was seen [41]. Tacrolimus was also not associated with steroid dose reduction [42].

Rituximab is a chimeric mouse and human monoclonal antibody against CD20, which reduces CD20+ B cells and thus antibody production. The utility of rituximab has been suggested mostly by retrospective studies including a systematic review [43, 44]. A prospective trial in MuSK myasthenia also showed improvement on the primary outcome, which was a combination of clinical status and dosing of other immunosuppressants [45]. A phase 2 clinical trial on AchR-positive myasthenia gravis has been completed and results soon to be published [46]. A treatment protocol has not been established, but the most common regimen is 375 mg/m$^2$ weekly for 4 weeks, followed by optional re-dosing every 4–6 months.

Eculizumab is a humanized monoclonal antibody that prevents the formation of the membrane attack complex by binding to the terminal complement protein C5. A phase 3 clinical trial showed benefit on secondary analysis [47]. Eculizumab was FDA approved for acetylcholine receptor antibody-positive myasthenia gravis in 2017. It is administered weekly for the first 5 weeks and then every 2 weeks. Meningococcal vaccination is required prior to treatment initiation.

#### Immunomodulating Therapy

There are two forms of more rapid-acting immunomodulating therapy commonly used in MG. Intravenous immunoglobulin (IVIg) and plasma exchange (PLEX) are commonly used for rapid control of myasthenia gravis and for myasthenic crisis. A randomized placebo-controlled trial showed that 2 g/kg of IVIg leads to improvement in patients with worsening weakness, but the most benefit was seen in patients with severe disease [48]. There are no trials of PLEX against placebo, but no difference was found between PLEX and IVIg in myasthenic crisis [49]. More recently an open-label trial suggested that immunoglobulin administered subcutaneously was beneficial in mild to moderate worsening symptoms [50]. IVIg and PLEX are also frequently used as a pretreatment prior to elective surgeries (i.e., thymectomy) to help reduce the risk of perioperative worsening and myasthenic crisis.

#### Thymectomy

As discussed previously, the thymus is believed to play a central role in MG. If there is a thymoma, thymectomy is always indicated for tumor removal regardless of age. Even though thymectomy has been a common practice in AchR antibody-positive generalized, non-thymomatous MG, its benefit was not confirmed due to conflicting results of prior retrospective studies until recently. A randomized, placebo-controlled trial, performed in patients between 18 and 65 years of age, showed that thymectomy resulted in better clinical status, lower prednisone dose, lower non-steroidal immunosuppression, and less myasthenic crisis admissions [51]. There are no trials comparing the surgical techniques. Thymectomy can be performed with a transternal approach or video-assisted thoracoscopy, and though there are no guidelines for thymectomy procedures, the latter is the favored approach by most. There is no evidence for the role of thymectomy in seronegative myasthenia or myasthenia associated with antibodies against MuSK, cortactin, or LRP4.

## Lambert-Eaton Myasthenic Syndrome

### Epidemiology

Lambert-Eaton myasthenic syndrome (LEMS) is a rare neuromuscular junction disorder, with an estimated worldwide incidence of 0.48 per million and prevalence of 2.5–3.3 per million [52]. Approximately 47–62% of all LEMS cases are paraneoplastic (often due to small-cell lung cancer [SCLC]), while the remainder is caused by a non-tumor, autoimmune disease [53]. The median age of onset of cancer-associated LEMS is 60 years with a slight male predominance (59–

70%) [54]. Cigarette smoking is a risk factor for LEMS, which is reflective of its strong association with SCLC. In non-tumor LEMS (NT-LEMS), the median age of onset is 50 years but may also be seen in all ages with a bimodal peak at 35 years and a larger peak at 60 years [53–55]. There is a slight female predominance at 52% [53]. A personal and/or family history of autoimmune disease is a risk factor for NT-LEMS [56].

## Pathophysiology

In both paraneoplastic and NT-LEMS, autoantibodies are formed against the P/Q-type VGCC via cross-linking and downregulation, thereby decreasing the amount of acetylcholine released at the neuromuscular junction [57]. The reduction of the number of active zones has also been implicated in the action of the autoantibodies [58]. The downregulation of VGCC by targeted antibodies also occurs at muscarinic terminals causing the autonomic symptoms seen in LEMS [59]. In addition to IgG antibodies targeted to P/Q-type VGCCs, IgG can target N-type VGCC in LEMS [60]. N-type VGCCs play a role in the neurotransmitter release from nerve terminals in the autonomic system [61].

In SCLC-LEMS, the inciting event is believed to be antibodies formed against VGCC expressed on SCLC tumors [62]. SCLC expresses three types of VGCC: N, P/Q, and L [63]. Of patients with LEMS, 10–15% are seronegative for the P/Q-type VGCC antibody [53]. In these cases, it is hypothesized that P/Q-type VGCC antibodies are present in undetectable levels or that there are antibodies to a different VGCC epitope or molecule that result in a similar phenotype [57]. Other antibodies have recently been implicated in LEMS, such as antibodies to SOX1 (immunogenic antigen of SCLC) and synaptotagmin (synaptic vesicle protein) [57].

The exact immunopathological trigger in NT-LEMS is unknown, but there is an increased association with the HLA B8-DR3 haplotype [64]. There is also an association with underlying autoimmune diseases in NT-LEMS, such as autoimmune thyroid disease, diabetes mellitus, rheumatoid arthritis, and systemic lupus erythematosus [56].

## Clinical Manifestations

Similar to MG, muscle weakness in LEMS is fatigable; however, unlike MG, proximal muscle weakness of the lower extremities is a common presenting symptom. The clinical triad of proximal weakness, autonomic symptoms, and areflexia is commonly seen. Additional presenting symptoms include generalized weakness (12%), arm weakness (2%), muscle ache or stiffness (12%), and autonomic dysfunction (6%) [65]. Symptoms are typically gradual in onset but may appear subacutely. The pattern of progression of weakness is often proximal to distal, caudal to cranial, and, lastly, spreading to the oculobulbar region. Isolated ocular weakness is not expected, unlike in myasthenia gravis [53, 59]; and although patients have reported diplopia, clinical findings of ophthalmoplegia are rare [55, 65].

Autonomic symptoms are present in 80–96% of patients and include xerostomia, dry eyes, erectile dysfunction, constipation, micturition abnormalities, abnormalities of sweating, orthostatic hypotension, and blurred vision [65, 66]. Cerebellar ataxia, although rare, may be seen in paraneoplastic LEMS, due to antibodies to P/Q--type VGCC in the cerebellum [54].

On examination, Lambert's sign can be elicited, whereby, for a short time during strength testing, muscle power increases. This reflects the increased accumulation of calcium during sustained contraction stimulating more than usual release of acetylcholine [59]. Diminished or absent reflexes also can be seen on examination. Similar to the Lambert's sign, the phenomenon of transient return of reflexes can be seen post-exercise.

## Neoplastic Associations

LEMS may exist as a paraneoplastic syndrome with malignancy association or as an autoimmune process in the absence of malignancy. Of patients with LEMS, 47–62% have paraneoplastic LEMS, with SCLC being the most common associated cancer [53]. Other cancers that have been associated with paraneoplastic LEMS include lung adenocarcinoma, breast cancer, prostate carcinoma, lymphoma, and leukemia [67]. Diagnosis of LEMS often precedes the diagnosis of underlying malignancy, and the development of LEMS in paraneoplastic LEMS is due to antigenic expression of VGCC on the tumor leading to antibody formation [68]. SCLC significantly impacts the prognosis of LEMS, as survival is poor in patients with SCLC. However, patients with concomitant SCLC and LEMS have longer survival than SCLC in patients without LEMS. This likely represents lead-time bias with the earlier detection and diagnosis of SCLC [68]. The prognosis of NT-LEMS, although variable, is very different from SCLC-related LEMS as the life expectancy in NT-LEMS is normal [53].

Given the differences in prognosis between paraneoplastic LEMS and NT-LEMS, Titulaer and colleagues developed a prediction score (DELTA-P score) to help identify LEMS patients at a greatest risk of having underlying SCLC based on the presence of bulbar symptoms (dysarthria/dysphagia "D"), erectile dysfunction (E), weight loss (L), tobacco use at diagnosis (T), age at onset ≥50 (A), and Karnofsky performance score <70 (P). A score of 0–1 excludes SCLC, and a score of 3–6 should prompt the physician to screen for underlying malignancy aggressively as the risk of SCLC rises to 100% with a score of 6 [69]. Vigilant screening for underlying malignancy is recommended in patients with newly diagnosed LEMS. Screening is done via CT chest and fluorodeoxyglucose (FDG) positron emission tomography (FDG-PET) if CT chest is negative. Based on the DELTA-P score of 0–1, screening should be repeated in 6 months and may cease thereafter if negative. Screening should be done every 6 months for 2 years with a score of 2. A score of 3 or greater should prompt initial re-screening in 3 months and then every 6 months for 2 years [69, 70].

## Diagnosis

Diagnosis for LEMS is based on history and physical examination and is supported by serologic testing and specific electrodiagnostic findings.

### Serology

Detection of a high serum titer of P/Q-type VGCC (>1.00 nmol/L) antibodies is highly specific of both paraneoplastic and NT-LEMS. The absence of P/Q-type VGCC antibodies does not exclude the diagnosis [60]. Autoantibodies to P/Q-type VGCC are detected in approximately 90% of patients with LEMS, while approximately 30% of LEMS patients have antibodies to N-type VGCC; often antibodies to both VGCC types overlap [71, 72]. Furthermore, the presence of SOX-1 antibodies is suggestive of SCLC-LEMS. Although antibodies against synaptotagmin and M1 muscarinic AchR have been implicated in LEMS, they are not of diagnostic value [54].

### Electrodiagnostics

The electrodiagnostic features of LEMS span multiple modalities including motor and sensory nerve conduction studies, repetitive nerve stimulation (RNS), and single-fiber EMG.

At rest, the baseline motor amplitudes are reduced. During low-frequency (2–5 Hz) stimulation in RNS, there is decrement in the motor amplitudes [29]. In contrast to MG, in LEMS the decrement is sustained throughout all trains given the presynaptic dysfunction of acetylcholine release. During high-frequency RNS (20–50 Hz) or after brief maximal isometric muscle contraction (a mimicker of high-frequency RNS), there is marked increment in amplitude as intracellular calcium is increased in the presynaptic nerve terminal overcoming the presynaptic defect in acetylcholine release [29]. A greater than 100% increment is diagnostic of LEMS, but studies have shown that even 60% increment is just as specific for the diagnosis of LEMS [73]. Post-exercise testing is preferred to high-frequency RNS as the latter is painful. Single-fiber EMG shows increased jitter or blocking as well as possible myopathic motor units on routine EMG.

## Treatment of Lambert-Eaton Myasthenic Syndrome

The treatment of LEMS depends on the etiology. Symptomatic therapy includes 3,4-diaminopyridine, pyridostigmine, and guanidine. Immune therapy is indicated in autoimmune LEMS, but in paraneoplastic LEMS, treatment of tumor is the therapy of choice.

Symptomatic therapy includes the use of 3,4-diaminopyridine (3,4-DAP), which reversibly blocks the presynaptic voltage-gated potassium channels; this prolongs the duration of the presynaptic action potential and increases calcium entry. There is also evidence that it also directly targets the beta subunit of the VGCC [62]. The drug is well tolerated with less central side effects (such as seizures) as it crosses the blood-brain barrier less than its counterpart 4-aminopyridine [62]. Randomized, placebo-controlled trials have shown the benefits of 3,4-DAP in both motor and autonomic symptoms of LEMS [74–76]. Pyridostigmine also has been used with some symptomatic benefit, though a small clinical trial did not find any benefit with pyridostigmine alone or in addition of 3,4-DAP when compared to 3,4-DAP alone [77]. Guanidine, like 3,4-DAP, blocks the presynaptic potassium channels. An open-label study of nine patients suggested that low-dose guanidine combined with pyridostigmine is beneficial for symptomatic control [78]. Gastrointestinal side effects are usually the limiting factors for tolerability. There are also case reports of renal failure and myelosuppression [79, 80].

Because of the rarity of the condition, there is very little research on immunotherapy for it. A randomized, placebo-controlled, crossover trial with IVIg in nine patients showed reduced antibody titers and short-term improvement in strength over an 8-week period [81]. There are reports that PLEX is associated with clinical improvement [82]. Data on long-term immune therapy with steroids and azathioprine comes mostly from retrospective data with evidence to suggest it helps control both cancer and non-cancer-associated LEMS [83, 84]. There are two reported cases of rituximab-responsive LEMS as well [85].

## Immune Checkpoint Inhibitors and Neuromuscular Junction Disorders

Immune checkpoint inhibitors (ICPIs) have revolutionized the treatment of cancer such as Hodgkin's lymphoma, metastatic melanoma, renal cell carcinoma, and non-small cell lung cancer. Unfortunately, they have increasingly been implicated in triggering autoimmune disease. ICPIs are currently approved by the FDA in three forms: cytoplasmic T-lymphocyte-associated antigen-4 (CTLA-4: ipilimumab), programed cell death-1 (PD-1: pembrolizumab, nivolumab, cemiplimab), and programed cell death ligand-1 (PD-L1: avelumab, atezolizumab, durvalumab) [86]. Although neurologic manifestations are relatively rare in patients who have received ICPIs (1–3% of patients), neuromuscular diseases (MG, myositis, chronic and acute inflammatory demyelinating polyneuropathies) are the most common autoimmune neurologic manifestation [87]. ICPIs have been documented to cause de novo myasthenia gravis and exacerbate existing or premorbid myasthenia gravis [88]. Specific features such as bulbar and respiratory weakness appear to be more common in ICPI-related myasthenia gravis [87]. Acetylcholine receptor antibodies are detectable in two-thirds of the cases [87]. To date, cases of ICPI-related LEMS have not been reported [87]. Depending on the severity, the treatment of ICPI-related MG includes steroids, IVIg, PLEX, rituximab, and discontinuation of the checkpoint inhibitor [86]. Further peripheral nervous system and central nervous system complications of ICPIs are further discussed in Chap. 25.

## Conclusion

Autoimmune neuromuscular junction disorders include myasthenia gravis and Lambert-Eaton myasthenic syndrome. In both, autoantibodies target different proteins of the neuromuscular junction effectively impairing transmission. Sometimes antibodies are not detected, making diagnosis more difficult in seronegative cases. Both disorders have characteristic clinical findings, course, and neoplastic associations. Immune therapy is the mainstay of treatment, but tumor removal is key when found, particularly in Lambert-Eaton myasthenic syndrome. There are many options for the treatment of both conditions, but management should be personalized depending on severity, risk factors, and side effects.

## References


1. Nishimune H, Shigemoto K. Practical anatomy of the neuromuscular junction in health and disease. Neurol Clin. 2018;36(2):231–40.
2. Kawakami Y, Ito M, Hirayama M, Sahashi K, Ohkawara B, Masuda A, et al. Anti-MuSK autoantibodies block binding of collagen Q to MuSK. Neurology. 2011;77(20):1819–26.
3. Carr AS, Cardwell CR, McCarron PO, McConville J. A systematic review of population based epidemiological studies in Myasthenia Gravis. BMC Neurol. 2010;10:46.
4. Peragallo JH. Pediatric myasthenia gravis. Semin Pediatr Neurol. 2017;24(2):116–21.
5. Heldal AT, Owe JF, Gilhus NE, Romi F. Seropositive myasthenia gravis: a nationwide epidemiologic study. Neurology. 2009;73(2):150–1.
6. Querol L, Illa I. Myasthenia gravis and the neuromuscular junction. Curr Opin Neurol. 2013;26(5):459–65.
7. Lindstrom JM, Engel AG, Seybold ME, Lennon VA, Lambert EH. Pathological mechanisms in experimental autoimmune myasthenia gravis. II. Passive transfer of experimental autoimmune myasthenia gravis in rats with anti-acetylcholine receptor antibodies. J Exp Med. 1976;144(3):739–53.
8. Hoch W, McConville J, Helms S, Newsom-Davis J, Melms A, Vincent A. Auto-antibodies to the receptor tyrosine kinase MuSK in patients with myasthenia gravis without acetylcholine receptor antibodies. Nat Med. 2001;7(3):365–8.
9. Richman DP, Nishi K, Morell SW, Chang JM, Ferns MJ, Wollmann RL, et al. Acute severe animal model of anti-muscle-specific kinase myasthenia: combined postsynaptic and presynaptic changes. Arch Neurol. 2012;69(4):453–60.
10. Zhang B, Tzartos JS, Belimezi M, Ragheb S, Bealmear B, Lewis RA, et al. Autoantibodies to lipoprotein-related protein 4 in patients with double-seronegative myasthenia gravis. Arch Neurol. 2012;69(4):445–51.
11. Gallardo E, Martinez-Hernandez E, Titulaer MJ, Huijbers MG, Martinez MA, Ramos A, et al. Cortactin autoantibodies in myasthenia gravis. Autoimmun Rev. 2014;13(10):1003–7.
12. Berrih-Aknin S, Le Panse R. Myasthenia gravis: a comprehensive review of immune dysregulation and etiological mechanisms. J Autoimmun. 2014;52:90–100.
13. Hehir MK, Silvestri NJ. Generalized myasthenia gravis: classification, clinical presentation, natural history, and epidemiology. Neurol Clin. 2018;36(2):253–60.
14. Klimiec E, Quirke M, Leite MI, Hilton-Jones D. Thymus imaging in myasthenia gravis: the relevance in clinical practice. Muscle Nerve. 2018;58:153.
15. Richards J, Howard JF Jr. Seronegative myasthenia gravis associated with malignant thymoma. Neuromuscul Disord. 2017;27(5):417–8.
16. Keesey JC. Clinical evaluation and management of myasthenia gravis. Muscle Nerve. 2004;29(4):484–505.
17. Nicolle MW. Myasthenia Gravis and Lambert-Eaton Myasthenic Syndrome. Continuum (Minneap Minn). 2016;22(6, Muscle and Neuromuscular Junction Disorders):1978–2005.
18. Sih M, Soliven B, Mathenia N, Jacobsen J, Rezania K. Head-drop: a frequent feature of late-onset myasthenia gravis. Muscle Nerve. 2017;56(3):441–4.
19. Silvestri NJ, Wolfe GI. Myasthenia gravis. Semin Neurol. 2012;32(3):215–26.
20. Wendell LC, Levine JM. Myasthenic crisis. Neurohospitalist. 2011;1(1):16–22.
21. Mehrizi M, Fontem RF, Gearhart TR, Pascuzzi RM. Medications and myasthenia gravis (a reference for health care professionals). Myasthenia Gravis Foundation of America. 2012.
22. Thieben MJ, Blacker DJ, Liu PY, Harper CM Jr, Wijdicks EF. Pulmonary function tests and blood gases in worsening myasthenia gravis. Muscle Nerve. 2005;32(5):664–7.
23. Pascuzzi RM. The edrophonium test. Semin Neurol. 2003;23(1):83–8.
24. Sethi KD, Rivner MH, Swift TR. Ice pack test for myasthenia gravis. Neurology. 1987;37(8):1383–5.
25. Oh SJ, Kim DE, Kuruoglu R, Bradley RJ, Dwyer D. Diagnostic sensitivity of the laboratory tests in myasthenia gravis. Muscle Nerve. 1992;15(6):720–4.
26. Pasnoor M, Dimachkie MM, Farmakidis C, Barohn RJ. Diagnosis of myasthenia gravis. Neurol Clin. 2018;36(2):261–74.
27. Baslo MB, Deymeer F, Serdaroglu P, Parman Y, Ozdemir C, Cuttini M. Decrement pattern in Lambert-Eaton myasthenic syndrome is different from myasthenia gravis. Neuromuscul Disord. 2006;16(7):454–8.

28. Bromberg MB, Scott DM. Single fiber EMG reference values: reformatted in tabular form. AD HOC Committee of the AAEM Single Fiber Special Interest Group. Muscle Nerve. 1994;17(7):820–1.
29. Preston DC. Neuromuscular junction disorders. Electromyography and neuromuscular disorders. 3rd ed. Philadelphia: Elsevier; 2013.
30. Walker M. Treatment of myasthenia gravis with Physostigmine. Lancet. 1934;223:1200.
31. Osserman KE, Teng P, Kaplan LI. Studies in myasthenia gravis; preliminary report on therapy with mestinon bromide. J Am Med Assoc. 1954;155(11):961–5.
32. Schwab RS, Timberlake WH. Pyridostigmin (mestinon) in the treatment of myasthenia gravis. N Engl J Med. 1954;251(7):271–2.
33. Jenkins RB. Treatment of myasthenia gravis with prednisone. Lancet. 1972;1(7754):765–7.
34. Benatar M, McDermott MP, Sanders DB, Wolfe GI, Barohn RJ, Nowak RJ, et al. Efficacy of prednisone for the treatment of ocular myasthenia (EPITOME): a randomized, controlled trial. Muscle Nerve. 2016;53(3):363–9.
35. Palace J, Newsom-Davis J, Lecky B. A randomized double-blind trial of prednisolone alone or with azathioprine in myasthenia gravis. Myasthenia Gravis Study Group. Neurology. 1998;50(6):1778–83.
36. Chaudhry V, Cornblath DR, Griffin JW, O'Brien R, Drachman DB. Mycophenolate mofetil: a safe and promising immunosuppressant in neuromuscular diseases. Neurology. 2001;56(1):94–6.
37. Ciafaloni E, Massey JM, Tucker-Lipscomb B, Sanders DB. Mycophenolate mofetil for myasthenia gravis: an open-label pilot study. Neurology. 2001;56(1):97–9.
38. Sanders DB, Hart IK, Mantegazza R, Shukla SS, Siddiqi ZA, De Baets MH, et al. An international, phase III, randomized trial of mycophenolate mofetil in myasthenia gravis. Neurology. 2008;71(6):400–6.
39. Heckmann JM, Rawoot A, Bateman K, Renison R, Badri M. A single-blinded trial of methotrexate versus azathioprine as steroid-sparing agents in generalized myasthenia gravis. BMC Neurol. 2011;11:97.
40. Pasnoor M, He J, Herbelin L, Burns TM, Nations S, Bril V, et al. A randomized controlled trial of methotrexate for patients with generalized myasthenia gravis. Neurology. 2016;87(1):57–64.
41. Tindall RS, Phillips JT, Rollins JA, Wells L, Hall K. A clinical therapeutic trial of cyclosporine in myasthenia gravis. Ann N Y Acad Sci. 1993;681:539–51.
42. Yoshikawa H, Kiuchi T, Saida T, Takamori M. Randomised, double-blind, placebo-controlled study of tacrolimus in myasthenia gravis. J Neurol Neurosurg Psychiatry. 2011;82(9):970–7.
43. Tandan R, Hehir MK 2nd, Waheed W, Howard DB. Rituximab treatment of myasthenia gravis: a systematic review. Muscle Nerve. 2017;56(2):185–96.
44. Collongues N, Casez O, Lacour A, Tranchant C, Vermersch P, de Seze J, et al. Rituximab in refractory and non-refractory myasthenia: a retrospective multicenter study. Muscle Nerve. 2012;46(5):687–91.
45. Hehir MK, Hobson-Webb LD, Benatar M, Barnett C, Silvestri NJ, Howard JF Jr, et al. Rituximab as treatment for anti-MuSK myasthenia gravis: Multicenter blinded prospective review. Neurology. 2017;89(10):1069–77.
46. Nowak R. BeatMG: phase II trial of rituximab in Myasthenia Gravis. Available from: https://clinicaltrials.gov/ct2/show/NCT02110706
47. Howard JF Jr, Utsugisawa K, Benatar M, Murai H, Barohn RJ, Illa I, et al. Safety and efficacy of eculizumab in anti-acetylcholine receptor antibody-positive refractory generalised myasthenia gravis (REGAIN): a phase 3, randomised, double-blind, placebo-controlled, multicentre study. Lancet Neurol. 2017;16(12):976–86.
48. Zinman L, Ng E, Bril V. IV immunoglobulin in patients with myasthenia gravis: a randomized controlled trial. Neurology. 2007;68(11):837–41.
49. Gajdos P, Chevret S, Clair B, Tranchant C, Chastang C. Clinical trial of plasma exchange and high-dose intravenous immunoglobulin in myasthenia gravis. Myasthenia Gravis Clinical Study Group. Ann Neurol. 1997;41(6):789–96.
50. Beecher G, Anderson D, Siddiqi ZA. Subcutaneous immunoglobulin in myasthenia gravis exacerbation: a prospective, open-label trial. Neurology. 2017;89(11):1135–41.
51. Wolfe GI, Kaminski HJ, Cutter GR. Randomized trial of thymectomy in myasthenia gravis. N Engl J Med. 2016;375(20):2006–7.
52. Abenroth DC, Smith AG, Greenlee JE, Austin SD, Clardy SL. Lambert-Eaton myasthenic syndrome: epidemiology and therapeutic response in the national veterans affairs population. Muscle Nerve. 2017;56(3):421–6.
53. Titulaer MJ, Lang B, Verschuuren JJ. Lambert-Eaton myasthenic syndrome: from clinical characteristics to therapeutic strategies. Lancet Neurol. 2011;10(12):1098–107.
54. Schoser B, Eymard B, Datt J, Mantegazza R. Lambert-Eaton myasthenic syndrome (LEMS): a rare autoimmune presynaptic disorder often associated with cancer. J Neurol. 2017;264(9):1854–63.
55. Kesner VG, Oh SJ, Dimachkie MM, Barohn RJ. Lambert-Eaton Myasthenic syndrome. Neurol Clin. 2018;36(2):379–94.
56. Wirtz PW, Bradshaw J, Wintzen AR, Verschuuren JJ. Associated autoimmune diseases in patients with the Lambert-Eaton myasthenic syndrome and their families. J Neurol. 2004;251(10):1255–9.
57. Takamori M. Lambert-Eaton myasthenic syndrome: search for alternative autoimmune targets and possible compensatory mechanisms based on presynaptic calcium homeostasis. J Neuroimmunol. 2008;201–202:145–52.
58. Tarr TB, Wipf P, Meriney SD. Synaptic pathophysiology and treatment of Lambert-Eaton Myasthenic syndrome. Mol Neurobiol. 2015;52(1):456–63.
59. Mareska M, Gutmann L. Lambert-Eaton myasthenic syndrome. Semin Neurol. 2004;24(2):149–53.

60. Zalewski NL, Lennon VA, Lachance DH, Klein CJ, Pittock SJ, McKeon A. P/Q- and N-type calcium-channel antibodies: oncological, neurological, and serological accompaniments. Muscle Nerve. 2016;54(2):220–7.
61. Hirning LD, Fox AP, McCleskey EW, Olivera BM, Thayer SA, Miller RJ, et al. Dominant role of N-type Ca2+ channels in evoked release of nor-epinephrine from sympathetic neurons. Science. 1988;239(4835):57–61.
62. Hulsbrink R, Hashemolhosseini S. Lambert-Eaton myasthenic syndrome - diagnosis, pathogenesis and therapy. Clin Neurophysiol. 2014;125(12):2328–36.
63. Lang B, Newsom-Davis J. Immunopathology of the Lambert-Eaton myasthenic syndrome. Springer Semin Immunopathol. 1995;17(1):3–15.
64. Wirtz PW, Willcox N, van der Slik AR, Lang B, Maddison P, Koeleman BP, et al. HLA and smoking in prediction and prognosis of small cell lung cancer in autoimmune Lambert-Eaton myasthenic syndrome. J Neuroimmunol. 2005;159(1–2):230–7.
65. O'Neill JH, Murray NM, Newsom-Davis J. The Lambert-Eaton myasthenic syndrome. A review of 50 cases. Brain. 1988;111(Pt 3):577–96.
66. Waterman SA. Autonomic dysfunction in Lambert-Eaton myasthenic syndrome. Clin Auton Res. 2001;11(3):145–54.
67. Wirtz PW, Smallegange TM, Wintzen AR, Verschuuren JJ. Differences in clinical features between the Lambert-Eaton myasthenic syndrome with and without cancer: an analysis of 227 published cases. Clin Neurol Neurosurg. 2002;104(4):359–63.
68. Lipka AF, Verschuuren JJ, Titulaer MJ. SOX1 antibodies in Lambert-Eaton myasthenic syndrome and screening for small cell lung carcinoma. Ann N Y Acad Sci. 2012;1275:70–7.
69. Titulaer MJ, Maddison P, Sont JK, Wirtz PW, Hilton-Jones D, Klooster R, et al. Clinical Dutch-English Lambert-Eaton Myasthenic syndrome (LEMS) tumor association prediction score accurately predicts small-cell lung cancer in the LEMS. J Clin Oncol. 2011;29(7):902–8.
70. Briggs SE, Gozzard P, Talbot DC. The association between Lambert-Eaton myasthenic syndrome and small cell lung carcinoma. Immunotargets Ther. 2013;2:31–7.
71. O'Suilleabhain P, Low PA, Lennon VA. Autonomic dysfunction in the Lambert-Eaton myasthenic syndrome: serologic and clinical correlates. Neurology. 1998;50(1):88–93.
72. Iwasa K, Pinto A, Vincent A, Lang B. LEMS IgG binds to extracellular determinants on N-type voltage-gated calcium channels, but does not reduce VGCC expression. Ann N Y Acad Sci. 2003;998:196–9.
73. Oh SJ, Kurokawa K, Claussen GC, Ryan HF Jr. Electrophysiological diagnostic criteria of Lambert-Eaton myasthenic syndrome. Muscle Nerve. 2005;32(4):515–20.
74. McEvoy KM, Windebank AJ, Daube JR, Low PA. 3,4-Diaminopyridine in the treatment of Lambert-Eaton myasthenic syndrome. N Engl J Med. 1989;321(23):1567–71.
75. Sanders DB, Massey JM, Sanders LL, Edwards LJ. A randomized trial of 3,4-diaminopyridine in Lambert-Eaton myasthenic syndrome. Neurology. 2000;54(3):603–7.
76. Oh SJ, Claussen GG, Hatanaka Y, Morgan MB. 3,4-Diaminopyridine is more effective than placebo in a randomized, double-blind, cross-over drug study in LEMS. Muscle Nerve. 2009;40(5):795–800.
77. Wirtz PW, Verschuuren JJ, van Dijk JG, de Kam ML, Schoemaker RC, van Hasselt JG, et al. Efficacy of 3,4-diaminopyridine and pyridostigmine in the treatment of Lambert-Eaton myasthenic syndrome: a randomized, double-blind, placebo-controlled, crossover study. Clin Pharmacol Ther. 2009;86(1):44–8.
78. Oh SJ, Kim DS, Head TC, Claussen GC. Low-dose guanidine and pyridostigmine: relatively safe and effective long-term symptomatic therapy in Lambert-Eaton myasthenic syndrome. Muscle Nerve. 1997;20(9):1146–52.
79. Oh SJ, Kim KW. Guanidine hydrochloride in the Eaton-Lambert syndrome. Electrophysiologic improvement. Neurology. 1973;23(10):1084–90.
80. Blumhardt LD, Joekes AM, Marshall J, Philalithis PE. Guanidine treatment and impaired renal function in the Eaton-Lambert syndrome. Br Med J. 1977;1(6066):946–7.
81. Bain PG, Motomura M, Newsom-Davis J, Misbah SA, Chapel HM, Lee ML, et al. Effects of intravenous immunoglobulin on muscle weakness and calcium-channel autoantibodies in the Lambert-Eaton myasthenic syndrome. Neurology. 1996;47(3):678–83.
82. Newsom-Davis J, Murray NM. Plasma exchange and immunosuppressive drug treatment in the Lambert-Eaton myasthenic syndrome. Neurology. 1984;34(4):480–5.
83. Chalk CH, Murray NM, Newsom-Davis J, O'Neill JH, Spiro SG. Response of the Lambert-Eaton myasthenic syndrome to treatment of associated small-cell lung carcinoma. Neurology. 1990;40(10):1552–6.
84. Tim RW, Massey JM, Sanders DB. Lambert-Eaton myasthenic syndrome (LEMS). Clinical and electrodiagnostic features and response to therapy in 59 patients. Ann N Y Acad Sci. 1998;841:823–6.
85. Maddison P, McConville J, Farrugia ME, Davies N, Rose M, Norwood F, et al. The use of rituximab in myasthenia gravis and Lambert-Eaton myasthenic syndrome. J Neurol Neurosurg Psychiatry. 2011;82(6):671–3.
86. Kolb NA, Trevino CR, Waheed W, Sobhani F, Landry KK, Thomas AA, et al. Neuromuscular complications of immune checkpoint inhibitor therapy. Muscle Nerve. 2018;58(1):10–22.
87. Kao JC, Brickshawana A, Liewluck T. Neuromuscular complications of programmed cell death-1 (PD-1) inhibitors. Curr Neurol Neurosci Rep. 2018;18(10):63.
88. Dalakas MC. Neurological complications of immune checkpoint inhibitors: what happens when you 'take the brakes off' the immune system. Ther Adv Neurol Disord. 2018;11:1756286418799864.

# Immune and Inflammatory Myopathies

# 20

Andrew R. Findlay and Robert C. Bucelli

**Key Points**

1. Immune and inflammatory myopathies (IIM) are characterized by weakness with immune or inflammatory changes on muscle biopsy and commonly have extra-muscular manifestations.
2. Unique clinical features, autoantibodies, and histopathological patterns are used to phenotypically categorize patients and predict treatment response and prognosis.
3. The most commonly agreed-upon criteria recognize four main categories: dermatomyositis, sporadic inclusion body myositis, antisynthetase syndrome, and immune-mediated necrotizing myopathy.
4. Excluding sporadic inclusion body myositis, IIM treatment still largely relies on empirical use of corticosteroids and steroid-sparing agents.

A. R. Findlay (✉) · R. C. Bucelli
Department of Neurology, Neuromuscular Section, Barnes Jewish Hospital, Washington University School of Medicine in Saint Louis, Saint Louis, MO, USA
e-mail: arfindlay@wustl.edu

## Introduction

Immune and inflammatory myopathies (IIM), collectively known as myositis, are characterized by progressive weakness and inflammatory cellular infiltrates within skeletal muscle. Damage to specific tissues within skeletal muscle, such as connective tissue or blood vessels, may cause syndromes involving multiple organ systems other than muscle, including skin, lungs, and joints. IIM subtypes have historically been defined by clinical and histopathological differences and traditionally were classified as polymyositis (PM) or dermatomyositis (DM). Progress has been made in revising Bohan and Peter's original diagnostic criteria from 1975 in order to more accurately align clinical, autoantibody, and histopathological data with prognosis and response to treatment [1, 2]. PM has been overvalued [3, 4], and pathologic criteria isolated two new subgroups, previously referred to as PM, including sporadic inclusion body myositis (sIBM) [1, 2] and immune-mediated necrotizing myopathy (IMNM) [2]. These classification approaches, however, define overlapping entities. For example, antisynthetase syndrome is often classified as DM, as PM, or as an overlap syndrome [5–7]. Myopathology [8] and autoantibodies can help define subgroups of patients in terms of clinical or pathologic phenotypes, prognosis, and response to treatment [9–13]. The most up-to-date, and commonly accepted, classification



A. L. Piquet, E. Alvarez (eds.), *Neuroimmunology*, https://doi.org/10.1007/978-3-030-61883-4_20

criteria will be used in this chapter and eliminates PM as a distinct entity recognizing only DM, IMNM, sIBM, and antisynthetase syndrome [14]. PM does not represent a subgroup of patients, and use of this term should probably be discontinued [14]. It is important to recognize that all classification criteria have their drawbacks. A major issue with the criteria used for this chapter is that it fails to account for other forms of IIM, such as brachio-cervical inflammatory myopathy, focal myositis, eosinophilic myositis, and granulomatous myopathies. This chapter describes current knowledge of the epidemiology, clinical characteristics, diagnostic evaluation, classification, pathogenesis, treatment, and prognosis of IIM.

## Epidemiology

Most epidemiologic studies have used criteria that fail to distinguish sIBM, IMNM, and antisynthetase syndrome (discussed previously), causing inaccuracies in incidence and prevalence studies and making it easier to analyze data for IIM as a collective group. Incidence rates for IIM range between 4.27 and 7.89 per 100,000 person years, and prevalence ranges from 9.54 to 32.74 cases per 100,000 individuals [15–17]. sIBM prevalence has been reported as 9.3 per million [18]. Using recent classifications, of all IIMs, sIBM accounts for 29.6%, IMNM 35%, DM 20%, and antisynthetase syndrome 15.4% [14]. DM, IMNM, and antisynthetase syndromes occur more frequently in females [19]. DM may affect children and adults [19], whereas sIBM is seen more commonly in male patients over the age of 50 [20]. Mean annual medical costs and number of ambulatory visits, specialty visits, and inpatient hospital stays are significantly higher among subjects with IIM compared to matched controls [21].

## Clinical Features

### Dermatomyositis

Dermatomyositis may present with subacute onset progressive proximal weakness, cutaneous manifestations, or both [22]. The deltoids tend to be more severely affected [14]. Some patients may present with only skin changes and are considered to have hypomyopathic or amyopathic forms of disease [23]. Others may present with isolated muscle weakness and never develop rash or only develop rash months later [24]. Juvenile patients may present initially with a febrile illness [24]. Pathognomonic skin features include violaceous periorbital edema (heliotrope rash) and papular lesions on the extensor surfaces of metacarpophalangeal and interphalangeal joints, Gottron's papules (Fig. 20.1a). Other findings may include an erythematous rash over extensor surfaces of limbs (Gottron's sign), over the neck and chest (V sign), and over the back of the neck and shoulders (shawl sign), limb edema, alopecia, skin ulcers, calcinosis, and panniculitis (Fig. 20.1b–f) [14]. Lesions may be photosensitive and pruritic [25]. Juvenile patients more commonly develop cutaneous calcinosis (30–70% of juvenile cases and 10% of adult cases) over pressure points [25, 26]. Myalgias may also be present [27].

### Immune-Mediated Necrotizing Myopathy

IMNM is typically characterized by rapid progression of severe proximal weakness, with prominent involvement of the psoas muscles and exceptionally high creatine kinase (CK). Toxic or drug-induced etiologies, as well as some hereditary myopathies (e.g., limb-girdle muscular dystrophy), may appear similar to IMNM and should be ruled out [28–30]. Patients may have mild myalgias or have no muscle pain whatsoever [31]. Extra-muscular manifestations are generally mild if they occur [32–34].

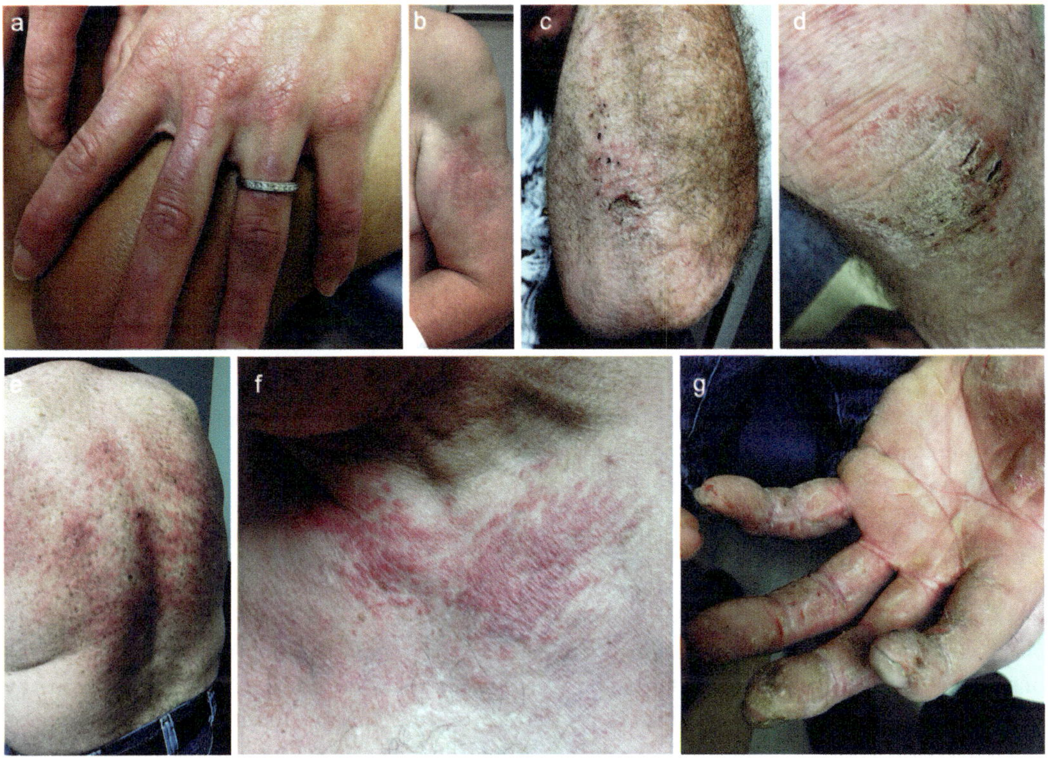


**Fig. 20.1** Cutaneous manifestations of immune and inflammatory myopathies (IIM). Gottron's papules (**a**), Gottron's sign, an erythematous rash over extensor surfaces such as elbows (**b**, **c**) or knees (**d**), shawl sign (**e**), V sign (**f**), mechanics hands (**g**)

## Antisynthetase Syndromes

Overlap myositis occurs when a patient has an autoimmune myopathy associated with other autoimmune diseases such as systemic lupus erythematosus, rheumatoid arthritis, Sjögren's syndrome, or systemic sclerosis [35]. Antisynthetase syndrome, with autoantibodies targeting aminoacyl tRNA synthetases, is the most representative form of overlap myositis [36]. Patients may present with a combination of inflammatory myopathy, interstitial lung disease (ILD), arthritis, Raynaud syndrome, fever, or hyperkeratotic finger lesions called mechanic's hands (Fig. 20.1g) [37]. Antisynthetase syndrome may also cause skin rashes similar to dermatomyositis [37]. Myopathic features include proximal weakness similar to DM, although some patients may have no weakness at all and clinical manifestations of muscle disease may be limited to myalgias in isolation.

## sIBM

sIBM often presents slowly with progression over 5–8 years before affected patients come to medical attention [18, 38]. Characteristic findings include asymmetric wasting and weakness of the wrist flexors, deep finger flexors, and quadriceps muscles (Fig. 20.2a, b) [24]. Tibialis anterior weakness, dysphagia, and mild facial weakness may also be present [39–42]. In a study of 57 patients with sIBM, the initial presenting symptoms were quadriceps weakness (79%), finger weakness (12%), foot drop (7%), and dysphagia

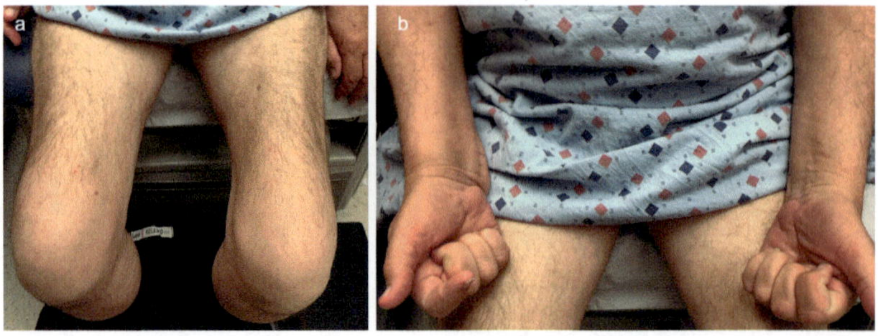


**Fig. 20.2** Clinical features of sporadic inclusion body myositis (sIBM). Atrophy and wasting of quadriceps muscles (**a**) and asymmetrical wrist and finger flexor weakness causing impaired ability to make a fist, worse on patient's right side (**b**)

(1.8%) [43]. Asymmetric involvement was very common (82%), with the patient's non-dominant side commonly being more severely affected [43]. There may be evidence of a generalized sensory peripheral neuropathy on clinical exam [44]. Up to 15% of sIBM patients have a coexisting autoimmune disorder or condition with altered immune function [45]. Sporadic IBM is not associated with heart disease [39] or an increased risk of malignancy [46]. Primary respiratory failure is rare; however, progressive dysphagia may occur and may lead to aspiration [40, 41].

## Diagnostic Evaluation

### Elevated Muscle Enzymes

Serum CK levels are a sensitive measure of muscle disease activity in IIM [24]. They do not correlate well with disease activity when comparing different patients, but they can reflect changes in disease activity within an individual patient. Levels are typically highest in IMNM (2300 U/L–7000 U/L) and lowest in sIBM (160 U/L–793 U/L) [14]. Aldolase levels may also be prominently elevated, presumably from intramuscular connective tissue damage. For example, antisynthetase syndromes with perimysial pathology may have isolated aldolase elevation [42]. Other muscle enzymes, including myoglobin, lactate dehydrogenase, aspartate aminotransferase, and alanine aminotransferase, may also be elevated. Patients taking hepatotoxic steroid-sparing agents, such as azathioprine (AZA) or methotrexate, may develop elevated transaminases. The liver enzyme gamma-glutamyl transpeptidase (GGT) can aid in differentiating liver damage in IIM patients, as it is not released by damaged muscle [47].

### Electrodiagnostics

Electromyography (EMG) typically reveals an irritable myopathic pattern characterized by increased insertional and spontaneous activity (fibrillation potentials, positive sharp waves, and occasionally complex repetitive discharges), polyphasic motor unit action potentials (MUAPs) with small duration and low amplitude, and early MUAP recruitment. sIBM patients may have evidence of neuropathy on nerve conduction studies and mixed myopathic and neurogenic changes on EMG [44].

### Muscle Imaging

Muscle MRI in sIBM patients demonstrates severe involvement of the anterior compartment of the thigh and forearm [48]. DM, IMNM, and antisynthetase syndrome patients often have a nonspecific pattern with hyperintensities on intramuscular T2-weighted magnetic resonance imaging (MRI) scans [48, 49]. Some recommend using muscle MRI to select the site of muscle biopsy [50]. Caution should be used with this approach as neurogenic changes from denervation appear similar to changes related to myositis on MRI. MRI also cannot distinguish between IIM and hereditary myopathies [51].

## Antibodies

A screen for autoantibodies is common in the evaluation of patients with IIM or suspected IIM. However, their role in the pathophysiology of IIM is unclear. Some may be directly involved in pathophysiology and others simply an epiphenomenon. Antibodies are categorized as myositis-associated autoantibodies (MAAs) or myositis-specific autoantibodies (MSAs). MSAs are found predominantly in the serum of patients with IIM, but are not 100% specific for IIM [52, 53]. MAAs are primarily encountered in other connective tissue diseases and occasionally found in patients with IIM [52, 53]. MSAs can help classify homogenous phenotypic subsets of patients and help predict the degree of muscle, skin, and lung involvement, as well as risk of an associated malignancy (Table 20.1) [14, 29, 32, 33, 41, 52, 54–63]. Recent classification schemes suggest MSAs are crucial for accurate categorization of IIM [14].

**Dermatomyositis** Approximately 70% of patients with DM have a dermatomyositis-specific autoantibody [52], many associated with a unique clinical phenotype (Table 20.1) [56–61]. Autoantibodies against Mi2, a nuclear antigen, are associated with classic DM characteristics, severe skin manifestations, proximal weakness, and a lower risk of associated malignancy relative to DM associated with other MSAs [56]. DM patients with autoantibodies against nuclear matrix protein NXP2 are more likely to present with both proximal and distal muscle weakness, subcutaneous edema, and dysphagia and are more prone to develop calcinosis [58]. Patients with anti-NXP2 or anti-transcription intermediary factor (TIF)-1 autoantibodies are associated with increased risk of malignancy within 3 years of diagnosis. Accordingly, comprehensive cancer screening or positron emission tomography-computed tomography (PET-CT) scans are particularly important for these patients [57, 58, 64, 65]. DM patients with antibodies against

**Table 20.1** Myositis-specific autoantibodies

| Myositis-specific autoantibody | Phenotype features |
|---|---|
| *Antisynthetase syndrome* | |
| Anti-histidyl-tRNA synthetase (anti-Jo-1) | 90% muscle involvement; 50–75% interstitial lung disease (ILD) [54] |
| Anti-threonyl-tRNA synthetase (anti-PL-7) | 44% muscle involvement; 80% ILD [55] |
| Anti-alanyl-tRNA synthetase (anti-PL-12) | 50% muscle involvement; 90% ILD [54] |
| Anti-glycyl-tRNA synthetase (anti-EJ) | |
| Anti-isoleucyl-tRNA synthetase (anti-OJ) | |
| Anti-asparaginyl-tRNA synthetase (anti-KS) | |
| Anti-tyrosyl-tRNA synthetase (anti-Ha) | |
| Anti-phenylalanyl-tRNA synthetase (anti-Zo) | |
| *Dermatomyositis* | |
| Anti-Mi-2 | Severe skin manifestations, good response to treatment, less risk of malignancy relative to other forms of dermatomyositis [56] |
| Anti-transcriptional intermediary factor 1+ (anti-TIF-1) | Adults: Increased risk of malignancy. Children: Severe cutaneous involvement [57] |
| Anti-nuclear matrix protein 2 (anti-NXP2) | Increased risk of calcinosis. Increased risk of malignancy in adults [57, 58] |
| Anti-melanoma differentiation-associated protein 5 (anti-MDA5) | Skin ulcerations, palmar papules, and severe ILD syndrome [59, 60] |
| Anti-small ubiquitin-like modifier 1 (SAE) | Skin manifestations before muscle; dysphagia [61] |
| *Immune-mediated necrotizing myopathy* | |
| Anti-signal recognition particle (anti-SRP) | Severe weakness. Difficult to treat. ILD more common than anti-HMGCR. Onset is most common in autumn [32, 33, 62] |
| Anti-3-hydroxy-3-methylglutaryl coenzyme A reductase (HMGCR) | Can cause pure muscle involvement or antisynthetase syndrome-like picture. Increased risk of malignancy vs. anti-signal recognition particle (SRP) [29, 62, 63] |

small ubiquitin-like modifier activating enzyme (SAE) or melanoma differentiation-associated gene 5 (MDA5) typically have more cutaneous than muscle involvement [59–61, 66]. MDA5 patients are commonly hypomyopathic or amyopathic and may develop ulcers on the palmar surface of their hands and a rapidly progressive form of ILD [59, 60, 66]. IIM patients suspected to have interstitial lung disease should initially be evaluated and monitored using pulmonary function tests (carbon monoxide diffusion and inspiratory and expiratory pressures) and high-resolution CT scans.

**IMNM** Several autoantibodies associated with IMNM have been identified, each with specific characteristics and clinical outcomes (Table 20.1) [29, 32, 33, 62, 63]. These include anti-signal recognition particle (SRP) and anti-3-hydroxy-3-methylglutaryl coenzyme A reductase (HMGCR) autoantibodies. Patients with SRP or HMGCR antibodies often share several features, including high CK, and an aggressive refractory disease course in some patients [62]. The IMNM classification does not perfectly overlap with all patients with SRP or HMGCR antibodies. Only two-thirds of IMNM patients are reported to have antibodies to SRP or HMGCR, and around 20% of patients with SRP or HMGCR antibodies do not have key histopathology characteristics of IMNM [33, 63, 67]. Only two-thirds of HMGCR patients have necrosis or regeneration, and one-third have lymphocytic infiltrates [68]. Approximately 60% of these patients will have prominent perimysial pathology, and as high as 37% will have systemic features such as ILD and skin rash, features more commonly seen with overlap or antisynthetase syndromes [68]. The association between statin usage and increased risk of developing IMNM associated with HMGCR antibodies is a subject of ongoing debate [68]. However, it is clear that some patients do have a form of disease triggered by exposure to statins, likely from feedback mechanisms that lead to increased HMGCR expression in muscle tissue [29]. Patients with HMGCR antibodies might have an increased risk of malignancy [69]. Some patients may even present with a slowly progressive disease course and be misdiagnosed with limb-girdle muscular dystrophy [29]. SRP patients tend to have more severe weakness than HMGCR patients [32, 62]. In addition to necrosis and regeneration, SRP muscle pathology demonstrates prominent endomysial fibrosis and capillary pathology [70]. SRP patients may be at greater risk for developing interstitial lung disease and possibly cardiac involvement when compared to patients with HMGCR antibodies [62, 71]. If cardiac involvement is suspected, an electrocardiogram (ECG) and echocardiogram should be performed. Seronegative IMNM is thought to be associated with increased risk of malignancy, female predominance, frequent occurrence of associated connective tissue disorders, and increased risk of extra-muscular disease activity [69, 72].

**Antisynthetase Syndrome** Autoantibodies against histidyl (anti-Jo-1)-, threonyl (anti-PL7)-, and alanyl (anti-PL12)-tRNA synthetases are the most common [36, 73]. About 90% of patients with anti-Jo-1 autoantibodies have an inflammatory myopathy, while approximately 50% of patients with anti-PL12 autoantibodies present with interstitial lung disease but no muscle involvement [36]. Muscle weakness tends to be more severe in patients with anti-Jo-1 autoantibodies, while lung involvement is more severe in patients with anti-PL7 and anti-PL12 autoantibodies [36, 73].

**Sporadic IBM** Autoantibodies against cytosolic 5′-nucleotidase 1A (NT5C1A) are present in 30–60% of patients with sIBM. NT5C1A autoantibodies are not specific for sIBM, as they are found in 15–20% of patients with DM, 10% of patients with systemic lupus erythematosus, and 12% of patients with Sjögren's syndrome [74–77]. SIBM patients with NT5C1A antibodies are more commonly female, have greater motor and functional disability, and have more prominent bulbar, facial, and respiratory involvement [78].

## Histopathology

Muscle biopsy is a valuable diagnostic tool in patients suspected to have an IIM. The key pathological characteristics of IIM initially recognized by Bohan and Peter criteria were degeneration, regeneration, necrosis, and interstitial mononuclear infiltrates. Inflammatory cell infiltrates are not specific to IIM as they can be seen in muscular dystrophies such as dysferlinopathy, calpainopathy, facioscapulohumeral muscular dystrophy, metabolic myopathies following rhabdomyolysis, granulomatous disorders, myasthenia gravis, vasculitis, and lymphoma, among other disorders [24, 79–84]. Muscle biopsies from patients with DM, IMNM, sIBM, and antisynthetase syndrome are known to have many unique pathological features, suggesting different pathophysiological mechanisms exist for each [8, 82–84]. Each IIM affects specific regions or tissues within skeletal muscle, including connective tissue, blood vessels, and muscle fibers (Fig. 20.3). Pathology can also predict lung involvement, risk of malignancy, and response to immunomodulatory treatment.

**Dermatomyositis** "Perifascicular atrophy" is the classic feature described in DM (Fig. 20.4a). Some claim this finding is very specific for DM; however, there are several inconsistences [14, 41, 85]. For example, some patients will have prominent perifascicular necrosis instead of atrophy, and others may have minimal inflammatory infiltrates and prominent necrosis similar to IMNM [41, 86]. The classic dermatomyositis clinicopathological picture may therefore be more accurately characterized as dermatomyositis with vascular pathology (DM-VP) [87]. DM-VP biopsies demonstrate a perifascicular myopathy with muscle fiber atrophy, reduced cytochrome oxidase staining, and increased MHC class 1 expression (Fig. 20.4a, b) [8]. The vascular pathology is characterized by abnormal, damaged endomysial capillaries with alkaline phosphatase staining, C5b-9 deposition, and lymphocytic foci surrounding larger vessels in vascular perimysium (Fig. 20.4c, d) [8].

A clinically different subset of patients that are often included under the umbrella categorization of DM have damage to perimysial connective tissue and perifascicular muscle fiber pathology that is often mistaken for DM-VP [8, 88]. Biopsies demonstrate perimysial connective tissue pathology including fragmentation, acid phosphatase-positive histiocytic cells, and alkaline phosphatase staining of the perimysium (Fig. 20.5a–c). Muscle fiber pathology includes necrosis and regeneration, more prominent in

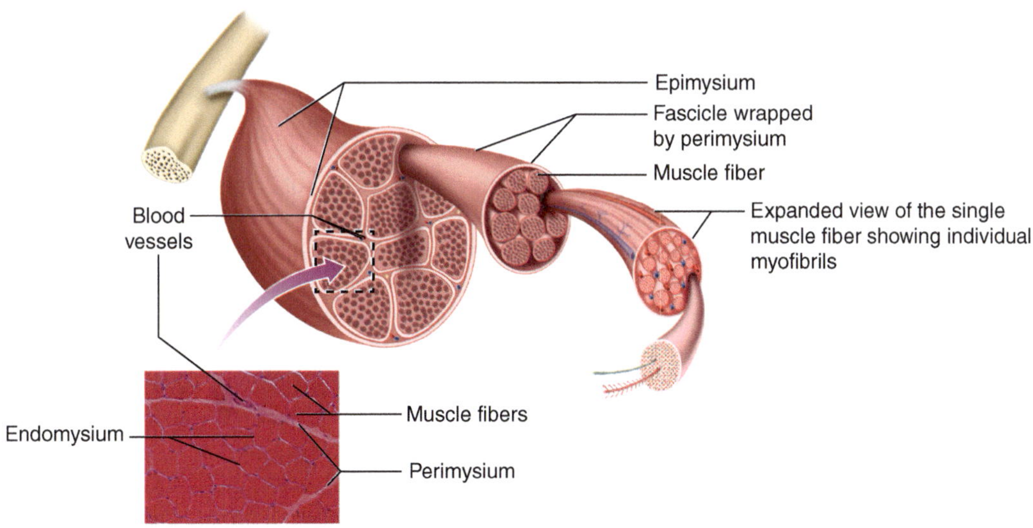


**Fig. 20.3** Muscle anatomy. Individual muscle fibers are surrounded by the endomysium, which contains capillaries. Muscle fibers are grouped into fascicles, which are separated by the perimysium. Perimysial connective tissue may contain vasculature. The epimysial connective tissue envelops all fascicles within the muscle

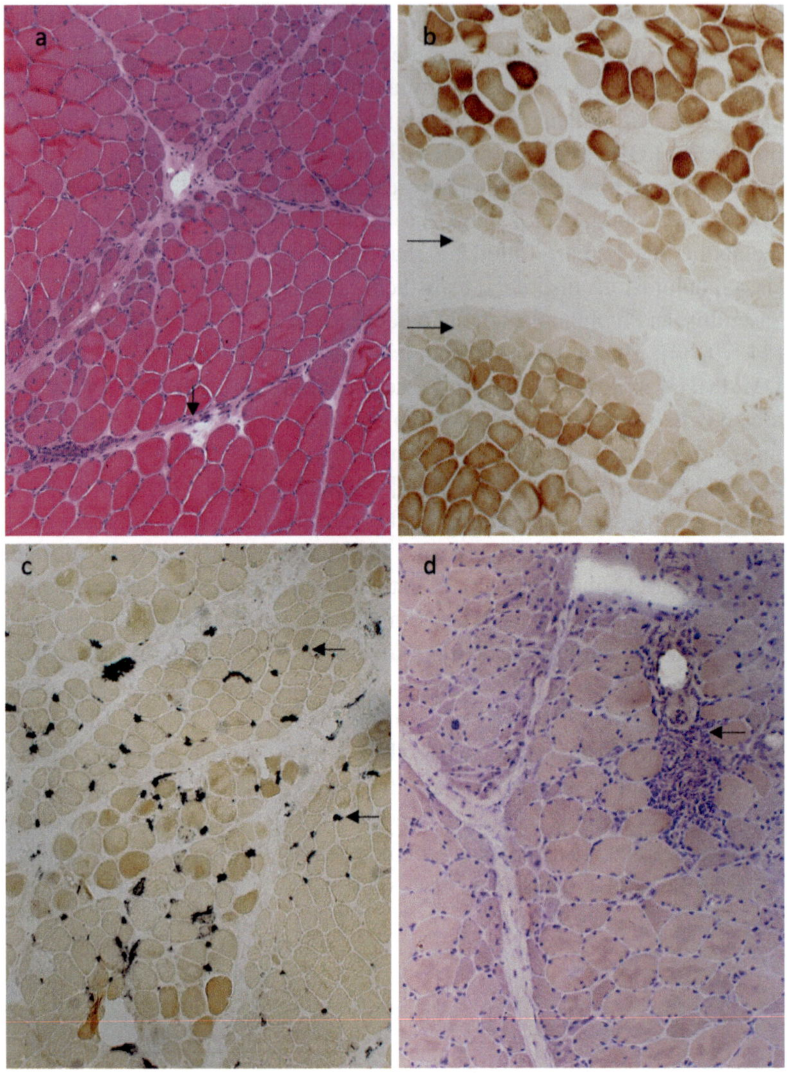


**Fig. 20.4** Dermatomyositis with vascular pathology histopathology. H&E demonstrating perifascicular atrophy. Note absence of atrophy adjacent to vascular perimysium (arrow) (**a**). Perifascicular pattern of reduced cytochrome oxidase staining (arrows) (**b**). Alkaline phosphatase highlights enlarged, abnormal endomysial capillaries (**c**). Perivascular lymphocytic infiltrates around intermediate-sized vessels (arrow), distant from muscle fiber atrophy (**d**)

regions neighboring the perimysium (Fig. 20.5d) [8, 88]. These disorders have been termed immune myopathies with perimysial pathology (IMPP) [89]. When compared to DM-VP, IMPP is associated with the clinical picture of antisynthetase syndrome with increased risk of ILD, Raynaud phenomenon, mechanic's hands, inflammatory arthritis, and a higher CK level. IMPP also predicts a sustained response to immunomodulatory therapy and is less frequently associated with a concurrent malignancy [88]. Because of this, IMPP patients require regular screening for ILD. While IMPP patients may have MSAs such as anti-Jo-1 or HMGCR, the large percentage of patients without MSAs emphasizes the important role of myopathology in identifying patients at higher risk of severe comorbid conditions such as ILD.

Regional ischemic immune myopathy (RIIM) is another distinctive pathologic group observed in dermatomyopathy patients and is likely caused by ischemia in border zones between damaged intermediate-sized perimysial blood vessels [90]. Histopathology reveals an unusual pattern of regional muscle fiber necrosis and regeneration with capillary loss in border zones between intermediate-sized perimysial vessels, vascular pathology with damaged walls of intermediate-

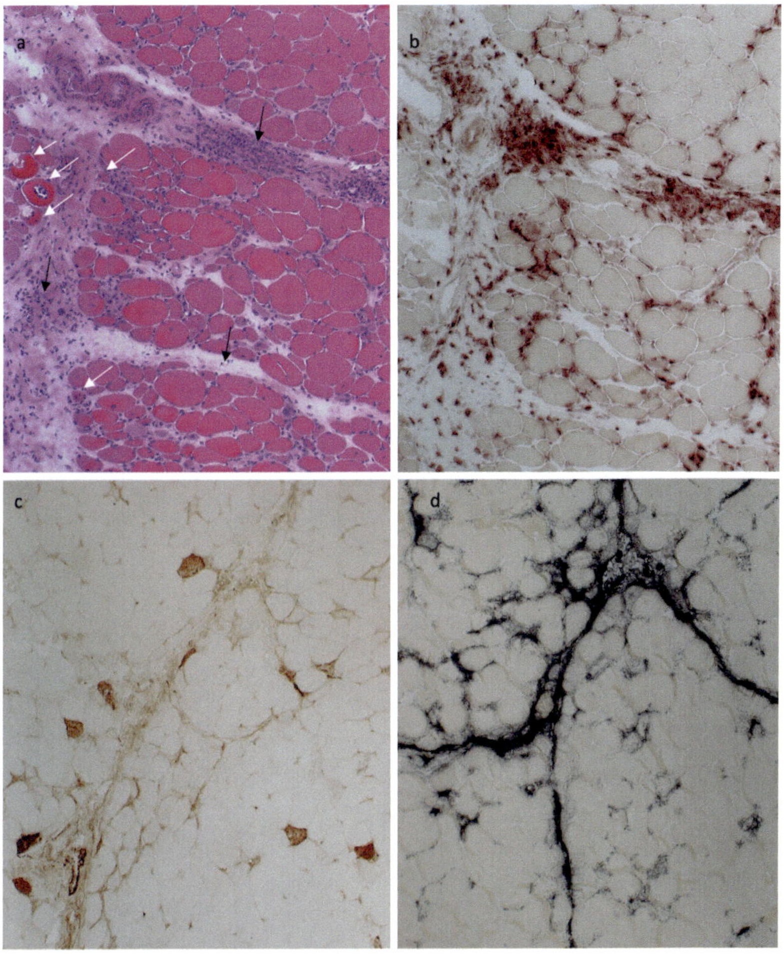


**Fig. 20.5** Immune myopathy with perimysial pathology. Perifascicular pattern of necrosis and regenerating fibers (white arrows), with widened, pale, cellular perimysium (dark arrows) (**a**). Acid phosphatase-positive histiocytic cells occupying the perimysium (**b**). Alkaline phosphatase stains the perimysium (**c**). C5b9 stains the perimysium and cytoplasm of necrotic fibers (**d**)

sized perimysial veins, and connective tissue with expression of the ischemia marker carbonic anhydrase IX but no mononuclear inflammatory foci [90].

**IMNM** The term "necrotizing" may be misleading and imply the whole muscle is necrotic. Immune myopathy "with myofiber necrosis" may be more accurate indicating single myofibers are undergoing necrosis. Regardless, IMNM biopsies typically demonstrate scattered necrotic muscle fibers, although these may be rare or completely absent. Different stages of necrosis/myophagocytosis and regeneration should also be identified [91]. Lymphocytic infiltrates are minimal, if present at all [91]. Sarcolemmal MHC class 1 expression may be seen on non-necrotic and non-regenerating fibers but is often less robust than that seen in other IIMs (Fig. 20.6a–d) [62, 91]. Patchy $C_{5b-9}$ deposition may be seen. Anti-SRP myopathies more commonly have prominent endomysial fibrosis and enlarged capillaries (Fig. 20.6d–f) [62, 70]. Anti-HMGCR myopathies frequently have perimysial pathology and nuclear abnormalities (Fig. 20.6g–j) [62, 88]. It should be noted that muscle fiber necrosis by itself is not useful for subclassifying IIM [91]. Many different myopathic disorders have prominent muscle fiber necrosis with variable patterns. For example, IMPPs have prominent necrotic fibers near the perimysium [88]. Brachio-cervical inflammatory myopathy has randomly scattered necrotic fibers [92]. Regional ischemic immune myopathy (RIIM) has necrosis of muscle fibers in border zones between vessels [90]. Hereditary and other types of acquired myopathies may also have abundant scattered necrosis.

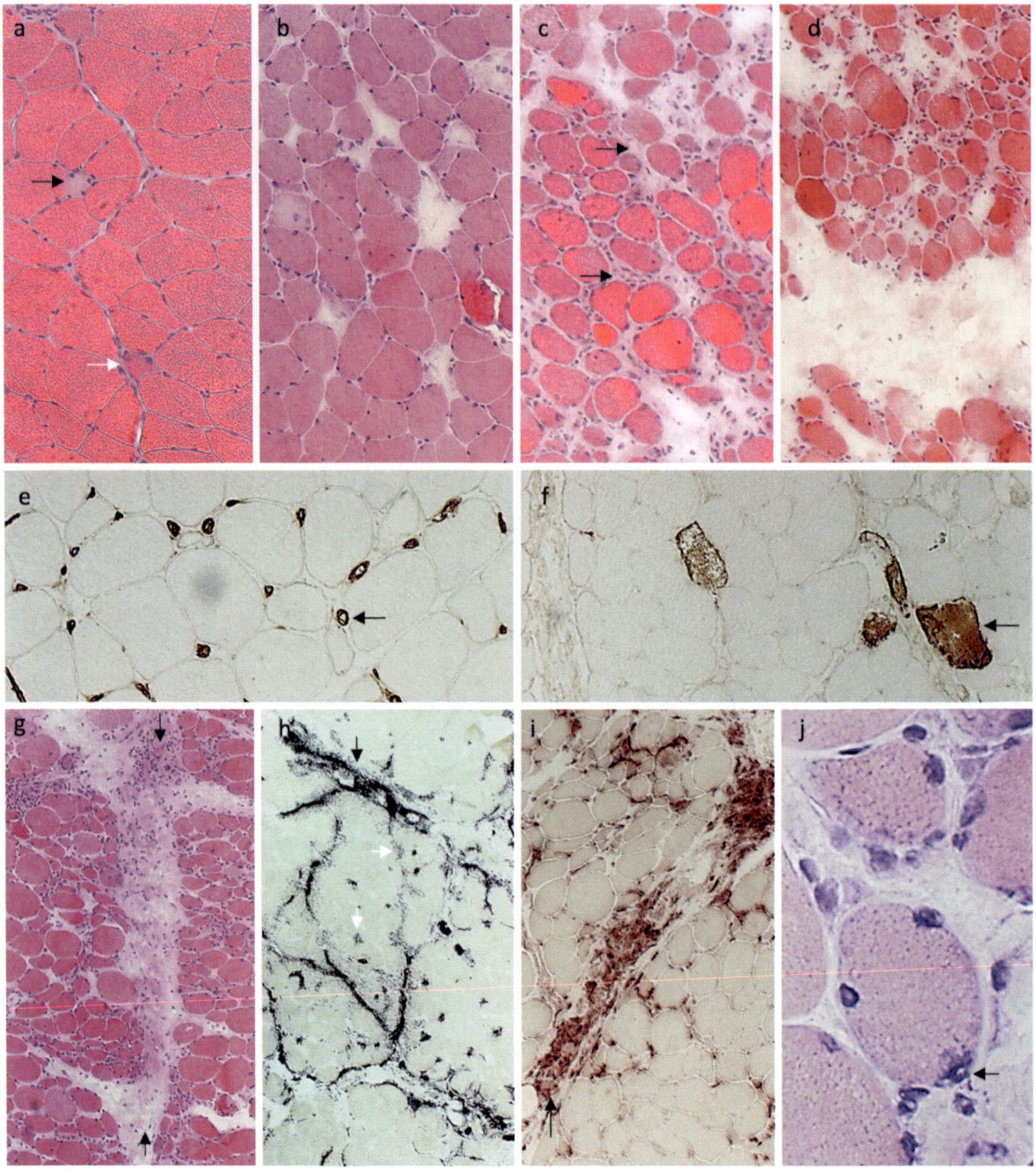


**Fig. 20.6** Anti-signal recognition particle (SRP) myopathy and anti-3-hydroxy-3-methylglutaryl coenzyme A reductase (HMGCR) myopathy. Varying degrees of pathology seen in anti-SRP myopathy (**a-d**). Early pathology with scattered necrotic (dark arrow) and regenerating fibers (white arrow) (**a**). Later in disease with mild pathology (**b**), intermediate pathology with moderately increased endomysial connective tissue (arrows) (**c**), and severe pathology with prominently increased connective tissue (**d**). Ulex staining highlights enlarged capillaries (arrow) (**e**). C5b-9 stains the sarcoplasm of necrotic fibers (arrow) (**f**). Anti-HMGCR myopathy more commonly demonstrates immune myopathies with perimysial pathology (IMPP) pathology associated with necrosis. H&E with widened, fragmented, and cellular perimysium with fatty replacement (arrows) (**g**). Alkaline phosphatase highlights the perimysium (dark arrow) and sarcoplasm of immature fibers (white arrows) (**h**). Acid phosphatase highlights histiocytic cells within the perimysium (arrow) (**i**). Congo red staining illustrating nuclear pathology with irregular shapes and clear centers (arrow) (**j**)

**Antisynthetase Syndrome** Muscle biopsies most commonly demonstrate an IMPP pattern with damaged, fragmented perimysium with adjacent perifascicular myofiber necrosis (Fig. 20.5) [8, 88, 93]. The key clinical difference between DM and antisynthetase syndrome patients with IMPP is simply the presence or absence of antisynthetase autoantibodies [88, 93]. Some antisynthetase syndrome patients may have more widespread necrosis and regeneration [93]. On electron microscopy, nuclear actin aggregation may be seen [94].

## Sporadic IBM

Muscle biopsies from patients with sIBM demonstrate a coexistence of mononuclear inflammatory cells and protein aggregation. Specifically pathology reveals an IIM with vacuoles, aggregates, and mitochondrial pathology (Fig. 20.7a–i) [8]. This combination of findings has been abbreviated as IM-VAMP [8]. Atrophic fibers are often grouped and may have a neurogenic appearance [95]. The inflammatory infiltrate is located within the endomysium and composed of CD8 T cells that surround and invade non-necrotic fibers [1, 96]. MHC class 1 is often expressed on the sarcolemma. Vacuoles contain granular basophilic debris and are immuno-reactive for markers of autophagy, amyloid, and aggregation-prone proteins such as TAR DNA-binding protein 43 and phosphorylated neurofilament [1, 8, 97–101]. Aggregates are visualized on H&E as eosinophilic inclusions and may also be highlighted using AMPDA or SMI-31. These tubulo-filamentous inclusions may be seen on electron microscopy and gave rise to the name inclusion body myositis [102].

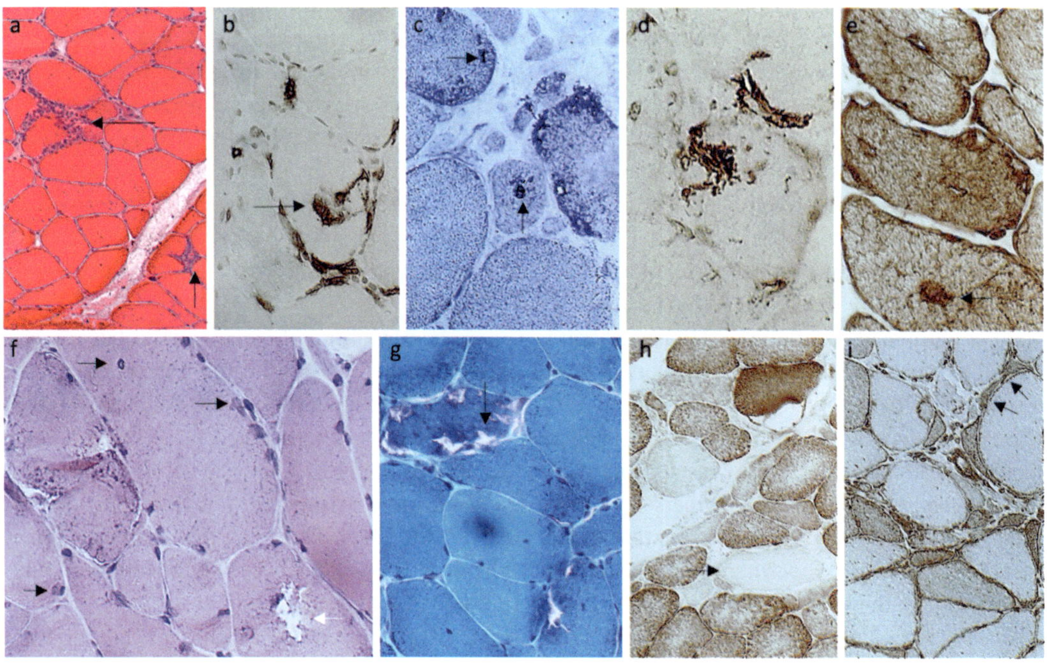


**Fig. 20.7** Sporadic inclusion body myositis (sIBM). H&E demonstrating several key features of sIBM including fiber size variability and endomysial inflammatory cell infiltrates invading non-necrotic fibers (arrows) (**a**). Focal invasion of non-necrotic fiber by CD8-positive cells (arrow) (**b**). Aggregates demonstrated by AMPDA (arrows) (**c**), LC-3 (**d**), and desmin staining (arrow) (**e**). Congo red staining illustrating nuclear pathology with clear centers and irregular borders (black arrows) (**f**). Rimmed vacuoles on Congo red staining (white arrow) (**f**) and Gomori trichrome (arrow) (**g**). Cytochrome oxidase-negative fibers (arrow) (**h**). MHC-1 upregulation on sarcolemma (arrows) (**i**)

Mitochondrial pathology manifests as scattered cytochrome oxidase-negative fibers. Abnormal myonuclei are also seen [8, 103]. Muscle from individual sIBM patients may show one or all of these features. Individual muscle fibers typically show only one of these features [8].

**Others** Many other IIM exist and are not included or well categorized by current classification schemes. These disorders are best understood based on their histopathological patterns.

Routine muscle pathology in brachio-cervical inflammatory myopathy (BCIM, also referred to as B-cell inflammatory myopathy) can be similar to that seen in sIBM with focal invasion by inflammatory cells that are commonly endomysial and perivascular [92]. Perimysial connective tissue staining for alkaline phosphatase; foci of B-cell inflammatory infiltrates (CD20 positive), often associated with ectopic lymphoid structures (ELS); or prominent endomysial C5b-9 complement deposition aid in distinguishing BCIM from other forms of IIM. BCIM syndromes frequently overlap with other immune disorders including myasthenia gravis and rheumatoid arthritis and preferentially involve the proximal arms and posterior neck [92].

Histiocytic inflammatory myopathies have focal collections of cells located in the endomysium or perimysium. Acid phosphatase, esterase, and CD68 stains label cells in the centers of these histiocytic inflammatory foci [8]. In contrast, these stains label only 10–30% of cells in focal mononuclear cell collections. Muscle fiber damage appears as replacement of fibers by histiocytic cells and endomysial connective tissue. The best described histiocytic syndromes are granulomatous myopathies, some of which are associated with sarcoidosis [8, 104]. Histiocytic foci and granulomas in muscle can occur without myopathy in systemic sarcoidosis or vasculitic lesions [8, 104]. Collections of histiocytic cells are also found in macrophagic myofasciitis (MMF) and inflammatory myopathy with abundant macrophages (IMAMs) [8, 105–107]. MMF and IMAM may be related to immunizations and be clinically silent [108]. Therefore, identifying IMAM histiocytic cell collections should not preclude further search for alternative causes of weakness.

## Pathogenesis

The clinical and histopathological distinctions between IIMs suggest different pathogenic processes underlie each, but the precise mechanisms leading to tissue injury are poorly defined.

**Dermatomyositis** The precise mechanisms responsible for DM are unknown. Several different DM models have been proposed. One model focuses on a central role for type 1 interferons (IFN) causing capillary, myofiber, and connective tissue injury [109–111]. Alternatively, DM myofiber injury may result from an antibody and complement-mediated microangiopathy [8, 112, 113], and the resulting hypoxia triggers IFN production [114]. The pathogenic role of myositis-specific autoantibodies in DM is uncertain [11, 41]. DM in some patients is a paraneoplastic syndrome associated with cancer through unknown mechanisms. Future studies of DM pathogenesis should avoid lumping together distinct clinicopathological groups (DM-VP, IMPP, RIIM), which likely have different pathomechanisms.

A combination of genetic risk factors and exposure to environmental factors may be required to trigger DM. Certain class 2 HLA alleles have been implicated in dermatomyositis pathogenesis [115]. Exposure to ultraviolet light is also a known risk factor for developing dermatomyositis [116]. However, the majority of people with known genetic risk factors and high ultraviolet light exposure never develop DM. Mutations in *TIF1* genes in tumors from patients with DM positive for anti-TIF1 autoantibodies have been reported [117]. Once a patient has developed DM, it is unclear what mechanisms maintain muscle damage and weakness.

**IMNM** The mechanisms underlying this condition are unknown. Despite the lack of substantial immune cell invasion of muscle, this condition can respond to immunosuppressive therapies, suggesting it is immune-mediated. While statins are known to cause rhabdomyolysis, their association with anti-HMGCR myopathy is not clear, and the condition should not be called a statin myopathy [29, 68]. That stated, there is evidence that HMGCR is expressed by muscle fibers, particularly regenerating fibers, and that antigen expression is increased by statin exposure (via feedback mechanisms similar to those present in hepatocytes) [29]. Accordingly, in patients with anti-HMGCR antibodies, exposure to statins could lead to increased antigen expression and further, immune-mediated, muscle damage. Class 2 HLA-allele DRB1*08:03 is associated with anti-SRP myopathy, and DRB1*11:01 is an immunogenic risk factor for anti-HMGCR myopathy [118]. Some have proposed anti-SRP and anti-HMGCR antibodies are directly pathogenic [67]; however, these antibodies were unable to induce necrosis in vitro [119], indicating further studies are required.

**Antisynthetase Syndrome** Little is known about what triggers and maintains autoimmunity in antisynthetase syndrome. A pathogenic role for these antibodies remains unproven. Mouse models of myositis induced by immunization with histidyl-tRNA synthetase are not dependent on the development of antibody responses [120]. Instead, they are thought to be mediated by innate immune mechanisms or by the action of histidyl-tRNA synthetase as a chemokine [120].

**Sporadic IBM** Pathologic features of sIBM can be divided into two categories: inflammatory changes and myodegenerative pathologies [8]. These two pathologies have led to pathomechanistic speculation as to whether sIBM is a primary inflammatory, or a primary degenerative, myopathy.

Several lines of evidence suggest that unlike other IIM, sIBM is a primary degenerative myopathy. Rimmed vacuoles are immuno-reactive for autophagic markers such as LC3 suggesting they are autophagic in origin [97–100]. Inclusions are also immuno-reactive for aggregate-prone proteins including amyloid precursor protein, phosphorylated neurofilament, and TDP-43 [99, 121–124]. Rimmed vacuoles may also be found in hereditary inclusion body myopathies or protein aggregate myopathies. Dominantly inherited mutations in the ubiquitin adaptor valosin containing protein (VCP) cause a multisystem degenerative syndrome manifesting with IBM, Paget's disease of bone (PDB), motor neuron disease, and fronto-temporal dementia [125]. Rare variants in SQSTM1 have also been identified in patients with a similar phenotype [126]. Both SQSTM1 and VCP accumulate in sIBM patient muscle, often within or adjacent to rimmed vacuoles [100, 125]. FYCO1, similar to SQSTM1, is an autophagic adaptor protein that binds autophagosomes and facilitates their maturation to acidic lysosomes along microtubules [127]. FYCO1 is a strong marker of rimmed vacuoles, and disease-associated variants impair autophagosome binding in skeletal muscle suggesting they may disrupt autophagic degradation [128]. FYCO1 variants are statistically overrepresented in sIBM patients compared to controls and may serve as risk alleles [128]. These studies support that the degeneration in sIBM patient muscle may be due to a more global disruption in protein degradation pathways, and future treatment strategies aimed at improving protein degradation or protein aggregates may be therapeutic for sIBM. As proof of concept, mice expressing pathogenic VCP mutations were treated with a small molecule, arimoclomol, that enhances the heat shock response. This causes a coordinated upregulation of protein chaperones to facilitate proper folding or degradation of misfolded proteins [129]. Arimoclomol reduced both ubiquitin and TDP-43 pathology and increased forelimb grip strength. These data were supported by a phase II clinical trial in 16 sIBM patients [129].

A number of observations have also strongly implicated autoimmunity as a central pathologic mechanism in sIBM. For example, the invasion of myofibers by cytotoxic CD8+ T cells is a prominent feature in muscle biopsies from sIBM patients [96, 130, 131]. Both oligoclonal and polyclonal expansions of T cells exist within muscle from sIBM patients and support the idea that there is a continuous antigen-driven inflammatory process in sIBM [132]. Many sIBM patients have abnormal clonal expansions of circulating granular lymphocytes that express CD57, a marker of persistent antigenic stimulation that defines a population of T cells with increased cytotoxic potential and resistance to apoptosis [133]. In fact, most sIBM patients meet criteria for T-cell large granular lymphocytic leukemia (T-LGL) [133]. In sIBM, muscle is invaded by the CD8+ CD57+ lymphocytes, which contain cytotoxic granules, analogous to T-LGL where these same cells invade the bone marrow, spleen, and liver. These findings suggest persistent antigenic stimulation of T cells precipitates a neoplastic-like disorder, with cytotoxic T cells invading muscle and circulating in the blood [134].

Dense inflammatory collections consistent with ELS have also been identified in sIBM muscle [135]. Clonally related B cells and plasma cells within these intramuscular lymphoid structures suggest antigen-stimulated maturation of antibody-producing plasma cells occurs locally within sIBM muscle. These findings led to the discovery of autoantibodies targeting cytosolic 5′-nucleotidase 1A (NT5C1A) [136, 137], an enzyme that catalyzes the hydrolysis of adenosine monophosphate to adenosine and inorganic phosphate. NT5C1A is aberrantly localized to perinuclear regions and vacuole rims in sIBM skeletal muscle cells [138]. Whether the abnormal distribution of NT5C1A plays a role in triggering an autoimmune response in sIBM has not been determined.

Another interesting point regarding sIBM pathogenesis is the sIBM-like syndrome that develops in human immunodeficiency virus (HIV)-positive patients. Initially these patients may present, at a younger age of onset, with very high CK levels and proximal weakness that may improve with treatment. However, all patients eventually develop features most consistent with inclusion body myositis, including finger and wrist flexor weakness, rimmed vacuoles, or anti-NT5C1A autoantibodies [139].

## Classification

There have been many attempts to establish classification and diagnostic criteria for IIMs. Bohan and Peter proposed their system to establish clear guidelines for diagnosis and classification of PM and DM [140]. These criteria are too inclusive, allowing patients with various muscular dystrophies to be diagnosed with IIM [79], and they are unable to distinguish sIBM, IMNM, antisynthetase syndrome, and DM. Many other classification schemes have been proposed, all attempting to improve the homogeneity of diagnostic categories, so treatment and prognosis may be evaluated accurately. No universally accepted classification system currently exists. IIMs such as BCIM, focal myositis, and others are distinct and well characterized clinically and pathologically, yet are not recognized by current classification schemes [92, 141, 142].

Clinical-serologic associations have helped to more accurately categorize patients and predict risk of malignancy or ILD; however, the utility of classification schemes based on MSAs is limited [10]. Many MSAs lack specificity for a distinct syndrome [14, 52, 53]. MSAs also lack sensitivity as many IIM patients are seronegative [14, 52, 53].

Other classification schemes have placed more emphasis on muscle pathology and facilitated the initial distinction between sIBM and PM [22, 143]. The importance of histopathologic criteria was demonstrated by a retrospective follow-up study of 165 IIM patients that suggested the diagnosis of PM is rare and actually includes a heterogeneous group of disorders [4].

In 2003, two new distinct pathologic entities were proposed at a consensus conference of the European Neuromuscular Centre (ENMC), IMNM and nonspecific myositis, which included patients with nonspecific perimysial/perivascular infiltrates, but without biopsy features diagnostic of DM or PM [144]. In 2011, another classification system was proposed based solely on myopathology that avoided inconsistencies of other clinical classification systems [8]. It utilizes pathologic characteristics, types of muscle fiber damage, and tissues involved to subclassify IIMs. It defined six new pathologic classes: IMPP (seen in antisynthetase syndrome or "DM" cases with ILD), myovasculopathies (seen in dermatomyopathies such as DM-VP and RIIM), immune polymyopathies (such as anti-SRP and HMGCR myopathies), immune myopathies with endomysial pathology (seen in BCIM), histiocytic inflammatory myopathy (seen in granulomatous disorders, MMF, and IMAM), and IM-VAMP (seen in sIBM) [8, 68, 70, 87–90, 92, 99, 104].

While this system provides consistent and inclusive classification, such specialized myopathological techniques are not widely available. In addition, accurate interpretation of specimens is also problematic [14]. This is evident in the most recent classification scheme, which proposes using only clinical findings and MSAs while excluding histopathology [14]. They note this system may be used to determine what type of IIM a patient has, not if a patient has IIM. Based on phenotypic, biological, and immunologic data, four clusters (DM, IBM, IMNM, antisynthetase syndrome) were identified. They developed a simplified decisional tree with 78.4% correct estimation of their self-defined clusters using three variables: DM rash, antisynthetase syndrome antibodies, and finger flexor scores of 3 or less on the Medical Research Council (MRC) scale [14]. Many problems result from this oversimplification. By ignoring histopathology, many antisynthetase antibody-negative patients are miscategorized. This includes seronegative IMPP patients who are still at increased risk of ILD [88]. This scheme also improperly classifies 35% of sIBM patients as IMNM and 8.7% of IMNM patients as sIBM [14]. This leads to a very problematic situation of incorrectly initiating or withholding immunosuppression in the setting of not having a biopsy to guide further management. While many aspects of these criteria are not ideal, they have been useful in eliminating polymyositis as a diagnostic entity.

## Treatment

Treatment for IIMs remains challenging. The absence of standardized treatment guidelines is reflective of their low prevalence, phenotypic heterogeneity, and suboptimal classification systems. Currently, treatment requires a multidisciplinary approach managed by experienced clinicians.

**IIMs Other Than sIBM** The shortage of adequate randomized trials has resulted in treatment strategies relying on historical clinical practice, case series, and expert opinion.

Glucocorticoids are first-line treatment, but side effects (weight gain, osteoporosis, hypertension, diabetes) limit their use as a monotherapy. At initial presentation, intravenous methylprednisolone (IVMP) is typically given at 1 gram daily for 3–5 days depending on severity. More conservative approaches will initiate prednisone at starting doses of 0.5–1 mg/kg/day at a maximum of 100 mg/day. Some will maintain daily prednisone for 4–6 weeks and then taper. We utilize pulse dose steroids to minimize side effects [145–147], starting at 1 gram/week for 1–2 months, followed by 1 gram every 2 weeks for another 1–2 months, at which time patients are reassessed. Further tapering is facilitated by slowly increasing time between doses or reducing total dose and guided by repeat clinical examinations.

Other immunosuppressive and immunomodulatory drugs commonly used for IIMs include methotrexate, azathioprine, mycophenolate mofetil, cyclosporine, tacrolimus, intravenous immunoglobulin (IVIg), rituximab, and cyclophosphamide (Table 20.2). Certain clinical settings guide the selection of different drugs. Methotrexate is useful as a steroid-sparing agent for muscle and joint disease when relatively quick onset (months) is desired, but may cause lung toxicity and should be avoided in patients with ILD [148]. Azathioprine is useful in patients

**Table 20.2** Immunomodulatory treatments for immune and inflammatory myopathies (IIM)

| Drug | Indications | Dose | Side effects | Monitoring |
|---|---|---|---|---|
| Corticosteroids | Severe cases, all manifestations | 1 g/day for 3–5 d and then daily prednisone or intermittent dosing: 1 g/week for 1 month, 1 g/every other week for 2 months. Taper further via slow dose or frequency reduction | Hypertension, weight gain, hyperglycemia, osteoporosis, cataracts, infection, insomnia | Weight, blood pressure, serum glucose, bone density, cataracts |
| | All patients, all manifestations | Daily: 0.5–1 mg/kg/day. Intermittent: 3.5–7 mg/kg/week. Taper: After 3–6 months or clinical improvement. Reduce by 5 mg every 2–6 weeks | | |
| Azathioprine | Steroid sparing. Muscle involvement | 2–3 mg/kg/day | Myelosuppression, hepatotoxicity, malignancy, teratogenicity, alopecia, flu-like hypersensitivity reaction | Thiopurine methyltransferase enzyme activity before initiation, CBC, and CMP |
| Methotrexate | Steroid sparing. Muscle involvement. Avoid in ILD | 7.5 mg/week for 2 weeks, titrate to maximum 25 mg/week in 2.5 mg increments IM/SQ administration may have more efficacy than PO | Hepatotoxicity, myelosuppression, alopecia, pneumonitis, teratogenicity, malignancy, renal insufficiency | Weekly CBC and CMP for 1 month, monthly for 6 months, every 3 months thereafter |
| Cyclosporine | Steroid sparing. Skin involvement and ILD | 3–5 mg/kg/day | Hypertension, nephrotoxicity, hepatotoxicity, myelosuppression | Blood pressure, CBC, CMP, cyclosporine troughs with goal 50–150 ng/ml |
| Tacrolimus | Steroid sparing. ILD | 0.06 mg/kg/day | Hypertension, hepatotoxicity, nephrotoxicity, hirsutism, tremor, teratogenicity | Blood pressure, CMP, tacrolimus troughs with goal 2–9 ng/ml |
| Mycophenolate mofetil | ILD | 2–3 g/day in divided doses | Myelosuppression, nausea, diarrhea, hypertension | Blood pressure, CBC |
| Cyclophosphamide | ILD | IV: 0.7–1 g/M$^2$ for 1 d/month for 5–6 months Oral: 10–15 mg/kg per month for 6–12 months | Vomiting, alopecia, hemorrhagic cystitis, myelosuppression, malignancy, infertility | Urinalysis, monthly CBC |
| IVIg | Dysphagia and severe disease refractory to other treatments | 2 g/kg over 2–5 days and then 0.4–2 g/kg every 4–6 weeks | Hypotension, arrhythmia, diaphoresis, flushing, nephrotoxicity, headache, aseptic meningitis, anaphylaxis, thrombosis | Heart rate, blood pressure, kidney function |
| Rituximab | Severe IIM, rapidly progressive ILD | 375 mg/M$^2$ weekly for 2 weeks and then every 10 weeks for 2 years | Infusion reaction, infection, progressive multifocal leukoencephalopathy | CD19 count, quantitative immunoglobulins, CBC, and BMP |

*CBC* complete blood count, *CMP* comprehensive metabolic panel, *ILD* interstitial lung disease, *IM* intramuscular, *SQ* subcutaneous, *PO* per oral, *IV* intravenous, *IVIg* intravenous immunoglobulin, *BMP* basic metabolic panel with normal thiopurine methyltransferase activity for long-term immunosuppression when rapid onset is not necessary [149]. Mycophenolate mofetil, cyclosporine, and tacrolimus may be useful for ILD refractory to corticosteroids [150–152]. Cyclosporine and tacrolimus have been used for skin manifestations in DM [150, 151]. Cyclophosphamide may be used in patients with more severe ILD who do not respond to steroids; however, it is associated with more adverse events including infertility [153]. IVIg has shown efficacy in a randomized controlled trial and in a retrospective study for the management of dermatomyositis [154, 155]. IVIg and methotrexate are also effective for anti-HMGCR myopathy [91, 156]. Subcutaneous immunoglobulins (SCIg) may be an alternative to intravenous administration, but reports of SCIg use in IIM are quite limited. Rituximab, a monoclonal antibody targeting CD20 on B lymphocytes, was assessed in refractory DM and polymyositis [157]. While the rituximab arm of this study failed to meet the investigator defined primary endpoint, there were clear benefits to rituximab use in this patient population in that 83% of subjects receiving rituximab met the trial definition of improvement, a criteria generated from measures including muscle strength, muscle enzyme testing, and qualitative disease severity scales [157]. It also appears to be efficacious in patients with antisynthetase syndrome, with or without ILD, and in patients with anti-Mi2, anti-SRP, and anti-HMGCR antibodies [32, 91, 158].

The treatment strategy for juvenile DM is similar to adults [159]. The initial prednisone dose is 2 mg/kg, and methotrexate is the main steroid-sparing agent, although azathioprine, cyclosporine, and tacrolimus have been used. IVIg is the preferred agent for refractory cases. Rituximab is increasingly utilized, and cyclophosphamide is used for severe or life-threatening cases [159].

Evidence is conflicting regarding the use of anti-tumor necrosis factor agents in IIM [160–162]. In fact, exposure to anti-TNF drugs has been reported as a precipitant for IIMs in the literature. Abatacept, a fusion protein that inhibits T-cell co-stimulation, showed benefit by reducing disease activity in a pilot study of 20 IIM patients [163]. Case reports have noted efficacy in IIM for tofacitinib [164] and ruxolitinib (Janus kinase inhibitors) [165], tocilizumab (IL-6 antagonist) [166], anakinra (IL-1 antagonist) [167], and alemtuzumab (anti-CD52) [168]; however, confirmatory studies are required.

**Sporadic IBM** In contrast to other IIMs, no pharmacological therapy has been shown to be effective for sIBM. Treatment of this form of myositis remains largely supportive. Immunosuppressive drugs, such as corticosteroids, azathioprine, methotrexate, or etanercept, have not shown efficacy in sIBM [134, 169]. Alemtuzumab showed a trend toward a reduction of biomarkers in a pilot study that was not confirmed in a subsequent study [170]. Bimagrumab [171] (a monoclonal antibody that blocks the myostatin pathway) and follistatin [172] (myostatin inhibitor locally delivered using an adeno-associated virus) improved thigh muscle volume and performance on the 6-minute walk test but did not significantly improve muscle strength. Rapamycin, also known as sirolimus, improved performance on the 6-minute walk test but did not improve quadriceps strength [173]. Oxandrolone and simvastatin were also not effective [174, 175]. A randomized controlled trial (NCT02483845) investigating natalizumab, an FDA-approved therapy for multiple sclerosis that prevents T-cell egression out of vasculature, is ongoing. A large randomized controlled trial of arimoclomol is ongoing (NCT02753530).

## Management

**Physical Exercise** Physical exercise and rehabilitation programs under the supervision of a physical therapist are safe in all types of IIM and are generally recommended to increase strength and reduce disability [176, 177].

**Skin Disease** Patients with skin manifestations should use sunscreen and avoid UV rays. Topical steroids and tacrolimus have been used [178]. Hydroxychloroquine, an antimalarial drug, also is also commonly used for cutaneous manifestations.

**Calcinosis** Calcinosis commonly fails to respond to immunosuppressive and immunomodulatory therapies. Diltiazem may help [179]. Abatacept and sodium thiosulfate, a calcium chelator, improved calcinosis in a case report [180]. Surgical excision is an option [159].

**Dysphagia** Dysphagia may occur in all subtypes of IIM and is particularly common in sIBM. IVIg may improve swallowing in sIBM and other forms of IIM [181–183]. Cricopharyngeomyotomy, pharyngoesophageal dilation, and injection of botulinum toxin may be used when dysphagia results from failure of upper esophageal sphincter relaxation [184–186].

**Treatment of Associated ILD** Patients with even mild ILD should be intensively treated from onset with glucocorticoids and a second-line immunosuppressant agent (tacrolimus or mycophenolate mofetil). When ILD progression is detected, immediate, intensive treatment should be initiated. This includes methylprednisolone pulses along with a second-line immunosuppressant (tacrolimus, cyclophosphamide, or rituximab). Other treatments to consider include two courses of polymyxin in 24 hours, daily plasmapheresis for 3 days followed by every other day for a total of seven sessions, and IVIg after each plasmapheresis session [41, 187, 188]. Lung transplantation may be considered as a last-resort treatment [188].

## Conclusions and Future Directions

Currently, four main types of inflammatory myopathies are recognized: dermatomyositis, immune-mediated necrotizing myopathy, sporadic inclusion body myositis, and antisynthetase syndrome. The ongoing controversy regarding classification of IIMs will likely only be resolved through a deeper understanding of pathogenesis. Improved alignment of clinical, laboratory, and histopathologic data will facilitate the development of more efficacious treatments.

## References


1. Griggs RC, Askanas V, DiMauro S, Engel A, Karpati G, Mendell JR, et al. Inclusion body myositis and myopathies. Ann Neurol. 1995;38(5):705–13.
2. Hoogendijk JE, Amato AA, Neuromuscular BL. 119th ENMC international workshop: trial design in adult idiopathic inflammatory myopathies, with the exception of inclusion body myositis. Neuromuscul Disord. 2004;14(5):337–45.
3. Amato AA, Griggs RC. Unicorns, dragons, polymyositis, and other mythological beasts. Neurology. 2003;61(3):288–9.
4. van der Meulen MFG, Bronner IM, Hoogendijk JE, Burger H, van Venrooij WJ, Voskuyl AE, et al. Polymyositis: an overdiagnosed entity. Neurology. 2003;61(3):316–21.
5. Matsushita T, Hasegawa M, Fujimoto M, Hamaguchi Y, Komura K, Hirano T, et al. Clinical evaluation of anti-aminoacyl tRNA synthetase antibodies in Japanese patients with dermatomyositis. J Rheumatol. 2007;34(5):1012–8.
6. Hamaguchi Y, Fujimoto M, Matsushita T, Kaji K, Komura K, Hasegawa M, et al. Common and distinct clinical features in adult patients with anti-aminoacyl-tRNA synthetase antibodies: heterogeneity within the syndrome. PLoS One. 2013;8(4):e60442.
7. Marguerie C, Bunn CC, Beynon HL, Bernstein RM, Hughes JM, So AK, et al. Polymyositis, pulmonary fibrosis and autoantibodies to aminoacyl-tRNA synthetase enzymes. Q J Med. 1990;77(282):1019–38.
8. Pestronk A. Acquired immune and inflammatory myopathies: pathologic classification. Curr Opin Rheumatol. 2011;23(6):595–604.
9. Troyanov Y, Targoff IN, Tremblay J-L, Goulet J-R, Raymond Y, Senécal J-L. Novel classification of idiopathic inflammatory myopathies based on overlap syndrome features and autoantibodies: analysis of 100 French Canadian patients. Medicine. 2005;84(4):231–49.
10. Love LA, Leff RL, Fraser DD, Targoff IN, Dalakas M, Plotz PH, et al. A new approach to the classification of idiopathic inflammatory myopathy: myositis-specific autoantibodies define useful homogeneous patient groups. Medicine. 1991;70(6):360–74.
11. Mammen AL. Autoimmune myopathies: autoantibodies, phenotypes and pathogenesis. Nat Rev Neurol. 2011;7(6):343–54.

12. Allenbach Y, Benveniste O. Apport des auto-anticorps au cours des myopathies auto-immunes. Rev Neurol. 2013;169(8–9):656–62.
13. Benveniste O, Stenzel W, Allenbach Y. Advances in serological diagnostics of inflammatory myopathies. Curr Opin Neurol. 2016;29(5):662–73.
14. Mariampillai K, Granger B, Amelin D, Guiguet M, Hachulla E, Maurier F, et al. Development of a new classification system for idiopathic inflammatory myopathies based on clinical manifestations and myositis-specific autoantibodies. JAMA Neurol. 2018;75(12):1528–37.
15. Smoyer-Tomic KE, Amato AA, Fernandes AW. Incidence and prevalence of idiopathic inflammatory myopathies among commercially insured, Medicare supplemental insured, and Medicaid enrolled populations: an administrative claims analysis. BMC Musculoskelet Disord. 2012;13(1):103.
16. Bernatsky S, Joseph L, Pineau CA, Bélisle P, Boivin JF, Banerjee D, et al. Estimating the prevalence of polymyositis and dermatomyositis from administrative data: age, sex and regional differences. Ann Rheum Dis. 2009;68(7):1192–6.
17. Furst DE, Amato AA, Iorga ŞR, Gajria K, Fernandes AW. Epidemiology of adult idiopathic inflammatory myopathies in a U.S. managed care plan. Muscle Nerve. 2012;45(5):676–83.
18. Needham M, Corbett A, Day T, Christiansen F, Fabian V, Mastaglia FL. Prevalence of sporadic inclusion body myositis and factors contributing to delayed diagnosis. J Clin Neurosci. 2008;15(12):1350–3.
19. Prieto S, Grau JM. The geoepidemiology of autoimmune muscle disease. Autoimmun Rev. 2010;9(5):A330–4.
20. Mastaglia FL, Phillips BA. Idiopathic inflammatory myopathies: epidemiology, classification, and diagnostic criteria. Rheum Dis Clin N Am. 2002;28(4):723–41.
21. Furst DE, Amato AA, Iorga ŞR, Bancroft T, Fernandes AW. Medical costs and health-care resource use in patients with inflammatory myopathies in an insured population. Muscle Nerve. 2012;46(4):496–505.
22. Dalakas MC. Polymyositis, dermatomyositis, and inclusion-body myositis. N Engl J Med. 1991;325(21):1487–98.
23. Lundberg IE, Tjärnlund A, Bottai M, Werth VP, Pilkington C, de Visser M, et al. 2017 European League Against Rheumatism/American College of Rheumatology Classification Criteria for adult and juvenile idiopathic inflammatory myopathies and their major subgroups. Arthritis Rheumatol. 2017;69(12):2271–82.
24. Amato AA, Barohn RJ. Evaluation and treatment of inflammatory myopathies. J Neurol Neurosurg Psychiatry. 2009;80(10):1060–8.
25. Khan S, Christopher-Stine L. Polymyositis, dermatomyositis, and autoimmune necrotizing myopathy: clinical features. Rheum Dis Clin N Am. 2011;37(2):143–58.
26. Orlow SJ, Watsky KL. Skin and bones. II. J Am Acad Dermatol. 1991;25(3):447–62.
27. Shirani Z, Kucenic MJ, Carroll CL, Fleischer AB, Feldman SR, Yosipovitch G, et al. Pruritus in adult dermatomyositis. Clin Exp Dermatol. 2004;29(3):273–6.
28. Schneider I, Stoltenburg G, Deschauer M, Winterholler M, Hanisch F. Limb girdle muscular dystrophy type 2L presenting as necrotizing myopathy. Acta Myol. 2014;33(1):19–21.
29. Mohassel P, Mammen AL. Anti-HMGCR myopathy. J Neuromuscul Dis. 2018;5(1):11–20.
30. Mohassel P, Landon-Cardinal O, Foley AR, Donkervoort S, Pak KS, Wahl C, et al. Anti-HMGCR myopathy may resemble limb-girdle muscular dystrophy. Neurol Neuroimmunol Neuroinflamm. 2019;6(1):e523.
31. Rider LG, Miller FW. Deciphering the clinical presentations, pathogenesis, and treatment of the idiopathic inflammatory myopathies. JAMA. 2011;305(2):183–90.
32. Fernandez IP, Parks C, Werner JL, Albayda J, Paik JJ, Danoff SK, et al. Longitudinal course of disease in a large cohort of myositis patients with autoantibodies recognizing the signal recognition particle. Arthritis Care Res (Hoboken). 2017;69(2):263–70.
33. Suzuki S, Nishikawa A, Kuwana M, Nishimura H, Watanabe Y, Nakahara J, et al. Inflammatory myopathy with anti-signal recognition particle antibodies: case series of 100 patients. Orphanet J Rare Dis. 2015;10(1):61.
34. Tiniakou E, Pinal-Fernandez I, Lloyd TE, Albayda J, Paik J, Werner JL, et al. More severe disease and slower recovery in younger patients with anti-3-hydroxy-3-methylglutaryl-coenzyme a reductase-associated autoimmune myopathy. Rheumatology (Oxford). 2017;56(5):787–94.
35. Clements PJ, Furst DE, Campion DS, Bohan A, Harris R, Levy J, et al. Muscle disease in progressive systemic sclerosis: diagnostic and therapeutic considerations. Arthritis Rheum. 1978;21(1):62–71.
36. Pinal-Fernandez I, Casal-Dominguez M, Huapaya JA, Albayda J, Paik JJ, Johnson C, et al. A longitudinal cohort study of the anti-synthetase syndrome: increased severity of interstitial lung disease in black patients and patients with anti-PL7 and anti-PL12 autoantibodies. Rheumatology (Oxford). 2017;56(6):999–1007.
37. Dalakas MC. Inflammatory muscle diseases. N Engl J Med. 2015;373(4):393–4.
38. Needham M, Mastaglia FL. Inclusion body myositis: current pathogenetic concepts and diagnostic and therapeutic approaches. Lancet Neurol. 2007;6(7):620–31.
39. Cox FM, Delgado V, Verschuuren JJ, Ballieux BE, Bax JJ, Wintzen AR, et al. The heart in sporadic inclusion body myositis: a study in 51 patients. J Neurol. 2009;257(3):447–51.
40. Voermans NC, Vaneker M, Hengstman GJD, Laak ter HJ, Zimmerman C, Schelhaas HJ, et al. Primary

respiratory failure in inclusion body myositis. Neurology. 2004;63(11):2191–2.
41. Selva-O'Callaghan A, Pinal-Fernandez I, Trallero-Araguás E, Milisenda JC, Grau-Junyent JM, Mammen AL. Classification and management of adult inflammatory myopathies. Lancet Neurol. 2018;17(9):816–28.
42. Nozaki K, Pestronk A. High aldolase with normal creatine kinase in serum predicts a myopathy with perimysial pathology. J Neurol Neurosurg Psychiatry. 2009;80(8):904–8.
43. Needham M, James I, Corbett A, Day T, Christiansen F, Phillips B, et al. Sporadic inclusion body myositis: phenotypic variability and influence of HLA-DR3 in a cohort of 57 Australian cases. J Neurol Neurosurg Psychiatry. 2008;79(9):1056–60.
44. Amato AA, Gronseth GS, Jackson CE, Wolfe GI, Katz JS, Bryan WW, et al. Inclusion body myositis: clinical and pathological boundaries. Ann Neurol. 1996;40(4):581–6.
45. Koffman BM, Rugiero M, Dalakas MC. Immune-mediated conditions and antibodies associated with sporadic inclusion body myositis. Muscle Nerve. 1998;21(1):115–7.
46. Lotz BP, Engel AG, Nishino H, Stevens JC, Litchy WJ. Inclusion body myositis observations in 40 patients. Brain. 1989;112(3):727–47.
47. Rosales XQ, Chu M-L, Shilling C, Wall C, Pastores GM, Mendell JR. Fidelity of gamma-glutamyl transferase (GGT) in differentiating skeletal muscle from liver damage. J Child Neurol. 2008;23(7):748–51.
48. Tasca G, Monforte M, De Fino C, Kley RA, Ricci E, Mirabella M. Magnetic resonance imaging pattern recognition in sporadic inclusion-body myositis. Muscle Nerve. 2015;52(6):956–62.
49. Pinal-Fernandez I, Casal-Dominguez M, Carrino JA, Lahouti AH, Basharat P, Albayda J, et al. Thigh muscle MRI in immune-mediated necrotising myopathy: extensive oedema, early muscle damage and role of anti-SRP autoantibodies as a marker of severity. Ann Rheum Dis. 2017;76(4):681–7.
50. Van De Vlekkert J, Maas M, Hoogendijk JE, de Visser M, Van Schaik IN. Combining MRI and muscle biopsy improves diagnostic accuracy in subacute-onset idiopathic inflammatory myopathy. Muscle Nerve. 2015;51(2):253–8.
51. Theodorou DJ, Theodorou SJ, Kakitsubata Y. Skeletal muscle disease: patterns of MRI appearances. Br J Radiol. 2014;85(1020):e1298–308.
52. Betteridge Z, McHugh N. Myositis-specific autoantibodies: an important tool to support diagnosis of myositis. J Intern Med. 2016;280(1):8–23.
53. Targoff IN. Myositis specific autoantibodies. Curr Rheumatol Rep. 2006;8(3):196–203.
54. Kalluri M, Sahn SA, Oddis CV, Gharib SL, Christopher-Stine L, Danoff SK, et al. Clinical profile of anti-PL-12 autoantibody: cohort study and review of the literature. Chest. 2009;135(6):1550–6.
55. Hervier B, Devilliers H, Stanciu R, Meyer A, Uzunhan Y, Masseau A, et al. Hierarchical cluster and survival analyses of antisynthetase syndrome: phenotype and outcome are correlated with anti-tRNA synthetase antibody specificity. Autoimmun Rev. 2012;12(2):210–7.
56. Ghirardello A, Zampieri S, Iaccarino L, Tarricone E, Bendo R, Gambari PF, et al. Anti-Mi-2 antibodies. Autoimmunity. 2005;38(1):79–83.
57. Fiorentino DF, Chung LS, Christopher-Stine L, Zaba L, Li S, Mammen AL, et al. Most patients with cancer-associated dermatomyositis have antibodies to nuclear matrix protein NXP-2 or transcription intermediary factor 1γ. Arthritis Rheum. 2013;65(11):2954–62.
58. Albayda J, Pinal-Fernandez I, Huang W, Parks C, Paik J, Casciola-Rosen L, et al. Antinuclear matrix protein 2 autoantibodies and edema, muscle disease, and malignancy risk in dermatomyositis patients. Arthritis Care Res (Hoboken). 2017;69(11):1771–6.
59. Labrador-Horrillo M, Martinez MA, Selva-O'Callaghan A, Trallero-Araguás E, Balada E, Vilardell-Tarrés M, et al. Anti-MDA5 antibodies in a large Mediterranean population of adults with Dermatomyositis. J Immunol Res. 2014;2014(5):1–8.
60. Narang NS, Casciola-Rosen L, Li S, Chung L, Fiorentino DF. Cutaneous ulceration in dermatomyositis: association with anti-melanoma differentiation-associated gene 5 antibodies and interstitial lung disease. Arthritis Care Res (Hoboken). 2015;67(5):667–72.
61. Ge Y, Lu X, Shu X, Peng Q, Wang G. Clinical characteristics of anti-SAE antibodies in Chinese patients with dermatomyositis in comparison with different patient cohorts. Sci Rep. 2016;7(1):188.
62. Watanabe Y, Uruha A, Suzuki S, Nakahara J, Hamanaka K, Takayama K, et al. Clinical features and prognosis in anti-SRP and anti-HMGCR necrotising myopathy. J Neurol Neurosurg Psychiatry. 2016;87(10):1038–44.
63. Mammen AL, Chung T, Christopher-Stine L, Rosen P, Rosen A, Doering KR, et al. Autoantibodies against 3-hydroxy-3-methylglutaryl-coenzyme A reductase in patients with statin-associated autoimmune myopathy. Arthritis Rheum. 2011;63(3):713–21.
64. Trallero-Araguás E, Rodrigo-Pendás JÁ, Selva-O'Callaghan A, Martínez-Gómez X, Bosch X, Labrador-Horrillo M, et al. Usefulness of anti-p155 autoantibody for diagnosing cancer-associated dermatomyositis: a systematic review and meta-analysis. Arthritis Rheum. 2012;64(2):523–32.
65. Selva-O'Callaghan A, Grau JM, Gámez-Cenzano C, Vidaller-Palacín A, Martínez-Gómez X, Trallero-Araguás E, et al. Conventional cancer screening versus PET/CT in dermatomyositis/polymyositis. Am J Med. 2010;123(6):558–62.
66. Sato S, Hoshino K, Satoh T, Fujita T, Kawakami Y, Fujita T, et al. RNA helicase encoded by melanoma differentiation-associated gene 5 is a major autoantigen in patients with clinically amyopathic dermatomyositis: association with rapidly pro-

gressive interstitial lung disease. Arthritis Rheum. 2009;60(7):2193–200.
67. Allenbach Y, Arouche-Delaperche L, Preusse C, Radbruch H, Butler-Browne G, Champtiaux N, et al. Necrosis in anti-SRP+ and anti-HMGCR+myopathies: role of autoantibodies and complement. Neurology. 2018;90(6):e507–17.
68. Alshehri A, Choksi R, Bucelli R, Pestronk A. Myopathy with anti-HMGCR antibodies: perimysium and myofiber pathology. Neurol Neuroimmunol Neuroinflamm. 2015;2(4):e124.
69. Allenbach Y, Keraen J, Bouvier A-M, Jooste V, Champtiaux N, Hervier B, et al. High risk of cancer in autoimmune necrotizing myopathies: usefulness of myositis specific antibody. Brain. 2016;139(Pt 8):2131–5.
70. Miller T, Al-Lozi MT, Lopate G, Pestronk A. Myopathy with antibodies to the signal recognition particle: clinical and pathological features. J Neurol Neurosurg Psychiatry. 2002;73(4):420–8.
71. Targoff IN, Johnson AE, Miller FW. Antibody to signal recognition particle in polymyositis. Arthritis Rheum. 1990;33(9):1361–70.
72. Lim J, Rietveld A, De Bleecker JL, Badrising UA, Saris CGJ, van der Kooi AJ, et al. Seronegative patients form a distinctive subgroup of immune-mediated necrotizing myopathy. Neurol Neuroimmunol Neuroinflamm. 2019;6(1):e513.
73. Trallero-Araguás E, Grau-Junyent JM, Labirua-Iturburu A, García-Hernández FJ, Monteagudo-Jiménez M, Fraile-Rodriguez G, et al. Clinical manifestations and long-term outcome of anti-Jo1 antisynthetase patients in a large cohort of Spanish patients from the GEAS-IIM group. Semin Arthritis Rheum. 2016;46(2):225–31.
74. Lloyd TE, Christopher-Stine L, Pinal-Fernandez I, Tiniakou E, Petri M, Baer A, et al. Cytosolic 5′-Nucleotidase 1A as a target of circulating autoantibodies in autoimmune diseases. Arthritis Care Res (Hoboken). 2015;68(1):66–71.
75. Herbert MK, Stammen-Vogelzangs J, Verbeek MM, Rietveld A, Lundberg IE, Chinoy H, et al. Disease specificity of autoantibodies to cytosolic 5′-nucleotidase 1A in sporadic inclusion body myositis versus known autoimmune diseases. Ann Rheum Dis. 2016;75(4):696–701.
76. Muro Y, Nakanishi H, Katsuno M, Kono M, Akiyama M. Prevalence of anti-NT5C1A antibodies in Japanese patients with autoimmune rheumatic diseases in comparison with other patient cohorts. Clin Chim Acta. 2017;472:1–4.
77. Lilleker JB, Rietveld A, Pye SR, Mariampillai K, Benveniste O, Peeters MTJ, et al. Cytosolic 5′-nucleotidase 1A autoantibody profile and clinical characteristics in inclusion body myositis. Ann Rheum Dis. 2017;76(5):862–8.
78. Goyal NA, Cash TM, Alam U, Enam S, Tierney P, Araujo N, et al. Seropositivity for NT5c1A antibody in sporadic inclusion body myositis predicts more severe motor, bulbar and respiratory involvement. J Neurol Neurosurg Psychiatry. 2016;87(4):373–8.
79. Gallardo E, García RR, De Luna N, Pou A, Brown RH, Illa I. Inflammation in dysferlin myopathy: Immunohistochemical characterization of 13 patients. Neurology. 2001;57(11):2136–8.
80. Krahn M, De Munain AL, Streichenberger N, Bernard R, Pécheux C, Testard H, et al. CAPN3 mutations in patients with idiopathic eosinophilic myositis. Ann Neurol. 2006;59(6):905–11.
81. Arahata K, Ishihara T, Fukunaga H, Orimo S, Lee JH, Goto K, et al. Inflammatory response in facioscapulohumeral muscular dystrophy (FSHD): Immunocytochemical and genetic analyses. Muscle Nerve. 1995;18(S13):S56–66.
82. Maeshima S, Koike H, Noda S, Noda T, Nakanishi H, Iijima M, et al. Clinicopathological features of sarcoidosis manifesting as generalized chronic myopathy. J Neurol. 2015;262(4):1035–45.
83. Fenichel GM, Shy GM. Muscle biopsy experience in myasthenia gravis. Arch Neurol. 1963;9(3): 237–43.
84. Tahsili-Fahadan P, Rashidi A, Cimino PJ, Bucelli RC, Keyrouz SG. Neurologic manifestations of intravascular large B-cell lymphoma. Neurol Clin Pract. 2016;6(1):55–60.
85. Suárez-Calvet X, Gallardo E, Pinal-Fernandez I, De Luna N, Lleixà C, Díaz-Manera J, et al. RIG-I expression in perifascicular myofibers is a reliable biomarker of dermatomyositis. Arthritis Res Ther. 2017;19(1):174.
86. Pinal-Fernandez I, Casciola-Rosen LA, Christopher-Stine L, Corse AM, Mammen AL. The prevalence of individual histopathologic features varies according to autoantibody status in muscle biopsies from patients with dermatomyositis. J Rheumatol. 2015;42(8):1448–54.
87. Pestronk A, Schmidt RE, Choksi R. Vascular pathology in dermatomyositis and anatomic relations to myopathology. Muscle Nerve. 2010;42(1):53–61.
88. Bucelli RC, Pestronk A. Immune myopathies with perimysial pathology: clinical and laboratory features. Neurol Neuroimmunol Neuroinflamm. 2018;5(2):e434.
89. Mozaffar T, Pestronk A. Myopathy with anti-Jo-1 antibodies: pathology in perimysium and neighbouring muscle fibres. J Neurol Neurosurg Psychiatry. 2000;68(4):472–8.
90. Cai C, Alshehri A, Choksi R, Pestronk A. Regional ischemic immune myopathy: a paraneoplastic dermatomyopathy. J Neuropathol Exp Neurol. 2014;73(12):1126–33.
91. Allenbach Y, Mammen AL, Benveniste O, Stenzel W. Immune-Mediated Necrotizing Myopathies Working Group. 224th ENMC International Workshop: Clinico-sero-pathological classification of immune-mediated necrotizing myopathies Zandvoort, The Netherlands, 14–16 October 2016. Neuromuscul Disord. 2018;28(1):87–99.
92. Pestronk A, Kos K, Lopate G, Al-Lozi MT. Brachio-cervical inflammatory myopathies: clinical, immune, and myopathologic features. Arthritis Rheum. 2006;54(5):1687–96.

93. Noguchi E, Uruha A, Suzuki S, Hamanaka K, Ohnuki Y, Tsugawa J, et al. Skeletal muscle involvement in antisynthetase syndrome. JAMA Neurol. 2017;74(8):992–9.
94. Stenzel W, Preuße C, Allenbach Y, Pehl D, Junckerstorff R, Heppner FL, et al. Nuclear actin aggregation is a hallmark of anti-synthetase syndrome-induced dysimmune myopathy. Neurology. 2015;84(13):1346–54.
95. Mitsui T, Kunishige M, Kuroda Y, Kaji R. Heterogeneous muscular involvement in inclusion body myositis. Lab Med. 2010;41(7):393–6.
96. Arahata K, Engel AG. Monoclonal antibody analysis of mononuclear cells in myopathies. I: quantitation of subsets according to diagnosis and sites of accumulation and demonstration and counts of muscle fibers invaded by T cells. Ann Neurol. 1984;16(2):193–208.
97. Hiniker A, Daniels BH, Lee HS, Margeta M. Comparative utility of LC3, p62 and TDP-43 immunohistochemistry in differentiation of inclusion body myositis from polymyositis and related inflammatory myopathies. Acta Neuropathol Commun. 2013;1(1):29.
98. Lünemann JD, Schmidt J, Schmid D, Barthel K, Wrede A, Dalakas MC, et al. β-Amyloid is a substrate of autophagy in sporadic inclusion body myositis. Ann Neurol. 2007;61(5):476–83.
99. Temiz P, Weihl CC, Pestronk A. Inflammatory myopathies with mitochondrial pathology and protein aggregates. J Neurol Sci. 2009;278(1–2):25–9.
100. Nogalska A, Terracciano C, D'Agostino C, Engel WK, Askanas V. p62/SQSTM1 is overexpressed and prominently accumulated in inclusions of sporadic inclusion-body myositis muscle fibers, and can help differentiating it from polymyositis and dermatomyositis. Acta Neuropathol. 2009;118(3):407–13.
101. Mendell JR, Sahenk Z, Gales T, Paul L. Amyloid filaments in inclusion body myositis: novel findings provide insight into nature of filaments. arch neurol. 1991;48(12):1229–34.
102. Dalakas M. Muscle biopsy findings in inflammatory myopathies. Rheum Dis Clin N Am. 2002;28(4):779–98, vi.
103. Greenberg S, Pinkus J, Amato A. Nuclear membrane proteins are present within rimmed vacuoles in inclusion-body myositis. Muscle Nerve. 2006;34(4):406–16.
104. Mozaffar T, Lopate G, Pestronk A. Clinical correlates of granulomas in muscle. J Neurol. 1998;245(8):519–24.
105. Israeli E, Agmon-levin N, Blank M, Shoenfeld Y. Macrophagic myofasciitis a vaccine (alum) autoimmune-related disease. Clin Rev Allergy Immunol. 2011;41(2):163–8.
106. Hara S, Henmi T, Kawakami A, Fujikawa K, Mukae H, Ishimatsu Y, et al. Clinical, serologic and magnetic resonance imaging of 3 cases of inflammatory myopathy with abundant macrophages in the japanese population. Rheumatol Int. 2013;33(4):1059–64.
107. Bassez G, Authier F-J, Lechapt-Zalcman E, Delfau-Larue M, Plonquet A, Coquet M, et al. Inflammatory myopathy with abundant macrophages (imam): a condition sharing similarities with cytophagic histiocytic panniculitis and distinct from macrophagic myofasciitis. J Neuropathol Exp Neurol. 2003;62(5):464–74.
108. Iach B, Cupler E. Macrophagic myofasciitis in children is a localized reaction to vaccination. J Child Neurol. 2007;23(6):614–9.
109. Greenberg SA. Dermatomyositis and type 1 interferons. Curr Rheumatol Rep. 2010;12(3):198–203.
110. Greenberg SA. Proposed immunologic models of the inflammatory myopathies and potential therapeutic implications. Neurology. 2007;69(21):2008–19.
111. Greenberg SA. Type 1 interferons and myositis. Arthritis Res Ther. 2010;12 Suppl 1(Suppl 1):S4.
112. Dalakas MC. Inflammatory disorders of muscle: progress in polymyositis, dermatomyositis and inclusion body myositis. Curr Opin Neurol. 2004;17(5):561–7.
113. Dalakas MC, Hohlfeld R. Polymyositis and dermatomyositis. Lancet. 2003;362(9388):971–82.
114. De Luna N, Suárez-Calvet X, Lleixà C, Díaz-Manera J, Olivé M, Illa I, et al. Hypoxia triggers IFN-I production in muscle: implications in dermatomyositis. Sci Rep. 2017;7(1):8595.
115. Miller FW, Chen W, O'Hanlon TP, Cooper RG, Vencovsky J, Rider LG, et al. Genome-wide association study identifies HLA 8.1 ancestral haplotype alleles as major genetic risk factors for myositis phenotypes. Genes Immun. 2015;16(7):470–80.
116. Mamyrova G, Rider LG, Ehrlich A, Jones O, Pachman LM, Nickeson R, et al. Environmental factors associated with disease flare in juvenile and adult dermatomyositis. Rheumatology (Oxford). 2017;56(8):1342–7.
117. Pinal-Fernandez I, Ferrer-Fabregas B, Trallero-Araguás E, Balada E, Martinez MA, Milisenda JC, et al. Tumour TIF1 mutations and loss of heterozygosity related to cancer-associated myositis. Rheumatology (Oxford). 2018;57(2):388–96.
118. Ohnuki Y, Suzuki S, Shiina T, Uruha A, Watanabe Y, Suzuki S, et al. HLA-DRB1 alleles in immune-mediated necrotizing myopathy. Neurology. 2016;87(18):1954–5.
119. Arouche-Delaperche L, Allenbach Y, Amelin D, Preusse C, Mouly V, Mauhin W, et al. Pathogenic role of anti-signal recognition protein and anti-3-Hydroxy-3-methylglutaryl-CoA reductase antibodies in necrotizing myopathies: Myofiber atrophy and impairment of muscle regeneration in necrotizing autoimmune myopathies. Ann Neurol. 2017;81(4):538–48.
120. Soejima M, Kang EH, Gu X, Katsumata Y, Clemens PR, Ascherman DP. Role of innate immunity in a murine model of histidyl–transfer RNA synthetase (Jo-1)–mediated myositis. Arthritis Rheum. 2011;63(2):479–87.

121. Askanas V, Engel WK, Alvarez RB. Light and electron microscopic localization of beta-amyloid protein in muscle biopsies of patients with inclusion-body myositis. Am J Pathol. 1992;141(1):31–6.
122. Mirabella M, Alvarez RB, Bilak M, Engel WK, Askanas V. Difference in expression of phosphorylated tau epitopes between sporadic inclusion-body myositis and hereditary inclusion-body myopathies. J Neuropathol Exp Neurol. 1996;55(7):774–86.
123. Salajegheh M, Pinkus JL, Taylor JP, Amato AA, Nazareno R, Baloh RH, et al. Sarcoplasmic redistribution of nuclear TDP-43 in inclusion body myositis. Muscle Nerve. 2009;40(1):19–31.
124. Weihl CC, Temiz P, Miller SE, Watts G, Smith C, Forman M, et al. TDP-43 accumulation in inclusion body myopathy muscle suggests a common pathogenic mechanism with frontotemporal dementia. J Neurol Neurosurg Psychiatry. 2008;79(10):1186–9.
125. Watts GDJ, Wymer J, Kovach MJ, Mehta SG, Mumm S, Darvish D, et al. Inclusion body myopathy associated with Paget disease of bone and frontotemporal dementia is caused by mutant valosin-containing protein. Nat Genet. 2004;36(4):377–81.
126. Bucelli RC, Arhzaouy K, Pestronk A, Pittman SK, Rojas L, Sue CM, et al. SQSTM1 splice site mutation in distal myopathy with rimmed vacuoles. Neurology. 2015;85(8):665–74.
127. Pankiv S, Johansen T. FYCO1: linking autophagosomes to microtubule plus end-directing molecular motors. Autophagy. 2010;6(4):550–2.
128. Güttsches AK, Brady S, Krause K, Maerkens A, Uszkoreit J, Eisenacher M, et al. Proteomics of rimmed vacuoles define new risk allele in inclusion body myositis. Ann Neurol. 2017;81(2):227–39.
129. Ahmed M, Machado PM, Miller A, Spicer C, Herbelin L, He J, et al. Targeting protein homeostasis in sporadic inclusion body myositis. Sci Transl Med. 2016;8(331):331ra41–1.
130. Arahata K, Engel AG. Monoclonal antibody analysis of mononuclear cells in myopathies. IV: cell-mediated cytotoxicity and muscle fiber necrosis. Ann Neurol. 1988;23(2):168–73.
131. Arahata K, Engel AG. Monoclonal antibody analysis of mononuclear cells in myopathies. V: identification and quantitation of T8+ cytotoxic and T8+ suppressor cells. Ann Neurol. 1988;23(5):493–9.
132. Fyhr I-M, Moslemi A-R, Lindberg C, Oldfors A. T cell receptor β-chain repertoire in inclusion body myositis. J Neuroimmunol. 1998;91(1–2):129–34.
133. Greenberg SA, Pinkus JL, Amato AA, Kristensen T, Dorfman DM. Association of inclusion body myositis with T cell large granular lymphocytic leukaemia. Brain. 2016;139(5):1348–60.
134. Weihl CC, Mammen AL. Sporadic inclusion body myositis – a myodegenerative disease or an inflammatory myopathy. Neuropathol Appl Neurobiol. 2017;43(1):82–91.
135. Salajegheh M, Pinkus JL, Amato AA, Morehouse C, Jallal B, Yao Y, et al. Permissive environment for B-cell maturation in myositis muscle in the absence of B-cell follicles. Muscle Nerve. 2010;42(4):576–83.
136. Salajegheh M, Lam T, Greenberg SA. Autoantibodies against a 43 KDa muscle protein in inclusion body myositis. PLoS One. 2011;6(5):e20266.
137. Pluk H, van Hoeve BJA, van Dooren SHJ, Stammen-Vogelzangs J, van der Heijden A, Schelhaas HJ, et al. Autoantibodies to cytosolic 5′-nucleotidase 1A in inclusion body myositis. Ann Neurol. 2013;73(3):397–407.
138. Larman HB, Salajegheh M, Nazareno R, Lam T, Sauld J, Steen H, et al. Cytosolic 5′-nucleotidase 1A autoimmunity in sporadic inclusion body myositis. Ann Neurol. 2013;73(3):408–18.
139. Lloyd TE, Pinal-Fernandez I, Michelle EH, Christopher-Stine L, Pak K, Sacktor N, et al. Overlapping features of polymyositis and inclusion body myositis in HIV-infected patients. Neurology. 2017;88(15):1454–60.
140. Bohan A, Peter JB. Polymyositis and dermatomyositis. N Engl J Med. 1975;292(7):344–7.
141. Smith AG, Urbanits S, Blaivas M, Grisold W, Russell JW. Clinical and pathologic features of focal myositis. Muscle Nerve. 2000;23(10):1569–75.
142. Rojana-Udomsart A, Fabian V, Hollingsworth PN, Walters SE, Zilko PJ, Mastaglia FL. Paraspinal and scapular myopathy associated with scleroderma. J Clin Neuromuscul Dis. 2010;11(4):213–22.
143. Carpenter S, Karpati G, Heller I, Eisen A. Inclusion body myositis: a distinct variety of idiopathic inflammatory myopathy. Neurology. 1978;28(1):8–17.
144. Hoogendijk JE, Amato AA, Lecky BR, Choy EH, Lundberg IE, Rose MR, et al. 119th ENMC international workshop: trial design in adult idiopathic inflammatory myopathies, with the exception of inclusion body myositis, 10-12 October 2003, Naarden, The Netherlands. Neuromuscul Disord. 2004;14(5):337–45.
145. Connolly AM, Schierbecker J, Renna R, Florence J. High dose weekly oral prednisone improves strength in boys with Duchenne muscular dystrophy. Neuromuscul Disord. 2002;12(10):917–25.
146. Escolar DM, Hache LP, Clemens PR, Cnaan A, McDonald CM, Viswanathan V, et al. Randomized, blinded trial of weekend vs daily prednisone in Duchenne muscular dystrophy. Neurology. 2011;77(5):444–52.
147. Matsubara S, Sawa Y, Takamori M, Yokoyama H, Kida H. Pulsed intravenous methylprednisolone combined with oral steroids as the initial treatment of inflammatory myopathies. J Neurol Neurosurg Psychiatry. 1994;57(8):1008.
148. Vencovsky J, Jarosová K, Machácek S, Studýnková J, Kafková J, Bartůnková J, et al. Cyclosporine A versus methotrexate in the treatment of polymyositis and dermatomyositis. Scand J Rheumatol. 2000;29(2):95–102.

149. Dalakas MC. Inflammatory myopathies: management of steroid resistance. Curr Opin Neurol. 2011;24(5):457–62.
150. Go DJ, Park JK, Kang EH, Kwon HM, Lee YJ, Song Y-W, et al. Survival benefit associated with early cyclosporine treatment for dermatomyositis-associated interstitial lung disease. Rheumatol Int. 2016;36(1):125–31.
151. Labirua-Iturburu A, Selva-O'Callaghan A, Martínez-Gómez X, Trallero-Araguás E, Labrador-Horrillo M, Vilardell-Tarrés M. Calcineurin inhibitors in a cohort of patients with antisynthetase-associated interstitial lung disease. Clin Exp Rheumatol. 2013;31(3):436–9.
152. Hanaoka H, Iida H, Kiyokawa T, Takakuwa Y, Kawahata K. Mycophenolate mofetil treatment with or without a calcineurin inhibitor in resistant inflammatory myopathy. Clin Rheumatol. 2018;372:1734–6.
153. Barnes H, Holland AE, Westall GP, Goh NS, Glaspole IN. Cyclophosphamide for connective tissue disease-associated interstitial lung disease. Cochrane Database Syst Rev. 2018;1(12):CD010908.
154. Kampylafka EI, Kosmidis ML, Panagiotakos DB, Dalakas M, Moutsopoulos HM, Tzioufas AG. The effect of intravenous immunoglobulin (IVIG) treatment on patients with dermatomyositis: a 4-year follow-up study. Clin Exp Rheumatol. 2012;30(3):397–401.
155. Dalakas MC, Illa I, Dambrosia JM, Soueidan SA, Stein DP, Otero C, et al. A controlled trial of high-dose intravenous immune globulin infusions as treatment for dermatomyositis. N Engl J Med. 1993;329(27):1993–2000.
156. Mammen AL, Tiniakou E. Intravenous immune globulin for statin-triggered autoimmune myopathy. N Engl J Med. 2015;373(17):1680–2.
157. Oddis CV, Reed AM, Aggarwal R, Rider LG, Ascherman DP, Levesque MC, et al. Rituximab in the treatment of refractory adult and juvenile dermatomyositis and adult polymyositis: a randomized, placebo-phase trial. Arthritis Rheum. 2013;65(2):314–24.
158. Aggarwal R, Bandos A, Reed AM, Ascherman DP, Barohn RJ, Feldman BM, et al. Predictors of clinical improvement in rituximab-treated refractory adult and juvenile dermatomyositis and adult polymyositis. Arthritis Rheumatol. 2014;66(3):740–9.
159. Malik A, Hayat G, Kalia JS, Guzman MA. Idiopathic inflammatory myopathies: clinical approach and management. Front Neurol. 2016;7(21):1487.
160. Chen D, Wang X-B, Zhou Y, Zhu X-C. Efficacy of infliximab in the treatment for dermatomyositis with acute interstitial pneumonia: a study of fourteen cases and literature review. Rheumatol Int. 2013;33(10):2455–8.
161. Schiffenbauer A, Garg M, Castro C, Pokrovnichka A, Joe G, Shrader J, et al. A randomized, double-blind, placebo-controlled trial of infliximab in refractory polymyositis and dermatomyositis. Semin Arthritis Rheum. 2018;47(6):858–64.
162. Muscle Study Group. A randomized, pilot trial of etanercept in dermatomyositis. Ann Neurol. 2011;70(3):427–36.
163. Kerola AM, Kauppi MJ. Abatacept as a successful therapy for myositis—a case-based review. Clin Rheumatol. 2014;34(3):609–12.
164. Kurtzman DJB, Wright NA, Lin J, Femia AN, Merola JF, Patel M, et al. Tofacitinib citrate for refractory cutaneous dermatomyositis: an alternative treatment. JAMA Dermatol. 2016;152(8):944–5.
165. Hornung T, Janzen V, Heidgen F-J, Wolf D, Bieber T, Wenzel J. Remission of recalcitrant dermatomyositis treated with ruxolitinib. N Engl J Med. 2014;371(26):2537–8.
166. Narazaki M, Hagihara K, Shima Y, Ogata A, Kishimoto T, Tanaka T. Therapeutic effect of tocilizumab on two patients with polymyositis. Rheumatology (Oxford). 2011;50(7):1344–6.
167. Zong M, Dorph C, Dastmalchi M, Alexanderson H, Pieper J, Amoudruz P, et al. Anakinra treatment in patients with refractory inflammatory myopathies and possible predictive response biomarkers: a mechanistic study with 12 months follow-up. Ann Rheum Dis. 2014;73(5):913–20.
168. Ruck T, Bittner S, Kuhlmann T, Wiendl H, Meuth SG. Long-term efficacy of alemtuzumab in polymyositis. Rheumatology (Oxford). 2015;54(3):560–2.
169. Benveniste O, Guiguet M, Freebody J, Dubourg O, Squier W, Maisonobe T, et al. Long-term observational study of sporadic inclusion body myositis. Brain. 2011;134(Pt 11):3176–84.
170. Schmidt K, Kleinschnitz K, Rakocevic G, Dalakas MC, Schmidt J. Molecular treatment effects of alemtuzumab in skeletal muscles of patients with IBM. BMC Neurol. 2016;16(1):48.
171. Amato AA, Sivakumar K, Goyal N, David WS, Salajegheh M, Praestgaard J, et al. Treatment of sporadic inclusion body myositis with bimagrumab. Neurology. 2014;83(24):2239–46.
172. Mendell JR, Sahenk Z, Al-Zaidy S, Rodino-Klapac LR, Lowes LP, Alfano LN, et al. Follistatin gene therapy for sporadic inclusion body myositis improves functional outcomes. Mol Ther. 2017;25(4):870–9.
173. Lilleker JB, Bukhari M, Chinoy H. Rapamycin for inclusion body myositis: targeting non-inflammatory mechanisms. Rheumatology (Oxford). 2018;129:611.
174. Rutkove SB, Parker RA, Nardin RA, Connolly CE, Felice KJ, Raynor EM. A pilot randomized trial of oxandrolone in inclusion body myositis. Neurology. 2002;58(7):1081–7.
175. Sancricca C, Mora M, Ricci E, Tonali PA, Mantegazza R, Mirabella M. Pilot trial of simvastatin in the treatment of sporadic inclusion-body myositis. Neurol Sci. 2011;32(5):841–7.
176. Munters LA, Dastmalchi M, Andgren V, Emilson C, Bergegård J, Regardt M, et al. Improvement in

health and possible reduction in disease activity using endurance exercise in patients with established polymyositis and dermatomyositis: a multicenter randomized controlled trial with a 1-year open extension follow up. Arthritis Care Res (Hoboken). 2013;65(12):1959–68.

177. Alexanderson H, Munters LA, Dastmalchi M, Loell I, Heimburger M, Opava CH, et al. Resistive home exercise in patients with recent-onset polymyositis and dermatomyositis – a randomized controlled single-blinded study with a 2-year follow up. J Rheumatol. 2014;41(6):1124–32.
178. Yoshimasu T, Ohtani T, Sakamoto T, Oshima A, Furukawa F. Topical FK506 (tacrolimus) therapy for facial erythematous lesions of cutaneous lupus erythematosus and dermatomyositis. Eur J Dermatol. 2002;12(1):50–2.
179. Balin SJ, Wetter DA, Andersen LK, Davis MDP. Calcinosis cutis occurring in association with autoimmune connective tissue disease: the Mayo Clinic experience with 78 patients, 1996–2009. Arch Dermatol. 2012;148(4):455–62.
180. Arabshahi B, Silverman RA, Jones OY, Rider LG. Abatacept and sodium thiosulfate for treatment of recalcitrant juvenile dermatomyositis complicated by ulceration and calcinosis. J Pediatr. 2012;160(3):520–2.
181. Walter MC, Lochmüller H, Toepfer M, Schlotter B, Reilich P, Schröder M, et al. High-dose immunoglobulin therapy in sporadic inclusion body myositis: a double-blind, placebo-controlled study. J Neurol. 2000;247(1):22–8.
182. Dalakas MC, Sonies B, Dambrosia J, Sekul E, Cupler E, Sivakumar K. Treatment of inclusion-body myositis with IVIg: a double-blind, placebo-controlled study. Neurology. 1997;48(3):712–6.
183. Marie I, Menard JF, Hatron PY, Hachulla E, Mouthon L, Tiev K, et al. Intravenous immunoglobulins for steroid-refractory esophageal involvement related to polymyositis and dermatomyositis: a series of 73 patients. Arthritis Care Res (Hoboken). 2010;62(12):1748–55.
184. Oh TH, Brumfield KA, Hoskin TL, Stolp KA, Murray JA, Bassford JR. Dysphagia in inflammatory myopathy: clinical characteristics, treatment strategies, and outcome in 62 patients. Mayo Clin Proc. 2007;82(4):441–7.
185. Langdon PC, Mulcahy K, Shepherd KL, Low VH, Mastaglia FL. Pharyngeal dysphagia in inflammatory muscle diseases resulting from impaired suprahyoid musculature. Dysphagia. 2012;27(3): 408–17.
186. Liu LW, Tarnopolsky M, Armstrong D. Injection of botulinum toxin a to upper esophageal sphincter for oropharyngeal dysphagia in two patients with inclusion body myositis. Can J Gastroenterol. 2004;18(6):397–9.
187. Teruya A, Kawamura K, Ichikado K, Sato S, Yasuda Y, Yoshioka M. Successful polymyxin B Hemoperfusion treatment associated with serial reduction of serum anti-CADM-140/MDA5 antibody levels in rapidly progressive interstitial lung disease with amyopathic dermatomyositis. Chest. 2013;144(6):1934–6.
188. Morisset J, Johnson C, Rich E, Collard HR, Lee JS. Management of myositis-related interstitial lung disease. Chest. 2016;150(5):1118–28.

# Autoimmune Autonomic Disorders 21

Elisabeth Golden, Kyle Blackburn,
and Steven Vernino

**Key Points**

1. There is a complex bi-directional relationship between the autonomic nervous system and the immune system; autonomic dysfunction can occur due to or in association with autoimmunity.
2. Autoimmune autonomic ganglionopathy is characterized by pan-autonomic failure, associated with pathogenic ganglionic acetylcholine receptor antibodies in 50% of cases and often responsive to antibody-targeted immunotherapy.
3. Autonomic dysfunction can be paraneoplastic in origin or a secondary feature of other autoimmune neurologic conditions such as Lambert–Eaton myasthenic syndrome, Guillain–Barré syndrome, and various autoimmune encephalitides.
4. Postural orthostatic tachycardia syndrome (POTS) is associated with a variety of comorbid autoimmune disorders and autoantibodies; however, an autoimmune cause for POTS is not established.

## Introduction

Disorders of the autonomic nervous system (ANS) are commonly attributed to toxic/metabolic causes (such as diabetes) or neurodegenerative disorders (especially α[alpha]-synucleinopathies), but there is also a close relationship between the autonomic nervous system and the immune system. In some cases, autonomic dysfunction can occur due to or in association with autoimmunity. A few autoimmune disorders specifically target the autonomic nervous system, other autoimmune diseases have prominent autonomic manifestations, and some common forms of dysautonomia have features of autoimmunity or inflammation, but the associations are less well defined (Table 21.1).

**Table 21.1** Autoimmune autonomic disorders

| |
|---|
| *Autonomic disorders with definite autoimmune etiology* |
| Autoimmune autonomic ganglionopathy |
| Paraneoplastic autonomic/enteric neuropathy |
| Lambert–Eaton myasthenic syndrome |
| *Autonomic conditions with possible autoimmune etiology* |
| Immune-mediated sensory and autonomic neuropathies |
| Postural orthostatic tachycardia syndrome |
| *Other autoimmune disorders with prominent autonomic features* |
| Guillain–Barré syndrome |
| NMDA-R encephalitis |
| LGI1 and CASPR2 ab encephalitis (including Morvan syndrome) |
| Anti-DPPX-associated encephalitis |
| Sjögren's syndrome |

E. Golden · K. Blackburn · S. Vernino (✉)
Department of Neurology, University of Texas Southwestern Medical Center, Dallas, TX, USA
e-mail: steven.vernino@utsouthwestern.edu



A. L. Piquet, E. Alvarez (eds.), *Neuroimmunology*, https://doi.org/10.1007/978-3-030-61883-4_21

## Function and Evaluation of the Autonomic Nervous System

The autonomic nervous system (ANS) controls the automatic functions of the body that are not under conscious control. The opposing actions of the sympathetic ("fight or flight") and parasympathetic ("rest and digest") systems maintain homeostasis as they control such diverse bodily functions as lacrimation, salivation, pupillary function, heart rate, blood pressure, digestion, bladder and sexual function, and sweating. A distributed network in the brain (including the hypothalamus, insula, amygdala, and multiple brainstem nuclei) integrates afferent input and exerts central control over autonomic efferents. The peripheral ANS consists of preganglionic neurons in the brainstem and spinal cord that project to neurons in peripheral autonomic ganglia whose unmyelinated postganglionic axons innervate target organs throughout the body. Synapses occur at the autonomic ganglia, where acetylcholine (ACh) acts on ganglionic nicotinic acetylcholine receptors (AChRs). At the level of the target organs (e.g., eye, heart, skin, and bladder), parasympathetic nerves release ACh to act on muscarinic AChRs, while sympathetic nerves use norepinephrine to act on alpha- and beta-adrenergic receptors as well as ACh to act on muscarinic AChR at sweat glands (Fig. 21.1).

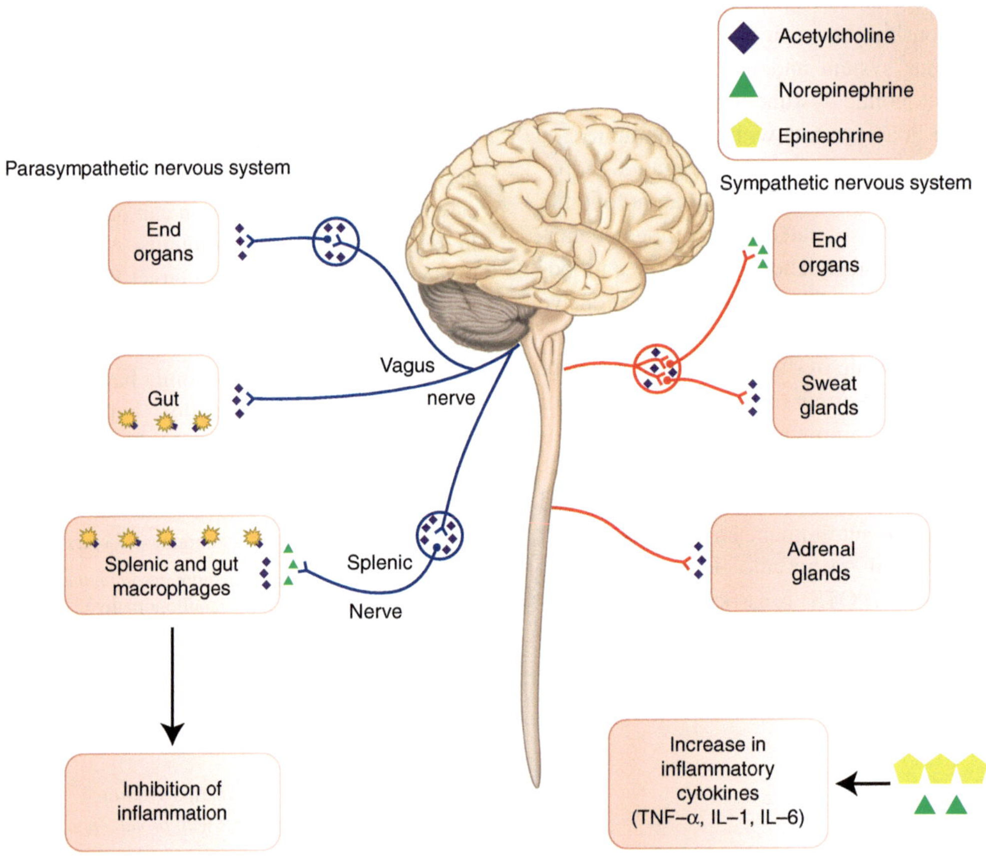


**Fig. 21.1** Interaction of the autonomic and immune system. The parasympathetic and sympathetic nerves innervate multiple target end-organs throughout the body (such as heart, eye, blood vessels, and sweat glands). The parasympathetic nervous system via the vagus nerve innervates the gut where local cholinergic transmission regulates gut motility as well as immune activity. The vagus nerve also mediates an anti-inflammatory pathway through projections to the spleen via the splenic nerve (with both adrenergic and cholinergic connections). Macrophages in the gut and spleen are downregulated via the action of acetylcholine on nicotinic AChR (α[alpha]7-type). Conversely, activation of the sympathetic nervous system is associated with increased levels of catecholamines (norepinephrine and epinephrine), which act to increase inflammation and levels of pro-inflammatory cytokines

Patients with autonomic disorders may present with a variety of symptoms, including fatigue, orthostatic lightheadedness, light sensitivity, dry eyes and dry mouth, early satiety, constipation or diarrhea, urinary retention or incontinence, and sexual dysfunction. Some of these symptoms are vague and not immediately suggestive of an autonomic disorder. A careful history and physical examination is the first step in evaluation and must include assessment of distal sensation, measurement of blood pressure and heart rate during orthostatic stress (standing), and assessment of the pupillary light reflex. Laboratory autonomic testing can provide additional objective information. The standard battery of tests [1] includes the following: (1) assessment of sympathetic sudomotor function via the Quantitative Sudomotor Axon Reflex Test (QSART) or thermoregulatory sweat test, (2) assessment of cardiovagal (parasympathetic) function via heart rate response to paced deep breathing and the Valsalva maneuver, (3) assessment of sympathetic adrenergic function via the blood pressure response to Valsalva maneuver, and (4) assessment of sympathetic adrenergic function via heart rate and blood pressure response to head-up tilt.

There is a complex bi-directional relationship between the autonomic nervous system and the immune system. The vagus nerve controls heart rate, gastrointestinal motility, and many other visceral functions, but also mediates a neuro-inflammatory reflex that controls immune responses and inflammation during infection or tissue injury (Fig. 21.1). Specifically, efferent vagus nerve activity suppresses pro-inflammatory cytokine levels in animal models through cholinergic innervation of spleen and other immune organs [2]. Adrenergic (sympathetic) signaling can increase production of pro-inflammatory cytokines such as interleukin-6 (IL-6). Conversely, the autonomic nervous system can be a target for autoimmune disorders.

## Autoimmune Autonomic Ganglionopathy with Ganglionic AChR Antibodies

The classic example of an autoimmune autonomic disorder is autoimmune autonomic ganglionopathy (AAG). The first patient with "pure pan-dysautonomia with recovery" was described in 1969 [3, 4], and the first large series of patients with purported immune-mediated failure of the autonomic nervous system was published in 1994 as "idiopathic autonomic neuropathy" [5]. In more recent years, this condition has been well characterized [6].

About 50% of patients with AAG are seropositive for antibodies against the AChR found at the autonomic ganglia [7, 8]. The ganglionic AChR (composed of 2α[alpha]3 subunits and 3ß[beta] subunits) is highly homologous to, but antigenically distinct from, the muscle AChR at the neuromuscular junction (α[alpha]1-type) [9]. Antibodies in AAG bind specifically to the ganglionic (α[alpha]3-type) AChR [10]. The pathogenicity of ganglionic AChR antibodies in AAG has been well established. Rabbits actively immunized against the α(alpha)3 subunit develop diffuse autonomic failure with gastrointestinal hypomotility, urinary retention, and impaired pupillary light reflex [11, 12], and mice injected with immunoglobulin G (IgG) from affected rabbits or humans develop a similar, though self-limited, phenotype of autonomic failure [13]. Transient neonatal AAG due to passive transfer of maternal antibodies has also been reported [14]. In patients with AAG, ganglionic AChR antibody levels correlate with severity of disease [7, 15, 16]. A decrease in antibody level is often seen during spontaneous clinical recovery [7] or response to treatment [17, 18], further supporting the pathogenicity of these antibodies.

The disease may follow an acute to subacute or a chronic, insidious course [15]. Those who present subacutely often report an antecedent illness, followed by the development of orthostatic hypotension (OH) and prominent cholinergic failure (dry mouth, dry eyes, impaired pupillary constriction, upper gastrointestinal symptoms, and urinary retention). Patients with prominent cholinergic symptoms tend to have higher antibody levels than those without these symptoms. Antibody levels also correlate with the severity of orthostatic hypotension [16].

Patients with AAG may also follow a chronic, insidious course [15]. They may present with the insidious onset of orthostatic hypotension, clinically indistinguishable from pure autonomic failure, or with prominent cholinergic symptoms. Antibody levels do not reliably distinguish between subacute and chronic presentations [15, 16].

Seropositive patients have a unique finding of premature pupillary redilation with prolonged light stimulus [19]. This "pupillary fatigue" (thought to be caused by impaired synaptic transmission at the ciliary ganglia) is hypothesized to be analogous to the muscle fatigue of myasthenia gravis (due to disrupted synaptic transmission at the neuromuscular junction). In the rabbit model of AAG, the severity of pupillary dysfunction correlated with antibody level [20].

Classically, about a quarter of patients with AAG describe paresthesias [5, 8], although a recent study from Japan reported a 46% prevalence of sensory disturbance [21]. However, objective evidence of sensory or motor impairment is not a feature of this disease. Central nervous system (CNS) manifestations have been reported in a minority of patients [21–24], and in fact, in one small study, cognitive impairment was independent of the orthostatic hypotension and improved with immunomodulation [25]. Cases of AAG reported in Japan have also noted psychiatric symptoms, coughing episodes, and endocrine dysfunction [8, 21, 26]. Cerebrospinal fluid (CSF) protein may be mildly elevated, generally without pleocytosis. Spontaneous recovery may occur in about one-third of patients but is typically incomplete [5, 15].

Antibody-targeted therapies such as intravenous immunoglobulin (IVIg) and plasma exchange (PLEX) are considered first-line treatments for AAG and have shown benefit in case reports and series [6, 27]. One case report documented a close relationship between antibody level and clinical and laboratory evidence of autonomic impairment during repeated courses of PLEX [18]. Additional benefit may be achieved with the combination of PLEX or IVIg and oral immunosuppressant therapy [17, 27, 28]. Rituximab has also been used in an effort to deplete the pathogenic antibodies, with good results in several case reports [29–32].

Although high levels of ganglionic acetylcholine receptor (gAChR) antibody (≥1.00 nmol/L [22]) are fairly specific for AAG, low-level antibodies may be seen in a variety of scenarios. In patients with limited autonomic symptoms such as isolated orthostatic intolerance or gastrointestinal dysmotility [7, 15, 16, 22, 33, 34], the clinical significance of antibody positivity is often not clear (see further discussion in the section on postural orthostatic tachycardia syndrome [POTS]). Low levels of antibody positivity may also be seen in a small percentage of controls (3–5% of healthy and "other disease" controls depending on the assay methods [35]). Low levels of gAChR antibodies have been reported in cases of small and large fiber polyneuropathy [22, 36], neurodegenerative disease, and particularly in a variety of neurological and non-neurological autoimmune diseases [36], including lupus, celiac disease [37], Guillain–Barré syndrome [38], and Sjögren's syndrome (SS) [39, 40], which may suggest that these low antibody levels may be non-specifically associated with general autoimmunity rather than with a specific disease process. A history of prior or active malignancy has been documented in a minority of antibody-positive patients [22, 36]. The classic associations are thymoma [41, 42] and small-cell lung cancer [7], but other studies have documented a wide variety of lung and other neoplasms [9, 22, 36, 43]. In more recent studies, gAChR antibody levels <0.2 nmol/L were found to have poor specificity, and so these low-level results should be interpreted with caution, particularly when the clinical features are not suggestive of an autoimmune autonomic disorder [22, 36].

## Paraneoplastic Autonomic and Enteric Neuropathies

Autonomic neuropathy or ganglionopathy can also occur as a paraneoplastic syndrome in the setting of known or occult malignancy. Paraneoplastic autonomic neuropathy can manifest as diffuse autonomic failure similar to AAG, with orthostatic hypotension, anhidrosis, dry mouth, and gastrointestinal dysmotility. Autonomic involvement can occur in isolation or in conjunction with other nervous system manifestations, such as sensory neuropathy or limbic encephalitis [44, 45]. The onset of autonomic symptoms can vary from an acute presentation to an insidious progression over several months.

In some cases, involvement is restricted to the enteric nervous system, with prominent symptoms of gut dysmotility in the absence of other autonomic features. The classical syndrome is paraneoplastic enteric neuropathy (gastroparesis) associated with small-cell lung carcinoma. These patients often present with severe nausea, anorexia, weight loss, and constipation, leading to initial evaluation by a gastroenterologist. Motility studies reveal impaired esophageal motility, delayed gastric emptying, and intestinal hypomotility [45, 46].

Multiple antibodies have been identified in cases of paraneoplastic autonomic neuropathy. The gAChR can be found in up to 30% of paraneoplastic autonomic neuropathies, usually in association with thymoma [22]. Other antibodies, such as ANNA 1 (anti-Hu) or CRMP5 (CV2), have been detected in cases of autonomic and enteric neuropathies and are strongly associated with malignancy, most commonly with small-cell lung carcinoma [47–49].

## Seronegative Autoimmune Autonomic Neuropathy

About half of patients who present with idiopathic subacute autonomic failure do not have detectable paraneoplastic or gAChR antibodies [7]. These cases are often referred to as "seronegative AAG," although in many cases the features are more consistent with autonomic neuropathy rather than ganglionopathy. Seropositive and seronegative cases have been compared [50]. Seropositive cases are more likely to have prominent cholinergic symptoms. Seronegative cases may respond to immunotherapy, supporting an autoimmune pathogenesis [28, 51].

A recent series of six cases with seronegative autoimmune autonomic failure highlighted some distinctive features [52, 53]. Rather than predominant cholinergic symptoms, as in seropositive patients, these patients all had severe sympathetic deficits, with varying levels of parasympathetic impairment. Premature pupillary redilation was not seen. Three patients also had severe sensory symptoms and neuropathic pain requiring multiple medications for control, which is not a feature of AAG. Trials of IVIg, PLEX, and even rituximab were largely ineffective; however, three patients showed remarkable response to high-dose intravenous steroids. The lack of response to antibody-targeted therapy but preserved response to steroids might indicate a cell-mediated or inflammatory disorder, and the disproportionate sympathetic failure and prominent sensory symptoms suggest a localization to the peripheral nerves rather than the autonomic ganglia.

Another presumed autoimmune autonomic disorder is described as acute autonomic and sensory neuropathy, first reported by Colan et al. in 1980 [54]. Much of the literature consists of case reports heterogeneous in clinical presentation, extent of sensory involvement and autonomic failure, and response to treatment. One larger series by Koike et al. [55] described diffuse autonomic failure and small fiber sensory deficits with neuropathic pain. Over half progressed to develop sensory ataxia, and several of these had dorsal column hyperintensities on spinal magnetic resonance imaging (MRI). About two-thirds had an identified antecedent event. Treatment trials with IVIg, PLEX, and steroids yielded variable responses. It is unclear if there is a common pathophysiology to these sensory and autonomic neuropathies, and the relationship with other forms of presumed immune-mediated small fiber neuropathies warrants further investigation [56].

## Autoimmunity in Postural Orthostatic Tachycardia Syndrome

Postural orthostatic tachycardia syndrome (POTS) is considered the most common disorder of the autonomic nervous system [57]. This condition predominantly affects young women (female:male ratio 4.5:1, average age 15–25 years) [58] and carries substantial morbidity. The hallmark of POTS is orthostatic intolerance, with symptoms such as lightheadedness, cognitive difficulties, palpitations, tremulousness, and dependent acrocyanosis. Objectively, the diagnosis is made in symptomatic patients based on sustained heart rate increment of ≥30 beats per minute within 10 min of standing or head-up tilt, in the absence of orthostatic hypotension [59]. There are also more pervasive symptoms including fatigue, unrefreshing sleep, and memory complaints, as well as a high incidence of comorbid chronic pain disorders, headaches, and gastrointestinal disorders [58, 60]. A number of different phenotypes have been described (e.g., neuropathic, hyperadrenergic, hypovolemic, and mast cell activation) [60], likely reflecting heterogeneous pathophysiology.

Several features may suggest an autoimmune etiology for POTS in a subset of patients. The demographics of POTS patients (primarily young females) puts them at higher risk of autoimmunity, and the diverse associated symptoms may also suggest an autoimmune disorder. Symptoms may appear after stressors such as surgery or viral infection (28% in a review of 152 POTS patients at the Mayo clinic) and may present acutely or subacutely (26% in the same study) [61]. There have been multiple reports of POTS onset after vaccination, though causality has not been established [62]. POTS patients often have a personal or family history of other autoimmune diseases. Sixteen percent of patients reported a personal diagnosis of autoimmune disease in a 3300-patient online survey [63], and 20% in a 100-patient retrospective chart review [64], including Hashimoto's thyroiditis, antiphospholipid syndrome, rheumatoid arthritis, celiac disease, lupus, and Sjögren's syndrome. Rare cases of POTS representing the initial manifestation of antiphospholipid syndrome [65] or Sjögren's syndrome [66] have been reported.

Furthermore, POTS patients have been found to have a high prevalence of a variety of autoantibodies. In the aforementioned chart review [64], 31% had positive autoantibodies including antinuclear antibodies (ANA; 25% of all patients), antiphospholipid, tissue transglutaminase, anti-SSB, anti-double-stranded DNA, and atypical antineutrophil cytoplasmic antibody (ANCA). A study of 33 patients tested for thyroid, neural, and muscular autoantibodies found positivity in one-third, most commonly thyroid [67]. Small pilot studies have demonstrated autoimmunoreactive IgGs against cardiac lipid-raft–associated proteins [68] and a variety of other cardiac membrane proteins [69].

Autoantibodies more directly pertinent to the autonomic nervous system have been investigated in POTS. Antibody positivity has been noted against alpha- and beta-adrenergic receptors [70], angiotensin II type 1 receptors [71], and muscarinic AChR [72]. The prevalence of ganglionic AChR antibodies in POTS has been repeatedly studied. Prevalence has been reported from 5% to 29% [7, 61, 73–76], generally at low levels, and in two of the larger studies [73, 74], the prevalence did not differ between POTS patients and healthy controls. Ultimately, despite the high prevalence of autoantibodies in POTS, it remains unclear whether these autoantibodies contribute to the underlying pathophysiology of the disease or are simply markers of immune dysregulation. Unless another confirmed autoimmune disorder is being treated, immunotherapy is not currently recommended for POTS outside the context of clinical trials.

Aside from the presence of specific autoimmune conditions or autoantibodies, there is also some evidence that POTS may itself be a pro-inflammatory state. In a study of POTS patients, those with documented sympathetic upregulation and parasympathetic withdrawal were found to have significantly higher levels of IL-6 [77]. Upregulation of inflammatory cytokines could potentially account for some of the symptoms commonly associated with POTS but not directly related to postural tachycardia (fatigue, brain fog, etc.). Vagus nerve stimulation is being considered as a potential immunomodulatory therapy for POTS [78].

## Other Autoimmune Conditions with Associated Autonomic Dysfunction

### Lambert–Eaton Myasthenic Syndrome

Lambert–Eaton myasthenic syndrome (LEMS) is a presynaptic neuromuscular junction disorder that is associated with antibodies against P/Q-type voltage calcium channels and with small-cell carcinoma. Patients typically present with symmetric proximal weakness, more pronounced in the lower extremities, and areflexia. The majority of patients with LEMS also develop symptoms of cholinergic autonomic impairment, commonly dry mouth, reduced heart rate variability, and erectile dysfunction [79]. Treatment typically involves management of malignancy and immunotherapy. Treatment with 3,4-diaminopyridine facilitates acetylcholine release from the presynaptic terminal and has been shown in randomized trials to improve both motor and autonomic manifestations of LEMS [80].

## Voltage-Gated Potassium Channel Complex Disorders

The voltage-gated potassium channel complex refers to a group of peripheral nerve or synaptic proteins, which may be targeted in a diverse set of autoimmune neurological disorders. Specifically, antibodies directed against leucine-rich, glioma-inactivated 1 (LGI1) and contactin-associated protein-like 2 (CASPR2) have been implicated in patients with encephalopathies and peripheral nerve hyperexcitability disorders. In a large cohort of LGI1 and CASPR2 antibody-positive patients, autonomic symptoms were reported in approximately 25% of cases. On autonomic testing, orthostatic hypotension and changes in sweating were the most common abnormalities. Gastrointestinal symptoms were reported in a few patients [81].

More recently, neurocardiac issues were described in patients with LGI1 and CASPR2 antibodies. Episodic bradycardia and sinus pauses were reported in three patients with LGI1 antibodies prior to the onset of encephalitis symptoms. All patients were diagnosed with sick sinus syndrome and required pacemakers in the months prior to the onset of other neurological symptoms [82]. In a Mayo Clinic cohort, two patients with CASPR2 antibodies had bradycardia and recurrent asystole requiring pacemakers. In addition, two patients (one LGI1 positive and one CASPR2 positive) died from sudden cardiac arrest [81].

Morvan syndrome is a characteristic disorder with antibodies to CASPR2 or LGI1, often associated with thymoma. Defining features include prominent encephalopathy, insomnia, signs of peripheral nerve hyperexcitability (fasciculations, myokymia, and neuromyotonia), and autonomic hyperactivity. Commonly encountered autonomic features include hyperhidrosis, sialorrhea, urinary difficulties, and tachycardia. Autonomic testing can demonstrate impaired cardiovagal and cardiovascular function, and evidence of impaired temperature regulation with either anhidrosis or hyperhidrosis [83]. See Chap. 12 for more details regarding encephalitis presentations with LGI1 and CASPR2.

## DPPX Antibody-Associated Encephalopathy

Antibodies to dipeptidyl-peptidase-like protein 6 (DPPX), a regulatory protein of the Kv4.2 potassium channel, define a unique syndrome with features of CNS hyperexcitability (e.g., tremor, myoclonus, hyperekplexia, and rigidity) and encephalopathy. Severe diarrhea is a hallmark of the disorder, often preceding other neurological manifestations (confusion, seizures, and psychosis) by several months. Other gastrointestinal manifestations include profound weight loss, gastroparesis, and constipation [84–86]. Urinary dysfunction and ventricular arrhythmias, including one young patient with unexplained cardiac arrest, have also been described [86]. Sera containing DPPX antibodies has been shown to increase neuronal activity in the myenteric plexus, lending support to pathogenic effect of the antibodies on enteric autonomic neurons [87].

## NMDA Receptor Antibody Encephalitis

N-methyl-D-aspartate (NMDA) receptor encephalitis is characterized by the subacute development of prominent psychosis, seizures, dyskinesias, and alterations in consciousness. Autonomic instability is a common feature in later stages of the disease, frequently presenting with dysrhythmias, temperature dysregulation, and labile blood pressures. These symptoms tend to improve with treatment, and persistent autonomic symptoms following recovery have not been reported [88]. See Chap. 12 for more details regarding NMDA receptor encephalitis.

## Guillain–Barré Syndrome

Autonomic instability is a common feature of Guillain–Barré syndrome and can represent a major cause of mortality. Symptoms such as blood pressure fluctuations, resting tachycardia, and ileus occur in the acute phase of illness and are more frequent in patients with severe motor deficits. Improvement in these symptoms tends to occur simultaneously with motor recovery, though residual autonomic symptoms are common [89, 90].

## Sjögren's Syndrome

Sjögren's syndrome (SS) is a common systemic autoimmune disease characterized by dysfunction of exocrine glands that results in symptoms of dry mouth and dry eyes. Neurological symptoms, most commonly paresthesias and numbness, are fairly common in SS (prevalence estimates vary from 20% to 50%). Objective evidence of small fiber or large fiber sensory neuropathy may present in symptomatic or asymptomatic patients [91]. Autonomic symptoms are reported in approximately half of SS patients and are more frequent in those with neuropathy [92]. Formal autonomic studies do not show a consistent pattern; a mixture of sympathetic and parasympathetic abnormalities, gastrointestinal dysmotility, POTS, and pupillary abnormalities have been reported [66]. Generalized autonomic failure has been rarely reported in patients with biopsy-proven SS [92].

While the underlying pathophysiology is poorly understood, evidence suggests peripheral nervous system involvement in SS may be T-cell mediated. Autopsies have demonstrated CD8 T-cell infiltration in the sympathetic ganglion of patients with SS and prominent orthostatic hypotension [93]. Treatment with immunotherapy is often ineffective, though small uncontrolled series have shown improvement in autonomic symptoms and testing following treatment with intravenous immunoglobulin [66].

## Conclusion

Severe, subacute autonomic failure can be the hallmark of autoimmune neurological disorders, and autonomic manifestations are found in several other antibody-mediated neurological conditions. The role of autoimmunity in other autonomic conditions, such as POTS, remains unclear. In patients with autonomic dysfunction, consideration of autoimmune causes is warranted as well as the use of immunotherapy in select cases.

## References

1. Low P, Sletten D. Laboratory evaluation of autonomic failure. In: Low P, Benarroch E, editors. Clinical autonomic disorders. 3rd ed. Baltimore: Lippincott Williams & Wilkins; 2008. p. 130–63.
2. Pavlov VA, Tracey KJ. The vagus nerve and the inflammatory reflex-linking immunity and metabolism. Nat Rev Endocrinol. 2012;8(12):743–54.
3. Young RR, Asbury AK, Adams RD, Corbett JL. Pure pan-dysautonomia with recovery. Trans Am Neurol Assoc. 1969;94:355–7.
4. Young RR, Asbury AK, Corbett JL, Adams RD. Pure pan-dysautonomia with recovery. Description and discussion of diagnostic criteria. Brain. 1975;98(4):613–36.
5. Suarez GA, Fealey RD, Camilleri M, Low PA. Idiopathic autonomic neuropathy: clinical, neurophysiologic, and follow-up studies on 27 patients. Neurology. 1994;44(9):1675–82.
6. Muppidi S, Vernino S. Autoimmune autonomic failure. Handb Clin Neurol. 2013;117:321–7.
7. Vernino S, Low PA, Fealey RD, Stewart JD, Farrugia G, Lennon VA. Autoantibodies to ganglionic acetylcholine receptors in autoimmune autonomic neuropathies. N Engl J Med. 2000;343(12):847–55.
8. Nakane S, Higuchi O, Koga M, Kanda T, Murata K, Suzuki T, et al. Clinical features of autoimmune autonomic ganglionopathy and the detection of subunit-specific autoantibodies to the ganglionic acetylcholine receptor in Japanese patients. PLoS One. 2015;10(3):e0118312.

9. Vernino S, Adamski J, Kryzer TJ, Fealey RD, Lennon VA. Neuronal nicotinic ACh receptor antibody in subacute autonomic neuropathy and cancer-related syndromes. Neurology. 1998;50(6):1806–13.
10. Vernino S, Lindstrom J, Hopkins S, Wang Z, Low PA, Muscle Study Group. Characterization of ganglionic acetylcholine receptor autoantibodies. J Neuroimmunol. 2008;197(1):63–9.
11. Lennon VA, Ermilov LG, Szurszewski JH, Vernino S. Immunization with neuronal nicotinic acetylcholine receptor induces neurological autoimmune disease. J Clin Invest. 2003;111(6):907–13.
12. Vernino S, Low PA, Lennon VA. Experimental autoimmune autonomic neuropathy. J Neurophysiol. 2003;90(3):2053–9.
13. Vernino S, Ermilov LG, Sha L, Szurszewski JH, Low PA, Lennon VA. Passive transfer of autoimmune autonomic neuropathy to mice. J Neurosci. 2004;24(32):7037–42.
14. Baker SK, Chow BM, Vernino SA. Transient neonatal autoimmune autonomic ganglionopathy. Neurol Neuroimmunol Neuroinflamm. 2014;1(3):e35.
15. Klein CM, Vernino S, Lennon VA, Sandroni P, Fealey RD, Benrud-Larson L, et al. The spectrum of autoimmune autonomic neuropathies. Ann Neurol. 2003;53(6):752–8.
16. Gibbons CH, Freeman R. Antibody titers predict clinical features of autoimmune autonomic ganglionopathy. Auton Neurosci. 2009;146(1–2):8–12.
17. Gibbons CH, Vernino SA, Freeman R. Combined immunomodulatory therapy in autoimmune autonomic ganglionopathy. Arch Neurol. 2008;65(2):213–7.
18. Schroeder C, Vernino S, Birkenfeld AL, Tank J, Heusser K, Lipp A, et al. Plasma exchange for primary autoimmune autonomic failure. N Engl J Med. 2005;353(15):1585–90.
19. Muppidi S, Scribner M, Gibbons CH, Adams-Huet B, Spaeth EB, Vernino S. A unique manifestation of pupillary fatigue in autoimmune autonomic ganglionopathy. Arch Neurol. 2012;69(5):644–8.
20. Mukherjee S, Vernino S. Dysfunction of the pupillary light reflex in experimental autoimmune autonomic ganglionopathy. Auton Neurosci. 2007;137(1–2):19–26.
21. Nakane S, Mukaino A, Maeda Y, Higuchi O, Matsuo H, Ando Y. Extra-autonomic manifestations in autoimmune autonomic ganglionopathy: a Japanese survey. J Neurol Neurosurg Psychiatry. 2017;88(4):367–8.
22. McKeon A, Lennon VA, Lachance DH, Fealey RD, Pittock SJ. Ganglionic acetylcholine receptor autoantibody: oncological, neurological, and serological accompaniments. Arch Neurol. 2009;66(6):735–41.
23. Baker SK, Morillo C, Vernino S. Autoimmune autonomic ganglionopathy with late-onset encephalopathy. Auton Neurosci. 2009;146(1–2):29–32.
24. Kuki I, Kawawaki H, Okazaki S, Hattori Y, Horino A, Higuchi O, et al. Autoimmune autonomic ganglionopathy in a pediatric patient presenting with acute encephalitis. Brain Dev. 2016;38(6):605–8.
25. Gibbons CH, Centi J, Vernino S, Freeman R. Autoimmune autonomic ganglionopathy with reversible cognitive impairment. Arch Neurol. 2012;69(4):461–6.
26. Hayashi M, Ishii Y. A Japanese case of autoimmune autonomic ganglionopathy (AAG) and a review of AAG cases in Japan. Auton Neurosci. 2009;146(1–2):26–8.
27. Iodice V, Kimpinski K, Vernino S, Sandroni P, Low PA. Immunotherapy for autoimmune autonomic ganglionopathy. Auton Neurosci. 2009;146(1–2):22–5.
28. Iodice V, Kimpinski K, Vernino S, Sandroni P, Fealey RD, Low PA. Efficacy of immunotherapy in seropositive and seronegative putative autoimmune autonomic ganglionopathy. Neurology. 2009;72(23):2002–8.
29. Gupta A, Harris S, Vernino S, Naina HV. Rituximab-based therapy and long-term control of autoimmune autonomic ganglionopathy. Clin Auton Res. 2015;25(4):255–8.
30. Imrich R, Vernino S, Eldadah BA, Holmes C, Goldstein DS. Autoimmune autonomic ganglionopathy: treatment by plasma exchanges and rituximab. Clin Auton Res. 2009;19(4):259–62.
31. Hollenbeck R, Black BK, Peltier AC, Biaggioni I, Robertson D, Winton EF, et al. Long-term treatment with rituximab of autoimmune autonomic ganglionopathy in a patient with lymphoma. Arch Neurol. 2011;68(3):372–5.
32. Benizri S, Agmon-Levin N, Kitrey ND, Carter D, Goshen E, Sharabi Y. Clinical problem solving: a tobacco merchant who can't spit. Isr Med Assoc J. 2017;19(12):786–91.
33. Sandroni P, Low PA. Other autonomic neuropathies associated with ganglionic antibody. Auton Neurosci. 2009;146(1–2):13–7.
34. Mukaino A, Minami H, Isomoto H, Hamamoto H, Ihara E, Maeda Y, et al. Anti-ganglionic AChR antibodies in Japanese patients with motility disorders. J Gastroenterol. 2018;53:1227.
35. Lang K, Pruss H. Frequencies of neuronal autoantibodies in healthy controls: Estimation of disease specificity. Neurol Neuroimmunol Neuroinflamm. 2017;4(5):e386.
36. Li Y, Jammoul A, Mente K, Li J, Shields RW, Vernino S, et al. Clinical experience of seropositive ganglionic acetylcholine receptor antibody in a tertiary neurology referral center. Muscle Nerve. 2015;52(3):386–91.
37. Briani C, Doria A, Ruggero S, Toffanin E, Luca M, Albergoni MP, et al. Antibodies to muscle and ganglionic acetylcholine receptors (AchR) in celiac disease. Autoimmunity. 2008;41(1):100–4.
38. Nakane S, Higuchi O, Hamada Y, Maeda Y, Mukaino A, Sakai W, et al. Ganglionic acetylcholine receptor autoantibodies in patients with Guillain–Barre syndrome. J Neuroimmunol. 2016;295–296:54–9.
39. Kondo T, Inoue H, Usui T, Mimori T, Tomimoto H, Vernino S, et al. Autoimmune autonomic ganglionopathy with Sjogren's syndrome: significance of ganglionic acetylcholine receptor antibody and therapeutic approach. Auton Neurosci. 2009;146(1–2):33–5.

40. Mukaino A, Nakane S, Higuchi O, Nakamura H, Miyagi T, Shiroma K, et al. Insights from the ganglionic acetylcholine receptor autoantibodies in patients with Sjogren's syndrome. Mod Rheumatol. 2016;26(5):708–15.
41. Vernino S, Lennon VA. Autoantibody profiles and neurological correlations of thymoma. Clin Auton Res. 2004;10(21):7270–5.
42. Zekeridou A, McKeon A, Lennon VA. Frequency of synaptic autoantibody accompaniments and neurological manifestations of thymoma. JAMA Neurol. 2016;73(7):853–9.
43. Nakae Y, Hyuga M, Terada Y, Kishimoto W, Fukunaga A, Tabata S, et al. Multiple myeloma presenting with autoimmune autonomic ganglionopathy. Intern Med. 2017;56(24):3347–51.
44. McKeon A, Benarroch EE. Autoimmune autonomic disorders. Handb Clin Neurol. 2016;133:405–16.
45. Sharp L, Vernino S. Paraneoplastic neuromuscular disorders. Muscle Nerve. 2012;46(6):841–50.
46. Dhamija R, Tan KM, Pittock SJ, Foxx-Orenstein A, Benarroch E, Lennon VA. Serologic profiles aiding the diagnosis of autoimmune gastrointestinal dysmotility. Clin Gastroenterol Hepatol. 2008;6(9):988–92.
47. Lucchinetti CF, Kimmel DW, Lennon VA. Paraneoplastic and oncologic profiles of patients seropositive for type 1 antineuronal nuclear autoantibodies. Neurology. 1998;50(3):652–7.
48. Vernino S, Eggenberger ER, Rogers LR, Lennon VA. Paraneoplastic neurological autoimmunity associated with ANNA-1 autoantibody and thymoma. Neurology. 2002;59(6):929–32.
49. Yu Z, Kryzer TJ, Griesmann GE, Kim K, Benarroch EE, Lennon VA. CRMP-5 neuronal autoantibody: marker of lung cancer and thymoma-related autoimmunity. Ann Neurol. 2001;49(2):146–54.
50. Sandroni P, Vernino S, Klein CM, Lennon VA, Benrud-Larson L, Sletten D, et al. Idiopathic autonomic neuropathy: comparison of cases seropositive and seronegative for ganglionic acetylcholine receptor antibody. Arch Neurol. 2004;61(1):44–8.
51. Tijero B, Del Pino R, Perez-Concha T, Acera MA, Gabilondo I, Berganzo K, et al. Seronegative and seropositive autoimmune autonomic ganglionopathy (AAG): same clinical picture, same response to immunotherapy. J Neuroimmunol. 2018;319:68–70.
52. Golden EP, Bryarly MA, Vernino S. Seronegative autoimmune autonomic neuropathy: a distinct clinical entity. Clin Auton Res. 2018;28(1):115–23.
53. Goldstein DS, Holmes C, Sullivan P, Donadio V, Isonaka R, Zhong E, et al. Autoimmunity-associated autonomic failure with sympathetic denervation. Clin Auton Res. 2017;27(1):57–62.
54. Colan RV, Snead OC, Oh SJ, Kashlan MB. Acute autonomic and sensory neuropathy. Ann Neurol. 1980;8(4):441–4.
55. Koike H, Atsuta N, Adachi H, Iijima M, Katsuno M, Yasuda T, et al. Clinicopathological features of acute autonomic and sensory neuropathy. Brain. 2010;133(10):2881–96.
56. Muppidi S. Autoimmune autonomic neuropathies: time to look beyond autoimmune autonomic ganglionopathy. Clin Auton Res. 2018;28(1):7–8.
57. Schondorf R, Benoit J, Wein T, Phaneuf D. Orthostatic intolerance in the chronic fatigue syndrome. J Auton Nerv Syst. 1999;75(2–3):192–201.
58. Benarroch EE. Postural tachycardia syndrome: a heterogeneous and multifactorial disorder. Mayo Clin Proc. 2012;87(12):1214–25.
59. Freeman R, Wieling W, Axelrod FB, Benditt DG, Benarroch E, Biaggioni I, et al. Consensus statement on the definition of orthostatic hypotension, neurally mediated syncope and the postural tachycardia syndrome. Auton Neurosci. 2011;161(1–2):46–8.
60. Raj SR. Postural tachycardia syndrome (POTS). Circulation. 2013;127(23):2336–42.
61. Thieben MJ, Sandroni P, Sletten DM, Benrud-Larson LM, Fealey RD, Vernino S, et al. Postural orthostatic tachycardia syndrome: the Mayo clinic experience. Mayo Clin Proc. 2007;82(3):308–13.
62. Butts BN, Fischer PR, Mack KJ. Human papillomavirus vaccine and postural orthostatic tachycardia syndrome: a review of current literature. J Child Neurol. 2017;32(11):956–65.
63. Raj SR, Stiles LE, Shaw BH, Green EA, Dorminy CA, Shibao CA, et al. The face of postural tachycardia syndrome: a cross-sectional community-based survey. Heart Rhythm. 2016;13(5):S328.
64. Blitshteyn S. Autoimmune markers and autoimmune disorders in patients with postural tachycardia syndrome (POTS). Lupus. 2015;24(13):1364–9.
65. Schofield JR. Autonomic neuropathy-in its many guises-as the initial manifestation of the antiphospholipid syndrome. Immunol Res. 2017;65(2):532–42.
66. Goodman BP, Crepeau A, Dhawan PS, Khoury JA, Harris LA. Spectrum of autonomic nervous system impairment in Sjogren syndrome. Neurologist. 2017;22(4):127–30.
67. Singer W, Klein C, Low P, Lennon V. Autoantibodies in the postural tachycardia syndrome (P1.272). Neurology. 2015;84(14 Supplement):s1–s196.
68. Wang XL, Ling TY, Charlesworth MC, Figueroa JJ, Low P, Shen WK, et al. Autoimmunoreactive IgGs against cardiac lipid raft-associated proteins in patients with postural orthostatic tachycardia syndrome. Transl Res. 2013;162(1):34–44.
69. Wang XL, Chai Q, Charlesworth MC, Figueroa JJ, Low P, Shen WK, et al. Autoimmunoreactive IgGs from patients with postural orthostatic tachycardia syndrome. Proteomics Clin Appl. 2012;6(11–12):615–25.
70. Li H, Yu X, Liles C, Khan M, Vanderlinde-Wood M, Galloway A, et al. Autoimmune basis for postural tachycardia syndrome. J Am Heart Assoc. 2014;3(1):e000755.
71. Yu X, Li H, Murphy TA, Nuss Z, Liles J, Liles C, et al. Angiotensin II type 1 receptor autoantibodies in postural tachycardia syndrome. JAHA. 2018;7(8):e008351.

72. Dubey D, Hopkins S, Vernino S. M1 and M2 muscarinic receptor antibodies among patients with postural orthostatic tachycardia syndrome: potential disease biomarker. J Clin Neuromuscul Dis. 2016;17:179S–9.
73. Vernino S, Hopkins S, Okamoto L, Black B, Dorminy C, Paranjape S, et al. Prevalence of ganglionic AChR antibodies in postural tachycardia syndrome (POTS) (P1.276). Neurology. 2015;84(14 Supplement):277–88.
74. Vernino S, Bryarly M, Hopkins S, Okamoto LE, Black BK, Paranjape SY, et al. The prevalence of ganglionic AchR antibodies in postural tachycardia syndrome (POTS). Clin Auton Res. 2016;26:328.
75. Watari M, Nakane S, Mukaino A, Nakajima M, Mori Y, Maeda Y, et al. Autoimmune postural orthostatic tachycardia syndrome. Ann Clin Transl Neurol. 2018;5(4):486–92.
76. Li J, Zhang Q, Liao Y, Zhang C, Hao H, Du J. The value of acetylcholine receptor antibody in children with postural tachycardia syndrome. Pediatr Cardiol. 2015;36(1):165–70.
77. Okamoto LE, Raj SR, Gamboa A, Shibao CA, Arnold AC, Garland EM, et al. Sympathetic activation is associated with increased IL-6, but not CRP in the absence of obesity: lessons from postural tachycardia syndrome and obesity. Am J Physiol Heart Circ Physiol. 2015;309(12):H2098–107.
78. Vernino S, Stiles LE. Autoimmunity in postural orthostatic tachycardia syndrome: current understanding. Auton Neurosci. 2018;215:78.
79. O'Suilleabhain P, Low PA, Lennon VA. Autonomic dysfunction in the Lambert–Eaton myasthenic syndrome: serologic and clinical correlates. Neurology. 1998;50(1):88–93.
80. Wirtz PW, Verschuuren JJ, van Dijk JG, de Kam ML, Schoemaker RC, van Hasselt JG, et al. Efficacy of 3,4-diaminopyridine and pyridostigmine in the treatment of Lambert–Eaton myasthenic syndrome: a randomized, double-blind, placebo-controlled, crossover study. Clin Pharmacol Ther. 2009;86(1):44–8.
81. Gadoth A, Pittock SJ, Dubey D, McKeon A, Britton JW, Schmeling JE, et al. Expanded phenotypes and outcomes among 256 LGI1/CASPR2-IgG-positive patients. Ann Neurol. 2017;82(1):79–92.
82. Naasan G, Irani SR, Bettcher BM, Geschwind MD, Gelfand JM. Episodic bradycardia as neurocardiac prodrome to voltage-gated potassium channel complex/leucine-rich, glioma inactivated 1 antibody encephalitis. JAMA Neurol. 2014;71(10):1300–4.
83. Josephs KA, Silber MH, Fealey RD, Nippoldt TB, Auger RG, Vernino S. Neurophysiologic studies in Morvan syndrome. J Clin Neurophysiol. 2004;21(6):440–5.
84. Boronat A, Gelfand JM, Gresa-Arribas N, Jeong HY, Walsh M, Roberts K, et al. Encephalitis and antibodies to dipeptidyl-peptidase-like protein-6, a subunit of Kv4.2 potassium channels. Ann Neurol. 2013;73(1):120–8.
85. Hara M, Arino H, Petit-Pedrol M, Sabater L, Titulaer MJ, Martinez-Hernandez E, et al. DPPX antibody-associated encephalitis: main syndrome and antibody effects. Neurology. 2017;88(14):1340–8.
86. Tobin WO, Lennon VA, Komorowski L, Probst C, Clardy SL, Aksamit AJ, et al. DPPX potassium channel antibody: frequency, clinical accompaniments, and outcomes in 20 patients. Neurology. 2014;83(20):1797–803.
87. Piepgras J, Holtje M, Michel K, Li Q, Otto C, Drenckhahn C, et al. Anti-DPPX encephalitis: pathogenic effects of antibodies on gut and brain neurons. Neurology. 2015;85(10):890–7.
88. Dalmau J, Gleichman AJ, Hughes EG, Rossi JE, Peng X, Lai M, et al. Anti-NMDA-receptor encephalitis: case series and analysis of the effects of antibodies. Lancet Neurol. 2008;7(12):1091–8.
89. van Doorn PA, Ruts L, Jacobs BC. Clinical features, pathogenesis, and treatment of Guillain–Barre syndrome. Lancet Neurol. 2008;7(10):939–50.
90. Low PA, Vernino S, Suarez G. Autonomic dysfunction in peripheral nerve disease. Muscle Nerve. 2003;27(6):646–61.
91. Griffin JW, Cornblath DR, Alexander E, Campbell J, Low PA, Bird S, et al. Ataxic sensory neuropathy and dorsal root ganglionitis associated with Sjogren's syndrome. Ann Neurol. 1990;27(3):304–15.
92. Mori K, Iijima M, Koike H, Hattori N, Tanaka F, Watanabe H, et al. The wide spectrum of clinical manifestations in Sjogren's syndrome-associated neuropathy. Brain. 2005;128(Pt 11):2518–34.
93. Kawagashira Y, Koike H, Fujioka Y, Hashimoto R, Tomita M, Morozumi S, et al. Differential, size-dependent sensory neuron involvement in the painful and ataxic forms of primary Sjogren's syndrome-associated neuropathy. J Neurol Sci. 2012;319(1–2):139–46.

# Part V

# Other CNS Inflammatory Diseases

# Neurorheumatology

# 22

Johannes Pulst-Korenberg
and Shamik Bhattacharyya

**Key Points**

1. Neuropsychiatric lupus is a broad term encompassing inflammatory and non-inflammatory mechanisms of neurologic injury. Many common symptoms in lupus such as headache are not causally associated with disease activity.
2. Rheumatoid arthritis commonly affects the C1–C2 atlantoaxial joint, and progressive disease can cause myelopathy or lower brainstem dysfunction.
3. Myelitis in Sjögren's syndrome is commonly associated with presence of anti-aquaporin 4 antibody and is treated similarly to neuromyelitis optica.
4. Parenchymal neuro-Behçet's disease often has brainstem involvement and can be misdiagnosed as multiple sclerosis on initial imaging.
5. Immunoglobulin G4 (IgG4)-related disease usually does not cause isolated parenchymal brain or spinal cord disease and is more associated with orbitopathy, pachymeningitis, or pituitary gland and stalk disease.

## Introduction

Systemic autoimmune diseases commonly affect the nervous system, and there are many more patients with systemic autoimmune diseases with neurological complaints than there are patients with primary neuro-inflammatory diseases. The interface between rheumatology and neurology is a growing field informed by better understanding of clinical symptoms, imaging findings, and laboratory markers. Here we review the neurological syndromes associated with systemic lupus erythematosus (SLE), rheumatoid arthritis (RA), Sjögren's syndrome (SS), Behçet's disease (BD) and immunoglobulin G4 (IgG4)-related disease.

## Systemic Lupus Erythematosus

Systemic lupus erythematosus (SLE) is a chronic, multisystemic disease of unclear origin associated with immune-mediated injury to multiple

J. Pulst-Korenberg
Department of Neurology, Keck School of Medicine, University of Southern California, Los Angeles, CA, USA

S. Bhattacharyya (✉)
Department of Neurology, Brigham and Women's Hospital, Boston, MA, USA
e-mail: sbhattacharyya3@bwh.harvard.edu



A. L. Piquet, E. Alvarez (eds.), *Neuroimmunology*, https://doi.org/10.1007/978-3-030-61883-4_22

organs. The disease is heterogenous clinically with many predisposing factors. SLE is more common in women indicating a hormonal effect and/or genetic differences on disease pathogenesis. The concordance rate for monozygotic twins is about 24%, suggesting a genetic risk [1]. However, the genetic predisposition is polygenic, and multiple genes in the inflammatory pathway have been identified in genome-wide association studies (GWAS). Environmental triggers likely play a role in disease pathogenesis as well. Exposure to medications such as isoniazid and hydralazine can generate lupus-like clinical symptoms with autoantibodies and improve with cessation of the drug. Epstein–Barr virus is hypothesized to be a possible triggering factor, and there is interest in the role of the gut microbiome.

SLE is marked by autoantibodies against nuclear antigens and immune complexes formed by autoantibodies and cellular debris. Some of the clinical manifestations are directly from immune complex deposition such as in nephritis. Others are from pathogenic autoantibodies such as antiphospholipid syndrome mediated by antiphospholipid antibodies. Still others result from vasculitis and accelerated vasculopathy. Hence, the disease mechanisms are broad and varying among patients. Clinically, this heterogeneity of disease activity is reflected in the breadth of clinical signs and symptoms present in SLE. Many have constitutional symptoms such as fatigue, periodic fevers, arthralgias, myalgias, and weight loss. There are a range of other disease manifestations such as rash (malar or discoid rash or photosensitivity), oral ulcers, arthritis, renal disease, serositis, and cytopenias. On laboratory testing, the most common abnormality is the presence of anti-nuclear antibody (ANA) present in more than 95% of patients with SLE. Other laboratory abnormalities include the presence of anti-dsDNA, anti-Sm, and antiphospholipid antibodies. Low complement levels are frequent with active disease as well.

SLE is primarily diagnosed clinically based on clinical judgment. Multiple classification schemes have been formulated over the years including the widely used American College of Rheumatology and more recently the Systemic Lupus International Collaborating Clinics (SLICC) criteria [2]. These are meant for classification and not for the diagnosis of individual patients, but can provide a helpful guide. Diagnosis is typically approached based on the clinical symptoms and investigations and excluding competing diagnoses such as RA, systemic sclerosis, systemic vasculitis, fibromyalgia, and malignancies.

Neuropsychiatric symptoms are common in SLE, estimated to be present in about 56% of patients [3]. Historically, many terms were used to describe neurologic involvement in SLE including lupus cerebritis or lupoid sclerosis. The current preferred term is neuropsychiatric systemic lupus erythematosus (NPSLE). To standardize nomenclature and reporting, the American College of Rheumatology developed 19 standard NPSLE syndromes with diagnostic criteria and exclusion (Table 22.1) [4]. However, these criteria were not meant to imply causation but rather to set reporting standards. The most frequent syndromes encountered are headache, mood disorder, cognitive dysfunction, seizures, and stroke. An important task for a clinician encountering these syndromes is to first deter-

**Table 22.1** Neuropsychiatric syndromes proposed by American College of Rheumatology

| |
|---|
| 1. Aseptic meningitis |
| 2. Cerebrovascular disease |
| 3. Demyelinating syndrome (central nervous system) |
| 4. Headache |
| 5. Movement disorder |
| 6. Myelopathy |
| 7. Seizure disorder |
| 8. Acute confusional state |
| 9. Anxiety disorder |
| 10. Cognitive dysfunction |
| 11. Mood disorder |
| 12. Psychosis |
| 13. Acute inflammatory demyelinating polyradiculoneuropathy |
| 14. Autonomic disorder |
| 15. Mononeuropathy, single/multiplex |
| 16. Myasthenia gravis |
| 17. Cranial neuropathy |
| 18. Plexopathy |
| 19. Polyneuropathy |

Reprinted with permission from Liang et al. [4]

mine whether the symptoms are causally attributable to SLE or as is often common, caused by other etiologies.

Headache, for example, is the most commonly associated neuropsychiatric symptom in SLE, but in isolation without other neurologic findings, headache is unlikely to be causally linked. The pooled estimate of headache in patients in SLE is about 57%, which is similar to the general population [5]. The quality of the headache (migraine or tension type) is not associated with SLE, and headache itself generally has not correlated with systemic disease activity. Hence, in general, headache in SLE should be treated symptomatically and investigated routinely as in other patients. This conclusion does not imply that headache in SLE is *never* dangerous. For example, headache in a patient with SLE and antiphospholipid antibodies may signal the beginning of central venous sinus thrombosis (Fig. 22.1).

The causes of mood disorder in SLE are varied. Depression is common in SLE, though the relative contribution of the disease itself versus competing factors such as chronic illness and background genetic risk is unclear. Elevated disease activity modestly increases the odds of major depression by about 10% [6]. Depression in SLE generally is treated symptomatically because of the multifactorial etiology. In contrast, psychosis is more closely causally linked to SLE. In a large inception cohort, psychosis was a rare event occurring in 1.5% of patients but was primarily attributable to SLE when it occurred. Most psychotic episodes occurred early within 3 years of the diagnosis of SLE with increased risk in men and those of African ancestry [7]. When considering psychosis in SLE, steroid-induced psychosis has to be initially excluded. There are no laboratory or imaging studies that can confirm the diagnosis of SLE psychosis. Cerebrospinal fluid (CSF) studies are not consistently abnormal, and magnetic resonance imaging (MRI) of the brain does not have specific findings. The anti-Ribosomal P antibody had been proposed to be helpful in this regard. However, in further studies, the antibody lacks sensitivity (about 27%) and has moderate specificity (about 80%) [8]. Hence, the presence or absence of the anti-Ribosomal P antibody has a limited role in the diagnosis of mood disorder associated with SLE.

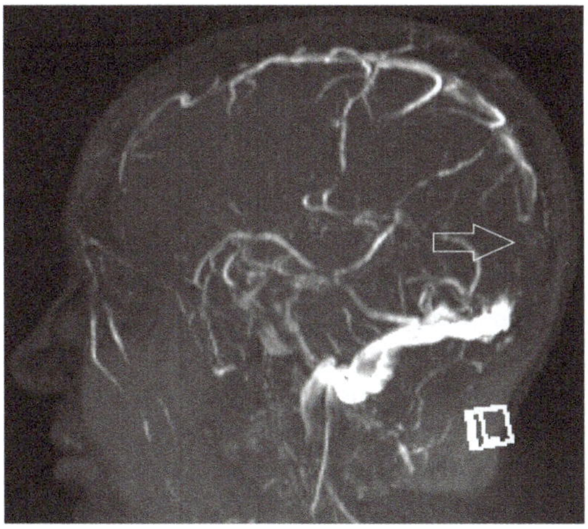

**Fig. 22.1** Magnetic resonance (MR) venogram showing superior sagittal sinus thrombosis (arrow) in a patient with systemic lupus erythematosus (SLE) presenting with worsening headache

Seizures in SLE are often multifactorial. They can be provoked by factors such as electrolyte disturbance, dialysis disequilibrium, uremia, or ischemic stroke. For seizures attributed to SLE, the majority occur around the time of SLE diagnosis [9]. Most seizures attributed to SLE are fortunately single episodes and do not result in epilepsy. No autoantibodies are known to be specific for seizure in SLE. Therapeutically, provoking factors including posterior reversible encephalopathy syndrome should be first sought for. Otherwise, first-time seizures in SLE should be treated similarly to the general population in which many patients with normal imaging and electroencephalogram (EEG) do not necessarily need antiepileptic therapy after the first seizure.

Stroke, on the other hand, is a major cause of morbidity in SLE. In a prospective series of 1000 patients with SLE, stroke accounted for about 12% of deaths [10]. There is an overall twofold increase risk of stroke in patients with SLE compared to the general population. The pathogenesis of the stroke in SLE is complex. Most notably, "lupus vasculitis" is *not* the most common cause as shown by multiple pathological series [11, 12]. Small vessel vasculopathy is the most frequent neuropathological finding without signs of vasculitis. This is seen on MRI as increased burden of white matter T2 changes. Even when controlling for concurrent risk factors such as hypertension or diabetes mellitus, SLE itself confers an increased risk of accruing these white matter changes. How best to respond clinically to these changes is unclear. Stroke in SLE can also occur from cardioembolism such as Libman–Sacks endocarditis (sterile valvular lesions), infective endocarditis (predisposed from frequent immune suppression), and paradoxical embolism. Microembolism to the brain might be more prevalent than appreciated as shown by recent transcranial Doppler studies demonstrating microembolic signals even in apparently asymptomatic patients with SLE [13]. Antiphospholipid syndrome is another major cause of ischemic stroke in SLE. In patients with ischemic stroke with SLE, antiphospholipid antibodies are found in as many as two-thirds of patients. The antiphospholipid syndrome can also cause Libman–Sacks endocarditis and small vessel injury. Atherosclerosis is also accelerated in SLE compared to the general population partly from comorbidities such as renal dysfunction, hypertension, and corticosteroid exposure and partly from long-term SLE itself. Finally, the occurrence of cerebral vasculitis in SLE is quite rare compared to the other mechanisms discussed previously. There have been individual case reports of pathologically proven vasculitis with stroke in patients with SLE [14], but diagnosis of SLE vasculitis without pathological confirmation based on imaging and angiographic evidence should be believed with caution.

Therapeutically, there are no clear guidelines for treatment of stroke in SLE. Acutely, ischemic stroke is treated similarly to the general population with consideration of thrombectomy and thrombolysis in patients who present early with significant deficits. In addition to usual head and vascular imaging, patients with SLE and ischemic stroke should be tested for antiphospholipid antibodies. Patients may also need more extensive heart imaging based on suspicion for sterile and infective endocarditis. Note that because of increased rate of atherosclerosis in SLE, many may have beading on angiographic imaging of the head without having vasculitis. Secondary prevention is usually with antiplatelet therapy except for antiphospholipid syndrome in which anticoagulation is used.

Neuro-inflammatory disorders can certainly also occur with SLE. For example, SLE has been associated with myelitis. A significant number of these patients have anti-aquaporin 4 (AQP4) antibodies and are treated similarly to those with neuromyelitis optica spectrum disorder (NMOSD; see Chap. 15). However, there are cases of SLE myelitis without other known pathogenic antibodies. Testing for AQP4 and myelin oligodendrocyte glycoprotein (MOG) antibodies are important in the diagnostic workup of these patients as treatment paradigms may be different. SLE can also infrequently cause optic neuritis, encephalitis, and hypophysitis [15, 16]. Because these are individually rare events, less is known about the pathogenesis including what are the appropriate antibodies to check, and larger ongoing cohort studies will provide more information in the future. When the mechanism is suspected to be inflammatory, the treatment is typically with steroids and immunosuppression.

## Rheumatoid Arthritis

Rheumatoid arthritis (RA) is a common disease estimated to affect as many as 0.5–1% of adults in the United States and is characterized by chronic synovial inflammation and erosive arthritis. The most typical clinical presentation is arthritis of the small joints causing prolonged morning stiffness. The disease is diagnosed by combination of clinical symptoms and laboratory markers. Generally, acute-phase reactants such as C-reactive protein (CRP) and erythrocyte sedimentation rate (ESR) are elevated during disease activity. Serologically, rheumatoid factor and anti-citrullinated peptide antibody are present in about 60–70% [17]. The anti-citrullinated peptide antibody is more specific for the disease.

RA rarely affects the central nervous system (CNS) primarily, and what is known is based on case reports or small case series. Rheumatoid meningitis is characterized pathologically by a combination of lymphocytic infiltration of the meninges, rheumatoid nodules, and intracranial vasculitis. Clinically, patients can present with focal symptoms referable to underlying parenchymal dysfunction, seizures, stroke-like episodes, cranial neuropathies, and headaches. Cerebrospinal fluid studies are variable with pleocytosis, mildly elevated protein, and normal to low glucose. Although the majority of patients have long-standing seropositive disease, it can present even if the patient's systemic symptoms are otherwise well controlled or as the presenting manifestation. Imaging with MRI can variably show a combination of pachymeningeal and leptomeningeal gadolinium enhancement (Fig. 22.2). Historically, reported successful treatments have included corticosteroids with or without adjunctive methotrexate or cyclophosphamide, although more recently rituximab has also shown efficacy [18, 19].

In contrast to the rarity of direct CNS involvement, rheumatoid arthritis commonly affects the cervical spine and can cause secondary spondylotic myelopathy, radiculopathy, or rarely vertebral artery vasculopathy. The joints most affected cause C1–C2 atlantoaxial subluxation (AAS), vertical subluxation of the axis (VS) or "superior migration of the odontoid," and C3-C7 subaxial subluxation (SAS). In some cases, large rheumatoid nodules themselves may cause compressive myelopathy. The pooled frequency of any cervical spine abnormality is estimated at 45% [20]. Patients typically present with C2 radiculopathy (pain in the occipital region) with atlantoaxial subluxation, and in compressive myelopathy, early symptoms are paresthesia in the hands and gait instability (Fig. 22.3).

Although the treatment of RA has improved with time, a 2015 meta-analysis comparing data from the 1970s to 2000s estimates that the frequency of myelopathy has remained constant at 5% [20]. The prevalence of atlantoaxial subluxation appears to have decreased from 36% to 24%, but the combined frequency of other forms of subluxation appears constant at 24%. There is observational evidence that more intensive disease-modifying therapy can slow the development of new cervical spine lesions, although it may not prevent progression of preexisting ones. When patients develop myelopathy, therapy is generally with surgical stabilization. Surgical outcomes vary based on type of subluxation and underlying disease control. In a

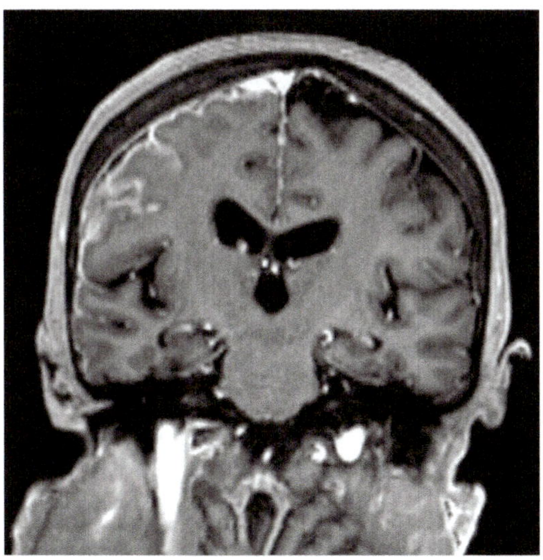

**Fig. 22.2** Coronal post-gadolinium contrast magnetic resonance imaging (MRI) with enhancement in the pachymeningeal and leptomeningeal areas over the right hemisphere

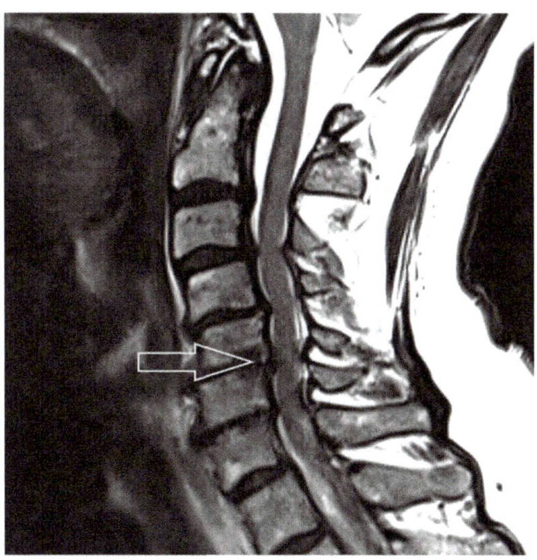

**Fig. 22.3** Sagittal view of magnetic resonance imaging (MRI) of cervical spine (T2 sequence) demonstrating subaxial subluxation with spinal cord compression (arrow) in a patient with rheumatoid arthritis

large retrospective single-center study of 118 seropositive RA patients, the best outcomes were seen in patients with atlantoaxial subluxation [21]. Patients with subaxial subluxation did the poorest, but also had worse pre-intervention neurological status.

Peripheral neuropathy in RA can result from entrapment due to degenerative bony changes, synovial inflammation, drug toxicity, vasculitis, amyloidosis, or rheumatoid nodules. In a large study of 108 patients, only 21% complained of neuropathic symptoms although 57% had electrophysiologic evidence of neuropathy [22]. Of these, the most common neuropathy was sensory or sensorimotor axonal (85.5%), followed by demyelinating (15%). Sural nerve biopsy pathology was variable, possibly reflecting the various etiologies of neuropathy in RA. Studies conflict as to whether disease duration is a risk factor. Carpal tunnel syndrome is the most common peripheral neuropathy in RA, and a recent meta-analysis estimated the adjusted odds ratio at 1.96 for patients with RA relative to the general population [23]. Less common entrapment neuropathies include posterior interosseous neuropathy at the arcade of Frohse and ulnar neuropathy at the cubital tunnel. Autonomic neuropathy has also been associated with RA. Unlike other RA neuropathies, there is less clear evidence that RA disease activity is linked to autonomic dysfunction. Entrapment neuropathy is most effectively treated with surgical decompression. Disease-modifying anti-rheumatic drugs (DMARDs) and biologics have some evidence for efficacy in RA-associated autonomic neuropathy [24].

The clinical presentation of multiple mononeuropathies should raise suspicion for a peripheral nervous system (PNS) vasculitis. RA-associated vasculitis is histologically identical to polyarteritis nodosa and generally occurs with long-standing disease. With effective treatment of RA, the incidence of vasculitis has decreased substantially [25]. When the vasculitis is present systemically, the CNS may also be affected in the form of ischemic strokes in addition to neuropathy and usually skin changes. Treatment is generally by therapy for RA.

## Sjögren's Syndrome

Sjögren's syndrome (SS) is characterized by focal lymphocytic infiltration of exocrine glands, especially the lacrimal and salivary glands resulting in the typical sicca syndrome of dry eyes and dry mouth. SS is called primary when occurring by itself or secondary when occurring in the context of other autoimmune diseases such as SLE. The disease is diagnosed by the combination of symptoms of dry eyes/dry mouth, objective evidence of glandular dysfunction, and supportive laboratory testing. While not specific, anti-SSA/Ro and/or anti-SSB/La are associated with SS and present in the majority of patients. There have been multiple classification criteria developed over the years, most recently by European League Against Rheumatism (EULAR), which created a scoring scheme and reports sensitivity and specificity over 90% compared to expert opinion [26].

In addition to the sicca syndrome (dry eyes/dry mouth), SS has many extraglandular manifestations. These can be broadly grouped into periepithelial involvement (bronchiolitis, interstitial nephritis, and bronchiolitis) and those with extra-epithelial pathology (glomerulonephritis, vasculitis, neuropathy, and fatigue). Neurologically, SS can affect both the CNS and PNS. Frequently, patients with SS complain of debilitating fatigue and cognitive slowing. The causes of the symptoms have been unclear, and investigations including MRI and cerebrospinal fluid analysis are generally normal. Overall, patients with SS have a higher burden of white matter T2 hyperintensities on brain MRI compared to the general population [27]. The clinical relevance of these findings is unclear. SS can also cause more focal symptoms including recurrent demyelinating lesions affecting the spinal cord, brain, and optic nerves. Emerging data suggest that many of these patients, especially those with spinal cord lesions, harbor AQP4 antibodies found in NMOSD (Fig. 22.4) [28]. However, there are also patients with inflammatory CNS lesions with SS who do not have NMOSD.

Peripheral neuropathy is the more recognized neurological manifestation of SS. There are myr-

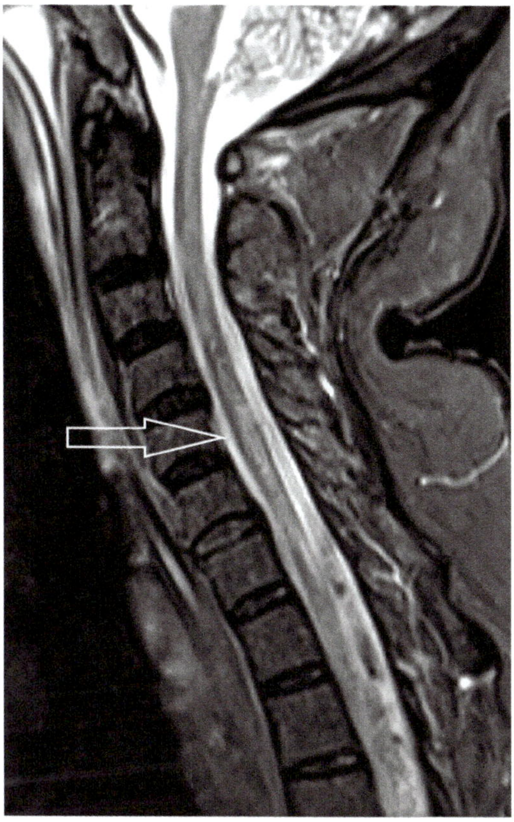

**Fig. 22.4** Sjögren's syndrome associated longitudinally extensive myelitis demonstrated in sagittal view of cervical spine magnetic resonance imaging (MRI) as T2 hyperintensity with cord edema (arrow). The patient tested positive for aquaporin 4 (AQP4) antibody, thus diagnosis consistent with neuromyelitis optica spectrum disorder (NMOSD), which can often coexist with Sjögren's and/or systemic lupus erythematosus (SLE)

iad forms of peripheral neuropathy described, which occur in up to 64% of patients: sensory ataxic, sensory axonal, pure motor, sensorimotor, multiple mononeuropathies, multiple cranial neuropathies, trigeminal, autonomic, and radiculoneuropathy [29]. Peripheral neuropathy frequently precedes sicca symptoms and diagnosis of SS. The different types of neuropathy do not share the same underlying pathology. For example, ganglioneuropathy is associated with lymphocytic infiltration, whereas vasculitis and perivascular cell invasion are associated with multiple mononeuropathies. These entities are usually distinguishable by clinical history, Electromyography (EMG), and sometimes MRI findings, although skin biopsy may be needed to diagnose isolated small fiber sensory neuropathy.

One entity of particular neurologic relevance is the sensory ataxic neuronopathy, also called "ganglionopathy." It is a relatively rare neurologic phenomenon characterized by the rapid onset of sensory ataxia, vibratory and proprioceptive loss, and loss of reflexes potentially with a non-length-dependent pattern of sensory loss. This has a narrow differential diagnosis, and so Sjögren's should be strongly considered as an etiology if the appropriate signs present. Treatment for the aforementioned neuropathies varies greatly, as the pathology and severity have a large range. Evidence is largely at the level of case reports and by clinical experience; at best modest improvement should be expected in most cases. Sensory ataxic neuronopathy has been treated with infliximab, plasmapheresis, mycophenolate mofetil, intravenous immunoglobulin (IVIg), and rituximab [30–33]. Interpretation of the reported treatment series is limited by lack of knowledge of untreated history of sensory neuronopathy, limited time of follow-up, and non-uniform neurologic assessment.

## Neuro-Behçet's Disease

Behçet's disease (BD) is a multisystemic inflammatory disorder of unknown etiology. It is characterized pathologically by a perivasculitis without fibrinoid necrosis or frank vasculitis, although there is no pathognomonic diagnostic feature. The incidence varies greatly by geographic region, favoring the area formerly known as the Silk Road from Japan to the Mediterranean. Incidence is as high as 80–370 cases per 100,000 in Turkey to as low as 1 per 15,000 to 1 per 500,000 in North America [34]. Neurologic involvement occurs in approximately 9.4% of BD and is the first manifestation in only 6% of patients [35]. The typical patient is between ages 20 and 40, with a 2.8:1 ratio between men and women.

Diagnostic criteria were put forth in 1990 by the International Study Group for Behçet's Disease, including recurrent oral ulceration and any two of the following: recurrent genital ulceration, eye lesions, skin lesions, or a positive pathergy test. In 2013, diagnostic criteria were revised by another group involving a weighted sum of syndromes (2 points for ocular lesions, genital aphthosis, and oral aphthosis; 1 point for skin, neurological, and vascular changes; and 1 point for pathergy). A score greater than or equal to 4 achieved 95% sensitivity and 91% specificity [36]. In 2014, diagnostic criteria were put forth by the International Neuro Behçet's Advisory Group for the diagnosis of Definite or Probable Neuro-Behçet's Disease (NBD), requiring that the patient either satisfy criteria for BD, have a characteristic neurologic syndrome, or both for Definite disease.

The neurologic manifestations of BD fall into one of two main types: (1) an inflammatory, usually multifocal, parenchymal disorder and (2) a vascular variant characterized by cerebral venous sinus thrombosis (CVST) and intracranial hypertension. Headache is the most common neurologic symptom in BD, and it is estimated that only ~10% of BD patients with headache have neurologic involvement. The headache may flare with systemic disease activity without nervous system disease. Parenchymal NBD can involve the brainstem, cerebrum, optic nerves, spinal cord, or multifocal involving any of the previous areas. Clinically, it can mimic an atypical multiple sclerosis (MS) picture, with relapses or a slowly progressive course. Radiographically, however, there is substantially more gray matter involvement, brainstem atrophy, and mottled or diffuse enhancement (Fig. 22.5). In one large longitudinal study, 31% had a monophasic course, 27% relapsing-remitting, and 20% with a progressive course. New T1 black holes were common (35%), 29% had increasing lesion size, 47% had new enhancing lesions, and 12% demonstrated persistent enhancement despite therapy. Interestingly, 71% of patients fulfilled the McDonald criteria for MS by the end of the study [37].

Non-parenchymal manifestations include venous sinus thrombosis, intracranial hypertension, meningitis, aneurysm, and rarely arterial thrombosis. A large meta-analysis estimates the incidence of CVST at 3/1000 person-years, with a fivefold increase in patients with neurologic

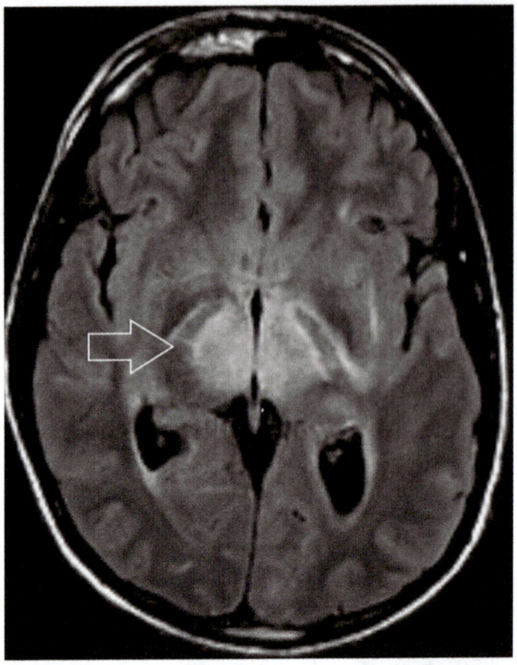

**Fig. 22.5** Axial fluid-attenuated inversion recovery (FLAIR) magnetic resonance imaging (MRI) sequence demonstrating bi-thalamic T2 hyperintensity (arrow) in a patient with parenchymal Behçet's disease

involvement [38]. The most common clinical scenario is one of the gradual onset (77%) of an intracranial hypertension syndrome, with headache (92–100%), papilledema (63–89%), cranial nerve VI palsy (8–57%), and nausea/vomiting (19%). Less common were focal deficits, pyramidal signs, or seizure. The most common sites of involvement were the superior sagittal (64%) and transverse sinuses (61%), followed by sigmoid (9%) and straight (4%). Unlike in parenchymal involvement, cerebrospinal fluid can be normal with the exception of an elevated opening pressure. There is little involvement of the PNS in NBD.

Inflammatory markers can be elevated in the acute stages of NBD but have limited diagnostic utility. Human Leukocyte Antigen HLA-B51 is clearly associated with BD depending on ethnicity. On cerebrospinal fluid analysis, a neutrophilic or lymphocytic pleocytosis is usually present, protein is increased, and oligoclonal bands are usually absent. Increased CSF interleukin 6 (IL-6) levels have been found in patients with acute inflammatory NBD as compared to other CNS inflammatory diseases (multiple sclerosis and subacute sclerosing panencephalitis) and may be useful in trending in patients with otherwise benign CSF studies [39]. However, this finding is not considered a fully sensitive biomarker.

In 2018, the European League Against Rheumatism (EULAR) issued updated treatment guidelines for Behçet's disease [40]. For parenchymal disease, pulse doses of intravenous steroids should be administered followed by a taper over months in combination with immunosuppression such as with azathioprine. Tumor necrosis factor (TNF)-alpha therapy should be considered first-line for severe neurologic manifestations or for refractory disease [41]. Mycophenolate mofetil, cyclophosphamide, and methotrexate have historically also been used but did not receive a recommendation. Recently, three refractory cases have been treated with tocilizumab, and anti-IL6 directed antibody, to good effect [42]. Notably, cyclosporine should be avoided due to meta-analyses showing paradoxically increased nervous system involvement [43]. For CVST, intravenous steroids and taper should also be initiated, and anticoagulation is generally used to treat the thrombus.

## IgG4-Related Disease

Immunoglobulin G4-related disease (IgG4-RD) is an immune-mediated multisystemic disease with pathological hallmarks of storiform fibrosis, obliterative phlebitis, and lymphoplasmacytic infiltrate concentrated with IgG4 positive plasma cells. Since its recognition, IgG4-RD has been the unifying explanation for a remarkably diverse number of conditions including autoimmune pancreatitis, periaortitis, retroperitoneal fibrosis, and eosinophilic angiocentric fibrosis [44]. The true prevalence of IgG4-RD is uncertain because of lack of recognition of the unique pathology and emerging awareness of the diagnosis. In contrast to most other autoimmune diseases, IgG4-RD more commonly affects men than women, especially in the middle-age to elderly age groups [45]. In the head and neck region, the balance between the sexes is more even unlike IgG4-RD systemically [45]. While the underlying cause of IgG4-RD is unclear, the IgG4 is not pathological by itself; rather, it is more a marker of disease. In fact, secreted IgG4 cannot cross-link antigens to create immune complexes and binds poorly to Fc receptors [44]. T cells and antigen-presenting B cells are likely important in disease pathogenesis, though the precise sequence of disease initiation and propagation are still under investigation.

IgG4-RD is preferably diagnosed by tissue histopathology. There is no imaging or serological marker that is very sensitive or specific to obviate the need for obtaining tissue. Common competing diagnoses such as sarcoidosis or granulomatosis with polyangiitis or neoplasms can all have similar imaging appearance but with divergent treatments, and firm early diagnosis is preferred. High-serum IgG4 concentration is common in IgG4-RD, but it is not sensitive or specific for the disease and as such cannot be relied on to exclude or diagnose IgG4-RD especially in patients with single organ disease. For example, diseases such as Sjögren's syndrome, chronic sinusitis, leukemia, and recurrent pneumonia can all cause elevated serum IgG4 levels [46]. The ratio between IgG4 to IgG1 (greater than 0.24) might provide better diagnostic specificity [47]. On the other hand, somewhere between 3% and 30% of patients with IgG4-RD will have normal serum IgG4 levels [48]. Longitudinally, there may be a role for tracking IgG4 levels to assess for disease remission and recurrence [48].

Neurologically, IgG4-RD rarely causes isolated parenchymal central nervous system disease. In the head and neck region, IgG4-RD is usually associated with orbitopathy, pachymeningitis, and pituitary gland and stalk disease. IgG4-related orbitopathy causes a chronic, progressive, and painless syndrome marked by periorbital swelling, extraocular movement abnormalities, and optic neuropathy. The periorbital swelling can be caused either by infiltration of the orbital structures or by lacrimal gland inflammation. On imaging, MRI is more sensitive than computed tomography (CT) for defining the extent of the disease, and fat-suppressed orbital images on MRI are important for complete orbital evaluation. IgG4-RD is visualized as either localized masses or enlargement of the lacrimal gland, extraocular muscles, or optic sheath. These lesions are typically well defined with T2 iso- to hypo-intensity and show homogenous enhancement with intravenous gadolinium [49].

Hypertrophic pachymeningitis is a slowly progressive disease marked by thickening of the pachymeninges. Neurological symptoms are generally caused by site of involvement and neural/vascular compression. In the head, the most frequent symptoms are headache in combination with cranial nerve palsies [50]. However, parenchymal dysfunction can also occur including seizures, cognitive dysfunction, or weakness/numbness in the extremities. Pachymeningitis can extend into the spine as well and cause radicular symptoms [51]. On imaging, there is pachymeningeal thickening with enhancement on MRI and can be focal (such as over one convexity or skull base) or be diffuse [50]. Similar to hypertrophic pachymeningitis, pituitary lesions present with slow progression of hypopituitarism or diabetes insipidus and found on imaging to have enlargement of the pituitary with thickened stalk. Often, there is concurrent involvement of the meninges.

For treatment of IgG4-RD, glucocorticoids are first-line therapy for inducing remission often starting at prednisone equivalent of 30–40 mg/day for 2–4 weeks and then gradually tapering over 3–6 months [48]. Although effective, glucocorticoids alone often do not sufficiently control the disease or maintain remission when tapered. For example, in a French cohort study of IgG4-RD, only 30% were able to stop steroids [52]. Long-term steroids are also poorly tolerated by older patients usually afflicted with IgG4-RD. Multiple agents have been used as steroid-sparing agents including azathioprine, mycophenolate mofetil, methotrexate, and cyclophosphamide; however, there are no prospective trials to guide use of these agents or to assess additional efficacy over glucocorticoids alone [48]. There is better data for rituximab use. In a single arm, open label, prospective trial of rituximab, the majority of patients treated with rituximab alone without glucocorticoids were able to achieve and maintain remission [53]. There have been no further blinded, placebo-controlled trials for rituximab for IgG4-RD, but existing observational data and clinical experience argue strongly for its efficacy. An area of uncertainty is how long to treat patients before tapering or stopping therapy. In the absence of stronger data, this is generally approached on a case-by-case basis guided by prior disease history and patient preference.

## Conclusion

There is wide variability in the neurological manifestations of systemic autoimmune diseases, and even within individual diseases, mechanisms can vary. Hence, rather than applying blanket recommendations for "lupus cerebritis," treatment should be tailored to understanding the mechanism of neurological disease whenever possible.

## References

1. Deapen D, Escalante A, Weinrib L, Horwitz D, Bachman B, Roy-Burman P, et al. A revised estimate of twin concordance in systemic lupus erythematosus. Arthritis Rheum. 1992;35(3):311–8.
2. Petri M, Orbai A-M, Alarcón GS, Gordon C, Merrill JT, Fortin PR, et al. Derivation and validation of the systemic lupus international collaborating clinics classification criteria for systemic lupus erythematosus. Arthritis Rheum. 2012;64(8):2677–86.
3. Unterman A, Nolte JES, Boaz M, Abady M, Shoenfeld Y, Zandman-Goddard G. Neuropsychiatric syndromes in systemic lupus erythematosus: a meta-analysis. Semin Arthritis Rheum. 2011;41(1):1–11.
4. Liang MH, Corzillius M, Bae SC, Lew RA, Fortin PR, Gordon C, Isenberg D, Alarcón GS, Straaton KV, Denburg J, Denburg S. The American College of Rheumatology nomenclature and case definitions for neuropsychiatric lupus syndromes. Arthritis Rheum. 1999;42(4):599–608.
5. Mitsikostas DD, Sfikakis PP, Goadsby PJ. A meta-analysis for headache in systemic lupus erythematosus: the evidence and the myth. Brain. 2004;127(Pt 5):1200–9.
6. Bachen EA, Chesney MA, Criswell LA. Prevalence of mood and anxiety disorders in women with systemic lupus erythematosus. Arthritis Rheum. 2009;61(6):822–9.
7. Hanly JG, Li Q, Su L, Urowitz MB, Gordon C, Bae S-C, et al. Psychosis in systemic lupus erythematosus:

results from an International Inception Cohort Study. Arthritis Rheum. 2019;71(2):281–9.
8. Karassa FB, Afeltra A, Ambrozic A, Chang D-M, De Keyser F, Doria A, et al. Accuracy of anti-ribosomal P protein antibody testing for the diagnosis of neuropsychiatric systemic lupus erythematosus: an international meta-analysis. Arthritis Rheum. 2006;54(1):312–24.
9. Hanly JG, Urowitz MB, Su L, Gordon C, Bae S-C, Sanchez-Guerrero J, et al. Seizure disorders in systemic lupus erythematosus results from an international, prospective, inception cohort study. Ann Rheum Dis. 2012;71(9):1502–9.
10. Cervera R, Khamashta MA, Font J, Sebastiani GD, Gil A, Lavilla P, et al. Morbidity and mortality in systemic lupus erythematosus during a 10-year period: a comparison of early and late manifestations in a cohort of 1,000 patients. Medicine (Baltimore). 2003;82(5):299–308.
11. Johnson RT, Richardson EP. The neurological manifestations of systemic lupus erythematosus. Medicine (Baltimore). 1968;47(4):337–69.
12. Devinsky O, Petito CK, Alonso DR. Clinical and neuropathological findings in systemic lupus erythematosus: the role of vasculitis, heart emboli, and thrombotic thrombocytopenic purpura. Ann Neurol. 1988;23(4):380–4.
13. Baizabal-Carvallo JF, Samson Y. Microembolic signals in systemic lupus erythematosus and other cerebral small vessel diseases. J Neurol. 2010;257(4):503–8.
14. Goel D, Reddy SR, Sundaram C, Prayaga AK, Rajasekhar L, Narsimulu G. Active necrotizing cerebral vasculitis in systemic lupus erythematosus. Neuropathology. 2007;27(6):561–5.
15. de Andrade FA, Guimarães Moreira Balbi G, Bortoloti de Azevedo LG, Provenzano Sá G, Vieira de Moraes Junior H, Mendes Klumb E, Abramino Levy R, et al. Neuro-ophthalmologic manifestations in systemic lupus erythematosus. Lupus. 2017;26(5):522–8.
16. Yang N, Bhattacharyya S, Weinblatt M. Refractory syndrome of inappropriate secretion of antidiuretic hormone in systemic lupus erythematosus-associated hypophysitis. J Rheumatol. 2017;44(4):541–2.
17. Whiting PF, Smidt N, Sterne JAC, Harbord R, Burton A, Burke M, et al. Systematic review: accuracy of anti-citrullinated peptide antibodies for diagnosing rheumatoid arthritis. Ann Intern Med. 2010;152(7):456–64; W155–166.
18. Harrison NS, Kishore S, Majithia V. Rheumatoid meningitis: successful remission with rituximab. BMJ Case Rep. 2018;11(1):e226642.
19. Schuster S, Braass H, Iking-Konert C, Schnoor U, Matschke J, Gerloff C, et al. Rheumatoid meningitis: a rare cause of aseptic meningitis with frequently stroke-like episodes. Neurol Clin Pract. 2018;8(5):451–5.
20. Zhang T, Pope J. Cervical spine involvement in rheumatoid arthritis over time: results from a meta-analysis. Arthritis Res Ther. 2015;17:148.
21. Miyamoto H, Sumi M, Uno K. Outcome of surgery for rheumatoid cervical spine at one institute over three decades. Spine J. 2013;13(11):1477–84.
22. Biswas M, Chatterjee A, Ghosh SK, Dasgupta S, Ghosh K, Ganguly PK. Prevalence, types, clinical associations, and determinants of peripheral neuropathy in rheumatoid patients. Ann Indian Acad Neurol. 2011;14(3):194–7.
23. Shiri R. Arthritis as a risk factor for carpal tunnel syndrome: a meta-analysis. Scand J Rheumatol. 2016;45(5):339–46.
24. Syngle A, Verma I, Krishan P, Garg N, Syngle V. Disease-modifying anti-rheumatic drugs improve autonomic neuropathy in arthritis: DIANA study. Clin Rheumatol. 2015;34(7):1233–41.
25. Watts RA, Mooney J, Lane SE, Scott DGI. Rheumatoid vasculitis: becoming extinct? Rheumatology (Oxford). 2004;43(7):920–3.
26. Shiboski CH, Shiboski SC, Seror R, Criswell LA, Labetoulle M, Lietman TM, et al. 2016 American College of Rheumatology/European League Against Rheumatism classification criteria for primary Sjögren's syndrome: a consensus and data-driven methodology involving three international patient cohorts. Ann Rheum Dis. 2017;76(1):9–16.
27. Morgen K, McFarland HF, Pillemer SR. Central nervous system disease in primary Sjogrens syndrome: the role of magnetic resonance imaging. Semin Arthritis Rheum. 2004;34(3):623–30.
28. Kahlenberg JM. Neuromyelitis optica spectrum disorder as an initial presentation of primary Sjögren's syndrome. Semin Arthritis Rheum. 2011;40(4):343–8.
29. Pavlakis PP, Alexopoulos H, Kosmidis ML, Mamali I, Moutsopoulos HM, Tzioufas AG, et al. Peripheral neuropathies in Sjögren's syndrome: a critical update on clinical features and pathogenetic mechanisms. J Autoimmun. 2012;39(1–2):27–33.
30. Pereira PR, Viala K, Maisonobe T, Haroche J, Mathian A, Hié M, et al. Sjögren sensory neuronopathy (Sjögren Ganglionopathy): long-term outcome and treatment response in a series of 13 cases. Medicine (Baltimore). 2016;95(19):e3632.
31. Caroyer J-M, Manto MU, Steinfeld SD. Severe sensory neuronopathy responsive to infliximab in primary Sjögren's syndrome. Neurology. 2002;59(7):1113–4.
32. Gorson KC, Natarajan N, Ropper AH, Weinstein R. Rituximab treatment in patients with IVIg-dependent immune polyneuropathy: a prospective pilot trial. Muscle Nerve. 2007;35(1):66–9.
33. Chen WH, Yeh JH, Chiu HC. Plasmapheresis in the treatment of ataxic sensory neuropathy associated with Sjögren's syndrome. Eur Neurol. 2001;45(4):270–4.
34. Yurdakul S, Hamuryudan V, Yazici H. Behçet syndrome. Curr Opin Rheumatol. 2004;16(1):38–42.
35. Al-Araji A, Kidd DP. Neuro-Behçet's disease: epidemiology, clinical characteristics, and management. Lancet Neurol. 2009;8(2):192–204.
36. International Team for the Revision of the International Criteria for Behçet's Disease (ITR-ICBD). The

International Criteria for Behçet's Disease (ICBD): a collaborative study of 27 countries on the sensitivity and specificity of the new criteria. J Eur Acad Dermatol Venereol. 2014;28(3):338–47.
37. Borhani Haghighi A, Sarhadi S, Farahangiz S. MRI findings of neuro-Behcet's disease. Clin Rheumatol. 2011;30(6):765–70.
38. Aguiar de Sousa D, Mestre T, Ferro JM. Cerebral venous thrombosis in Behçet's disease: a systematic review. J Neurol. 2011;258(5):719–27.
39. Akman-Demir G, Tüzün E, Içöz S, Yeşilot N, Yentür SP, Kürtüncü M, et al. Interleukin-6 in neuro-Behçet's disease: association with disease subsets and long-term outcome. Cytokine. 2008;44(3):373–6.
40. Hatemi G, Christensen R, Bang D, Bodaghi B, Celik AF, Fortune F, et al. 2018 update of the EULAR recommendations for the management of Behçet's syndrome. Ann Rheum Dis. 2018;77(6):808–18.
41. Vallet H, Riviere S, Sanna A, Deroux A, Moulis G, Addimanda O, et al. Efficacy of anti-TNF alpha in severe and/or refractory Behçet's disease: multicenter study of 124 patients. J Autoimmun. 2015;62:67–74.
42. Addimanda O, Pipitone N, Pazzola G, Salvarani C. Tocilizumab for severe refractory neuro-Behçet: three cases IL-6 blockade in neuro-Behçet. Semin Arthritis Rheum. 2015;44(4):472–5.
43. Kötter I, Günaydin I, Batra M, Vonthein R, Stübiger N, Fierlbeck G, et al. CNS involvement occurs more frequently in patients with Behçet's disease under cyclosporin A (CSA) than under other medications--results of a retrospective analysis of 117 cases. Clin Rheumatol. 2006;25(4):482–6.
44. Kamisawa T, Zen Y, Pillai S, Stone JH. IgG4-related disease. Lancet. 2015;385(9976):1460–71.
45. Zen Y, Nakanuma Y. IgG4-related disease: a cross-sectional study of 114 cases. Am J Surg Pathol. 2010;34(12):1812–9.
46. Carruthers MN, Khosroshahi A, Augustin T, Deshpande V, Stone JH. The diagnostic utility of serum IgG4 concentrations in IgG4-related disease. Ann Rheum Dis. 2015;74(1):14–8.
47. Boonstra K, Culver EL, de Buy Wenniger LM, van Heerde MJ, van Erpecum KJ, Poen AC, et al. Serum immunoglobulin G4 and immunoglobulin G1 for distinguishing immunoglobulin G4-associated cholangitis from primary sclerosing cholangitis. Hepatology. 2014;59(5):1954–63.
48. Khosroshahi A, Wallace ZS, Crowe JL, Akamizu T, Azumi A, Carruthers MN, et al. International consensus guidance statement on the management and treatment of IgG4-related disease. Arthritis Rheum. 2015;67(7):1688–99.
49. Katsura M, Mori H, Kunimatsu A, Sasaki H, Abe O, Machida T, et al. Radiological features of IgG4-related disease in the head, neck, and brain. Neuroradiology. 2012;54(8):873–82.
50. Lu LX, Della-Torre E, Stone JH, Clark SW. IgG4-related hypertrophic pachymeningitis: clinical features, diagnostic criteria, and treatment. JAMA Neurol. 2014;71(6):785–93.
51. Williams MM, Mashaly H, Puduvalli VK, Jin M, Mendel E. Immunoglobulin G4-related disease mimicking an epidural spinal cord tumor: case report. J Neurosurg Spine. 2017;26(1):76–80.
52. Ebbo M, Daniel L, Pavic M, Sève P, Hamidou M, Andres E, et al. IgG4-related systemic disease: features and treatment response in a French cohort: results of a multicenter registry. Medicine (Baltimore). 2012;91(1):49–56.
53. Carruthers MN, Topazian MD, Khosroshahi A, Witzig TE, Wallace ZS, Hart PA, et al. Rituximab for IgG4-related disease: a prospective, open-label trial. Ann Rheum Dis. 2015;74(6):1171–7.

# Neurosarcoidosis 23

James E. Eaton and Siddharama Pawate

**Key Points**

1. Sarcoidosis is a multi-system granulomatous disorder that is presumably immune-mediated, reflecting the response to as-yet unidentified antigen(s).
2. Sarcoidosis affects the nervous system in 5% of cases with asymptomatic involvement in another 1 in 5 patients. Over half of patients diagnosed with neurosarcoidosis do not carry a prior diagnosis of sarcoid.
3. Clinical presentation can vary and central nervous system (CNS) presentations can be classified as (1) cranial neuropathies, (2) meningeal, (3) brain parenchymal, and (4) spinal cord. In addition, peripheral neuropathy and myopathy are also seen.
4. Biopsy is key for diagnosis and determines the level of certainty. Thus "definite neurosarcoidosis" requires nervous system biopsy and "probable neurosarcoidosis" requires biopsy from extraneural tissue. Without biopsy, "possible neurosarcoidosis" can be diagnosed if the clinical picture is compatible and other causes are ruled out.
5. Glucocorticoids are the cornerstone of treatment, especially in the acute stage, while "steroid-sparing agents" such as methotrexate, mycophenolate mofetil, or azathioprine are used for prolonged therapy to minimize steroid toxicity. Infliximab appears to be efficacious when these agents are inadequate.

J. E. Eaton
Department of Neurology, Vanderbilt University Medical Center, Neurology Clinic, Nashville, TN, USA

S. Pawate (✉)
Department of Neurology, Vanderbilt University Medical Center, Nashville, TN, USA
e-mail: siddharama.pawate@vumc.org

## Introduction

Sarcoidosis is a multi-system disorder characterized by the formation of non-caseating granulomas in affected tissues. Granulomas are thought to represent the immune response to unidentified antigens, designed to confine the antigen and protect the surrounding tissue [1, 2]. However, this process damages the surrounding tissue; this is especially of consequence when sarcoid granulomas occur in critical organs such as the central nervous system, heart, and the eye. Neurological involvement due to sarcoidosis is termed neurosarcoidosis. Winkler in 1905 first identified neu-



A. L. Piquet, E. Alvarez (eds.), *Neuroimmunology*, https://doi.org/10.1007/978-3-030-61883-4_23

rosarcoidosis by observing non-caseating granulomas in peripheral nerves [3]. Over the following decades it was increasingly recognized that sarcoid could affect any part of the central and peripheral nervous system. Colover in 1948 compiled 118 cases of sarcoidosis with neurologic involvement. His review was the first to compile the different presentations of neurosarcoidosis and described the high tendency for cranial nerve involvement [4].

The diagnosis of neurosarcoidosis has remained difficult. Clinical presentation is varied. Imaging and cerebrospinal fluid (CSF) tests are useful to demonstrate inflammation but are not specific. Biomarkers to aid in diagnosis and disease monitoring have remained elusive. The diagnosis of sarcoidosis rests on the demonstration of granulomas in affected organs by biopsy, but this is not always feasible in neurosarcoidosis, especially in "isolated neurosarcoidosis," which is seen in 10–17% of neurosarcoid cases, where the disease is confined to the nervous system without involvement of other organs. Immunosuppression is the cornerstone of treatment. Patients are typically treated with corticosteroids. "Steroid-sparing agents" such as methotrexate, mycophenolate mofetil, and azathioprine are utilized for longer-term treatment. Newer treatment options, especially anti-tumor necrosis factor (TNF) agents, appear to provide better outcomes in patients refractory to other agents. We will discuss the advances in our understanding of neurosarcoidosis in this chapter.

## Epidemiology

Sarcoidosis is a ubiquitous disease, affecting all ethnicities and genders to varying degrees. There is significant geographical variation. Spain, for example, has a reported incidence as low as 0.42 per 100,000 while the Nordic countries range from 7 to 24 per 100,000 [5]. In general women are affected more than men. In the United States, young African American women are most frequently affected [6]. While epidemiologic information is incomplete, more recent data suggests 2 peak incidences of sarcoidosis: ages 25–29 and 65–69. A study using insurance claims and OptumInsight between 2009 and 2013 included ~15% of US residents and found half of initial diagnosis of sarcoidosis occurring in patients older than 55 [7].

The incidence of neurologic manifestations in sarcoidosis varies. Reports of neurologic symptoms in patients range from 5% to 20% [8, 9]. Notably neurologic manifestations can be the presenting symptom in a significant portion of patients. Fritz et al. compiled all the case series from 1980 to 2016 and found that among patients that were diagnosed with neurosarcoidosis, neurological symptoms were the initial presentation in 69% [10].

## Pathophysiology

Granuloma formation in sarcoidosis is presumably incited by an antigen that the immune system does not completely clear. The majority of patients develop pulmonary disease. As a result, most of the research in sarcoid granuloma formation has been focused on the lungs [11, 12].

Sarcoid granulomas are composed of epithelioid macrophages, activated monocytes, T and B lymphocytes, multinucleated giant cells, and fibroblasts. They are organized into two regions: the core and the crust. The core is a collection of tight clusters of macrophages, giant cells, and epithelioid cells. The crust is a collection of lymphocytes, predominantly T cells [11, 12]. In early granuloma formation, the aggregating macrophages are converted into epithelioid cells. The epithelioid cells then organize into a cluster that makes up an immature granuloma. Activated T cells will accumulate at sites of inflammation. This leads to increased activation and expression of cytokines and chemokines, especially TNF-alpha produced by T cells and macrophages, that lead to the formation of mature granulomas [13]. Fibroblast proliferation with collagen deposition leads to fibrosis [14].

## Clinical Features

There is a wide range of clinical presentations in neurosarcoidosis. Our understanding of clinical manifestations of neurosarcoidosis is derived mostly from retrospective studies performed at single tertiary care centers. Clinical manifestations depend on the neuroanatomical substrate infiltrated by granulomas. These manifestations can be classified as (1) cranial neuropathies, (2) meningeal, (3) brain parenchymal, (4) pituitary-hypothalamic, (5) spinal cord, (6) peripheral neuropathy, and (7) myopathy. Less commonly patients can manifest with cerebrovascular complications such as ischemic strokes or hemorrhages. The recent compilation by Fritz et al. [10] found the frequency of the presentations of patients as cranial neuropathy (53%), headache (32%), optic neuritis (21%), myelopathy (18%), peripheral nerve (17%), seizure (12%), meningitis (10%), and pituitary-hypothalamic dysfunction (8%). Other recent studies have reflected similar patterns of disease manifestations [15–18].

Manifestations of neurosarcoidosis in pediatric population are less well studied. Seizures are the most frequently encountered presenting symptom in children [19, 20]. As in adults, patients can also develop encephalopathy, transverse myelitis, and cranial neuropathies.

## Cranial Neuropathy

Cranial neuropathies are the most common finding in neurosarcoidosis. They can develop from direct nerve involvement, leptomeningeal encasement of nerves or nerve roots, or granuloma formation in the brainstem. Cranial nerve involvement may also be indirect as in CN VI palsy due to increased intracranial pressure from sarcoid meningitis.

The facial nerve is the most commonly affected cranial nerve, followed by optic nerve [10]. Facial nerve involvement may be unilateral, but can be bilateral in a third of those involved, either concurrently or sequentially [21]. Optic nerve involvement is one of the more frequently encountered findings as well and presents clinically with blurring of vision and visual field defects [22–24]. Vision loss can be rarely due to extrinsic compression of the optic nerve from granulomatous inflammation [23]. Sensorineural hearing loss and/or vertigo due to vestibulocochlear nerve involvement can occur and may be intermittent; bilateral sensorineural hearing loss should prompt an evaluation for neurosarcoidosis [25]. CN V involvement may manifest as facial numbness or, less commonly, trigeminal neuralgia. Other cranial nerves are involved less often, with CN XI and XII being the least common.

The differential diagnosis for individual cranial neuropathies is extensive [26, 27]. Neurosarcoidosis rises higher in the differential in patients with multiple cranial neuropathies or cranial neuropathies as part of more widespread symptoms such as seizures or myelopathy. Leptomeningeal disease, either infectious or carcinomatous, can present with multiple cranial neuropathies. Other infections such as Lyme disease can be considered with multiple cranial neuropathies. Individual nerves can have unique differential diagnosis considerations. Optic neuritis, for example, has an extensive differential diagnosis [28] including demyelinating disease (e.g., multiple sclerosis, neuromyelitis optica), autoimmune disorders (e.g., Sjögren's syndrome, temporal arteritis), neoplasm (lymphoma, glioma, metastatic disease), viral infections (e.g., herpes simplex virus, varicella zoster virus, cytomegalovirus), and bacterial (syphilis, tuberculosis, *Bartonella*), among many others. The differential for individual cranial neuropathies is beyond the scope of this chapter. Inflammatory ophthalmologic disease is discussed further in Chap. 28.

Outcomes in sarcoid cranial neuropathies tend to be favorable, with complete recovery occurring in the majority of patients. Optic neuropathy has been considered to have poorer outcomes, but it should be noted that these reports were published before anti-TNF therapies began to be widely used to treat neurosarcoidosis [23, 29].

## Meningeal Disease

The most common meningeal manifestation of neurosarcoidosis is “aseptic meningitis,” presenting with headaches, fever, and neck rigidity. Neurosarcoid meningitis usually has a benign course and a good outcome, but a subset of patients can develop chronic meningitis [21, 30]. Brain magnetic resonance imaging (MRI) usually shows leptomeningeal enhancement, especially in the posterior fossa. Headaches and hydrocephalus can result from diffuse meningeal involvement. Other manifestations include single or multiple cranial neuropathies and ataxia. Cerebrospinal fluid typically shows pleocytosis that is lymphocytic and monocytic, elevated protein and, in more severe cases, low CSF glucose (less than half the simultaneous serum glucose value). Contrasted MRI studies are necessary to clearly identify meningeal involvement. Patients with continued hydrocephalus despite appropriate immunomodulatory therapy may require ventriculoperitoneal shunting.

Involvement of the dura, or pachymeningitis, is less commonly seen and usually presents with headaches and cranial neuropathies [31].

Chronic meningitis can be secondary to a variety of causes [32]. Infections, especially mycobacterial and fungal, are considerations. Patients on chronic immunosuppression for sarcoidosis can develop such infectious meningitides. Other less common considerations would include neuro-Behçet's and immunoglobulin 4 (IgG4)-related disease, the latter being a common cause of pachymeningitis [31]. Leptomeningeal carcinomatosis is often a primary concern in patients presenting in this manner. Meningeal involvement is often feasible to biopsy and should be performed in patients with an unclear diagnosis.

## Parenchymal

Granulomas can develop in any part of the brain, including white matter, cortex, deep gray matter, pituitary, and hypothalamus and cause a wide range of symptoms [23]. Headaches and seizures are a frequent presentation. Focal deficits such as hemiplegia, ataxia, apraxia, and visual field may occur. Parenchymal involvement in the brainstem may manifest as cranial neuropathies. Mass lesions in the ventricular system can lead to obstructive hydrocephalus.

Imaging modalities will highlight these lesions well, but there are not pathognomonic findings to clearly differentiate sarcoid lesions from other lesions. There are typically T2/fluid-attenuated inversion recovery (FLAIR) signal changes and varying degrees of contrast enhancement. The differential is broad and often these granulomas are not distinguishable from neoplasms without tissue diagnosis. Other diagnostic considerations would include abscess, progressive multifocal leukoencephalopathy, tumefactive demyelinating lesions, and other space-occupying lesions.

Pituitary-hypothalamic infiltration by sarcoid granulomas can manifest with panhypopituitarism, diabetes insipidus, or amenorrhea-galactorrhea, among other symptoms of pituitary dysfunction [33, 34].

## Cerebrovascular Manifestations

There is a small but growing body of literature regarding vascular events attributable to neurosarcoidosis. Small arterial and venous involvement has been observed in pathology studies. Interestingly the small arterial perforators are the most frequently affected, with large vessel involvement much less frequent [35]. A recent review by Bathla et al. collected the English language cases of cerebrovascular manifestations of sarcoid until 2017. They found 13 cases of ischemic stroke, 16 cases of hemorrhagic stroke, and 5 venous sinus thromboses [36].

These patients will generally develop small unusual infarcts or hemorrhages. Vascular manifestations of sarcoid are unusual and there are not many clues for diagnosis. Primary CNS angiitis, infectious vasculitis, amyloid angiopathy, and other systemic small- or medium-vessel vasculitides should be considered in the differential as well.

## Myelopathy

Spinal cord involvement occurs in 12–24% of patients and can be particularly devastating, given the clinically eloquent role of spinal cord in neurological functioning [37, 38]. Transverse myelitis and longitudinally extensive transverse myelitis can develop. Meningeal disease can also lead to compressive myelopathies. Patients can have a range of symptoms such as mixed motor and sensory findings and possible bowel or bladder dysfunction, and some may have back pain.

Longitudinally extensive transverse myelitis is defined as spinal cord involvement spanning three or more vertebral segments. Inflammatory disorders are a common etiology with neurosarcoidosis being an important consideration. Sohn et al. collected 29 cases of spinal cord neurosarcoidosis [38]. Most patients had intramedullary disease (22/27) and about half had meningeal involvement (13/27). Longitudinally extensive transverse myelitis was seen in 17/22 of patients and the longest lesion was 9 vertebral levels [38]. A similar study was done by Durel where they had 20 patients with spine involvement. Longitudinally extensive lesions were present in 75% of patients. The cervical and thoracic spine was most frequently involved. Three patients had involvement of the entire cord [39]. The differential diagnosis of longitudinally extensive transverse myelitis includes neuromyelitis optica spectrum disorders (NMOSD) (see Chaps. 15 and 27), Behçet's disease, and acute disseminated encephalomyelitis (ADEM) and other demyelinating syndromes [40].

## Peripheral Neuropathy

Neurosarcoid large fiber neuropathy usually manifests as (1) axonal distal sensorimotor polyneuropathy with stocking-glove sensory symptoms manifesting as pain and paresthesias or (2) an asymmetric polyradiculoneuropathy in which proximal and distal nerve segments are equally affected in a non-length-dependent distribution. Less common presentations include mononeuritis multiplex, multifocal motor neuropathy with conduction block, mononeuropathy, and plexopathy. Electromyography/nerve conduction study (EMG/NCS) usually shows an axonal loss; demyelination is less common [41–44].

Small fiber neuropathy, with predominant complaint of neuropathic pain, numbness, and allodynia, is often seen in patients with sarcoidosis. It may be non-length dependent. Autonomic small fibers may be involved, resulting in dysautonomia and abnormal tilt-table testing [45–49]. Small fiber neuropathy has been proposed to represent a "para-neurosarcoidosis," reflecting the fact that it is not a direct consequence of granulomas in small fibers but a distant effect, presumed to be caused by inflammatory cytokines [50].

## Myopathy

Sarcoid myopathy is usually asymptomatic or may present with palpable nodules in muscles that are painless. Strength is not affected and serum creatine kinase (CK) is usually normal in these cases. Myopathy may also present as a polymyositis, with weakness affecting proximal muscles and elevated serum CK [51–53].

## Diagnosis and Workup

Neurosarcoidosis patients may be conceptualized in three groups:

1. Patients with known, active systemic sarcoidosis presenting with neurological symptoms. In these patients neurosarcoidosis is usually high in the differential, and if the initial workup points to neurosarcoidosis, a therapeutic trial with high-dose steroids may be justified.
2. Patients with a history of systemic sarcoidosis that is not active, presenting with neurological problems. Neurosarcoidosis is usually high in the differential for these patients as well.
3. Patients with no prior history of systemic sarcoidosis. These are the patients in whom there may be a delay and diagnosis requires a high index of suspicion [54].

A detailed history, physical exam, and neurologic exam are vital to correct diagnosis. Sarcoidosis is often multi-systemic. A thorough exam is necessary to guide appropriate diagnostic testing.

Imaging provides a noninvasive way to look for structural evidence of granulomatous involvement. Typically, patients will require brain and/or spine imaging. Contrasted MRI scans (Fig. 23.1) provide a high resolution of detail and can highlight sarcoid lesions [55]. Fels and colleagues looked at a group of patients with MRI imaging and neurosarcoidosis to try to characterize typical patterns. They found that parenchymal lesions were T1 isointense and T2 hyperintense, more clearly seen on FLAIR sequences, and exhibited multifocal enhancement. Leptomeninges may appear normal on T1 or T2 sequences, but a contrasted T1 study usually shows linear or nodular meningeal enhancement [56]. Shah et al. also

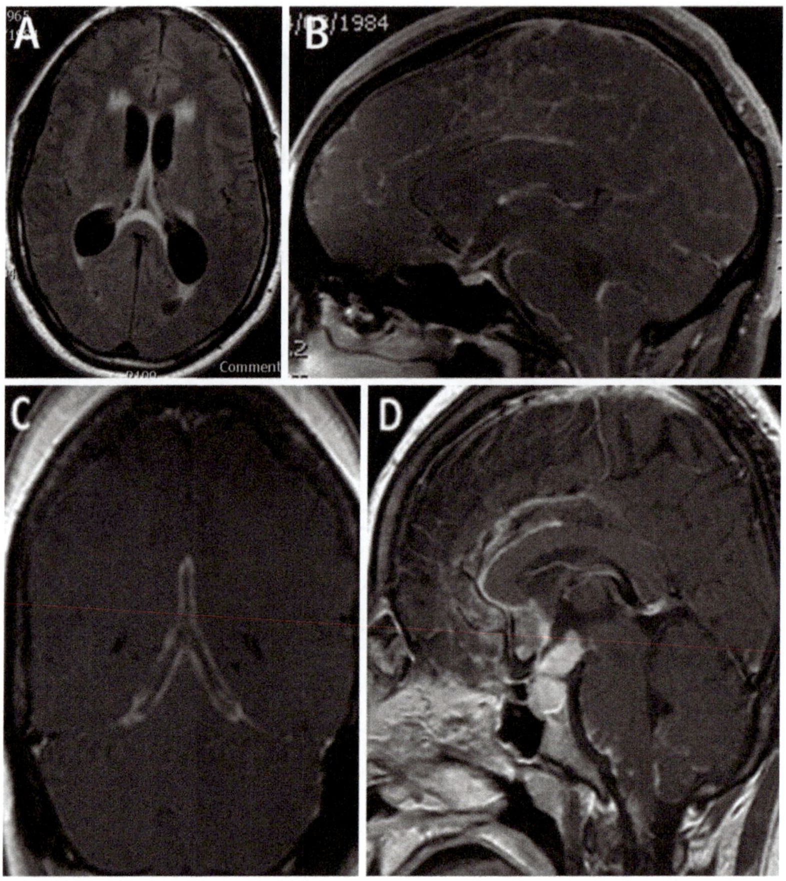


**Fig. 23.1** Imaging features suggestive of neurosarcoidosis. This figure shows brain and spinal cord imaging in patients with neurological problems in the context of biopsy-proven systemic sarcoidosis. Panels (**a**) and (**b**): Leptomeningeal involvement leading to hydrocephalus. Panel (**a**) shows the axial T2-FLAIR image. There is enlargement of the lateral ventricles with periventricular edema. Panel (**b**) shows the postcontrast T1 image showing the enlargement of 3rd and 4th ventricles as well, and extensive nodular leptomeningeal enhancement. Note the postcontrast enhancement of the meninges along the anterior surface of the brainstem extending into the cervical spinal cord. Panel (**c**): Dural enhancement due to sarcoid pachymeningitis. Panel (**d**): Pituitary-hypothalamic involvement in a patient presenting with panhypopituitarism. Note also the leptomeningeal involvement in the frontal lobe sulci, as well as along the anterior surface of the brainstem. Panels (**e**) and (**f**): A contrast-enhancing lesion in the left temporal lobe (**e**), in a patient with seizures that was FDG-avid on PET scan (**f**)

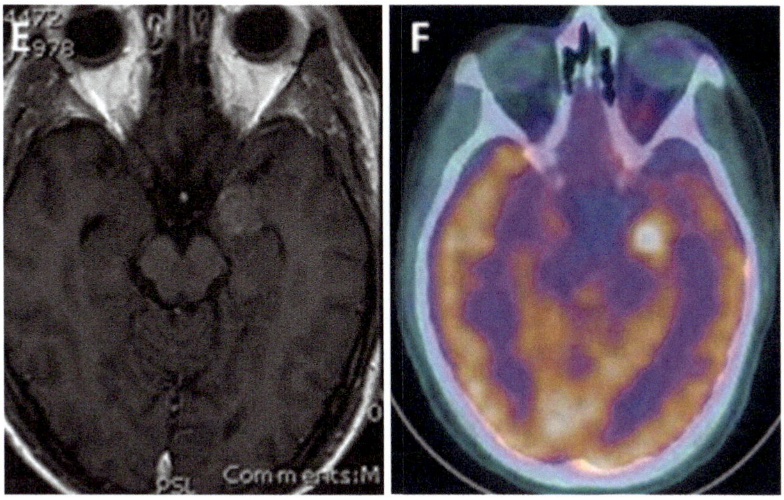


**Fig. 23.1** (continued)

demonstrated that spine lesions are typically T2 hyperintense as well [57]. For those with contra-indications to MRI, computed tomography (CT) is an option but has lower sensitivity and specificity. Iodinated contrast CT scans can sometimes highlight more lesions and meningeal involvement [56].

Systemic imaging is helpful to establish extra-neural sarcoidosis and provide possible targets for biopsy. Chest X-ray was historically the key for diagnosis of pulmonary sarcoidosis but chest CT is more sensitive [58]. More extensive body imaging may be useful, such as positron emission tomography (PET). This can help identify lesions amenable for biopsy. Gallium-67 scintigraphy was utilized for decades for evaluation of possible sarcoidosis as well as lymphomas and occult infections. PET imaging has largely supplanted gallium, but in facilities where PET imaging is not available, gallium scintigraphy may be an option [59].

Serologic and CSF testing is often utilized. CSF typically shows pleocytosis and increased protein as nonspecific markers of inflammation and, less commonly, decreased glucose [50]. Serum and CSF angiotensin-converting enzyme (ACE) levels can be elevated in sarcoidosis, but lack sensitivity and specificity. A retrospective study found a sensitivity and specificity of CSF-ACE of 66.7% and 67.3%, respectively [60]. Most authorities do not recommend relying on these levels for neurosarcoidosis diagnosis [50]. Soluble interleukin 2 receptor (sIL-2R) is found to be elevated in situations with CD4+ T cell activation, as in granulomatous disease, but does not have clear diagnostic utility at this time [50].

Ultimately, histopathology is necessary for confident diagnosis. This was reflected in the recently issued consensus definition and diagnostic criteria issued by the Neurosarcoidosis Consortium Consensus Group. They proposed criteria to stratify the diagnosis into definite, probable, and possible neurosarcoidosis (Table 23.1) [50].

It is also important to evaluate for other conditions that can cause a wide range of symptoms. Testing varies between patients depending on the clinical picture. Conditions that can present similarly to sarcoidosis would include CNS neoplasms such as gliomas or lymphoma, demyelinating disease, and atypical infections such as fungi, tuberculosis, or syphilis.

## Treatment Considerations

Immunosuppression is the basis of treatment for sarcoidosis. The treatment of neurosarcoidosis is challenging for several reasons. There are no randomized clinical trials to guide therapy. The clinical course varies significantly and there are no tools available to predict it. There are significant side effects with immunosuppressive therapy

**Table 23.1** Criteria for the diagnosis of neurosarcoidosis as proposed by the Neurosarcoidosis Consortium Consensus Group

| | |
|---|---|
| Definite | Clinical presentation and diagnostic evaluation suggests neurosarcoidosis as defined by the clinical manifestations and MRI, CSF, and/or EMG/NCS findings typical of granulomatous inflammation of the nervous system and rigorous exclusion of other causes<br>Nervous system pathology is consistent with neurosarcoidosis<br>Extraneural sarcoidosis is evident<br>No extraneural sarcoidosis is evident (i.e., isolated CNS sarcoidosis) |
| Probable | Clinical presentation and diagnostic evaluation suggests neurosarcoidosis as defined by the clinical manifestations and MRI, CSF, and/or EMG/NCS findings typical of granulomatous inflammation of the nervous system and rigorous exclusion of other causes<br>There is pathologic confirmation of systemic granulomatous disease consistent with sarcoidosis |
| Possible | Clinical presentation and diagnostic evaluation suggests neurosarcoidosis as defined by the clinical manifestations and MRI, CSF, and/or EMG/NCS findings typical of granulomatous inflammation of the nervous system and rigorous exclusion of other causes<br>There is no pathologic confirmation of granulomatous disease |

Reprinted with permission from Stern et al. [50]

*MRI* magnetic resonance imaging, *CSF* cerebrospinal fluid, *EMG* electromyography, *NCS* nerve conduction study, *CNS* central nervous system

including opportunistic infections that become more likely with longer duration of treatment. The duration of treatment that is necessary is not clear and symptoms can recur after cessation of treatment. Controlled trials to compare treatment modalities for neurosarcoidosis are lacking. Due to these features, the treatment of neurosarcoidosis is based primarily on clinical experience and small clinical case series.

For treatment planning, Lower and Weiss classified neurosarcoidosis into [61]:

1. Mild (isolated CN VII or aseptic meningitis), with usually monophasic course. In these patients, a course of steroids may be adequate.
2. Moderate and severe disease is based on initial clinical presentation and initial response to treatment, with parenchymal, leptomeningeal, or spinal cord disease, which tend to have a chronic course and a need for prolonged immunosuppression.

Steroids have been used to treat neurosarcoidosis for decades and are typically first line. The clinical response to glucocorticoids is usually favorable. Stern and colleagues achieved remission in 79% on appropriate glucocorticoid therapy [62].

The addition or substitution of other immunosuppressants is warranted in patients with intolerance to steroids, moderate/severe clinical course, or progression of disease on steroids. The medications that have been used include methotrexate, mycophenolate mofetil, and azathioprine [23] and cyclophosphamide [63]. There is sparse data comparing the efficacy of these agents. A comparative study of methotrexate and mycophenolate mofetil had 32 and 14 patients, respectively. Patients on mycophenolate had higher rates of relapse compared to patients on methotrexate (78.6% compared to 46.8%). Adverse events were higher in the methotrexate group [64]. This suggests that methotrexate may be the more efficacious agent for patients who can tolerate it.

As noted in the pathophysiology section above, TNF alpha plays an important role in the formation and perpetuation of granulomas. Infliximab, a monoclonal antibody against TNF, has recently emerged as an option to treat sarcoidosis refractory to other agents. A double-blind placebo study was done with infliximab in steroid-refractory pulmonary sarcoidosis. They found improvement in radiographic evidence of disease burden and vital capacity over a 24-week period but did not find improvement in exercise tolerance or dyspnea scores [65]. Two recent retrospective studies have looked at infliximab in neurosarcoidosis. A multicenter study from the United States looked at infliximab use in 66 patients. In a majority of patients, infliximab was combined with an oral agent (mycophenolate or methotrexate) with the aim of obtaining synergistic effects, as well as the reduction of formation of anti-infliximab antibodies that reduce the effectiveness of infliximab. More than

80% of patients had favorable response clinically and radiographically [66]. The other study from France had similar results [67]. In both studies, approximately half of patients relapsed after discontinuation of infliximab, indicating that infliximab suppresses disease activity but may not result in its eradication in the period that they were treated for. Another anti-TNF agent, adalimumab, has much more limited data in the form of isolated case reports [68, 69].

Rituximab is a monoclonal antibody directed against CD20+. It has been utilized for multiple autoimmune disease and hematologic malignancies. It has also been used in isolated cases successfully for neurosarcoidosis [70].

Mass lesions in the spine or brain in emergent situations have been treated using radiation [71]. Patients who have developed hydrocephalus could potentially require ventriculoperitoneal shunting.

There is no formalized treatment algorithm available. Patients may require high-dose steroids, such as intravenous (IV) methylprednisolone, for initial presentations or acute exacerbations. Mild cases such as isolated facial nerve palsy or aseptic meningitis may be treated with a prolonged oral prednisone taper. When longer-term therapy is anticipated, methotrexate is a reasonable next choice in most patients. Other steroid-sparing agents are worth consideration such as mycophenolate or azathioprine. Infliximab is an option in cases refractory to other agents [66, 67].

Table 23.2 summarizes the available therapeutic agents used in the treatment of neurosarcoidosis.

**Table 23.2** Medications for the treatment of neurosarcoidosis (see text for references)

| Medications | Typical dosing | Clinical considerations |
|---|---|---|
| Glucocorticoids | Acute presentations: 1000 mg IV methylprednisolone for 3–5 days<br>Chronic use: titration depends on the severity of symptoms. Usually 1 mg/kg/day at the beginning and titrated to the lowest possible dose | Relatively fast effect<br>Dose- and duration-related side effects: hyperglycemia/diabetes, bruising, hair thinning, cushingoid appearance |
| Methotrexate | 5–20 mg once weekly | Relatively well tolerated<br>Easy administration<br>GI and dermatologic side effects can be limiting<br>Bone marrow suppression<br>Give at least 1 mg of folic acid daily<br>Cannot be used in pregnancy |
| Azathioprine | 50–200 mg/day | Thiopurine-S-methyltransferase (TMPT) testing prior to initiation. If normal it does not need a slow titration<br>Bone marrow suppression, anorexia, and pancreatitis can develop |
| Mycophenolate mofetil | 1000–1500 mg twice a day | Often well tolerated<br>GI side effects, tremor, and bone marrow suppression can develop |
| Cyclophosphamide | 500–750 mg/m$^2$ (meter squared body surface area) if there is normal renal function for initial dose – subsequent doses typically every 2–4 weeks; depending on cell counts, it may need to be reduced if patients develop leukopenia | Effective for prompt immunosuppression for patients that fail steroids<br>Risk of infertility<br>Ideally would be discontinued and switched to a less toxic medication when disease stabilizes, typically 6 months |
| Infliximab | IV infusion 5 mg/kg of ideal body weight. Loading dose given at weeks 0, 2, and 6. Maintenance every 4–8 weeks after | Evidence shows it works in steroid-refractory cases<br>Infusion therapy<br>TB screening before<br>Can cause acute hepatitis |
| Adalimumab | 40 mg subcutaneous every other week | Limited data, but theoretically should be as effective as infliximab<br>TB screening |

*IV* intravenous, *GI* gastrointestinal, *TB* tuberculosis

## Conclusion

Neurosarcoidosis remains a difficult disease from identification and diagnosis to management. The etiology of the disease is not clear and diagnosis remains challenging. More consistent biomarkers are needed to improve diagnostic certainty, especially in cases where neural biopsy is not feasible. There are multiple treatment options available, but there have been limited head-to-head comparisons and no prospective studies. Further studies are needed to provide clearer directions in treatment decisions.

## References


1. Iannuzzi MC, Rybicki BA, Teirstein AS. Sarcoidosis. N Engl J Med. 2007;357(21):2153–65.
2. Valeyre D, Prasse A, Nunes H, Uzunhan Y, Brillet PY, Muller-Quernheim J. Sarcoidosis. Lancet. 2014;383(9923):1155–67.
3. Winkler M. Beitrag zur Frage der Sarkoidose (Boeck). Arch Dermatol Syphilis. 1905;77:3–24.
4. Colover J. Sarcoidosis with involvement of the nervous system. Brain. 1948;71(Pt. 4):451–75.
5. Lopez-Campos JL, Rodriguez-Becerra E, Neumosur Task Group of the Registry of Interstitial Lung Diseases (RENIA). Incidence of interstitial lung diseases in the south of Spain 1998-2000: the RENIA study. Eur J Epidemiol. 2004;19(2):155–61.
6. Rybicki BA, Major M, Popovich J Jr, Maliarik MJ, Iannuzzi MC. Racial differences in sarcoidosis incidence: a 5-year study in a health maintenance organization. Am J Epidemiol. 1997;145(3):234–41.
7. Baughman RP, Field S, Costabel U, Crystal RG, Culver DA, Drent M, et al. Sarcoidosis in America. Analysis based on health care use. Ann Am Thorac Soc. 2016;13(8):1244–52.
8. Nozaki K, Judson MA. Neurosarcoidosis. Curr Treat Options Neurol. 2013;15(4):492–504.
9. Segal BM. Neurosarcoidosis: diagnostic approaches and therapeutic strategies. Curr Opin Neurol. 2013;26(3):307–13.
10. Fritz D, van de Beek D, Brouwer MC. Clinical features, treatment and outcome in neurosarcoidosis: systematic review and meta-analysis. BMC Neurol. 2016;16(1):220.
11. Chen ES, Moller DR. Etiologies of sarcoidosis. Clin Rev Allergy Immunol. 2015;49(1):6–18.
12. Sakthivel P, Bruder D. Mechanism of granuloma formation in sarcoidosis. Curr Opin Hematol. 2017;24(1):59–65.
13. Iannuzzi MC, Fontana JR. Sarcoidosis: clinical presentation, immunopathogenesis, and therapeutics. JAMA. 2011;305(4):391–9.
14. Patterson KC, Chen ES. The pathogenesis of pulmonary sarcoidosis and implications for treatment. Chest. 2018;153(6):1432–42.
15. Cacao G, Branco A, Meireles M, Alves JE, Mateus A, Silva AM, et al. Neurosarcoidosis according to Zajicek and Scolding criteria: 15 probable and definite cases, their treatment and outcomes. J Neurol Sci. 2017;379:84–8.
16. Kidd DP. Sarcoidosis of the central nervous system: clinical features, imaging, and CSF results. J Neurol. 2018;265(8):1906–15.
17. Leonhard SE, Fritz D, Eftimov F, van der Kooi AJ, van de Beek D, Brouwer MC. Neurosarcoidosis in a tertiary referral center: a cross-sectional cohort study. Medicine (Baltimore). 2016;95(14):e3277.
18. Ungprasert P, Crowson CS, Matteson EL. Characteristics and long-term outcome of neurosarcoidosis: a population-based study from 1976 to 2013. Neuroepidemiology. 2017;48(3–4):87–94.
19. Baumann RJ, Robertson WC Jr. Neurosarcoid presents differently in children than in adults. Pediatrics. 2003;112(6 Pt 1):e480–6.
20. Rao R, Dimitriades VR, Weimer M, Sandlin C. Neurosarcoidosis in pediatric patients: a case report and review of isolated and systemic neurosarcoidosis. Pediatr Neurol. 2016;63:45–52.
21. Stern BJ, Krumholz A, Johns C, Scott P, Nissim J. Sarcoidosis and its neurological manifestations. Arch Neurol. 1985;42(9):909–17.
22. Kidd DP, Burton BJ, Graham EM, Plant GT. Optic neuropathy associated with systemic sarcoidosis. Neurol Neuroimmunol Neuroinflamm. 2016;3(5):e270.
23. Pawate S, Moses H, Sriram S. Presentations and outcomes of neurosarcoidosis: a study of 54 cases. QJM. 2009;102(7):449–60.
24. Terushkin V, Stern BJ, Judson MA, Hagiwara M, Pramanik B, Sanchez M, et al. Neurosarcoidosis: presentations and management. Neurologist. 2010;16(1):2–15.
25. Neurosarcoidosis AAJ. Continuum. Neurologic manifestations of systemic disease. 14th ed. Minneapolis: American Academy of Neurology; 2013. p. 181–96.
26. Carlson ML, White JR Jr, Espahbodi M, Haynes DS, Driscoll CL, Aksamit AJ, et al. Cranial base manifestations of neurosarcoidosis: a review of 305 patients. Otol Neurotol. 2015;36(1):156–66.
27. Fritz D, Voortman M, van de Beek D, Drent M, Brouwer MC. Many faces of neurosarcoidosis: from chronic meningitis to myelopathy. Curr Opin Pulm Med. 2017;23(5):439–46.
28. Toosy AT, Mason DF, Miller DH. Optic neuritis. Lancet Neurol. 2014;13(1):83–99.
29. Koczman JJ, Rouleau J, Gaunt M, Kardon RH, Wall M, Lee AG. Neuro-ophthalmic sarcoidosis: the University of Iowa experience. Semin Ophthalmol. 2008;23(3):157–68.
30. Gullapalli D, Phillips LH. Neurosarcoidosis. Curr Neurol Neurosci Rep. 2004;4(6):441–7.
31. Hahn LD, Fulbright R, Baehring JM. Hypertrophic pachymeningitis. J Neurol Sci. 2016;367:278–83.

32. Thakur KT, Wilson MR. Chronic meningitis. Continuum (Minneap Minn). 2018;24(5, Neuroinfectious Disease):1298–326.
33. Murialdo G, Tamagno G. Endocrine aspects of neurosarcoidosis. J Endocrinol Investig. 2002;25(7):650–62.
34. Porter N, Beynon HL, Randeva HS. Endocrine and reproductive manifestations of sarcoidosis. QJM. 2003;96(8):553–61.
35. Brown MM, Thompson AJ, Wedzicha JA, Swash M. Sarcoidosis presenting with stroke. Stroke. 1989;20(3):400–5.
36. Bathla G, Watal P, Gupta S, Nagpal P, Mohan S, Moritani T. Cerebrovascular manifestations of neurosarcoidosis: an underrecognized aspect of the imaging spectrum. AJNR Am J Neuroradiol. 2018;39(7):1194–200.
37. Bradley DA, Lower EE, Baughman RP. Diagnosis and management of spinal cord sarcoidosis. Sarcoidosis Vasc Diffuse Lung Dis. 2006;23(1):58–65.
38. Sohn M, Culver DA, Judson MA, Scott TF, Tavee J, Nozaki K. Spinal cord neurosarcoidosis. Am J Med Sci. 2014;347(3):195–8.
39. Durel CA, Marignier R, Maucort-Boulch D, Iwaz J, Berthoux E, Ruivard M, et al. Clinical features and prognostic factors of spinal cord sarcoidosis: a multicenter observational study of 20 BIOPSY-PROVEN patients. J Neurol. 2016;263(5):981–90.
40. Tobin WO, Weinshenker BG, Lucchinetti CF. Longitudinally extensive transverse myelitis. Curr Opin Neurol. 2014;27(3):279–89.
41. Burns TM, Dyck PJ, Aksamit AJ, Dyck PJ. The natural history and long-term outcome of 57 limb sarcoidosis neuropathy cases. J Neurol Sci. 2006;244(1–2):77–87.
42. Said G, Lacroix C, Plante-Bordeneuve V, Le Page L, Pico F, Presles O, et al. Nerve granulomas and vasculitis in sarcoid peripheral neuropathy: a clinicopathological study of 11 patients. Brain. 2002;125(Pt 2):264–75.
43. Vital A, Lagueny A, Ferrer X, Louiset P, Canron MH, Vital C. Sarcoid neuropathy: clinico-pathological study of 4 new cases and review of the literature. Clin Neuropathol. 2008;27(2):96–105.
44. Vital A, Vital C. Clinical neuropathology practice guide 3-2014: combined nerve and muscle biopsy in the diagnostic workup of neuropathy – the Bordeaux experience. Clin Neuropathol. 2014;33(3):172–8.
45. Hoitsma E, Faber CG, van Santen-Hoeufft M, De Vries J, Reulen JP, Drent M. Improvement of small fiber neuropathy in a sarcoidosis patient after treatment with infliximab. Sarcoidosis Vasc Diffuse Lung Dis. 2006;23(1):73–7.
46. Hoitsma E, Marziniak M, Faber CG, Reulen JP, Sommer C, De Baets M, et al. Small fibre neuropathy in sarcoidosis. Lancet. 2002;359(9323):2085–6.
47. Hoitsma E, Reulen JP, de Baets M, Drent M, Spaans F, Faber CG. Small fiber neuropathy: a common and important clinical disorder. J Neurol Sci. 2004;227(1):119–30.
48. Tavee J, Culver D. Sarcoidosis and small-fiber neuropathy. Curr Pain Headache Rep. 2011;15(3):201–6.
49. Tavee JO, Karwa K, Ahmed Z, Thompson N, Parambil J, Culver DA. Sarcoidosis-associated small fiber neuropathy in a large cohort: clinical aspects and response to IVIG and anti-TNF alpha treatment. Respir Med. 2017;126:135.
50. Stern BJ, Royal W 3rd, Gelfand JM, Clifford DB, Tavee J, Pawate S, et al. Definition and consensus diagnostic criteria for neurosarcoidosis: from the neurosarcoidosis consortium consensus group. JAMA Neurol. 2018;75(12):1546–53.
51. Androdias G, Maillet D, Marignier R, Pinede L, Confavreux C, Broussolle C, et al. Mycophenolate mofetil may be effective in CNS sarcoidosis but not in sarcoid myopathy. Neurology. 2011;76(13):1168–72.
52. Berger C, Sommer C, Meinck HM. Isolated sarcoid myopathy. Muscle Nerve. 2002;26(4):553–6.
53. Fayad F, Liote F, Berenbaum F, Orcel P, Bardin T. Muscle involvement in sarcoidosis: a retrospective and followup studies. J Rheumatol. 2006;33(1):98–103.
54. Hoitsma E, Faber CG, Drent M, Sharma OP. Neurosarcoidosis: a clinical dilemma. Lancet Neurol. 2004;3(7):397–407.
55. Stjepanovic MI, Vucinic VM, Jovanovic D, Mijajlovic M, Trifunovic VS, Ivanov JV. Radiological presentation of neurosarcoidosis. Med Pregl. 2014;67(1–2):24–7.
56. Fels C, Riegel A, Javaheripour-Otto K, Obenauer S. Neurosarcoidosis: findings in MRI. Clin Imaging. 2004;28(3):166–9.
57. Shah R, Roberson GH, Cure JK. Correlation of MR imaging findings and clinical manifestations in neurosarcoidosis. AJNR Am J Neuroradiol. 2009;30(5):953–61.
58. Greco FG, Spagnolo P, Muri M, Paladini I, Chizzolini F, Piciucchi S, et al. The value of chest radiograph and computed tomography in pulmonary sarcoidosis. Sarcoidosis Vasc Diffuse Lung Dis. 2014;31(2):108–16.
59. Treglia G, Taralli S, Giordano A. Emerging role of whole-body 18F-fluorodeoxyglucose positron emission tomography as a marker of disease activity in patients with sarcoidosis: a systematic review. Sarcoidosis Vasc Diffuse Lung Dis. 2011;28(2):87–94.
60. Bridel C, Courvoisier DS, Vuilleumier N, Lalive PH. Cerebrospinal fluid angiotensin-converting enzyme for diagnosis of neurosarcoidosis. J Neuroimmunol. 2015;285:1–3.
61. Lower EE, Weiss KL. Neurosarcoidosis. Clin Chest Med. 2008;29(3):475–92, ix.
62. Stern BJ, Aksamit A, Clifford D, Scott TF, Neurosarcoidosis Study G. Neurologic presentations of sarcoidosis. Neurol Clin. 2010;28(1):185–98.
63. Doty JD, Mazur JE, Judson MA. Treatment of corticosteroid-resistant neurosarcoidosis with a

short-course cyclophosphamide regimen. Chest. 2003;124(5):2023–6.
64. Bitoun S, Bouvry D, Borie R, Mahevas M, Sacre K, Haroche J, et al. Treatment of neurosarcoidosis: a comparative study of methotrexate and mycophenolate mofetil. Neurology. 2016;87(24):2517–21.
65. Rossman MD, Newman LS, Baughman RP, Teirstein A, Weinberger SE, Miller W, et al. A double-blinded, randomized, placebo-controlled trial of infliximab in subjects with active pulmonary sarcoidosis. Sarcoidosis Vasc Diffuse Lung Dis. 2006;23(3):201–8.
66. Gelfand JM, Bradshaw MJ, Stern BJ, Clifford DB, Wang Y, Cho TA, et al. Infliximab for the treatment of CNS sarcoidosis: a multi-institutional series. Neurology. 2017;89(20):2092–100.
67. Cohen Aubart F, Bouvry D, Galanaud D, Dehais C, Mathey G, Psimaras D, et al. Long-term outcomes of refractory neurosarcoidosis treated with infliximab. J Neurol. 2017;264(5):891–7.
68. Marnane M, Lynch T, Scott J, Stack J, Kelly PJ. Steroid-unresponsive neurosarcoidosis successfully treated with adalimumab. J Neurol. 2009;256(1):139–40.
69. Metyas S, Tawadrous M, Yeter KC, Arkfeld DG. Neurosarcoidosis mimicking multiple sclerosis successfully treated with methotrexate and adalimumab. Int J Rheum Dis. 2014;17(2):214–6.
70. Bomprezzi R, Pati S, Chansakul C, Vollmer T. A case of neurosarcoidosis successfully treated with rituximab. Neurology. 2010;75(6):568–70.
71. Motta M, Alongi F, Bolognesi A, Cozzarini C, Di Muzio N, Fazio F. Remission of refractory neurosarcoidosis treated with brain radiotherapy: a case report and a literature review. Neurologist. 2008;14(2):120–4.

# Central Nervous System Vasculitis

# 24

Christine M. Gill, Amanda L. Piquet,
and Tracey A. Cho

**Key Points**

1. Vasculitis of the nervous system can occur in isolation, known as primary angiitis of the central nervous system (PACNS), or in the setting of systemic vasculitis.
2. Systemic vasculitides are a heterogeneous group of disorders often related to connective tissue disorders, underlying rheumatological disease, or infection.
3. Clinically, PACNS often presents with subacute, progressive symptoms of headache, cognitive impairment, and/or focal neurological deficits.
4. CNS vasculitis can have significant morbidity and mortality; therefore, a high level of clinical suspicion, prompt diagnosis, and treatment is essential.
5. Currently available ancillary tests lack sensitivity and specificity for PACNS, making diagnosis challenging.
6. There are no formal guidelines for the treatment of PACNS; however, chronic immunotherapy is often required with commonly used agents including azathioprine, methotrexate, leflunomide, mycophenolate mofetil, and rituximab.

## Introduction

Vasculitis is a term that describes a heterogenous group of related conditions in which there is inflammation involving the walls of blood vessels of variable size and distribution. Vasculitides are generally classified according to the Chapel Hill Consensus Conference definitions based on the size of vessels involved [1]. Vasculitides may further be classified by histopathologic findings or according to underlying etiology. Vasculitis of any vessel size may involve the peripheral and/or central nervous systems resulting in a myriad of neurologic symptoms. Nervous system vasculitis confined to either the central nervous system

C. M. Gill · T. A. Cho (✉)
Department of Neurology, University of Iowa Carver College of Medicine, University of Iowa Hospitals and Clinics, Iowa City, IA, USA
e-mail: tracey-cho@uiowa.edu

A. L. Piquet
Department of Neurology, Neuroimmunology, Neuroinfectious disease and Neurohospitalist Sections, Neuro-Infectious Diseases Group, University of Colorado School of Medicine, Aurora, CO, USA


A. L. Piquet, E. Alvarez (eds.), *Neuroimmunology*, https://doi.org/10.1007/978-3-030-61883-4_24

(CNS) or peripheral nervous system (PNS) is termed primary angiitis of the CNS or isolated PNS vasculitis, respectively. Secondary nervous system vasculitis describes CNS and/or PNS vasculitis occurring in the setting of systemic vasculitis, connective tissue disease, infection, or other miscellaneous conditions. The primary focus of this chapter will be CNS vasculitis.

CNS vasculitis is a diagnostically challenging condition owing to its variable, non-specific presentation and the limitations of currently available diagnostic studies. A high level of clinical suspicion is necessary to ensure prompt recognition of possible CNS vasculitis. Timely and thorough evaluation of suspected cases of CNS vasculitis is essential and is largely focused on ruling out mimics and determining the underlying etiology. Treatment of CNS vasculitis may vary slightly based on the underlying etiology, but generally involves long-term immunosuppression. Historically, CNS vasculitis was considered nearly universally fatal. In recent years, mortality from CNS vasculitis has significantly decreased due to increased recognition, advances in diagnostic studies, and more aggressive treatment. Despite these improvements, significant morbidity and mortality may occur if diagnosis and treatment are delayed. Physicians, particularly neurologists, must therefore be vigilant in recognizing cases of possible CNS vasculitis.

## Primary Angiitis of the Central Nervous System (PACNS)

### Epidemiology

Primary angiitis of the central nervous system (PACNS) is a rare, single-organ vasculitis affecting small and medium vessels of the brain, spinal cord, and/or meninges. Although the first clinical descriptions of cerebral vasculitis were published by Harbitz in 1922 [2], PACNS was not recognized as a separate nosologic entity until 1959 when Cravioto and Feign described several cases of non-infectious granulomatous angiitis confined to the CNS [3].

The true incidence of PACNS is unknown but has been estimated at 2.4 cases per 1,000,000 person-years based on a single case series of 101 patients [4]. Classically, PACNS is considered to have a male predilection, though some cohorts report nearly equal rates of male and female involvement. The median age of onset is approximately 50 years, but PACNS can occur at any age [4].

### Etiology and Pathogenesis

Much remains unknown about the cause of PACNS. Histopathologic studies implicating memory T cells in the pathogenesis of PACNS [5] suggest that PACNS may stem from an antigen-specific immune response in the walls of cerebral vessels [6]. The specific trigger or triggers of this response are unknown; however, infectious agents including varicella-zoster virus (VZV) [7], mycoplasma [8, 9], human immunodeficiency virus (HIV) [10, 11], and West Nile virus [12] have been speculated to be possible triggers. It is important to note that VZV may also cause a direct infectious vasculitis that is nosologically distinct from PACNS. This infectious/parainfectious vasculitis should not be confused with theories positing that VZV is a trigger for the development of PACNS.

Amyloid beta deposition has also been proposed as a potential trigger. This hypothesis is supported by transgenic mouse models of cerebral amyloid angiopathy in whom an inflammatory response to vascular amyloid ranging from minimal to severe granulomatous vasculitis has been observed [13]. Further, an association between cerebral amyloid angiopathy and PACNS has been recognized in humans, though amyloid beta deposition is not present in all cases of PACNS. PACNS associated with amyloid deposition in cerebral vessel walls, termed amyloid beta-related angiitis (ABRA), is now considered a distinct subtype of PACNS [14].

Several germline mutations have been associated with PACNS, and cases of PACNS in persons with defined monogenic immune dysregulation syndromes are increasingly reported, suggesting that genetic factors may play a role in the development of PACNS [15–19]. A small European study found a significantly higher frequency of the HLA-A*69 haplotype in those with biopsy-proven PACNS as compared to controls, further suggesting that genetic factors may affect susceptibility to developing PACNS [20]. Unfortunately the low incidence of disease limits gene association studies, and the genetic factors in PACNS remain poorly understood. The role of other potential etiologic factors is unclear and has not been adequately studied.

The pathogenesis of PACNS is also not fully understood, but it is generally accepted that it involves an inflammatory immune response resulting in vessel wall damage. Three distinct histopathologic patterns are seen: granulomatous, necrotizing, and lymphocytic. Of these, granulomatous is most commonly seen in adults. The damaged vessel ultimately becomes narrowed, occluded, and/or thrombosed resulting in hypoperfusion of the territory supplied by the affected vessel. Aneurysm, hemorrhage, and space-occupying lesions with mass effect may also occur and result in abnormal perfusion.

## Clinical Presentation

PACNS typically presents insidiously with slowly progressive symptoms, though it may present acutely. Symptoms are non-specific and vary based on the regions of the CNS involved. Subacute, progressive headache is the most commonly reported symptom, seen in approximately 60% of patients (range 43–100% in various cohorts) [21]. "Thunderclap headache" is virtually never seen and if reported should prompt evaluation for mimics of PACNS.

Cognitive impairment is the second most common manifestation of PACNS, reported in up to 50% of patients [4, 6]. Other neuropsychiatric manifestations include behavioral disturbance, personality change, and mood disturbance [4, 22].

Transient or persistent focal neurologic deficits are also common, often correlating with stroke on imaging. Ataxia, seizure, cranial neuropathies, and intracerebral hemorrhage are less common. Spinal cord involvement, predominately thoracic, with resultant myelopathy is implicated in approximately 5% of patients with PACNS, though isolated spinal cord involvement is exceedingly rare [23].

## Subtypes of PACNS

Several subtypes of PACNS have been described based on differences in presentation, neuroimaging findings, and prognosis. These subtypes have not been validated to represent distinct clinical or pathophysiological entities, but may provide a clinically useful framework when considering cases of PACNS [4]. Subtypes of PACNS include the following:

- *Angiography-positive PACNS* affects medium to large cerebral vessels, resulting in abnormal angiographic findings suggesting vasculitis. Focal neurologic deficits and seizure tend to be more common in this subtype.
- *Angiography-negative, biopsy-positive PACNS* affects only very small arteries or arterioles with diameters below the resolution of angiography. Cognitive dysfunction, aphasia, and other so-called "diffuse" neurologic deficits are common in this subtype. Magnetic resonance imaging (MRI) often shows enhancement of the meninges and/or parenchymal lesions. Response to treatment is generally favorable with good outcomes [24].
- *PACNS with prominent leptomeningeal enhancement* often presents acutely with cognitive dysfunction. Angiography is frequently negative. Prognosis is overall favorable, and most respond to treatment with normalization of MRI [25].
- *Amyloid-β(beta)-related angiitis (ABRA)* is a very rare form of CNS vasculitis occurring in a small percentage of patients with cerebral amyloid angiopathy (CAA). ABRA is thought to result from an abnormal immune response to amyloid-β(beta) depositions in vessel walls. The mean age of onset is 67 years, meaning that patients with ABRA tend to be younger than those with noninflammatory CAA (mean

77 years) but older than those with classic PACNS (mean 45–50 years). The most common presentation is altered mental status with or without hallucinations. Headache and focal neurologic deficits are also common. Cerebrospinal fluid (CSF) protein is elevated in the majority. MRI is universally abnormal with T2/fluid-attenuated inversion recovery (FLAIR) hyperintensities with minimal gadolinium enhancement (see Fig. 24.1). Most will also have multiple microhemorrhages in the cortico-subcortical junction, a feature not commonly seen in PACNS. Infrequently, mass-like lesions may be present and may be confused for malignancy. Angiography is frequently normal. Though typical cases may be diagnosed based on clinical and imaging features, biopsy remains the gold standard for diagnosis of ABRA. Histopathology reveals angiodestructive inflammation with abundant Aβ(beta) in affected vessels (though scant within brain parenchyma). Most patients responding favorably to immunosuppressive therapy with complete remission and normalization of MRI [14, 26–28].

- *Rapidly progressive PACNS* affects multiple, bilateral large cerebral vessels resulting in significant vascular abnormalities that are often seen on angiography. MRI shows multiple, bilateral infarcts that may correlate with focal neurologic deficits. As the name suggests, patients with this subtype often deteriorate quickly and have a poorer response to immunosuppressive treatment. Fatality is high in this subtype [29].
- *PACNS with a solitary tumor-like mass lesion* occurs in around 4% of cases. Presenting symptoms tend to correlate with the location and size of the lesion. Excision of the lesion may be curative in some; however, favorable

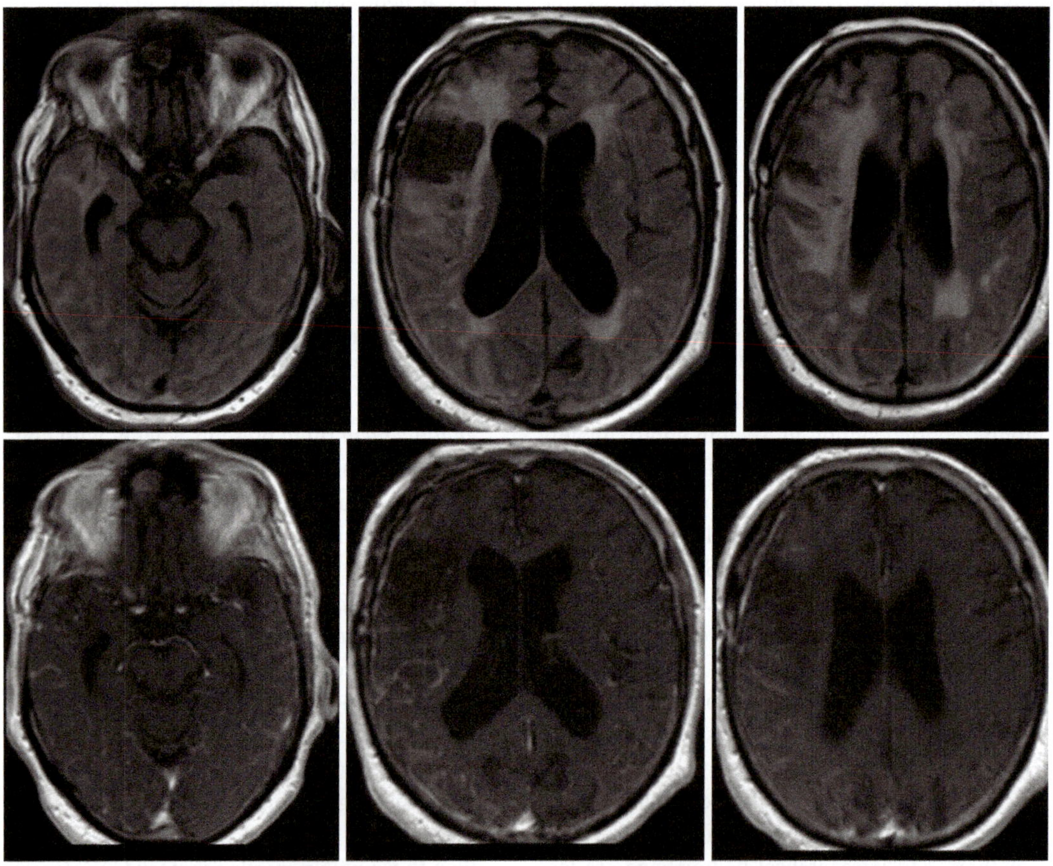

**Fig. 24.1** Magnetic resonance imaging (MRI) brain in amyloid-β(beta)-related angiitis (ABRA). Top row: Fluid-attenuated inversion recovery (FLAIR) images from patient with biopsy-proven ABRA. Bottom row: T1 post-contrasted images from the same patient. (Image courtesy of Dr. Amanda Piquet)

**Table 24.1** Diagnostic Criteria for Adult PACNS

| |
|---|
| 1. History or clinical finding of an acquired neurologic deficit that is unexplained after thorough evaluation |
| 2. Classic angiographic or histopathologic features of angiitis within the CNS |
| 3. No evidence of systemic vasculitis or any other condition that could produce the angiographic or histopathologic findings |

Adapted from Calabrese and Mallek [32]

outcomes with aggressive immunotherapy have also been reported [30].

- *PACNS with intracranial hemorrhage* is reported in up to 12% of cases. Patients with hemorrhage are less likely to experience cognitive dysfunction, persistent neurologic deficit, or stroke on MRI as compared to those without hemorrhage [31].

## Diagnosis

Diagnostic criteria for adult PACNS were proposed by Calabrese and Mallek in 1988 (Table 24.1) [32], and though not validated, these criteria are still widely used in the evaluation of suspected PACNS. PACNS is essentially a diagnosis of exclusion and can only be made once a comprehensive and systematic evaluation for causes of secondary CNS vasculitis and other mimics of PACNS has been performed.

## Childhood PACNS (cPACNS)

PACNS is also seen in the pediatric population, termed childhood PACNS (cPACNS). Similar to adult PACNS, little is known about the epidemiology and pathogenesis of cPACNS. As in adults, both sexes are affected. Unlike adults, the distribution of vessel involvement differs between the sexes, with large and medium vessel involvement being more common in boys [33] and small vessel involvement being more common in girls [34]. Like adults, children with PACNS present with a variety of non-specific focal and/or diffuse neurologic deficits. Three major subtypes of cPACNS are recognized and include the following [35]:

- *Non-progressive cPACNS* is characterized by unilateral involvement of the distal internal carotid and/or proximal middle cerebral arteries [36, 37]. The majority present with acute focal neurologic deficits; concomitant cognitive or behavioral changes are uncommon. CSF analysis is normal in more than 50%; however, an otherwise unexplained increase in opening pressure may be seen [35]. MRI shows unilateral ischemic lesions within large vessel territories. Angiography is frequently abnormal [36]. The course is generally monophasic with an overall favorable prognosis.
- *Progressive cPACNS* also involves medium to large cerebral vessels; however, involvement is not strictly unilateral. The vast majority will present with headache in addition to focal and diffuse neurologic deficits. CSF is mildly inflammatory with pleocytosis and/or elevated protein. MRI shows ischemic lesions in multiple vascular territories. Angiography is often abnormal [35, 38].
- *Small vessel cPACNS* may present acutely with a rapidly progressive meningitis-like illness or may present more insidiously with cognitive dysfunction, headache, and/or focal seizures. CSF is abnormal in more than 90%. MRI shows inflammatory lesions of the subcortical white matter and cortical gray matter. By definition, angiography is negative [36, 39].

Unlike adult PACNS, the most common histopathologic pattern seen in pediatric biopsies is lymphocytic. Granulomatous or necrotizing patterns are exceedingly rare, and beta amyloid deposition is not seen [40, 41].

The diagnostic approach in cPACNS is similar to that in adult PACNS with focus on ruling out other forms of childhood inflammatory brain disease, causes of secondary CNS vasculitis, and other mimics of cPACNS.

Treatment of cPACNS is extrapolated from adult PACNS literature, and the latter will be discussed in greater detail later in this chapter.

## Secondary CNS Vasculitis (SCNSV)

In evaluating cases of possible PACNS, causes of secondary CNS vasculitis must be considered. Secondary CNS vasculitis is not confined to the CNS and typically occurs in the setting of systemic vasculitis, underlying connective tissue disease, or infection. Symptoms of SCNSV and PACNS are nearly indistinguishable; however, specific ancillary tests are available for many causes of SCNSV. As a result, those with SCNSV tend to be diagnosed earlier than those with PACNS (mean time to diagnosis of 8 months vs. 27 months, respectively) [22]. The major categories of SCNSV are discussed below.

## CNS Vasculitis Secondary to Primary Systemic Vasculitis

Primary systemic vasculitis is reported to involve the CNS in approximately one-fourth of cases and should be considered when evaluating cases of suspected CNS vasculitis. Often the diagnosis of systemic vasculitis has been established before CNS involvement occurs or may be inferred from biopsy and/or angiography of a non-CNS site. Important considerations in this category are as follows:

### Behçet Disease

Behçet disease is a relapsing variable vessel vasculitis that is characterized by recurrent oral and genital ulcers and ocular inflammation (i.e., anterior uveitis). Behçet disease predominately affects persons aged 20–40 years. Men are affected twice as often as women. Diagnosis of Behçet disease is based on published consensus criteria [42]. Neurologic involvement occurs in 10–40% of patients with Behçet disease, on average 5 years after the onset of mucocutaneous and ocular manifestations. Neuro-Behçet disease may be classified as parenchymal or non-parenchymal. The parenchymal form is more common and severe, presenting with meningoencephalitis, seizure, and/or myelitis. Parenchymal lesions may present with symptoms mimicking PACNS. CSF is frequently inflammatory with elevated protein and/or pleocytosis (often neutrophilic); generally, protein and/or white blood cells (WBC) are more elevated in neuro-Behçet disease than in PACNS. Elevated immunoglobulin G (IgG) index may be seen in 70%. Oligoclonal bands may also be seen, though are typically transient. Imaging often shows enhancing lesions with predilection for the corticospinal tracts and brainstem and may help to distinguish Behçet disease from PACNS [43].

### Anti-neutrophil Cytoplasmic Antibody (ANCA)-Associated Vasculitides

The anti-neutrophil cytoplasmic antibody (ANCA)-associated vasculitides (AAV) are primary small vessel vasculitides characterized by the presence of ANCA. The ANCA-associated vasculitides include granulomatosis with polyangiitis (GPA, formerly Wegener's granulomatosis), eosinophilic granulomatosis with polyangiitis (EGPA, formerly Churg-Strauss syndrome), and microscopic polyangiitis (MPA). The global incidence of AAV is estimated at 0.9–10.3 per million depending on the cohort examined [44–47]. Neurologic involvement in AAV is not uncommon, most frequently involving the PNS. Less than 15% will have CNS involvement [48], typically presenting with small-medium vessel vasculitis of the brain and/or spinal cord invasion or compression of CNS structures by granulomatous pathology in adjacent structures, or de novo granulomatous lesions in the CNS [49]. Headache, stroke, hemorrhage, and encephalopathy are common presenting symptoms. Constitutional symptoms are common. Additionally, symptoms attributable to involvement of the ears, nose, throat, lung, and/or kidneys are frequently reported (Table 24.2) [50–54].

ANCA serologic testing is crucial in the evaluation of possible AAV. Unfortunately, the sensitivity of ANCA is limited and may be negative in

**Table 24.2** ANCA-Associated Vasculitides

| | |
|---|---|
| GPA [50, 51] | Upper and lower respiratory tract, kidneys. CNS involvement with parenchymal or dural lesions in 4–11%. |
| EGPA [52, 53] | Allergic prodrome (asthma, nasal polyps) -> eosinophilic phase (peripheral eosinophilia, organ involvement) -> vasculitic phase. CNS involvement in 6–10%. ANCA positive in roughly 1/3 (MPO serotype predominant). |
| MPA | Glomerulonephritis, skin manifestations, and mononeuritis multiplex are classic. CNS involvement in 37–72% with hemorrhage, ischemic stroke, and/or pachymeningitis [54]. |

*GPA* granulomatosis with polyangiitis, *EGPA* eosinophilic granulomatosis with polyangiitis, *CNS* central nervous system, *ANCA* antineutrophil cytoplasmic antibodies, *MPO* myeloperoxidase, *MPA* microscopic polyangiitis

up to 50% of pathologically confirmed cases of GPA and MPA [55]. ANCA positivity typically falls into two serotypes: PR3- and MPO-ANCA. The distinction between these serotypes is important as differences in epidemiology, clinical presentation, and prognosis have been recognized. The rate of neurologic involvement appears to be similar between the two serotypes [49, 56]. CSF analysis is sensitive but not specific, most commonly showing mild lymphocytic pleocytosis with elevated protein.

Additional systemic vasculitides associated with CNS vasculitis are summarized in Table 24.3 [57–64].

**Table 24.3** Systemic vasculitides associated with CNS vasculitis

| | |
|---|---|
| *Large vessel vasculitis* | |
| Giant cell arteritis/ temporal arteritis [57] | Large vessel vasculitis affecting adults over age 50. Headache, fever, malaise, myalgia, and anorexia are common initial symptoms. Vision loss due to ischemic optic neuropathy is a primary concern. Coexisting polymyalgia rheumatica in 40%. ESR markedly elevated |
| Takayasu disease [58, 59] | Granulomatous vasculitis of aorta and its major branches primarily affecting patients under age 40. F:M ratio 10:1. Cranial nerve palsies (isolated or multiple), stroke, and/ or ischemic optic neuropathy |
| *Medium vessel vasculitis* | |
| Polyarteritis nodosa (PAN) [58, 59] | Systemic medium vessel necrotizing vasculitis. Onset age 40–50. May be idiopathic or associated with chronic infection (e.g., hepatitis B). CNS involvement in 20–40%; occurs later in disease course. Cognitive dysfunction, atypical persistent headache, visual disturbance, and/or seizures. Intracranial microaneurysms and intracranial hemorrhage may be seen |
| Kawasaki disease [60] | Predominately affects children under 5 years of age. Encephalopathy, seizures, stroke, intracranial hemorrhage, ataxia, and/or cranial nerve palsy |
| *Small vessel vasculitis* | |
| ANCA-associated vasculitis | See text |
| Cryoglobulinemic vasculitis [58, 61] | Vasculitis with cryoglobulin immune deposits affecting small vessels (capillaries, venules, arterioles). Classic triad of purpura, weakness, and arthralgia. Hepatitis C infection in ~80%; B-cell lymphoproliferative disorders, autoimmune disease, and other infection are other causes |
| IgA vasculitis (formerly Henoch-Schönlein purpura) [59] | Typically young people under age 20. CNS involvement with headache, altered consciousness, seizure, focal neurologic deficits |
| *Variable vessel vasculitis* | |
| Behçet syndrome | See text |
| Cogan syndrome [62–64] | Poorly understood systemic variable vessel vasculitis. Non-syphilitic interstitial keratitis and Meniere's disease-like symptoms are classic. Atypical types with other forms of ocular inflammation and/or long interval between development of audiovestibular and ocular symptoms. May be associated with RA or inflammatory bowel disease |

*ESR* erythrocyte sedimentation rate, *ANCA* anti-neutrophil cytoplasmic antibody, *IgA* immunoglobulin A, *RA* rheumatoid arthritis

## CNS Vasculitis Secondary to Connective Tissue Disease or Other Systemic Rheumatic Disease

The connective tissue diseases are a group of chronic autoimmune inflammatory disorders primarily targeting muscle, joints, and skin. Neurologic involvement in connective tissue disease is not uncommon, most often affecting the PNS. CNS manifestations may reflect non-vasculitic complications or, more rarely, true CNS vasculitis. CNS manifestations due to rheumatic diseases may occur with or without systemic symptoms. Other systemic manifestations may aid in the diagnosis; however, CNS manifestations may occur independent of systemic symptoms. Laboratory tests may also help to distinguish these conditions from PACNS [65].

Rheumatoid arthritis (RA), systemic lupus erythematosus (SLE), Sjögren's syndrome (SS), and systemic sclerosis are among the most likely to involve the CNS. Sarcoidosis may also be complicated by CNS vasculitis and can be difficult to distinguish from PACNS, even histologically. Other systemic rheumatologic conditions (inflammatory bowel disease, dermatomyositis, etc.) may also be complicated by CNS involvement; however, further discussion of these conditions is beyond the scope of this chapter. Dermatomyositis and other immune-mediated myopathies are discussed in Chap. 20. Neurological syndromes, beyond vasculitis, associated with rheumatological disease are further discussed in Chap. 22. Connective tissue/rheumatic diseases that are associated with CNS vasculitis are listed in Table 24.4 [57, 65–78].

**Table 24.4** Connective tissue/rheumatologic diseases associated with CNS vasculitis

| | |
|---|---|
| Systemic lupus erythematosus (SLE) | Multisystemic autoimmune disease with heterogenous clinical manifestations that overlap with several other autoimmune diseases. CNS involvement may occur at any time during the disease course and may be the initial presentation [57, 66, 67]. Common CNS manifestations include headache, cerebrovascular disease, seizure, psychosis, cognitive dysfunction, and chorea. CNS vasculitis is rare and may be confused with cerebral vasculopathy, which is more commonly seen. The development of vasculitis involving any organ system is associated with longer disease duration, younger age at disease onset, and male sex. Vasculitis often develops during a "lupus flare" [68]. ANA, anti-Smith, and/or anti-dsDNA antibodies are frequently present. C3 and C4 may be reduced. CSF is inflammatory in the majority. MRI may be normal in up to 40%. Of those with abnormal MRI, findings consistent with small vessel disease are most common [69] |
| Sjögren's syndrome (SS) | Chronic inflammation and destruction of lacrimal and salivary glands result in hallmark symptoms of dry eye (keratoconjunctivitis sicca) and dry mouth (xerostomia). Neurologic involvement, when it occurs, frequently precedes sicca symptoms [57, 70]. PNS involvement (sensory ganglionopathy, polyneuropathy) is most common; CNS complications include stroke, seizure, trigeminal neuralgia, aseptic meningoencephalitis, and transverse myelitis. ANA, RF, SSA, and/or SSB may be present, although one review noted that SSA and SSB antibodies are less common in those with neurologic involvement as compared to those without [65]. Imaging findings are similar to other forms of vasculitis; however, the presence of enlarged lacrimal and/or salivary glands may be suggestive of Sjögren's syndrome [71] |
| Rheumatoid arthritis (RA) | Most common inflammatory arthritis<br>Extra-articular manifestations associated with disease severity, disease duration, the presence of autoantibodies, and the presence of comorbidities [72, 73]. The incidence of neurologic involvement estimated at 11% (PNS >> CNS) [74]. CNS involvement is rare with fewer than ten pathologically confirmed cases of rheumatoid vasculitis with CNS involvement reported in the literature. Symptoms are similar to other forms of vasculitis and may occur in the absence of systemic disease activity [57] |
| Systemic sclerosis | Widespread microvasculopathy with multiorgan fibrosis (skin, heart, lungs, gastrointestinal). ANA, anti-Scl-70, anti-centromere, anti-RNA-polymerase III, anti-U3-RNP, anti-U1-RNP, and anti-Pm/Scl may be detected. Vasculitis may be associated with ANCA or mixed cryoglobulinemia [75] |

**Table 24.4** (continued)

| | |
|---|---|
| Mixed connective tissue disease | Overlap syndrome associated with anti-U1RNP antibody. Anti-Smith and anti-dsDNA negative |
| Dermatomyositis | Vasculitis very rare. Usually in conjunction with vasculitis in other sites (skin, retina, and/or lung). CNS involvement usually shortly after diagnosis [76, 77] |
| Sarcoidosis | Idiopathic, multisystemic autoimmune disorder characterized by noncaseating granulomas. Lung, eye, lymph node, and skin involvement common. CNS involvement rare, though may be initial presentation. Isolated neurosarcoidosis also described. CNS vasculitis rare; may occur with systemic vasculitis or as isolated cerebral vasculitis. CSF is inflammatory and hypoglycorrhachia may be present (may be confused for infection or malignancy). Serum angiotensin-converting enzyme (ACE) elevated in some. CSF ACE may be elevated but sensitivity and specificity are poor [78] |
| Relapsing polychondritis | – |
| Inflammatory bowel disease | – |
| Non-vasculitic autoimmune inflammatory meningoencephalitis[a] | – |
| Antiphospholipid syndrome[a] | – |

*CNS* central nervous system, *ANA* antinuclear antibody, *CSF* cerebrospinal fluid, *MRI* magnetic resonance imaging, *PNS* peripheral nervous system, *RF* rheumatoid factor, *ANCA* anti-neutrophil cytoplasmic antibody
[a]Conditions causing CNS manifestations mimicking vasculitis without true vasculitis

## CNS Vasculitis Secondary to Infection

Several infections are known to be associated with CNS vasculitis. Infection may cause vessel damage directly (infection of the endothelium with resultant inflammatory response) or indirectly (triggering local immune response with spread of inflammation to the vessel wall). Any infection that causes meningitis at the base of the brain can potentially lead to local inflammation involving the circle of Willis and its branches. Various bacterial, viral, fungal, parasitic, and rickettsial infections have been associated with the development of CNS vasculitis. Ruling out infection is paramount, as unwarranted immunosuppression (treatment for other forms of vasculitis) may impair immune clearance of microbes in patients with active infection. Positive culture, serology, and/or molecular testing for pathogens associated with CNS vasculitis establishes this diagnosis. Infections causing subacute or chronic meningitis are the most likely to mimic CNS vasculitis.

Of the myriad infections associated with CNS vasculitis, varicella-zoster virus (VZV) is of particular interest and has been associated with a range of vascular complications. Primary infection with VZV causes the childhood disease chickenpox; this disease has become much less common after the introduction of widespread vaccination in 1995. Once cleared after primary infection, VZV remains latent in sensory ganglia. The most common neurological complication of VZV reactivation is herpes zoster (shingles); however, latent VZV infection can affect nearly any location along the neuraxis, including cerebral blood vessels [79–81]. Classically, zoster of the ophthalmic division of the trigeminal nerve is followed by contralateral hemiparesis due to vasculopathy of the MCA. More recently, with HIV and other causes of immunocompromise, a more diffuse small vessel vasculopathy has been recognized. VZV vasculopathy/vasculitis may occur without preceding zoster, and a high degree of suspicion is required to identify VZV as the etiology in these cases. Imaging most often demonstrates ischemic or hemorrhagic stroke in deep structures or at the gray-white junction. Post-contrast vessel wall enhancement may be seen. Spinal cord infarction, subarachnoid hemorrhage, aneurysm, and arterial dissection are less common [82, 83]. CSF analysis reveals a mononuclear pleocytosis and may demonstrate an increased number of red blood cells. In VZV-associated CNS vasculitis, the sensitivity of CSF VZV polymerase chain reaction (PCR) is only 30%; therefore, diagnosis often relies on the detection of intrathecal synthesis of VZV-specific IgG [84–86]. See Table 24.5 [87–94].

## Other Causes of Secondary CNS Vasculitis

### Drug-Induced

Generally in the setting of drug use, neurologic deficits are due to vasospasm rather than a true

**Table 24.5** Infections associated with CNS vasculitis

| | |
|---|---|
| **Bacterial** | |
| Acute bacterial meningitis (*S. pneumoniae*, *N. meningitidis*) | Vascular complications in up to 20%. Acutely, vasculitis, vasospasm, and/or mycotic aneurysm. Delayed or chronic vasculopathy may occur [87, 88] |
| Bacterial endocarditis | MRI and angiographic findings may be indistinguishable from PACNS. Positive blood cultures and vegetations on echocardiogram help to distinguish [89] |
| Tuberculosis (TB) | Basilar meningitis leads to vasculitis of vessels at the base of the brain. MRI shows basilar meningitis ± hydrocephalus. CSF with pleocytosis, elevated protein, normal to low glucose. Diagnosis by detection of TB in CSF (stain, culture, and/or PCR) [90, 91] |
| Syphilis/*Treponema pallidum* | Acquired or congenital. CNS involvement in up to 10% of untreated patients. Meningovascular syphilis in ~40–60% of neurosyphilis; usually occurs months to years after primary infection. Imaging shows smooth or beaded segmental narrowing of involved vessels ± gummas in brain parenchyma. Serum treponemal antibody universally positive; RPR frequently positive. CSF is inflammatory. CSF VDRL is confirmatory but relatively insensitive. All should be tested for concurrent HIV [92] |
| Other spirochetal (Lyme/*B. burgdorferi*, leptospira) | Exceedingly rare, scant evidence |
| *Mycoplasma pneumoniae* | Exceedingly rare |
| *Bartonella* | Cat-scratch disease |
| *T. whipplei* | Whipple's disease |
| **Viral** | |
| Herpes viruses (VZV, EBV, CMV, HSV 1 and 2) | Evidence for HSV, EBV, CMV limited to single patient reports. CMV seen in the immunocompromised |
| Hepatitis B virus (HBV), hepatitis C virus (HCV) | HCV may cause vasculitis independently. HBV and HCV may also cause CNS vasculitis as a result of associated autoimmune phenomena such as cryoglobulinemia or PAN |
| Retrovirus (HIV, HTLV-1) | Increased risk of stroke in HIV, though several possible mechanisms. CNS vasculitis exceedingly rare; often related to secondary opportunistic infection |
| Parvovirus B19 | – |
| Enterovirus | – |
| Arboviruses (West Nile virus, Zika virus) | – |
| **Fungal** | |
| Cryptococcus | Most common cause of fungal meningitis. Vascular involvement in up to 32%. Diagnosis with CSF cryptococcal antigen, India ink stain, and/or culture [88] |
| Aspergillus | Most common invasive mold infection worldwide. Immunocompromised or immunocompetent patients. <50% of CSF cultures positive. CSF galactomannan emerging as biomarker for CNS aspergillosis. Non-CNS tissue biopsy (e.g., lungs) often required. MRI may show ring-enhancing lesions, meningeal enhancement, ischemic stroke, or hemorrhage [88] |
| Coccidioides | Soil-based fungus endemic to American Southwest and Latin America. Basilar meningitis with local vasculitis; may have chronic vasculopathy. CSF may show eosinophilic pleocytosis. CSF culture positive in approximately one-third. Complement fixation in serum and CSF positive in ~40% [93] |
| Mucormycosis | – |
| Candida | – |
| Histoplasmosis | – |

**Table 24.5** (continued)

| | |
|---|---|
| **Parasitic** | |
| Cysticercosis/*Taenia solium* | Infection with larval stage of pork tapeworm. Most common CNS parasite. Accounts for 10% of stroke in endemic areas due to subarachnoid neurocysticercosis with spread of inflammation to vessels at the base of the brain. MRI shows lenticulostriate lacunar infarcts and/or hemorrhage. Angiography positive in ~50% (MCA and PCA most often affected) [94] |
| Toxoplasmosis/*Toxoplasma gondii* | – |
| *Plasmodium falciparum* | – |
| *Schistosoma mansoni* | – |
| **Rickettsial** | |
| *Rickettsia rickettsii* | – |
| Scrub typhus | – |

*Abbreviations*: *CNS* central nervous system, *MRI* magnetic resonance imaging, *PACNS* primary angiitis of the central nervous system, *CSF* cerebrospinal fluid, *PCR* polymerase chain reaction, *RPR* rapid plasma regain, *VDRL* Venereal Disease Research Laboratory test, *HIV* human immunodeficiency virus, *HSV* herpes simplex virus, *VZV* varicella-zoster virus, *EBV* Epstein-Barr virus, *CMV* cytomegalovirus, *PAN* polyarteritis nodosa, *HTLV* human T-lymphotropic virus, *MCA* middle cerebral artery, *PCA* posterior cerebral artery

vasculitis; however, rare histologically documented cases of vasculitis have been reported [95]. Amphetamine, methamphetamine, cocaine, and related drugs have been implicated in multi-organ arteritis involving the CNS [96]. Neurologic symptoms tend to occur in close temporal correlation with drug use. CSF, MRI, and angiography are frequently abnormal.

### Radiation-Induced

Radiation-induced CNS vasculopathy/vasculitis generally evolves very slowly with ischemia occurring years or even decades after irradiation. Small arteries and capillaries are most commonly affected. Rarely, large vessel injury may occur resulting in occlusive vasculopathy. MRI demonstrates thickening and prominent enhancement of affected large vessel walls; vessel wall enhancement may help to distinguish this condition from Moyamoya disease [97].

## Mimics of CNS Vasculitis

In addition to the aforementioned causes of secondary CNS vasculitis, there are a multitude of non-vasculitic conditions that may be confused for CNS vasculitis. Ruling out these mimics is essential in the evaluation of patients with possible CNS vasculitis and may greatly impact management and prognosis. The main differential considerations vary somewhat based on age. Noninflammatory vasculopathies and vasospasm are primary differential diagnoses of angiography-positive cPACNS. The main differential diagnoses of angiography-negative/small vessel vasculitis in children are non-vasculitic inflammatory brain diseases/encephalitis, antibody-mediated inflammatory brain diseases, and primary demyelinating diseases.

In adults, the most important mimics are atherosclerotic disease and vasospasm (i.e., reversible cerebral vasoconstriction syndrome). Embolic phenomena, autoimmune encephalitis, malignancy, genetic conditions, and various noninflammatory vasculopathies may also be considered in the appropriate clinical circumstances.

## Reversible Cerebral Vasoconstriction Syndrome

Reversible cerebral vasoconstriction syndrome (RCVS) is an important differential consideration in any patient presenting with possible CNS vasculitis. Classically, RCVS is a disease of middle-aged women (mean age of onset ~45 years). The vast majority present with "thunderclap" headache with or without focal neurologic deficits. This is in contrast to the more insidious headache associated with CNS vasculitis. Over half of cases occur in the postpartum period or after exposure to vasoactive drugs (cocaine, 3,4-methylenedioxy-methamphetamine [MDMA], amphetamines, triptans, selective serotonin reuptake inhibitors [SSRIs], etc.). Serum inflammatory markers are normal. CSF shows mild elevation in protein with minimal pleocytosis. In contrast to CNS vasculitis, MRI in RCVS is often normal, though infarcts and/or hemorrhage may be seen. Angiography shows multifocal, segmental cerebral artery vasoconstriction that can appear identical to findings seen in CNS vasculitis. As a rule, angiographic abnormalities in RCVS show complete or near-complete resolution on repeat examination within 12 weeks of onset, and this may help distinguish RCVS from CNS vasculitis. The course is typically monophasic, though may recur in ~5%. Management is largely supportive. Calcium channel blockers are often used, though evidence of their efficacy in RCVS is lacking [98–100]. See Table 24.6 [21].

**Table 24.6** Mimics of CNS vasculitis

| | |
|---|---|
| Noninflammatory vasculopathies | Intracranial atherosclerosis |
| | Chronic hypertensive vasculopathy |
| | Intracranial dissection |
| | Fibromuscular dysplasia |
| | Moyamoya syndrome |
| Genetic cerebral vasculopathies | Connective tissue vasculopathy (e.g., Marfan syndrome, Ehlers-Danlos syndrome) |
| | Mitochondrial diseases (e.g., CADASIL, MELAS) |
| | Metabolic vasculopathies (Fabry disease, McArdle disease[a]) |
| Cerebral vasculopathy syndromes | Susac syndrome (encephalopathy, hearing loss, branch retinal artery occlusion) |
| | Kohlmeier-Degos disease (retinocerebral vasculopathy with cerebral leukodystrophy) |
| | Hereditary endotheliopathy with retinopathy, nephropathy, and stroke |
| | Sneddon syndrome (progressive noninflammatory cerebral arteriopathy and livedo reticularis) |
| | PHACES (posterior fossa malformations, hemangiomas, arterial anomalies, cardiac defects, eye abnormalities, sternal cleft, and supraumbilical raphe syndromes)[a] |
| Vasospasm | RCVS (reversible cerebral vasoconstriction syndrome) |
| | Migraine |
| | Vasospasm secondary to hypertension |
| | Drug or medication induced |
| | Ion channelopathies (e.g., familial hemiplegic migraine) |
| Other conditions associated with vasculopathy | Hemoglobinopathies (e.g., sickle cell) |
| | Thrombotic/embolic phenomenon<br>Coagulopathies<br>Antiphospholipid antibody syndrome<br>Myxoma and other cardiac tumors<br>Atheroembolism |
| | Malignancy-associated vasculopathy<br>Angiotropic and intravascular lymphoproliferative disorders<br>Graft-versus-host disease<br>Intravascular lymphoma |
| Non-vasculitic inflammatory brain diseases | ADEM (acute disseminated encephalomyelitis)[a] |
| | Antibody-mediated inflammatory brain disease/autoimmune encephalitis |
| | Progressive multifocal leukoencephalopathy |
| | Posterior reversible encephalopathy syndrome (PRES) |
| | T-cell associated parenchymal inflammatory brain disease (Rasmussen's encephalitis)[a] |
| | Necrotizing and hemorrhagic inflammatory brain disease[a] |
| | FIRES (febrile infection-related epilepsy syndrome)[a] |

Adapted from Ref. [21]

*CADASIL* cerebral autosomal dominant arteriopathy with subcortical infarcts and leukoencephalopathy, *MELAS* mitochondrial encephalomyopathy, lactic acidosis, and stroke-like episodes

[a]Primarily pediatric concerns

## Diagnostic Approach in CNS Vasculitis

Given the protean manifestations of CNS vasculitis, a high level of clinical suspicion is required for prompt recognition of patients with possible CNS vasculitis. Byram et al. recommend that evaluation for possible CNS vasculitis be performed in the following circumstances [101]:

- Strokes of different ages in multiple vascular territories with inflammatory CSF profile
- Stroke in the young without cardiovascular/stroke risk factors
- Chronic meningitis without evidence of infectious or neoplastic etiology
- Subacute to chronic headache with cognitive dysfunction
- Combination of otherwise unexplained focal and diffuse neurologic symptoms

The evaluation of suspected CNS vasculitis begins with a thorough clinical history and examination with particular focus on signs/symptoms that may suggest a systemic process. Additionally, the patient's age, sex, ethnicity, and family history may help in narrowing the differential.

All patients presenting with cerebral ischemia should first undergo appropriate stroke workup. In patients presenting with significant encephalopathy, evaluation for toxic/metabolic disturbances should take precedence.

## Laboratory Testing

An initial step in evaluating suspected CNS vasculitis includes laboratory testing for markers of inflammation, markers of systemic disease, and markers of non-vasculitic inflammatory brain diseases. At present, there are no blood markers with sufficient sensitivity or specificity to diagnose PACNS; however, results of these tests may be useful in evaluating causes of secondary vasculitis or other mimics of PACNS. More targeted testing for specific rheumatic or other autoimmune conditions may be considered based on clinical suspicion. Most patients should be tested for HIV status. Otherwise, cultures, serologies, and serum molecular testing for common and epidemiologically appropriate infectious agents should be considered depending on the clinical context.

## CSF Analysis

CSF analysis should be performed in all patients who do not have a contraindication to lumbar puncture. CSF analysis in CNS vasculitis reveals at least one abnormality in the vast majority (80–95%) of patients, giving CSF analysis a high (but not 100%) negative predictive value for ruling out CNS vasculitis [101]. CSF in PACNS most often shows a lymphocyte predominant pleocytosis and/or elevated protein with normal glucose. Oligoclonal bands and/or elevated IgG index may be seen in a minority [4, 102, 103].

## Imaging

Neuroimaging studies are essential in the evaluation of possible CNS vasculitis. While imaging findings alone are not adequately sensitive or specific to distinguish PACNS from other causes of CNS vasculitis, some imaging findings may be suggestive of particular etiologies. For example, a nasal mass with destruction of the nasal septum may suggest GPA, while enlarged lacrimal and salivary glands may suggest Sjögren's syndrome.

### MRI

MRI is abnormal in 95–100% of patients with CNS vasculitis. Most commonly, MRI demonstrates multiple ischemic lesions of varying ages that are frequently bilateral and supratentorial. Non-specific cortical and/or subcortical lesions, hemorrhage, and leptomeningeal enhancement (focal or diffuse) may be seen. The presence of leptomeningeal enhancement may help to distinguish CNS vasculitis from non-vasculitic inflammatory conditions; however, it does not help to distinguish PACNS from secondary forms of CNS vasculitis [71, 104, 105].

## MRA

Contrast-enhanced magnetic resonance angiography (MRA) can be useful in evaluating for vessel abnormalities in CNS vasculitis but is not as sensitive as conventional angiography in detecting lesions of the posterior circulation or distal vasculature [4, 37]. More recently, 3T MRI vessel wall imaging has been found to be sensitive and specific for CNS vasculitis. Classically, smooth circumferential concentric wall thickening with diffuse gadolinium enhancement of the vessel wall is seen [106, 107]. Vessel wall enhancement appears to resolve with immunosuppressive therapy, suggesting that 3T MRI vessel wall imaging may also be useful in monitoring disease activity [108].

## CT/CTA

CT is less sensitive than MRI for evaluating brain parenchyma, apart from detecting acute hemorrhage. Computed tomography angiography (CTA) offers comparable to slightly better resolution of large- and medium-size intracranial arteries compared to MRA, but does not significantly improve sensitivity in patients with normal MRA [71].

## Digital Subtraction Angiography (DSA)

Digital subtraction angiography (DSA, also known as conventional angiography) has better resolution than MRA or CTA for detecting changes in small arteries. The classic angiographic finding in CNS vasculitis is alternating areas of stenosis and dilation referred to as "beading" in at least two separate vascular distributions (see Fig. 24.2). Aneurysms, collateral circulation, isolated regions of vessel narrowing in multiple branches, and multiple abrupt vessel occlusions have also been reported. Angiographic findings in CNS vasculitis may be indistinguishable from a range of noninflammatory vasculopathies [109]. The sensitivity of DSA ranges from 40% to 90% with a reported specificity of ~30% [110]. DSA is unlikely to add additional diagnostic value in patients with clearly abnormal MRA; however, the utility of DSA in patients with normal MRA is a topic of debate [4, 110]. Nonetheless, DSA remains a core part of the evaluation of possible PACNS, and the sensitivity and specificity remain dependent to some extent on the experience and judgment of the operator.

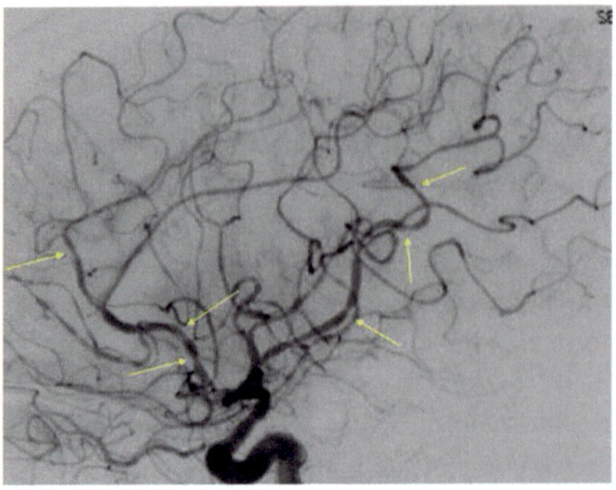

**Fig. 24.2** Digital subtraction angiography from a patient with primary angiitis of the central nervous system (PACNS). Arrows denote areas of focal stenosis involving the anterior and middle cerebral arteries. (Image courtesy of Dr. Amanda Piquet)

## Other Imaging Modalities

Color duplex ultrasonography may be used in evaluating large vessel, particularly extra-cranial, vasculitis and can provide information about both the arterial wall and the flow within the vessel. This modality may be particularly useful in the evaluation of giant cell arteritis (GCA). Detection of a "halo sign" (hypoechoic edematous wall swelling) in the temporal arteries has a higher sensitivity than biopsy for GCA, and specificity can reach 100% if bilateral halo sign is identified [111].

Other imaging modalities have been studied and may have some utility in the evaluation of CNS vasculitis; however, use of these modalities is not standard practice at this time. Positron emission tomography (PET)/CT may be useful in cases of large vessel vasculitis when clinical and other imaging findings are non-specific and may have some utility in monitoring disease activity [71, 112]. Single-photon emission computed tomography (SPECT) imaging may detect regional CNS perfusion abnormalities and may be able to provide information about the pathophysiology in cerebral vasculitides beyond the resolution of MRA, CTA, and DSA [113, 114].

## Brain Biopsy

Brain biopsy remains the gold standard in the diagnosis of PACNS despite its limited sensitivity. The diagnostic yield of brain biopsy for suspected PACNS is typically estimated at more than 50%. Although PACNS may not be confirmed, biopsy may still help to identify alternate diagnoses such as mimics or secondary causes of CNS vasculitis. In order to maximize yield, abnormal areas on MRI should be targeted when possible, and full-thickness biopsy including white matter, cortex, leptomeninges, and dura should be obtained [40]. The morbidity rate of biopsy is estimated at 10–16% [115, 116].

Classically, pathology in PACNS demonstrates an angiocentric transmural and/or perivascular inflammatory infiltrate with reactive endothelial cells. Importantly, viral inclusions, neuronophagia, parenchymal inflammation, and demyelination are not seen. Amyloid deposition may be seen in adults; however, it is not seen in children [14, 40].

Three distinct histologic subtypes are seen in PACNS: (1) granulomatous (segmental granulomatous inflammation with multinucleated giant cells), (2) lymphocytic (lymphocytic infiltrate with occasional plasma cells), and (3) necrotizing (transmural fibrinoid necrosis; associated with intracranial hemorrhage) [4, 6]. Granulomatous inflammation is most common in adult PACNS. In cPACNS, a lymphocytic pattern is present in the vast majority with granulomatous or necrotizing patterns rarely seen.

## Treatment

Treatment of PACNS involves both an acute/induction phase and a maintenance phase. There are no validated guidelines for the treatment of PACNS; however, the authors use the European League Against Rheumatism (EULAR) recommendations for the management of primary small and medium vessel vasculitis as a guide for treatment of PACNS [117]. High-dose glucocorticoids are the mainstay of treatment in the acute phase. There are no consensus guidelines regarding glucocorticoid regimens in PACNS; however, patients typically receive a short 3–5-day course of IV methylprednisolone (or equivalent) followed by a high-dose oral glucocorticoid with a prolonged taper over several months (Table 24.7).

Historically, cyclophosphamide has frequently been used in conjunction with glucocorticoids in the remission induction phase (Table 24.7). When cyclophosphamide is used, the typical duration is 3–6 months. Rituximab is also becoming increasingly used in the management of PACNS (extrapolated from data in primary systemic small-medium vessel vasculitides). Intravenous immunoglobulin (IVIg) or plasma exchange may be used in certain circumstances.

Following this induction phase, maintenance therapy is initiated. Azathioprine, methotrexate, leflunomide, and mycophenolate mofetil are the most commonly used agents; however, the biologic agent rituximab is gaining popularity as maintenance therapy for PACNS. Other biologic agents such as infliximab may be considered in

**Table 24.7** Suggested treatment of PACNS using glucocorticoids and cyclophosphamide

| | | |
|---|---|---|
| Glucocorticoid | IV methylprednisolone 15 mg/kg daily | May be used as initial 3–5-day pulse and then followed by PO prednisone |
| | PO prednisone 1 mg/kg/day (up to 80 mg/day) | No validated taper, though generally very slow taper over several months |
| Cyclophosphamide | IV 15 mg/kg (max 1200 mg) every 2 weeks for first 3 pulses; every 3 weeks for the next 3–6 pulses | Antiemetic therapy should be given.<br>2-Mercaptoethanesulfonate sodium (mesna) should be given to prevent bladder toxicity. Close monitoring with CBC (baseline, day 7, day 10, day 27, and following each dose), urinalysis every 3–6 months, and BUN/creatinine every 2 weeks is required |
| | PO 2 mg/kg/day (max 200 mg/day) | Close laboratory monitoring as above |

*IV* intravenous, *PO* per oral, *CBC* complete blood count, *BUN* blood urea nitrogen refractory cases, though evidence of their efficacy is limited.

The treatment of non-infectious secondary CNS vasculitis is dependent to some extent on the underlying etiology and is largely based on EULAR guidelines [117–120]. Infectious CNS vasculitis is generally treated with appropriate antimicrobial agents with or without adjuvant glucocorticoids.

All patients receiving high-dose glucocorticoids should receive calcium and vitamin D supplementation to protect against glucocorticoid-induced osteoporosis. Patients receiving high-dose glucocorticoid with other immunosuppressive agents should also be given prophylaxis against *Pneumocystis* pneumonia.

Close follow-up is essential in all patients being treated for CNS vasculitis with appropriate clinical and laboratory monitoring for potential adverse effects of therapy. Response to treatment is monitored with serial neurologic examinations, serial neuroimaging studies, and periodic reassessment of symptoms.

## Prognosis

The prognosis in CNS vasculitis varies depending on the underlying etiology. Among those with PACNS, approximately 85% have a favorable response to therapy. Approximately one-third will experience relapse. Mortality in PACNS is estimated as high as 15% [102]. Prognosis is secondary; CNS vasculitis varies; however, those with non-infectious secondary CNS vasculitis seem to have a lower relapse-free survival rate than those with PACNS [22].

## Conclusion

PACNS is a rare single-organ vasculitis that, by definition, is confined to the CNS. The clinical symptoms, laboratory results, and imaging findings in PACNS are all non-specific, overlapping with numerous other conditions. The diagnosis of PACNS, therefore, relies heavily on ruling out other conditions that may present similarly to PACNS including CNS involvement systemic vasculitis, infection, malignancy, and myriad noninflammatory vasculopathies. The treatment in PACNS is extrapolated from treatment of systemic vasculitides and involves use of immunosuppressive medications. High rates of morbidity and mortality can occur with PACNS, particularly if diagnosis and treatment are delayed. Physicians, particularly neurologists, must maintain a high level of clinical suspicion in order to recognize symptoms of PACNS and initiate appropriate evaluation and prompt treatment.

## References

1. Jennette JC, Falk RJ, Bacon PA, Basu N, Cid MC, Ferrario F, et al. 2012 revised international Chapel Hill consensus conference nomenclature of vasculitides. Arthritis Rheum. 2013;65(1):1–11.
2. Harbitz F. Unknown forms of arteritis with special reference to their relation to syphilitic arteritis and periarteritis nodosa. Am J Med Sci. 1922;163(2):250–71.
3. Cravioto H, Feigin I. Noninfectious granulomatous angiitis with a predilection for the nervous system. Neurology. 1959;9:599–609.
4. Salvarani C, Brown RD Jr, Calamia KT, Christianson TJ, Weigand SD, Miller DV, et al. Primary central nervous system vasculitis: analysis of 101 patients. Ann Neurol. 2007;62(5):442–51.
5. Iwase T, Ojika K, Mitake S, Katada E, Katano H, Mase M, et al. Involvement of CD45RO+ T lymphocyte infiltration in a patient with primary angiitis of the central nervous system restricted to small vessels. Eur Neurol. 2001;45(3):184–5.
6. Salvarani C, Brown RD Jr, Hunder GG. Adult primary central nervous system vasculitis. Lancet. 2012;380(9843):767–77.
7. Nagel MA, Cohrs RJ, Mahalingam R, Wellish MC, Forghani B, Schiller A, et al. The varicella zoster virus vasculopathies: clinical, CSF, imaging, and virologic features. Neurology. 2008;70(11):853–60.
8. Thomas L, Davidson M, McCluskey RT. Studies of PPLO infection. I. The production of cerebral polyarteritis by Mycoplasma gallisepticum in Turkeys; the neurotoxic property of the Mycoplasma. J Exp Med. 1966;123(5):897–912.
9. Arthur G, Margolis G. Mycoplasma-like structures in granulomatous angiitis of the central nervous system. Case reports with light and electron microscopic studies. Arch Pathol Lab Med. 1977;101(7):382–7.
10. Nogueras C, Sala M, Sasal M, Vinas J, Garcia N, Bella MR, et al. Recurrent stroke as a manifestation of primary angiitis of the central nervous system in a patient infected with human immunodeficiency virus. Arch Neurol. 2002;59(3):468–73.

11. Kossorotoff M, Touze E, Godon-Hardy S, Serre I, Mateus C, Mas JL, et al. Cerebral vasculopathy with aneurysm formation in HIV-infected young adults. Neurology. 2006;66(7):1121–2.
12. Alexander JJ, Lasky AS, Graf WD. Stroke associated with central nervous system vasculitis after West Nile virus infection. J Child Neurol. 2006;21(7):623–5.
13. Winkler DT, Bondolfi L, Herzig MC, Jann L, Calhoun ME, Wiederhold KH, et al. Spontaneous hemorrhagic stroke in a mouse model of cerebral amyloid angiopathy. J Neurosci. 2001;21(5):1619–27.
14. Salvarani C, Brown RD Jr, Calamia KT, Christianson TJ, Huston J 3rd, Meschia JF, et al. Primary central nervous system vasculitis: comparison of patients with and without cerebral amyloid angiopathy. Rheumatology (Oxford). 2008;47(11):1671–7.
15. Zhou Q, Yang D, Ombrello AK, Zavialov AV, Toro C, Zavialov AV, et al. Early-onset stroke and vasculopathy associated with mutations in ADA2. N Engl J Med. 2014;370(10):911–20.
16. Garg N, Kasapcopur O, Foster J 2nd, Barut K, Tekin A, Kizilkilic O, et al. Novel adenosine deaminase 2 mutations in a child with a fatal vasculopathy. Eur J Pediatr. 2014;173(6):827–30.
17. AlKhater SA. CNS vasculitis and stroke as a complication of DOCK8 deficiency: a case report. BMC Neurol. 2016;16:54.
18. Crow YJ, Casanova JL. STING-associated vasculopathy with onset in infancy--a new interferonopathy. N Engl J Med. 2014;371(6):568–71.
19. Moshous D, Feyen O, Lankisch P, Schwarz K, Schaper J, Schneider M, et al. Primary necrotizing lymphocytic central nervous system vasculitis due to perforin deficiency in a four-year-old girl. Arthritis Rheum. 2007;56(3):995–9.
20. Kraemer M, Becker J, Horn PA, Schwitalla JC, Keyvani K, Metz I, et al. Association of primary central nervous system vasculitis with the presence of specific human leucocyte antigen gene variant. Clin Neurol Neurosurg. 2017;160:137–41.
21. Twilt M, Benseler SM. Central nervous system vasculitis in adults and children. Handb Clin Neurol. 2016;133:283–300.
22. Vera-Lastra O, Sepulveda-Delgado J, Cruz-Dominguez Mdel P, Medina G, Casarrubias-Ramirez M, Molina-Carrion LE, et al. Primary and secondary central nervous system vasculitis: clinical manifestations, laboratory findings, neuroimaging, and treatment analysis. Clin Rheumatol. 2015;34(4):729–38.
23. Salvarani C, Brown RD Jr, Calamia KT, Christianson TJ, Huston J 3rd, Meschia JF, et al. Primary CNS vasculitis with spinal cord involvement. Neurology. 2008;70(24 Pt 2):2394–400.
24. Salvarani C, Brown RD Jr, Calamia KT, Christianson TJ, Huston J 3rd, Meschia JF, et al. Angiography-negative primary central nervous system vasculitis: a syndrome involving small cerebral vessels. Medicine (Baltimore). 2008;87(5):264–71.
25. Salvarani C, Brown RD Jr, Calamia KT, Christianson TJ, Huston J 3rd, Meschia JF, et al. Primary central nervous system vasculitis with prominent leptomeningeal enhancement: a subset with a benign outcome. Arthritis Rheum. 2008;58(2):595–603.
26. Moussaddy A, Levy A, Strbian D, Sundararajan S, Berthelet F, Lanthier S. Inflammatory cerebral amyloid angiopathy, amyloid-beta-related angiitis, and primary angiitis of the central nervous system: similarities and differences. Stroke. 2015;46(9):e210–3.
27. Danve A, Grafe M, Deodhar A. Amyloid beta-related angiitis--a case report and comprehensive review of literature of 94 cases. Semin Arthritis Rheum. 2014;44(1):86–92.
28. Scolding NJ, Joseph F, Kirby PA, Mazanti I, Gray F, Mikol J, et al. Abeta-related angiitis: primary angiitis of the central nervous system associated with cerebral amyloid angiopathy. Brain. 2005;128(Pt 3):500–15.
29. Salvarani C, Brown RD Jr, Calamia KT, Christianson TJ, Huston J 3rd, Meschia JF, et al. Rapidly progressive primary central nervous system vasculitis. Rheumatology (Oxford). 2011;50(2):349–58.
30. Molloy ES, Singhal AB, Calabrese LH. Tumour-like mass lesion: an under-recognised presentation of primary angiitis of the central nervous system. Ann Rheum Dis. 2008;67(12):1732–5.
31. Salvarani C, Brown RD Jr, Calamia KT, Christianson TJ, Huston J 3rd, Meschia JF, et al. Primary central nervous system vasculitis presenting with intracranial hemorrhage. Arthritis Rheum. 2011;63(11):3598–606.
32. Calabrese LH, Mallek JA. Primary angiitis of the central nervous system. Report of 8 new cases, review of the literature, and proposal for diagnostic criteria. Medicine (Baltimore). 1988;67(1):20–39.
33. Golomb MR, Fullerton HJ, Nowak-Gottl U, Deveber G, International Pediatric Stroke Study Group. Male predominance in childhood ischemic stroke: findings from the international pediatric stroke study. Stroke. 2009;40(1):52–7.
34. Benseler SM, deVeber G, Hawkins C, Schneider R, Tyrrell PN, Aviv RI, et al. Angiography-negative primary central nervous system vasculitis in children: a newly recognized inflammatory central nervous system disease. Arthritis Rheum. 2005;52(7):2159–67.
35. Benseler SM, Silverman E, Aviv RI, Schneider R, Armstrong D, Tyrrell PN, et al. Primary central nervous system vasculitis in children. Arthritis Rheum. 2006;54(4):1291–7.
36. Aviv RI, Benseler SM, Silverman ED, Tyrrell PN, Deveber G, Tsang LM, et al. MR imaging and angiography of primary CNS vasculitis of childhood. AJNR Am J Neuroradiol. 2006;27(1):192–9.
37. Eleftheriou D, Cox T, Saunders D, Klein NJ, Brogan PA, Ganesan V. Investigation of childhood central nervous system vasculitis: magnetic resonance angiography versus catheter cerebral angiography. Dev Med Child Neurol. 2010;52(9):863–7.
38. Aviv RI, Benseler SM, DeVeber G, Silverman ED, Tyrrell PN, Tsang LM, et al. Angiography of primary central nervous system angiitis of childhood: conventional angiography versus magnetic resonance angi-

ography at presentation. AJNR Am J Neuroradiol. 2007;28(1):9–15.
39. Hutchinson C, Elbers J, Halliday W, Branson H, Laughlin S, Armstrong D, et al. Treatment of small vessel primary CNS vasculitis in children: an open-label cohort study. Lancet Neurol. 2010;9(11):1078–84.
40. Elbers J, Halliday W, Hawkins C, Hutchinson C, Benseler SM. Brain biopsy in children with primary small-vessel central nervous system vasculitis. Ann Neurol. 2010;68(5):602–10.
41. Lanthier S. Primary angiitis of the central nervous system in children: 10 cases proven by biopsy. J Rheumatol. 2002;29(7):1575–6.
42. International Team for the Revision of the International Criteria for Behcet's D. The International Criteria for Behcet's Disease (ICBD): a collaborative study of 27 countries on the sensitivity and specificity of the new criteria. J Eur Acad Dermatol Venereol. 2014;28(3):338–47.
43. Berlit P, Kramer M, Group ftC. Primary angiitis of the CNS (PACNS) and Behcet disease. Neurol Res Pract. 2019;1(1):10.
44. Watts RA, Lane SE, Bentham G, Scott DG. Epidemiology of systemic vasculitis: a ten-year study in the United Kingdom. Arthritis Rheum. 2000;43(2):414–9.
45. Ormerod AS, Cook MC. Epidemiology of primary systemic vasculitis in the Australian Capital Territory and South-Eastern New South Wales. Intern Med J. 2008;38(11):816–23.
46. Mahr AD. Epidemiological features of Wegener's granulomatosis and microscopic polyangiitis: two diseases or one 'anti-neutrophil cytoplasm antibodies-associated vasculitis' entity? APMIS Suppl. 2009;127:41–7.
47. Younger DS. Epidemiology of the Vasculitides. Neurol Clin. 2019;37(2):201–17.
48. Graf J. Central nervous system disease in antineutrophil cytoplasmic antibodies-associated vasculitis. Rheum Dis Clin N Am. 2017;43(4):573–8.
49. Zheng Y, Zhang Y, Cai M, Lai N, Chen Z, Ding M. Central nervous system involvement in ANCA-associated vasculitis: what neurologists need to know. Front Neurol. 2018;9:1166.
50. De Luna G, Terrier B, Kaminsky P, Le Quellec A, Maurier F, Solans R, et al. Central nervous system involvement of granulomatosis with polyangiitis: clinical-radiological presentation distinguishes different outcomes. Rheumatology (Oxford). 2015;54(3):424–32.
51. Holle JU, Gross WL. Neurological involvement in Wegener's granulomatosis. Curr Opin Rheumatol. 2011;23(1):7–11.
52. Greco A, Rizzo MI, De Virgilio A, Gallo A, Fusconi M, Ruoppolo G, et al. Churg-Strauss syndrome. Autoimmun Rev. 2015;14(4):341–8.
53. Andre R, Cottin V, Saraux JL, Blaison G, Bienvenu B, Cathebras P, et al. Central nervous system involvement in eosinophilic granulomatosis with polyangiitis (Churg-Strauss): report of 26 patients and review of the literature. Autoimmun Rev. 2017;16(9):963–9.
54. Ku BD, Shin HY. Multiple bilateral non-hemorrhagic cerebral infarctions associated with microscopic polyangiitis. Clin Neurol Neurosurg. 2009;111(10):904–6.
55. Bossuyt X, Cohen Tervaert JW, Arimura Y, Blockmans D, Flores-Suarez LF, Guillevin L, et al. Position paper: revised 2017 international consensus on testing of ANCAs in granulomatosis with polyangiitis and microscopic polyangiitis. Nat Rev Rheumatol. 2017;13(11):683–92.
56. Hilhorst M, van Paassen P, Tervaert JW, Limburg Renal Registry. Proteinase 3-ANCA Vasculitis versus Myeloperoxidase-ANCA Vasculitis. J Am Soc Nephrol. 2015;26(10):2314–27.
57. Bougea A, Anagnostou E, Spandideas N, Triantafyllou N, Kararizou E. An update of neurological manifestations of vasculitides and connective tissue diseases: a literature review. Einstein (Sao Paulo). 2015;13(4):627–35.
58. Nadeau SE. Neurologic manifestations of systemic vasculitis. Neurol Clin. 2002;20(1):123–50, vi.
59. Zhang S, Yuan D, Tan G. Neurological involvement in primary systemic vasculitis. Front Neurol. 2019;10:430.
60. Alves NR, Magalhaes CM, Almeida Rde F, Santos RC, Gandolfi L, Pratesi R. Prospective study of Kawasaki disease complications: review of 115 cases. Rev Assoc Med Bras (1992). 2011;57(3):295–300.
61. Terrier B, Cacoub P. Cryoglobulinemia vasculitis: an update. Curr Opin Rheumatol. 2013;25(1):10–8.
62. Kessel A, Vadasz Z, Toubi E. Cogan syndrome--pathogenesis, clinical variants and treatment approaches. Autoimmun Rev. 2014;13(4–5):351–4.
63. Jancatova D, Zelenik K, Kominek P, Matousek P. Atypical Cogan's syndrome: a case report and summary of current treatment options. Int J Pediatr Otorhinolaryngol. 2015;79(3):428–31.
64. Tirelli G, Tomietto P, Quatela E, Perrino F, Nicastro L, Cattin L, et al. Sudden hearing loss and Crohn disease: when Cogan syndrome must be suspected. Am J Otolaryngol. 2015;36(4):590–7.
65. Bougea A, Anagnostou E, Konstantinos G, George P, Triantafyllou N, Kararizou E. A systematic review of peripheral and central nervous system involvement of rheumatoid arthritis, systemic lupus erythematosus, primary Sjogren's syndrome, and associated immunological profiles. Int J Chronic Dis. 2015;2015:910352.
66. Unterman A, Nolte JE, Boaz M, Abady M, Shoenfeld Y, Zandman-Goddard G. Neuropsychiatric syndromes in systemic lupus erythematosus: a meta-analysis. Semin Arthritis Rheum. 2011;41(1):1–11.
67. Popescu A, Kao AH. Neuropsychiatric systemic lupus erythematosus. Curr Neuropharmacol. 2011;9(3):449–57.
68. Barile-Fabris L, Hernandez-Cabrera MF, Barragan-Garfias JA. Vasculitis in systemic lupus erythematosus. Curr Rheumatol Rep. 2014;16(9):440.
69. Sarbu N, Alobeidi F, Toledano P, Espinosa G, Giles I, Rahman A, et al. Brain abnormalities in newly diagnosed neuropsychiatric lupus: systematic MRI approach and correlation with clinical and laboratory

data in a large multicenter cohort. Autoimmun Rev. 2015;14(2):153–9.

70. Berkowitz AL, Samuels MA. The neurology of Sjogren's syndrome and the rheumatology of peripheral neuropathy and myelitis. Pract Neurol. 2014;14(1):14–22.
71. Abdel Razek AA, Alvarez H, Bagg S, Refaat S, Castillo M. Imaging spectrum of CNS vasculitis. Radiographics. 2014;34(4):873–94.
72. Prete M, Racanelli V, Digiglio L, Vacca A, Dammacco F, Perosa F. Extra-articular manifestations of rheumatoid arthritis: an update. Autoimmun Rev. 2011;11(2):123–31.
73. Makol A, Crowson CS, Wetter DA, Sokumbi O, Matteson EL, Warrington KJ. Vasculitis associated with rheumatoid arthritis: a case-control study. Rheumatology (Oxford). 2014;53(5):890–9.
74. Joaquim AF, Appenzeller S. Neuropsychiatric manifestations in rheumatoid arthritis. Autoimmun Rev. 2015;14(12):1116–22.
75. Quemeneur T, Mouthon L, Cacoub P, Meyer O, Michon-Pasturel U, Vanhille P, et al. Systemic vasculitis during the course of systemic sclerosis: report of 12 cases and review of the literature. Medicine (Baltimore). 2013;92(1):1–9.
76. Elst EF, Kamphuis SS, Prakken BJ, Wulffraat NM, van der Net J, Peters AC, et al. Case report: severe central nervous system involvement in juvenile dermatomyositis. J Rheumatol. 2003;30(9):2059–63.
77. Ramanan AV, Sawhney S, Murray KJ. Central nervous system complications in two cases of juvenile onset dermatomyositis. Rheumatology (Oxford). 2001;40(11):1293–8.
78. Macedo PJ, da Silveira VC, Ramos LT, Nobrega FR, Vasconcellos LF. Isolated central nervous system vasculitis as a manifestation of neurosarcoidosis. J Stroke Cerebrovasc Dis. 2016;25(6):e89–92.
79. Erskine N, Tran H, Levin L, Ulbricht C, Fingeroth J, Kiefe C, et al. A systematic review and meta-analysis on herpes zoster and the risk of cardiac and cerebrovascular events. PLoS One. 2017;12(7):e0181565.
80. Zhang Y, Luo G, Huang Y, Yu Q, Wang L, Li K. Risk of stroke/transient ischemic attack or myocardial infarction with herpes zoster: a systematic review and meta-analysis. J Stroke Cerebrovasc Dis. 2017;26(8):1807–16.
81. Mineyko A, Kirton A. Mechanisms of pediatric cerebral arteriopathy: an inflammatory debate. Pediatr Neurol. 2013;48(1):14–23.
82. Gomes LJ. The role of imaging in the diagnosis of central nervous system vasculitis. Curr Allergy Asthma Rep. 2010;10(3):163–70.
83. Gilden D, Cohrs RJ, Mahalingam R, Nagel MA. Varicella zoster virus vasculopathies: diverse clinical manifestations, laboratory features, pathogenesis, and treatment. Lancet Neurol. 2009;8(8):731–40.
84. Grahn A, Studahl M. Varicella-zoster virus infections of the central nervous system – prognosis, diagnostics and treatment. J Infect. 2015;71(3):281–93.
85. Nagel MA, Jones D, Wyborny A. Varicella zoster virus vasculopathy: the expanding clinical spectrum and pathogenesis. J Neuroimmunol. 2017;308:112–7.
86. Nagel MA, Bubak AN. Varicella zoster virus vasculopathy. J Infect Dis. 2018;218(suppl_2):S107–S12.
87. Klein M, Koedel U, Pfefferkorn T, Zeller G, Woehrl B, Pfister HW. Arterial cerebrovascular complications in 94 adults with acute bacterial meningitis. Crit Care. 2011;15(6):R281.
88. Younger DS, Coyle PK. Central nervous system vasculitis due to infection. Neurol Clin. 2019;37(2):441–63.
89. Berlit P. Isolated angiitis of the CNS and bacterial endocarditis: similarities and differences. J Neurol. 2009;256(5):792–5.
90. Lammie GA, Hewlett RH, Schoeman JF, Donald PR. Tuberculous cerebrovascular disease: a review. J Infect. 2009;59(3):156–66.
91. Thwaites GE. Advances in the diagnosis and treatment of tuberculous meningitis. Curr Opin Neurol. 2013;26(3):295–300.
92. Cordato DJ, Djekic S, Taneja SR, Maley M, Beran RG, Cappelen-Smith C, et al. Prevalence of positive syphilis serology and meningovascular neurosyphilis in patients admitted with stroke and TIA from a culturally diverse population (2005–09). J Clin Neurosci. 2013;20(7):943–7.
93. Williams PL, Johnson R, Pappagianis D, Einstein H, Slager U, Koster FT, et al. Vasculitic and encephalitic complications associated with Coccidioides immitis infection of the central nervous system in humans: report of 10 cases and review. Clin Infect Dis. 1992;14(3):673–82.
94. Marquez JM, Arauz A. Cerebrovascular complications of neurocysticercosis. Neurologist. 2012;18(1):17–22.
95. Calabrese LH, Duna GF. Drug-induced vasculitis. Curr Opin Rheumatol. 1996;8(1):34–40.
96. Younger DS. Central nervous system vasculitis due to substance abuse. Neurol Clin. 2019;37(2):425–40.
97. Aoki S, Hayashi N, Abe O, Shirouzu I, Ishigame K, Okubo T, et al. Radiation-induced arteritis: thickened wall with prominent enhancement on cranial MR images report of five cases and comparison with 18 cases of Moyamoya disease. Radiology. 2002;223(3):683–8.
98. Dutra LA, de Souza AW, Grinberg-Dias G, Barsottini OG, Appenzeller S. Central nervous system vasculitis in adults: an update. Autoimmun Rev. 2017;16(2):123–31.
99. de Boysson H, Parienti JJ, Mawet J, Arquizan C, Boulouis G, Burcin C, et al. Primary angiitis of the CNS and reversible cerebral vasoconstriction syndrome: a comparative study. Neurology. 2018;91(16):e1468–e78.
100. Ducros A. Reversible cerebral vasoconstriction syndrome. Handb Clin Neurol. 2014;121:1725–41.

101. Byram K, Hajj-Ali RA, Calabrese L. CNS vasculitis: an approach to differential diagnosis and management. Curr Rheumatol Rep. 2018;20(7):37.
102. Salvarani C, Brown RD Jr, Christianson TJ, Huston J 3rd, Giannini C, Miller DV, et al. Adult primary central nervous system vasculitis treatment and course: analysis of one hundred sixty-three patients. Arthritis Rheumatol. 2015;67(6):1637–45.
103. de Boysson H, Zuber M, Naggara O, Neau JP, Gray F, Bousser MG, et al. Primary angiitis of the central nervous system: description of the first fifty-two adults enrolled in the French cohort of patients with primary vasculitis of the central nervous system. Arthritis Rheumatol. 2014;66(5):1315–26.
104. Verhey LH, Branson HM, Shroff MM, Callen DJ, Sled JG, Narayanan S, et al. MRI parameters for prediction of multiple sclerosis diagnosis in children with acute CNS demyelination: a prospective national cohort study. Lancet Neurol. 2011;10(12):1065–73.
105. Boulouis G, de Boysson H, Zuber M, Guillevin L, Meary E, Costalat V, et al. Primary angiitis of the central nervous system: magnetic resonance imaging spectrum of parenchymal, meningeal, and vascular lesions at baseline. Stroke. 2017;48(5):1248–55.
106. Swartz RH, Bhuta SS, Farb RI, Agid R, Willinsky RA, Terbrugge KG, et al. Intracranial arterial wall imaging using high-resolution 3-tesla contrast-enhanced MRI. Neurology. 2009;72(7):627–34.
107. Mandell DM, Matouk CC, Farb RI, Krings T, Agid R, terBrugge K, et al. Vessel wall MRI to differentiate between reversible cerebral vasoconstriction syndrome and central nervous system vasculitis: preliminary results. Stroke. 2012;43(3):860–2.
108. Mineyko A, Narendran A, Fritzler ML, Wei XC, Schmeling H, Kirton A. Inflammatory biomarkers of pediatric focal cerebral arteriopathy. Neurology. 2012;79(13):1406–8.
109. Haroon M, Molloy E, Farrell M, Alraqi S. Central nervous system vasculitis: all that glitters is not gold. J Rheumatol. 2012;39(3):662–3.
110. Chen SH, Sur S, Sedighim S, Kassi A, Yavagal D, Peterson EC, et al. Utility of diagnostic cerebral angiography in the management of suspected central nervous system vasculitis. J Clin Neurosci. 2019;64:98–100.
111. Monti S, Floris A, Ponte C, Schmidt WA, Diamantopoulos AP, Pereira C, et al. The use of ultrasound to assess giant cell arteritis: review of the current evidence and practical guide for the rheumatologist. Rheumatology (Oxford). 2018;57(2):227–35.
112. Pacheco Castellanos Mdel C, Minguez Vega M, Martinez Caballero A, Bernabeu Gonzalvez MP. Early diagnosis of large vessel vasculitis: usefulness of positron emission tomography with computed tomography. Reumatol Clin. 2013;9(1):65–8.
113. Yuh WT, Ueda T, Maley JE, Quets JP, White M, Hahn PY, et al. Diagnosis of microvasculopathy in CNS vasculitis: value of perfusion and diffusion imaging. J Magn Reson Imaging. 1999;10(3):310–3.
114. Masdeu JC, Arbizu J, Toledo J, Valero M. SPECT and PET in neurology. Neurologia. 2006;21(5):219–25.
115. Bai HX, Zou Y, Lee AM, Lancaster E, Yang L. Diagnostic value and safety of brain biopsy in patients with cryptogenic neurological disease: a systematic review and meta-analysis of 831 cases. Neurosurgery. 2015;77(2):283–95; discussion 95.
116. Torres J, Loomis C, Cucchiara B, Smith M, Messe S. Diagnostic yield and safety of brain biopsy for suspected primary central nervous system angiitis. Stroke. 2016;47(8):2127–9.
117. Mukhtyar C, Guillevin L, Cid MC, Dasgupta B, de Groot K, Gross W, et al. EULAR recommendations for the management of primary small and medium vessel vasculitis. Ann Rheum Dis. 2009;68(3):310–7.
118. Mukhtyar C, Guillevin L, Cid MC, Dasgupta B, de Groot K, Gross W, et al. EULAR recommendations for the management of large vessel vasculitis. Ann Rheum Dis. 2009;68(3):318–23.
119. Yates M, Watts RA, Bajema IM, Cid MC, Crestani B, Hauser T, et al. EULAR/ERA-EDTA recommendations for the management of ANCA-associated vasculitis. Ann Rheum Dis. 2016;75(9):1583–94.
120. Hellmich B, Agueda A, Monti S, Buttgereit F, de Boysson H, Brouwer E, et al. 2018 update of the EULAR recommendations for the management of large vessel vasculitis. Ann Rheum Dis. 2020;79(1):19–30.

# Infections, Immunodeficiency, and Complications of Immunomodulatory Therapies in Neuroimmunology

# 25

Amanda L. Piquet, Daniel M. Pastula, J. David Beckham, and Kenneth L. Tyler

**Key Points**

1. A dysfunctional adaptive immune response can lead to increased susceptibility to infection, malignancy, or autoimmunity.
2. Viral central nervous system (CNS) infections have been associated with development of CNS antibody-mediated autoimmune disease with one well-described entity being anti-N-methyl-D-aspartate receptor (NMDAR) antibody-associated encephalitis following herpes simplex virus encephalitis.
3. In addition to viral triggers, dysregulation of the immune system can occur in the setting of primary immunodeficiency (e.g., combined variable immunodeficiency), medically induced immunodeficiency (e.g., hypogammaglobulinemia with rituximab), or medically induced dysregulation (e.g., novel oncological therapies with immune checkpoint inhibitors) and can result in secondary autoimmune and neurological complications.
4. Complications of immunotherapy often used in multiple sclerosis (MS) and other autoimmune neurological disease include progressive multifocal leukoencephalopathy, secondary hypogammaglobulinemia with B-cell-depleting agents, and secondary autoimmunity with alemtuzumab.
5. Immune checkpoint inhibitors are novel treatments for advanced or refractory cancers, but can have serious peripheral and central nervous system complications.

A. L. Piquet (✉)
Department of Neurology, Section of Neuroimmunology, Neuro-Infectious Diseases Group, University of Colorado School of Medicine, Aurora, CO, USA
e-mail: Amanda.Piquet@CUAnschutz.edu

D. M. Pastula
Department of Neurology, Medicine (Infectious Diseases), and Epidemiology, Neuro-Infectious Diseases Group, University of Colorado School of Medicine, Aurora, CO, USA

J. D. Beckham
Departments of Medicine, Neurology, and Immunology & Microbiology, Neuro-Infectious Diseases Group, University of Colorado School of Medicine, Rocky Mountain Regional VA Medical Center, Aurora, CO, USA

K. L. Tyler
Department of Medicine and Immunology-Microbiology, Department of Neurology, Neuro-Infectious Diseases Group, University of Colorado School of Medicine, University of Colorado Hospital and Rocky Mountain Regional VA Medical Center, Aurora, CO, USA



A. L. Piquet, E. Alvarez (eds.), *Neuroimmunology*, https://doi.org/10.1007/978-3-030-61883-4_25

## Introduction

In the presentation of neurological syndromes (e.g., encephalopathy/encephalitis, meningitis, and/or myelopathy/myelitis), the differential diagnosis is often broad and includes various infectious, inflammatory, autoimmune, vascular, and neoplastic etiologies. There are numerous inflammatory and autoimmune etiologies alone. However, evaluating a patient for an underlying infection and/or immunodeficiency is also a critical aspect of the workup. The first part of this chapter will focus on the association of viral infections as triggers of autoimmunity and the dysregulation of the immune system in the development of autoimmune disorders. One specific example of this is the development of anti-N-methyl-D-aspartate receptor (NMDAR) antibody-associated encephalitis following herpes simplex virus encephalitis (HSVE) in a subset of patients.

In additional to viral triggers, dysregulation of the immune system by either primary immunodeficiency (e.g., combined variable immunodeficiency [CVID]), medically induced immunodeficiency (e.g., hypogammaglobulinemia with rituximab), or medically induced dysregulation (e.g., novel oncological therapies with immune checkpoint inhibitors) can result in secondary autoimmune and neurological complications. Furthermore, patients on immunosuppressive therapy for the treatment of underlying autoimmune or inflammatory diseases are at increased risk for opportunistic infections (e.g., progressive multifocal leukoencephalopathy [PML] caused by John Cunningham [JC] virus after natalizumab use). The second part of this chapter will focus on immunodeficiency in neurological disease and notable side effects of these immunomodulatory therapies.

## Infectious Triggers in Autoimmune and Inflammatory Diseases

Viral central nervous system (CNS) infections have been associated with development of CNS antibody-mediated autoimmune disease. One notable example is HSVE leading to the development of antibody-mediated autoimmune encephalitis (AE), particularly NMDAR encephalitis (discussed in Chap. 12). The intertwined relationship between infection and inflammatory diseases can lead to the development of post- or para-infectious syndromes often associated with an antecedent or concomitant infection, respectively. A classic example is acute disseminated encephalomyelitis (ADEM) [1, 2]. ADEM typically presents as a monophasic illness often preceded by a viral prodrome or rarely by immunization [3]; this disease is further discussed in Chap. 29. Other forms of post-infectious encephalitis include acute hemorrhagic leukoencephalitis (a very rare form of ADEM) and Bickerstaff's brainstem encephalitis (a rare variant of the anti-Gq1b antibody syndrome characterized by acute ophthalmoplegia, ataxia, and encephalopathy). Common pathogens implicated in post- or para-infectious encephalitis include varicella-zoster virus (VZV), influenza virus, coxsackievirus, Epstein-Barr virus (EBV), and mycoplasma, as well as many others [1].

There are multiple mechanisms by which infection can lead to autoimmunity including molecular mimicry, bystander activation, epitope spreading, or presentation of cryptic antigens [4]. When a pathogen has similar proteins to a self-antigen, then the pathogen can act as a "mimic," with the immune response to the pathogen cross-reacting with the related self-antigens in a process known as molecular mimicry [5]. B and T cells are activated in response to the microbial antigen and lead to damage of self-tissues [6]. Molecular mimicry has been implicated for many immune-medicated neurological syndromes, including Sydenham chorea secondary to *Streptococcus pyogenes* infection [7], Guillain-Barré syndrome in the setting of *Campylobacter jejuni* [8], and cross-reactivity between EBV and myelin basic protein [9] (the relationship between EBV and multiple sclerosis is further discussed in Chap. 10).

In the setting of Sydenham chorea, a rare clinical manifestation following streptococcal infection, antibodies to the basal ganglia in the brain, and molecular mimicry between basal ganglia and *Streptococcus pyogenes*-derived proteins has been suggested [10] as a potential mechanism. However, the exact mechanism remains elusive. Animal model studies demonstrated abnormal movement and behavioral disorders in mice primed with *S. pyogenes* [11]. These mice had antibody deposits in the brain as well as serum antibody cross-reactive to several regions of the brain [11]. Additionally, rabbits immunized with streptococcal M protein from *S. pyogenes* developed antibodies that cross-reacted with several human brain proteins, demonstrating potential cross-reactive epitopes [12].

On the other hand, the mechanism of bystander activation occurs when viral infections lead to significant activation of antigen-presenting cells (APCs). These activated APCs then potentially activate autoreactive T cells, which can initiate autoimmune disease [13]. Additionally, viral-specific T cells might also initiate bystander activation via destruction and cell death. Viral infected cells can release cytokines such as tumor necrosis factor (TNF), lymphotoxin, and nitric oxide, which can lead to killing of unaffected neighboring cells. This mechanism has been supported by the experimental animal model in non-obese diabetic mice for type 1 diabetes [14].

## Evidence of HSVE-Induced Autoimmunity

The relationship between NMDAR antibodies after HSVE was first identified through case observations [15, 16]. In 2014, five patients (four children and one adult) with HSVE were prospectively identified who presented with neurological relapse associated with positive cerebrospinal fluid (CSF) anti-NMDAR-antibodies [17]. Relapse occurred at a median of 24 days after the onset of HSVE, with the children demonstrating choreoathetosis and mental status changes and the adult demonstrating abnormal behavior and personality. There was no evidence of CSF HSV polymerase chain reaction (PCR) positivity to indicate HSVE recurrence, but all five patients had evidence of NMDAR antibody synthesis 1–4 weeks post-HSVE preceding the neurological relapse. The NMDAR antibody was not present during the initial HSVE infection. Apart from one patient who improved spontaneously, the other four patients experienced improvement after a course of immunotherapy. The authors also retrospectively studied the presence of autoantibodies against neuronal cell-surface antigens in archived serum and CSF of 34 HSVE patients. Twelve of 34 (35%) of these patients were found to have antibodies to neuronal cell-surface antigens, 2 developed NMDAR antibodies, 9 developed antibodies against unknown antigens, and 1 developed antibodies to both the NMDAR and unknown antigens. These findings suggest that the presence of autoantibodies increased over time following infection with HSVE and that the initial HSV infection may lead to release of multiple antigens that can serve as potential targets for development of autoimmunity [17].

In a large multicenter study, a prospective cohort of 53 patients with HSVE, 27% developed symptoms of AE within 3 months of completing a 14–21-day course of acyclovir treatment [18]. The median time on onset after HSVE was 32 days, and it was noted that children 4 years and younger developed choreoathetosis, while older children and adults developed primarily psychiatric manifestations [18]. This particular study demonstrated that development of AE after HSVE is frequent and supports prior literature that the clinical presentation tended to vary with age.

It appears that HSVE is a robust trigger for anti-neuronal autoimmunity [17, 19]. While it is plausible to consider molecular mimicry as a possible mechanism, no HSV epitope cross-reacting with neuronal antigens or the NMDAR has yet been identified. Alternatively, other mechanisms such as the "bystander effect" has been postulated as HSV induces an intense inflammatory response in limbic structures leading to necrosis and the presentation of locally and abundantly expressed NMDAR epitopes to the immune system and initiating an autoimmune response [20]. This is supported by the concomitant presence of neuronal autoantibodies in the setting of herpes infections not isolated to HSVE, but also demonstrated with EBV, VZV, and human herpes virus 6 (HHV-6) [19, 21]. However, this finding with other herpes infections is much more limited in the literature and isolated to a few case reports.

The immune response leading to the development of NMDAR autoantibodies and subsequent development of NMDAR encephalitis may not be due to the presence of the HSV specifically, but rather to the virus-induced immune response, and development of synaptic autoimmunity may be secondary to the inflammation and neuronal tissue damage. This mechanism is further supported by the demonstration of autoantibodies against other neuronal antigens and not exclusively against NMDAR [17]. With HSVE often affecting limbic structures, which have a rich expression of NMDA receptors, it is possible that the inflammation and damage leading to subsequent release of otherwise immune-privileged NMDAR and other neuronal antigens could induce a broad autoimmune response to these potential epitopes. This anatomic affinity of HSV may also explain why similar development of NMDAR encephalitis has not been documented following other neurological disease, such as stroke or bacterial meningitis, where there is neuronal damage but involvement of the limbic system is uncommon.

Furthermore, this relationship between NMDAR encephalitis and HSV may not be limited to CNS HSV involvement. A retrospective study of 39 NMDAR encephalitis pediatric patients demonstrated that in the setting of non-encephalitic infection (e.g., oral or skin lesions using confirmatory testing HSV-1 immunoglobulin [Ig] G in the serum), there appears to be a meaningful association between HSV and development of anti-NMDAR encephalitis compared to age-matched controls [22]. The relationship between NMDAR antibodies and HSV may provide a clue to the pathophysiology of AE in certain patients. However, the exact mechanism remains unknown, and why certain patients are more vulnerable compared to others in the development of AE is unclear. Abnormalities of immune regulation and genetic susceptibility may play roles.

## Immunodeficiency and Autoimmunity

A dysfunctional adaptive immune response can lead to increased susceptibility to infection, malignancy, or autoimmunity. CVID, which is the most common severe primary antibody deficiency, is a known disease model for this association to underlying autoimmunity. In CVID, a compromised humoral immunity results in low levels of IgG, IgA, and/or IgM. Phenotypes of this disease include patients who present with both acute and chronic infections, as well as a wide range of inflammatory and autoimmune diseases [23]. Diagnostic criteria for CVID include a low IgG level (IgG < 700 mg/dL), often with associated low IgA or IgM, a history of infections (typically bacterial), and a poor or absent response to pneumococcal vaccine challenge [24].

Given that patients with CVID cannot make pathogen-specific antibodies, infections tend to be the main clinical feature. When patients present with neurological syndromes of meningoencephalitis in the setting of known CVID, ruling out an underlying infection becomes paramount. Likewise, CVID should be considered in the first presentation of atypical neurologic infections. Common CNS infections seen in CVID include enteroviral meningitis as well as other viral infections (e.g., adenovirus, cytomegalovirus virus [CMV], HSV, VZV, and West Nile virus [WNV]), bacterial infection with *Streptococcus pneumoniae*, certain fungal infections including histoplasmosis and cryptococcus, and toxoplasmosis infection. However, there are innumerable pathogens known to cause infectious encephalitis, and in the setting of underlying immunodeficiency, identifying the pathogen can pose a diagnostic challenge. Diagnostics with high-throughput next-generation sequencing (NGS) platforms could be advantageous as they may provide an advantage over the traditional approach of testing each individual suspected pathogen using specific PCR assays [25]. This approach has proven useful in other forms of immunodeficiency and has facilitated diagnosis of a wide variety of infections including neuroleptospirosis [26] and astrovirus encephalitis [27, 28]. Patients with an unknown immune status who present with an atypical CNS infection should have a full immunodeficiency workup to rule out possible unrecognized CVID or other acquired immunodeficiency, such as human immunodeficiency virus (HIV) or less commonly idiopathic CD4 deficiency. Beyond unusual infections, there have been several clinical manifestations identified in and associated with CVID, including chronic lung disease, diffuse granulomatous disease, enteropathy with or without associated nutrient deficiency, lymphoid hyperplasia, endocrine dysfunction, and malignancy [29].

### Autoimmune Disease in CVID

Autoimmunity occurs in approximately 25–30% of patients with CVID [29], and autoimmune complications identified include neutropenia, vitiligo, Sjögren's syndrome/sicca syndrome, autoimmune thyroiditis, diabetes mellitus, pernicious anemia, systemic lupus erythematosus (SLE), antiphospholipid syndrome, uveitis, juvenile rheumatoid arthritis, lichen planus, vasculitis, and psoriasis [23]. Granulomatous disease has also been reported, most commonly in the form of granulomatous lymphocytic interstitial lung disease (GLILD).

### Granulomatous Disease in CVID

In CVID, 8–22% of patients develop granulomatous disease or "atypical sarcoid-like" lesions [29–32]. Many times, there can be multi-organ system involvement, and patients can have biopsy-proven granulomatous disease long before CVID is recognized [29, 33]. While CVID-associated granulomatous disease is still not commonly recognized by physicians [33], understanding of neurological involvement and complications is even less common [34, 35]. It is common for CVID to be confused with sarcoidosis as granulomas in CVID are similarly noncaseating. There are no current guidelines for the diagnosis and treatment of noncaseating granulomatous disease. However, patients presumptively diagnosed with sarcoidosis should have serum immunoglobulins measured as part of their evaluation before the initiation of immunotherapy in order to rule out CVID as an etiology of the granulomatous disease. Neurosarcoidosis is discussed further in Chap. 23.

## Complications of Immunomodulatory Therapies: Infections and Immune-Related

### Progressive Multifocal Leukoencephalopathy

Progressive multifocal leukoencephalopathy (PML) is a rare, often fatal, demyelinating disease seen almost exclusively in immunocompromised hosts caused by the infection of oligodendrocytes with the JC virus. The infection of oligodendrocytes causes cell lysis leading to neurological impairment including progressive cognitive and motor dysfunction [36]. JC virus infection can also lead to non-PML forms of CNS infection including encephalitis and a cerebellar granule cell neuronopathy [37]. PML occurs predominantly in patients with HIV infections or those with underlying neoplastic disease and organ transplants undergoing immune suppression. With the use of immunomodulatory treatments in the field of neuroimmunology, PML has been well recognized in the MS population, most commonly associated with the use of natalizumab. Natalizumab, approved for the treatment of relapsing-remitting MS (RRMS) in 2004, is a monoclonal antibody to α(alpha)4β(beta)1 integrin that blocks the trafficking of lymphocytes across the blood-brain barrier via the vascular cell adhesion molecule-1 (VCAM-1) [38]. After three cases of PML (two of which were fatal), natalizumab was removed from the market in February 2005. After the development of risk mitigation strategies using anti-JC virus antibody screening, it was released back on market in 2006. The initial prevalence of natalizumab-associated PML in MS was estimated to be 1 in 1000 [39]. It is now recognized that duration of exposure to natalizumab, prior exposure to immunosuppressive medications, and higher JC virus antibody index are key factors increasing the overall risk of PML. In patients treated with natalizumab for more than 24 months, with prior immunosuppressive therapy, and with an elevated JC virus antibody index, the risk of PML can be as high as 1 in 70 [36, 40, 41]. Patients without JC virus antibodies appear to be at extremely low risk of developing PML (<1 in 10,000), indicating that disease likely develops from reactivation of latent virus in previously infected individuals. Regular monitoring of JC virus antibody titers for seroconversion or increase in antibody index may identify patients requiring drug discontinuation and consideration of alternate therapies. Regular magnetic resonance imaging (MRI) monitoring may also identify patients with clinically asymptomatic disease. Studies are underway to examine whether extended dosing intervals of natalizumab (e.g., every 5 weeks instead of the standard monthly regimen) reduce the risk of PML without compromising therapeutic efficacy [42]. A wide variety of alternative disease-modifying therapies, including those with similar efficacy including rituximab and ocrelizumab, are now available for patients with positive JC virus serostatus or elevated JCV antibody indices.

PML risk is not exclusively confined to natalizumab. Cases of PML associated with other disease-modifying therapies including fingolimod, dimethyl fumarate, teriflunomide, rituximab, and ocrelizumab have been reported, although the risk is significantly less than natalizumab. In cases of PML seen with dimethyl fumarate, associated lymphopenia appears to be a contributing risk factor [43].

Brain MRI is essential in the diagnosis of PML. Although no single feature is considered pathognomonic for PML, typical features include hyperintense lesions on T2-weighted and fluid-attenuated inversion recovery (FLAIR) sequences, hypointense lesions on T1-weighted images, and lack of gadolinium enhancement. However, as many of 15% of patients with HIV-associated PML and 40% of natalizumab-associated PML may have gadolinium enhancement at the time of diagnosis [44]. The lesions are often but not invariably multifocal as the name implies. The frontal lobes and parieto-occipital regions are most commonly affected [44]. Definitive diagnosis is made by JC virus PCR positivity in the CSF or by brain biopsy demonstrating viral antigen or DNA in the presence of consistent neuropathological features including viral inclusions seen in the oligodendroglial nuclei. See Fig. 25.1.

## Secondary Hypogammaglobulinemia with B-Cell-Depleting Therapies

The use of anti-CD20 monoclonal antibodies (mAb) has greatly increased over the last several years for autoimmune neurological diseases and multiple sclerosis (MS). Rituximab is a chimeric anti-CD20 mAb that depletes precursor, and eventually mature, antibody-producing plasma cells. Ocrelizumab, which is the first US Food and Drug Administration (FDA) anti-CD20 therapy approved for MS in 2017, is a fully humanized anti-CD20 mAb. Additionally, there are other CD20 mAbs under investigation for MS. For neuromyelitis optic spectrum disease (NMOSD) (discussed further in Chap. 15), there recently completed clinical trial using a similar mechanism of B-cell depletion with an anti-CD19 mAb, known as inebilizumab [45]. With the emergence of these new drugs, it is likely that the use of B-cell-depleting mAbs will continue to grow.

The CD20 antigen is an epitope that is expressed on the B-cell lineage cells with the exception of plasma cells, while the CD19 antigen target is also expressed on B cells as well as plasmablasts and plasma cells. All CD20-positive B cells express CD19, but not all CD19-positive cells express CD20. Given this, anti-CD19 mAb therapy is thought to have enhanced antibody-dependent cell-mediated cytotoxic effects beyond the anti-CD20 mAb therapy. Regardless, both mechanisms cause overall B-cell suppression and can lead to secondary hypogammaglobulinemia. Hypogammaglobulinemia is an immune disorder leading to the reduction of gamma globulin classes such as IgG, IgM, and

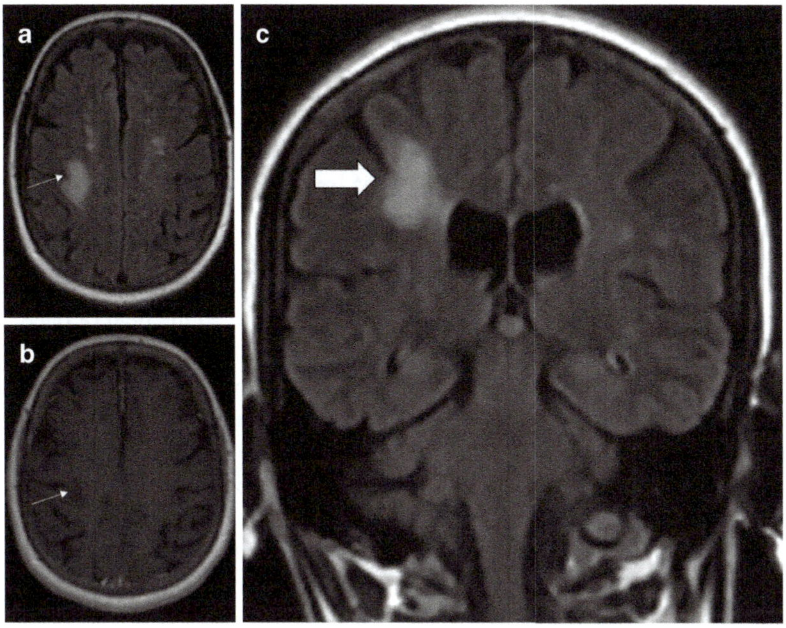


**Fig. 25.1** Magnetic resonance imaging (MRI) of a 58-year-old bone marrow transplant (BMT) recipient for acute myeloid leukemia who developed progressive multifocal leukoencephalopathy (PML), confirmed by JC virus polymerase chain reaction (PCR) testing in the cerebrospinal fluid (CSF). The patient was severely immunocompromised after treatment with multiple immunotherapies including rituximab, methotrexate, and tacrolimus following her BMT; laboratory testing included CD4+ count of 100, immunoglobulin G (IgG) level of 585, and CD20 counts suppressed at <1. MRI demonstrated T2 hyperintense signal change on fluid-attenuated inversion recovery (FLAIR) sequence (panel A, thin white arrow) without associated contrast enhancement (panel B) on T1 post-contrast imaging; however, there is an associated T1 hypointense lesion (thin white arrow). Coronal FLAIR (panel C) demonstrates a white matter lesion with involvement of the subcortical U-fibers (thick arrow). (MR images courtesy of John Corboy, MD)

IgA. Similar to CVID (discussed earlier), this can lead to a variety of infectious complications. This can include CNS viral infections including those caused by enterovirus, adenovirus, CMV, HSV, VZV, and WNV. Additionally, common systemic, recurrent infections encountered include sinopulmonary infections (pneumonia, chronic otitis media, sinusitis, bronchitis), bacterial infections, or parasitic gastroenteritis. Risk of infectious complications is associated with low pre-treatment IgG levels [46]. Thus, current recommendations include testing patients' antibody titers prior to treatment with anti-B-cell therapy [46, 47] and considering alternative immunotherapy in those patients with hypogammaglobulinemia. If hypogammaglobulinemia is noted after rituximab therapy, intravenous immunoglobulin G (IVIg) can be used as replacement therapy [48].

In a cohort of 50 individuals with NMOSD treated over an extended period with rituximab, 5 developed serious infections in the setting of antibody deficiency, and hypogammaglobulinemia was documented in 64% of patients [49]. Moreover, in a cohort study of 4479 patients receiving rituximab for various neurological and systemic autoimmune disorders, most patients (85.4%) did not have immunoglobulins checked before rituximab therapy. However, if patients did have hypogammaglobulinemia noted prior to initiation of rituximab, there was a dramatic increase in severe infections leading to hospitalization [46]. These studies emphasize the need for proper screening and monitoring of patients on B-cell-depleting agents to help minimize the risk of secondary infectious complications due to underlying immunodeficiency.

## Autoimmunity with Alemtuzumab

Alemtuzumab is a humanized anti-CD52 monoclonal antibody used to treat refractory and aggressive RRMS. While a highly efficacious therapy, there are several side effects ranging from infusion reactions, secondary autoimmune disease, malignancies, and infections. Secondary B-cell autoimmune disease with alemtuzumab often peaks 2–3 years after treatment initiation and had been reported in about 50% of MS patients within 5–7 years of treatment [50]. While thyroid disease is the most frequently encountered autoimmune adverse event, other adverse events have included immune thyrombocytopenia (ITP) and nephropathies [51]. In the CAMMS223 trial, a fatality associated with ITP led to the implementation of a strict safety monitoring program to help with early detection of autoimmune events [52].

It has been hypothesized that after rapid depletion of both B- and T-cell populations expressing anti-CD52, subsequent repopulation of B cells in the absence of T-cell regulation allows for the development of autoimmunity [53]. It has also been postulated that this tends to occur more frequently in those individuals with genetic susceptibility for autoimmunity, such as patients with MS compared to those circumstances in which alemtuzumab is used in the treatment of malignancies [53–55]. Additional studies are needed to better understand this mechanism to help mitigate the effect of alemtuzumab on secondary autoimmunity.

## Neurological Complications of Immune Checkpoint Inhibitors

Immune checkpoint inhibitors (ICPIs) are novel treatments for advanced or refractory cancers and have transformed the field of oncology. It is now apparent that these drugs can also cause immune-related side effects. Examples of ICPIs include monoclonal antibodies targeting the cytotoxic T lymphocyte-associated antigen 4 (CTLA4) and the programmed death-1 receptor (PD-1) or its ligand (PD-L1). Immune-related adverse events (irAEs) affect the colon, liver, lungs, pituitary, thyroid, heart, and nervous system. Cardiac and neurological events can be prominent, with a 2018 meta-analysis of 112 trials involving 19,217 patients demonstrating fatal neurological events in 6% (11/193) treated with ipilimumab, 15% (50/333) with anti-programed death-1/programed death ligand-1 (PD-1/PD-L1), and 7% (8/87) with combination therapy [56]. Neurological events include myasthenia gravis, encephalitis, meningitis, polyradiculitis, cranial polyneuropathy, myositis, and acute and chronic inflammatory demyelinating polyneuropathy [57]. Examples of specific drugs and their associated neurological complications are listed in Table 25.1.

**Table 25.1** Immune checkpoint inhibitors with US Food and Drug Administration (FDA) approval for the treatment of cancer and associated neurological syndromes

| Class of ICPI | Clinical indication | Neurological irAE reported |
|---|---|---|
| *Anti-PD-1 mAbs* | | |
| Nivolumab | Metastatic melanoma, metastatic squamous non-small cell lung cancer, metastatic urothelial carcinoma | Encephalitis, AIDP, CIPD, myasthenic syndromes, myositis |
| Pembrolizumab | Metastatic melanoma, metastatic non-small cell lung cancer, head and neck squamous cell carcinoma, pediatric refractory classic Hodgkin's lymphoma | Encephalitis, AIPD, CIPD, cranial neuropathies, myasthenic syndromes, myositis |
| *Anti-CTLA4 mAbs* | | |
| Ipilimumab | Metastatic melanoma, small cell lung cancer, bladder cancer | Aseptic meningitis, encephalitis, AIDP, CIPD, cranial neuropathies, myasthenic syndromes, myositis |

*ICPI* immune checkpoint inhibitor, *irAE* immune-related adverse events, *PD-1* programmed death-1 receptor, *mAbs* monoclonal antibodies, *AIDP* acute inflammatory demyelinating polyneuropathy, *CIPD* chronic inflammatory demyelinating polyneuropathy, *CTLA4* cytotoxic T lymphocyte-associated antigen 4

Neurological irAEs involving the central nervous system are rare, but encephalitis and meningitis are the best characterized toxicities [58]. While encephalitis has been described in several patents on ICPIs, one case reported the development of NMDAR antibodies after treatment of metastatic cancer with nivolumab and ipilimumab [59]. While the causality could not be proven in this particular case, there is a suggestion based on the time course that the immune checkpoint inhibition may have unmasked or accelerated preexisting autoimmune reactions targeting neuronal epitopes. Patients with known MS may be at risk for relapse when treated with ICPs [60, 61], and there has been a case report of fatal CNS demyelination in a patient without a history of MS [62].

Neurological iAEs involving the peripheral nervous system include a wide range of presentations including mild-to-moderate peripheral neuropathies, myasthenic syndrome, to severe or fatal cases of acute inflammatory demyelinating polyneuropathy (AIDP) and myositis [58]. Exacerbation of existing myasthenia gravis (MG) has been reported with anti-PD1 [63–65], while de novo presentations of MG are estimated to occur in 0.1–0.2% of patients treated with ICIs [58, 66–68].

Often neurological adverse events occur shortly after initiation of ICPIs, with a mean onset of 8 weeks [57]. There is no validated therapeutic approach to the management of these events, but often the treatment approach includes discontinuation of the ICPI and prompt initiation of corticosteroids. Given the broad spectrum of neurological complications, expedited diagnosis can be challenging but paramount as rapid initiation of corticosteroids for treatment are mandatory to prevent clinical deterioration and sometimes fatal outcomes.

## Conclusion

The intricacies of the immune system are complex. While there certainly appears to be clinical evidence supporting a role for infection-induced autoimmunity, evidence is stronger for some diseases than for others. The complexity and heterogeneity of individual immune systems has made this area difficult to study and has resulted in controversial evidence at times. Further research with animal models to clarify potential mechanisms of disease induction, as well as studying predisposing genetic markers in patients, are needed to better understand the pathogenesis of autoimmunity. Moreover, our understanding of the immune system and the interplay with the CNS continues to evolve, especially in the face of our expanding knowledge and experience with immunomodulatory therapies.

## References


1. Hynson JL, Kornberg AJ, Coleman LT, Shield L, Harvey AS, Kean MJ. Clinical and neuroradiologic features of acute disseminated encephalomyelitis in children. Neurology. 2001;56(10):1308–12.
2. Kipps A, Dick G, Moodie JW. Measles and the central nervous system. Lancet. 1983;2(8364):1406–10.
3. Karussis D, Petrou P. The spectrum of post-vaccination inflammatory CNS demyelinating syndromes. Autoimmun Rev. 2014;13(3):215–24.
4. Ercolini AM, Miller SD. The role of infections in autoimmune disease. Clin Exp Immunol. 2009;155(1):1–15.
5. Fujinami RS, Oldstone MB, Wroblewska Z, Frankel ME, Koprowski H. Molecular mimicry in virus infection: crossreaction of measles virus phosphoprotein or of herpes simplex virus protein with human intermediate filaments. Proc Natl Acad Sci U S A. 1983;80(8):2346–50.
6. Cusick MF, Libbey JE, Fujinami RS. Molecular mimicry as a mechanism of autoimmune disease. Clin Rev Allergy Immunol. 2012;42(1):102–11.
7. Kirvan CA, Swedo SE, Heuser JS, Cunningham MW. Mimicry and autoantibody-mediated neuronal cell signaling in Sydenham chorea. Nat Med. 2003;9(7):914–20.
8. Ang CW, Jacobs BC, Laman JD. The Guillain-Barré syndrome: a true case of molecular mimicry. Trends Immunol. 2004;25(2):61–6.
9. Mameli G, Cocco E, Frau J, Marrosu MG, Sechi LA. Epstein Barr Virus and Mycobacterium avium subsp. paratuberculosis peptides are recognized in sera and cerebrospinal fluid of MS patients. Sci Rep. 2016;6:22401.
10. Husby G, van de Rijn I, Zabriskie JB, Abdin ZH, Williams RC. Antibodies reacting with cytoplasm of subthalamic and caudate nuclei neurons in chorea and acute rheumatic fever. J Exp Med. 1976;144(4):1094–110.
11. Hoffman KL, Hornig M, Yaddanapudi K, Jabado O, Lipkin WI. A murine model for neuropsychiatric

disorders associated with group a beta-hemolytic streptococcal infection. J Neurosci. 2004;24(7):1780–91.
12. Bronze MS, Dale JB. Epitopes of streptococcal m proteins that evoke antibodies that cross-react with human brain. J Immunol. 1993;151(5):2820–8.
13. Fujinami RS, von Herrath MG, Christen U, Whitton JL. Molecular mimicry, bystander activation, or viral persistence: infections and autoimmune disease. Clin Microbiol Rev. 2006;19(1):80–94.
14. Pearson JA, Wong FS, Wen L. The importance of the Non Obese Diabetic (NOD) mouse model in autoimmune diabetes. J Autoimmun. 2016;66:76–88.
15. Armangue T, Titulaer MJ, Málaga I, Bataller L, Gabilondo I, Graus F, et al. Pediatric anti-N-methyl-D-aspartate receptor encephalitis-clinical analysis and novel findings in a series of 20 patients. J Pediatr. 2013;162(4):850–6.e2.
16. Leypoldt F, Titulaer MJ, Aguilar E, Walther J, Bönstrup M, Havemeister S, et al. Herpes simplex virus-1 encephalitis can trigger anti-NMDA receptor encephalitis: case report. Neurology. 2013;81(18):1637–9.
17. Armangue T, Leypoldt F, Málaga I, Raspall-Chaure M, Marti I, Nichter C, et al. Herpes simplex virus encephalitis is a trigger of brain autoimmunity. Ann Neurol. 2014;75(2):317–23.
18. Armangue T, Spatola M, Vlagea A, Mattozzi S, Cárceles-Cordon M, Martinez-Heras E, et al. Frequency, symptoms, risk factors, and outcomes of autoimmune encephalitis after herpes simplex encephalitis: a prospective observational study and retrospective analysis. Lancet Neurol. 2018;17(9):760–72.
19. Linnoila JJ, Binnicker MJ, Majed M, Klein CJ, McKeon A. CSF herpes virus and autoantibody profiles in the evaluation of encephalitis. Neurol Neuroimmunol Neuroinflamm. 2016;3(4):e245.
20. Höftberger R, Armangue T, Leypoldt F, Graus F, Dalmau J. Clinical neuropathology practice guide 4-2013: post-herpes simplex encephalitis: N-methyl-D-aspartate receptor antibodies are part of the problem. Clin Neuropathol. 2013;32(4):251–4.
21. Schäbitz WR, Rogalewski A, Hagemeister C, Bien CG. VZV brainstem encephalitis triggers NMDA receptor immunoreaction. Neurology. 2014;83(24):2309–11.
22. Salovin A, Glanzman J, Roslin K, Armangue T, Lynch DR, Panzer JA. Anti-NMDA receptor encephalitis and nonencephalitic HSV-1 infection. Neurol Neuroimmunol Neuroinflamm. 2018;5(4):e458.
23. Jolles S. The variable in common variable immunodeficiency: a disease of complex phenotypes. J Allergy Clin Immunol Pract. 2013;1(6):545–56; quiz 57.
24. Conley ME, Notarangelo LD, Etzioni A. Diagnostic criteria for primary immunodeficiencies. Representing PAGID (Pan-American Group for Immunodeficiency) and ESID (European Society for Immunodeficiencies). Clin Immunol. 1999;93(3):190–7.
25. Schubert RD, Wilson MR. A tale of two approaches: how metagenomics and proteomics are shaping the future of encephalitis diagnostics. Curr Opin Neurol. 2015;28(3):283–7.
26. Wilson MR, Naccache SN, Samayoa E, Biagtan M, Bashir H, Yu G, et al. Actionable diagnosis of neuroleptospirosis by next-generation sequencing. N Engl J Med. 2014;370(25):2408–17.
27. Quan PL, Wagner TA, Briese T, Torgerson TR, Hornig M, Tashmukhamedova A, et al. Astrovirus encephalitis in boy with X-linked agammaglobulinemia. Emerg Infect Dis. 2010;16(6):918–25.
28. Naccache SN, Peggs KS, Mattes FM, Phadke R, Garson JA, Grant P, et al. Diagnosis of neuroinvasive astrovirus infection in an immunocompromised adult with encephalitis by unbiased next-generation sequencing. Clin Infect Dis. 2015;60(6):919–23.
29. Bonilla FA, Barlan I, Chapel H, Costa-Carvalho BT, Cunningham-Rundles C, de la Morena MT, et al. International Consensus Document (ICON): common variable immunodeficiency disorders. J Allergy Clin Immunol Pract. 2016;4(1):38–59.
30. Boursiquot JN, Gérard L, Malphettes M, Fieschi C, Galicier L, Boutboul D, et al. Granulomatous disease in CVID: retrospective analysis of clinical characteristics and treatment efficacy in a cohort of 59 patients. J Clin Immunol. 2013;33(1):84–95.
31. Morimoto Y, Routes JM. Granulomatous disease in common variable immunodeficiency. Curr Allergy Asthma Rep. 2005;5(5):370–5.
32. Ardeniz O, Cunningham-Rundles C. Granulomatous disease in common variable immunodeficiency. Clin Immunol. 2009;133(2):198–207.
33. Arnold DF, Wiggins J, Cunningham-Rundles C, Misbah SA, Chapel HM. Granulomatous disease: distinguishing primary antibody disease from sarcoidosis. Clin Immunol. 2008;128(1):18–22.
34. Misbah SA, Spickett GP, Esiri MM, Hughes JT, Matthews WB, Thompson RA, et al. Recurrent intracranial granulomata presenting as space-occupying lesions in a patient with common variable immunodeficiency. Postgrad Med J. 1992;68(799):359–62.
35. Nguyen JT, Green A, Wilson MR, DeRisi JL, Gundling K. Neurologic complications of common variable immunodeficiency. J Clin Immunol. 2016;36(8):793–800.
36. Major EO, Yousry TA, Clifford DB. Pathogenesis of progressive multifocal leukoencephalopathy and risks associated with treatments for multiple sclerosis: a decade of lessons learned. Lancet Neurol. 2018;17(5):467–80.
37. Miskin DP, Koralnik IJ. Novel syndromes associated with JC virus infection of neurons and meningeal cells: no longer a gray area. Curr Opin Neurol. 2015;28(3):288–94.
38. O'Connor PW, Goodman A, Willmer-Hulme AJ, Libonati MA, Metz L, Murray RS, et al. Randomized multicenter trial of natalizumab in acute MS relapses: clinical and MRI effects. Neurology. 2004;62(11):2038–43.
39. Yousry TA, Major EO, Ryschkewitsch C, Fahle G, Fischer S, Hou J, et al. Evaluation of patients treated with natalizumab for progressive multifocal leukoencephalopathy. N Engl J Med. 2006;354(9):924–33.

40. Borchardt J, Berger JR. Re-evaluating the incidence of natalizumab-associated progressive multifocal leukoencephalopathy. Mult Scler Relat Disord. 2016;8:145–50.
41. Berger JR, Fox RJ. Reassessing the risk of natalizumab-associated PML. J Neurovirol. 2016;22(4):533–5.
42. Zhovtis Ryerson L, Frohman TC, Foley J, Kister I, Weinstock-Guttman B, Tornatore C, et al. Extended interval dosing of natalizumab in multiple sclerosis. J Neurol Neurosurg Psychiatry. 2016;87(8):885–9.
43. Longbrake EE, Ramsbottom MJ, Cantoni C, Ghezzi L, Cross AH, Piccio L. Dimethyl fumarate selectively reduces memory T cells in multiple sclerosis patients. Mult Scler. 2016;22(8):1061–70.
44. Berger JR, Aksamit AJ, Clifford DB, Davis L, Koralnik IJ, Sejvar JJ, et al. PML diagnostic criteria: consensus statement from the AAN neuroinfectious disease section. Neurology. 2013;80(15):1430–8.
45. Cree BA, Lamb S, Morgan K, Chen A, Waubant E, Genain C. An open label study of the effects of rituximab in neuromyelitis optica. Neurology. 2005;64(7):1270–2.
46. Barmettler S, Ong M, Farmer JR, Choi H, Walter J. Association of immunoglobulin levels, infectious risk, and mortality with rituximab and hypogammaglobulinemia. JAMA Netw Open. 2018;1(7):e184169.
47. Roberts DM, Jones RB, Smith RM, Alberici F, Kumaratne DS, Burns S, et al. Rituximab-associated hypogammaglobulinemia: incidence, predictors and outcomes in patients with multi-system autoimmune disease. J Autoimmun. 2015;57:60–5.
48. Barmettler S, Price C. Continuing IGG replacement therapy for hypogammaglobulinemia after rituximab--for how long? J Allergy Clin Immunol. 2015;136(5):1407–9.
49. Tallantyre EC, Whittam DH, Jolles S, Paling D, Constantinesecu C, Robertson NP, et al. Secondary antibody deficiency: a complication of anti-CD20 therapy for neuroinflammation. J Neurol. 2018;265(5):1115–22.
50. Tuohy O, Costelloe L, Hill-Cawthorne G, Bjornson I, Harding K, Robertson N, et al. Alemtuzumab treatment of multiple sclerosis: long-term safety and efficacy. J Neurol Neurosurg Psychiatry. 2015;86(2):208–15.
51. Devonshire V, Phillips R, Wass H, Da Roza G, Senior P. Monitoring and management of autoimmunity in multiple sclerosis patients treated with alemtuzumab: practical recommendations. J Neurol. 2018;265(11):2494–505.
52. Coles AJ, Fox E, Vladic A, Gazda SK, Brinar V, Selmaj KW, et al. Alemtuzumab more effective than interferon β-1a at 5-year follow-up of camms223 clinical trial. Neurology. 2012;78(14):1069–78.
53. Baker D, Herrod SS, Alvarez-Gonzalez C, Giovannoni G, Schmierer K. Interpreting lymphocyte reconstitution data from the pivotal phase 3 trials of alemtuzumab. JAMA Neurol. 2017;74(8):961–9.
54. Marrie RA, Reider N, Cohen J, Stuve O, Sorensen PS, Cutter G, et al. A systematic review of the incidence and prevalence of autoimmune disease in multiple sclerosis. Mult Scler. 2015;21(3):282–93.
55. Demko S, Summers J, Keegan P, Pazdur R. FDA drug approval summary: alemtuzumab as single-agent treatment for B-cell chronic lymphocytic leukemia. Oncologist. 2008;13(2):167–74.
56. Wang DY, Salem JE, Cohen JV, Chandra S, Menzer C, Ye F, et al. Fatal toxic effects associated with immune checkpoint inhibitors: a systematic review and meta-analysis. JAMA Oncol. 2018;4(12):1721–8.
57. Fellner A, Makranz C, Lotem M, Bokstein F, Taliansky A, Rosenberg S, et al. Neurologic complications of immune checkpoint inhibitors. J Neuro-Oncol. 2018;137(3):601–9.
58. Touat M, Talmasov D, Ricard D, Psimaras D. Neurological toxicities associated with immune-checkpoint inhibitors. Curr Opin Neurol. 2017;30(6):659–68.
59. Williams TJ, Benavides DR, Patrice KA, Dalmau JO, de Ávila AL, Le DT, et al. Association of autoimmune encephalitis with combined immune checkpoint inhibitor treatment for metastatic cancer. JAMA Neurol. 2016;73(8):928–33.
60. Gettings EJ, Hackett CT, Scott TF. Severe relapse in a multiple sclerosis patient associated with ipilimumab treatment of melanoma. Mult Scler. 2015;21(5):670.
61. Gerdes LA, Held K, Beltrán E, Berking C, Prinz JC, Junker A, et al. CTLA4 as immunological checkpoint in the development of multiple sclerosis. Ann Neurol. 2016;80(2):294–300.
62. Maurice C, Schneider R, Kiehl TR, Bavi P, Roehrl MH, Mason WP, et al. Subacute CNS demyelination after treatment with nivolumab for melanoma. Cancer Immunol Res. 2015;3(12):1299–302.
63. Lau KH, Kumar A, Yang IH, Nowak RJ. Exacerbation of myasthenia gravis in a patient with melanoma treated with pembrolizumab. Muscle Nerve. 2016;54(1):157–61.
64. Zhu J, Li Y. Myasthenia gravis exacerbation associated with pembrolizumab. Muscle Nerve. 2016;54(3):506–7.
65. Phadke SD, Ghabour R, Swick BL, Swenson A, Milhem M, Zakharia Y. Pembrolizumab therapy triggering an exacerbation of preexisting autoimmune disease: a report of 2 patient cases. J Investig Med High Impact Case Rep. 2016;4(4):2324709616674316.
66. Suzuki S, Ishikawa N, Konoeda F, Seki N, Fukushima S, Takahashi K, et al. Nivolumab-related myasthenia gravis with myositis and myocarditis in Japan. Neurology. 2017;89(11):1127–34.
67. Shirai T, Sano T, Kamijo F, Saito N, Miyake T, Kodaira M, et al. Acetylcholine receptor binding antibody-associated myasthenia gravis and rhabdomyolysis induced by nivolumab in a patient with melanoma. Jpn J Clin Oncol. 2016;46(1):86–8.
68. Johnson DB, Saranga-Perry V, Lavin PJ, Burnette WB, Clark SW, Uskavitch DR, et al. Myasthenia gravis induced by ipilimumab in patients with metastatic melanoma. J Clin Oncol. 2015;33(33):e122–4.

# Oncological Mimics in Inflammatory CNS Disease

# 26

Alexander C. Mohler and Douglas E. Ney

**Key Points**

1. Oncological conditions involving the central nervous system (CNS) are serious, often difficult to diagnosis, and are associated with unique treatment strategies.
2. Histiocytic disorders are a rare, heterogeneous class of diseases that have the propensity for both CNS and systemic involvement.
3. Activating mutations in the mitogen-activated protein kinase (MAPK) pathway are found in a significant number of histiocytic disorders indicating a clonal, neoplastic origin with potential for targeted treatments.
4. Primary central nervous system lymphoma is a rare extranodal form of non-Hodgkin lymphoma involving the CNS without systemic involvement. Though highly aggressive, it is a potentially curable malignancy.
5. Though lacking formal diagnostic criteria, chronic lymphocytic inflammation with pontine perivascular enhancement responsive to steroids (CLIPPERS) is typically associated with characteristic clinical, imaging, and pathologic findings typically involving the brain stem. Central to the diagnosis is the exquisite sensitivity to steroids, but CLIPPERS has been associated with the subsequent development of malignancy.

## Introduction

Evaluation of inflammatory disease within the central nervous system (CNS) can be a diagnostic challenge given diverse presentations of patients. Extensive diagnostic evaluation is often unrevealing and can lead to further uncertainty. Malignancy or "premalignant" conditions within the CNS can mimic inflammatory disease, thus making these entities important for consideration when evaluating for inflammatory conditions. While rare, disease such as histiocytic disorders and primary central nervous system lymphoma (PCNSL) occurs with enough frequency to be considered when presented with a potential CNS

A. C. Mohler
Department of Neurology, Rhode Island Hospital and the Miriam Hospital, Warren Alpert Medical School of Brown University, Providence, RI, USA

D. E. Ney (✉)
Departments of Neurology and Neurosurgery, University of Colorado School of Medicine, Aurora, CO, USA
e-mail: Douglas.Ney@cuanschutz.edu



A. L. Piquet, E. Alvarez (eds.), *Neuroimmunology*, https://doi.org/10.1007/978-3-030-61883-4_26

inflammatory process. Entities such as chronic lymphocytic inflammation with pontine perivascular enhancement responsive to steroids (CLIPPERS) exhibit common characteristics suggestive of the diagnosis and may represent early manifestations of malignancy. Metastatic disease to the brain or CNS effects from systemic malignancy are important conditions to consider but are outside the scope of this chapter. Similarly, neurologic paraneoplastic conditions are discussed in Chap. 16 in this book. Herein, we review common oncological mimics of CNS inflammatory disease including histiocytic disorders, PCNSL, and CLIPPERS.

## Histiocytic Disorders

Histiocytic disorders are a rare heterogeneous class of diseases of varied clinical course and prognoses. The World Health Organization Classification of Tumors of the Nervous System subclassifies histiocytic disorders into several different entities including Langerhans cell histiocytosis (LCH), Erdheim-Chester disease (ECD), Rosai-Dorfman-Destombes disease (RDD), juvenile xanthogranuloma (JXA), histiocytic sarcoma (HS), and indeterminate cell histiocytosis (ICH) [1]. These conditions are essentially malignancies of histiocytes or tissue macrophages that can present with isolated or multifocal tissue infiltration both intracranially and systemically. Since histiocytes are part of the mononuclear phagocyte system, they can present like and often be mistaken for immunologic conditions. Indeed, while the pathophysiology of these disorders has long been debated to be autoimmune, recent discovery of activating mutations in the MAPK and phosphatidylinositol 3-kinase/protein kinase B (PI3K-AKT) pathways involving a large proportion of patients has cemented that these are likely clonal and neoplastic in origin [2].

### Langerhans Cell Histiocytosis

LCH is characterized by the proliferation of CD1a-positive and CD207-positive histiocytes [3]. Initially thought to be related to epidermal Langerhans cells, transcriptional profiling of Langerhans histiocytosis suggests greater similarity to bone marrow-derived monocyte and dendritic cell precursors [4]. The pathogenesis is not entirely clear, and there is some suggestion that it may result from an abnormal inflammatory response as well as secondary malignant transformation [5]. Biopsy for pathologic confirmation is recommended in all cases, particularly for patients requiring systemic therapy.

LCH typically occurs in childhood with an incidence of 0.5 cases per 100,000 individuals under age 15 and 0.1 cases per 100,000 in patients older than 15 years [6, 7]. This disorder can affect any organ but most frequently affects the bone, skin, and pituitary gland. Central nervous system involvement occurs in about 20–50% of cases [8] and most commonly occurs via direct extension through the calvarium but can involve the meninges or less commonly present as intraparenchymal masses [9]. A wide variety of neurologic symptoms can occur depending on the location of the lesion; diabetes insipidus is the most common neurologic manifestation and occurs in 15–50% of patients with CNS involvement [10].

Two major types of CNS lesions have been described: tumor lesions and degenerative lesions. Tumor lesions are space-occupying lesions that avidly enhance on magnetic resonance imaging (MRI) (Fig. 26.1) and are associated with clinical signs and symptoms related to tumor infiltration including acute to subacute focal neurologic deficits, cranial nerve palsies, increased intracranial pressure, seizures, or behavior change. In contrast, neurodegenerative lesions are accompanied by slowly progressive cerebellar dysfunction, cognitive impairment, or a pyramidal syndrome. Degenerative lesions are atrophic on MRI scan, do not typically enhance, and are associated with high T2 signal in the white matter [11, 12].

Consensus guidelines exist for the diagnosis, workup, and treatment of LCH [13]. However, the evidence to support any one treatment over another is scant. The discovery that slightly over half of all cases harbor oncogenic *BRAF* mutations [14] has opened the door to targeted treatments.

## Erdheim-Chester Disease

Erdheim-Chester disease (ECD) is a rare non-Langerhans cell histiocytosis. Fewer than 800 cases have been reported, but recognition of this disease is improving [15, 16]. The mean age at diagnosis is approximately 55 years with a male predominance (3:1) [15]. Diagnosis is based on histopathologic demonstration of foamy or eosinophilic histiocytes positive for CD68 and CD163 and negative for CD1a, CD207, and S100 (Fig. 26.2a, b) [15].

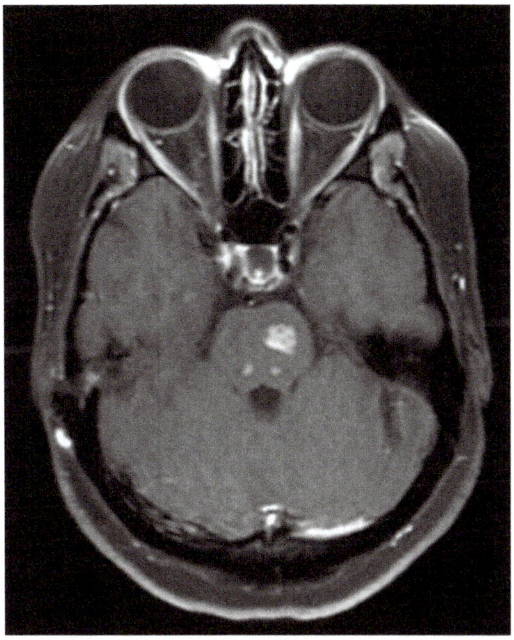

**Fig. 26.1** Magnetic resonance imaging of a 28-year-old presenting with dysarthria and hemiparesis. Biopsy showed Langerhans cell histiocytosis

ECD is a heterogenous multisystem disease that can affect any organ including the CNS. Commonly, the presenting symptom is bone pain due to sclerotic lesions of the long bones, which occur in 96% of reported cases [17]. Cardiac involvement is also quite common and is a source of increased mortality [18]. Close to half of patients with ECD have neurologic manifestations. When the CNS is involved, presenting symptoms typically involve cerebellar (41%) or pyramidal symptoms (45%) [19]. However, a wide variety of neurologic manifestations can occur depending on the location of CNS involvement. Signs and symptoms can include seizures, headaches, neurocognitive deficits, weakness, and numbness [19]. Much like LCH, diabetes insipidus occurs in about 25% of patients, although other endocrinopathies have been reported as well [16]. CNS involvement has also been identified as a poor prognostic factor [20].

Radiographic manifestations of ECD typically include an intraparenchymal predominance, a meningeal predominance, or a combination of both. Intraparenchymal lesions tend to be enhancing T2 hyperintense abnormalities on MRI (Fig. 26.3). Intraparenchymal lesions may be mistaken for demyelinating disease in many cases as lesions often may not appear to be space occupying. Lesions that involve the meninges can appear as dural thickening and may resemble other intracranial meningeal-based tumors like meningiomas [21].

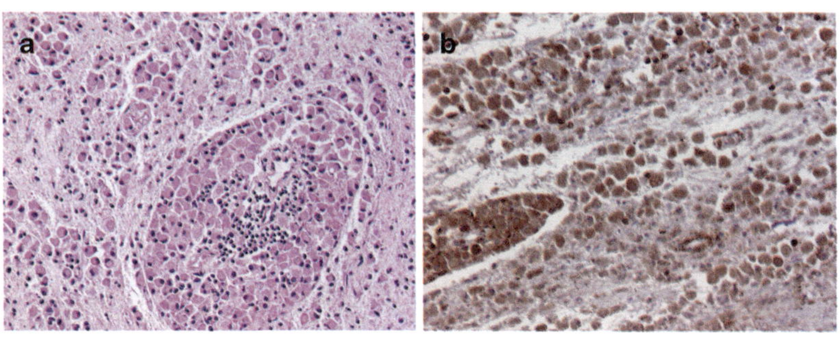


**Fig. 26.2** Photomicrograph of brain tissue from a patient with Erdheim-Chester disease. (**a**) Seen are variably foamy histiocytes with cytologically normal nuclei. (**b**) Histiocytes strongly staining for CD68. Notably, normal macrophages seen in conditions such as infarcts of demyelinating disorders are also CD68 positive

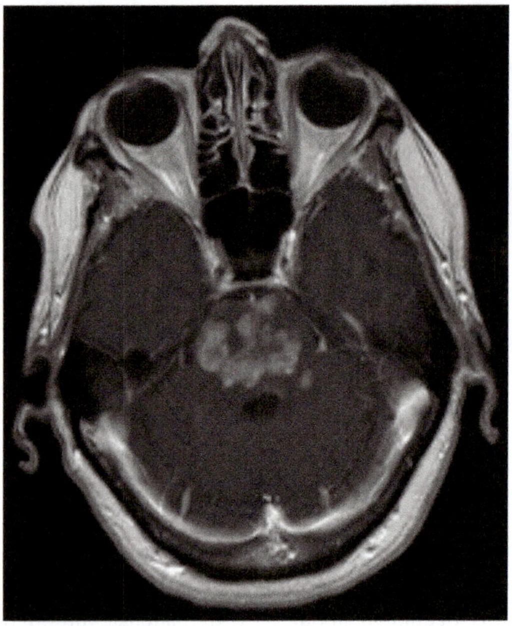

**Fig. 26.3** Magnetic resonance imaging of a 70-year-old woman with progressive brain stem symptoms. Biopsy showed Erdheim-Chester disease with a *BRAF V600E* mutation

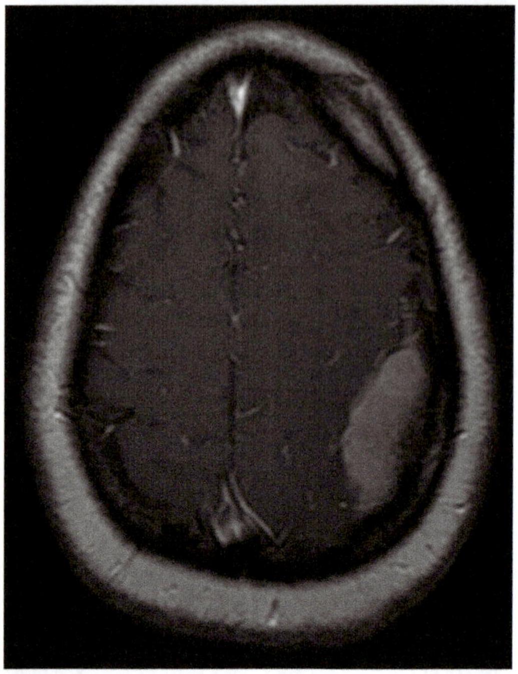

**Fig. 26.4** Magnetic resonance imaging of a 42-year-old African-American patient with biopsy-proven Rosai-Dorfman-Destombes disease

Many reported regimens have been used with varied success for the treatment of ECD [15]. However, much like LCH, many cases of ECD exhibit *BRAF* mutations and can respond successfully to medications targeting this mutational pathway [22].

## Rosai-Dorfman-Destombes Disease

Rosai-Dorfman-Destombes disease (RDD) is considered a non-Langerhans cell histiocytosis much like ECD. As compared to ECD, it has a lower mean age of 21 years and is more common in males and patients of African descent [23, 24]. Histopathologic features include accumulation of CD68-positive and CD1a-negative histiocytes; however, unlike ECD, RDD histiocytes are S100 positive [23]. Unlike both LCH and ECD, RDD is not typically associated with *BRAF* mutations [25].

Classically, patients present with systemic findings of bilateral massive lymphadenopathy, but extranodal disease may occur in up to 43% of cases [26]. Extranodal sites may include the nasal sinuses, skin, bone, liver, mediastinum, and CNS. In the CNS, RDD typically presents with dural-based disease that may occur in almost any location, including the spinal cord, with a predilection for the cerebral convexities (Fig. 26.4). It may also be associated with intraparenchymal lesions which involve the brain stem and cerebellum, although this is less common [23].

Outcomes for RDD are typically favorable in patients with nodal or cutaneous disease. Patients with kidney, liver, or lower respiratory tract disease seem to have a worse prognosis and may warrant more aggressive therapies [23]. CNS disease can be fatal and can have significant associated morbidity. In these cases, aggressive therapies may also be warranted. While multiple treatment approaches have been attempted, typically in case reports, no standard treatment recommendations exist.

## Juvenile Xanthogranuloma

Juvenile xanthogranuloma (JXG) typically presents in very young patients with a mean age of approximately 2 years and is typically a self-limited disease [27]. Cutaneous papules or nodules present typically on the face or head and neck but can involve any skin surface. Lesion number ranges from a few to hundreds. The condition is typically self-limiting with gradual resolution over months to years. Systemic JXG can occur in 5–10% of patients; CNS involvement can also occur [28]. In most cases when the CNS is involved, it presents in the brain with a much smaller percentage of spinal cord involvement. As is the case in most of the histiocytic disorders, neurologic symptoms at presentation are typically based on lesion location. Although CNS involvement in this condition is rare, it is associated with a high degree of morbidity and mortality in this otherwise benign condition [28].

### Histiocytic Sarcoma

This condition is a rare aggressive neoplasm involving malignant histiocytes. Primarily, this condition is reported in case reports, which have described cases involving the parenchyma, meninges, and cavernous sinus. Radiologically, CNS involvement often mimics demyelinating disease, lymphoma, or even glioma. Due to the extreme rarity of this condition, it is often mistaken for these other entities [29].

## Primary Central Nervous System Lymphoma

Primary central nervous system lymphoma (PCNSL) is an uncommon extranodal non-Hodgkin lymphoma involving the brain, spinal cord, leptomeninges, or eyes without evidence of systemic involvement. Though highly aggressive, it is a potentially curable malignancy. Up to 90% of patients will respond to appropriate treatment, but half of patients will relapse within 2 years, and prognosis after recurrence is poor [30]. PCNSL is rare and accounts for only 4% of all tumors involving the CNS and 4–6% of all extranodal lymphomas [31, 32]. The median age of diagnosis is 65 years, and the incidence in the elderly population is rising [33, 34]. PCNSL also occurs in the setting of immunosuppression such as human immunodeficiency virus/acquired immunodeficiency syndrome (HIV/AIDS), post-transplant immunosuppression, or congenital immunodeficiency.

Two widely recognized prognostic models were developed to predict outcomes in patients with PCNSL. The International Extranodal Lymphoma Study Group identified age, Eastern Cooperative Oncology Group (ECOG) status, serum lactic dehydrogenase (LDH), cerebrospinal fluid (CSF) protein concentration, and presence of deep brain structures as independent predictors of survival [35]. The Memorial Sloan Kettering Cancer Center model is a simplified model using only Karnofsky performance status (KPS) and age as predictors of outcome [36]. It is also worth noting that delays in treatment have been associated with poorer outcomes [37]. Thus, prompt diagnosis and initiation of treatment is of utmost importance.

PCNSL can present in any compartment of the CNS. As such, a wide variety of symptoms can occur in these patients including focal neurologic deficits, seizures, increased intracranial pressure, and cognitive difficulties, which can progress rapidly over weeks. Most cases of PCNSL present with lesions involving the brain parenchyma, primarily in the supratentorial compartment. Lesions are often periventricular and involve the deep white matter or corpus callosum. PCNSL can appear as a solitary lesion or multifocal disease. The frontal lobe white matter is thought to be the most common location. A hemispheric location, as opposed to deep structures, is seen in up to 38% of patients [38]. Rarely, lesions develop in the cerebellum or spinal cord. Leptomeningeal disease is seen in approximately 20% of cases of PCNSL but can be present in close to two-thirds of secondary CNS lymphoma [39, 40]. Isolated leptomeningeal involvement is unusual.

Computed tomography (CT) imaging typically reveals iso- or hypo-dense lesions. Contrast enhancement is quite common and is present in nearly all cases of CNS lymphoma [40]. On MRI, PCNSL typically presents as sharply demarcated

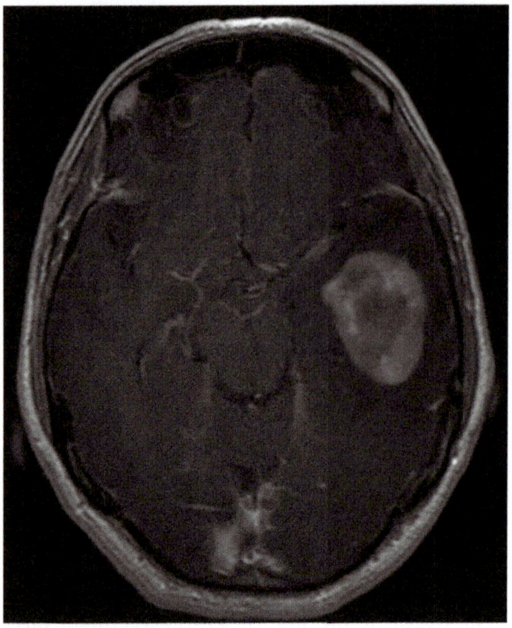

**Fig. 26.5** Magnetic resonance imaging of a 72-year-old man who presented with confusion and word-finding difficulties. Biopsy showed diffuse large B-cell lymphoma. Negative systemic imaging confirmed primary central nervous system lymphoma

homogeneous enhancement that is iso- to hypointense on pre-contrast T1-weighted sequences (Fig. 26.5). Due to the dense cellularity, there is often diffusion restriction on diffusion-weighted imaging (Fig. 26.6a–d) [41]. It is very uncommon for CNS lymphoma patients to present without enhancement. However, in the setting of immunosuppression such as HIV/AIDS, enhancement may be more heterogeneous, exhibit central necrosis, or be absent [42, 43]. Symptoms concerning for PCNSL should prompt urgent imaging, preferably with MRI.

The diagnosis of primary CNS lymphoma is dependent on pathologic examination and thus requires the tumor sampling in most cases. Stereotactic biopsy is the procedure of choice. Corticosteroids should be avoided prior to biopsy as they are lymphotoxic and can obscure pathology results [44, 45]. Greater than 90% of cases are diffuse large B-cell lymphoma (DLBCL), with the remainder consisting of T-cell, Burkitt, or low-grade lymphomas [46]. Microscopically, these lesions tend to be densely populated with numerous small cells with large central, pleomorphic nuclei. Perivascular aggregates with invasion into surrounding parenchyma are common. Malignant lymphoma cells exhibit expression of pan-B-cell markers including CD19, CD20, and CD79a. MUM1 is nearly always positive, BCL6 positivity is seen in half of cases, and BCL-2 is variably expressed [46]. These characteristics suggest that the majority of PCNSL most closely resembles a postgerminal center or activated B-cell (ABC) immunophenotype [47]. While subtyping may have prognostic implications, it does not affect treatment.

Although most patients presenting with lymphoma of the CNS are thought to have primary disease, systemic spread to the CNS does occur. Differentiating between primary and secondary CNS lymphoma is important as the treatment approach differs. Workup to determine the extent of spread throughout the CNS is imperative as well. Standard serum evaluation for patients with PCNSL should include a complete blood count, metabolic panel with liver function tests, hepatitis serologies, LDH, and HIV testing. Baseline evaluation with a CT scan of the chest, abdomen, and pelvis or preferably a positron emission tomography (PET) scan is required to identify potential systemic disease. A bone marrow biopsy is also warranted [48]. In men, testicular ultrasound to rule out concomitant testicular lymphoma is warranted as up to 15% of patients with testicular lymphomas metastasize to the brain [49]. Further, a thorough ophthalmologic evaluation including slit lamp examination is always necessary as the eye can serve as a reservoir of disease and potentially reseed the CNS following treatment. To assess for leptomeningeal involvement, a lumbar puncture should also be done if there is no risk for herniation. CSF should be sent for flow cytometry, cytology, and IgH rearrangement in addition to basic CSF studies [50]. Documented involvement of any of these compartments should prompt end of treatment reassessment to ensure disease is eradicated.

Treatment of PCNSL consists of induction therapy with the goal of achieving a complete response (CR) or elimination of gross disease. This is followed by consolidation to eliminate

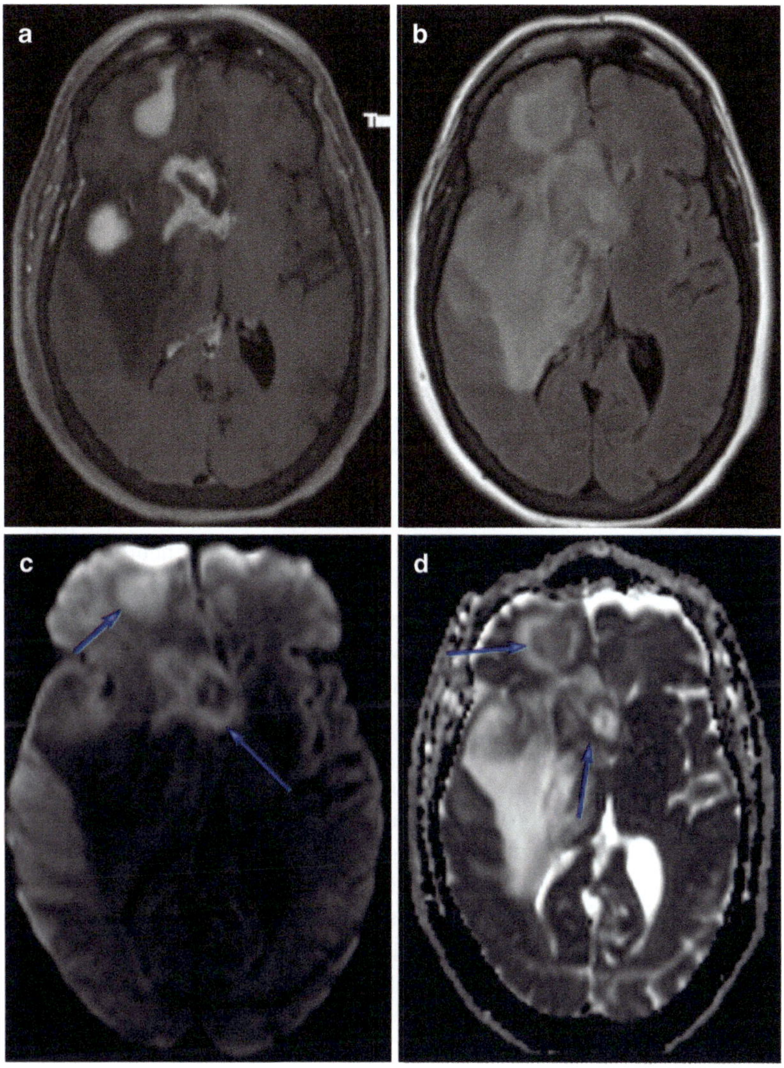


**Fig. 26.6** Magnetic resonance imaging of a 66-year-old woman presenting with 2 weeks of confusion and found to have multifocal enhancing lesions (**a**). T2 fluid-attenuated inversion recovery (FLAIR)-weighted imaging (**b**) showed significant edema associated with the enhancing lesions. Diffusion-weighted imaging (**c**) and apparent diffusion coefficient imaging (**d**) showed mild diffusion restriction. Biopsy proved diffuse large B-cell lymphoma

microscopic disease and maintain remission. Due to the rarity of primary CNS lymphoma and lack of phase three randomized trials, a standardized treatment regimen does not exist. Untreated, this disease is almost universally fatal within the first few months of symptom onset. However, with aggressive treatment, improved response rates and prolonged disease control are achievable.

Traditionally, whole-brain radiation therapy (WBRT) was used with overall response rates of 90% and extension of survival beyond a year [51, 52]. Treatment with radiotherapy is not curative, however, and overall survival with WBRT alone is only 12–18 months [51, 53]. Ultimately, the role of WBRT is debated. When combined with chemotherapy, neurotoxicity rates are high and does not likely prolong overall survival [54, 55].

Chemotherapy regimens used to treat systemic lymphoma have limited effectiveness in treating CNS disease [39]. This is likely due to limited blood-brain barrier penetration of most chemotherapies used to treat systemic lymphoma. In general, the chemotherapeutic approach to PCNSL has trended toward multi-agent chemotherapeutic regimens; however, there is no standard first-line regimen. High-dose methotrexate (MTX) is considered the backbone of therapy for PCNSL. MTX, when given at relatively high doses (>1.5 gm/m$^2$) and as a rapid infusion, results in tumoricidal concentrations with the brain and CSF [56, 57]. Single-agent MTX has been given at doses as high as 8 gm/m$^2$ [58]. The addition of high-dose cytarabine to MTX has resulted in improved response rates and progression-free survival but with increased rates of hematological toxicity [59]. Rituximab, an anti-CD20 monoclonal antibody, has been incorporated into many varied chemotherapy regimens with noted improvement in response rates and overall survival [60–64]. It has become common practice to incorporate rituximab into MTX-containing regimens. Common combination regimens include MTX typically with an alkylating agent with or without rituximab.

High-dose MTX is relatively well tolerated and specific toxicities vary based on regimen. Aside from the potential delayed neurotoxicity (particularly when combined with WBRT), an additional limitation is the need for hospitalization during administration. Common systemic side effects include renal toxicity, hepatitis, myelosuppression, and mucositis. Treatment is administered with aggressive hydration, urine alkalinization, and leucovorin to prevent systemic organ damage. With appropriate supportive care, MTX-based regimens are safe to administer, even in older patients with medical comorbidities. However, impaired renal function (creatinine clearance <30 mL/min) is a contraindication for MTX.

Consolidative therapies following complete response to induction vary depending on physician preference, and no standard exists. Reduced-dose WBRT has been used as a consolidative approach with encouraging results and low rates of neurotoxicity, but the small numbers of patients and limited follow-up limit the applicability [65, 66]. There has been increasing interest in pursuing high-dose chemotherapy with autologous stem cell transplantation with curative intent as a consolidative option. In younger otherwise healthy patients, this may be a good option, and early studies suggest an improvement in survival with the use of stem cell transplantation for consolidation [67, 68]. Certain patients may not be able to tolerate the rigors of high-dose chemotherapy given in preparation for transplantation. These patients may rely on other chemotherapeutic approaches for consolidation [37]. Maintenance rituximab has been used as a strategy to maintain remission, and trials using maintenance strategies are ongoing [69].

## Primary CNS Lymphoma in the Elderly

More than half of patients with PCNSL are 60 years of age or greater, and age is an independent predictor of survival [35, 36]. Moreover, the risk of neurotoxicity is highest in the elderly, and treatment may be complicated by multiple comorbidities. Thus, the optimal treatment for this group has yet to be defined. However, a large meta-analysis of elderly patients with newly diagnosed PCNSL showed that 73% of patients received a MTX (median dose of 3 g/m$^2$)-containing regimen, and this was associated with improved survival, particularly when combined with an oral alkylating agent. Interestingly, more aggressive chemotherapy regimens were not associated with superior outcomes [70]. It is generally agreed that WBRT should be avoided and multi-agent MTX-based regimens be considered for first-line treatment.

## Primary CNS Lymphoma in the Immunocompromised

The epidemic of HIV in the 1980s and 1990s led to an increase in CNS lymphoproliferative disorders including CNS lymphomas. Prognosis in these patients was initially quite poor [71]. PCNSL in these patients is thought to be driven by Epstein-Barr virus (EBV) infection. The advent of antiretroviral therapies has significantly reduced the incidence of this subtype of CNS lymphoma in patients with HIV-related lymphoproliferative disorders [72]. WBRT is an option for treating patients with this condition and is especially useful in patients with significant comorbidities. In general, treatment of the cause of the immunosuppression has improved survival in these patients, thought to be due to immune recovery [73]. Aside from WBRT, other approaches include the use of MTX alone or in combination with additional chemotherapies concurrently with antiretroviral management. A recent study suggested that a combination of zidovudine, ganciclovir, and rituximab without MTX produced complete and durable responses in these patients [74].

## Subtypes of CNS Lymphoma

There are many subtypes of PCNSL, which are differentiated by unusual areas of disease involvement or uncommon cellular type. Primary ocular lymphoma is uncommon and is felt to be a subtype of PCNSL as opposed to systemic disease. Due to the blood-ocular barrier, this condition is treated in similar fashion to CNS lymphoma with methotrexate- or cytarabine-based regimens. Primary leptomeningeal lymphoma is also an uncommon presentation of CNS disease but treated in a similar fashion.

Hodgkin lymphoma can rarely involve the brain as well as lymphomas of the T-cell type [75, 76]. Low-grade lymphomas such as marginal zone lymphoma or mucosa-associated lymphoid tissue (MALT) lymphoma can involve the CNS; however, dural or extra-axial involvement is much more typical of these lower-grade lesions [77]. Intravascular lymphoma is a subtype of lymphoma that has typical CNS involvement but is considered disease with systemic involvement. This condition involves invasion of malignant B cells into the lumens and vessel walls of smaller blood vessels. It is often rapidly fatal, can affect multiple organs, and causes ischemic injury. Intracranially, MR imaging may be variable but often shows infarct-like white matter lesions in vascular territories with variable patterns of parenchymal and meningeal enhancement (Fig. 26.7a, b) [78].

## Lymphomatoid Granulomatosis

Lymphomatoid granulomatosis is an uncommon lymphoproliferative disorder that primarily involves the lungs but in rare occasions can involve other sites including the CNS. In even rarer instances, it can involve the CNS in isola-

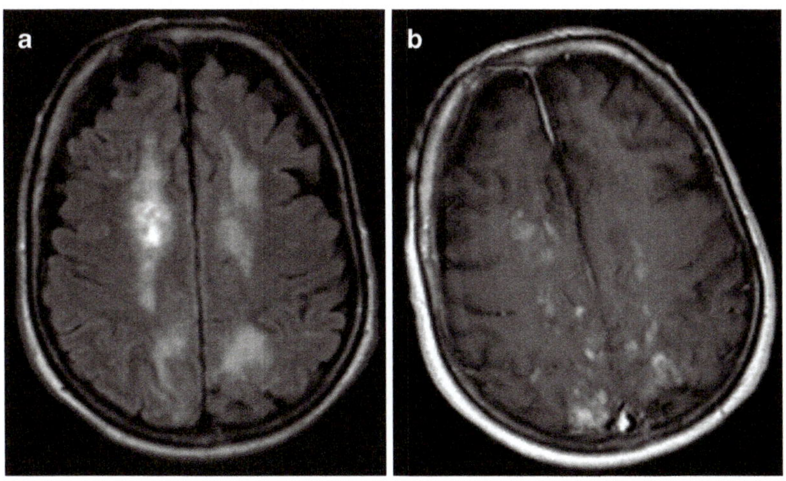


**Fig. 26.7** Magnetic resonance imaging of a 48-year-old woman presenting with diffuse white matter disease. T2 fluid-attenuated inversion recovery (FLAIR)-weighted imaging (**a**) showed diffuse areas of high signal within the white matter. T1-weighted post-contrast images (**b**) showed variable patchy enhancement. Biopsy demonstrated malignant lymphoid cells aggregated within the vasculature consistent with intravascular lymphoma

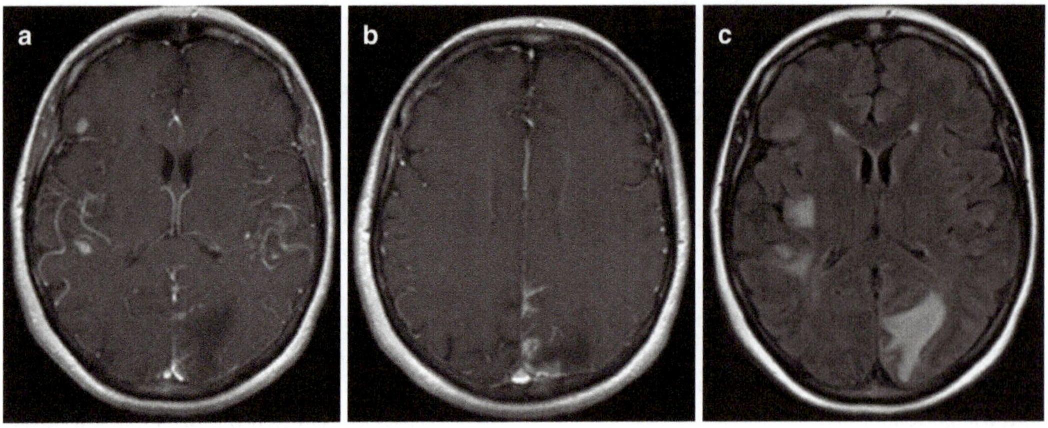


**Fig. 26.8** Patient with biopsy-proven lymphomatoid granulomatosis. T1 post-contrast imaging (**a**, **b**) shows multifocal enhancement throughout both cerebral hemispheres. T2 fluid-attenuated inversion recovery (FLAIR) imaging (**c**) with areas of surrounding edema

tion. In cases of primary CNS lymphomatoid granulomatosis, this disease can present with both mass-like lesions and more diffuse-infiltrating lesions, the former being more common [79]. There is no classic presentation seen on MRI or CT imaging of the brain, necessitating evaluation for more common conditions. Linear or punctate enhancement is slightly more specific for this condition and is thought to represent the angiocentric nature of this condition (Fig. 26.8a–c) [79, 80].

Histopathologically, this condition is characterized by angiocentric lymphoid aggregates that can invade and destroy blood vessels. This can lead to infarcted tissue, both brain and tumor alike. The infiltrate is generally a mixture of CD20-positive large, monoclonal, neoplastic B cells with a significant T-cell population without clonality. The B-cell population is of uncertain malignant potential and may not make up a large percentage of cells within the infiltrate. Pathogenesis is unclear; however, it is generally thought to be an EBV-driven process, particularly in cases with systemic presentations [81, 82]. The World Health Organization separates this condition into three grades: grade 1 is the least aggressive; grade 3 is the most aggressive with large atypical CD20+ B cells and extensive necrosis [82].

There is no standard therapy for this condition, and treatment depends on extent of involvement and grade of disease. For more benign-appearing presentations, steroids alone may be enough to treat this condition. More aggressive disease may warrant treatment similar to PCNSL with systemic chemotherapy. In a review of 22 cases of isolated CNS lymphomatoid granulomatosis, treatment included steroids alone, radiotherapy alone, chemotherapies including cyclophosphamide, cisplatin, cytarabine, methotrexate, or a combination of these treatments [79].

## Chronic Lymphocytic Inflammation with Pontine Perivascular Enhancement Responsive to Steroids (CLIPPERS)

CLIPPERS is a CNS inflammatory disorder of unknown etiology predominantly affecting the brain stem. It is characterized by punctate, curvilinear gadolinium-enhancing lesions “peppering” the pons and cerebellum and is noted to have exquisite sensitivity to steroids [83]. While associated with characteristic clinical, imaging, and pathologic findings, the lack of formal diagnostic criteria and biomarkers often leads to uncertainty regarding the diagnosis.

Given the predilection for involvement of the brain stem, common presenting symptoms include dysarthria, diplopia, gait ataxia, vertigo, sensory changes of the face, and cognitive impairment [84]. Symptoms typically progress over weeks to months and rapid progression is unusual. Typical age of onset is in midlife, but a wide range has been reported [83, 84]. While clinical symptoms may be suggestive, none make the diagnosis. Further, no specific laboratory abnormalities are characteristic of CLIPPERS, and laboratory testing should be used to exclude alternative diagnoses. No pathologic hallmark has been identified in CLIPPERS, and the primary reason for biopsy is to exclude other diagnoses. Typical histologic findings include perivascular inflammation with CD3-reactive T-lymphocytes (predominantly CD4 positive), activated microglia, CD68-positive histiocytes, and occasionally CD20-positive lymphocytes [83, 84]. Diagnostic criteria using clinical, radiological, and pathologic criteria have been proposed [84].

Central to the diagnosis of CLIPPERS is the overall radiographic and clinical responsiveness to steroids. The natural history seems to be that of a relapsing-remitting disorder, and some patients require long-term immunosuppression [84]. The occurrence of suspected CLIPPERS has been associated with malignancy, and it has been postulated that CLIPPERS represents a premalignant state [85–89]. While likely that CLIPPERS represents a spectrum of entities, it is clear that long-term follow-up is necessary to determine the course of the disease as well as potential development of other diseases.

## Conclusion

While rare, oncological disorders occur within the CNS with enough frequency to warrant consideration when evaluating patients for potential inflammatory disease. Varied presentations and diagnostic uncertainty can complicate evaluation and treatment. However, conditions such as histiocytic disorders, PCNSL, or CLIPPERS represent specific diseases with unique therapeutic strategies. As our understanding of these diseases evolves, outcomes are improving making early and accurate diagnosis imperative.

## References

1. Louis DN, Ohgaki H, Wiestler OD, Cavenee WK. (eds.): WHO Classification of Tumours of the Central Nervous System. IARC: Lyon 2007.
2. Durham BH, Diamond EL, Abdel-Wahab O. Histiocytic neoplasms in the era of personalized genomic medicine. Curr Opin Hematol. 2016;23(4):416–25.
3. Collin M, Bigley V, McClain KL, Allen CE. Cell(s) of origin of Langerhans cell histiocytosis. Hematol Oncol Clin North Am. 2015;29(5):825–38.
4. Allen CE, Li L, Peters TL, Leung HC, Yu A, Man TK, et al. Cell-specific gene expression in Langerhans cell histiocytosis lesions reveals a distinct profile compared with epidermal Langerhans cells. J Immunol. 2010;184(8):4557–67.
5. Delprat C, Arico M. Blood spotlight on Langerhans cell histiocytosis. Blood. 2014;124(6):867–72.
6. Guyot-Goubin A, Donadieu J, Barkaoui M, Bellec S, Thomas C, Clavel J. Descriptive epidemiology of childhood Langerhans cell histiocytosis in France, 2000–2004. Pediatr Blood Cancer. 2008;51(1):71–5.
7. Stalemark H, Laurencikas E, Karis J, Gavhed D, Fadeel B, Henter JI. Incidence of Langerhans cell histiocytosis in children: a population-based study. Pediatr Blood Cancer. 2008;51(1):76–81.
8. Laurencikas E, Gavhed D, Stalemark H, van't Hooft I, Prayer D, Grois N, et al. Incidence and pattern of radiological central nervous system Langerhans cell histiocytosis in children: a population based study. Pediatr Blood Cancer. 2011;56(2):250–7.
9. Porto L, Schoning S, Hattingen E, Sorensen J, Jurcoane A, Lehrnbecher T. Central nervous system imaging in childhood Langerhans cell histiocytosis – a reference center analysis. Radiol Oncol. 2015;49(3):242–9.
10. Grois N, Flucher-Wolfram B, Heitger A, Mostbeck GH, Hofmann J, Gadner H. Diabetes insipidus in Langerhans cell histiocytosis: results from the DAL-HX 83 study. Med Pediatr Oncol. 1995;24(4):248–56.
11. Le Guennec L, Decaix C, Donadieu J, Santiago-Ribeiro M, Martin-Duverneuil N, Levy R, et al. The cognitive spectrum in neurodegenerative Langerhans cell histiocytosis. J Neurol. 2014;261(8):1537–43.
12. Nanduri VR, Lillywhite L, Chapman C, Parry L, Pritchard J, Vargha-Khadem F. Cognitive outcome of long-term survivors of multisystem Langerhans cell histiocytosis: a single-institution, cross-sectional study. J Clin Oncol. 2003;21(15):2961–7.
13. Haupt R, Minkov M, Astigarraga I, Schafer E, Nanduri V, Jubran R, et al. Langerhans Cell Histiocytosis

(LCH): guidelines for diagnosis, clinical work-up, and treatment for patients till the age of 18 years. Pediatr Blood Cancer. 2013;60(2):175–84.
14. Badalian-Very G, Vergilio JA, Degar BA, MacConaill LE, Brandner B, Calicchio ML, et al. Recurrent BRAF mutations in Langerhans cell histiocytosis. Blood. 2010;116(11):1919–23.
15. Diamond EL, Dagna L, Hyman DM, Cavalli G, Janku F, Estrada-Veras J, et al. Consensus guidelines for the diagnosis and clinical management of Erdheim-Chester disease. Blood. 2014;124(4):483–92.
16. Haroche J, Arnaud L, Cohen-Aubart F, Hervier B, Charlotte F, Emile JF, et al. Erdheim-Chester disease. Curr Rheumatol Rep. 2014;16(4):412.
17. Mazor RD, Manevich-Mazor M, Shoenfeld Y. Erdheim-Chester disease: a comprehensive review of the literature. Orphanet J Rare Dis. 2013;8:137.
18. Haroche J, Amoura Z, Dion E, Wechsler B, Costedoat-Chalumeau N, Cacoub P, et al. Cardiovascular involvement, an overlooked feature of Erdheim-Chester disease: report of 6 new cases and a literature review. Medicine (Baltimore). 2004;83(6):371–92.
19. Lachenal F, Cotton F, Desmurs-Clavel H, Haroche J, Taillia H, Magy N, et al. Neurological manifestations and neuroradiological presentation of Erdheim-Chester disease: report of 6 cases and systematic review of the literature. J Neurol. 2006;253(10):1267–77.
20. Arnaud L, Hervier B, Neel A, Hamidou MA, Kahn JE, Wechsler B, et al. CNS involvement and treatment with interferon-alpha are independent prognostic factors in Erdheim-Chester disease: a multicenter survival analysis of 53 patients. Blood. 2011;117(10):2778–82.
21. Parks NE, Goyal G, Go RS, Mandrekar J, Tobin WO. Neuroradiologic manifestations of Erdheim-Chester disease. Neurol Clin Pract. 2018;8(1):15–20.
22. Haroche J, Cohen-Aubart F, Emile JF, Maksud P, Drier A, Toledano D, et al. Reproducible and sustained efficacy of targeted therapy with vemurafenib in patients with BRAF(V600E)-mutated Erdheim-Chester disease. J Clin Oncol. 2015;33(5):411–8.
23. Abla O, Jacobsen E, Picarsic J, Krenova Z, Jaffe R, Emile JF, et al. Consensus recommendations for the diagnosis and clinical management of Rosai-Dorfman-Destombes disease. Blood. 2018;131(26):2877–90.
24. Mahzoni P, Zavareh MH, Bagheri M, Hani N, Moqtader B. Intracranial rosai-dorfman disease. J Res Med Sci. 2012;17(3):304–7.
25. Haroche J, Charlotte F, Arnaud L, von Deimling A, Helias-Rodzewicz Z, Hervier B, et al. High prevalence of BRAF V600E mutations in Erdheim-Chester disease but not in other non-Langerhans cell histiocytoses. Blood. 2012;120(13):2700–3.
26. Foucar E, Rosai J, Dorfman R. Sinus histiocytosis with massive lymphadenopathy (Rosai-Dorfman disease): review of the entity. Semin Diagn Pathol. 1990;7(1):19–73.
27. Janssen D, Harms D. Juvenile xanthogranuloma in childhood and adolescence: a clinicopathologic study of 129 patients from the kiel pediatric tumor registry. Am J Surg Pathol. 2005;29(1):21–8.
28. Haroche J, Abla O. Uncommon histiocytic disorders: Rosai-Dorfman, juvenile xanthogranuloma, and Erdheim-Chester disease. Hematology Am Soc Hematol Educ Program. 2015;2015:571–8.
29. So H, Kim SA, Yoon DH, Khang SK, Hwang J, Suh CH, et al. Primary histiocytic sarcoma of the central nervous system. Cancer Res Treat. 2015;47(2):322–8.
30. DeAngelis LM, Seiferheld W, Schold SC, Fisher B, Schultz CJ, Radiation Therapy Oncology Group Study. Combination chemotherapy and radiotherapy for primary central nervous system lymphoma: Radiation Therapy Oncology Group Study 93–10. J Clin Oncol. 2002;20(24):4643–8.
31. Ostrom QT, Gittleman H, Liao P, Vecchione-Koval T, Wolinsky Y, Kruchko C, et al. CBTRUS statistical report: primary brain and other central nervous system tumors diagnosed in the United States in 2010–2014. Neuro Oncol. 2017;19(suppl_5):v1–v88.
32. Villano JL, Koshy M, Shaikh H, Dolecek TA, McCarthy BJ. Age, gender, and racial differences in incidence and survival in primary CNS lymphoma. Br J Cancer. 2011;105(9):1414–8.
33. Eloranta S, Branvall E, Celsing F, Papworth K, Ljungqvist M, Enblad G, et al. Increasing incidence of primary central nervous system lymphoma but no improvement in survival in Sweden 2000–2013. Eur J Haematol. 2018;100(1):61–8.
34. O'Neill BP, Decker PA, Tieu C, Cerhan JR. The changing incidence of primary central nervous system lymphoma is driven primarily by the changing incidence in young and middle-aged men and differs from time trends in systemic diffuse large B-cell non-Hodgkin's lymphoma. Am J Hematol. 2013;88(12):997–1000.
35. Ferreri AJ, Blay JY, Reni M, Pasini F, Spina M, Ambrosetti A, et al. Prognostic scoring system for primary CNS lymphomas: the International Extranodal Lymphoma Study Group experience. J Clin Oncol. 2003;21(2):266–72.
36. Abrey LE, Ben-Porat L, Panageas KS, Yahalom J, Berkey B, Curran W, et al. Primary central nervous system lymphoma: the Memorial Sloan-Kettering Cancer Center prognostic model. J Clin Oncol. 2006;24(36):5711–5.
37. Rubenstein JL, Hsi ED, Johnson JL, Jung SH, Nakashima MO, Grant B, et al. Intensive chemotherapy and immunotherapy in patients with newly diagnosed primary CNS lymphoma: CALGB 50202 (Alliance 50202). J Clin Oncol. 2013;31(25):3061–8.
38. Gerstner ER, Batchelor TT. Primary central nervous system lymphoma. Arch Neurol. 2010;67(3):291–7.
39. Fischer L, Martus P, Weller M, Klasen HA, Rohden B, Roth A, et al. Meningeal dissemination in primary CNS lymphoma: prospective evaluation of 282 patients. Neurology. 2008;71(14):1102–8.
40. Haldorsen IS, Espeland A, Larsson EM. Central nervous system lymphoma: characteristic findings on traditional and advanced imaging. AJNR Am J Neuroradiol. 2011;32(6):984–92.

41. Zacharia TT, Law M, Naidich TP, Leeds NE. Central nervous system lymphoma characterization by diffusion-weighted imaging and MR spectroscopy. J Neuroimaging. 2008;18(4):411–7.
42. Bathla G, Hegde A. Lymphomatous involvement of the central nervous system. Clin Radiol. 2016;71(6):602–9.
43. Koubska E, Weichet J, Malikova H. Central nervous system lymphoma: a morphological MRI study. Neuro Endocrinol Lett. 2016;37(4):318–24.
44. Gametchu B. Glucocorticoid receptor-like antigen in lymphoma cell membranes: correlation to cell lysis. Science. 1987;236(4800):456–61.
45. Weller M. Glucocorticoid treatment of primary CNS lymphoma. J Neuro-Oncol. 1999;43(3):237–9.
46. Giannini C, Dogan A, Salomao DR. CNS lymphoma: a practical diagnostic approach. J Neuropathol Exp Neurol. 2014;73(6):478–94.
47. Hans CP, Weisenburger DD, Greiner TC, Gascoyne RD, Delabie J, Ott G, et al. Confirmation of the molecular classification of diffuse large B-cell lymphoma by immunohistochemistry using a tissue microarray. Blood. 2004;103(1):275–82.
48. Abrey LE, Batchelor TT, Ferreri AJ, Gospodarowicz M, Pulczynski EJ, Zucca E, et al. Report of an international workshop to standardize baseline evaluation and response criteria for primary CNS lymphoma. J Clin Oncol. 2005;23(22):5034–43.
49. Zucca E, Conconi A, Mughal TI, Sarris AH, Seymour JF, Vitolo U, et al. Patterns of outcome and prognostic factors in primary large-cell lymphoma of the testis in a survey by the International Extranodal Lymphoma Study Group. J Clin Oncol. 2003;21(1):20–7.
50. Ekstein D, Ben-Yehuda D, Slyusarevsky E, Lossos A, Linetsky E, Siegal T. CSF analysis of IGH gene rearrangement in CNS lymphoma: relationship to the disease course. J Neurol Sci. 2006;247(1):39–46.
51. Nelson DF, Martz KL, Bonner H, Nelson JS, Newall J, Kerman HD, et al. Non-Hodgkin's lymphoma of the brain: can high dose, large volume radiation therapy improve survival? Report on a prospective trial by the Radiation Therapy Oncology Group (RTOG): RTOG 8315. Int J Radiat Oncol Biol Phys. 1992;23(1):9–17.
52. Taguchi S, Hamamoto Y, Fujii T, Ochi T, Harada H, Ohnishi T, et al. Prognosis of primary central nervous system lymphoma treated with radiotherapy alone. Jpn J Radiol. 2012;30(10):806–10.
53. Shibamoto Y, Ogino H, Hasegawa M, Suzuki K, Nishio M, Fujii T, et al. Results of radiation monotherapy for primary central nervous system lymphoma in the 1990s. Int J Radiat Oncol Biol Phys. 2005;62(3):809–13.
54. Gavrilovic IT, Hormigo A, Yahalom J, DeAngelis LM, Abrey LE. Long-term follow-up of high-dose methotrexate-based therapy with and without whole brain irradiation for newly diagnosed primary CNS lymphoma. J Clin Oncol. 2006;24(28):4570–4.
55. Thiel E, Korfel A, Martus P, Kanz L, Griesinger F, Rauch M, et al. High-dose methotrexate with or without whole brain radiotherapy for primary CNS lymphoma (G-PCNSL-SG-1): a phase 3, randomised, non-inferiority trial. Lancet Oncol. 2010;11(11):1036–47.
56. Hiraga S, Arita N, Ohnishi T, Kohmura E, Yamamoto K, Oku Y, et al. Rapid infusion of high-dose methotrexate resulting in enhanced penetration into cerebrospinal fluid and intensified tumor response in primary central nervous system lymphomas. J Neurosurg. 1999;91(2):221–30.
57. Shapiro WR, Young DF, Mehta BM. Methotrexate: distribution in cerebrospinal fluid after intravenous, ventricular and lumbar injections. N Engl J Med. 1975;293(4):161–6.
58. Batchelor T, Carson K, O'Neill A, Grossman SA, Alavi J, New P, et al. Treatment of primary CNS lymphoma with methotrexate and deferred radiotherapy: a report of NABTT 96-07. J Clin Oncol. 2003;21(6):1044–9.
59. Ferreri AJ, Reni M, Foppoli M, Martelli M, Pangalis GA, Frezzato M, et al. High-dose cytarabine plus high-dose methotrexate versus high-dose methotrexate alone in patients with primary CNS lymphoma: a randomised phase 2 trial. Lancet. 2009;374(9700):1512–20.
60. Ferreri AJ, Cwynarski K, Pulczynski E, Ponzoni M, Deckert M, Politi LS, et al. Chemoimmunotherapy with methotrexate, cytarabine, thiotepa, and rituximab (MATRix regimen) in patients with primary CNS lymphoma: results of the first randomisation of the International Extranodal Lymphoma Study Group-32 (IELSG32) phase 2 trial. Lancet Haematol. 2016;3(5):e217–27.
61. Gregory G, Arumugaswamy A, Leung T, Chan KL, Abikhair M, Tam C, et al. Rituximab is associated with improved survival for aggressive B cell CNS lymphoma. Neuro-Oncology. 2013;15(8):1068–73.
62. Holdhoff M, Ambady P, Abdelaziz A, Sarai G, Bonekamp D, Blakeley J, et al. High-dose methotrexate with or without rituximab in newly diagnosed primary CNS lymphoma. Neurology. 2014;83(3):235–9.
63. Kansara R, Shenkier TN, Connors JM, Sehn LH, Savage KJ, Gerrie AS, et al. Rituximab with high-dose methotrexate in primary central nervous system lymphoma. Am J Hematol. 2015;90(12):1149–54.
64. Mocikova H, Pytlik R, Sykorova A, Janikova A, Prochazka V, Vokurka S, et al. Role of rituximab in treatment of patients with primary central nervous system lymphoma: a retrospective analysis of the Czech Lymphoma Study Group Registry. Leuk Lymphoma. 2016;57(12):2777–83.
65. Glass J, Won M, Schultz CJ, Brat D, Bartlett NL, Suh JH, et al. Phase I and II study of induction chemotherapy with methotrexate, rituximab, and temozolomide, followed by whole-brain radiotherapy and postirradiation temozolomide for primary CNS lymphoma: NRG Oncology RTOG 0227. J Clin Oncol. 2016;34(14):1620–5.
66. Morris PG, Correa DD, Yahalom J, Raizer JJ, Schiff D, Grant B, et al. Rituximab, methotrexate, procar-

bazine, and vincristine followed by consolidation reduced-dose whole-brain radiotherapy and cytarabine in newly diagnosed primary CNS lymphoma: final results and long-term outcome. J Clin Oncol. 2013;31(31):3971–9.
67. Illerhaus G, Kasenda B, Ihorst G, Egerer G, Lamprecht M, Keller U, et al. High-dose chemotherapy with autologous haemopoietic stem cell transplantation for newly diagnosed primary CNS lymphoma: a prospective, single-arm, phase 2 trial. Lancet Haematol. 2016;3(8):e388–97.
68. Omuro A, Correa DD, DeAngelis LM, Moskowitz CH, Matasar MJ, Kaley TJ, et al. R-MPV followed by high-dose chemotherapy with tbc and autologous stem-cell transplant for newly diagnosed primary CNS lymphoma. Blood. 2015;125(9):1403–10.
69. Ney DE, Abrey LE. Maintenance therapy for central nervous system lymphoma with rituximab. Leuk Lymphoma. 2009;50(9):1548–51.
70. Kasenda B, Ferreri AJ, Marturano E, Forst D, Bromberg J, Ghesquieres H, et al. First-line treatment and outcome of elderly patients with primary central nervous system lymphoma (PCNSL)--a systematic review and individual patient data meta-analysis. Ann Oncol. 2015;26(7):1305–13.
71. Pipkin S, Scheer S, Okeigwe I, Schwarcz S, Harris DH, Hessol NA. The effect of HAART and calendar period on Kaposi's sarcoma and non-Hodgkin lymphoma: results of a match between an AIDS and cancer registry. AIDS. 2011;25(4):463–71.
72. Collaboration of Observational HIVERESG, Bohlius J, Schmidlin K, Costagliola D, Fatkenheuer G, May M, et al. Incidence and risk factors of HIV-related non-Hodgkin's lymphoma in the era of combination antiretroviral therapy: a European Multicohort Study. Antivir Ther. 2009;14(8):1065–74.
73. Hoffmann C, Tabrizian S, Wolf E, Eggers C, Stoehr A, Plettenberg A, et al. Survival of AIDS patients with primary central nervous system lymphoma is dramatically improved by HAART-induced immune recovery. AIDS. 2001;15(16):2119–27.
74. Dugan JP, Haverkos BM, Villagomez L, Martin LK, Lustberg M, Patton J, et al. Complete and durable responses in primary central nervous system posttransplant lymphoproliferative disorder with zidovudine, ganciclovir, rituximab, and dexamethasone. Clin Cancer Res. 2018;24(14):3273–81.
75. Martinez DL, Gujrati M, Geoffroy F, Tsung AJ. Isolated CNS Hodgkin's lymphoma: implications for tissue diagnosis. CNS Oncol. 2014;3(6):383–7.
76. Menon MP, Nicolae A, Meeker H, Raffeld M, Xi L, Jegalian AG, et al. Primary CNS T-cell lymphomas: a clinical, morphologic, immunophenotypic, and molecular analysis. Am J Surg Pathol. 2015;39(12):1719–29.
77. Matmati K, Matmati N, Hannun YA, Rumboldt Z, Patel S, Lazarchick J, et al. Dural MALT lymphoma with disseminated disease. Hematol Rep. 2010;2(1):e10.
78. Slone HW, Blake JJ, Shah R, Guttikonda S, Bourekas EC. CT and MRI findings of intracranial lymphoma. AJR Am J Roentgenol. 2005;184(5):1679–85.
79. Lucantoni C, De Bonis P, Doglietto F, Esposito G, Larocca LM, Mangiola A, et al. Primary cerebral lymphomatoid granulomatosis: report of four cases and literature review. J Neuro-Oncol. 2009;94(2):235–42.
80. Tateishi U, Terae S, Ogata A, Sawamura Y, Suzuki Y, Abe S, et al. MR imaging of the brain in lymphomatoid granulomatosis. AJNR Am J Neuroradiol. 2001;22(7):1283–90.
81. Demetrick DJ, Hamilton MG, Curry B, Tranmer BI. Epstein-barr virus-associated primary B-cell lymphoproliferative disorder of the cerebellum in an immune competent man. Cancer. 1992;70(2):519–28.
82. Roschewski M, Wilson WH. Lymphomatoid granulomatosis. Cancer J. 2012;18(5):469–74.
83. Pittock SJ, Debruyne J, Krecke KN, Giannini C, van den Ameele J, De Herdt V, et al. Chronic lymphocytic inflammation with pontine perivascular enhancement responsive to steroids (CLIPPERS). Brain. 2010;133(9):2626–34.
84. Tobin WO, Guo Y, Krecke KN, Parisi JE, Lucchinetti CF, Pittock SJ, et al. Diagnostic criteria for chronic lymphocytic inflammation with pontine perivascular enhancement responsive to steroids (CLIPPERS). Brain. 2017;140(9):2415–25.
85. De Graaff HJ, Wattjes MP, Rozemuller-Kwakkel AJ, Petzold A, Killestein J. Fatal B-cell lymphoma following chronic lymphocytic inflammation with pontine perivascular enhancement responsive to steroids. JAMA Neurol. 2013;70(7):915–8.
86. Jones JL, Dean AF, Antoun N, Scoffings DJ, Burnet NG, Coles AJ. 'Radiologically compatible clippers' may conceal a number of pathologies. Brain. 2011;134(Pt 8):e187.
87. Limousin N, Praline J, Motica O, Cottier JP, Rousselot-Denis C, Mokhtari K, et al. Brain biopsy is required in steroid-resistant patients with chronic lymphocytic inflammation with pontine perivascular enhancement responsive to steroids (CLIPPERS). J Neuro-Oncol. 2012;107(1):223–4.
88. Lin AW, Das S, Fraser JA, Ang LC, Florendo-Cumbermack A, Jenkins ME, et al. Emergence of primary CNS lymphoma in a patient with findings of CLIPPERS. Can J Neurol Sci. 2014;41(4):528–9.
89. Taieb G, Uro-Coste E, Clanet M, Lassmann H, Benouaich-Amiel A, Laurent C, et al. A central nervous system B-cell lymphoma arising two years after initial diagnosis of CLIPPERS. J Neurol Sci. 2014;344(1–2):224–6.

# Part VI

# Clinical Approaches in Neuroimmunology

# Clinical Approach to Autoimmune Myelitis and Myelopathy

# 27

Cynthia Wang and Benjamin Greenberg

**Patient Vignette 1**

A 19-year-old young woman with no significant past medical history presents with 2 days of worsening bilateral leg weakness and gait instability. Magnetic resonance imaging (MRI) of the spine showed abnormal T2/fluid-attenuated inversion recovery (FLAIR) signal with patchy enhancement of the thoracic cord extending over 4 vertebral segments and affecting both central gray and white matter. MRI of the brain was normal. Cerebrospinal fluid (CSF) studies revealed 6 nucleated cells, normal protein, glucose, 0 oligoclonal bands, normal immunoglobulin G (IgG) index, and negative aquaporin-4(AQP4)/neuromyelitis optica (NMO) IgG. She was diagnosed with acute transverse myelitis and started on intravenous (IV) corticosteroids with marked improvement in her leg weakness. At her follow-up appointment 2 months later, her gait has returned to normal, but she notes right eye pain and blurry vision. On exam, she has a right relative afferent pupillary defect and right optic nerve margins are indistinct. MRI of the brain demonstrates right optic neuritis. Anti-myelin oligodendrocyte antibodies returned positive.

This case illustrates a possible clinical presentation for transverse myelitis and how neuroimaging and laboratory studies help guide diagnosis. Features of the patient's presentation that argue against multiple sclerosis (MS) include longitudinally extensive spinal cord lesion, normal MRI brain, absence of oligoclonal bands, and normal IgG index. Neuromyelitis optica spectrum disorder (NMOSD) should be considered and can be assessed with very good sensitivity and specificity through a cell-based assay for aquaporin-4 antibodies. It is now also possible to test for myelin oligodendrocyte glycoprotein (MOG) antibodies, which often overlaps phenotypically with AQP-4 related NMOSD. Idiopathic acute transverse myelitis is often diagnosed once the above etiologies have been excluded.

## Introduction

Autoimmune myelitis encompasses a heterogeneous set of clinical syndromes with differing etiologies and disease mechanisms. While inflammation can be present in infectious myelitis and non-immune-mediated myelopathies, autoimmune myelitis refers to spinal cord injury resulting solely from a pathological immune response. In some cases, autoimmune myelitis can exist in isolation, sometimes called idiopathic or post-infectious acute transverse myelitis, or as

C. Wang
Department of Pediatrics and Department of Neurology and Neurotherapeutics, Children's Health Dallas, University of Texas Southwestern, Dallas, TX, USA

B. Greenberg (✉)
Transverse Myelitis Program, Department of Neurology and Neurotherapeutics, University of Texas Southwestern, Dallas, TX, USA
e-mail: Benjamin.greenberg@utsouthwestern.edu



A. L. Piquet, E. Alvarez (eds.), *Neuroimmunology*, https://doi.org/10.1007/978-3-030-61883-4_27

a manifestation of relapsing syndrome, such as multiple sclerosis (MS) or neuromyelitis optica spectrum disorder (NMOSD).

The initial approach to an individual presenting with myelopathic symptoms is to recognize the syndrome, confirm the diagnosis with laboratory and imaging studies, and provide timely treatments that reduce ongoing inflammation. Acute management of the various forms of autoimmune myelitis is similar, but notably differs from infectious, metabolic, compressive, and vascular myelopathies. Once autoimmune myelitis is established as the likely diagnosis, it is essential to evaluate for an underlying cause, particularly identifying relapsing immune-mediated conditions that require long-term preventative treatment.

This chapter provides an overview of the current understanding of autoimmune myelitis including the various etiologies, pathogenesis, clinical manifestations, diagnostic workup, differential diagnosis, acute treatment, long-term management strategies, and prognosis. As autoimmune myelitis has the potential to cause significant motor, sensory, and autonomic deficits and enduring functional impairment, it is essential for clinicians to appropriately manage autoimmune myelitis in order to improve outcomes for affected individuals.

## Pathology and Immunopathogenesis

Although clinical presentations of autoimmune myelitis overlap, the immunopathogenesis may be distinct depending on the underlying cause. Although myelitis has largely been conceptualized as a demyelinating condition, multiple mechanisms may be present, affecting both white and gray matter, neuronal and glia function, and myelin and axonal integrity.

Histopathology of idiopathic transverse myelitis is not well characterized since biopsy specimens are rarely obtained. In aggressive forms of the disease associated with MRI appearance of necrosis, spinal cord lesions can show gliosis, reactive lymphocytes, and demyelination [1, 2]. Focal infiltration by monocytes and lymphocytes into spinal cord and perivascular spaces, along with astrogial and microglial activation, has also been reported [3]. Several studies have examined CSF cytokine levels and identified elevations in individuals with transverse myelitis, including interleukin-6 (IL-6) levels, which was associated with disease severity and subsequent disability [4, 5].

In neuromyelitis optica, aquaporin-4 immunoglobulin G (AQP4-IgG) targets the dominant water channel protein in the central nervous system, which is expressed on astrocyte foot processes at the blood-brain barrier, subpial, and subependymal regions. These antibodies are known to be pathogenic, leading to antibody-dependent cellular cytotoxicity. Pathology of these lesions demonstrates perivascular immunoglobulin deposition and complement activation leading to AQP4 loss [6]. Myelin and neuronal damage occurs secondary to the primary astrocytopathy and immune activation [7].

In contrast, the pathological hallmark of multiple sclerosis is demyelination with variable preservation of axons. Histology of MS lesions demonstrates immunopathological heterogeneity including T-cell and/or antibody-mediated gliosis and inflammation, affecting both white and gray matter [8]. Intrathecal antibody production, as demonstrated by the presence of CSF-specific oligoclonal bands and elevated IgG index, is supportive of a multiple sclerosis diagnosis [9, 10].

## Demographics, Clinical Features, and Epidemiology

Although idiopathic autoimmune myelitis is relatively rare, it is difficult to quantify its incidence and prevalence as myelitis stems from various causes. Demographic features such as gender and age predilection also differ based on the underlying etiology. The relative contributions from idiopathic myelitis versus disease-associated myelitis are difficult to discern as previous studies have examined heterogeneous populations with varying length of follow-up. In one publication, 354 patients with transverse myelitis (TM) were studied with approximately 64% of cases representing idiopathic TM and 36% of cases representing disease-associated myelitis [11]. Other reports indicate a greater contribution from secondary TM representing a majority of cases (85–70%) [12, 13].

For idiopathic transverse myelitis, the incidence is estimated to be 1–8 per million people per year [14]. It can affect males and females and all ages and ethnic backgrounds without a strong bias for any group. Some reports indicate a bimodal age distribution with increased incidence in patients between 10–19 years and 30–39 years [15]. In children, idiopathic myelitis has been more frequently reported in children under 5 and over 10 years [16, 17]. In contrast, transverse myelitis as a part of multiple sclerosis typically affects women (female to male ratio of 1.4–2.3:1) women with mean age of onset of 28–31 years [18]. Myelitis as part of AQP4-IgG positive NMOSD tends to have an even more pronounced female predilection (female to male ratio of 5–10:1), with a mean age of onset of 32–41 years [19, 20].

## Clinical Features

Myelopathies often manifest with common symptoms at onset regardless of etiology. Historically, acute transverse myelitis was defined to involve bilateral extremities and a clear sensory level by the Transverse Myelitis Consortium Working Group [16], but the disease can involve varying degrees of weakness and sensory change to the neck, trunk, and extremities depending on the location and extent of spinal cord inflammation. Motor symptoms often include a rapidly progressive paraparesis or quadriparesis. Sensory symptoms can involve pain, numbness, and dysthesias/paresthesias. In the acute setting, an individual may have flaccid tone and diminished reflexes, but spasticity usually develops over days to weeks in upper motor neuron-predominant immune-mediated myelitis. Autonomic nervous system dysfunction occurs below the level of the lesion and can manifest as urinary and bowel incontinence and/or retention, and sexual dysfunction. The time course from symptom onset to nadir usually ranges from 4 hours to 21 days [16], though a progression over several days is most typical.

## Diagnostic Approach

Since no single diagnostic test has perfect sensitivity or specificity for immune-mediated myelitis, a clinician must use clinical, radiographical, and laboratory studies to establish a diagnosis. At initial presentation, clinical findings may not be adequate to conclusively determine if symptoms localize to the central nervous system (CNS) or peripheral nervous system (PNS). Likewise, a clinical history may not be sufficient to determine if an acute myelopathy is related to inflammation.

In the acute setting, myelopathies can cause flaccid tone and diminished reflexes potentially leading to diagnostic confusion with Guillain-Barré syndrome. However, features such as a sensory level or urinary retention/incontinence argue against a peripheral etiology. An acute to subacute time course of symptom onset can support an immune etiology. However, myelopathies with hyperacute or chronic presentations prompt consideration of alternative non-immune etiologies such as vascular and metabolic myelopathies, respectively.

Regardless of suspected diagnosis, any presentations concerning for acute myelopathy should be treated as medical emergencies. Medical literature notes that compressive myelopathies lead to irreversible injury if not correctly managed, and at the bedside a clinician cannot determine a cause of a myelopathy, thus, urgent neuroimaging is indicated in all cases. CSF analysis is also critical in evaluating myelopathic patients. However, inability to complete these studies should not delay empirical initiation of IV corticosteroids as these are unlikely to be contraindicated regardless of etiology [21].

## Diagnostic Workup

### Magnetic Resonance Imaging

The typical MRI characteristics of spinal cord usually demonstrate abnormal T2/FLAIR signal affecting one or more spinal segments often with associated edema. Idiopathic transverse myelitis typically leads to central lesions affecting two-thirds or more of the cross section of the cord, extends over more than two segments, and has predilection for the thoracic cord [22]. In contrast, MS-related myelitis is associated with partial, dorsolateral lesions less than two segments in length. NMOSD is classically associated with long segment myelitis and may be associated with T1 hypointensity, cord enhancement, and cord expansion. Gadolinium contrast enhancement may be present with inflammatory myelopathies. Obtaining complete spinal cord imaging is recommended as lower extremity weakness may still originate from the cervical spine. Children are more likely to have more longitudinally extensive myelitis; thus, the extent of a lesion is not as useful in pediatric TM to differentiate between idiopathic TM, MS, or AQP4-IgG or myelin oligodendrocyte glycoprotein (MOG) antibody syndromes [23]. Figures 27.1a–h and 27.2a–f provide examples of typical MRI findings in different forms of myelitis.

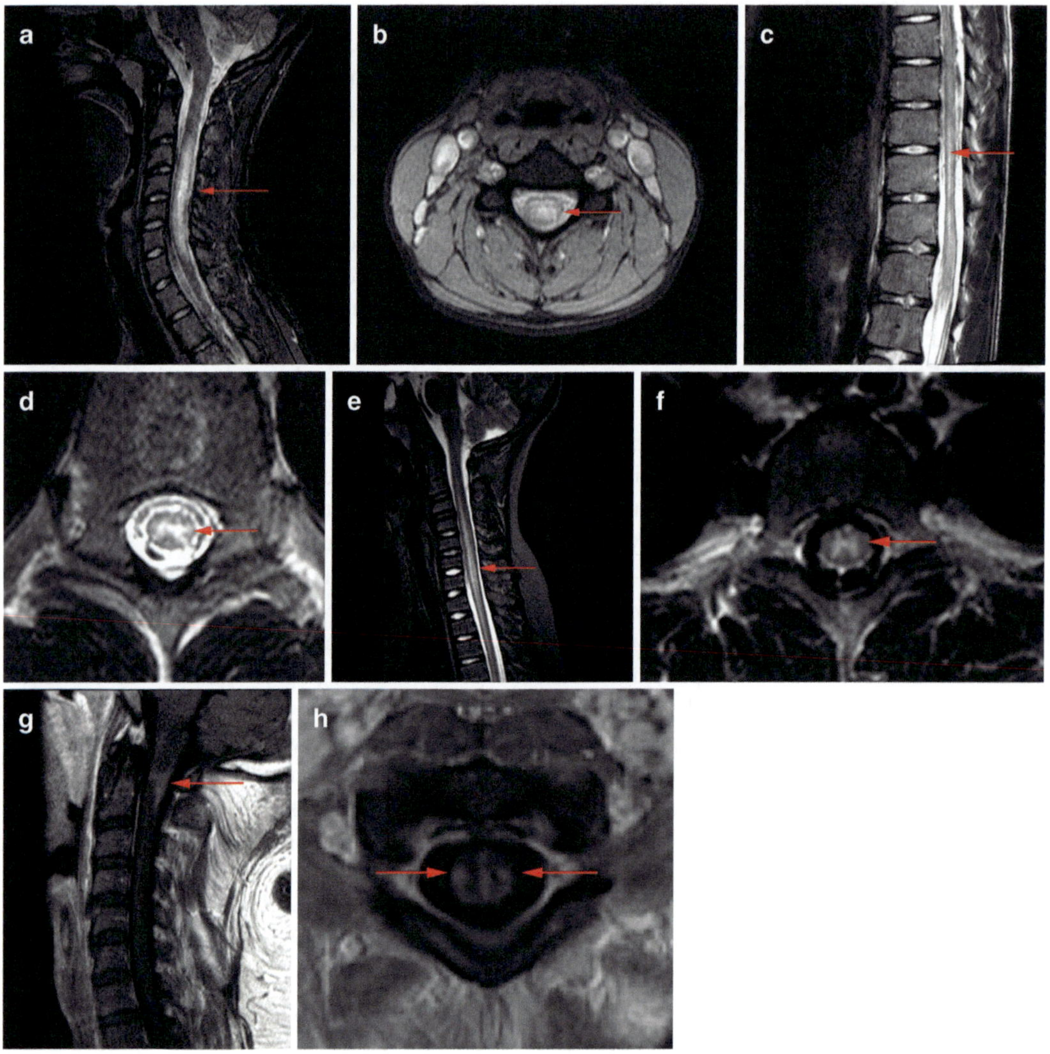


**Fig. 27.1** Magnetic resonance imaging (MRI) T2 sagittal and axial images of patients with: (**a, b**) AQP4 IgG positive NMOSD, (**c, d**) MOG IgG positive myelitis, (**e, f**) acute flaccid myelitis (AFM), (**g, h**) neurosarcoidosis. Abbreviations: AQP4 aquaporin-4, IgG immunoglobulin G, NMOSD neuromyelitis optica spectrum disorder, MOG myelin oligodendrocyte glycoprotein

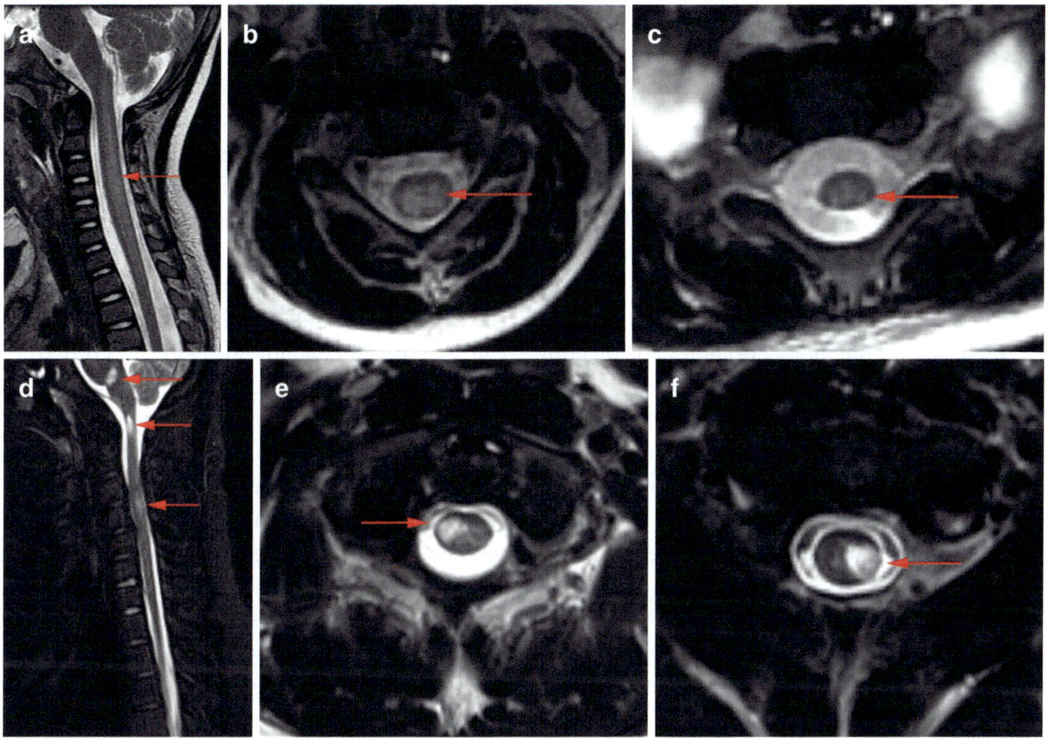


**Fig. 27.2** Magnetic resonance imaging (MRI) T2 sagittal and axial images of patients with: (**a**–**c**) idiopathic transverse myelitis, (**d**–**f**) multiple sclerosis

## Cerebrospinal Fluid Analysis

CSF is abnormal in about 50% of the cases with lymphocytic pleocytosis, elevated protein, and typically normal glucose. The presence of unmatched oligoclonal bands in CSF compared to serum is suggestive of multiple sclerosis. In particular, one large meta-analysis found that detection of oligoclonal bands has a sensitivity of 88% and specificity of 86% of identifying individuals with MS [9]. Anti-AQP4 IgG can be assessed in CSF, though this is slightly less sensitive compared to serum testing.

## Serum Laboratory Testing

Anti-AQP4 IgG can be detected with excellent sensitivity and specificity in the serum. In one study aquaporin-4 antibody assays, enzyme-linked immunosorbent assays (ELISAs) were found to have mean sensitivity and specificity of 64% and 85.9% versus 75.7% and 99.8% for cell-based assays [24]. Commercial testing for anti-MOG antibodies through cell-based assays is also increasingly available. Paraneoplastic antibody panels may be useful, particularly screening for collapsing response-mediator protein-5 (CRMP-5) and amphiphysin antibodies, which have been associated with myelitis [25, 26]. Rheumatologic workup including assessing for anti-nuclear antibodies (ANAs), anti-Smith, anti-phospholipid antibodies, SS-A, and SS-B should be considered. Evaluating vitamin B12, copper, and vitamin E levels can identify nutritional/metabolic myelopathies. 25-hydroxy vitamin D level is often low in the general population and is important to optimize in autoimmune conditions such as MS in which higher levels of vitamin D may confer greater protection against relapses [27].

Concurrent acute illness and fever should prompt consideration of infectious myelitis, which may be related to direct injury from a pathogen. Infectious myelitis and parainfectious myelitis have been reported with viruses such as West Nile virus (WNV), varicella zoster virus (VZV), herpes simplex virus (HSV), human T-lymphotropic virus (HTLV)-1, human immunodeficiency virus (HIV), and Zika virus (although HTLV and HIV are associated with chronic progressive forms of myelopathy) [28]. Bacterial infections including mycoplasma, Lyme, syphilis, and *Listeria monocytogenes* have also been detected in association with myelitis [17]. However, infectious etiologies for myelitis are rare relative to immune-mediated causes, and corticosteroids have been used as adjunctive therapy for some viral and bacterial infections of the CNS [29]. Thus, empiric therapy for suspected immune-mediated myelitis generally should not be delayed while investigating for infection.

## Other Studies

Additional imaging studies can be considered based on clinical suspicion. For instance, computed tomography (CT) of chest, abdomen, and pelvis should be obtained to evaluate for malignancy if there is concern for a paraneoplastic myelitis. Chest imaging to identify systemic sarcoidosis may include CT chest, gallium scan, and fluorodeoxyglucose (FDG)-positron emission tomography (PET) scan (chest X-ray has low sensitivity) [30]. Visual evoked potentials and optical coherence tomography (OCT) may establish past history of optic neuritis suggestive of MS or NMOSD.

## Acute Treatment

There have not been any large-scale prospective randomized clinical trials demonstrating the relative efficacy of acute treatments used for myelitis. Based on American Academy of Neurology (AAN) guidelines, only class IV evidence exists to support corticosteroid use. However, given little risk of harm [31], steroids are generally administered if an inflammatory myelitis is a part of the differential diagnosis. High-dose IV corticosteroids are typically first-line treatments based on expert opinion and consensus. It is typically dosed at 1000 mg (1 gm) methylprednisilone daily or dexamethasone 200 mg daily for 3–5 days in adults and 30 mg/kg (up to 1000 mg) methylprednisilone daily for same duration in children.

Therapeutic plasma exchange (PLEX) is an effective treatment for autoimmune myelitis, particularly those that are humorally mediated such as with AQP4 and MOG antibody syndromes. Case series in pediatric patients have identified PLEX as a safe and effective therapy [32]. This can be utilized concurrently with IV corticosteroids for severe relapses or following steroid-unresponsive relapses. Indications for PLEX may include significant motor impairment/inability to ambulate and bilateral and/or severe vision loss. A typical course of treatment is 1.1–1.5 plasma volume exchanges every other day for 5–7 sessions.

In one retrospective study, patients with idiopathic TM who were treated with corticosteroids plus PLEX had twice as much improvement on the Expanded Disability Status Scale (EDSS) compared to those treated with corticosteroids alone [33]. There is even stronger evidence and a rational approach to use PLEX for NMO IgG-associated myelitis, given presumed mechanism of removing pathogenic antibodies and complement. In NMOSD patients, early initiation of PLEX appears to improve outcomes [34]. Thus, PLEX is often administered concurrently with IV corticosteroids in individuals with severe symptoms and prior attacks poorly responsive to corticosteroids [35–37]. Likewise, PLEX has been shown to be effective for MS-related myelitis as well, particularly in those with pattern II lesions distinguished by prominent immunoglobulin deposition and complement activation [38].

Other acute therapies include intravenous immunoglobulin (IVIg), which has been with some success following poor response to corticosteroids, but it was not recommended based on AAN guidelines [39]. Intravenous cyclophosphamide given as a 800–1200 mg/m$^2$ pulsed dose can also be quite effective, particularly in cases of severe, longitudinally extensive myelitis and myelitis secondary to connective tissue disorders such as systemic lupus erythematosus (SLE).

## Forms of Autoimmune Myelitis

Acute myelitis can be a consequence of recurrent demyelinating and systemic inflammatory conditions. Acquired central nervous system autoimmune syndromes include multiple sclerosis, neuromyelitis optica spectrum disorder associated with AQP4 antibodies, and MOG antibody-associated encephalomyelitis. Here we will review the most common forms of autoimmune myelitis and some of the distinguishing features of each. Table 27.1 presents characteristic MRI, CSF, and other findings in forms of autoimmune myelitis. Table 27.2 reviews features of myelopathies in which inflammation is not the principal disease mechanism.

## Myelitis Secondary to Multiple Sclerosis

Up to 85% of individuals with multiple sclerosis have spinal cord involvement at some point in their disease, and 20–40% may experience myelitis as their first clinical event [40]. Spinal cord lesions related to multiple sclerosis are typically short (1–2 vertebral segments) and involve only part of the cross section of the cord (typically dorsal). Sometimes multiple small lesions, or "skip lesions," can take on a confluent appearance, resembling NMOSD. During the acute relapse, the degree of cord swelling is lower compared to NMOSD. Following myelitis, atrophy typically occurs at the affected locations giving a thin appearance to the spinal cord.

**Table 27.1** Characteristic magnetic resonance imaging (MRI), cerebrospinal fluid (CSF), and other findings in forms of autoimmune myelitis

| Diagnosis | MRI spine features | CSF features | Other tests |
|---|---|---|---|
| Multiple sclerosis | Oval-shaped lesions, predilection for cervical and posterior cord | Positive oligoclonal bands (OCBs), increased IgG index | MRI of the brain may show juxtacortical and periventricular lesions |
| Aquaporin-4 (AQP4) Ab NMOSD | Longitudinally extensive transverse myelitis (LETM) | OCBs rare, positive aquaporin-4 IgG | Positive serum AQP4 IgG |
| MOG Ab myelitis | LETM, gray and white matter | OCBs rare, IgG index may be elevated | Positive serum MOG IgG |
| Neurosarcoidosis | LETM, favors cervical and thoracic cord, patchy and/or leptomeningeal enhancement, "Trident sign" | Lymphocytic pleocytosis, low glucose, OCB rare | Chest CT |
| Systemic lupus erythematosus | LETM, cord swelling, gray and white matter | Pleocytosis, OCBs rare | ANA, dsDNA, anti-Smith antibodies |
| Idiopathic transverse myelitis | LETM, thoracic cord, affects 2/3 of cross section of cord | Increased protein, OCBs can be seen | Rule out infectious and other inflammatory causes of TM |
| Acute flaccid myelitis | Gray matter predominant, anterior horns, "owl's eye" | Enterovirus PCR (rarely positive) | Respiratory enterovirus PCR |

*Abbreviations: IgG* immunoglobulin G, *NMOSD* neuromyelitis optica spectrum disorder, *MOG* myelin oligodendrocyte glycoprotein, *CT* computed tomography, *TM* transverse myelitis, *PCR* polymerase chain reaction

**Table 27.2** Causes of myelopathy

| Diagnosis | MRI spine features | Other tests |
|---|---|---|
| Subacute combined degeneration | Dorsal column and pyramidal tract involvement | Vitamin B12, copper, vitamin E, |
| Tabes dorsalis (syphilitic myelopathy) | Dorsal columns, dorsal roots | Treponemal antibody |
| Anterior spinal artery occlusion | Abnormal signal in central spinal cord, no enhancement | Spinal angiogram, MR angiography |
| Compression/spondylotic | "Pancake"-like gadolinium contrast enhancement | MRI, CT, CT myelogram |

*MR* magnetic resonance, *MRI* magnetic resonance imaging, *CT* computed tomography

In addition to the characteristics of the spinal cord imaging, brain imaging can also be very helpful in establishing a diagnosis of multiple sclerosis. Demonstration of two or more brain lesions in the typical periventricular, juxtacortical, and brainstem locations is highly suggestive for MS and provides evidence of disease dissemination in space. In addition, the presence of intrathecal immunoglobulin synthesis (elevated IgG index) and unique CSF oligoclonal bands (compared to serum) provides additional support for multiple sclerosis [41]. With the 2017 revision of the McDonald criteria, oligoclonal bands serve as a surrogate for dissemination in time. Transverse myelitis can be considered in the broader category of clinically isolated syndrome (CIS) when a person is suspected to have multiple sclerosis but has not met diagnostic criteria.

## Myelitis Secondary to Neuromyelitis Optica Associated with AQP4 Antibodies

Prior to the discovery of the NMO IgG, neuromyelitis optica was chiefly distinguished from multiple sclerosis due to long segment spinal cord and optic nerve involvement with relative sparing of the brain. NMO-related myelitis is still typically regarded to involve three or more vertebral segments; however, shorter NMOSD lesions have been reported [42]. After its molecular target was identified as aquaporin-4, a number of other syndromes including brainstem and brain involvement are now accepted to be part of the NMO disease spectrum. Given the variability in phenotypes and high morbidity associated with the disease without treatment, testing all patients with history of transverse myelitis for anti-AQP4 antibodies is recommended.

Radiographically, NMOSD myelitis manifests as a centrally located lesion that is hyperintense on T2/FLAIR sequences (Fig. 27.1a, b) and sometimes hypointense on T1 imaging in the most severely affected areas. The cervical spine and the thoracic spine are commonly involved. Within the brain, the dorsal medulla (area postrema) and diencephalon are frequent targets of inflammation leading to clinical symptoms of persistent vomiting and hiccups, and hypothermia and hypersomnolence, respectively.

NMOSD relapses can be very debilitating and are often associated with poor recovery. In addition to corticosteroids, PLEX should be considered early in the acute setting for individuals with attacks leading to vision loss and motor impairments. One clinical attack with positive anti-AQP4 antibodies is sufficient to make the diagnosis of NMOSD and to initiate chronic immunosuppressive treatment. Commonly used therapies include rituximab, mycophenolate mofetil, azathioprine, and prednisone; in retrospective studies, comparing these drugs in NMO, rituximab was the most effective option, becoming the mainstay first-line therapy [43]. Although none of these treatments are approved by the US Food and Drug Administration (FDA); as of 2020, however, there are now three drugs with FDA approval in NMOSD including eculizumab, inebilizumab, and satralizumab (discussed in Chap. 17).

## Myelitis Secondary to Anti-Myelin Oligodendrocyte Glycoprotein (MOG) Syndrome

Our understanding of MOG antibody-related syndromes has advanced greatly in recent years. Historically, poor sensitivity and specificity of MOG assays led to doubt about the clinical relevance of the antibodies. With improved cell-based assay testing methods, researchers began to find anti-MOG seropositivity in pediatric populations with histories of acute disseminated encephalomyelitis (ADEM), optic neuritis, and/or transverse myelitis. Other syndromes including confluent white matter abnormalities resembling a leukodystrophy and cortical encephalitis with seizures have also been reported [44, 45].

As some of these individuals satisfy criteria for NMOSD and both are antibody-mediated, some groups consider MOG syndromes a type of seronegative NMOSD. However, others argue that the molecular target, epidemiology, and clinical syndromes are distinct and that MOG antibody disease should to be classified as a separate entity from NMOSD [46]. In particular, MOG antibody disease differs from AQP4-associated disease in epidemiological features including younger age of onset, less predilection for the female sex, and higher frequency in Caucasian populations. MOG antibody disease also has a stronger association with optic neuritis, especially bilateral optic neuritis. Longitudinally extensive myelitis can occur with MOG antibody disease though it has a higher frequency of conus medullaris involvement compared to AQP4 NMOSD and is less likely to result in cord necrosis and cavitation [47]. Presently, it is unclear to what extent MOG antibody diseases cause relapsing demyelination, though persistence of MOG antibodies beyond 6–12 months is associated with increased risk of relapse [48].

## Myelitis Secondary to Systemic Inflammatory Disorders/Connective Tissue Disorders

Systemic inflammatory disorders including systemic lupus erythematosus (SLE) and Sjögren's syndrome have been associated with longitudinally extensive transverse myelitis. In these situations, myelitis may represent co-existing presence of AQP-4 antibodies rather than being a direct result of the rheumatologic disorder [49]. However, SLE can be associated with a gray matter myelitis, which likely represents a distinct entity [50]. Treatments such as cyclophosphamide may confer distinct benefit in SLE-related myelitis [51].

## Myelitis Secondary to Sarcoidosis

Only 5–10% of individuals with sarcoidosis present with neurological dysfunction, and even a smaller portion develop isolated myelitis. In contrast to other inflammatory myelopathies, myelitis secondary to sarcoidosis typically has a more indolent course. Distinguishing neurological features include cranial nerve involvement and meningeal enhancement. In particular, central canal and dorsal-subpial enhancement on MRI, the so-called "trident sign," has been reported to be a sensitive finding for neurosarcoidosis [52, 53]. Neurosarcoidosis is further discussed in Chap. 23.

## Idiopathic Transverse Myelitis

The diagnosis of idiopathic transverse myelitis is made when no definitive etiology is identified despite an appropriate and thorough investigation. It is hypothesized to be a parainfectious or postinfectious immune reaction as there is a history of a recent respiratory, gastrointestinal illness in 30–60% of cases [54]. It is regarded as a monophasic autoimmune process and only 10–30% of individuals go on to develop multiple sclerosis in the absence of brain lesions and intrathecal unique oligoclonal bands [41]. Interestingly, in a study of patients evaluated at the Mayo Clinic who had prior diagnosis of idiopathic TM, an alternative specific myelopathy diagnosis was found in 69.9% of 226 patients. Clinically isolated syndrome/multiple sclerosis and vascular myelopathy represented the two most commonly identified etiologies [55]. Although idiopathic TM is not considered to be heritable, one report of two sisters with monophasic TM and a shared rare missense mutation affecting endosomal sorting complex protein suggests the existence of genetic susceptibility [56].

Myelitis of unknown etiology accompanied by multifocal brain lesions and encephalopathy is consistent with acute disseminated encephalomyelitis (ADEM). With increasing availability of commercial anti-MOG antibody testing, future studies may be able to delineate the frequency of anti-MOG seropositivity in this population.

## Acute Flaccid Myelitis

Although not considered to be an autoimmune myelitis, acute flaccid myelitis (AFM) is an increasingly recognized syndrome in children presenting with flaccid weakness. Increased incidence of the disease was observed in summer and fall of 2014 and 2016, coinciding temporally and geographically with a rise in enterovirus infections. Children typically present following a viral respiratory illness with rapidly progressive, often asymmetric, flaccid weakness. Sensation and bowel and bladder function are relatively preserved. Radiographically, it corresponds to hyperintense signal in the central gray matter of the spinal cord, particularly the anterior horns and sometimes the brainstem, involving cranial nerve nuclei [57].

Although its clinical and radiological phenotype is similar to poliomyelitis, the etiology for AFM is believed to be enterovirus D68 as this infection was confirmed in a portion of AFM cases from nasopharyngeal specimens. CSF is almost universally negative for pathogens, notably enterovirus polymerase chain reaction (PCR). Enterovirus A71 has also been implicated in other cohorts of AFM patients [58]. Further prospective epidemiological studies are needed to establish a causal relationship between specific enterovirus subtypes and AFM [59].

Motor recovery from AFM is slower and less complete compared to autoimmune myelitis. While all individuals in a Colorado cohort demonstrated functional gains, the majority of children continued to have persistent motor deficits at 1 year [60]. There is no known specific treatment for AFM, and differing paradigms of intervention have been controversial. To date, there is no specific contraindication to immunotherapy, but its use has not been recommended by some public health practitioners out of concern for worsening underlying infections.

**Patient Vignette 2**

A 24-year-old man develops headache, nausea, and vomiting after weight lifting. Within 20 minutes, he also notices bilateral leg weakness and numbness. Emergency Medical Services (EMS) is contacted and brings him to an emergency department for evaluation. While undergoing CT head, he has urinary incontinence and develops weakness and numbness in his arms. He exhibits respiratory distress with hypercarbia and is intubated for airway protection 2 hours after the onset of symptoms. MRI of the spine shows abnormal T2/FLAIR signal from C2 to 5 without enhancement, affecting primarily the anterior spinal cord gray matter. MRI of the brain shows a few punctate areas of abnormal T2/FLAIR signal, which restrict diffusion. CSF studies demonstrate normal indices with two nucleated cells and protein 42 mg/dL. He is started on IV corticosteroids due to concern for acute disseminated encephalomyelitis but has no improvement with the treatment. Infectious testing is negative for West Nile Virus and enterovirus. He undergoes a spinal angiogram that reveals irregularity in the lumen of the vertebral arteries.

Vascular myelopathies and autoimmune myelitis can present similarly and can be challenging to differentiate. However, the timing from symptom onset to nadir of weakness can be helpful in distinguishing these etiologies. In particular, vascular myelopathies can have a hyperacute progression versus the acute/subacute pattern of immune-mediated and inflammatory myelitis. Other features of the case that argue against autoimmune myelitis include non-inflammatory CSF, pattern of spinal cord involvement (i.e., anterior spinal cord), and lack of improvement with immunotherapy. Treatments for acute transverse myelitis such as corticosteroids and IVIg often do harm or improve vascular myelopathies, while therapeutic plasma exchange can potentially lead to worsening of symptoms due to fluctuations in blood pressure and spinal cord perfusion. Acute flaccid myelitis is an alternate consideration with spinal cord involvement largely restricted to anterior horns and central gray matter of spinal cord. Cases have primarily been reported in children during late summer-fall season. There is no acute therapy that has been proven to be beneficial for this condition, which is thought to be related to direct viral injury of spinal cord.

## Prognosis/Long-Term Management

To date, there have been no published large natural history studies of idiopathic transverse myelitis. Although most individuals who experience myelitis have some degree of recovery, up to 40% have residual deficits that interfere with activities of daily living and impact quality of life [61]. The rate of improvement is typically most marked in the first 3–6 months following the onset of symptoms and continues with rehabilitation therapy over many years. Poor outcomes are associated with rapid onset of symptoms and complete paraplegia with spinal shock. Recurrence of TM is associated with systemic autoimmune diseases, and it is rare for people with idiopathic TM to have another inflammatory episode.

For patients determined to have myelitis secondary to an autoimmune condition, chronic immunomodulatory therapy should be considered to minimize the risk of recurrent disease. Given the severity and poor recovery from NMOSD-related attacks and uniformly relapsing disease course, identification of anti-AQP4 antibodies or clinical attack meeting Wingerchuk diagnostic criteria for NMOSD should prompt initiation of therapies. Commonly used, though not FDA-approved treatments, include rituximab, mycophenolate mofetil, or azathioprine [20]. A growing armamentarium of multiple sclerosis drugs with varying routes of administration, efficacy, and safety profiles are available. Anti-tumor necrosis factor (TNF) alpha antagonists and mycophenolate mofetil can be effective in sarcoidosis [62].

## Conclusion and Future Directions

Our understanding of the etiologies of myelitis has grown significantly in recent decades with discoveries of the molecular targets of autoimmunity in a subset of patients. What was previously regarded as a demyelinating process is now known to involve varying degrees of white and gray matter involvement depending on underlying pathobiology. Research targeted at understanding mechanisms of autoimmune myelopathies will lead to improvements in the future treatment of individuals afflicted with these diseases. Furthermore, small-molecule-based, monoclonal-antibody-based, and cell-based therapies to repair damage within the spinal cord are underway and hold promise for improving outcomes in patients [63–65].

## References

1. Katz JD, Ropper AH. Progressive necrotic myelopathy: clinical course in 9 patients. Arch Neurol. 2000;57(3):355–61.
2. Mirich DR, Kucharczyk W, Keller MA, Deck J. Subacute necrotizing myelopathy: Mr imaging in four pathologically proved cases. AJNR Am J Neuroradiol. 1991;12(6):1077–83.
3. Kaplin AI, Krishnan C, Deshpande DM, Pardo CA, Kerr DA. Diagnosis and management of acute myelopathies. Neurologist. 2005;11(1):2–18.
4. Kaplin AI, Deshpande DM, Scott E, Krishnan C, Carmen JS, Shats I, et al. Il-6 induces regionally selective spinal cord injury in patients with the neuroinflammatory disorder transverse myelitis. J Clin Invest. 2005;115(10):2731–41.
5. Dixit P, Garg RK, Malhotra HS, Jain A, Verma R, Sharma PK, et al. Cytokines and matrix metalloproteinases in the cerebrospinal fluid of patients with acute transverse myelitis: an outcome analysis. Inflamm Res. 2016;65(2):125–32.
6. Guo Y, Weigand SD, Popescu BF, Lennon VA, Parisi JE, Pittock SJ, et al. Pathogenic implications of cerebrospinal fluid barrier pathology in neuromyelitis optica. Acta Neuropathol. 2017;133(4):597–612.
7. Lucchinetti CF, Guo Y, Popescu BF, Fujihara K, Itoyama Y, Misu T. The pathology of an autoimmune astrocytopathy: lessons learned from neuromyelitis optica. Brain Pathol. 2014;24(1):83–97.
8. Popescu BF, Pirko I, Lucchinetti CF. Pathology of multiple sclerosis: where do we stand? Continuum (Minneap Minn). 2013;19(4 Multiple Sclerosis):901–21.
9. Dobson R, Ramagopalan S, Davis A, Giovannoni G. Cerebrospinal fluid oligoclonal bands in multiple sclerosis and clinically isolated syndromes: a meta-analysis of prevalence, prognosis and effect of latitude. J Neurol Neurosurg Psychiatry. 2013;84(8):909–14.
10. Stangel M, Fredrikson S, Meinl E, Petzold A, Stuve O, Tumani H. The utility of cerebrospinal fluid analysis in patients with multiple sclerosis. Nat Rev Neurol. 2013;9(5):267–76.
11. Krishnan C, Kaplin AI, Pardo CA, Kerr DA, Keswani SC. Demyelinating disorders: update on transverse myelitis. Curr Neurol Neurosci Rep. 2006;6(3):236–43.
12. de Seze J, Lanctin C, Lebrun C, Malikova I, Papeix C, Wiertlewski S, et al. Idiopathic acute transverse myelitis: application of the recent diagnostic criteria. Neurology. 2005;65(12):1950–3.
13. Beh SC, Greenberg BM, Frohman T, Frohman EM. Transverse myelitis. Neurol Clin. 2013;31(1):79–138.
14. Berman M, Feldman S, Alter M, Zilber N, Kahana E. Acute transverse myelitis: incidence and etiologic considerations. Neurology. 1981;31(8):966–71.

15. Jeffery DR, Mandler RN, Davis LE. Transverse myelitis. Retrospective analysis of 33 cases, with differentiation of cases associated with multiple sclerosis and parainfectious events. Arch Neurol. 1993;50(5):532–5.
16. Transverse Myelitis Consortium Working G. Proposed diagnostic criteria and nosology of acute transverse myelitis. Neurology. 2002;59(4):499–505.
17. Absoud M, Greenberg BM, Lim M, Lotze T, Thomas T, Deiva K. Pediatric transverse myelitis. Neurology. 2016;87(9 Suppl 2):S46–52.
18. Alonso A, Hernan MA. Temporal trends in the incidence of multiple sclerosis: a systematic review. Neurology. 2008;71(2):129–35.
19. Wingerchuk DM. Neuromyelitis optica: effect of gender. J Neurol Sci. 2009;286(1–2):18–23.
20. Mealy MA, Wingerchuk DM, Greenberg BM, Levy M. Epidemiology of neuromyelitis optica in the United States: a multicenter analysis. Arch Neurol. 2012;69(9):1176–80.
21. Loblaw DA, Mitera G, Ford M, Laperriere NJ. A 2011 updated systematic review and clinical practice guideline for the management of malignant extradural spinal cord compression. Int J Radiat Oncol Biol Phys. 2012;84(2):312–7.
22. Goh C, Desmond PM, Phal PM. Mri in transverse myelitis. J Magn Reson Imaging. 2014;40(6):1267–79.
23. Krupp LB, Tardieu M, Amato MP, Banwell B, Chitnis T, Dale RC, et al. International pediatric multiple sclerosis study group criteria for pediatric multiple sclerosis and immune-mediated central nervous system demyelinating disorders: revisions to the 2007 definitions. Mult Scler. 2013;19(10):1261–7.
24. Waters PJ, Pittock SJ, Bennett JL, Jarius S, Weinshenker BG, Wingerchuk DM. Evaluation of aquaporin-4 antibody assays. Clin Exp Neuroimmunol. 2014;5(3):290–303.
25. Keegan BM, Pittock SJ, Lennon VA. Autoimmune myelopathy associated with collapsin response-mediator protein-5 immunoglobulin g. Ann Neurol. 2008;63(4):531–4.
26. Virani S, Tan M, Abraham J. Transverse myelitis: Amphiphysin autoimmunity paraneoplastic syndrome in a woman with breast cancer. Clin Adv Hematol Oncol. 2009;7(3):180–2.
27. Ascherio A, Munger KL, White R, Kochert K, Simon KC, Polman CH, et al. Vitamin d as an early predictor of multiple sclerosis activity and progression. JAMA Neurol. 2014;71(3):306–14.
28. Zukor K, Wang H, Siddharthan V, Julander JG, Morrey JD. Zika virus-induced acute myelitis and motor deficits in adult interferon alphabeta/gamma receptor knockout mice. J Neurovirol. 2018;24(3):273–90.
29. Fitch MT, van de Beek D. Drug insight: steroids in cns infectious diseases–new indications for an old therapy. Nat Clin Pract Neurol. 2008;4(2):97–104.
30. Silva M, Nunes H, Valeyre D, Sverzellati N. Imaging of sarcoidosis. Clin Rev Allergy Immunol. 2015;49(1):45–53.
31. Scott TF, Frohman EM, De Seze J, Gronseth GS, Weinshenker BG, Therapeutics, et al. Evidence-based guideline: clinical evaluation and treatment of transverse myelitis: report of the therapeutics and technology assessment subcommittee of the american academy of neurology. Neurology. 2011;77(24):2128–34.
32. Noland DK, Greenberg BM. Safety and efficacy of plasma exchange in pediatric transverse myelitis. Neurol Clin Pract. 2018;8(4):327–30.
33. Greenberg BM, Thomas KP, Krishnan C, Kaplin AI, Calabresi PA, Kerr DA. Idiopathic transverse myelitis: corticosteroids, plasma exchange, or cyclophosphamide. Neurology. 2007;68(19):1614–7.
34. Trebst C, Jarius S, Berthele A, Paul F, Schippling S, Wildemann B, et al. Update on the diagnosis and treatment of neuromyelitis optica: recommendations of the neuromyelitis optica study group (nemos). J Neurol. 2014;261(1):1–16.
35. Wang KC, Wang SJ, Lee CL, Chen SY, Tsai CP. The rescue effect of plasma exchange for neuromyelitis optica. J Clin Neurosci. 2011;18(1):43–6.
36. Bonnan M, Cabre P. Improvement to baseline after plasma exchange in spinal attacks associated with neuromyelitis optica. Mult Scler J Exp Transl Clin. 2015;1:2055217315622794.
37. Bonnan M, Cabre P. Plasma exchange in severe attacks of neuromyelitis optica. Mult Scler Int. 2012;2012:787630.
38. Keegan M, Konig F, McClelland R, Bruck W, Morales Y, Bitsch A, et al. Relation between humoral pathological changes in multiple sclerosis and response to therapeutic plasma exchange. Lancet. 2005;366(9485):579–82.
39. Pavlou E, Gkampeta A, Kouskouras K, Evangeliou A, Athanasiadou-Piperopoulou F. Idiopathic acute transverse myelitis: complete recovery after intravenous immunoglobulin. Hippokratia. 2012;16(3):283–5.
40. Ikuta F, Zimmerman HM. Distribution of plaques in seventy autopsy cases of multiple sclerosis in the United States. Neurology. 1976;26(6 PT 2):26–8.
41. Scott TF, Kassab SL, Singh S. Acute partial transverse myelitis with normal cerebral magnetic resonance imaging: transition rate to clinically definite multiple sclerosis. Mult Scler. 2005;11(4):373–7.
42. Flanagan EP, Weinshenker BG, Krecke KN, Lennon VA, Lucchinetti CF, McKeon A, et al. Short myelitis lesions in aquaporin-4-igg-positive neuromyelitis optica spectrum disorders. JAMA Neurol. 2015;72(1):81–7.
43. Mealy MA, Wingerchuk DM, Palace J, Greenberg BM, Levy M. Comparison of relapse and treatment failure rates among patients with neuromyelitis optica: Multicenter study of treatment efficacy. JAMA Neurol. 2014;71(3):324–30.
44. Hacohen Y, Rossor T, Mankad K, Chong W, Lux A, Wassmer E, et al. 'Leukodystrophy-like' phenotype in children with myelin oligodendrocyte glycoprotein antibody-associated disease. Dev Med Child Neurol. 2018;60(4):417–23.

45. Ogawa R, Nakashima I, Takahashi T, Kaneko K, Akaishi T, Takai Y, et al. Mog antibody-positive, benign, unilateral, cerebral cortical encephalitis with epilepsy. Neurol Neuroimmunol Neuroinflamm. 2017;4(2):e322.
46. Weber MS, Derfuss T, Bruck W. Anti-myelin oligodendrocyte glycoprotein antibody-associated central nervous system demyelination-a novel disease entity? JAMA Neurol. 2018;75(8):909–10.
47. Kitley J, Leite MI, Kuker W, Quaghebeur G, George J, Waters P, et al. Longitudinally extensive transverse myelitis with and without aquaporin 4 antibodies. JAMA Neurol. 2013;70(11):1375–81.
48. Hacohen Y, Wong YY, Lechner C, Jurynczyk M, Wright S, Konuskan B, et al. Disease course and treatment responses in children with relapsing myelin oligodendrocyte glycoprotein antibody-associated disease. JAMA Neurol. 2018;75(4):478–87.
49. Pittock SJ, Lennon VA, de Seze J, Vermersch P, Homburger HA, Wingerchuk DM, et al. Neuromyelitis optica and non organ-specific autoimmunity. Arch Neurol. 2008;65(1):78–83.
50. Birnbaum J, Petri M, Thompson R, Izbudak I, Kerr D. Distinct subtypes of myelitis in systemic lupus erythematosus. Arthritis Rheum. 2009;60(11):3378–87.
51. Barile L, Lavalle C. Transverse myelitis in systemic lupus erythematosus–the effect of iv pulse methylprednisolone and cyclophosphamide. J Rheumatol. 1992;19(3):370–2.
52. Zalewski NL, Krecke KN, Weinshenker BG, Aksamit AJ, Conway BL, McKeon A, et al. Central canal enhancement and the trident sign in spinal cord sarcoidosis. Neurology. 2016;87(7):743–4.
53. Jolliffe EA, Keegan BM, Flanagan EP. Trident sign trumps aquaporin-4-igg elisa in diagnostic value in a case of longitudinally extensive transverse myelitis. Mult Scler Relat Disord. 2018;23:7–8.
54. Knebusch M, Strassburg HM, Reiners K. Acute transverse myelitis in childhood: nine cases and review of the literature. Dev Med Child Neurol. 1998;40(9):631–9.
55. Zalewski NL, Flanagan EP, Keegan BM. Evaluation of idiopathic transverse myelitis revealing specific myelopathy diagnoses. Neurology. 2018;90(2):e96–e102.
56. Mealy MA, Nam TS, Pardo SJ, Pardo CA, Sobreira NL, Avramopoulos D, et al. Familial monophasic acute transverse myelitis due to the pathogenic variant in vps37a. Neurol Genet. 2018;4(1):e213.
57. Messacar K, Schreiner TL, Van Haren K, Yang M, Glaser CA, Tyler KL, et al. Acute flaccid myelitis: a clinical review of us cases 2012–2015. Ann Neurol. 2016;80(3):326–38.
58. Teoh HL, Mohammad SS, Britton PN, Kandula T, Lorentzos MS, Booy R, et al. Clinical characteristics and functional motor outcomes of enterovirus 71 neurological disease in children. JAMA Neurol. 2016;73(3):300–7.
59. Messacar K, Asturias EJ, Hixon AM, Van Leer-Buter C, HGM N, Tyler KL, et al. Enterovirus d68 and acute flaccid myelitis-evaluating the evidence for causality. Lancet Infect Dis. 2018;18(8):e239–e47.
60. Martin JA, Messacar K, Yang ML, Maloney JA, Lindwall J, Carry T, et al. Outcomes of Colorado children with acute flaccid myelitis at 1 year. Neurology. 2017;89(2):129–37.
61. Pidcock FS, Krishnan C, Crawford TO, Salorio CF, Trovato M, Kerr DA. Acute transverse myelitis in childhood: center-based analysis of 47 cases. Neurology. 2007;68(18):1474–80.
62. Moravan M, Segal BM. Treatment of cns sarcoidosis with infliximab and mycophenolate mofetil. Neurology. 2009;72(4):337–40.
63. Ruggieri S, Tortorella C, Gasperini C. Anti lingo 1 (opicinumab) a new monoclonal antibody tested in relapsing remitting multiple sclerosis. Expert Rev Neurother. 2017;17(11):1081–9.
64. Wooliscroft L, Silbermann E, Cameron M, Bourdette D. Approaches to remyelination therapies in multiple sclerosis. Curr Treat Options Neurol. 2019;21(7):34.
65. Katoh H, Yokota K, Fehlings MG. Regeneration of spinal cord connectivity through stem cell transplantation and biomaterial scaffolds. Front Cell Neurosci. 2019;13:248.

# Clinical Approach to Autoimmune and Inflammatory Ophthalmologic Disease

# 28

Megan Esch and Shiv Saidha

**Patient Vignette**

A 22-year-old Caucasian female presents with a 5-day history of dull pain behind the left eye, particularly worse with movement. She reports a "fog" covering the central vision. Her visual acuity (VA) is 20/20 in the right eye, 20/60 in the left eye. The fundoscopic exam is unremarkable, but swinging flashlight testing demonstrates paradoxical dilation of left pupil on direct application of light (relative afferent pupillary defect, RAPD). She completes 12/12 Ishihara color plates correctly in the right eye and 9/12 in the left eye.

A magnetic resonance imaging (MRI) scan of the brain and orbits is shown in Fig. 28.1. Post-contrast coronal and axial views demonstrate a short segment, enhancing lesion anterior to the chiasm of the left optic nerve (arrows). There were at least two additional enhancing brain lesions seen (*) and at least three non-enhancing T2 hyperintense lesions with morphology and characteristics, including location, consistent with demyelinating disease.

She received 3 days of intravenous (IV) methylprednisolone with complete recovery of vision 1 week following onset. At follow-up visit 12 weeks later, although her VA had improved to 20/30 in the left eye, color deficits persisted. Optical coherence tomography (OCT) is completed at that time (Fig. 28.2). This spectral-domain OCT demonstrates mild retinal nerve fiber layer (RNFL) thinning in the clinically unaffected right eye (mean thickness: 80 um), and marked RNFL thinning in the left eye (mean thickness: 58 um), with predominant involvement of the temporal RNFL quadrant. This case represents a typical course of inflammatory, demyelinating acute optic neuritis (AON) as frequently occurs during the course of multiple sclerosis (MS). There was fast, nearly complete recovery of visual acuity following a short course of steroids, although with subtle persistent deficits in color vision.

## Introduction

In discussion of immunological disorders of the central nervous system (CNS), special consideration of ophthalmologic dysfunction and the anterior visual system is warranted. Affliction of the afferent visual system, in particular acute optic neuritis (AON), is commonly seen in multiple sclerosis (MS) and other inflammatory autoimmune conditions. AON is an inflammatory condition of the optic nerve and is often considered a "forme fruste" of MS. The clinical presentation of AON is variable, though most commonly involves pain and impairment or loss of vision in

M. Esch
Department of Neurology, Geisinger Medical Center, Danville, PA, USA

S. Saidha (✉)
Department of Neurology, Johns Hopkins University, Baltimore, MD, USA
e-mail: ssaidha2@jhmi.edu





A. L. Piquet, E. Alvarez (eds.), *Neuroimmunology*, https://doi.org/10.1007/978-3-030-61883-4_28

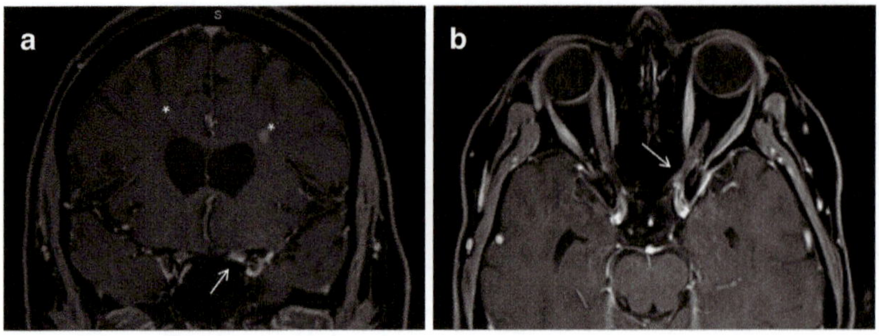


**Fig. 28.1** Magnetic resonance imaging (MRI): T1 post-contrast coronal (**a**) and axial (**b**) scans demonstrate a short segment, enhancing lesion anterior to the optic chiasm within the left optic nerve (arrows). There were two additional enhancing lesions (*) and at least three non-enhancing T2 hyperintense lesions (not shown) with morphology and characteristics, including location, consistent with demyelinating disease

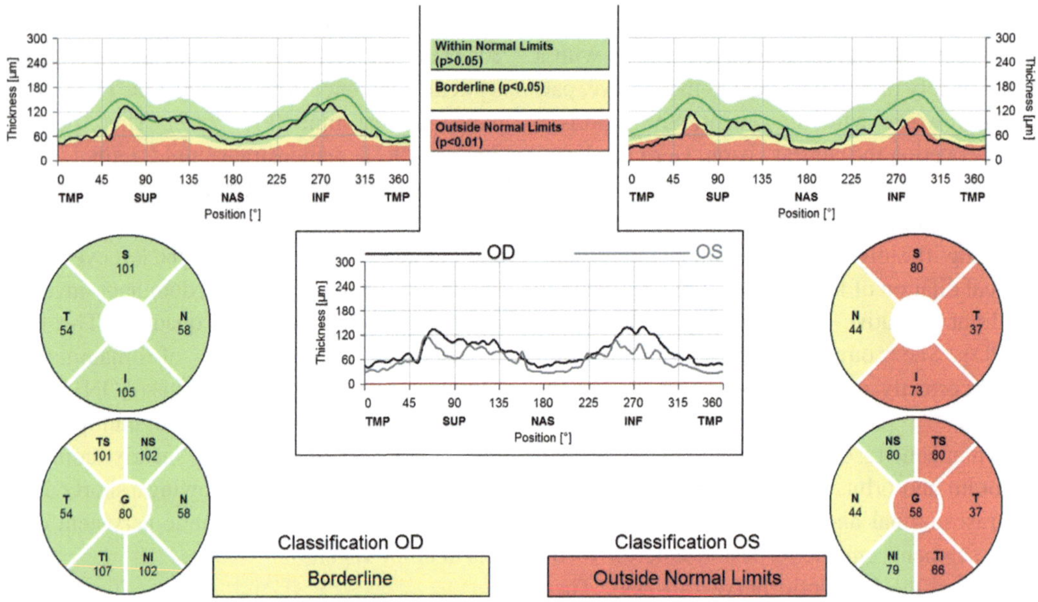


**Fig. 28.2** Optical coherence tomography (OCT): This spectral-domain OCT demonstrates borderline reduction in average retinal nerve fiber layer (RNFL) thickness in the clinically unaffected right eye (mean thickness: 80 um), potentially representing subclinical optical neuropathy, and marked reduction (<1st percentile) in average RNFL thickness in the left eye (mean thickness: 58 um), with predominant involvement of the temporal RNFL quadrant, consistent with prior optic neuritis

the affected eye. Severity, duration, and resolution of symptoms depend on underlying disease pathology. In the Western world, AON is most commonly seen in the setting of demyelinating conditions such as MS. The spectrum of conditions associated with AON, however, is broad and includes non-MS autoimmune and demyelinating diseases, infections, granulomatous disease, paraneoplastic processes, and rarely, hereditary diseases. Identifying potential "red-flags" or atypical features of AON is important in the consideration of alternative diagnoses to MS as the underlying mechanism. While diagnosis of AON is primarily clinical, utilization of magnetic resonance imaging (MRI) and electrophysiological studies has historically aided AON diagnosis. Utilization of more advanced imaging techniques, such as optical coherence tomography (OCT), has allowed for more specific analyses and monitoring of axonal and neuronal degeneration following AON. The treatment of classic demyelinating, including idiopathic, forms of AON has primarily focused on the use of corticosteroids, although this approach may not actually be supported by evidence. Recovery from AON is largely dependent on the severity and underlying pathology of the AON. Potential neuroprotective and remyelinating strategies (in MS) are in development, and clinical trials are ongoing. In this chapter, we aim to address the most common clinical symptomatology associated with inflammatory and immunological features of the afferent visual system, notable features on imaging and ancillary testing that may aid in the diagnosis of these disorders, as well as current and potential future treatment options.

## Epidemiology

The worldwide incidence of unilateral AON is between 0.94 [1] and 2.18 per 100,000 persons per year. The rates of AON are similar across geographically distinct populations; per 100,000 rates are reported as 1.6 in Japan, [2], 1.6 in Croatia [3], 1 in the United Kingdom (UK) [4], and 1.46 in Sweden [5], although one study in the United States (US) found the incidence of AON to be as high as 5 per 100,000 [6]. Women of northern European ancestry have up to a threefold higher incidence of AON than men [1, 5]. Similar risk factors exist for AON as multiple sclerosis. Seasonal variances demonstrate AON is most likely to occur during Spring [5, 7] and at higher latitudes [8]. There is suggestion of a genetic predisposition in patients who have had infectious mononucleosis with high titers of Epstein–Barr virus and positive *HLA-DRB1*1501* status [9].

While the incidence of unilateral AON is similar across ethnicities [10], the clinical characteristics of AON among populations may differ. AON in Asian populations tends to have a lower association with MS (6%), than in Caucasian populations [2], and a higher incidence of bilateral or recurrent presentation. African American populations tend to present with a more severe form of AON at onset and may experience poorer recovery than their Caucasian American counterparts [11]. The global incidence of AON is 1–5/100,000 persons per year. There is a threefold predilection for women than men. Underlying diseases associated with AON differ globally.

## Clinical Presentation and Evaluation of AON

The Optic Neuritis Study Group contributed to the fundamental understanding of clinical presentation and treatment of AON through the large, randomized, placebo-controlled Optic Neuritis Treatment Trial (ONTT) conducted in the 1990s [12].

The classic presentation of AON is subacute, painful, monocular loss of vision. Pain in this disorder is typically described as dull or aching, located behind the eye or temporally, and may be confused initially with headache. More than 90% of patients enrolled in the ONTT presented with ocular pain with eye movement. Periocular pain is seen most commonly in patients with retrobulbar involvement of the optic nerve; anterior optic nerve involvement is less frequently associated with painful symptoms and predicts a favorable prognosis regarding the final risk of developing MS. Lesions of the optic chiasm are more likely to present with painless bitemporal hemianopia. Pain in AON is thought to be due to irritation of the inflamed nerve with the tension of motion. The duration of painful symptoms is often for just a few days and may precede, or develop concurrently, with visual loss. Visual dysfunction may last for weeks at a time, typically reaching peak symptoms at about 2 weeks post-onset [12]. The visual loss is often reported as a "blurring" or "fog" over the affected eye, and only in very severe cases will patients experience complete loss of vision with just retention of light or motion perception. Typically, there is central (or centrocecal) vision impairment, although arcuate defects, hemianopsias, and global visual distortion can occur. The severity of visual loss on high-contrast visual acuity (HCVA) examination is variable. Two-thirds of patients in the ONTT demonstrated HCVA worse than 20/40, with half of those patients demonstrating HCVA worse than 20/200 [13]. Abnormalities in color perception, dyschromatopsias, are also described: patients report "faded" or "dim" colors, particularly of the red–green spectrum. While 35% of patients demonstrate HCVA better than 20/40 at diagnosis, 88% of patients show abnormalities of color vision as measured by Ishihara plates [13]. Ishihara plates are pseudoisochromatic colored plates containing different-sized and colored dots. The dots are configured to showcase a number that is easily depicted by patients with normal red–green color vision, but with difficulty (or not at all) by those with color vision deficits. Notably, shift in color defects may be seen as AON evolves; in the acute phase, blue/yellow defects may predominate, while red–green color defects might persist more chronically following AON [14]. Patients may also report positive visual phenomena (phosphenes) with eye movement or auditory stimuli; these are known as Moore's lightning streaks and may even present following recovery of visual acuity. A visual illusion by which objects in linear motion are perceived as following an ellipsoid trajectory is known as the Pulfrich phenomenon and is likely related to inter-eye asymmetries in depth perception; this effect may be ameliorated by placing neutral eye filters over the unaffected eye [15]. Pain is a presenting feature of AON in approximately 90% of patients. Vision loss tends to be central. Associated symptoms include color vision loss and visual field defects; these abnormalities can persist indefinitely beyond recovery of HCVA.

Abnormalities of visual dysfunction may be seen disproportionate to loss/recovery of HCVA. Although 1 year following AON, 90% of patients recover to an HCVA of 20/40 or better, regardless of treatment group, abnormalities of contrast visual acuity, color vision, and visual field deficits persist [16–18]. Low-contrast vision (contrast sensitivity) and low-contrast letter acuity (LCLA) have been shown to be more sensitive measures of visual dysfunction in optic neuropathy than HCVA [19–21]. Contrast sensitivity measures the perception of progressively dimming shades of same-sized gray letters on a retroluminated background, while LCLA utilizes the same shade of light gray letter becoming progressively smaller on the light background of a Sloan chart [22]. Sloan charts alter the degree of contrast so that eyes are tested as high as 100%-contrast on a lit background, and as low as 1.25%-contrast. At a contrast of 1.25%, the "lightest" shade of gray, letters reflect only 1.25%-contrast as compared to the background. Sabadia et al. [23] demonstrated that even with average binocular HCVA scores of 20/20 in patients with prior monocular AON, as compared to healthy controls, LCLA is reduced by as much as 45% on 1.25% Sloan charts. These are not used as measurements of visual dysfunction in the typical office setting, but may provide more information regarding the degree of visual loss in patients who have experienced prior clinical or subclinical optic neuropathies. Six months post-AON, rates of abnormal contrast sensitivity exceed abnormalities of HCVA by more than double [24]. These persistent changes are associated with reductions in quality of life in MS patients and have important implications in overall health-related quality of life measures [25–27].

Importantly for the clinician, initial fundoscopic examination may be normal (normal in 64% of patients with AON). Pallor of the optic disk may take weeks to months to develop and is not typically seen in the acute phase (Fig. 28.3). A relative afferent pupillary defect (RAPD) is usually observed in the affected eye and is often present in most patients from onset; this can be elicited by swinging a light between the pupils of affected and unaffected eyes and observing reduction of constriction, or frank dilation, of the affected pupil. Absence of an RAPD may indicate an alternative mechanism than AON for vision loss, including retinal diseases or, alternatively, bilateral optic nerve pathology. Optic disk swelling (papillitis) on fundoscopic exam is seen in only 35% of AON cases [12]. If swelling, retinal hemorrhages, or exudates are seen acutely, this should prompt consideration of an alternative diagnosis. Fundoscopic exam is normal in 64% of patients with AON. RAPD is present in most patients from onset, although it may resolve after the acute phase. Presence of optic disk swelling on initial exam should prompt consideration of an alternative diagnosis.

Lumbar puncture may be performed in patients who have imaging findings that are sus-

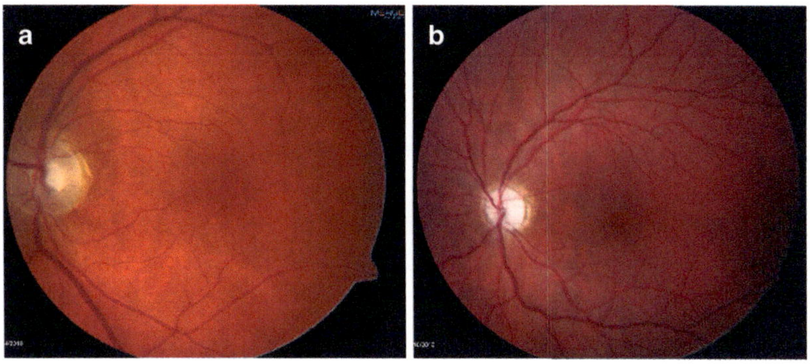


**Fig. 28.3** Fundus photography is seen demonstrating the left optic disk. A normal left optic disk with normal cup-to-disk ratio (0.4) and crisp optic disk margins are seen in (**a**). (**b**) Twelve weeks following acute optic neuritis, the patient has developed significant pallor of the left optic disk

picious for MS or if the etiology of AON is unclear. Cerebrospinal fluid (CSF) profile may assist with differentiation of inflammatory, infectious, or paraneoplastic causes of optic neuropathy from potentially demyelinating etiologies. While the cellular profile of CSF is normal in 77% of patients with MS, mild CSF pleocytosis (<50 cell/uL) and elevated protein (<1.1 g/L) may be seen. Significant pleocytosis or protein elevation and hypoglycorrhachia suggest alternative inflammatory or infectious etiologies. Oligoclonal bands unique to the CSF may support a diagnosis of MS; these are present in approximately 88% of MS patients and demonstrate sensitivities of around 95% and specificities of around 86% [28]. The presence of identical or "mirror" CSF and serum bands may alternatively support a systemic autoimmune disorder. As with any ancillary testing, the clinical profile must corroborate the laboratory findings; the low positive predictive value of oligoclonal bands may prompt consideration of an alternative autoimmune, infectious, neoplastic, or vascular disorder in the correct clinical setting.

## Treatment of Optic Neuritis

### Corticosteroids

The ONTT demonstrated the efficacy of intravenous (IV) corticosteroids in hastening improvement in visual function following AON. Doses of 1 gram (g) of IV methylprednisolone for 3 days, given within 15 days of onset of symptoms, improved visual outcomes compared to placebo and low-dose oral prednisone at 15 days post-administration. This effect persisted at 6 months only for color vision and contrast sensitivity, but significance waned thereafter, and had disappeared by 1-year post-steroids [16, 17, 29]. The improvement was most robustly seen in patients with worse degrees of visual impairment as determined by HCVA, contrast acuity, color vision, and visual field testing. The ONTT findings suggest that patients with HCVA better than 20/40 may not benefit as significantly from IV steroids, while those with worse HCVA from onset (20/50 or worse) have higher likelihood of benefit. Further, while IV corticosteroids reduced the risk of development of MS for 2 years post-ON event, steroid administration did not change ultimate risk of developing MS.

The utility of oral corticosteroids in the treatment of ON has been even further debated. The ONTT found that low doses of oral prednisone compared to IV methylprednisolone and placebo were not only ineffective but also potentially increased the risk of further episodes of AON. More recent literature has suggested that treatment with bioequivalent dosing of oral prednisone (1250 mg) is not significantly different than treatment with 1 g IV methylprednisolone with regard to HCVA, LCLA, and P100 latency on visual evoked potential (VEP) [30] outcomes. Thus, high-dose oral corticosteroid administration may be a suitable alternative treatment in some patients. High-dose IV corticosteroids hasten visual recovery from AON when administered early; however, ultimate visual outcomes are unaffected by steroid administration. Low-dose oral corticosteroids might increase the risk of further AON. However, bioequivalent high-dose oral corticosteroids seem to be equivalent to IV corticosteroids.

## Intravenous Immunoglobulins and Plasma Exchange

There is a paucity of literature regarding the utility of nonsteroidal treatments in refractory cases of AON.

Intravenous immunoglobulin (IVIg) therapy is not felt to be beneficial in the treatment of steroid-refractory AON. Two early trials demonstrated no efficacy in improvement of residual visual deficits from prior AON. One trial demonstrated no improvement in chronic post-AON visual deficits upon administration of IVIg compared with placebo; the average time to receive IVIg was 4 years post-AON [31]. A second study demonstrated that patients who received IVIg within 4 weeks of symptom onset demonstrated no improvement in visual outcomes at 6 months compared to placebo, and there were similarly no differences in number of new or enhancing lesions on MRI, nor were there clinical relapses between groups [32]. A small, non-randomized prospective study of patients with AON who demonstrated HCVA of 20/400 or worse following treatment with corticosteroids, who then received IVIg, demonstrated significant improvement compared to patients who received corticosteroids alone [33]. No large, randomized controlled trial has been completed to validate this.

Plasma exchange (PLEX) might be useful in steroid-refractory cases of severe (VA 20/200 or worse) AON. However, no large, randomized, placebo-controlled trials of plasmapheresis in steroid-refractory AON currently exist. In an analysis of pooled data from four studies of 39 affected eyes with severe AON (<20/200), those patients who received PLEX early (median of 19 days post attack) demonstrated significant improvement of HCVA; the effect increased the earlier the treatment was given [34]. The latter finding is consistent with a previous study demonstrating the benefit of early administration of PLEX in improving steroid-refractory demyelinating central nervous system (CNS) deficits [35]. Additional case series have demonstrated similar benefits. A study of 41 affected eyes treated initially with corticosteroids, with residual HCVA of 20/200 or worse, received plasmapheresis; 56% of PLEX-treated patients demonstrated improvement to 20/40 or better [36]. In a study of 23 patients refractory to two rounds of IV corticosteroids treated subsequently with PLEX, 70% improved to HCVA of 20/50 or better [37]. Finally, an observational study of neuromyelitis optica (NMO) patients with AON compared outcomes of patients treated with corticosteroids alone ($n$ = 36) to those treated with corticosteroids followed by PLEX ($n$ = 16). Both groups demonstrated mean baseline HCVA of 20/400; those who received sequential corticosteroid and plasmapheresis treatment demonstrated mean final HCVA of 20/50, with 75% of those treated with PLEX demonstrating HCVA better than 20/40 (compared to only 39% of those treated with corticosteroids alone) [38].

Thus, while corticosteroids remain the recommended first-line treatment for AON, severe refractory cases (20/200 or worse) beyond 2–3 weeks following treatment may benefit from PLEX as second-line therapy. Steroid-refractory AON and severe visual loss (VA 20/200 or worse) may benefit from treatment with PLEX as second-line therapy.

## Causes of AON

### Multiple Sclerosis

AON occurs in up to 70% of MS patients at some point during their disease course and is the initial manifestation of MS in approximately 25% of cases [39]. In the ONTT, the overall probability of developing MS at 15 years following a single episode of AON was approximately 50%. This rate was higher (72%) in patients who had at least one brain MRI lesion consistent with demyelination at initial presentation, and lower (25%) in patients who demonstrated no brain lesions on baseline brain MRI. In the latter group, the risk of developing MS was highest in females and twice as likely in patients who demonstrated retrobulbar AON as opposed to those with anterior segment involvement. Lower risk of developing MS was associated with atypical features at onset, including severe papillitis, peripapillary hemorrhages or retinal exudates, significant light vision loss, and absence of periocular pain [29]. The frequency of abnormal white matter lesions on brain MRI at the time of AON presentation varies globally, with the lowest rates found in Japan (14%) [2] and highest rates found in the UK (77%) [40], underscoring the fact that while AON may be closely related to the future diagnosis of clinically definite MS (CDMS) in the Western world, this may not uniformly be the case across geographic regions. The ONTT demonstrated that while treatment with corticosteroids hastened AON symptom recovery, overall prognosis in terms of future conversion to MS was not significantly affected by the administration of steroids.

The importance of identifying and appropriately treating patients at high risk for developing CDMS following a single attack of AON has been previously elucidated. Early studies demonstrate that offering interferon-beta treatment to high-risk patients (those with a single clinical demyelinating event with ≥2 demyelinating MRI lesions) reduces the probability of conversion to CDMS over a 3-year period by more than a third [41–43]. In the Controlled High-Risk Avonex in Multiple Sclerosis Prevention Study (CHAMPS), patients with a first clinical demyelinating event (AON and other presentations), who had received IV corticosteroids, either received interferon-beta-1a intramuscularly once per week (Avonex) or placebo. At 3 years, the rate of conversion to CDMS was 35% in those patients who received interferon therapy, as compared to 50% in those who received placebo. A crossover-designed extension study used the same population of patients to determine whether delayed administration of interferon therapy following prior placebo treatment conferred the same benefit. This CHAMPIONS (Controlled High-Risk Avonex in Multiple Sclerosis Prevention Study in Ongoing Neurologic Surveillance) trial demonstrated that those patients who had crossed over from placebo to interferon therapy 2–3 years following enrollment had a higher probability of developing CDMS (35%) relative to those who had been on interferon therapy since the initiation of the original CHAMPS study (21%) 5 years following first clinical presentation [44]. These studies also demonstrated a reduction in volume and accumulation of T2-weighted lesions, as well as number of post-gadolinium enhancing lesions on MRI, in favor of treatment with a disease-modifying therapy relative to placebo [41–43].

### Alternative Diagnoses

Features of atypical vision loss should prompt investigation into alternative etiologies beyond MS. The presence of bilateral (either simultaneous or rapidly sequential) AON, severe visual loss refractory to corticosteroids and/or without significant improvement within 2–3 weeks, recurrent bouts of unilateral or bilateral AON, and/or significant disk edema or hemorrhages may indicate an alternative pathology to MS. Table 28.1 provides a list of conditions that can cause optic neuropathy, as well as a summary of their clinical, imaging, and laboratory features.

## Neuromyelitis Optica (NMO)

NMO is an inflammatory demyelinating condition, associated classically with severe bouts of AON and longitudinally extensive transverse myelitis (three or more vertebral segments in length) [45]. NMO is discussed further in Chap. 15. AON associated with NMO may be refractory to steroids and subsequently require PLEX [46]. In patients with isolated AON (i.e., lacking brain, brainstem, or spinal cord pathology), one study demonstrated that 56% of patients were seropositive for anti-aquaporin 4 (AQP4) antibodies [47].

**Table 28.1** Differential diagnosis of acute optic neuropathy

| | Important clinical features | Common imaging features | Laboratory/CSF studies |
|---|---|---|---|
| *Inflammatory* | | | |
| Idiopathic | Single episode of optic neuritis | Normal MRI of brain | – |
| Multiple sclerosis | Often unilateral, centrocecal defects, periocular pain in up to 90% | Classic multifocal periventricular and juxtacortical demyelinating lesions | Normal cellular profile in 77%<br>Oligoclonal bands present in about 88% |
| Neuromyelitis optica (NMO) | Severe optic neuritis, often bilateral or sequential<br>Longitudinally extensive transverse myelitis (LETM) or classic "brainstem" syndrome | Posterior optic nerve, chiasmal involvement<br>If LETM ≥ 3 segment spinal cord lesion<br>Non-specific deep white matter lesions | Serum anti-aquaporin 4 antibody<br>May see inflammatory CSF (pleiocytosis, elevated protein)<br>Oligoclonal bands typically absent (<25%) |
| Myelin oligodendrocyte glycoprotein (MOG) | Steroid responsive, recurrent bouts of optic neuritis<br>May experience myelitis, brainstem, or cerebellar encephalitis | MRI Brain may demonstrate non-specific deep white matter lesions | Serum anti-MOG antibody<br>May see inflammatory CSF (pleiocytosis, elevated protein)<br>Oligoclonal bands typically absent (<13%) |
| Chronic relapsing inflammatory optic neuropathy (CRION) | Steroid responsive, recurrent<br>No other neurologic features | Normal MRI of brain | |
| Arteritic ischemic optic neuropathy (AION) | Acute onset unilateral vision loss associated with temporal headache, myalgias, jaw claudication | – | High erythrocyte sedimentation rate (ESR)<br>Histopathology: panarteritis of the media layer of the vessel with CD4+ lymphocytes and macrophages |
| Sarcoidosis | Progressive onset, painless (75%), bilateral (31%)<br>Co-occurrence of uveitis, vitriitis, scleritis (36%) | May involve optic chiasm or leptomeninges<br>Optic nerve sheath enhancement | Often CSF pleiocytosis<br>If present, oligoclonal bands in CSF may mirror systemic bands |
| Systemic lupus erythematosus<br>Sjögren's syndrome<br>Rheumatoid arthritis<br>Wegener's granulomatosis | Often present with multi-organ involvement<br>Consider co-incidence of NMO<br>May be precipitated by use of TNF-alpha inhibitors<br>Pulmonary or sinus involvement | Typically normal MRI brain | Antinuclear autoantibodies for associated disease |
| *Infectious* | | | |
| Syphilis, HIV, tuberculosis | Identify risk factors, presence of other disease stigmata (e.g., chancre in syphilis) | – | Serologic markers to infectious agent |
| Lyme disease, *Bartonella henselae*, toxoplasmosis | Identify exposure: tick bites, cats in home<br>Fundoscopy: severe optic disk swelling, neuroretinitis (macular star), hemorrhages, and exudates | – | – |

**Table 28.1** (continued)

| | Important clinical features | Common imaging features | Laboratory/CSF studies |
|---|---|---|---|
| *Noninflammatory* | | | |
| Non-arteritic ischemic optic neuropathy (NAION) | Acute, painless, typically altitudinal vision loss | – | – |
| *Toxic and nutritional* | | | |
| Methanol poisoning | Bilateral vision loss, painless<br>History of alcohol abuse, exposure | Putaminal T2 hyperintensities | Anion gap metabolic acidosis |
| Vitamin B1, B12 | Gradual, painless loss of vision, often bilateral | May have posterior-spinal cord T2 hyperintensities | Low serum levels B1, B12 |
| *Hereditary* | | | |
| Leber's hereditary optic neuropathy (LHON) | Painless, progressive, bilateral, or sequential central vision loss<br>Maternal family history of early vision loss (<40 yrs)<br>Predominant in men (80%) | – | Mitochondrial DNA point mutations |
| Dominant optic atrophy | Bilateral. Starts first decade of life. May be associated with hearing loss, spastic paraplegia, peripheral neuropathy | – | Autosomal dominant |

*MRI* magnetic resonance imaging, *CSF* cerebrospinal fluid, *TNF* tumor necrosis factor, *HIV* human immunodeficiency virus

The discovery of this highly specific antibody has enabled targeted immunosuppressive treatment for patients who suffer from this severely disabling disease. In 2014, the Neuromyelitis Optic Study Group (NEMOS) provided updated recommendations regarding first-, second-, and third-line immunosuppressive therapies in NMO [48]. The goal of treatment in NMO is immune suppression. Until 2019, no US Food and Drug Administration (FDA)-approved immune therapy existed for the treatment of NMO. Ongoing studies have utilized biologics targeting the interleukin-6 and complement-dependent pathways in NMO. A placebo-controlled, double-blind, phase 3 trial utilizing eculizumab to target the complement-dependent pathway was completed in mid-2018, with the primary end point being time-to-first-attack following administration of drug. The results were overwhelmingly positive, with a relative risk reduction of time to first relapse of 94% in patients receiving eculizumab compared to those receiving placebo. Eculizumab was approved for use in patients with antibody-positive NMO spectrum disorder (NMOSD) as of June 2019 [49]. The frequency of eculizumab dosing, in addition to the cost of the medication, may prove to be pitfalls in its future clinical use; as of now, it is only approved for use in AQP4-antibody-positive patients. Two phase 3 trials of interleukin-6-targeting sartralizumab (SA237) as monotherapy (NCT02073279) or combined therapy (NCT02028884) have completed enrollment. Early results of sartralizumab as monotherapy in patients with antibody-positive or antibody-negative NMO demonstrated reduction in relapse rates of 79% in antibody-positive patients, and 34% in antibody-negative patients, compared to placebo [50]. A third agent, inebilizumab, targets CD19-B cells and has been found to reduce risk of relapse by 77% compared to placebo [51]. As of this writing, rituximab, azathioprine, and mycophenolate mofetil remain the most commonly utilized treatments, with rituximab demonstrating significant reductions in number of attacks per patient per year. Failure of rituximab in patients with antibody-positive disease is associated with repopulation of B cells [52, 53].

Ancillary studies help distinguish features of NMO and MS-related AON. MRI in NMO-AON demonstrates more frequent involvement of the posterior optic nerve and/or optic chiasm [54]. NMO-AON lesions may be "longitudinally extensive" (>17.6 mm); using cutoff criteria of 17.6 mm in ON lesion length, sensitivity and specificity for NMO are 76.9% and 80.8%, respectively [55]. OCT reflects the severity of AON in NMO patients as compared to MS. One study demonstrated that peripapillary retinal nerve fiber layer (RNFL) thickness is overall lower in eyes that have experienced AON of patients with NMO relative to MS (63.6 μ[mu]m vs. 88.3 μ[mu]m, $p < 0.0001$) [56]. In the same study, a single episode of AON was associated with 24 μ(mu)m greater loss of RNFL thickness in NMO than in MS; the authors suggested that RNFL loss of greater than 15 μ(mu)m following an AON event is more consistent with NMO as opposed to MS. Additionally, RNFL thinning in NMO preferentially affects the superior and inferior sectors, as opposed to the temporal sector in MS [57].

## Anti-myelin Oligodendrocyte Glycoprotein-Associated Disease

A recently discovered antibody-mediated demyelinating condition commonly presents with recurrent, steroid-responsive AON. Antibodies directed against myelin oligodendrocyte protein (MOG) were reported in four patients with recurrent neurologic symptoms previously diagnosed with NMO-IgG antibody-negative NMO spectrum disorder (NMOSD); all of these patients had good response to corticosteroids [58]. Subsequently, there has been heightened interest in reclassifying patients previously diagnosed with seronegative-NMOSD and determining clinical features that may distinguish the two diseases. A study of 50 MOG-IgG positive patients with 29 eyes affected by AON or recurrent AON demonstrated that compared to seropositive NMO patients, the former had higher rates of recurrent attacks (0.69 vs. 0.29 attacks/year, $p = 0.004$), but without significant differences in RNFL, composite ganglion cell and inner plexiform layer (GCIPL), or visual outcome measures between groups [59]. A large cohort study determined that AON is the most frequent presenting clinical phenotype (60.9%) in MOG-IgG-positive patients; bilateral AON was seen in 25% of the entire cohort. Presence of MOG-IgG confers lower risk of severe vision loss, better recovery from attacks, and lower risk of ultimate disability [60]. Maintenance therapy with steroids and immunosuppression is the mainstay of therapy to reduce the risk of relapses [61, 62]. MOG antibody disease is also discussed in Chap. 15.

An entity of relapsing bilateral AON without MRI brain findings has also been described. Chronic relapsing inflammatory optic neuropathy (CRION) is typically responsive to steroids and demonstrates normal brain MRI findings [63]. In some cases, patients may require immune suppression to reduce the frequency and recurrence of AON. Whether an overlap or relationship exists between MOG-related disorders and CRION has yet to be elucidated.

## Infectious Optic Neuritis

Although rare, inflammatory ocular disease may be related to infection. The presence of macular edema, exudates, fevers, and other systemic symptoms should prompt questioning pertaining to infectious risks. *Bartonella henselae* is responsible for "cat scratch disease," presents most commonly with a cat scratch, and may present with a pathognomonic "macular star" (neuroretinitis) on fundoscopy [64]. Few cases of AON following exposure to the tick-borne *Borrelia burgdorferi* have also been reported, and if clinical suspicion warrants (presence of classic skin rash or involvement of meninges on MRI), this should prompt serologic and/or CSF studies. Tuberculosis should be considered in endemic areas and high-risk populations. Additionally, high-risk sexual behavior, presence of syphilitic lesions, and any viral prodrome warrants testing for viral causes (human immunodeficiency virus [HIV], syphilis, cytomegalovirus [CMV], and toxoplasmosis) [65]. Presence of an infectious etiology should give caution to utilizing steroids in treatment of these forms of ocular disease, unless there is certainty that the infectious source has been appropriately treated and/or cleared.

## Optic Neuropathy Due to Systemic Disease

Involvement of multiple organ systems or previous diagnosis of a rheumatologic disorder should prompt a search for a single unifying diagnosis if possible.

Systemic lupus erythematosus (SLE), Sjögren's syndrome, and rheumatoid arthritis (RA) may be associated with neuro-ophthalmologic manifestations [66, 67]. Additionally, treatment of these conditions with anti-tumor necrosis factor (TNF) drugs may either induce de novo demyelination or unmask an already existing underlying demyelinating disorder [68]. Patients with SLE and lupus anticoagulant or antiphospholipid antibody syndrome may be at risk of ischemic or inflammatory demyelinating optic neuropathy. In patients who have systemic features of SLE, antinuclear antibodies and serum procoagulants should be part of the workup of visual loss. Patients with SLE may also experience ocular involvement (scleritis and uveitis). It should be noted that NMO patients may also harbor concurrent autoantibodies against alternative cellular antigens (in particular, Sjögren's syndrome-associated antibodies). One of the commonest initial manifestations of Sjögren's syndrome is sicca complex (xerophthalmia and xerostomia); "dry eyes" may cause a sensation of visual obscuration, but frank optic neuropathy may also occur and should be considered if clinical features of optic neuropathy are present. If AON is diagnosed or suspected in patients with systemic autoimmune diseases, testing for anti-NMO-IgG should be completed.

Paraneoplastic and neoplastic forms of vision loss are a heterogeneous group of disorders and are rare. These tend to present as painless, subacute to progressive vision loss. Paraneoplastic optic neuropathy may be seen with small cell lung cancer (SCLC), among numerous other cancers, and associated with Anti-Cv2 antibodies directed against collapsing-responsive mediator protein-5 (CRMP-5). Cancer-associated retinopathy (CAR) and melanoma-associated retinopathy (MAR) may present with subacute vision loss and photopsia, and the former can be associated with anti-recoverin, Ras-related GTPase (Rab6A), heat shock protein 26 (HSP26), and guanylyl cyclase-activating protein (GCAP) antibodies; one cohort of 173 patients demonstrated that just 50% of patients with CAR features and the aforementioned antibodies were diagnosed with an underlying neoplasm [69].

## Noninflammatory Causes of Optic Neuropathy

Non-arteritic and arteritic ischemic optic neuropathies (NAION and AION), respectively, may present as acute, painless onset of vision loss; these are commonly associated with altitudinal visual defects. In patients with vascular risk factors, acute onset vision loss should prompt consideration of NAION; recovery is typically poor. AION should be considered in elderly patients with temporal headache, jaw claudication, and systemic signs of myalgias and fatigue, features suggestive of giant cell (temporal) arteritis. This is a rheumatologic and ophthalmologic emergency, with high erythrocyte sedimentation rate (ESR) prompting swift initiation of steroids and temporal artery biopsy. Note that temporal artery biopsies may be falsely negative, as related to segmental involvement of the artery, creating "skip lesions"; increasing the length of the biopsied area may improve sensitivity, but if suspicion is high enough, angiography or positron emission tomography (PET) may be needed to increase diagnostic yield [70]. Family history of vision loss, particularly within the mother's lineage, should prompt consideration of mitochondrial heritable diseases; Leber's hereditary optic neuropathy (LHON) may present as simultaneous or sequential bilateral optic neuropathy and may be subacute to chronic in onset. Patients with history of alcohol abuse may be at risk for methanol- or ethylene-induced optic neuropathy (methanol and ethylene glycol may be cheaper, more easily accessed, and more potent than ethanol). Nutritional deficiencies, in particular vitamin B12, may also present with optic neuropathy. A long list of potentially optic neuropathy inducing drugs has been described in the treatment of tuberculosis, malaria, and neoplasms, among other conditions; if patients have been on chronic therapy for any of these conditions, a thorough review of medication list should be completed, although causality can be difficult to ascertain.

## Beyond the Optic Nerve

Ocular manifestations of inflammatory autoimmune conditions outside the optic nerve include the uvea and the retina, although a detailed overview of these processes is beyond the scope of this review.

Uveitis is associated with painful blurring of vision and is characteristically associated with reddening of the eye when involving the anterior chamber. Patients with involvement of the posterior chamber frequently describe "floaters." Autoimmune conditions causing anterior and/or posterior uveitis, which may have other neurologic and systemic manifestations, include SLE, Sjögren's syndrome, RA, sarcoidosis, granulomatosis with polyangiitis (Wegener's granulomatosis), and Behçet's disease [71], among others. These conditions may also present with vasculitis of the central nervous system and with optic neuropathy [72].

Pars planitis specifically refers to inflammatory cellular infiltration involving the pars plana and the vitreous; this is an "intermediate" uveitis and may be seen in approximately 15% of patients with MS [73]. On slit lamp examination, the presence of vitreous cells called "snowballs" and exudates called "snowbanks" are characteristic of pars planitis. Retinal neovascularization, cystoid macular edema, and retinal periphlebitis may also be seen, and the latter may be associated with more aggressive MS [74]. Uveitis is a vision-threatening condition that can lead to blindness if untreated; these findings should prompt urgent referral to ophthalmology. Treatment of uveitis depends on severity of symptoms and may require steroids and/or immunosuppressive agents.

Patients of Afro-Caribbean race presenting with signs of vision loss and systemic disease should prompt consideration for ocular sarcoidosis. Estimates of the prevalence of ocular involvement of sarcoidosis range from 12% to 80% [75]. Signs of intraocular granulomatous inflammation in sarcoidosis most commonly include anterior uveitis, panuveitis, vitritis, episcleritis/scleritis, and keratitis. AON is rarely associated with sarcoidosis, although it may occur in about 5% of patients with systemic sarcoidosis [76]. The onset of sarcoidosis-associated optic neuropathy may be acute or slowly progressive, is bilateral in nearly 31% of patients, and is painless in approximately 75%. Intraocular inflammation (most commonly anterior uveitis) coincides with optic neuritis in 36% of cases [77].

## Neuroimaging in AON

### Magnetic Resonance Imaging

When evaluating patients with an AON, ancillary testing can assist in confirming the diagnosis, evaluating the underlying cause, and helping to establish or determine the risk of MS. MRI is the most useful paraclinical tool in predicting the risk of progression to CDMS. As previously discussed, results of the ONTT demonstrated the 15-year risk of developing CDMS in patients who presented with AON and even just one classic demyelinating lesion on brain MRI to be 72%. In addition to brain imaging, obtaining MRI of the orbits allows for more precise assessment of the optic nerves. Orbital imaging may assist in distinguishing AON due to inflammatory versus noninflammatory causes (e.g., ischemic or compressive neuropathies). Kupersmith et al. demonstrated that inflammatory lesions of the optic nerve enhance on T1-weighted sequences in 94%

of acutely affected eyes; a range of location and segmental lengths of enhancement are seen. Length of the optic nerve enhancing lesion >17 mm and enhancement of the canalicular segment of the optic nerve are associated with poorer baseline visual function. However, visual recovery appears similar regardless of location or length of the enhancing segment, as well as independent of treatment with steroids or baseline visual function in the affected eye [78]. While length and location of an enhancing lesion of the optic nerve is not associated with overall visual recovery in AON, earlier studies utilizing T2-weighted short tau inversion recovery (STIR) sequences demonstrated that longer lesions and those in the canalicular segment were associated with poorer visual recovery over time [79].

MRI might also aid the quantification of axonal loss following AON. Numerous studies have extensively demonstrated that atrophy of the brain and spinal cord correlates with disability in MS [80, 81]; measure of atrophy is a surrogate marker for axonal and neuronal degeneration, the principal pathological substrates of disability in MS. Hickman et al. have utilized coronal imaging cuts on short echo fast fluid attenuated inversion recovery (sTE fFLAIR) sequences to measure cross-sectional area of the optic nerve. They demonstrated that reduced optic nerve area (atrophy) correlates with longer disease duration in patients with a single episode of AON [82]; further, continued atrophy 3 years following an initial AON correlates with worsened visual acuity [83]. Acutely, affected optic nerves demonstrate increased mean area as compared to healthy controls (16.1 $mm^2$ vs. 13.6 $mm^2$, mean difference = 2.5 $mm^2$, 95% CI 1.2–3.8, $p = 0.0003$), and this correlates with worse baseline visual function [84]. The acute increase in mean area of the optic nerve is primarily related to interstitial inflammatory edema, as well as other potential mechanisms including axonal congestion. Hickman then demonstrated that 1 year post-AON, mean area of the affected nerve declined and was significantly lower than the mean area in healthy controls (11.3 $mm^2$ vs. 13.1 $mm^2$, mean difference = −1.8 $mm^2$, 95% CI -0.5 to −3.2, $p = 0.008$). However, at 1 year, despite optic nerve atrophy, mean area did not correlate with visual function as measured by logMAR VA and Humphrey's visual field testing (HVF), as at baseline [84]. Thus, good recovery of function may occur despite underlying structural loss. This is in slight contradiction to the findings of Trip et al., who demonstrated that patients with poor visual recovery, more than 1 year following AON, demonstrate significantly reduced optic nerve area (30% decreased compared to controls), which correlates with reduced OCT-derived RNFL thickness [85].

## Magnetization Transfer and Diffusion-Weighted Imaging

More sophisticated neuroimaging techniques used primarily in research have allowed more detailed assessment of the tissue ultrastructure of the optic nerve following AON. Magnetic transfer imaging allows for measurement of the exchange of free protons with proton-bound macromolecules such as the protein and lipids found in myelin and axonal membranes. The ratio of proton exchange is the magnetization transfer ratio (MTR), which is reduced in demyelinated MS lesions compared to healthy white matter [86]. Reduced MTR has been demonstrated to inversely correlate with visual evoked potential latency [85, 87, 88], suggesting that MTR may be a relative surrogate of myelin, such as in the optic nerve. Hickman et al. [88] serially imaged AON affected, contralateral unaffected, and healthy optic nerves, and demonstrated that while baseline MTR were similar across all optic nerves assessed, AON affected nerves experienced a reduction in MTR over time following AON, with nadir reached at 8 months; subsequently, a slow increase in MTR was noted by 12 months, suggesting that MTR might also conversely capture remyelination, although this remains to be definitively elucidated in vivo. Finally, Trip et al. demonstrated an association of lesional MTR reduction with reduction in RNFL thickness (as measured with OCT), suggesting that MTR may also be affected by axonal loss and may not be exclusively myelin specific [85].

Diffusion-weighted imaging measures the movement of water molecules across tissues. Pathologic damage to tissue, including demyelination, disrupts this movement. Diffusion measurements have proven difficult in the optic nerve due to its mobility, although techniques utilizing fat and CSF suppression have allowed for improved resolution. Increased mean diffusivity (MD) or apparent diffusion coefficient (ADC) signals are reported in chronic optic nerve lesions and patients with poor AON recovery, correlating with visual acuity, and VEP amplitude and latency [89, 90]. Measurements of directional flow, fractional anisotropy (FA), are reduced across affected nerves and also correlate with VEP amplitude. In AON, reduced axial diffusivities at onset correlate with reduced contrast sensitivity and visual acuity at 1 and 3 months post-onset [91] and with worse contrast sensitivity, VA, VEP amplitude, latency, and RNFL thickness at 6 months [92]. In remote (>1 year) cases of AON, radial diffusivity and FA, but not axial diffusivity, correlate with CS, VA, and RNFL thickness on OCT. The authors of this study hypothesized that diffusion properties likely reflect disruption of axonal integrity in the remotely affected optic nerve.

Diffusion techniques are also utilized in tractography, which allows imaging of the white matter tracts from optic nerves and optic radiations to the visual cortices. These techniques have potential for understanding underlying structural connectivity and visual remapping or reorganization following an AON insult.

## Visual Evoked Potentials

Electrophysiology in AON was described as early as 1972, when delayed VEP latencies were found to correlate with clinical AON in 65% of patients [93]. Photic stimuli are applied to the eyes, and the response is recorded at the scalp over the primary visual cortex. This produces a waveform of compounded signals (normal and abnormal) along the afferent visual pathway, reflecting the integrity of the entire visual pathway. In AON, the P100 waveform (normally elicited approximately 100 ms after the stimulus) latency is prolonged, and amplitude is reduced [93]. Benefits of conventional VEP include its ease of use in-office by trained electrophysiology technicians and usefulness in detecting demyelination both clinically and subclinically. However, conventional VEP has limited localization, and the central and inferior fields predominate the recording. Multifocal VEP has eliminated some of this limitation by dividing the visual field into 60 equal sectors for individual evaluation [94]. Reduced waveform amplitudes on mfVEP are seen in 92% of patients with AON and possible MS, and in 100% of patients with AON previously diagnosed with MS [95]. Fraser et al. also demonstrate that 36% of patients with abnormal latencies go on to develop MS at 1 year, compared to no patients with normal latencies [96]. Reduction in amplitude on mfVEP correlates with RNFL thinning on OCT as early as 3 months post-AON, suggesting an indication for mfVEP as a marker of axonal loss [97].

## Optical Coherence Tomography

OCT was first utilized in MS in the late 1990s. OCT provides a fast, noninvasive, high-resolution technique by which to image the layers of the retina, without the confound of myelin, since the retina is unmyelinated under normal circumstances. OCT allows measurement of neuroaxonal degeneration within the retina. The RNFL is the innermost layer of the retina and comprises the axons of ganglion cell neurons (which are located in the ganglion cell layer beneath the RNFL). The axons within the RNFL coalesce at the optic disks to form the optic nerves and exit the eye posteriorly via the lamina cribrosa, beyond which they acquire myelin. Following AON, retrograde axonal degeneration ensues, resulting in thinning of the RNFL and consequently death of ganglion cell neurons from which RNFL axons arise.

In MS, RNFL thinning is observed following AON, regardless of visual recovery [98]. In the acute phase, similar to optic nerve area, the RNFL may demonstrate thickening related to inflammation. Subsequently, the RNFL begins to demonstrate thinning (as edema resolves and ongoing/ensuing axonal degeneration is captured), with a predilection for the temporal RNFL quadrant [99–101]. Using older time-domain OCT, it was found that AON results in approximately 20% RNFL loss [102, 103]. RNFL thinning on OCT correlates with reduced visual acuity, low-contrast letter acuity, color vision, visual field deficits, and visual quality of life measures [99, 101, 104–106]. Importantly, RNFL thinning is also demonstrated in MS eyes without a prior history of AON and correlates with visual function [107]. This likely relates to subclinical optic neuropathy. Notably, MS patients with likely subclinical optic neuropathy reflected by baseline RNFL measurements approximately <88 μ(mu)m have an up to fourfold increase in risk of future disability accumulation at 3–5 years [108]. Optic nerve pathology is virtually ubiquitous as part of the MS disease process, with up to 99% of MS patients exhibiting demyelinating plaques in their optic nerves at post-mortem [109]. Reduced RNFL thickness also correlates with MRI measures of brain atrophy [85, 110–112].

While the early focus of OCT was measures of RNFL and average macular thickness (a nonspecific measure of total macular thickness including all intervening retinal layers), modern spectral-domain OCT has much higher resolution (3–5 μ[mu]m) and reproducibility than older time-domain OCT, allowing discrete layers of the retina to be accurately and reliably quantified. In particular, GCIPL thickness may exhibit superior structure–function relationships in MS, as compared to RNFL thickness [113]. This may relate to superior reproducibility, but also lack of confound related to edema, as well as astrogliosis (which appears to principally occur in the RNFL) [114]. In contrast to the RNFL, the GCIPL does not demonstrate edema during AON, making it a feasible measure to accurately quantify AON-related neurodegeneration [115]. Patients with clinical and/or radiological evidence of MS-related disease activity also exhibit accelerated rates of GCIPL thinning compared to controls [116]. GCIPL atrophy mirrors whole brain and in particular gray matter atrophy across MS subtypes [117, 118] and also appears to be differentially modulated by different disease-modifying therapies [119]. OCT provides a fast, noninvasive, inexpensive in-office technique to measure axonal and neuronal degeneration in patients with MS. RNFL and GCIPL thinning may be seen in patients without prior AON. OCT measures correlate with brain atrophy on MRI. MS treatments may slow progression of neuroaxonal degeneration as measured by OCT.

## Neuroprotection/Neurorestoration in AON

Currently available treatments in MS are primarily aimed at reducing clinical relapses and overt radiographic inflammatory activity, and ultimately reducing the risk of future disability accrual. Potential therapies for neuroprotection and remyelination that could allow recovery of lost function are of great clinical interest and would meet a greatly unmet need not only in the areas of MS, AON, and related disorders but also neurology in general. The afferent visual pathway offers an opportune and structurally eloquent window for studying neurodegeneration and remyelination due to the relative ease with which clinical impairment, recovery, structural, and functional, including neurophysiologic, outcomes can be measured.

Among potential targets for therapies, HMG-CoA reductase inhibitors (statins) demonstrated a non-significant trend toward improvements in contrast sensitivity and VEP measures [120]. A phase II trial of erythropoietin given at 33,000 IU/day for 3 days following AON demonstrated significant reduction in RNFL thinning and VEP latency at 16 weeks post-AON, as compared to placebo, and a trend toward significance in improved visual outcome [121]. Phase 3 trials are ongoing. The antiepileptic agent phenytoin targets voltage-gated sodium channels and has demonstrated neuroprotection in preclinical animal models; compared with placebo, patients who received phenytoin demonstrated significantly less RNFL loss following AON, although no significant clinical visual outcomes were achieved [122]. Clemastine, a first-generation antihistamine that has demonstrated remyelinating properties in animal models, demonstrated significant reduction in VEP latency in a crossover-designed trial, but did not meet significance in other outcomes; fatigue was the most common side effect reported with the drug [123]. Finally, the monoclonal antibody opicinumab (anti-LINGO-1) promotes remyelination by binding a neural cell surface protein (LINGO) that blocks the differentiation of oligodendrocyte precursor cells to inhibit myelin. Phase 2 trials did not demonstrate significant differences in VEP latencies at 24 and 32 weeks post-treatment between anti-LINGO and placebo following AON in intention to treat analyses; per-protocol analyses favored anti-LINGO over placebo [124]. Further phase 2 trials of opicinumab in MS are ongoing.

In summary, while neuroprotective and remyelinating agents have demonstrated significant changes in structural and/or physiologic outcome measures (OCT and VEP), clinical efficacy of these drugs remains unclear. Nevertheless, the positive subclinical outcomes in phase 2 trials demonstrate the early potential of these agents to potentially influence neuroprotection and/or neurorepair.

## Conclusion

AON is most commonly associated with MS in the Western world. A thorough clinical history in conjunction with ancillary imaging and serological testing can help elucidate the underlying diagnosis of ophthalmologic dysfunction in patients with potential inflammatory, demyelinating disease. Early treatment in these instances can hasten visual recovery. Appropriate diagnosis of optic neuropathy in conjunction with ancillary studies allows for determination of the risk of developing CDMS. Advances in imaging techniques are promising with regard to the evaluation of underlying axonal and neuronal degeneration. Advances in understanding of AON and related disorders may help to guide therapeutic intervention for slowing disease progression and perhaps in repairing acute tissue damage.

## References

1. Wikstrom J. The epidemiology of optic neuritis in Finland. Acta Neurol Scand. 1975;52(3):196–206.
2. Wakakura M, Minei-Higa R, Oono S, Matsui Y, Tabuchi A, Kani K, et al. Baseline features of idiopathic optic neuritis as determined by a multicenter treatment trial in Japan. Optic Neuritis Treatment Trial Multicenter Cooperative Research Group (ONMRG). Jpn J Ophthalmol. 1999;43(2):127–32.
3. Loncarek K, Brajac I, Petricek I, Stalekar H, Cerovski B, Pokupe R. Epidemiology of monosymptomatic optic neuritis in Rijeka County, Croatia: meteorological aspects. Coll Antropol. 2005;29(1):309–13.
4. MacDonald BK, Cockerell OC, Sander JW, Shorvon SD. The incidence and lifetime prevalence of neurological disorders in a prospective community-based study in the UK. Brain. 2000;123(Pt 4):665–76.
5. Jin YP, de Pedro-Cuesta J, Söderström M, Stawiarz L, Link H. Incidence of optic neuritis in Stockholm, Sweden 1990–1995: I. Age, sex, birth, and ethnic-group related patterns. J Neurol Sci. 1998;159:107–14.
6. Rodriguez M, Siva A, Cross SA, O'Brien PC, Kurland LT. Optic neuritis: a population-based study in Olmsted County, Minnesota. Neurology. 1995;45(2):244–50.
7. Jin Y, de Pedro-Cuesta J, Söderström M, Stawiarz L, Link H. Seasonal patterns in optic neuritis and multiple sclerosis: a meta-analysis. J Neurol Sci. 2000;181(1–2):56–64.
8. Taylor BV, Lucas RM, Dear K, Kilpatrick TJ, Pender MP, van der Mei IA, et al. Latitudinal variation in incidence and type of first central nervous system demyelinating events. Mult Scler. 2010;16:398–405.
9. Lucas RM, Ponsonby AL, Dear K, Valery P, Pender MP, Burrows JM, et al. Current and past Epstein-Barr virus infection in risk of initial CNS demyelination. Neurology. 2011;77(4):371–9.
10. Wakakura M, Ishikawa S, Oono S, Tabuchi A, Kani K, Tazawa Y, et al. Incidence of acute idiopathic optic neuritis and its therapy in Japan: Optic Neuritis

Treatment Trial Multicenter Cooperative Research Group (ONMRG). Nippon Ganka Gakkai Zasshi. 1995;99(1):93–7.
11. Phillips P, Newman N, Lynn M. Optic neuritis in African Americans. Arch Neurol. 1998;55(2):186–92.
12. Optic Neuritis Study Group. The clinical profile of optic neuritis. Experience of the Optic Neuritis Treatment Trial. Arch Ophthalmol. 1991;109(12):1673–8.
13. Beck RW, Cleary PA, Anderson MM Jr, Keltner JL, Shults WT, Kaufman DI, et al. A randomized, controlled trial of corticosteroids in the treatment of acute optic neuritis. The Optic Neuritis Study Group. N Engl J Med. 1992;326:581–8.
14. Katz B. The dyschromatopsia of optic neuritis: a descriptive analysis of data from the optic neuritis treatment trial. Trans Am Ophthalmol Soc. 1995;93:685–708.
15. Sobhanian MJ, Agarwal R, Meltzer E, Kildebeck E, Frohman BS, Frohman AN, et al. Identification and treatment of the visual processing asymmetry in MS patients with optic neuritis: The Pulfrich phenomenon. J Neurolog Sci. 387:60–9.
16. Beck RW, Cleary PA. Optic neuritis treatment trial. One-year follow-up results. Arch Ophthalmol. 1993;111(6):773–5.
17. Optic Neuritis Study Group. Visual function more than 10 years after optic neuritis: experience of the Optic Neuritis Treatment Trial. Am J Ophthalmol. 2004;137(1):77–83.
18. Keltner J, Johnson C, Spurr J, Beck R, Optic Neuritis Study Group. Visual field profile of Optic Neuritis: one-year follow up in the Optic Neuritis Treatment Trial. Arch Ophthalmol. 1994;112:946–53.
19. Regan D, Raymond J, Ginsburg AP, Murray TJ. Contrast sensitivity, visual acuity, and the discrimination of Snellen letters in multiple sclerosis. Brain. 1981;104:333–50.
20. Regan D, Neima D. Low-contrast letter charts as a test of visual function. Ophthalmology. 1983;90:1192–200.
21. Kupersmith MJ, Nelson JI, Seiple WH, Carr RE, Weiss PA. The 20/20 eye in multiple sclerosis. Neurology. 1983;33:1015–20.
22. Balcer LJ, Raynowska J, Nolan R, Galetta SL, Kapoor R, Benedict R, et al.; Multiple Sclerosis Outcome Assessments Consortium. Validity of low-contrast letter acuity as a visual performance outcome measure for multiple sclerosis. Mult Scler J. 2017:23(5): 734–47.
23. Sabadia SB, Nolan RC, Galetta KM, Narayana KM, Wilson JA, Calabresi PA, et al. 20/40 or better visual acuity after optic neuritis: not as good as we once thought? J Neuro Ophth. 2016;36:369–76.
24. Trobe JD, Beck RW, Moke PS, Cleary PA. Contrast sensitivity and other vision tests in the optic neuritis treatment trial. Am J Ophthalmol. 1996;121(5):547–53.
25. Rudick R, Antel J, Confavreux C, Cutter G, Ellison G, Fischer J, et al. Clinical outcomes assessment in multiple sclerosis. Ann Neurol. 1996;40(3):469–79.
26. Mowry EM, Loguidice MJ, Daniels AB, Jacobs DA, Markowitz CE, Galetta SL, et al. Vision related quality of life in multiple sclerosis: correlation with new measures of low and high contrast letter acuity. J Neurol Neurosurg Psychiatry. 2009;80(7):767–72.
27. Balcer LJ, Miller DH, Reingold SC, Cohen JA. Vision and vision-related outcome measures in multiple sclerosis. Brain. 2015;138:11–27.
28. Freedman MS, Thompson EJ, Deisenhammer F, Giovannoni G, Grimsley G, Keir G, et al. Recommended standard cerebrospinal fluid analysis in the diagnosis of multiple sclerosis: a consensus statement. Arch Neurol. 2005;62(6):865–70.
29. Optic Neuritis Study Group. Multiple sclerosis risk after optic neuritis: final optic neuritis treatment trial follow-up. Arch Neurol. 2008;65:727–32.
30. Morrow SA, Fraser JA, Day C, Bowman D, Rosehart H, Kremenchutzky M, et al. Effect of treating acute optic neuritis with bioequivalent oral vs intravenous corticosteroids a randomized clinical trial. JAMA Neurol. 2018;75(6):690–6.
31. Noseworthy JH, O'Brien PC, Petterson TM, Weis J, Stevens L, Peterson WK, et al. A randomized trial of intravenous immunoglobulin in inflammatory demyelinating optic neuritis. Neurology. 2001;56:1514–22.
32. Roed HG, Langkilde A, Sellebjerg F, Lauritzen M, Bang P, Mørup A, et al. A double-blind, randomized trial of IV immunoglobulin treatment in acute optic neuritis. Neurology. 2005;64:804–10.
33. Tselis A, Perumal J, Caon C, Hreha S, Ching W, Din M, et al. Treatment of corticosteroid refractory optic neuritis in multiple sclerosis patients with intravenous immunoglobulin. Eur J Neurol. 2008;15:1163–7.
34. Bonnan M, Cbare P. Plasma exchange in severe attacks of neuromyelitis optica. Mult Scler Int. 2012;787630:1–9.
35. Keegan M, Pineda AA, McClelland RL, Darby CH, Rodriguez M, Weinshenker BG. Plasma exchange for severe attacks of CNS demyelination: predictors of response. Neurology. 2002;58:143–6.
36. Deschamps R, Gueguen A, Parquet N, Saheb S, Driss F, Mesnil M, et al. Plasma exchange response in 34 patients with severe optic neuritis. J Neurol. 2016;263(5):883–7.
37. Roesner S, Appel R, Gbadamosi J, Martin R, Heesen C. Treatment of steroid-unresponsive optic neuritis with plasma exchange. Acta Neurol Scand. 2012;126:103–8.
38. Merle H, Olindo S, Jeannin S, Valentino R, Mehdaoui H, Cabot F, et al. Treatment of optic neuritis by plasma exchange (add-on) in neuromyelitis optica. Arch Ophthalmol. 2012;130:858–62.
39. Toosy A, Mason D, Miller D. Optic neuritis. Lancet. 2014;13:83–99.

40. Swanton JK, Fernando K, Dalton CM, Miszkiel KA, Thompson AJ, Plant GT, et al. Is the frequency of abnormalities on magnetic resonance imaging an isolated optic neuritis, related to the prevalence of multiple sclerosis? A global comparison. J Neurol Neurosurg Psychiatry. 2006;77:1070–2.
41. CHAMPS Study Group. Interferon beta-1a for optic neuritis patients at high risk for multiple sclerosis. Am J Ophthalmol. 2001;132:463–71.
42. Jacobs LD, Beck RW, Simon JH, Kinkel RP, Brownscheidle CM, Murray TJ, et al. Intramuscular interferon beta-1a therapy initiated during a first demyelinating event in multiple sclerosis. CHAMPS Study Group. N Engl J Med. 2000;343:898–904.
43. Comi G, Filippi M, Barkhof F, Durelli L, Edan G, Fernández O, et al. Effect of early interferon treatment on conversion to definite multiple sclerosis: a randomised study. Lancet. 2001;357:576–82.
44. CHAMPIONS Study Group. IM interferon B-1a delays definite multiple sclerosis 5 years after a first demyelinating event. Neurology. 2006;66:678.
45. Wingerchuk DM, Lennon VA, Lucchinetti CF, Pittock SJ, Weinshenker BG. The spectrum of neuromyelitis optica. Lancet Neurol. 2007;6:805–15.
46. Weinshenker BG, O'Brien PC, Petterson TM, Noseworthy JH, Lucchinetti CF, Dodick DW, et al. A randomized trial of plasma exchange in acute central nervous system inflammatory demyelinating disease. Ann Neurol. 1999;46(6):878–86.
47. Petzold A, Pittock S, Lennon V, Maggiore C, Weinshenker BG, Plant GT. Neuromyelitis optica-IgG (aquaporin-4) autoantibodies in immune mediated optic neuritis. J Neurol Neurosurg Psychiatry. 2010;81:109–11.
48. Trebst C, Jarius S, Berthele A, Paul F, Schippling S, Wildemann B, et al.; Neuromyelitis Optica Study Group (NEMOS). Update on the diagnosis and treatment of neuromyelitis optica: recommendations of the Neuromyelitis Optica Study Group (NEMOS). J Neurol. 2014; 261:1–16.
49. Pittock SJ, Berthele A, Fujihara K, Kim HJ, Levy M, Palace J, et al. Eculizumab in aquaporin-4-positive neuromyelitis optica spectrum disorder. N Engl J Med. 2019;381(7):614–25.
50. McNamara D. Satralizumab slashes relapse rate in neuromyelitis optica. Medscape. October 17, 2018. https://www.medscape.com/viewarticle/903574#:~:text=BERLIN%20%E2%80%94%20Adding%20an%20anti-interleukin,controlled%2C%20phase%203%20study%20found.
51. Greg E. Inebilizumab reduces the risk of NMOSD attacks. Neurol Rev. June 2019. https://www.neurologyreviews-digital.com/neurologyreviews/nr_june_2019/MobilePagedArticle.action?articleId=1491876#articleId1491876.
52. Pellkofer H, Krumbholz M, Berthele A, Hemmer B, Gerdes LA, Havla J, et al. Long-term follow-up of patients with neuromyelitis optica after repeated therapy with rituximab. Neurology. 2011;76((15):1310–5.
53. Cree BA, Lamb S, Morgan K, Chen A, Waubant E, Genain C. An open label study of the effects of rituximab in neuromyelitis optica. Neurology. 2005;64(7):1270–2.
54. Khanna S, Sharma A, Huecker J, Gordon M, Naismith RT, Van Stavern GP. Magnetic resonance imaging of optic neuritis in patients with neuromyelitis optica versus multiple sclerosis. J Neuroophthalmol. 2012;32(3):216–20.
55. Mealy M, Whetstone A, Orman G, Izbudak I, Calabresi PA, Levy M. Longitudinally extensive optic neuritis as an MRI biomarker distinguishers neuromyelitis optica from multiple sclerosis. J Neurol Sci. 2015;355(1–2):59–63.
56. Ratchford JN, Quigg ME, Conger A, Frohman T, Frohman E, Balcer LJ, et al. Optical coherence tomography helps differentiate neuromyelitis optica and MS optic neuropathies. Neurology. 2009;73:302–8.
57. Naismith RT, Tutlam NT, Xu J, Klawiter EC, Shepherd J, Trinkaus K, et al. Optical coherence tomography differs in neuromyelitis optica compared with multiple sclerosis. Neurology. 2009;72(12):1077–82.
58. Kitley J, Woodhall M, Waters P, Leite MI, Devenney E, Craig J, et al. Myelin-oligodendrocyte glycoprotein antibodies in adults with a neuromyelitis optica phenotype. Neurology. 79:1273–7.
59. Pache F, Zimmermann H, Mikolajczak J, Schumacher S, Lacheta A, Oertel FC, et al.; in cooperation with the Neuromyelitis Optica Study Group (NEMOS). MOG-IGG in NMO and related disorders: a multicenter study of 50 patients. Part 4: Afferent visual system damage after optic neuritis in MOG_IgG-seropositive versus AQP4-IgG-seropositive patients. J Neuroinflammation. 2016;13(1):282.
60. Cobo-Calvo A, Ruiz A, Maillart E, Audoin B, Zephir H, Bourre B, et al.; OFSEP and NOMADMUS Study Group. Clinical spectrum and prognostic value of CNS MOG autoimmunity in adults: The MOGADOR study. Neurology. 2018;90;e1858–69.
61. Ramanathan S, Mohammad S, Tantsis E, Nguyen TK, Merheb V, Fung VSC, et al.; Australasian and New Zealand MOG Study Group. Clinical course, therapeutic responses, and outcomes in relapsing MOG antibody-associated demyelination. J Neurol Neurosurg Psychiatry. 2018;89:127–37.
62. Jarius S, Ruprecht K, Kleiter I, Borisow N, Asgari N, Pitarokoili K, et al.; in cooperation with the Neuromyelitis Optica Study Group (NEMOS). MOG-IgG in NMO and related disorders: a multicenter study of 50 patients. Part 2: Epidemiology, clinical presentation, radiological and laboratory features, treatment responses, and long-term outcome. J Neuroinflammation. 2016;13(1):280.

63. Kidd D, Burton B, Plant GT, Graham EM. Chronic relapsing inflammatory optic neuropathy (CRION). Brain. 2003;126(Pt2):276–84.
64. Chi SL, Stinnett S, Eggenberger E, Foroozan R, Golnik K, Lee MS, et al. Clinical characteristics in 53 patients with cat scratch optic neuropathy. Ophthalmology. 2012;119:183–7.
65. de Seze J. Inflammatory optic neuritis: from multiple sclerosis to neuromyelitis optica. Neuroophthalmology. 2013;37(4):141–5.
66. Man BL, Mok CC, Fu YP. Neuroophthalmologic manifestations of systemic lupus erythematosus: a systematic review. Int J Rheum Di. 2014;17(5):494–501.
67. Tang W, Wei S. Primary Sjögren's syndrome related optic neuritis. Int J Ophthalmol. 2013;6(6):888–91.
68. Winthrop KL, Chen L, Fraunfelder FW, Ku JH, Varley CD, Suhler E, et al. Initiation of anti-TNF therapy and the risk of optic neuritis: from the safety assessment of biologic ThERapy (SABER) Study. Am J Ophthalmol. 2013;155(1):183–189.e1.
69. Yang S, Dizhoor A, Wilson DJ, Adamus G. GCAP1, Rab6, and HSP27: Novel Autoantibody Targets in Cancer-Associated Retinopathy and Autoimmune Retinopathy. Transl Vis Sci Technol. 2016;5(3):1.
70. Al-Louzi O, Hauptman H, Saidha S. Biopsy-negative PET-positive giant-cell arteritis. Neurology. 2014;83(18):1674–6.
71. Selmi C. Diagnosis and classification of autoimmune uveitis. Autoimmun Rev. 2014;13(4):591–4.
72. Akdal G, Toydemir H, Saactci A. Characteristics of optic neuropathy in Behçet disease. Neurol Neuroimmunol Neuroinflamm. 2018;5:e490.
73. Donaldson MJ, Pulido JS, Herman DC, Diehl N, Hodge D. Pars planitis: a 20-year study of incidence, clinical features, and outcomes. A J Ophthal. 2007;144:812–7.
74. Ortiz-Pérez S, Martínez-Lapiscina EH, Gabilondo I, Fraga-Pumar E, Martínez-Heras E, Saiz A, et al. Retinal periphlebitis is associated with multiple sclerosis severity. Neurology. 2013;81:877–81.
75. Baughman RP, Teirstein AS, Judson MA, Rossman MD, Yeager H Jr, Bresnitz EA, et al ; Case Control Etiologic Study of Sarcoidosis (ACCESS) research group. Clinical characteristics of patients in a case control study of sarcoidosis. Am J Respir Crit Care Med. 2001;164:1885–9.
76. Rush J. Retrobulbar optic neuropathy in sarcoidosis. Ann Ophthalmol. 1980;12(4):390–4.
77. Kidd DP, Burton BJ, Graham EM, Plant GT. Optic neuropath associated with systemic sarcoidosis. Neurol Neuroimmunol Neuroinflamm. 2016;3(5):e270.
78. Kupersmith MJ, Alban T, Zeiffer B, Lefton D. Contrast-enhanced MRI in acute optic neuritis: relationship to visual performance. Brain. 2002;125(4):812–22.
79. Miller DH, Newton MR, van der Poel JC, du Boulay EP, Halliday AM, Kendall BE, et al. Magnetic resonance imaging of the optic nerve in optic neuritis. Neurology. 1988;38:175–9.
80. Losseff NA, Webb SL, O'Riordan JI, Page R, Wang L, Barker GJ, et al. Spinal cord atrophy and disability in multiple sclerosis. A new reproducible and sensitive MRI method with potential to monitor disease progression. Brain. 1996;119(3):701–8.
81. Oh J, Saidha S, Chen M, Smith SA, Prince J, Jones C, et al. Spinal cord quantitative MRI discriminates between disability levels in multiple sclerosis. Neurology. 2013;80(6):540–7.
82. Hickman SJ, Brex PA, Brierley CM, Silver NC, Barker GJ, Scolding NJ, et al. Detection of optic nerve atrophy following a single episode of unilateral optic neuritis by MRI using a fat-saturated short-echo fast FLAIR sequence. Neuroradiology. 2001;43:123–8.
83. Hickman SJ, Brierley CM, Brex PA, MacManus DG, Scolding NJ, Compston DA, et al. Continuing optic nerve atrophy following optic neuritis: a serial MRI study. Mult Scler. 2002;8:339–42.
84. Hickman SJ, Toosy AT, Jones SJ, Altmann DR, Miszkiel KA, MacManus DG, et al. A serial MRI study following optic nerve mean area in acute optic neuritis. Brain. 2004;127:2498–505.
85. Trip SA, Schlottmann PG, Jones SJ, Li WY, Garway-Heath DF, Thompson AJ, et al. Optic nerve magnetization transfer imaging and measures of axonal loss and demyelination in optic neuritis. Mult Scler. 2007;13:875–9.
86. Kolappan M, Henderson AP, Jenkins TM, Wheeler-Kingshott CA, Plant GT, Thompson AJ, et al. Assessing structure and function of the afferent visual pathway in multiple sclerosis and associated optic neuritis. J Neurol. 2009;256:305–19.
87. Thorpe JW, Barker GJ, Jones SJ, Moseley I, Losseff N, MacManus DG, et al. Magnetisation transfer ratios and transverse magnetization decay curves in optic neuritis: correlation with clinical findings and electrophysiology. J Neurol Neurosurg Psychiatry. 1995;59:487–92.
88. Hickman SJ, Toosy AT, Jones SJ, Altmann DR, Miszkiel KA, MacManus DG, et al. Serial magnetization transfer imaging in acute optic neuritis. Brain. 2004;127(Pt 3):692–700.
89. Hickman SJ, Wheeler-Kingshott CA, Jones SJ, Miszkiel KA, Barker GJ, Plant GT, et al. Optic nerve diffusion measurement from diffusion-weighted imaging in optic neuritis. AJNR Am J Neuroradiol. 2005;26(4):951–6.
90. Trip SA, Wheeler-Kingshott C, Jones SJ, Li WY, Barker GJ, Thompson AJ, et al. Optic nerve diffusion tensor imaging in optic neuritis. NeuroImage. 2006;30:498–505.
91. Naismith RT, Xu J, Tutlam NT, Snyder A, Benzinger T, Shimony J, et al. Disability in optic neuritis correlates with diffusion tensor-derived directional diffusivities. Neurology. 2009;72(7):589–94.

92. Naismith RT, Xu J, Tutlam NT, Lancia S, Trinkaus K, Song SK, et al. Diffusion tensor imaging in acute optic neuropathies: predictor of clinical outcomes. Arch Neurol. 2012;69(1):65–71.
93. Halliday AM, McDonald WI, Mushin J. Delayed visual evoked response in optic neuritis. Lancet. 1972;1(7758):982–5.
94. Hood DC, Odel JG, Winn BJ. The multifocal visual evoked potential. J Neuroophthalmol. 2003;23:279–89.
95. Fraser CL, Klistorner A, Graham SL, Garrick R, Billson FA, Grigg JR. Multifocal visual evoked potential analysis of inflammatory or demyelinating optic neuritis. Ophthalmology. 2006;113(2):323.e1–2.
96. Fraser C, Klistorner A, Graham S, Garrick R, Billson F, Grigg J. Multifocal visual evoked potential analysis: predicting progression to multiple sclerosis. Arch Neurol. 2006;63(6):847–50.
97. Alshowaeir D, Yannikas C, Garrick R, Van Der Walt A, Graham SL, Fraser C, et al. Multifocal VEP assessment of optic neuritis evolution. Clin Neurophysiol. 2015;126(8):1617–23.
98. Parisi V, Manni G, Spadaro M, Colacino G, Restuccia R, Marchi S, et al. Correlation between morphological and functional retinal impairment in multiple sclerosis patients. Invest Ophthalmol Vis Sci. 1999;40(11):2520–7.
99. Henderson AP, Trip SA, Schlottmann PG, Altmann DR, Garway-Heath DF, Plant GT, et al. An investigation of the retinal nerve fiber layer in progressive multiple sclerosis using optical coherence tomography. Brain. 2008;131(Pt 1):277–87.
100. Bock M, Brandt AU, Dörr J, Kraft H, Weinges-Evers N, Gaede G, et al. Patterns of retinal nerve fiber layer loss in multiple sclerosis patients with or without optic neuritis and glaucoma patients. Clin Neurol Neurosurg. 2010;112(8):647–52.
101. Costello F, Hodge W, Pan YI, Eggenberger E, Coupland S, Kardon RH. Tracking retinal nerve fiber layer loss after optic neuritis: a prospective study using optical coherence tomography. Mult Scler. 2008;14(7):893–905.
102. Costello F. Evaluating the use of optical coherence tomography in optic neuritis. Mult Scler Int. 2011;2011:148394.
103. Petzold A, de Boer JF, Schippling S, Vermersch P, Kardon R, Green A, et al. Optical coherence tomography in multiple sclerosis: a systematic review and meta-analysis. Lancet Neurol. 2010;9(9):921–32.
104. Bock M, Brandt AU, Kuchenbecker J, Dörr J, Pfueller CF, Weinges-Evers N, et al. Impairment of contrast visual acuity as a functional correlate of retinal nerve fibre layer thinning and total macular volume reduction in multiple sclerosis. Br J Ophthalmol. 2012;96(1):62–7.
105. Villoslada P, Cuneo A, Gelfand J, Hauser SL, Green A. Color vision is strongly associated with retinal thinning in multiple sclerosis. Mult Scler. 2012;18(7):991–9.
106. Walter SD, Ishikawa H, Galetta KM, Sakai RE, Feller DJ, Henderson SB, et al. Ganglion cell loss in relation to visual disability in multiple sclerosis. Ophthalmology. 2012;119(6):1250–7.
107. Fisher JB, Jacobs DA, Markowitz CE, Galetta SL, Volpe NJ, Nano-Schiavi ML, et al. Relation of visual function to retinal nerve fiber layer thickness in multiple sclerosis. Ophthalmology. 2006;113(2):324–32.
108. Martinez-Lapiscina EH, Arnow S, Wilson JA, Saidha S, Preiningerova JL, Oberwahrenbrock T, et al; IMSVISUAL consortium. Retinal thickness measured with optical coherence tomography and risk of disability worsening in multiple sclerosis: a cohort study. Lancet Neurol. 2016;15(6):574–84.
109. Ikuta F, Zimmerman HM. Distribution of plaques in seventy autopsy cases of multiple sclerosis in the United States. Neurology. 1976;26:26–8.
110. Gordon-Lipkin E, Chodkowski B, Reich DS, Smith SA, Pulicken M, Balcer LJ, et al. Retinal nerve fiber layer is associated with brain atrophy in multiple sclerosis. Neurology. 2007;69(16):1603–9.
111. Siger M, Dziegielewski K, Jasek L, Bieniek M, Nicpan A, Nawrocki J, et al. Optical coherence tomography in multiple sclerosis: thickness of the retinal nerve fiber layer as a potential measure of axonal loss and brain atrophy. J Neurol. 2008;255(10):1555–60.
112. Frohman EM, Dwyer MG, Frohman T, Cox JL, Salter A, Greenberg BM, et al. Relationship of optic nerve and brain conventional and non-conventional MRI measures and retinal nerve fiber layer thickness, as assessed by OCT and GDx: a pilot study. J Neurol Sci. 2009;282(1–2):96–105.
113. Saidha S, Syc SB, Durbin MK, Eckstein C, Oakley JD, Meyer SA, et al. Visual dysfunction in multiple sclerosis correlates better with optical coherence tomography derived estimates of macular ganglion cell layer thickness than peripapillary retinal nerve fiber layer thickness. Mult Scler. 2011;17(12):1449–63.
114. Saidha S, Syc SB, Ibrahim MA, Eckstein C, Warner CV, Farrell SK, et al. Primary retinal pathology in multiple sclerosis as detected by optical coherence tomography. Brain. 2011;134(Pt 2):518–33.
115. Syc SB, Saidha S, Newsome SD, Ratchford JN, Levy M, Ford E, et al. Optical coherence tomography segmentation reveals ganglion cell layer pathology after optic neuritis. Brain. 2011;135(Pt 2):521–33.
116. Ratchford JN, Saidha S, Sotirchos ES, Oh JA, Seigo MA, Eckstein C, et al. Active MS is associated with

accelerated retinal ganglion cell/inner plexiform layer thinning. Neurology. 2013;80(1):47–54.
117. Saidha S, Al-Louzi O, Ratchford JN, Bhargava P, Oh J, Newsome SD, et al. Optical coherence tomography reflects brain atrophy in multiple sclerosis: a four-year study. Ann Neurol. 2015;78(5):801–13.
118. Saidha S, Sotirchos ES, Oh J, Syc SB, Seigo MA, Shiee N, et al. Relationships between retinal axonal and neuronal measures and global central nervous system pathology in multiple sclerosis. JAMA Neurol. 2013;70(1):34–43.
119. Button J, Al-Louzi O, Lang A, Bhargava P, Newsome SD, Frohman T, et al. Disease-modifying therapies modulate retinal atrophy in multiple sclerosis: a retrospective study. Neurology. 2017;88(6):525–32.
120. Tsakiri A, Kallenbach K, Fuglø D, Wanscher B, Larsson H, Frederiksen J. Simvastatin improves final visual outcome in acute optic neuritis: a randomized study. Mult Scler. 2012;18(1):82–1.
121. Sühs KW, Hein K, Sättler MB, Görlitz A, Ciupka C, Scholz K, et al. A randomized, double-blind, phase 2 study of erythropoietin in optic neuritis. Ann Neurol. 2012;72(2):199–210.
122. Raftopoulos R, Hickman SJ, Toosy A, Sharrack B, Mallik S, Paling D, et al. Phenytoin for neuroprotection in patients with acute optic neuritis: a randomised, placebo-controlled, phase 2 trial. Lancet Neurol. 2016;15(3):259–69.
123. Green AJ, Gelfand JM, Cree BA, Bevan C, Boscardin WJ, Mei F, et al. Clemastine fumarate as a remyelinating therapy for multiple sclerosis (ReBUILD): a randomized, controlled, double-blind crossover trial. Lancet. 2015;390(10111):2481–9.
124. Cadavid D, Balcer L, Galetta S, Aktas O, Ziemssen T, Vanopdenbosch L, et al. Safety and efficacy of opicinumab in acute optic neuritis (RENEW): a randomised, placebo-controlled, phase 2 trial. Lancet. 2017;16(3):189–99.

# Clinical Approach to Pediatric Demyelinating Disease

# 29

Jonathan Douglas Santoro and Tanuja Chitnis

## Introduction

Demyelinating disorders in children are a complex and heterogeneous pathologic group. Differentiating between these disorders can often be difficult on initial evaluation as acute demyelinating events can be either monophasic (such as in acute demyelinating encephalomyelitis) or an initial presentation of a polyphasic/recurrent disorder, such as observed in multiple sclerosis (MS) or neuromyelitis optica spectrum disorder (NMOSD). Clinical history, physical examination, and neurodiagnostic studies are critical in the appraisal of demyelinating disease in children and are reviewed here.

J. D. Santoro (✉)
Pediatric Multiple Sclerosis and Demyelinating Disorders, Division of Pediatric Neurology, Department of Neurology, Massachusetts General Hospital, Department of Neurology, Harvard Medical School, Boston, MA, USA

Department of Pediatrics, Keck School of Medicine at USC, Los Angeles, CA, USA
e-mail: jdsantoro@chla.usc.edu

T. Chitnis
Pediatric Multiple Sclerosis and Demyelinating Disorders, Division of Pediatric Neurology, Department of Neurology, Massachusetts General Hospital, Department of Neurology, Harvard Medical School, Boston, MA, USA

## Acute Disseminated Encephalomyelitis

**Patient Vignette 1**

A 3-year-old, right-handed female presented to an urgent care facility for subacute onset of irritability and drowsiness over the last 12 hours. Her parents reported that between periods of drowsiness, the patient was aggressive (biting and hitting) and began slurring her speech 4 hours prior to seeking care. The patient's family noted that 7 days prior to the onset of symptoms, the patient had a mild upper respiratory tract infection that necessitated only supportive care.

On evaluation, neurologic examination demonstrated a somnolent and irritable child. Cranial nerve examination demonstrated no abnormalities nor was there meningismus present. There were some localizing features on the left including a motor exam with asymmetric movements, extensor plantar response, and ataxia with reaching for objects most prominent on the left. The patient was unable to walk and refused to stand independently.

Serum screening revealed normal chemistry, complete blood count (CBC), and liver function tests (LFTs). Lumbar puncture revealed a white blood cell (WBC) count of 14/mm$^3$ (range 0–5/mm$^3$) with 96% lymphocytes, protein of 51 mg/dL (range 20–45 mg/dL), and glucose of 55 mg/dL (range 40–80 mg/dL), and red blood cells were not present. The patient had a normal immunoglobulin G (IgG) index and no oligoclonal



A. L. Piquet, E. Alvarez (eds.), *Neuroimmunology*, https://doi.org/10.1007/978-3-030-61883-4_29

bands. Additional infectious testing was unremarkable.

Neuroimaging included magnetic resonance imaging (MRI) with and without contrast of the brain and C-spine. T2 imaging demonstrated multifocal areas of high signal in the deep gray matter, periventricular white matter bilaterally, and left midline cerebellum (Fig. 29.1). Diffusion weighted imaging (DWI) demonstrated minimal patchy peripheral restricted diffusion in the right thalamic lesion and the left cerebellar lesion. T1 with gadolinium demonstrated both prominent enhancement in cerebral hemispheres and scant enhancement in the left cerebellar lesion. Imaging of the orbits and cervical spine demonstrated no abnormalities, and magnetic resonance angiography (MRA) sequencing was similarly negative.

A diagnosis of acute disseminated encephalomyelitis (ADEM) was made. The patient received a 3-day pulse of intravenous (IV) methylprednisolone (MP) (30 mg/kg/d). The patient improved dramatically over 2 days, first with regard to slurred speech and ataxia and subsequently weakness and somnolence/irritability. The patient was able to ambulate fully and was at 95% of baseline per parents on day 3 of hospitalization. She was discharged home without further disease-modifying treatment.

Follow-up neuroimaging was obtained at 6 months after onset of symptoms and was stable. The patient was symptom-free with no relapses following diagnosis and has not developed any other immune or neurologic disease for a period of 4 years.

## Diagnostic Discussion

The clinical features of ADEM are variable and polyfocal at onset. Encephalopathy (defined as any alteration in mental state or behavior ranging from coma to irritability) is required for the diagnosis. Other neurologic features include acute hemiparesis, cerebellar ataxia, cranial neuropathies, optic neuritis (ON; identified by vision loss, pain with eye movement, and afferent pupillary defect), pyramidal dysfunction, and spinal cord dysfunction (known as transverse myelitis/myelopathy) [1–3]. Movement disorders, lan-

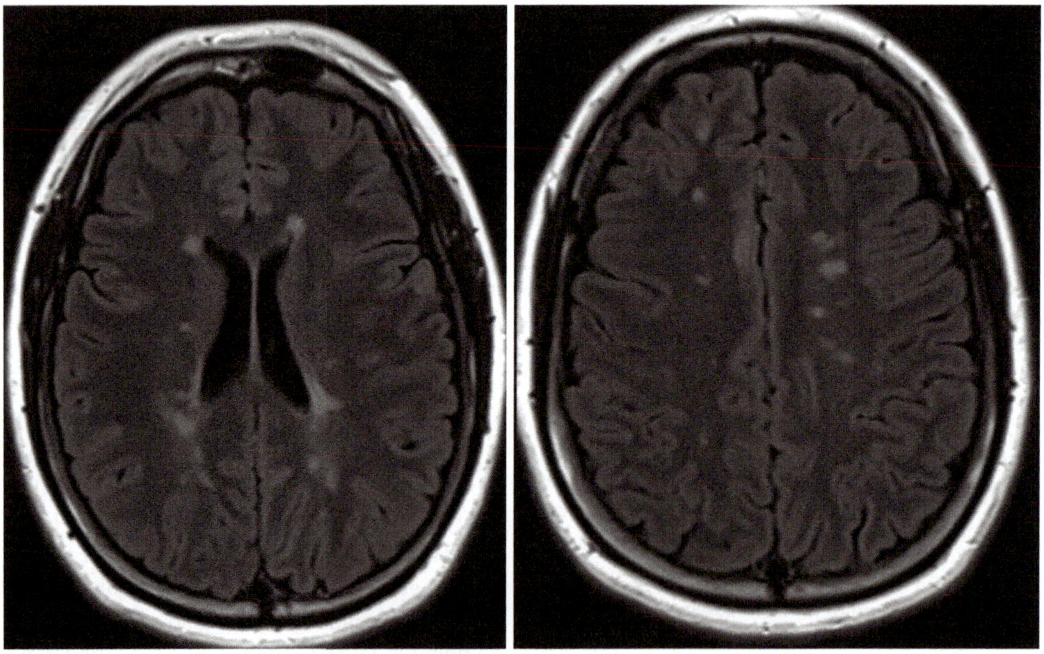

**Fig. 29.1** Axial T2 fluid-attenuated inversion recovery (FLAIR) sequence demonstrating multiple white matter lesions

guage disorders, and sensory disturbances are appreciated with less frequency. Associated non-neurologic findings include fever, meningismus, nausea/vomiting, and generalized weakness.

Preceding the onset of symptoms, up to 72% of patients will have a non-life-threatening viral illness 3–21 days prior [1, 2]. A minority of cases (5%) are associated with the administration of vaccines including, but not limited to, hepatitis B, influenza, diphtheria/pertussis/tetanus, measles/mumps/rubella, pneumococcus, and polio [4].

The diagnosis of ADEM is based on history, clinical, and radiologic findings although no one diagnostic test is considered confirmatory. The Pediatric Multiple Sclerosis Study Group generated criteria in 2007 [5] and 2012 [6] for the diagnosis of ADEM (Table 29.1).

Typically, clinical course follows a monophasic pattern with steady improvement after treatment back to neurologic baseline within the 3 months after the onset of symptoms. Most patients are clinically improved by 2–4 weeks after presentation [2].

The differential diagnosis in a child who presents with polyfocal neurologic abnormalities and encephalopathy is broad. Viral and bacterial meningitis/encephalitis must be ruled out, especially in the presence of fever and/or leukocytosis. Lumbar puncture is critical in the diagnostic workup. After ruling out infection and in the setting of white matter lesions on MRI of the brain, other inflammatory demyelinating disorders that should be considered include MS, neuromyelitis optica spectrum disorder (NMOSD), anti-myelin oligodendrocyte glycoprotein (MOG) antibody disease, autoimmune encephalitis, primary angiitis of the central nervous system (CNS), malignancy, mitochondrial disease, and leukodystrophies.

**Table 29.1** ADEM diagnostic criteria

| Clinical features (all required) | Characteristics of brain MRI lesions |
|---|---|
| A first polyfocal, clinical central nervous system event with presumed inflammatory demyelinating cause | Diffuse, poorly demarcated, large (>1–2 cm) lesions involving predominantly the cerebral white matter |
| Encephalopathy that cannot be explained by fever, systemic illness, or postictal symptoms | Deep gray matter lesions (e.g., involving the basal ganglia or thalamus) can be present |
| No new clinical and MRI findings emerge 3 months after the onset | T1 hypointense lesions in the white matter are rare |
| Brain MRI is abnormal during the acute (3-month) phase | |

*ADEM* acute demyelinating encephalomyelitis, *MRI* magnetic resonance imaging
Reprinted with permission from Krupp et al. [6]

Autoimmune encephalitis (see Chap. 12) may be difficult to differentiate from ADEM in children [7]. The most well-described entity, N-methyl-D-aspartic acid (NMDA) receptor antibody encephalitis, can be differentiated from ADEM by features such as the presence of movement disorders, epilepsy, waxing and waning behavioral and personality changes, autonomic instability, language disturbances, and MRI lesions (70% of MRIs can be normal in NMDA encephalitis in children) [8]. Other inflammatory and autoimmune conditions should be worked up if there are atypical clinical histories, neurologic examinations, or neuroimaging in the setting of subacute onset encephalopathy in children.

Relapsing and remitting disorders such as MS may initially present with symptoms consistent with ADEM [9, 10]. Distinguishing ADEM from a primary presentation of MS can be challenging. Clinical and neuroimaging features that may indicate a diagnosis of ADEM rather than MS are listed in Table 29.2 [4, 11–15]. In a retrospective study of neuroimaging in pediatric patients diagnosed with ADEM and then MS, Callen et al. identified three factors that predicted a higher likelihood of developing MS: ≥2 periventricular lesions, presence of T1 black holes, and absence of diffuse bilateral lesion restricted diffusion [12]. ADEM can present with both optic neuritis and myelitis, making it difficult in some circumstances to distinguish from NMOSD. Longitudinally extensive spinal lesions (>3 vertebrae), presence of aquaporin-4 (AQP4) antibody, and lack of clear encephalopathy may be helpful in differentiating these two syndromes.

Other etiologies that can mimic ADEM are best ruled out by history and laboratory workup. Malignancy, leukodystrophy, and primary angiitis of the nervous system tend to have more pro-

**Table 29.2** Indicators of a diagnosis of ADEM rather than MS

| Clinical [4, 11, 15] | Neuroimaging [11, 13, 15] |
|---|---|
| Ataxia (rare in MS) | Multiple, large, confluent lesions in asymmetric distributions throughout the white matter |
| Fever and meningismus | Lesions are poorly demarcated (MS lesions have more defined margins) |
| Prodromal viral illness or recent vaccination | Multiple lesions of the same age indicate ADEM rather than MS, which can have multiple lesions of various ages (active/inactive lesions) |
| Encephalopathy or polyfocal neurologic findings | Thalamic and deep grey lesions occur less frequently in MS |
| Absence of oligoclonal bands in CSF | |

*ADEM* acute demyelinating encephalomyelitis, *CSF* cerebrospinal fluid, *MS* multiple sclerosis

longed courses and neuroimaging that is specific to the etiology. Mitochondrial diseases can mimic many neurologic processes and can be triggered by infection and/or caloric insufficiency, but tend to have neuroimaging patterns, wherein symmetric deep gray matter lesions predominate.

Anti-MOG antibody-associated pathology is clinically heterogeneous and should be evaluated in all patients presenting with ADEM; this is discussed in more detail later in this chapter.

## Treatment of ADEM

The mainstay treatment of ADEM is IV methylprednisolone (IVMP), typically administered as 30 mg/kg/d (max 1000 mg/d) for 3–5 days [16]. Some centers will also provide an oral steroid taper for 4–6 weeks, but there is no data to support this practice. For refractory cases, intravenous immunoglobulin (IVIg) (typically administered as 0.4 g/kg for 5 days) and plasma exchange have been demonstrated to improve patients' clinical outcomes after failure of glucocorticoids, although the latter is limited to adult data [17, 18].

Prognosis associated with ADEM is excellent in most cases [19]. In cases that present with optic neuritis, this was a determinant of positive outcomes, although the opposite is observed in cases of ADEM-ON with positive anti-MOG antibody where visual outcomes are guarded [20]. Repeating neuroimaging, even in relatively straightforward cases 3–6 months after presentation, can be instrumental in risk stratification for patients with regard to developing MS at a later time point.

## Pediatric Multiple Sclerosis

### Patient Vignette 1, Part B

The same patient as presented previously does well off immunotherapy for a period of 4 years. In the absence of inflammatory triggers, she has a subsequent relapse at the age of 7 years old, consisting of right optic neuritis and right hemibody weakness. MRI at the time revealed new contrast-enhancing lesions in addition to prior T2 periventricular and juxtacortical lesions (Figs. 29.2 and 29.3a, b). Spinal imaging revealed a small enhancing lesion in the central cervical cord (C6-C7). Cerebrospinal fluid (CSF) revealed 14/mm$^3$ WBCs (range 0–5/mm$^3$), four oligoclonal bands, and an elevated IgG index of 1.10 (range 0–0.55). The patient was

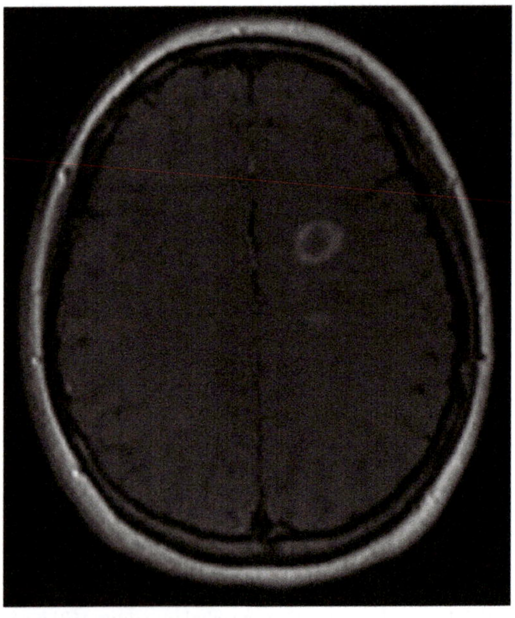

**Fig. 29.2** Axial T1 post-contrasted image demonstrating a characteristic ovoid, ring- enhancing lesion

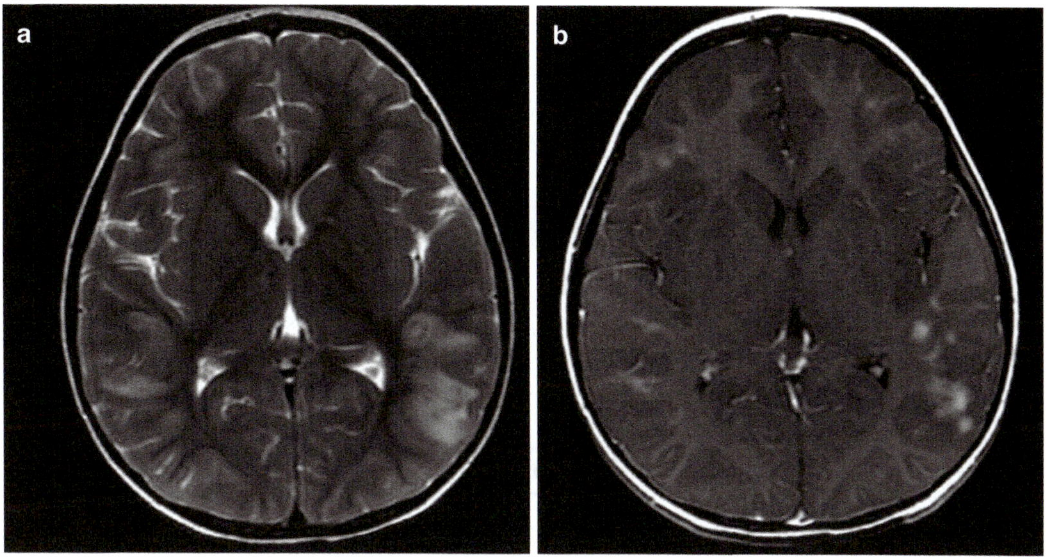


**Fig. 29.3** (**a**) Axial T2 demonstrate multiple large confluent hyperintensities in bilateral hemispheres. (**b**) Axial T1 post-contrast image demonstrating patchy gadolinium enhancement in both hemispheres

treated with IVMP (30 mg/kg/d) with improvement of symptoms over 5 days. The patient was diagnosed with pediatric onset MS in the setting of dissemination of disease in both time and space and she was started on interferon beta-1a.

The patient remained symptom-free on interferon monotherapy for 3 years with stable interval MRIs every 6 months, but at age 10, she had a third relapse consisting of right optic neuritis, right lower extremity weakness, and left hemibody sensory loss. MRI revealed an increase in the number and extent of patchy, confluent T2 fluid-attenuated inversion recovery (FLAIR), contrast-enhancing lesions of the cortex, and subcortical white matter. Characteristic "incomplete ring" contrast-enhancing lesions were present bilaterally. Brain MRA and spine MRI were normal at this time. She improved with IVMP and was discharged home on an oral taper. Immediately after finishing the steroid taper, her symptoms returned. She was readmitted again and a second 5-day course of IVMP (30 mg/kg/d) was administered, although this yielded minimal improvement. She was subsequently switched to plasmapheresis. Over the span of 2 weeks, she improved slowly. She was discharged at 80% of her neurologic baseline and continued to make a slow recovery. Anti-MOG antibody and AQP-4 antibodies have remained negative on repeat testing.

The patient was initiated on natalizumab (200 mg/month) following her prolonged hospitalization, although at her initial lab workup she was found to be John Cunningham (JC) virus positive. With these findings, the decision was made to use natalizumab for a period of 1 year and then consider alternative disease-modifying therapy (DMT). She did well for 1 year with no relapses and then was transitioned at 13 months to rituximab with no further relapses.

## Diagnostic Discussion

MS is an idiopathic inflammatory disorder characterized by demyelination of the CNS. The majority of those affected are adults, but pediatric presentations account for up to 5% of all cases [9]. Due to the infrequency of this disease and that time is often required for the diagnosis of this disorder, there can be a lag between diagnosis and the presenting neurologic event. In general, the first clinical presentation of MS can be varied depending on which portions of the CNS or spinal cord are involved. Clinical features that may be suggestive of a patient with an initial demyelinating event who will go on to develop MS include personal history of optic neuritis, absence of preceding infection, absence of encephalopathy, and absence of fever and/or meningismus.

Young children (<11 years) have been noted to have more heterogenous presentations of MS. This population is more likely to have preceding infection, optic nerve dysfunction, cognitive and behavioral issues, seizures, and brainstem/cerebellar findings on initial and subsequent evaluation in contrast to older children [10, 21]. This can make differentiation from other demyelinating events of the CNS such as encephalitis, meningitis, and ADEM considerably more difficult. Additionally, demyelinating entities such as ADEM are much more prevalent in younger children than MS, which can skew differential diagnoses during primary presentations.

Differentiation of disorders at the time of an initial demyelinating event based on clinical presentation is complex. Several factors involved in evaluating an initial demyelinating event include determining if the presentation is unifocal or polyfocal, what system within the neuroaxis is/are involved, and if encephalopathy is present (Fig. 29.4) [22]. Of note, classifying an event of

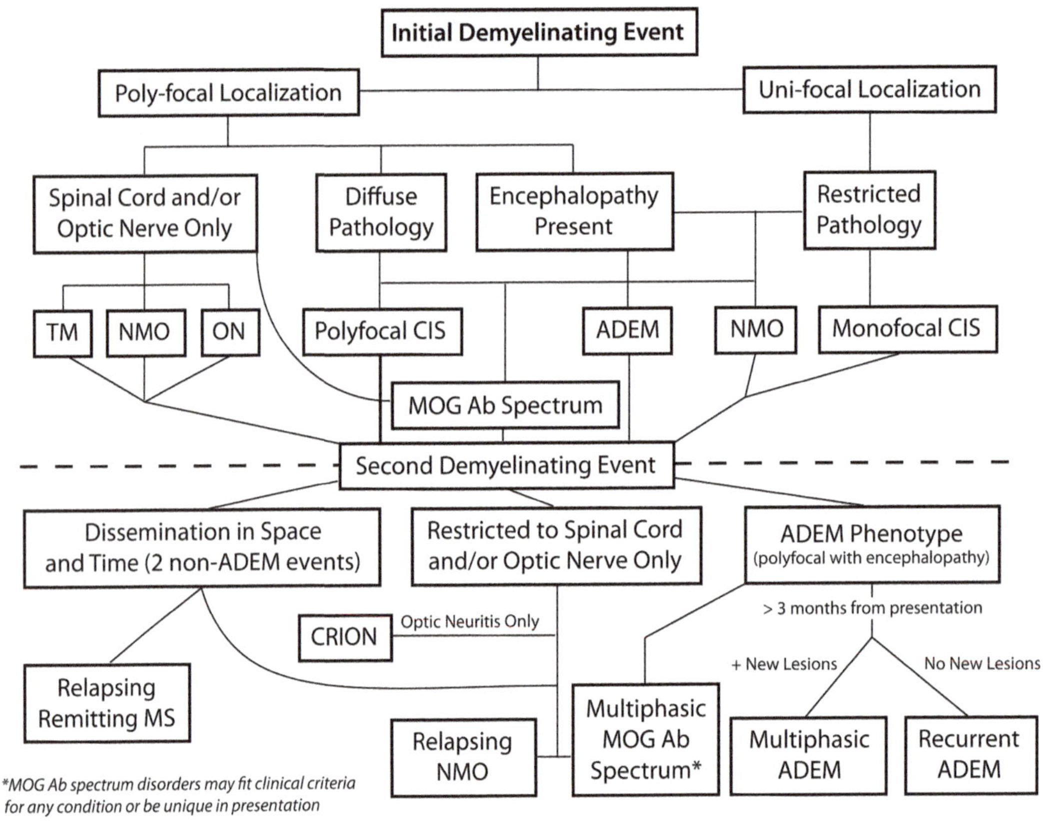


**Fig. 29.4** Evaluation of initial and secondary demyelinating events in children. Diffuse pathology: multiple demyelinating lesions in areas in the central nervous system (CNS), which may or may not be symptomatic
Polyfocal localization: indicates more than one nervous system location is involved with symptomatic presentations that include optic neuritis, transverse myelitis, encephalopathy, weakness, sensory deficits, and so on
Recurrent/multiphasic ADEM: term utilized for recrudescence of prior symptoms in the absence of new lesions as opposed to a relapsing course, which is considered multiphasic ADEM when encephalopathy is the primary phenotypic presentation at second demyelinating event
Restricted pathology: isolated demyelinating lesions in area of the CNS, which accounts for the symptoms at presentation
Unifocal localization: indicates only one of these entities is present in isolation (e.g., transverse myelitis)
*Abbreviations: ADEM* acute demyelinating encephalomyelitis, *CIS* clinically isolated syndrome, *CRION* chronic relapsing inflammatory optic neuropathy, *MOG* myelin oligodendrocyte glycoprotein, *MS* multiple sclerosis, *NMO* neuromyelitis optica, *ON* optic neuritis, *TM* transverse myelitis

initial demyelination can be helpful although should the clinical course become polyphasic, re-evaluation of the initial event in the context of the new secondary event is necessary. Secondary events should be stratified by if dissemination in time or space has occurred and if the disease remains limited to the spinal cord and/or optic nerve only on neuroimaging (Fig. 29.5). The concept of recurrent ADEM or multiphasic ADEM is being revised in the context of anti-MOG antibody spectrum disorders, and while both terms are still utilized, future research will likely elucidate the concept of relapsing ADEM in the setting of anti-MOG antibody disease.

MS is ultimately a clinical diagnosis, although neuroimaging is intimately involved in diagnostic practice. The diagnostic criteria for pediatric MS are based on the 2017 McDonald criteria (Table 29.3) [23]. Neuroimaging is an important ancillary study in the evaluation of demyelinating events in children. Abnormalities tend to be in four locations (periventricular, juxtacortical, infratentorial, and spinal cord) and as per the 2017 revised McDonald criteria, two of four locations are needed for diagnosis of MS [23]. Several studies examining imaging differences in pediatric MS have been completed. Callen et al. demonstrated that the criteria most useful for differentiating a first attack of MS from monophasic ADEM included absence of diffuse bilateral T2 pattern, presence of T1 "black holes," and ≥2 T2 periventricular lesions on initial imaging [12]. These findings yielded a positive predictive value of 95% and a negative predictive value of 79%. In a separate study, also by Callen et al., the group identified that having two of the following was predictive of a diagnosis of MS as opposed to other relapsing non-demyelinating disorders: >5 T2 lesions, >2 T2 periventricular lesions, and >1 T2 brainstem lesion [12]. Additionally, imaging involving the spinal cord in MS typically has lesions <3 vertebral segments in length, whereas NMOSD typically will be more extensive (and more classically associated with optic neuritis) [6]. Atypical neuroimaging features for MS include leptomeningeal enhancement (differential includes infection, neurosarcoidosis, vasculitis, hemophagocytic lymphohistiocytosis, or malignancy), multiple ring enhancing lesions (differential includes CNS tuberculosis, abscesses, toxoplasmosis, cysticercosis, fungal infection, or CNS lymphoma), increased size of lesions on serial imaging (differential includes CNS tumor, progressive multifocal leukoencephalopathy [PML], CNS lymphoma, or neurosarcoidosis), and hemorrhage (differential includes stroke, hemorrhagic ADEM variant, CNS vasculitis, or acute necrotizing encephalopathy) [24].

**Table 29.3** Diagnostic criteria for pediatric multiple sclerosis (MS)

| *Diagnostic criteria for pediatric multiple sclerosis* |
|---|
| Two or more non-encephalopathic CNS clinical events separated by more than 30 days, involving more than one area of the CNS |
| Single clinical event and MRI features rely on 2017 revised McDonald criteria for dissemination in time and space and in which a follow-up MRI shows at least one new enhancing or non-enhancing lesion consistent with dissemination in time criteria [23] |
| ADEM followed 3 months later by a non-encephalopathic clinical event with new lesions on brain MRI consistent with MS |
| A first, single, event that does not meet ADEM criteria and whose MRI findings are consistent with the 2017 revised McDonald criteria for dissemination in time and space |

*ADEM* acute demyelinating encephalomyelitis, *CNS* central nervous system, *MRI* magnetic resonance imaging

Although no definitive lab study is used in the diagnosis of MS, some biomarkers are used to support diagnosis. The 2017 revised McDonald criteria do not require additional lab studies when more than two attacks and more than two lesions on neuroimaging are present or in patients with more than two attacks implicating different CNS sites with at least one lesion on neuroimaging unless red flags are present. Patients with one attack can meet dissemination in space with testing of positive CSF-specific oligoclonal bands. The presence of oligoclonal bands is noted in up to 92% of children with definitive MS, although very young patients and those early in the disease course may be negative initially [25]. This can be useful in differentiating MS from other demyelinating entities such as NMOSD and ADEM that have significantly lower incidences of oligoclonal bands [15, 26].

Clinical, neuroimaging, and lab studies are of high yield in the assessment and diagnosis of MS, but mimics should also be considered during the initial workup and evaluation of the disease. Mimics to consider include bacteria (*Listeria monocytogenes*, *Brucella spp.*, *Borrelia burgdorferi, Bartonella henselae, Mycobacterium tuberculosis, Mycoplasma spp.*, and *Tropheryma whipplei*), viral infection (cytomegalovirus [CMV], varicella [VZV], enterovirus, herpes simplex virus [HSV], hepatitis C virus, human herpes virus type 6 [HHV-6], human immunodeficiency virus [HIV], human T-lymphotropic virus-1 [HTLV-1], and John Cunningham [JC] virus), fungus (*Histoplasma capsulatum*), and parasites/spirochetes (*Toxoplasma gondii*, *Taenia solium*, *Leptospira spp.*, and *Treponema pallidum*). Testing for these agents should be based on endemic patterns of disease, clinical history, and neuroimaging findings (Table 29.4).

**Table 29.4** Infectious mimics of pediatric multiple sclerosis (MS) and demyelinating disease

| Infectious agent | Clinical indicators of infection |
|---|---|
| **Bacterial infections** | |
| *Bartonella henselae* | Caused by scratches from cats/kittens. Associated with lymphadenopathy. |
| *Borrelia burgdorferi* | Caused by Ixodes tick bite in endemic areas of the United States. Associated with erythema chronicum migrans and "target lesions." Presentation may include multiple cranial neuropathies. |
| *Brucella spp.* | Transmitted through consumption of contaminated foods or close contact with livestock. Associated with cyclical fever and foul-smelling perspiration. Severe fatigue may be a presenting complaint. |
| *Listeria monocytogenes* | Transmitted through consumption of dairy and deli food. Associated with brainstem presentations (rhombencephalitis) and cranial neuropathy. |
| *Mycobacterium tuberculosis* | Common infectious agent spread by direct contact in endemic areas. HIV infection and immunosuppression may increase risk. Associated with brainstem presentations (rhombencephalitis), cranial neuropathies, and multiple ring enhancing lesions on neuroimaging (tuberculomas). |
| *Mycoplasma spp.* | Transmitted through respiratory contact. Associated with fatigue and mild–moderate respiratory complaints or pneumonia. |
| *Tropheryma whipplei* | Wide variety of symptoms but prominent GI complaints including abdominal pain, bloating sensation, constipation/diarrhea (mimicking irritable bowel syndrome), and low-grade fever. Patients may also have concurrent arthritis, uveitis, or spondylodiscitis. Very rare. |
| **Viral infections** | |
| CMV | Transmitted through direct contact or vertical transmission. Associated with periventricular lesions. |
| Enterovirus | Transmitted through direct contact. Typically induces mild illness but associated with acute flaccid myelitis of the upper spine (cervical). |
| Hepatitis C | Transmitted through sexual contact, IV drug usage, and vertical transmission. Variable symptoms based on duration of disease. |
| HHV-6 | Ubiquitous by age 2–3 years and cause of roseola in childhood. Reactivation most commonly seen in severely immunocompromised (stem cell transplant). Associated with encephalopathy, memory deficits, and seizures. |
| HIV/AIDS | Transmitted through IV drug use and sexual contact. Neurologic symptoms are variable and may include central and peripheral neurologic complaints. Increases risk for other opportunistic infections. |
| HSV 1/2 | Transmitted through direct contact or sexual contact. Associated with meningismus, fever, and seizures. Course is rapid and fatal if untreated. |
| HTLV-1 | Transmitted through sexual contact or breastfeeding and endemic in South America. Associated with progressive spastic paraparesis and may demonstrate atrophy of the thoracic/lumbar spinal cord on neuroimaging. |
| JC virus | Nearly ubiquitous virus. Associated with PML (diffuse and confluent lesions of the white matter) in persons on disease-modifying therapy or immunosuppression. Increased risk in persons already on natalizumab (Tysabri). |

**Table 29.4** (continued)

| Infectious agent | Clinical indicators of infection |
|---|---|
| VZV | Transmitted by direct contact. Virus typically indolent but reactivation may cause shingles, cranial neuropathy, or autonomic dysfunction. |
| **Fungal infections** | |
| *Histoplasma capsulatum* | Transmitted by direct contact with bat or bird excrement. Associated with pulmonary issues. Common among patients with HIV/AIDS and immunosuppressed. |
| *Parasitic infections* | |
| *Leptospira spp.* | Transmitted via contact with contaminated water, thus endemic to areas with high rainfall. Associated with renal failure and pulmonary hemorrhage. |
| *Treponema pallidum* | Transmitted through sexual contact. Symptoms vary depending on duration of infection but can include posterior column syndrome, Marcus Gunn pupil, optic neuritis, and encephalopathy/memory deficits. |
| *Taenia solium* | Transmitted fecal-oral. Associated with living in areas where pigs are present. Associated with epilepsy and neuroimaging may demonstrate multiple small cysticerci with or without scolex within. |
| *Toxoplasma gondii* | Transmitted through eating poorly cooked food but also associated with exposure to cats. Associated with influenza-like symptoms in acute phase with lymphadenopathy but may also cause encephalitis, retinitis, or seizures in acute or chronic phase. |

*AIDS* acquired immunodeficiency syndrome, *CMV* cytomegalovirus, *GI* gastrointestinal, *HHV-6* human herpes virus type 6, *HIV* human immunodeficiency virus, *HSV* herpes simplex virus, *HTLV-1* human T-lymphotropic virus-1, *IV* intravenous, *JC virus* John Cunningham virus, *PML* progressive multifocal leukoencephalopathy, *VZV* varicella

## Treatment Discussion

Treatment of pediatric MS is limited with regard to evidence-based interventions although many clinical trials, specifically in age groups <18-year-old, are underway. At this time there are no pediatric-specific disease-modifying therapies (DMT), with nearly all agents utilized being repurposed from adult clinical trials. Fingolimod is the only US Food and Drug Administration (FDA)-approved DMT for pediatric MS [27]. First-generation DMTs include interferon-beta and glatiramer acetate. Second-generation therapies include natalizumab, rituximab, ocrelizumab, and dimethyl fumarate, although consensus guidelines in the pediatric population are lacking. Medications with more serious adverse reactions such as teriflunomide, alemtuzumab, and cyclophosphamide are utilized less frequently and are often reserved for refractory cases. Therapy should be tailored to the patient with tolerability of side effects, compliance with modality of administration, and risk of immune suppression seriously considered [28]. General concepts in the approach to the treatment of MS in regard to DMT selection is further discussed in Chap. 31.

## Anti-Myelin Oligodendrocyte Glycoprotein (MOG) Antibody Spectrum Disorders

**Patient Vignette 2**

The patient is a 16-year-old, right-handed female who initially presented with subacute left optic neuritis with no precipitating factors. On evaluation, a left afferent pupillary defect was noted along with mild (grade 2) papilledema, but no extraocular muscle or other neurologic deficits. Neuroimaging at the time demonstrated gadolinium enhancement of the left optic nerve, but no evidence of other white matter lesions nor cervical spine pathology. The patient received a 3-day course of IVMP (1 g/d) with resolution of symptoms and improvements in her vision back to baseline.

Roughly 4 weeks later, the patient developed lower back pain, urinary incontinence, and bilateral lower extremity weakness. Imaging of the neuroaxis demonstrated T2 signal in the bilateral optic nerves and a large, longitudinally extensive (C4-T6) T2 lesion with patchy gadolinium enhancement (Fig. 29.5). Lumbar puncture at that time demonstrated leukocytosis with 218 WBCs $mm^3$ (range 0–5/$mm^3$), elevated protein of 67 mg/dL (range 20–45 mg/dL), and no oligoclonal

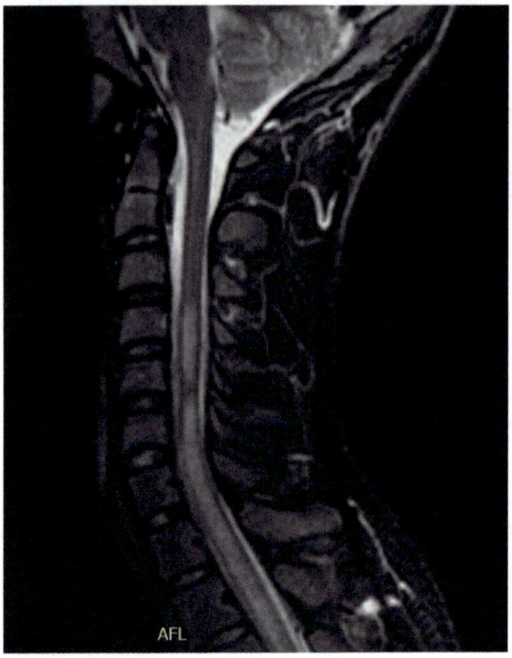


**Fig. 29.5** Sagittal T2 short tau inversion recovery (STIR) image of the cervical and thoracic spine demonstrating longitudinally extensive lesion of the spinal cord

bands. The patient was AQP4 antibody negative, but was noted to have positive anti-MOG antibodies with a titer of 1:100. The patient was treated with an additional course of 5 days of IVMP and subsequent 6-week steroid taper.

Four weeks into the patient's steroid taper, bilateral lower extremity weakness developed again, presenting as difficulty with ambulation. She was treated again with IVMP with minimal improvement. She then received a course of IVIg (2 g/kg over 3 days) followed by 375 mg/m$^2$ rituximab given over a 4-week period. Unfortunately, the patient developed worsening vision only a few weeks after this latter regimen and received another course of IVMP and IVIg. Given her refractory status to multiple immunomodulatory therapies, the decision was made to continue chronic IVIg. After developing an episode of transverse myelitis, a decision was made to switch to a combination of IVIg every 6 weeks and bi-annual rituximab infusions. The patient has accumulated disability over time, reaching a maximum Expanded Disability Status Scale (EDSS) of 6, although her most recent scoring is 2.

## Diagnostic Discussion

Myelin oligodendrocyte glycoprotein has been reported to be expressed entirely in the outermost lamellae of the myelin sheath within the CNS [29]. The location of expression of this protein is naturally a candidate for dysfunction in demyelinating disorders of the CNS given its unique expression pattern. An unanswered question at this time is whether antibodies against MOG are created in reaction to CNS inflammation or if they are directly pathogenic, causing inflammation directly.

Previous literature in both adult and pediatric patients has been controversial in that anti-MOG antibodies have been identified in MS, NMOSD, ADEM, multiphasic ADEM, and healthy controls, making its presence of uncertain clinical value. This is even more notable in children younger than 10 years of age with demyelinating events who tend to have higher rates of antiv-MOG antibody presence [30, 31]. In a recent review of 65 children with acute demyelinating events, 35% were anti-MOG antibody positive [32]. Interestingly, in this study presence of anti-MOG antibody had a high positive predictive value of 91% for non-MS-related disease. Additionally, in children with non-MS pathology (specifically ADEM), patients with anti-MOG antibody positivity were noted to have an increased frequency of large, bilateral, and widespread lesions including the brain and spinal cord, but ultimately had more favorable clinical outcomes compared to children who were anti-MOG antibody negative, although this same cohort of patient experiences polyphasic disease in up to 20% of cases [20].

Clinical presentations, as referenced prior, are extraordinarily heterogeneous within the anti-MOG antibody literature. In cases where patients have clinical symptoms similar to NMOSD, anti-MOG antibodies can be present in a minority of cases [33, 34]. These cases can be distinguished from "typical" presentations of NMOSD as they disproportionally affect men (as opposed to women in AQP4 positive disease), more likely involve the optic nerve than the spinal cord, have increased likelihood of bilateral optic neuritis, are more likely to be monophasic, have spinal lesions in lower portions of the spinal cord, and are associated with improved recovery [35]. The pathophysiology underlying anti-MOG antibody interactions in NMOSD and the clinical differences mentioned is unknown and seemingly counterintuitive to the general theory that more autoantibodies should produce a more severe phenotype, as would be the case with AQP4 antibodies seen in "classic" NMOSD.

When identified, the presence of anti-MOG antibodies poses a unique clinical conundrum in pediatric patients as monophasic and multiphasic forms have been reported. Monophasic illness has been associated with male gender and the absence of pathology involving the optic nerves [36]. Interestingly, these patients also have transient anti-MOG antibodies that are responsive to immune therapies. In contrast, patients who go on to have multiphasic disease (that can mimic MS, NMOSD, and multiphasic ADEM) tend to have a higher anti-MOG antibody titer at presentation, are refractory to intervention, and are older at presentation. Seroconversion to negative titers has been reported in both monophasic and relapsing disease. Although patients with continued seropositivity have been noted to have very high rates of polyphasic disease, up to 88% in children diagnosed with anti-MOG-associated ADEM [37–40]. While the idea of seroconversion is appealing, this has not been definitively linked to monophasic or more mild courses and longitudinal research will be needed to investigate these claims. As such, monitoring for signs that may indicate other diagnoses is critical as treatment and prognosis of anti-MOG antibody-associated disorders is complicated. Jarius et al. provided a list of "red flags" that should be considered before moving to definitive diagnosis and treatment and are provided in Table 29.5 [36].

**Table 29.5** "Red flags": conditions that should prompt physicians to challenge a positive test result (consider retesting the patient, ideally using an alternative, i.e., methodologically different cell-based assay; in case of doubt, consider seeking expert advice from a specialized center)

| |
|---|
| *Disease course*: |
| Chronic progressive disease (very rare in MOG-IgG-positive patients), including SPMS (especially SPMS without relapses) and PPMS |
| Sudden onset of symptoms, e.g., <4 hours from onset to maximum (consider ischemic cause), or continuous worsening of symptoms over weeks (consider tumor, sarcoidosis, etc.) |
| *MRI:* |
| Lesion adjacent to a lateral ventricle that is ovoid/round or associated with an inferior temporal lobe lesion or Dawson's finger-type lesion |
| Active brain MRI over time with silent increase in lesion burden between relapses (limited evidence) |
| *CSF:* |
| Bi- or tri-specific MRZ reaction (consider MS) |
| *Serology:* |
| MOG-IgG levels at or just barely above the assay-specific cut-off, especially (but not exclusively) if clinical picture is atypical |
| Positive MOG-IgM and/or MOG-IgA result with negative MOG-IgG (clinical significance unknown) |
| MOG-IgG positivity in the CSF but not in the serum (MOG-IgG is typically produced extrathecally) |
| AQP4-IgG/MOG-IgG "double-positive" test results (extremely rare; should prompt retesting for both antibodies) |
| *Others:* |
| Clinical or paraclinical findings suggesting diagnoses other than MOG-EM, NMOSD, or MS (e.g., neurotuberculosis, neuroborreliosis, neurosyphilis, neurosarcoidosis, Behçet syndrome, subacute combined degeneration of the spinal cord, Leber's hereditary optic neuropathy, vasculitis, CNS lymphoma, gliomatosis cerebri, paraneoplastic neurological disorders, PRES, PML, and evidence for CNS infection) |
| Combined central and peripheral demyelination (MOG is not expressed in the peripheral nervous system) |

*AQP4* aquaporin 4, *CNS* central nervous system, *CSF* cerebrospinal fluid, *Ig* immunoglobulin, *MOG* myelin oligodendrocyte glycoprotein, *MOG-EM* myelin oligodendrocyte glycoprotein encephalomyelitis, *MRI* magnetic resonance imaging, *MS* multiple sclerosis, *NMOSD* neuromyelitis optica spectrum disorder, *PML* progressive multifocal leukoencephalopathy, *PPMS* primary-progressive multiple sclerosis, *PRES* posterior reversible encephalopathy syndrome, *SPMS* secondary-progressive multiple sclerosis

Information regarding the treatment of anti-MOG antibody spectrum disorders is complex and limited in regard to both large-scale studies and long-term follow-up. Patients with anti-MOG antibody positivity and an acute demyelinating event respond to therapy with glucocorticoids [41]. Interestingly, these patients are vulnerable to relapses with steroid reduction and cessation prompting trials with multiple forms of immunotherapy. For this reason, prolonged and slow steroid tapers are often utilized anecdotally following initial anti-MOG antibody-associated demyelinating events. DMTs such as interferon beta and glatiramer acetate have been evaluated in patients with anti-MOG antibody positivity and are not associated with clinical improvement, resulting in no changes in annualized relapse rates or expanded disability status scale scores over time [42]. Other attempts at maintenance therapy, including azathioprine, mycophenolate mofetil, rituximab, and IVIg, have been associated with reduction in relapse in patients with relapsing anti-MOG antibody positive demyelinating disorders, although further large-scale studies are needed [42–46].

Prognosis is dependent on the systems involved, with patients who have optic neuritis and transverse myelitis having more accumulated disability than other presentations of the disorder.

## Conclusion

Demyelinating disorders in children are a complex and heterogeneous pathologic group. Although clear diagnostic criteria exist for MS, NMOSD, and ADEM, a broad differential and consideration of mimics of common demyelinating syndromes should be maintained based on the clinical, radiographic, and laboratory data available. In all demyelinating syndromes, anti-MOG antibody testing is warranted as data continues to emerge about the clinical and therapeutic implications of this autoantibody.

## References


1. Hynson J, Kornberg A, Coleman L, Shield L, Harvey A, Kean M. Clinical and neuroradiologic features of acute disseminated encephalomyelitis in children. Neurology. 2001;56(10):1308–12.
2. Murthy SN, Faden HS, Cohen ME, Bakshi R. Acute disseminated encephalomyelitis in children. Pediatrics. 2002;110(2 Pt 1):e21.
3. Tenembaum S, Chitnis T, Ness J, Hahn JS, International Pediatric MS Study Group. Acute disseminated encephalomyelitis. Neurology. 2007;68(16 Suppl 2):S23–36.
4. Stonehouse M, Gupte G, Wassmer E, Whitehouse WP. Acute disseminated encephalomyelitis: recognition in the hands of general paediatricians. Arch Dis Child. 2003;88(2):122–4.
5. Krupp LB, Banwell B, Tenembaum S, International Pediatric MS Study Group. Consensus definitions proposed for pediatric multiple sclerosis and related disorders. Neurology. 2007;68(16 Suppl 2):S7–12.
6. Krupp L, Tardieu M, Amato M, Banwell B, Chitnis T, Dale R, et al. International Pediatric Multiple Sclerosis Study Group criteria for pediatric multiple sclerosis and immune-mediated central nervous system demyelinating disorders: revisions to the 2007 definitions. Mult Scler J. 2013;19(10):1261–7.
7. Graus F, Titulaer MJ, Balu R, Benseler S, Bien CG, Cellucci, et al. A clinical approach to diagnosis of autoimmune encephalitis. Lancet Neurol. 2016;15(4):391–404.
8. Florance NR, Davis RL, Lam C, Szperka C, Zhou L, Ahmad S, et al. Anti-N-methyl-D-aspartate receptor (NMDAR) encephalitis in children and adolescents. Ann Neurol. 2009;66(1):11–8.
9. Renoux C, Vukusic S, Mikaeloff Y, Edan G, Clanet M, Dubois B, et al. Natural history of multiple sclerosis with childhood onset. NEJM. 2007;356(25):2603–13.
10. Banwell B, Ghezzi A, Bar-Or A, Mikaeloff Y, Tardieu M. Multiple sclerosis in children: clinical diagnosis, therapeutic strategies, and future directions. Lancet Neurol. 2007;6(10):887–902.
11. Castillo P, Woodruff B, Caselli R, Vernino S, Lucchinetti C, Swanson J, et al. Steroid-responsive encephalopathy associated with autoimmune thyroiditis. Arch Neurol. 2006;63(2):197–202.
12. Callen D, Shroff M, Branson H, Li D, Lotze T, Stephens D, et al. Role of MRI in the differentiation of ADEM from MS in children. Neurology. 2008;72(11):968–73.
13. Kesselring J, Miller DH, Robb SA, Kendall BE, Moseley IF, Kingsley D, et al. Acute disseminated encephalomyelitis. MRI findings and the distinction from multiple sclerosis. Brain. 1990;113(Pt 2):291–302.
14. Dale RC, de Sousa C, Chong WK, Cox TC, Harding B, Neville BG. Acute disseminated encephalomyeli-

tis, multiphasic disseminated encephalomyelitis and multiple sclerosis in children. Brain. 2000;123 Pt 12:2407–22.
15. Alper G, Heyman R, Wang L. Multiple sclerosis and acute disseminated encephalomyelitis diagnosed in children after long-term follow-up: comparison of presenting features. Dev Med Child Neurol. 2009;51(6):480–6.
16. Pohl D, Tenembaum S. Treatment of acute disseminated encephalomyelitis. Curr Treat Options Neurol. 2012;14:264–75.
17. Marchioni E, Marinou-Aktipi K, Uggett C, Bottanelli M, Pichiecchio A, Soragna D, et al. Effectiveness of intravenous immunoglobulin treatment in adult patients with steroid-resistant monophasic or recurrent acute disseminated encephalomyelitis. J Neurol. 2002;249(1):100–4.
18. Keegan M, Pineda AA, McClelland RL, Darby CH, Rodriguez M, Weinshenker BG. Plasma exchange for severe attacks of CNS demyelination: predictors of response. Neurology. 2002;58(1):143–6.
19. Tenembaum S, Chamoles N, Fejerman N. Acute disseminated encephalomyelitis: a long-term follow up study of 84 pediatric patients. Neurology. 2002;59(8):1224–31.
20. Baumann M, Sahin K, Lechner C, Hennes EM, Schanda K, Mader S, et al. Clinical and neuroradiologic differences of paediatric acute disseminating encephalomyelitis with and without antibodies to the myelin oligodendrocyte glycoprotein. J Neurol Neurosurg Psychiatry. 2015;86(3):265–72.
21. Mikaeloff Y, Caridade G, Assi S, Suissa S, Tardieu M. Prognostic factors for early severity in a childhood multiple sclerosis cohort. Pediatrics. 2006;118(3):1133–9.
22. Yeh E, Chitnis T, Krupp L, Ness J, Chabas D, Kuntz N, et al. Pediatric multiple sclerosis. Nat Rev Neurol. 2009;5(11):621–31.
23. Thompson A, Banwell B, Barkhof F, Carroll W, Coetzee T, Comi G, et al. Diagnosis of multiple sclerosis: 2017 revisions of the McDonald criteria. Lancet Neurol. 2018;17(2):162–73.
24. Tenembaum S. Pediatric multiple sclerosis. Neuroimaging Clin N Am. 2017;27(2):229–50.
25. Pohl D, Rostasy K, Reiber H, Hanefeld F. CSF characteristics in early-onset multiple sclerosis. Neurology. 2004;63(10):1966–7.
26. Juryńczyk M, Craner M, Palace J. Overlapping CNS inflammatory diseases: differentiating features of NMO and MS. J Neurol Neurosurg Psychiatry. 2014;86(1):20–5.
27. Chitnis T, Arnold DL, Banwell B, Bruck W, Ghezzi A, Giovannoni G, et al. Trial of Fingolimod versus interferon beta-1a in pediatric multiple sclerosis. N Engl J Med. 2018;379(11):1017–27.
28. Simone M, Chitnis T. Use of disease-modifying therapies in pediatric MS. Curr Treat Options Neurol. 2016;18(8)
29. Johns TG, Bernard CC. The structure and function of myelin oligodendrocyte glycoprotein. J Neurochem. 1999;72(1):1–9.
30. Brilot F, Dale RC, Selter RC, Grummel V, Kalluri SR, Aslam M, et al. Antibodies to native myelin oligodendrocyte glycoprotein in children with inflammatory demyelinating central nervous system disease. Ann Neurol. 2009;66(6):833–42.
31. McLaughlin KA, Chitnis T, Newcombe J, Franz B, Kennedy J, McArdel S, et al. Age-dependent B cell autoimmunity to myelin surface antigen in pediatric multiple sclerosis. J Immunol. 2009;183(6):4067–76.
32. Hacohen Y, Mankad K, Chong WK, Barkhof F, Vincent A, Lim M, et al. Diagnostic algorithm for relapsing acquired demyelinating syndromes in children. Neurology. 2017;89(3):269–78.
33. Kitley J, Woodhall M, Waters P, Leite MI, Devenney E, Craig J, et al. Myelin-oligodendrocyte glycoprotein antibodies in adults with neuromyelitis optica phenotype. Neurology. 2012;79(12):1273–7.
34. Sato DK, Callegara D, Lana-Peixoto MA, Waters PJ, de Haidar Jorge FM, Takahashi T, et al. Distinction between MOG antibody positive and AQP4 antibody positive NMO spectrum disorders. Neurology. 2014;82(6):474–81.
35. Kitley J, Waters P, Woodhall M, Leite MI, Murchison A, George J, et al. Neuromyelitis optica spectrum disorders with aquaporin-4 and myelin-oligodendrocyte glycoprotein antibodies: a comparative study. JAMA Neurol. 2014;71(3):276–83.
36. Jarius S, Paul F, Aktas O, Asgari N, Dale RC, de Seze J, Franciotta D, et al. MOG encephalomyelitis: international recommendations on diagnosis and antibody testing. J Neuro-Oncol. 2018;15(1):134.
37. Jarius S, Ruprecht K, Fleiter I, Borisow N, Asgari N, Pitarokoili K, et al. MOG-IgG in NMO and related disorders: a multicenter study of 50 patients. Part 2: epidemiology, clinical presentation, radiological and laboratory features, treatment responses, and long-term outcome. J Neuroinflammation. 2016;13(1):280.
38. Probstel AK, Dornmair K, Bittner R, Sperl P, Jenne D, Magalhaes S, et al. Antibodies to MOG are transient in childhood acute disseminated encephalomyelitis. Neurology. 2011;77(6):580–8.
39. Spadaro M, Gerdes LA, Krumbholz M, Ertl-Wagner B, Thaler FS, Schuh E, et al. Autoantibodies to MOG in a distinct subgroup of adult multiple sclerosis. Neurol Neuroimmunol Neuroinflamm. 2016;3(5):e257.
40. Lopez-Chiriboga AS, Majed M, Fryer J, Dubey D, McKeon A, Flanagan EP, et al. Association of MOG-IgG serostatus with relapse after acute disseminated encephalomyelitis and proposed diagnostic criteria for MOG-IgG-associated disorders. JAMA Neurol. 2018;75(11):1355–68.
41. Ramanathan S, Mohammad S, Tantsis E, Nguyen TK, Merheb V, Fung VSC, et al. Clinical course, therapeutic response and outcomes in relapsing

MOG antibody-associated demyelination. J Neurol Neurosurg Psychiatry. 2018;89(2):127–37.
42. Hacohen Y, Wong YY, Lechner C, Jurynczyk M, Wright S, Konuskan B, et al. Disease course and treatment responses in children with relapsing myelin oligodendrocyte glycoprotein antibody associated disease. JAMA Neurol. 2018;75(4):478–87.
43. Montcuquet A, Collongues N, Papeix C, Zephir H, Audoin B, Laplaud D, et al. Effectiveness of mycophenolate mofetil as first-line therapy in AQP4-IgG, MOG-IgG, and seronegative neuromyelitis optic spectrum disorders. Mult Scler. 2017;23(10):1377–84.
44. Reindl M, Jarius S, Rostasy K, Berger T. Myelin oligodendrocyte glycoprotein antibodies: how clinically useful are they? Curr Opin Neurol. 2017;30(3):295–301.
45. Dale RC, Tantsis EM, Merheb V, Kumaran RY, Sinmaz N, Pathmanandavel K, et al. Antibodies to MOG have a demyelination phenotype and affect oligodendrocyte cytoskeleton. Neurol Neuroimmunol Neuroinflamm. 2014;1(1):e12.
46. Ramanathan S, Prelog K, Barnes EH, Tantsis EM, Reddel SW, Henderson AP, et al. Radiological differentiation of optic neuritis with myelin oligodendrocyte glycoprotein antibodies, aquaporin-4 antibodies and multiple sclerosis. Mult Scler. 2016;22(4):470–82.

# Clinical Approach to Stiff Person Syndrome

# 30

Jonathan R. Galli and Stacey L. Clardy

## Patient Vignette 1

A 56-year-old woman reports muscle spasms that began in her proximal legs approximately 4 months prior to presentation. She explains that her spasms began in her right leg but progressed to involve both lower extremities the following month. The spasms initially occurred two to three times per month but subsequently increased to several times per week. She also notes progression of spasms into her upper and lower back. Spasms are triggered by loud noises. More recently, she describes increased tightness in her back and proximal legs even when spasms are not occurring. Her gait has become severely affected such that she is unable to walk without an assistive device; she has otherwise suffered falls, especially if she cannot place her hand against a wall for guidance.

On examination, her mental status is at her cognitive baseline, and no abnormalities are detected on brief cognitive screening. There are no cranial nerve abnormalities. Her motor examination demonstrates full strength in all of her extremities. She has increased tone in her proximal paraspinal muscles and lower extremities. Her reflexes are symmetrically brisk, graded at 3+ throughout all extremities, without Hoffman's or Babinski signs. Her sensory examination demonstrates normal testing of temperature, vibration, and sharp touch. The cerebellar examination is within normal limits. On gait examination, she has a slow, rigid-appearing gait, requiring extra effort to touch her heels fully to the ground, and she walks with a short stride, requiring a walker for assistance.

J. R. Galli · S. L. Clardy (✉)
Department of Neurology, University of Utah Health, Salt Lake City, UT, USA

Department of Neurology, George E. Wahlen Salt Lake City VHA, Salt Lake City, UT, USA
e-mail: Stacey.Clardy@hsc.utah.edu

## Introduction

Stiff person syndrome (SPS), also referred to as stiff man syndrome or Moersch-Woltman syndrome (in recognition of the Drs. Moersch and Woltman, the two Mayo Clinic neurologists who first described a series of affected male patients), was first described in 1956 in patients with primarily truncal and proximal muscle rigidity [1]. SPS is estimated to occur in about one to two per million people, though this is possibly an underestimation due to the historically limited availability of commercial testing of glycine receptor (GlyR) antibodies and also under-recognition of less severe or atypical presentations of the SPS spectrum of disease [2, 3]. Onset of disease is approximately 40 years old, with women more commonly affected than men [4, 5]. While more rare, pediatric onset of SPS has also been described [6].



A. L. Piquet, E. Alvarez (eds.), *Neuroimmunology*, https://doi.org/10.1007/978-3-030-61883-4_30

## Clinical Presentation

### Classic Stiff Person Syndrome

SPS was first described as a progressive, persistent, symmetrical rigidity of the axial, neck, and proximal limb muscles [1]. Patients often initially have stiffness and rigidity in the axial muscles, which may spread to the proximal limb muscles over weeks to months [7]. Intermittent, severe muscle spasms may also occur and may be triggered by volitional movement, startle, emotion, or tactile simulation [7]. As symptoms progress, volitional movement at the joint can become difficult due to muscle rigidity [7].

Examination of patients with SPS demonstrates normal strength and sensation [8]. Patients have significantly increased tone in the affected muscle groups, sometimes enough to make the muscles feel hardened to touch [1, 7]. Deep tendon reflexes may demonstrate hyperreflexia [8]. Spinal lordosis can occur, and the gait may appear rigid, described as a "wooden man" or "tin soldier" gait [7, 8].

Over time, several revisions have been suggested to create diagnostic criteria for SPS [7–9]. Currently, SPS diagnostic criteria based on work of Vasconcelos and Dalakas in 2003 include (1) simultaneous contraction of agonist and antagonist muscles leading to muscle stiffness, primarily in axial muscles, which is progressive over time; (2) episodic spasms in response to startle, emotion, volitional movement, or tactile stimulation; (3) simultaneous motor unit firing in agonist and antagonist muscle groups demonstrated on electromyography; (4) elevated serum glutamic acid decarboxylase (GAD65) autoantibodies; (5) symptomatic response to diazepam; and (6) absence of another neurologic diagnosis to explain muscle rigidity [9]. These criteria will likely require ongoing modification, as the clinical spectrum expands and as other role of other autoantibodies is clarified in patients.

See Table 30.1.

### Stiff Limb Syndrome

A variant of SPS has been described involving one or more limbs. Referred to as focal SPS or stiff limb syndrome, the symptoms most commonly

**Table 30.1** Clinical characteristics of stiff person syndrome (SPS)

| | |
|---|---|
| *Clinical findings* | |
| Progressive spasms and rigidity in muscles | |
| Co-contraction of agonist and antagonist muscle groups | |
| Exacerbation of spasm in response to startle or stimulation | |
| Symptomatic response to diazepam | |
| *Laboratory evaluation* | |
| Positive GAD65 autoantibodies in serum or CSF | |
| Alternatively, may have positive GlyR, amphiphysin, GABA$_A$ receptor, gephyrin autoantibodies | |
| CSF may demonstrate nonspecific pleocytosis and/or oligoclonal bands | |
| *Other diagnostic evaluation* | |
| EMG may demonstrate simultaneous motor unit firing in agonist and antagonist muscle groups | |
| Normal nerve conduction studies | |
| MRI may be used to rule out myelopathy or alternative causes of spasms and rigidity | |
| *Comparison of classic SPS, stiff limb syndrome, and PERM* | |
| Classic stiff person syndrome | Spasms occur primarily in axial and proximal muscle groups |
| Stiff limb syndrome | Focal involvement in limbs more distally, may progress to involve axial muscles over time |
| Progressive encephalomyelitis with rigidity and myoclonus (PERM) | Often have cranial nerve involvement, pyramidal signs, dysautonomia, seizures, and encephalopathy |

*CSF* cerebrospinal fluid, *EMG* electromyography, *GABA* γ (gamma)-aminobutyric acid, *GAD65* glutamic acid decarboxylase epitope 65, *GlyR* glycine receptor, *MRI* magnetic resonance imaging, *PERM* progressive encephalomyelitis with rigidity and myoclonus involve the legs [10, 11]. Spasms are painful and may occur with volitional movement, spontaneously, or with provocation, such as with reflex testing during the neurologic examination [12]. Patients with focal involvement tend to have fluctuating clinical courses [13]. These patients may also progress to a classic SPS phenotype, with greater involvement of axial musculature over time [5, 14]. GAD65 autoantibody positivity is seen in a large subset of patients with focal involvement, although seronegative cases may occur (and warrant further investigation to rule out mimics) [5, 13, 14]. Significant upper extremity or neck involvement is often seen in patients with amphiphysin autoantibodies as well as GlyR autoantibodies [5, 15, 16]. Similar to classic SPS, these patients may demonstrate co-contraction of agonist and antagonist muscle groups and/or continuous motor unit activation on electromyography (EMG). Unique segmented, hypersynchronous discharges may also occur with EMG in patients with stiff limb syndrome [12, 13]. These patients tend to require lower doses of benzodiazepines for symptomatic treatment, although they do not necessarily respond better to immunotherapy [5, 16].

## Progressive Encephalomyelitis with Rigidity and Myoclonus (PERM)

The term progressive encephalomyelitis with rigidity and myoclonus (PERM) is used to describe a more fulminant variant of SPS. Patients with PERM often present with cranial nerve findings prior to onset of spasms [13, 14]. In addition to the spasms, they may suffer a broader spectrum of neurologic symptoms including cognitive deficits, epilepsy, brain stem and cranial nerve dysfunction, pyramidal signs, sphincter dysfunction, sensory loss, and dysautonomia [13, 14, 16]. Most patients follow a pattern of rapid decline, leading to death within months to years [5, 13]. GAD65 and GlyR autoantibodies are commonly associated with PERM and can occur concomitantly in PERM patients [5, 16, 17]. Recent data suggests that patients with GlyR autoantibodies and PERM may respond better to immunotherapy than those with only GAD65 autoantibodies [16, 17].

## Serology

SPS was first associated with antibodies against glutamic acid decarboxylase in 1988 by Solimena et al. [18]. More recently, it was determined that the epitope 65 of glutamic acid was specifically bound by the autoantibody [19, 20]. Patients with SPS tend to have manyfold higher titers of GAD65 autoantibodies compared to patients exclusively diagnosed with insulin-dependent diabetes mellitus, and thus, the phrase "GAD65-associated neurologic autoimmunity" is useful to distinguish these neurologic conditions from the more common, isolated, insulin-dependent diabetes [20]. It is approximated that 60–80% of patients with SPS are GAD65 autoantibody seropositive [14]. While GAD65 autoantibodies serve as a marker supportive of SPS, there has not been any consistent correlation of GAD65 titers with disease severity [21]. GAD65 autoantibodies are also not specific for classical SPS and have been demonstrated at high titers in patients with cerebellar ataxia, limbic encephalitis, and epilepsy, as well as overlapping neurologic phenotypes (SPS "plus" syndromes) [10].

Other antibodies that have been associated in SPS include autoantibodies against amphiphysin, glycine receptor (GlyR), $\gamma$ (gamma)-aminobutyric acid $(GABA)_A$ receptor, gephyrin, and dipeptidyl peptidase-like protein 6 (DPPX) [22–26]. A subset of patients with SPS may also present without any clear causative autoantibody and warrant rigorous exclusion of other diagnoses [5]. GlyR antibodies may occur alone or with GAD65 antibodies, and patients with glycine receptor antibodies may have better clinical outcomes than those with GAD65 antibodies [16]. GlyR autoantibody testing should be strongly considered in GAD65-seronegative (or very low GAD65 titer positivity) PERM patients or in SPS patients with prominent positive visual symptoms [17, 27]. Amphiphysin in particular is an important consideration in GAD65-seronegative patients, given the overall higher occurrence of malignancy in these patients than in GAD65 patients [28]. Further, consideration for a paraneoplastic etiology of symptoms should be a consideration in all SPS patients regardless of antibody identification, and the treating physician should ensure that appropriate malignancy screening guidelines are up-to-date, and additionally query any specific symptoms suggestive of undiagnosed malignancy.

## Pathophysiology

The exact pathophysiological mechanism in SPS remains elusive. A leading theory is impairment of the GABA inhibition pathway, particularly within the spinal cord, leading to motor hyperactivity. This has been supported by the demonstration of exaggerated response to cutaneous or acoustic stimulation leading to spread of reflexes to additional muscle groups, which is thought to be secondary to neuronal hyperexcitability [29–32]. This disinhibition, particularly within the spinal cord, may be secondary to segmental interneuronitis [12]. Evidence of hyperexcitability within the motor cortex, however, suggests that there is disinhibition within the entire central nervous system [33]. The GABA inhibition pathway has been implicated in several studies that demonstrate reduced levels of GABA in both the brain and spinal cord of SPS patients [20, 34]. The autoantibodies associated with SPS bind antigens, which are found within the GABA inhibitory synapse, and also suggest this pathway is significantly involved in the underlying pathophysiology.

Substantial insight into the potential pathophysiology of SPS came with the discovery of the involvement of GAD65 autoantibodies in SPS [18]. The GAD65 autoantibody in patients with SPS binds two distinct regions of the GAD65 molecule, including the linear portion in the N-terminus [35, 36]. It has been hypothesized that the binding of the GAD65 molecule may inhibit GABA synthesis from glutamate, thus impairing inhibitory pathways [37]. Figure 30.1 demonstrates possible binding sites for the autoantibodies that have been implicated in SPS. The other associated autoantibodies in SPS also serve roles in the GABA inhibitory pathway, and while the exact pathophysiological mechanism of each unique autoantibody is not yet precisely understood, they do offer possible insight into the

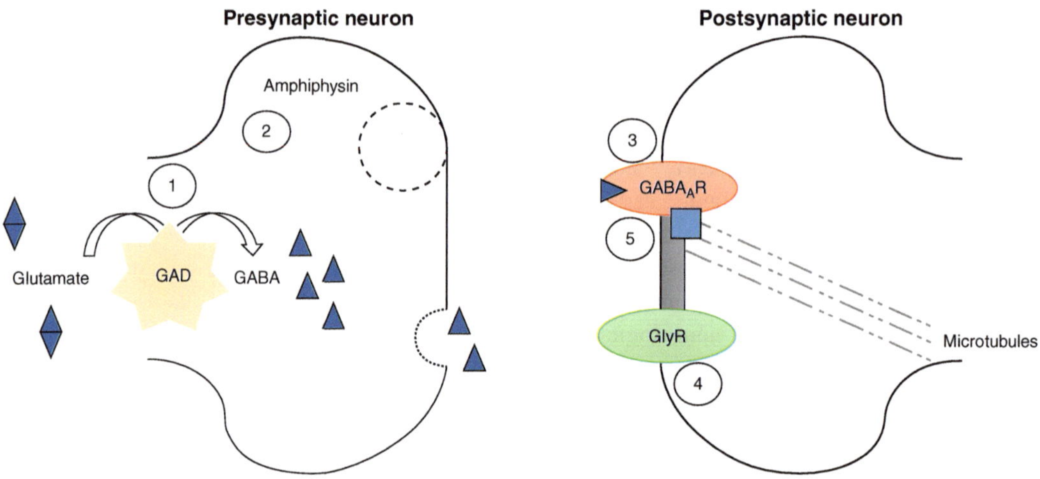


**Fig. 30.1** Potential binding sites for autoantibodies in the γ (gamma)-aminobutyric acid (GABA) inhibition pathway in stiff person syndrome. SPS is thought to be, in large part, secondary to pathology within the $GABA_A$ receptor synapse leading to neuronal hyperexcitability. (1) Glutamic acid decarboxylase (GAD) serves as the rate-limiting enzyme in the synthesis of glutamate to GABA, an inhibitory neurotransmitter. (2) Amphiphysin is involved in presynaptic endocytosis. (3) $GABA_A$ receptors ($GABA_A$ R) bind GABA postsynaptically and cause inhibition of the postsynaptic neuron. (4) Within the spinal cord, glycine binds postsynaptically to its receptor (GlyR) to cause inhibition of the postsynaptic neuron. (5) Gephyrin (shown as the gray box) and $GABA_A$ receptor-associated protein (shown as the blue box) are involved in anchoring and organizing $GABA_A$ receptor and GlyR

suspected increased excitability in SPS. $GABA_A$ receptors are essential in the inhibitory pathways within the central nervous system, and low titers of autoantibodies may occur concomitantly with GAD65 autoantibodies in patients with SPS [24, 38]. Glycine acts as an inhibitory neurotransmitter within the spinal cord, and autoantibodies against GlyR have also been associated with SPS, particularly PERM [39]. Autoantibodies against gephyrin and $GABA_A$ receptor-associated protein, both structural proteins involved in the GlyR and $GABA_A$ receptor plasma membrane organization, may also lead to a SPS phenotype [25, 40]. Amphiphysin is a protein involved in synaptic endocytosis, and mouse exposure to anti-amphiphysin antibodies can produce a similar phenotype as SPS, particularly affecting GABAergic neurons [41, 42].

The pathophysiological significance of GAD65 autoantibodies remains an area requiring additional investigation, particularly focusing on whether the specific epitope binding and/or the autoantibody titers in SPS are playing a direct role in disease activity [37, 43], especially in light of data suggesting that SPS-specific GAD65 autoantibodies in serum and cerebrospinal fluid (CSF) bind to different epitopes without phenotypical correlation [44], and that titers in serum and CSF do not correlate directly with disease severity [21]. Passive exposure to GAD65 human autoantibodies in mice does not necessarily induce symptoms, and GAD65 antibodies are seen in several other different neurologic diseases affecting other specific brain regions, including cerebellar ataxia and epilepsy, without an associated SPS phenotype. It has been suggested that B cells are likely involved in the GAD65-related neurologic disease process, given the intrathecal detection of GAD65 autoantibodies and the frequent presence of oligoclonal bands [44], but T cells are also thought to have a role in the maintenance of the ongoing pathologic immune response in SPS patients [45]. It has also been suggested that GAD65 molecule exposure to the immune system may initially occur during the exocytosis of GABA [46].

There are no pathognomonic findings reported in tissue studies from patients with SPS: autopsy studies have demonstrated chromatolysis, vacuolization, and microglial proliferation within anterior horn cells, particularly in the lumbar spine [47, 48], and T cell infiltration of cytotoxic CD8+ T cells may also be found [48]. Perivascular cuffing has been demonstrated in the spinal cord, as well as within the brain stem and cerebrum in a patient with GAD65-seropositive SPS [49]. Limited genetic studies do not help to further inform pathophysiology, as there is a paucity of data regarding the genetic risk in SPS, with one report of an association with human leukocyte antigen class II DQB1*0201 allele in patients with SPS [50] and another study noting higher rates of the HLA-DR haplotypes [4].

## Clinical Evaluation

Laboratory investigations of a patient with suspected SPS should include testing of GAD65 in both the serum and cerebrospinal fluid. Several commercial assays for testing GAD65 are available including radioimmunoassays, enzyme-linked immunosorbent assay, and enzymatic immunoprecipitation assay, with radioimmunoassays being the most sensitive and specific for GAD65 [51]. In patients with SPS, GAD65 serum titers are expected to be severalfold greater than the titers observed in type 1 diabetes, often five- to tenfold, but upper limits of assay detection and reporting in some of the commercial laboratories may limit the ability to determine the exact titer without requesting further dilution [36]. In patients who are seropositive, CSF GAD65 should also be evaluated to confirm intrathecal GAD65 antibody synthesis [20]. When GAD65 antibody testing is negative, consideration should be given to testing amphiphysin, GlyR, $GABA_A$ receptor, and/or gephyrin autoantibodies in the serum and CSF. Currently, testing of these additional autoantibodies may be limited to performance on a research basis in only a few labs internationally [16]. CSF findings are frequently nonspecific but may demonstrate leukocytosis or elevated protein [13, 16]. Oligoclonal bands may also be present in a subset of patients, although they are not specific for SPS [14].

EMG studies are important in the diagnostic evaluation for SPS, although not all patients with

SPS will demonstrate abnormalities, especially if tested while on treatment with symptomatic muscle-relaxing medications or if testing in unaffected muscles. Provocative measures may be required to induce spasm at the time of testing. Peripheral nerve conduction studies are expected to be normal [13]. EMG will classically demonstrate co-activation of agonist and antagonist muscle groups in muscles affected by spasms [1, 13, 52]. Motor unit potentials will have normal morphology but can be present in antagonist muscle groups during contraction of an agonist muscle group, demonstrating a lack of inhibition in opposing muscle groups [46]. These EMG findings improve with use of benzodiazepines, during sleep, or with general anesthesia [53]. Other electrophysiological findings include exaggerated brain stem reflexes, cutaneo-muscular reflexes, and exteroceptive reflexes elicited from cutaneous stimulation, although these are not routinely assessed for in a clinical setting [29–31]. In patients with more distal limb involvement, such as is seen in stiff limb syndrome, EMG may demonstrate segmented and synchronous discharges of motor units [13].

Magnetic resonance imaging (MRI) of both the brain and spine does not typically demonstrate any specific findings; abnormalities should raise suspicion of misdiagnosis or other concurrent abnormality (such as GABA-A receptor encephalitis). Imaging should be considered during diagnostic evaluation to rule out any intracranial or myelopathic process that may mimic SPS [46].

## Differential Diagnosis

Other causes of spasms and increased muscular tone should be considered during evaluation of SPS, and alternative causes or pyramidal or extrapyramidal disorders should be excluded. Common differential diagnostic considerations include genetic hyperekplexias; myelopathy secondary to infectious, inflammatory, or metabolic etiology; dystonia; Parkinson's disease and atypical parkinsonism; genetic myopathies such as Emery-Dreifuss or myotonic muscular dystrophy; neuromyotonia; hereditary spastic paraparesis; spinocerebellar degeneration; motor neuron disease; functional movement disorder; tetanus; serotonin syndrome; and neuroleptic malignant syndrome. Functional movement disorders or malingering should be considered quite cautiously, as SPS patients are at risk of treatment delay and complications if misdiagnosed with a functional movement disorder.

## Associated Conditions

There may be overlapping neurologic conditions with SPS. These include cerebellar ataxia, epilepsy, limbic encephalitis, myoclonus, parkinsonism, and peripheral neuropathy [5, 16]. Beyond the nervous system, patients with SPS also have an increased risk for other autoimmune diseases. Approximately 35–43% patients with GAD65-seropositive SPS have type 1 diabetes or latent autoimmune diabetes in adults (LADA). Conversely, the occurrence of SPS in type 1 diabetes is much lower, estimated at 1:10,000 [4, 5]. Type 1 diabetes or LADA can precede the symptoms of SPS by years [4, 5]. Other autoimmune conditions that have been associated with SPS include autoimmune thyroid disease, pernicious anemia, vitiligo, systemic lupus erythematosus, and celiac disease [5, 45, 46].

GAD65-associated SPS may occur rarely as a paraneoplastic condition, and at a minimum, standard malignancy screening should be pursued in all patients newly diagnosed with SPS. Screening may be expanded to include computed tomography imaging of the chest, abdomen, and pelvis or whole-body positron emission scan in atypical or high-risk patients. Amphiphysin-related SPS, in particular, is commonly association with breast cancer [22]. Other malignancies that have been associated with SPS include thyroid, colon, thymus, lung, renal, and lymphoma [45].

## Patient Vignette 2

A 45-year-old woman presents with spasms in her back and proximal lower extremities along with progressive difficulty walking over the last

3 years. Her past medical history is significant for Wegener's granulomatosis and autoimmune thyroiditis. On examination, she has increased muscle tone in her paraspinal muscles and proximal lower extremity muscles, along with spastic gait. Laboratory evaluation was notable for negative serum and CSF GAD65. Serum GlyR antibodies, however, were positive. CSF studies were otherwise normal. EMG demonstrated co-activation of muscles in agonist and antagonist muscle groups in her right lower extremity. MRI brain and spine were normal.

Symptomatic treatment with valium and baclofen provided only partial benefit. Immunotherapy with intravenous immunoglobulin (IVIg) marginally improved symptoms. Plasma exchange was started 3 months later with no benefit. Rituximab was initiated with no improvement in spasm frequency or ambulatory status. Ultimately, consideration was given to experimental autologous bone marrow transplant therapy, given her continued progression despite maximizing symptomatic therapy and failure of traditional immunotherapies.

## Treatment

The primary symptomatic treatment in patients with SPS is the use of antispasmodic agents. Diazepam, a $GABA_A$ receptor agonist, is an effective symptomatic therapy [8]. Benefit from diazepam or clonazepam can be substantial, and benzodiazepine therapy is considered first-line agent for symptomatic therapy [46]. There are no clinical trials to provide guidelines for dosing of benzodiazepines. Treatment should be initiated and cautiously up-titrated, taking into account both symptomatic relief and side effects, including respiratory depression and somnolence. Large doses may be necessary for effect [5]. If patients are prescribed concurrent opiates for another indication/condition, extreme caution must be undertaken to avoid accidental death, and discontinuation of opiates is strongly recommended; caregivers should be equipped with naloxone during the weaning of opiates if both medications must be continued concomitantly for a period of time. Additionally, polysomnography should be performed early in the course of treatment in appropriate patients to minimize the dangers of the use of sedating medications in untreated sleep apnea.

Baclofen, a $GABA_B$ agonist, was later described as an effective treatment, particularly when used together with a benzodiazepine [54, 55]. Baclofen may be dosed together with a benzodiazepine or as a monotherapy. There are no clinical trials to provide dosing instruction on baclofen; dosing should also be titrated to symptom relief and tolerance of side effects. Somnolence is the most common side effect with baclofen use, particularly when used along with benzodiazepines. Patients with classic SPS symptoms tend to need higher dosing than those with focal SPS [5]. Intrathecal baclofen is an option for patients who are not well controlled on oral medications or for those who do not tolerate higher doses of diazepam or baclofen. Outcomes in the use of intrathecal baclofen may vary, and some studies suggest it may be better tolerated than oral therapy [56, 57]. Risks of complications from catheter problems or pump malfunction should be seriously considered prior to placement [56, 57]. Intrathecal baclofen pump use in PERM patients may require higher doses and also increases the risk for dysautonomia [57]. Other medications that have been tried for symptomatic treatment include several antiepileptic medications, as well as propofol and dantrolene. Some have shown benefit in specific cases, although they are not widely used [58–63].

Immunotherapy has frequently been utilized in the treatment of SPS, given the presumed antibody-mediated disease mechanism, although large randomized control trials are currently lacking. Intravenous immunoglobulin has been most widely used, and small trials have shown benefit with improved ambulation, decreased falls, and increased ability to perform activities of daily living [64, 65]. IVIg is a relatively safe treatment in patients with SPS with most common side effects including headache, myalgias, and rash; more severe side effects include aseptic meningitis, thrombotic events, and renal tubular necrosis [65]. However, the benefit of IVIg is not always sustained long-term in patients and may not necessarily prevent disability [5, 64].

Plasma exchange has been used in SPS, with the hypothesized benefit relating to the filtration of immunoglobulins [66]. While no randomized control trials have been performed using plasma exchange in SPS, significant clinical improvement has been demonstrated in some patients with SPS in smaller studies [67]. Plasma exchange is generally well tolerated, with the most common adverse effects including catheter-associated infection and hypotension [66, 67].

Rituximab, an anti-CD20 monoclonal antibody, has been utilized for several autoantibody-mediated neurologic diseases including myasthenia gravis, neuromyelitis optica, and autoimmune encephalitis [68–70]. Small case series and case reports have reported benefit in SPS patients—notably who were refractory to other immunotherapy—but other larger, randomized trials have not shown benefit of rituximab treatment for SPS [71, 72].

Most recently, autologous stem cell transplantation is being pursued in treatment-refractory SPS patients in a few international centers. While large studies are currently lacking and greatly needed, this offers a promising therapy for particularly refractory and quickly progressive patients and may offer the potential of significant symptom improvement and perhaps clinical remission from disease [73].

## Conclusion and Future Directions

Understanding of SPS—particularly regarding epidemiology, pathophysiology, and effective treatments—remains sparse. Due to the rarity of diagnosed SPS, epidemiological data has been difficult to obtain, and larger population-based studies are necessary. Further studies to elucidate the pathophysiology of SPS can help to clarify the role the autoantibodies have in the disease, as well as offer better insight into treatment targets and strategies. Finally, rigorous randomized, multicenter controlled trials of specific treatments are crucial to improve patient outcomes and avoid unnecessary exposure to possibly ineffective and potentially dangerous immunotherapies.

## References


1. Moersch FP, Woltman HW. Progressive fluctuating muscular rigidity and spasm ("stiff-man" syndrome); report of a case and some observations in 13 other cases. Proc Staff Meet Mayo Clin. 1956;31(15):421–7.
2. Hadavi S, Noyce AJ, Leslie RD, Giovannoni G. Stiff person syndrome. Pract Neurol. 2011;11(5):272–82.
3. Galli JR, Austin SD, Greenlee JE, Clardy SL. Stiff person syndrome with anti-gad65 antibodies within the national veterans affairs health administration. Muscle Nerve. 2018;58(6):801–4.
4. Dalakas MC, Fujii M, Li M, McElroy B. The clinical spectrum of anti-gad antibody-positive patients with stiff-person syndrome. Neurology. 2000;55(10):1531–5.
5. McKeon A, Robinson MT, McEvoy KM, Matsumoto JY, Lennon VA, Ahlskog JE, et al. Stiff-man syndrome and variants: clinical course, treatments, and outcomes. Arch Neurol. 2012;69(2):230–8.
6. Clardy SL, Lennon VA, Dalmau J, Pittock SJ, Jones HR, Renaud DL, et al. Childhood onset of stiff-man syndrome. JAMA Neurol. 2013;70(12):1531–6.
7. Gordon EE, Januszko DM, Kaufman L. A critical survey of stiff-man syndrome. Am J Med. 1967;42(4):582–99.
8. Lorish TR, Thorsteinsson G, Howard FM. Stiff-man syndrome updated. Mayo Clin Proc. 1989;64(6):629–36.
9. Vasconcelos OM, Dalakas MC. Stiff-person syndrome. Curr Treat Options Neurol. 2003;5(1):79–90.
10. Saiz A, Graus F, Valldeoriola F, Valls-Solé J, Tolosa E. Stiff-leg syndrome: a focal form of stiff-man syndrome. Ann Neurol. 1998;43(3):400–3.
11. Thompson PD. The stiff-man syndrome and related disorders. Parkinsonism Relat Disord. 2001;8(2):147–53.
12. Brown P, Rothwell JC, Marsden CD. The stiff leg syndrome. J Neurol Neurosurg Psychiatry. 1997;62(1):31–7.
13. Barker RA, Revesz T, Thom M, Marsden CD, Brown P. Review of 23 patients affected by the stiff man syndrome: clinical subdivision into stiff trunk (man) syndrome, stiff limb syndrome, and progressive encephalomyelitis with rigidity. J Neurol Neurosurg Psychiatry. 1998;65(5):633–40.
14. Meinck HM, Thompson PD. Stiff man syndrome and related conditions. Mov Disord. 2002;17(5):853–66.
15. Murinson BB, Guarnaccia JB. Stiff-person syndrome with amphiphysin antibodies: distinctive features of a rare disease. Neurology. 2008;71(24):1955–8.
16. Martinez-Hernandez E, Ariño H, McKeon A, Iizuka T, Titulaer MJ, Simabukuro MM, et al. Clinical and immunologic investigations in patients with stiff-person spectrum disorder. JAMA Neurol. 2016;73(6):714–20.

17. Carvajal-González A, Leite MI, Waters P, Woodhall M, Coutinho E, Balint B, et al. Glycine receptor antibodies in perm and related syndromes: characteristics, clinical features and outcomes. Brain. 2014;137(Pt 8):2178–92.
18. Solimena M, Folli F, Denis-Donini S, Comi GC, Pozza G, De Camilli P, et al. Autoantibodies to glutamic acid decarboxylase in a patient with stiff-man syndrome, epilepsy, and type i diabetes mellitus. N Engl J Med. 1988;318(16):1012–20.
19. Li L, Hagopian WA, Brashear HR, Daniels T, Lernmark A. Identification of autoantibody epitopes of glutamic acid decarboxylase in stiff-man syndrome patients. J Immunol. 1994;152(2):930–4.
20. Dalakas MC, Li M, Fujii M, Jacobowitz DM. Stiff person syndrome: quantification, specificity, and intrathecal synthesis of gad65 antibodies. Neurology. 2001;57(5):780–4.
21. Rakocevic G, Raju R, Dalakas MC. Anti-glutamic acid decarboxylase antibodies in the serum and cerebrospinal fluid of patients with stiff-person syndrome: correlation with clinical severity. Arch Neurol. 2004;61(6):902–4.
22. Rosin L, DeCamilli P, Butler M, Solimena M, Schmitt HP, Morgenthaler N, et al. Stiff-man syndrome in a woman with breast cancer: an uncommon central nervous system paraneoplastic syndrome. Neurology. 1998;50(1):94–8.
23. Hutchinson M, Waters P, McHugh J, Gorman G, O'Riordan S, Connolly S, et al. Progressive encephalomyelitis, rigidity, and myoclonus: a novel glycine receptor antibody. Neurology. 2008;71(16):1291–2.
24. Petit-Pedrol M, Armangue T, Peng X, Bataller L, Cellucci T, Davis R, et al. Encephalitis with refractory seizures, status epilepticus, and antibodies to the gabaa receptor: a case series, characterisation of the antigen, and analysis of the effects of antibodies. Lancet Neurol. 2014;13(3):276–86.
25. Butler MH, Hayashi A, Ohkoshi N, Villmann C, Becker CM, Feng G, et al. Autoimmunity to gephyrin in stiff-man syndrome. Neuron. 2000;26(2):307–12.
26. Balint B, Jarius S, Nagel S, Haberkorn U, Probst C, Blöcker IM, et al. Progressive encephalomyelitis with rigidity and myoclonus: a new variant with dppx antibodies. Neurology. 2014;82(17):1521–8.
27. Piquet AL, Khan M, Warner JEA, Wicklund MP, Bennett JL, Leehey MA, et al. Novel clinical features of glycine receptor antibody syndrome: a series of 17 cases. Neurol Neuroimmunol Neuroinflamm. 2019;6(5):e592.
28. Graus F, Dalmau J. Paraneoplastic neurological syndromes. Curr Opin Neurol. 2012;25(6):795–801.
29. Khasani S, Becker K, Meinck HM. Hyperekplexia and stiff-man syndrome: abnormal brainstem reflexes suggest a physiological relationship. J Neurol Neurosurg Psychiatry. 2004;75(9):1265–9.
30. Mamoli B, Heiss WD, Maida E, Podreka I. Electrophysiological studies on the "stiff-man" syndrome. J Neurol. 1977;217(2):111–21.
31. Meinck HM, Ricker K, Conrad B. The stiff-man syndrome: new pathophysiological aspects from abnormal exteroceptive reflexes and the response to clomipramine, clonidine, and tizanidine. J Neurol Neurosurg Psychiatry. 1984;47(3):280–7.
32. Matsumoto JY, Caviness JN, McEvoy KM. The acoustic startle reflex in stiff-man syndrome. Neurology. 1994;44(10):1952–5.
33. Sandbrink F, Syed NA, Fujii MD, Dalakas MC, Floeter MK. Motor cortex excitability in stiff-person syndrome. Brain. 2000;123(Pt 11):2231–9.
34. Levy LM, Levy-Reis I, Fujii M, Dalakas MC. Brain gamma-aminobutyric acid changes in stiff-person syndrome. Arch Neurol. 2005;62(6):970–4.
35. Butler MH, Solimena M, Dirkx R, Hayday A, De Camilli P. Identification of a dominant epitope of glutamic acid decarboxylase (gad-65) recognized by autoantibodies in stiff-man syndrome. J Exp Med. 1993;178(6):2097–106.
36. Daw K, Ujihara N, Atkinson M, Powers AC. Glutamic acid decarboxylase autoantibodies in stiff-man syndrome and insulin-dependent diabetes mellitus exhibit similarities and differences in epitope recognition. J Immunol. 1996;156(2):818–25.
37. Dinkel K, Meinck HM, Jury KM, Karges W, Richter W. Inhibition of gamma-aminobutyric acid synthesis by glutamic acid decarboxylase autoantibodies in stiff-man syndrome. Ann Neurol. 1998;44(2):194–201.
38. Tretter V, Moss SJ. Gaba(a) receptor dynamics and constructing gabaergic synapses. Front Mol Neurosci. 2008;1:7.
39. Alexopoulos H, Dalakas MC. Immunology of stiff person syndrome and other gad-associated neurological disorders. Expert Rev Clin Immunol. 2013;9(11):1043–53.
40. Raju R, Rakocevic G, Chen Z, Hoehn G, Semino-Mora C, Shi W, et al. Autoimmunity to gabaa-receptor-associated protein in stiff-person syndrome. Brain. 2006;129(Pt 12):3270–6.
41. Wigge P, Vallis Y, McMahon HT. Inhibition of receptor-mediated endocytosis by the amphiphysin sh3 domain. Curr Biol. 1997;7(8):554–60.
42. Geis C, Weishaupt A, Hallermann S, Grünewald B, Wessig C, Wultsch T, et al. Stiff person syndrome-associated autoantibodies to amphiphysin mediate reduced gabaergic inhibition. Brain. 2010;133(11):3166–80.
43. Chang T, Alexopoulos H, McMenamin M, Carvajal-González A, Alexander SK, Deacon R, et al. Neuronal surface and glutamic acid decarboxylase autoantibodies in nonparaneoplastic stiff person syndrome. JAMA Neurol. 2013;70(9):1140–9.
44. Raju R, Foote J, Banga JP, Hall TR, Padoa CJ, Dalakas MC, et al. Analysis of gad65 autoantibodies in stiff-person syndrome patients. J Immunol. 2005;175(11):7755–62.
45. Baizabal-Carvallo JF, Jankovic J. Stiff-person syndrome: insights into a complex autoimmune disorder. J Neurol Neurosurg Psychiatry. 2015;86(8):840–8.

46. Rakocevic G, Floeter MK. Autoimmune stiff person syndrome and related myelopathies: understanding of electrophysiological and immunological processes. Muscle Nerve. 2012;45(5):623–34.
47. Saiz A, Mínguez A, Graus F, Marín C, Tolosa E, Cruz-Sánchez F. Stiff-man syndrome with vacuolar degeneration of anterior horn motor neurons. J Neurol. 1999;246(9):858–60.
48. Holmøy T, Skorstad G, Røste LS, Scheie D, Alvik K. Stiff person syndrome associated with lower motor neuron disease and infiltration of cytotoxic t cells in the spinal cord. Clin Neurol Neurosurg. 2009;111(8):708–12.
49. Warren JD, Scott G, Blumbergs PC, Thompson PD. Pathological evidence of encephalomyelitis in the stiff man syndrome with anti-gad antibodies. J Clin Neurosci. 2002;9(3):328–9.
50. Pugliese A, Solimena M, Awdeh ZL, Alper CA, Bugawan T, Erlich HA, et al. Association of hla-dqb1*0201 with stiff-man syndrome. J Clin Endocrinol Metab. 1993;77(6):1550–3.
51. Schmidli RS, Colman PG, Bonifacio E. Disease sensitivity and specificity of 52 assays for glutamic acid decarboxylase antibodies. The second international gadab workshop. Diabetes. 1995;44(6):636–40.
52. Armon C, McEvoy KM, Westmoreland BF, McManis PG. Clinical neurophysiologic studies in stiff-man syndrome: use of simultaneous video-electroencephalographic-surface electromyographic recording. Mayo Clin Proc. 1990;65(7):960–7.
53. Blum P, Jankovic J. Stiff-person syndrome: An autoimmune disease. Mov Disord. 1991;6(1):12–20.
54. Whelan JL. Baclofen in treatment of the 'stiff-man' syndrome. Arch Neurol. 1980;37(9):600–1.
55. Miller F, Korsvik H. Baclofen in the treatment of stiff-man syndrome. Ann Neurol. 1981;9(5):511–2.
56. Silbert PL, Matsumoto JY, McManis PG, Stolp-Smith KA, Elliott BA, McEvoy KM. Intrathecal baclofen therapy in stiff-man syndrome: a double-blind, placebo-controlled trial. Neurology. 1995;45(10):1893–7.
57. Stayer C, Tronnier V, Dressnandt J, Mauch E, Marquardt G, Rieke K, et al. Intrathecal baclofen therapy for stiff-man syndrome and progressive encephalomyelopathy with rigidity and myoclonus. Neurology. 1997;49(6):1591–7.
58. Spehlmann R, Norcross K, Rasmus SC, Schlageter NL. Improvement of stiff-man syndrome with sodium valproate. Neurology. 1981;31(9):1162–3.
59. Vermeij FH, van Doorn PA, Busch HF. Improvement of stiff-man syndrome with vigabatrin. Lancet. 1996;348(9027):612.
60. Rüegg SJ, Steck AJ, Fuhr P. Levetiracetam improves paroxysmal symptoms in a patient with stiff-person syndrome. Neurology. 2004;62(2):338.
61. Holmøy T. Long-term effect of gabapentin in stiff limb syndrome: a case report. Eur Neurol. 2007;58(4):251–2.
62. Hattan E, Angle MR, Chalk C. Unexpected benefit of propofol in stiff-person syndrome. Neurology. 2008;70(18):1641–2.
63. Schreiber AL, Vasudevan JM, Fetouh SK, Ankam NS, Hussain A, Rakocevic G. Atypical clinically diagnosed stiff-person syndrome response to dantrolene--a refractory case. Muscle Nerve. 2012;45(3):454–5.
64. Dalakas MC, Fujii M, Li M, Lutfi B, Kyhos J, McElroy B. High-dose intravenous immune globulin for stiff-person syndrome. N Engl J Med. 2001;345(26):1870–6.
65. Dalakas MC. The role of ivig in the treatment of patients with stiff person syndrome and other neurological diseases associated with anti-gad antibodies. J Neurol. 2005;252(Suppl 1):I19–25.
66. Pons-Estel GJ, Salerni GE, Serrano RM, Gomez-Puerta JA, Plasin MA, Aldasoro E, et al. Therapeutic plasma exchange for the management of refractory systemic autoimmune diseases: report of 31 cases and review of the literature. Autoimmun Rev. 2011;10(11):679–84.
67. Pagano MB, Murinson BB, Tobian AA, King KE. Efficacy of therapeutic plasma exchange for treatment of stiff-person syndrome. Transfusion. 2014;54(7):1851–6.
68. Tandan R, Hehir MK, Waheed W, Howard DB. Rituximab treatment of myasthenia gravis: a systematic review. Muscle Nerve. 2017;56(2):185–96.
69. Damato V, Evoli A, Iorio R. Efficacy and safety of rituximab therapy in neuromyelitis optica spectrum disorders: a systematic review and meta-analysis. JAMA Neurol. 2016;73(11):1342–8.
70. Gastaldi M, Thouin A, Vincent A. Antibody-mediated autoimmune encephalopathies and immunotherapies. Neurotherapeutics. 2016;13(1):147–62.
71. Baker MR, Das M, Isaacs J, Fawcett PR, Bates D. Treatment of stiff person syndrome with rituximab. J Neurol Neurosurg Psychiatry. 2005;76(7):999–1001.
72. Dalakas MC, Rakocevic G, Dambrosia JM, Alexopoulos H, McElroy B. A double-blind, placebo-controlled study of rituximab in patients with stiff person syndrome. Ann Neurol. 2017;82(2):271–7.
73. Sanders S, Bredeson C, Pringle CE, Martin L, Allan D, Bence-Bruckler I, et al. Autologous stem cell transplantation for stiff person syndrome: two cases from the Ottawa blood and marrow transplant program. JAMA Neurol. 2014;71(10):1296–9.

# Use of Disease-Modifying Therapies in Multiple Sclerosis

31

John R. Corboy and Robert H. Gross

## Patient Vignette

In 1999, a 36-year-old woman had a 3-week episode of numbness from the waist down, for which she did not seek medical attention and which cleared spontaneously. Three months later, she had blurred vision in the left eye with pain on lateral movement and was diagnosed clinically with optic neuritis. Brain and thoracic spine magnetic resonance imaging (MRI) scans revealed multifocal lesions in the periventricular region, in the right cerebellum, in the subcortical white matter, and at T10, with several brain lesions and the left optic nerve enhancing after administration of gadolinium (Gd). Spinal fluid examination revealed five oligoclonal bands unique to the central nervous system (CNS). She recovered incompletely after 3 days of intravenous (IV) methylprednisolone and continued to work and drive a car, noting exercise limitations (blurred vision), fatigue, and slowed cognitive processing. She was diagnosed with multiple sclerosis (MS) and began weekly intramuscular injections of interferon (IFN) β (beta)-1a, which she tolerated poorly due to continued flu-like side effects lasting 1 to 2 days after injection. She had two more relapses over the next 3 years resulting in persistent left leg weakness and imbalance with bladder urgency and switched in 2002 to subcutaneous IFN β (beta)-1a at a higher dose and greater frequency, which she tolerated poorly also. After another relapse in 2003, she switched to glatiramer acetate (GA) 20 mg daily, by subcutaneous injection, which she tolerated well. She had another relapse in late 2004 and considered switching to intravenous natalizumab (NTZ), but it was taken off the market in early 2005 before she started it. She continued GA, but after another relapse in 2009, she switched to NTZ every 4 weeks. She was no longer able to work due to fatigue, cognitive concerns, and leg weakness with imbalance. She had no new problems, but when the JC virus (JCV) serology test (STRATIFY) became available in 2013, she was noted to have been exposed to the virus, putting her at risk of development of progressive multifocal leukoencephalopathy (PML), so she switched to dimethyl fumarate (DMF) after her insurance refused to pay for rituximab (RTX) off-label. She remained relapse-free but began to note slow worsening of gait and cognitive function by 2015, at age 52. By 2018, she had been relapse-free and without magnetic resonance imaging (MRI) scan changes for 9 years but was using a cane intermittently and slowly worsening with gait and cognitive function. She asked her doctor if she should

J. R. Corboy (✉) · R. H. Gross
Department of Neurology, Rocky Mountain Multiple Sclerosis Center at the University of Colorado, Aurora, CO, USA
e-mail: John.corboy@ucdenver.edu



A. L. Piquet, E. Alvarez (eds.), *Neuroimmunology*, https://doi.org/10.1007/978-3-030-61883-4_31

switch to more effective therapy or, instead, consider a trial off her MS medication.

## Introduction

Drug development over the last 30 years has resulted in the regulatory approval of 21 disease-modifying therapies (DMTs) for MS in the United States, including multiple DMTs in the same class (interferons, fumarates, and sphingosine-1-phosphate modulators), multiple generics (two generic versions of GA, three generic versions of fingolimod), and several other drugs that are used off-label (RTX, mycophenolate mofetil, others). The first DMT has now been approved for pediatric use (fingolimod), and DMTs have been or are being tested in nearly all phases and phenotypes of the illness, including even those who have not yet had clinical symptoms suggestive of MS but whose MRIs are consistent with MS—radiologically isolated syndrome (RIS).

Given that patients may remain on a DMT for many years, their annual and cumulative costs are high, and potential risks of DMT use may be significant; maximum diagnostic certainty is extremely important. At the same time, however, evidence has accumulated that earliest possible treatment results in better outcomes. Along with unavailability of a reliable biomarker that convincingly separates those with relatively benign or severe prognoses at the earliest stages of the disease, this tension between early treatment and diagnostic certainty has helped fuel inconsistent approaches in the early use of DMTs in MS [1]. MRI markers such as the central vessel sign [2] and biomarkers such as neurofilament light [3] appear promising regarding enhancing diagnostic and perhaps prognostic certainty.

Much has been learned about the pathogenesis of MS (discussed in Part 2 of this book). The most cogent interpretation of available data suggests MS is both an inflammatory and a degenerative disorder of the central nervous system (CNS) [4]. The degenerative component begins early, even before overt symptoms suggestive of MS; persists throughout lifetime; and is linked pathologically to inflammation [5]. As people age, the clinical [6] and radiographic [7] manifestations of MS most closely linked to acute inflammation—relapses and gadolinium-enhancing (Gd+) lesions, respectively—diminish substantially. Consistent with this, autopsy studies reveal a markedly lower number of inflammatory white matter plaques in the aged MS population compared to younger patients with MS [5]. Older patients retain more diffuse background inflammation (as opposed to distinct white matter plaques) resulting in the accumulation of activated microglia, damaged mitochondria, and meningeal lymphocytic follicles that appear to be linked to underlying gray matter demyelination and atrophy [4, 8]. While there are many potential approaches to altering the underlying disease process in MS, presently available DMTs (PA-DMTs) all are immunotherapies of one form or another, directed primarily against the adaptive immune dysfunction seen most prominently in early MS lesions. It has been difficult to identify any potential neuroprotective therapies until very recently, and no PA-DMTs have yet been proven to remyelinate denuded axons or replace damaged CNS cells, either by external replacement or by inducement of available progenitor cells already residing in the CNS.

## Treatment Through Lifetime

The Patient Vignette illustrates several important aspects of MS natural history and treatment. MS phenotypes are perhaps best thought of as a continuum of phases through which a patient passes, or may pass, in their lifetime. Prior to overt MS symptoms, an unknown number of individuals will be found to have MRI evidence quite suggestive of MS: RIS [9, 10]. When MS actually "begins" remains difficult to ascertain. Based on the average number of lesions seen on the brain MRI scans of people in clinical trials [11] at the time of a first clinical relapse typical for MS, defined as clinically isolated syndrome (CIS), it is highly likely that most individuals ultimately diagnosed with MS will have lived through a prodrome [12] of non-overt symptomatology prior to the first clear clinical event. In addition to asymptomatic MRI lesions, this prodrome may consist of decreased activity levels [13], diminished cognitive function [14], loss of brain volume [15], subclinical motor impairment [16], elevated neurofilament light levels [17], and enhanced use of medical services [18]. With newer diagnostic criteria [19], many people are diagnosed with relapsing-remitting MS (RRMS) after a first demyelinating attack, in the context of appropriate MRI and/or cerebrospinal fluid (CSF) abnormalities. During the RRMS phase, patients will have a variable number of new, acute relapses; new MRI lesions in the optic nerve, brain, and spine; and varying degrees of recovery or disability related to these acute events. New relapses [6] and MRI lesions [7] diminish substantially with aging. Within 10 to 15 years after clinical onset of RRMS, many MS patients will begin to show signs of slow progression of symptoms with or without superimposed relapses, known as secondary progressive MS (SPMS) [19, 20]. Most relapses occur within 5 years (91.6%) after onset of progressive disease and/or before age 55 (95.2%) in those with SPMS [21]. Other individuals present with slow progression of MS without preceding clinical relapses, defined as primary progressive MS (PPMS) [20], typically with a similar time of progression onset as seen in SPMS. Ultimately, as people age, many will simply stop having relapses and significant MRI scan changes, with some continuing to progress in disability while others stabilize [22].

The PA-DMTs have been shown to have a variety of important positive treatment effects. To list them in detail for every DMT is beyond the scope of this chapter, but the reader is referred to the recent American Academy of Neurology guidelines on use of the DMTs where they are reviewed in detail [23]. Positive outcomes in randomized controlled trials (RCTs), generally of 6–24-month duration, include reduction of relapses, reduction in worsened MRI activity (including fewer new T1 and T2 lesions and slower brain atrophy rates), less accumulation of disability, less cognitive impairment, and better patient-reported outcomes (relatively less studied). All of the PA-DMTs are approved for RRMS, several (multiple IFNs, GA, teriflunomide, cladribine, and siponimod) are approved for CIS, one is being studied in RIS (DMF, NCT02739542), ocrelizumab is approved for PPMS, and mitoxantrone is approved for SPMS. In 2019, siponimod and cladribine have been approved also for "SPMS with activity," i.e., having both slow progression of symptoms and superimposed relapses and/or recent MRI changes based on the 2013 Lublin definition [24] (see Chap. 6). The verbiage from the US Food and Drug Administration (FDA) regarding these two latter approvals implies that patients with non-relapsing SPMS, who have worsening disability, are unlikely to benefit from these DMTs. The maximum age for inclusion criteria in clinical trials for all the FDA-approved DMTs is 55 with two exceptions, siponimod [25] (60 years old) and cladribine [26] (65 years old). Thus, there is only modest information regarding treatment suggestions for relatively older MS patients, especially those who either continue to slowly progress in spite of DMT use or those who have remained clinically and radiographically stable for prolonged periods while using a DMT. Given new demographic data concluding that 46% of adult US patients with MS are 55 or older [27], this lack of information in relatively older patients needs to be addressed. Finally, only fingolimod is approved for pediatric use [28]. Table 31.1 summarizes all the approved PA-DMTs.

Given the varying clinical and pathological manifestations of MS at different ages, it is not surprising that treatment with DMTs throughout these clinical phases may change based on markedly different principles and expectations. Treatment decisions in a young person with very early manifestations of MS are dominated by the desire to limit short-term relapses and long-term disability within a reasonable risk profile, based on a brief experience with the disease and limited availability of biomarkers to help guide understanding of their unique long-term prognosis. Pregnancy and breast-feeding issues may play a prominent role in treatment decisions [29]. In early middle age, the focus is on maintenance of a relapse-free and disability-stable status. While this view persists as people age further, the

**Table 31.1** Disease-modifying therapies in multiple sclerosis

| Drug | Dosing | Monitoring | Tolerability | Safety |
|---|---|---|---|---|
| Interferon beta-1a/1b | Variable; SC and IM | CBC, LFTs, and thyroid testing (before and every 3–6 months) | Injection site reactions, flu-like symptoms | Hematologic disturbances, LFT abnormalities, depression, skin necrosis |
| Glatiramer acetate | 20 mg SC daily, 40 mg SC three times weekly | None | Injection site reactions, lipoatrophy, postinjection syndrome (flushing, chest tightness, shortness of breath, anxiety) | Skin necrosis |
| Fingolimod<br>Siponimod | 0.5 mg PO daily<br>2.0 mg PO daily (after 5-day uptitration)<br>1.0 mg PO daily (after uptitration for those with CYP2C9*1/3* or 2*/3* genotype) | CBC and LFTs (before and every 3–6 months), VZV antibody (must be immune), EKG (before starting), macular evaluation (before and after starting), first dose monitoring of HR and BP (everyone on fingolimod and at-risk cardiac patients on siponimod), CYP2C9 variants (prior to siponimod) | Headache, back pain, GI distress (nausea, diarrhea, abdominal pain) | Bradycardia (with first dose); lymphopenia (expected); macular edema; LFT abnormalities; pulmonary function changes; infections including respiratory tract, fungal (e.g., cryptococcal meningitis), PML, and herpetic infections or reactivations (e.g., shingles); increased blood pressure; posterior reversible encephalopathy syndrome (PRES); fetal risk |
| Dimethyl fumarate<br>Diroximel fumarate | 120 mg (starting dose); 240 mg (standard dose) PO twice daily 231 mg (starting dose); 462 mg (standard dose) PO twice daily | CBC and LFTs (before and every 3–6 months) | Flushing, GI distress—less with diroximel fumarate (nausea, diarrhea, abdominal pain) | LFT abnormalities, lymphopenia, PML |
| Teriflunomide | 7 mg or 14 mg PO daily | CBC (before and every 3–6 months), LFTs (monthly for first 6 months and then every 3–6 months), pregnancy test, tuberculosis test | GI discomfort, alopecia, headache, paresthesia | Toxic epidermal necrolysis/Stevens-Johnson syndrome, teratogenicity, infection, interstitial lung disease, peripheral neuropathy, hypertension, pregnancy concerns |

| | | | | |
|---|---|---|---|---|
| Cladribine | 3.5 mg/kg PO divided into two yearly courses (1.75 mg/kg/course). Each course is given in two cycles 4–5 days each, with 23–27 days between cycles | CBC (before each course, 2 and 6 months after starting courses, and monthly if lymphocyte count < 200), LFTs (before each course), HIV, tuberculosis test, hepatitis B and C serologies, VZV antibody, pregnancy test | Headache, hypersensitivity, nausea, back and joint pain | Lymphopenia; malignancy (various types); infections including herpetic infections, pyelonephritis, tuberculosis, viral hepatitis, PML, and fungal infections; teratogenicity |
| Natalizumab | 300 mg IV every 4 weeks | CBC, LFTs, JCV antibody (every 3–6 months) | Headache, fatigue, allergic reactions | PML, anaphylaxis |
| Alemtuzumab | Cycle 1: 12 mg IV daily for 5 days<br>Cycle 2 (1 year later): 12 mg IV daily for 3 days | CBC, creatinine, urinalysis, and thyroid test (all monthly); VZV antibody (must be immune); hepatitis B and C serologies; HIV; tuberculosis test | Infusion reactions (tachycardia, BP decrease, fever, nausea, rash, flushing, headache) | Serious infusion reactions/anaphylaxis, infections including herpes infections, secondary autoimmune disease (thyroid, ITP, Goodpasture syndrome), malignancies, stroke |
| Ocrelizumab | Initial doses: 300 mg IV × 2 (2 weeks apart)<br>Subsequent doses: 600 mg IV every 6 months | CBC, CMP, lymphocyte subsets and immunoglobulin panel (all before and every 6 months), hepatitis B and C serologies, HIV, tuberculosis test | Infusion reactions (rash, flushing, itching, throat irritation) | Neutropenic fever, hypogammaglobulinemia, infections, hepatitis B reactivation |
| Mitoxantrone | 12 mg/m$^2$ IV every 3 months<br>Maximum cumulative lifetime dose: 140 mg/m$^2$ | CBC, CMP, pregnancy test, cardiac monitoring (baseline and periodically) with echocardiogram/MUGA scan/MRI | Infusion reactions, edema, GI distress, alopecia, amenorrhea | Neutropenic fever, cardiotoxicity, tissue necrosis with extravasation, leukemia, teratogenicity |

*BP* blood pressure, *CBC* complete blood count, *CMP* comprehensive metabolic panel, *EKG* electrocardiogram, *GI* gastrointestinal, *HIV* human immunodeficiency virus, *IM* intramuscular, *ITP* idiopathic thrombocytopenia purpura, *IV* intravenous, *JCV* John Cunningham virus, *LFT* liver function test, *MUGA* multigated acquisition, *MRI* magnetic resonance imaging, *PO* per os (oral), *PML* progressive multifocal leukoencephalopathy, *SC* subcutaneous, *VZV* varicella zoster virus

PA-DMTs have less obvious benefits [30, 31], and their risks may increase due to properties of the drug or the patient's own comorbidities [32]. Thus, the focus may shift to consideration of a de-escalation or discontinuation of PA-DMTs as the risk-benefit ratio evolves. Recently, the major neurological academies in the United States [23] and Europe [33] published revamped versions of guidelines for MS treatment, but processes in both academies resulted in reviews of randomized clinical trials and eschewed algorithmic approaches to disease modification. Both guidelines emphasized starting, monitoring, and switching therapies, and the American guideline had brief discussions of discontinuing DMT. Neither discussed the concepts of changing treatment goals during the lifetime of the patient or de-escalation, in any substantive fashion, the latter reflecting in part the lack of data available for review.

## Choosing a Disease-Modifying Therapy

The choice of an initial DMT should be made by a patient and their family in consultation with their medical provider after a complete discussion of the nature of MS and the choices available to alter its natural history. Adequate time should be reserved to answer the myriad questions about this complicated decision, which will be informed by disease and DMT information from a variety of sources. It will be based on a number of factors unique to that individual including the degree and perceived risk of the MS itself, patient comorbidities, and patient and doctor biases. In addition, drug aspects such as DMT efficacy, route of administration, short-term side effects, and long-term risks will be important. As of now, there are no reliable genetic or other biomarkers of response to specific DMTs that might guide treatment choice, though some baseline blood tests (JCV testing, CYP2C9 genotyping) might influence DMT selection. In a perfect world, insurance coverage would play no role in DMT decision-making, but in the United States, unfortunately, insurance intrusion is rampant, especially in the form of "step edits." These arbitrary, idiosyncratic, and ever-changing algorithms require failure (poorly and inconsistently defined) of one, or often several, less effective therapies before the insurance company will pay for more highly effective DMTs. These algorithms are nontransparent by nature, appear to be based solely on annual revisions in the costs of the various DMTs to the insurers, and are not based on any medically defensible principles of patient-centered care. Legislative efforts are underway in many states to limit, modify, or outlaw the use of step edits, which are similarly used in other disease states as well. It should be noted that physicians can help reduce costs [34] by using more cost-effective, but similar, approaches with comparable outcomes. Examples would include using generic GA or fingolimod, RTX (off-label) instead of ocrelizumab, or the less expensive between fingolimod and siponimod, i.e., drugs of the same class with similar mechanisms of action. Unfortunately, there is not enough clinical data for the generic forms of the various medications to determine if they are indeed clinically equivalent to the proprietary forms, as the FDA requires only that drug makers prove "chemical equivalency" for approval.

The timing for when to initiate a DMT varies throughout the United States and the world. Based on a number of RCTs (reviewed in references [23, 33]), in the United States, the standard of care has become to treat after a typical CIS when the MRI scans show lesions consistent with MS. While the PA-DMTs are extremely expensive [35] and medical-economic arguments have been made that they are not cost-effective given their present cost structure, the outcome data overwhelmingly demonstrate that early intervention has positive impacts in lowering level of relapses, MRI scan changes, and disability [23, 31, 33] and mortality [36] rates in young patients with typical inflammatory RRMS. While choosing to delay or never initiate a DMT may be reasonable in retrospect for some patients, until a highly reliable, early marker of benign prognosis is developed, making that decision with a newly diagnosed patient will lead to a highly uncertain outcome for many.

Our case demonstrates a typical scenario for someone diagnosed prior to 2004, when approved DMT choices were limited to interferons, GA, and mitoxantrone. Due to intolerance and ineffectiveness, switching between DMTs, even between interferons, was common. With the advent of natalizumab in 2004, for the first time, a clearly more efficacious drug was available, but it came with an initially poorly defined and potentially life-threatening risk of PML due to reactivation of the JCV. The concepts of risk stratification and mitigation became more relevant. DMT choice and sequencing, in essence, often reflected the history of the approval dates of the medications (with use of the older medications first).

With the development of greater number of DMT choices of varying routes of administration, effectiveness, and risk and noting that substantial numbers of MS patients still continue to develop disabling loss of neurological function in spite of early treatment, the initial choice of a DMT and the timing and sequence of switching DMTs have become more challenging [37]. There are two broad approaches, as below and as previously reviewed [38].

## Escalation Approach

The concept of escalation therapy is based on the principles that some significant number of MS patients may have a benign course not requiring more highly effective therapy and that it remains more appropriate to first use older, "tried-and-true" approaches that perhaps have less risk than more highly effective therapies. Even in 2020, this often still includes use of the injectable IFNs and GA. Proponents of this strategy point to the lack of long-term studies showing large enough differences in efficacy between DMTs that would warrant the greater risk for all or many from the outset. Conceptually, the patient lessens risk of the therapy but accepts a higher and poorly defined risk of new and disabling MS disease activity.

With the relatively more recent approval of multiple better-tolerated, oral DMTs, the concept of escalation therapy has shifted somewhat to using an oral choice first. Very few randomized controlled trials, either direct (teriflunomide vs. interferon [39]) or indirect (dimethyl fumarate vs. GA in a post hoc analysis [40]), have compared injectable to oral MS DMTs, but meta-analyses [41] and observational studies (reviewed in references [33, 42]) have been performed, leading many to conclude that the efficacy of the oral agents other than teriflunomide (i.e. fingolimod/siponimod, dimethyl fumarate/diroximel fumarate, and cladribine) exceeds that of the injectables and teriflunomide.

## Early Highly Effective Therapy Approach

In distinction to an escalation approach, more highly effective therapy from the outset may offer the largest number of MS patients the greatest likelihood of attaining lowest possible disability throughout their lifetime. The principles for this approach are that while studies vary regarding the degree and pace at which MS patients will accrue increasing levels of disability, a substantial majority of patients will develop unacceptable levels of disability in their lifetime. This is especially true if cognitive impairment and reduced employment are considered. Stated differently, "benign MS" is uncommon and, perhaps equally or more importantly, difficult to predict when initial treatment decisions are made. As noted previously, important and perhaps life-changing inflammatory and degenerative changes are already at work during the prodrome of MS; i.e., by the time practitioners first suggest a DMT for a patient, the disease may already have been active for several years. The risk of new inflammatory disease activity that is responsive to PA-DMTs is highest in young patients, just after onset, and diminishes as people age. Moreover, greater disease activity in the first few years after onset is associated with not only more short-term problems but also more rapid increase of disability and evolution into secondary progressive MS, where most disability accrues but where PA-DMTs are least effective. That is, the most important and effective time to treat aggressively is at the outset.

Both randomized, controlled clinical trials [43–45] and well-done observational studies [46, 47] convincingly show significant outcome differences between the older, less effective DMTs and the newer, more effective DMTs, especially the monoclonal antibodies alemtuzumab, ocrelizumab, and natalizumab. Indeed, the recently concluded fingolimod comparison to interferon β (beta)-1a in pediatric MS patients showed an 82% reduction in relapses for those treated with fingolimod [28]. Thus, earliest treatment amplified the efficacy distinctions between the two DMTs in younger patients. In a 2-year Swedish observational trial [47] of 494 RRMS patients receiving their first DMT, RTX was overall superior to IFNs, GA, oral DMTs, and NTZ in terms of likelihood of remaining on therapy at 2 years, relapse rate, and new neuroradiologic disease activity. It is true, however, that there are no long-term (greater than 2 years) controlled trials comparing low- and high-efficacy DMTs, but two such studies are presently underway (NCT03535298 and NCT03500328). In addition, it is false to state that a patient must opt between less effective, less risky choices and more effective but riskier DMTs. Risk mitigation strategies (e.g., not using NTZ in a patient who has been exposed to the JCV, avoiding fingolimod in a patient with diabetes or heart disease) can dramatically reduce important risks. For example, the serious adverse rate in the OPERA I and II studies [45] comparing ocrelizumab to interferon found only a 7% serious adverse event rate, which compares favorably to all other DMTs (reviewed in references [23, 33, 41]). Finally, comorbidities, which may limit the use of some of the more highly effective DMTs such as diabetes, hypertension, cancer, JCV, and seropositivity, increase with age [32]. It is likely safer to use the more highly effective DMTs in younger patients, although little age-specific data is available from controlled trials to fully ascertain this. Thus, in the "high-efficacy from the outset" construct, the majority of patients should be able to use a DMT with better outcomes and a reasonable risk profile. Stated more cynically, the reverse of the argument would be that deliberately using a less effective DMT at a time of highest need and most challenging prognostication simply guarantees substantially greater numbers of MS patients will develop life-altering disability with time.

High-efficacy therapy may be divided into two categories: (1) induction therapies of limited treatment duration and (2) less long-lasting high-efficacy therapies. The first, *induction therapy*, would seek to produce a permanent or near-permanent change in the underlying immune system. Some refer to this as "rebooting the immune system." Thus, after an induction period, the patient might de-escalate therapy to a less risky choice or theoretically not require any further DMT use. An example of induction therapy includes autologous hematopoietic stem cell transplantation (aHSCT), using pretransplantation conditioning regimens of varying intensities. Most early aHSCT trials were uncontrolled, open-label studies with relatively small numbers of participants of varying levels of inflammatory disease activity and levels of disability [48, 49]. One small controlled trial found aHSCT was superior to mitoxantrone [50], and a recent, larger RCT [51] in RRMS patients with active disease who were unresponsive to usual DMT compared a nonablative approach followed by aHSCT to usual MS therapy. It concluded that aHSCT was superior in this context. Overall, the data support the interpretation that younger individuals with active inflammatory disease unresponsive to approved DMTs may undergo a life-altering interruption in the previously negative trajectory of MS-related disability with this approach. Induction with mitoxantrone [52] and alemtuzumab [53] has also shown potential long-term benefits, with superior outcomes to IFNs for alemtuzumab. For all of these approaches, the known short- and long-term risks have limited their use to those with significant early and disabling disease activity, and potential delayed, unpredictable long-term effects on immune system function or cancer surveillance remain unclear. On May 1, 2019, after the collection of 39 stroke cases in those taking alemtuzumab, the European Medicines Agency (EMA) restricted (pending further review) use of alemtuzumab

[54] to those with highly active disease despite prior treatment with at least two DMTs or for whom all other DMTs are contraindicated. Mitoxantrone use has dropped precipitously since studies revealed elevated risks of delayed leukemias [55] after its use. Trials comparing aHSCT to alemtuzumab (NCT03477500), to standard of care with a variety of DMTs (NCT00273364), and to persistent therapy with alemtuzumab, ocrelizumab (OCR), or NTZ in RRMS patients who have failed at least one DMT (BEAT-MS, NCT04047628) either are underway or were set to begin in early 2020.

Alternatively, other highly effective therapies have more transient effects on immune system function. Thus, ongoing benefit may require more persistent treatment. In this context, a recent placebo-controlled trial of a brief induction with RTX (1000 mg twice over 2 weeks vs. placebo) followed by GA [56] showed only transient benefits of RTX. With persistent treatment, the requirement to continue to treat likely raises the lifetime treatment costs per patient but implies a greater degree of reversibility of immunosuppression. For example, a patient may successfully use NTZ for many years but then seroconvert to a JCV-positive state, increasing the risk of PML. That individual can stop the NTZ, and within 12–16 weeks, nearly all $\alpha$ (alpha)-4 integrin receptor occupancy will be substantially reduced [57], with the lymphocytes recovering apparently normal function, thus mitigating future PML risk. Similar protection from infectious complications may also occur upon cessation of other PA-DMTs, with treatment-induced lymphopenia resolving weeks after discontinuation of fingolimod and months after discontinuation of DMF and the anti-CD20 monoclonal antibodies OCR and RTX. Notably, both NTZ [57, 58] and fingolimod [59] have been associated with the greatest potential of severe rebound of disease activity, clinical or radiographic, after their discontinuation, suggesting minimal or no washout periods [60, 61] are most appropriate for those who had known active disease prior to starting these DMTs. It is not clear if rebound is a function of the disease biology of those who are using these DMTs or if rebound unmasks some feature unique to these particular medications. Both may be relevant.

As with our patient, switching DMT is very common. All available data suggest switching to a more highly effective DMT results in better outcomes, at least in the near term (see review in references [23, 33]). Over time, while maintaining DMT use or not, patients typically fall into two broad groups: (1) those who simply stop/nearly stop having new relapses and MRI scan changes over time and never enter a progressive phase and (2) those who enter a slowly progressive phase, with relapses similarly diminishing over time. Progression may seem to plateau for some in the progressive phase, and it is rare for patients to retain an active, relapsing state for many years [21].

## Intentional Treatment Discontinuation or De-escalation

Given the clinical, radiographic, and pathological changes over the lifetime of the patient, it is not surprising that subgroup analyses [30] of multiple phase III clinical trials with several DMTs of varying mechanisms of action have repeatedly shown greatest benefit in younger patients, especially those with recent relapses and/or enhancing MRI lesions on baseline scans. In addition, as described previously, discontinuation of DMTs in young, clinically and/or radiographically active MS patients may be associated with significant recurrence and even rebound of disease activity. The diminishing benefits of PA-DMTs as people age [30, 31] and the increase in comorbidities that may complicate the use of some DMTs in older individuals [32] argue it would be important to know if PA-DMTs may be deliberately de-escalated to safer choices or even discontinued as patients age. To answer these questions, however, there are few published studies. Indeed, it is unknown how many MS patients may de-escalate therapy as they age. In theory, when considering possible de-escalation, most individuals often will have already used one or more of the older, injectable therapies and escalated to a better therapy due to intolerance and/or ongoing disease activity in the past. De-escalation to an injectable in this context seems illogical or might simply be refused based on unwillingness to return to an injection therapy or a therapy that has already been proven ineffective in their case. Deliberate de-escalation to an oral agent such as DMF, teriflunomide, fingolimod, or siponimod might be considered more reasonable.

High-efficacy induction approaches are, by definition, succeeded by de-escalation to a less risky DMT or outright discontinuation of all DMTs at some point. Induction studies have been insufficient in length to conclude confidently that no further treatment is needed. In aHSCT studies [48, 49] with limited long-term follow-up, many patients have discontinued all DMTs. Among those using agents less likely to permanently alter the immune system, a number of database or single-center, observational studies or case series have reported analyses after DMT discontinuation [62–67]. Some have simply reported outcomes in those who have discontinued DMT and noted demographic or treatment features associated with risk of new disease activity or worsening. Others have done a similar analysis but compared patients prior to and after they stopped their DMT or to individuals in other databases. One group [62] has done a propensity-matched comparison of "stoppers and stayers." In the vast majority of these studies, patients stopped DMT for a panoply of reasons, including intolerance, perceived lack of efficacy, pregnancy, loss of insurance, and slow progression of the MS, and many had median ages in the 30s or 40s—what we would consider early for permanent discontinuation of DMTs in MS. Only two case series have looked explicitly at stopping exclusively because of lack of new disease activity [65, 66], and one of these looked only at 15 natalizumab patients with a mean age of 50 [66]. Two [65, 66] have focused on relatively older patients.

From these disparate studies, it can be concluded that variables associated with likelihood of recurrent relapse disease activity after DMT discontinuation include age, especially if under age 45; recent Gd-enhancing MRI lesions; and recent relapses. Stable middle-aged patients who discontinue natalizumab may have recurrence of inflammatory disease activity [66], consistent with prior studies showing significant recurrence, and even rebound, of activity after stopping natalizumab in younger patients. The FDA required a package insert change for fingolimod in 2018 noting a risk of severe disease activity in those discontinuing it, typically within 12 weeks of discontinuation, but no specific features were identified that defined those at greatest risk. The variables most consistently associated with worsening of disability after DMT discontinuation include older age, higher disability at time of discontinuation, and progressive disease prior to discontinuation. All of these are risks for progressive disability in general—with or without treatment. From these available studies, it is impossible to discern a specific patient and/or DMT profile that definitively predicts risk of relapse recurrence or disability worsening after DMT discontinuation. In the United States, a randomized (1:1), controlled, single-blind discontinuation study (NCT03073603) is underway, recruiting MS patients (relapsing or progressive) who are 55 or older, relapse-free for 5 years, and without new MRI scan disease activity for at least 3 years while continuously taking an approved DMT. Primary outcome measure will be new inflammatory disease activity, either relapse or any new/worsening brain MRI lesion(s). Secondary outcomes will include confirmed disability progression, other MRI measures, and several different patient-reported outcomes. A similar controlled trial is currently ongoing in France, enrolling only SPMS patients (NCT03653273), and a third is about to start in Denmark (Eva Strijbis, personal communication).

## Conclusion

To return to our Patient Vignette, DMT was initiated as soon as she became known to potential prescribers, but this might have been more effective had she been treated after her first attack of numbness in 1999. Her choices then reflected availability of relatively safe but modestly helpful injectable therapies. These were unable to halt more relapses or development of substantial disability. Over time, overt new inflammatory events diminished in the context of aging, use of more effective DMTs, and ongoing slow worsening of disability. Choices of initial treatment are now significantly more complicated, but our view is that the greatest number of individuals will have the highest likelihood of lifelong brain health [66, 67] and minimal disability if the default position is to use more highly effective therapy at the outset in the majority of newly diagnosed RRMS patients, i.e., the typical newly diagnosed individual. In this paradigm, high-efficacy therapy *is* optimization therapy for most. Stated differently, we would rather "overtreat" a small number of patients to maintain little to no disability in the majority of MS patients. At present, most practitioners would avoid induction therapy risks in treatment-naïve patients, rather saving this approach for those with demonstrated bad prognostic factors (multiple disabling early attacks on the background of significantly destructive MRI lesions while taking a DMT). Risk stratification with mitigation remains the cornerstone of highly effective DMT use. Of course, specific choices will be based on multiple factors and should always be based on patient-centered principles. It has been difficult to substantiate treatment benefit in someone 50 or older with long-standing MS whose disability is slowly progressing without superimposed relapses and MRI scan changes. While reasonable to consider an individual treatment discontinuation trial in this patient, the results cannot be easily predicted. De-escalation remains mostly unstudied, and no concrete recommendations can be made at this time. It is hoped that controlled discontinuation trials will begin to discern if it is safe and logical to discontinue PA-DMTs in those relatively older patients with no recent acute inflammatory CNS disease activity. Regardless of the results, however, there continues to be an urgent need for neuroprotective or regenerative approaches in MS that are not based on the general principle of alteration of the adaptive immune system, especially for those with significant fixed or worsening disability. This remains the greatest unmet need in MS treatment regimens.

## References


1. Solomon AJ, Corboy JR. The tension between early diagnosis and misdiagnosis of multiple sclerosis. Nat Rev Neurol. 2017;13(9):567–72.
2. Sinnecker T, Clarke MA, Meier D, Enzinger C, Calabrese M, De Stefano N, et al. MAGNIMS Study Group. Evaluation of the central vein sign as a diagnostic imaging biomarker in multiple sclerosis. JAMA Neurol. 2019;19:76(12).
3. Cantó E, Barro C, Zhao C, Caillier SJ, Michalak Z, Bove R, et al. Association between serum neurofilament light chain levels and long-term disease course among patients with multiple sclerosis followed up for 12 years. JAMA Neurol. 2019;76(11):1359–66.
4. Lassmann H. Multiple sclerosis pathology. Cold Spring Harb Perspect Med. 2018;8(3). pii: a028936.
5. Frischer JM, Bramow S, Dal-Bianco A, Lucchinetti CF, Rauschka H, Schmidbauer M, et al. The relation between inflammation and neurodegeneration in multiple sclerosis brains. Brain. 2009;132(Pt 5):1175–89.
6. Tremlett H, Zhao Y, Joseph J, Devonshire V. UBCMS Clinic Neurologists. Relapses in multiple sclerosis are age- and time-dependent. J Neurol Neurosurg Psychiatry. 2008;79:1368–74.
7. Tortorella C, Bellacosa A, Paolicelli D, Fuiani A, Di Monte E, Simone IL, et al. Age-related gadolinium-enhancement of MRI brain lesions in multiple sclerosis. J Neurol Sci. 2005;239(1):95–9.
8. Howell OW, Reeves CA, Nicholas R, Carassiti D, Radotra B, Gentleman SM, et al. Meningeal inflammation is widespread and linked to cortical pathology in multiple sclerosis. Brain. 2011;134:2755–71.
9. De Stefano N, Giorgio A, Tintoré M, Pia Amato M, Kappos L, Palace J, et al. MAGNIMS study group. Radiologically isolated syndrome or subclinical multiple sclerosis: MAGNIMS consensus recommendations. Mult Scler. 2018;24(2):214–21.
10. Okuda DT, Siva A, Kantarci O, Inglese M, Katz I, Tutuncu M, et al. Radiologically isolated syndrome: 5-year risk for an initial clinical event. PLoS One. 2014;9(3):e90509.
11. Comi G, Martinelli V, Rodegher M, Moiola L, Leocani L, Bajenaru O, et al. Effects of early treatment with glatiramer acetate in patients with clinically isolated syndrome. Mult Scler. 2013;19(8):1074–83.
12. Giovannoni G. The neurodegenerative prodrome in multiple sclerosis. Lancet Neurol. 2017;16(6):413–4.
13. Dorans KS, Massa J, Chitnis T, Ascherio A, Munger KL. Physical activity and the incidence of multiple sclerosis. Neurology. 2016;87(17):1770–6.
14. Menascu S, Stern M, Aloni R, Kalron A, Magalshvili D, Achiron A. Assessing cognitive performance in radiologically isolated syndrome. Mult Scler Relat Disord. 2019;32:70–3.
15. Rojas JI, Patrucco L, Míguez J, Besada C, Cristiano E. Brain atrophy in radiologically isolated syndromes. J Neuroimaging. 2015;25(1):68–71.

16. Bonzano L, Bove M, Sormani MP, Stromillo ML, Giorgio A, Amato MP, et al. Subclinical motor impairment assessed with an engineered glove correlates with magnetic resonance imaging tissue damage in radiologically isolated syndrome. Eur J Neurol. 2019;26(1):162–7.
17. Bjornevik K, Munger KL, Cortese M, Barro C, Healy BC, Niebuhr DW, et al. Serum neurofilament light chain levels in patients with presymptomatic multiple sclerosis. JAMA Neurol. 2019;77(1):58–64.
18. Marrie RA, Wijnands JMA, Kingwell E, Zhu F, Zhao Y, Ekuma O, et al. Higher health care use before a clinically isolated syndrome with or without subsequent MS. Mult Scler Relat Disord. 2019;35:42–9.
19. Thompson AJ, Banwell BL, Barkhof F, Carroll WM, Coetzee T, Comi G, et al. Diagnosis of multiple sclerosis: 2017 revisions of the McDonald criteria. Lancet Neurol. 2018;17(2):162–73.
20. Confavreux C, Vukusic S. Natural history of multiple sclerosis: a unifying concept. Brain. 2006;129(Pt 3):606–16.
21. Paz Soldán MM, Novotna M, Abou Zeid N, Kale N, Tutuncu M, Crusan DJ, et al. Relapses and disability accumulation in progressive multiple sclerosis. Neurology. 2015;84(1):81–8.
22. Lublin FD, Reingold SC, Cohen JA, Cutter GR, Sørensen PS, Thompson AJ, et al. Defining the clinical course of multiple sclerosis: the 2013 revisions. Neurology. 2014;83(3):278–86.
23. Rae-Grant A, Day GS, Marrie RA, Rabinstein A, Cree BAC, Gronseth GS, et al. Comprehensive systematic review summary: disease-modifying therapies for adults with multiple sclerosis: report of the guideline development, dissemination, and implementation Subcommittee of the American Academy of Neurology. Neurology. 2018;90(17):789–800.
24. Lublin FD, Reingold SC, Cohen JA, Cutter GR, Sørensen PS, Thompson AJ, et al. Defining the clinical course of multiple sclerosis: the 2013 revisions. Neurology. 2014;83(3):278–86.
25. Kappos L, Bar-Or A, Cree BAC, Fox RJ, Giovannoni G, Gold R, et al. EXPAND Clinical Investigators. Siponimod versus placebo in secondary progressive multiple sclerosis (EXPAND): a double-blind, randomised, phase 3 study. Lancet. 2018;391(10127):1263–73.
26. Giovannoni G, Comi G, Cook S, Rammohan K, Rieckmann P, Soelberg Sørensen P, Vermersch P, Chang P, Hamlett A, Musch B, Greenberg SJ; CLARITY Study Group. A placebo-controlled trial of oral cladribine for relapsing multiple sclerosis. N Engl J Med. 2010;362(5):416–26. https://doi.org/10.1056/NEJMoa0902533. Epub 2010 Jan 20.PMID: 20089960.
27. Wallin MT, Culpepper WJ, Campbell JD, Nelson LM, Langer-Gould A, Marrie RA, et al. US Multiple Sclerosis Prevalence Workgroup. The prevalence of MS in the United States: a population-based estimate using health claims data. Neurology. 2019;92(10):e1029–40.
28. Chitnis T, Arnold DL, Banwell B, Brück W, Ghezzi A, Giovannoni G, et al. PARADIGMS Study Group. Trial of Fingolimod versus interferon Beta-1a in pediatric multiple sclerosis. N Engl J Med. 2018;379(11):1017–27.
29. Langer-Gould AM. Pregnancy and family planning in multiple sclerosis. Continuum (Minneap Minn). 2019;25(3):773–92.
30. Signori A, Schiavetti I, Gallo F, Sormani MP. Subgroups of multiple sclerosis patients with larger treatment benefits: a meta-analysis of randomized trials. Eur J Neurol. 2015;22(6):960–6.
31. Weideman AM, Tapia-Maltos MA, Johnson K, Greenwood M, Bielekova B. Meta-analysis of the age-dependent efficacy of multiple sclerosis treatments. Front Neurol. 2017;8:577.
32. Marrie RA, Miller A, Sormani MP, Thompson A, Waubant E, Trojano M, et al. attendees of the International Workshop on Comorbidity in Multiple Sclerosis. The challenge of comorbidity in clinical trials for multiple sclerosis. Neurology. 2016;86(15):1437–45.
33. Montalban X, Gold R, Thompson AJ, Otero-Romero S, Amato MP, Chandraratna D, et al. ECTRIMS/EAN guideline on the pharmacological treatment of people with multiple sclerosis. Eur J Neurol. 2018;25(2):215–37.
34. Kister I, Corboy JR. Reducing costs while enhancing quality of care in MS. Neurology. 2016;87(15):1617–22.
35. Hartung DM, Bourdette DN, Ahmed S, Witham RH. The cost of multiple sclerosis drugs in the US and the pharmaceutical industry: too big to fail? Neurology. 2015;84:1–8.
36. Goodin DS, Reder AT, Ebers GC, Cutter G, Kremenchutzky M, Oger J, et al. Survival in MS: a randomized cohort study 21 years after the start of the pivotal IFNβ-1b trial. Neurology. 2012;78(17):1315–22.
37. Gross RH, Corboy JR. Monitoring, switching, and stopping multiple sclerosis disease-modifying therapies. Continuum (Minneap Minn). 2019;25(3):715–35.
38. Corboy JR, Weinshenker BG, Wingerchuk DM. Comment on 2018 American Academy of Neurology guidelines on disease-modifying therapies in MS. Neurology. 2018;90(24):1106–12.
39. Vermersch P, Czlonkowska A, Grimaldi LM, Confavreux C, Comi G, Kappos L, et al. TENERE Trial Group. Teriflunomide versus subcutaneous interferon beta-1a in patients with relapsing multiple sclerosis: a randomised, controlled phase 3 trial. Mult Scler. 2014;20(6):705–16.
40. Fox RJ, Miller DH, Phillips JT, Hutchinson M, Havrdova E, Kita M, et al. CONFIRM Study

Investigators. Placebo-controlled phase 3 study of oral BG-12 or glatiramer in multiple sclerosis. N Engl J Med. 2012;367(12):1087–97.
41. Li H, Hu F, Zhang Y, Li K. Comparative efficacy and acceptability of disease-modifying therapies in patients with relapsing-remitting multiple sclerosis: a systematic review and network meta-analysis. J Neurol. 2019. MubMed info is PMID: 31129710 https://doi.org/10.1007/s00415-019-09395-w.
42. Tur C, Kalincik T, Oh J, Sormani MP, Tintoré M, Butzkueven H, et al. Head-to-head drug comparisons in multiple sclerosis: urgent action needed. Neurology. 2019;93(18):793–809.
43. Cohen JA, Coles AJ, Arnold DL, Confavreux C, Fox EJ, Hartung HP, et al. CARE-MS I investigators. Alemtuzumab versus interferon beta 1a as first-line treatment for patients with relapsing-remitting multiple sclerosis: a randomised controlled phase 3 trial. Lancet. 2012;380(9856):1819–28.
44. Coles AJ, Twyman CL, Arnold DL, Cohen JA, Confavreux C, Fox EJ, et al. CARE-MS II investigators. Alemtuzumab for patients with relapsing multiple sclerosis after disease-modifying therapy: a randomised controlled phase 3 trial. Lancet. 2012;380(9856):1829–39.
45. Hauser SL, Bar-Or A, Comi G, Giovannoni G, Hartung HP, Hemmer B, et al. Ocrelizumab versus interferon beta-1a in relapsing multiple sclerosis. N Engl J Med. 2017;376:221–34.
46. Kalincik T, Brown JWL, Robertson N, Willis M, Scolding N, Rice CM, et al. MSBase Study Group. Treatment effectiveness of alemtuzumab compared with natalizumab, fingolimod, and interferon beta in relapsing-remitting multiple sclerosis: a cohort study. Lancet Neurol. 2017;16(4):271–81.
47. Granqvist M, Boremalm M, Poorghobad A, Svenningsson A, Salzer J, Frisell T, et al. Comparative effectiveness of rituximab and other initial treatment choices for multiple sclerosis. JAMA Neurol. 2018;75:320–7.
48. Muraro PA, Pasquini M, Atkins HL, Bowen JD, Farge D, Fassas A, et al. Multiple Sclerosis–Autologous Hematopoietic Stem Cell Transplantation (MS-AHSCT) Long-term Outcomes Study Group. Long-term outcomes after autologous hematopoietic stem cell transplantation for multiple sclerosis. JAMA Neurol. 2017;74(4):459–69.
49. Cohen JA, Baldassari LE, Atkins HL, Bowen JD, Bredeson C, Carpenter PA, et al. Autologous hematopoietic cell transplantation for treatment-refractory relapsing multiple sclerosis: position statement from the American Society for Blood and Marrow Transplantation. Biol Blood Marrow Transplant. 2019 May;25(5):845–54.
50. Mancardi GL, Sormani MP, Gualandi F, Saiz A, Carreras E, Merelli E, et al; ASTIMS Haemato-Neurological Collaborative Group, On behalf of the Autoimmune Disease Working Party (ADWP) of the European Group for Blood and Marrow Transplantation (EBMT); ASTIMS Haemato-Neurological Collaborative Group On behalf of the Autoimmune Disease Working Party ADWP of the European Group for Blood and Marrow Transplantation EBMT. Autologous hematopoietic stem cell transplantation in multiple sclerosis: a phase II trial. Neurology. 2015;84(10):981–8.
51. Burt RK, Balabanov R, Burman J, Sharrack B, Snowden JA, Oliveira MC, et al. Effect of nonmyeloablative hematopoietic stem cell transplantation vs continued disease-modifying therapy on disease progression in patients with relapsing-remitting multiple sclerosis: a randomized clinical trial. JAMA. 2019;321(2):165–74.
52. Edan G, Comi G, Le Page E, Leray E, Rocca MA. Filippi M; French–Italian Mitoxantrone Interferon-beta-1b Trial Group. Mitoxantrone prior to interferon beta-1b in aggressive relapsing multiple sclerosis: a 3-year randomised trial. J Neurol Neurosurg Psychiatry. 2011;82(12):1344–50.
53. Arroyo R, Bury DP, Guo JD, Margolin DH, Melanson M, Daizadeh N, et al. Impact of alemtuzumab on health-related quality of life over 6 years in CARE-MS II trial extension patients with relapsing-remitting multiple sclerosis. Mult Scler. 2019;30:1352458519849796.
54. European Medicines Agency 2019. Lemtrada. https://www.ema.europa.eu/en/medicines/human/referrals/lemtrada. Accessed 3 Jan 2020.
55. Buttmann M, Seuffert L, Mäder U, Toyka KV. Malignancies after mitoxantrone for multiple sclerosis: a retrospective cohort study. Neurology. 2016;86(23):2203–7.
56. Honce JM, Nair KV, Sillau S, Valdez B, Miravalle A, Alvarez E, Schreiner T, Corboy JR, Vollmer TL. Rituximab vs placebo induction prior to glatiramer acetate monotherapy in multiple sclerosis. Neurology. 2019;92(7):e723–32.
57. Derfuss T, Kovarik JM, Kappos L, Savelieva M, Chhabra R, Thakur A, et al. α4-integrin receptor desaturation and disease activity return after natalizumab cessation. Neurol Neuroimmunol Neuroinflamm. 2017;4(5):e388.
58. Fox RJ, Cree BA, De Sèze J, Gold R, Hartung HP, Jeffery D, et al. RESTORE. MS disease activity in RESTORE: a randomized 24-week natalizumab treatment interruption study. Neurology. 2014;82(17):1491–8.
59. Barry B, Erwin AA, Stevens J, Tornatore C. Fingolimod rebound: a review of the clinical experience and management considerations. Neurol Ther. 2019;8(2):241–50.
60. Vollmer B, Honce JM, Sillau S, Corboy JR, Vollmer T, Nair K, et al. The impact of very short transition times on switching from Natalizumab

to Fingolimod on imaging and clinical effectiveness outcomes in multiple sclerosis. J Neurol Sci. 2018;390:89–93.
61. Naegelin Y, Rasenack M, Andelova M, Von Felten S, Fischer-Barnicol B, Amann M, et al. Shortening the washout to 4 weeks when switching from natalizumab to fingolimod and risk of disease reactivation in multiple sclerosis. Mult Scler Relat Disord. 2018;25:14–20.
62. Kister I, Spelman T, Alroughani R, Lechner-Scott J, Duquette P, Grand'Maison F, et al. Discontinuing disease-modifying therapy in MS after a prolonged relapse-free period: a propensity score-matched study. J Neurol Neurosurg Psychiatry. 2016;87(10):1133–7.
63. Bsteh G, Feige J, Ehling R, Auer M, Hegen H, Di Pauli F, et al. Discontinuation of disease-modifying therapies in multiple sclerosis - clinical outcome and prognostic factors. Mult Scler. 2017;23(9):1241–8.
64. Bonenfant J, Bajeux E, Deburghgraeve V, Le Page E, Edan G, Kerbrat A. Can we stop immunomodulatory treatments in secondary progressive multiple sclerosis? Eur J Neurol. 2017;24(2):237–44.
65. Birnbaum G. Stopping disease-modifying therapy in nonrelapsing multiple sclerosis: experience from a clinical practice. Int J MS Care. 2017;19(1):11–4.
66. Fagius J, Feresiadou A, Larsson EM, Burman J. Discontinuation of disease modifying treatments in middle aged multiple sclerosis patients. First line drugs vs natalizumab. Mult Scler Relat Disord. 2017;12:82–7.
67. Kister I, Spelman T, Patti F, Duquette P, Trojano M, Izquierdo G, et al. Predictors of relapse and disability progression in MS patients who discontinue disease-modifying therapy. J Neurol Sci. 2018;391:72–6.

# Symptom Management in Multiple Sclerosis

# 32

Christopher Langston, Michelle Fabian, and Stephen Krieger

## Patient Vignette

A 35-year-old woman with multiple sclerosis (MS) returns to clinic with complaints of severe fatigue. For 3 years, her MS has been stable with no evidence of disease activity on disease-modifying therapy (DMT). History reveals that she has significant depression and insomnia, which is exacerbated by painful spasms at night, and that recently, she has been increasing her baclofen dosage during the day to treat her spastic gait.

She is referred for botulinum toxin injections to reduce her overall need for baclofen, and her baclofen dosing is rescheduled so that she takes less baclofen during the daytime and more before bedtime, which reduces her painful nighttime spasms and consequent awakenings. Sleep hygiene is reviewed with her. Concurrently, she is also referred to physical therapy for stretching and gait training, where she learns to use a cane, and she begins regular exercise for cardiovascular fitness. Her gait mechanics improve, and she is less worn down by routine activities. These measures collectively improve the quality of her sleep, reduce her daytime sleepiness, and increase her physical reserve.

Several months later, she feels physically better, but there is a residual degree of refractory fatigue and apathy that impairs her abilities at work. To treat her depression, she is started on bupropion, is referred for cognitive behavioral therapy, and is directed to an MS support group. After several months, she reports her depression has lifted; her fatigue is less pervasive, but she still finds herself "crashing" at work after lunch or while making dinner for her family. To relieve her fatigue and augment her depression treatment, she is started on amantadine and eventually methylphenidate once daily as needed. Over time, she is able to titrate down the methylphenidate and use it sparingly.

She finds these interventions helpful; while she still struggles with fatigue, it is no longer debilitating. In this way, the primary, secondary, and tertiary causes of her fatigue have been addressed with a multimodal approach that includes targeted pharmacotherapy, with attention to dose timing and drug de-escalation when feasible.

C. Langston (✉)
Department of Neurology, Albert Einstein College of Medicine, Bronx, NY, USA
e-mail: clangston@montefiore.org

M. Fabian · S. Krieger
Department of Neurology, Mount Sinai Hospital, Icahn School of Medicine at Mount Sinai, New York, NY, USA

## Introduction

The treatment of multiple sclerosis (MS) is directed toward improving each patient's quality of life, primarily by impeding illness progression and by promoting wellness. Treatments are



A. L. Piquet, E. Alvarez (eds.), *Neuroimmunology*, https://doi.org/10.1007/978-3-030-61883-4_32

distinguished by the time horizon of expected efficacy: *disease-modifying therapy* aims to minimize disability *in the future*, and *symptom management* aims to maximize function *in the present*. This chapter reviews the management of 11 common or important symptoms: gait dysfunction, spasticity, fatigue, visual impairment, tremor, cognitive impairment, emotional disorders, pain, bowel/bladder dysfunction, sexual dysfunction, and dysphagia/dysarthria.

MS symptoms vary depending on the affected areas of the central nervous system (CNS). Symptoms are classified by etiology: *primary* symptoms result directly from demyelination and axon loss, *secondary* symptoms result from chronic physical reactions to primary symptoms, and *tertiary* symptoms are the psychosocial effects of primary and secondary symptoms. Accordingly, a "single" symptom can have all three components. For example, "fatigue" has several layers: poorly functioning axonal connections can cause fatigue (primary) and an energy-inefficient gait, which saps endurance (secondary), leading to social isolation and depression, which consequently instills mental exhaustion (tertiary). Recognizing this three-tiered structure to any symptom is necessary to unwind the complex effects of MS.

Effective symptom management begins by taking a history that is comprehensive but targeted. The review of systems should be as complete as possible because MS affects each patient differently, symptoms evolve and fluctuate over time, and individual symptoms can have several related causes requiring independent attention. Moreover, MS can produce constellations of symptoms with self-reinforcing features such that overlooking one contributing factor misses an opportunity to break a vicious cycle [1]. A complete history can uncover "hidden" symptoms, such as sexual dysfunction [2], which patients are reluctant to bring up, and neglected symptoms, which are covert and insidious, especially when resilient patients adapt to them. For example, MS patients commonly report that their bathroom habits are "fine" because they have adapted to urinary urgency and occasional incontinence, even when this dysfunction negatively affects their lives. Surveys [3] of care providers and patients have shown that bowel/bladder/sexual dysfunction is difficult for both groups to broach. A comprehensive history that probes targeted areas can uncover these hidden, neglected, evolving, and clustering symptoms.

The history should prioritize symptoms according to risk of mortality and serious morbidity (e.g., trauma from fall, aspiration pneumonia from dysphagia, infection from skin breakdown, or suicide from depression) [4] and to their relevance to each patient's well-being. Compared to the feelings of their MS patients, neurologists often overemphasize the impact of physical dysfunction, and they tend to downplay vitality, general health, and body pain [5]. Moreover, symptoms that are common among MS patients are not necessarily bothersome or impactful. For example, fatigue, weakness, pain, and bladder symptoms are extremely prevalent and occur in more than half of MS patients, but patients may find them less distressing than their prevalence would suggest [6]. By contrast, tremor, sexual dysfunction, and dysarthria are infrequently reported but significantly bothersome to patients [6]. Some surveys have found the most bothersome MS symptoms are gait and balance dysfunction, vision problems, spasticity, and depression [6, 7].

Effective symptom management involves taking a broad approach to helping patients beyond pharmacotherapy alone. Just as the analysis of a particular symptom should be multifactorial and systemic, its management should be multimodal and multidisciplinary, potentially involving home visits, physical therapists, sexual therapists, cognitive psychologists, psychiatrists, urologists, and pain and rehabilitation specialists. Symptom management should evolve over the course of the disease. In newly diagnosed patients, promoting a feeling of self-efficacy in symptom management can substantially alter patients' outlook and quality of life [8]. Symptom management also goes beyond symptom therapy; it represents a critical opportunity to reassess polypharmacy, increase safety and prevent injury, maintain employment [9, 10], ameliorate financial anxieties [9], reduce social isolation and promote satisfying personal relationships, enable self-sufficiency with activities of daily living, and relieve caregiver burden. Although symptom management can be daunting, it is an exciting challenge to find creative solutions and promote hope.

An anticipatable concern is that symptom management can seem like a time-consuming, low-priority part of clinical encounters that distracts from disease modification. However, symptom management is central to excellent MS care. The American Academy of Neurology (AAN) has published 11 quality metrics for MS care, eight of which directly involve symptom management: fall risk, bladder infection, physical activity, fatigue, cognitive impairment, depression screening, depression outcomes, and overall quality of life [11]. Tracking these issues requires efficient data collection and documentation. Table 32.1 lists relevant clinical tools, some of which patients can complete in the waiting room prior to appointments [12–22]. Electronic medical record (EMR) templates and prompts can also promote completeness.

**Table 32.1** Some instruments for routine symptom management of multiple sclerosis (MS)

| Symptom | Instrument | Advantages | Disadvantages |
|---|---|---|---|
| *Quality of life* (including dysmobility, spasticity, fatigue, pain, bowel/bladder dysfunction, sexual dysfunction, dysphagia/dysarthria) | MSQoL-54 [12] | Extensively validated [13]<br>Internally consistent [13]<br>Test-retest reliable [13]<br>Sensitive to change [13]<br>Has fine-grained domains<br>Takes 20 minutes [13]<br>Many languages available [13] | Does not assess impact of vision loss [13]<br>Weighted to physical disability through inclusion of SF-36<br>Domains are not patient-specific<br>Many questions, which can be more time-consuming with some patients |
| | EQ-5D | Domains are patient-specific | Can miss impact of common, disease-specific symptoms |
| | MSIS-29 [14] | Test-rest reliable [13]<br>Sensitive to change [15, 16]<br>Includes physical and psychological domains [14] | Does not assess impact of vision loss and sexual dysfunction [13]<br>Domains are not fine-grained or patient-specific |
| *Cognition* | SDMT [17] | Sensitive to early cognitive decline in MS patients [17] | Not particularly specific to MS-related cognitive impairment |
| *Emotional disorders* | Emotional Lability Scale [18] | High sensitivity and specificity for PBA [18] | |
| | BDI [19] | Validated [19]<br>Internally consistent [19] | Scores may be influenced by other MS symptoms, [20] especially work difficulty and fatigue |
| | PHQ-2/PHQ-9 [20] | High sensitivity (PHQ-2) [21] and specificity (PHQ-9) [21] make it good for identifying patients with depression<br>Easily accessible through the public domain [21] | Scores may be influenced by other MS symptoms, esp. fatigue [22]<br>Does not distinguish mood frequency from severity<br>Unclear test-retest reliability and responsiveness to change render it less suitable for monitoring depression over time |
| | GHQ | Patient-administered<br>Generally valid, reliable, and internally consistent | Interpretation is complicated by different scoring methods and by versions with varying numbers of questions |

*Abbreviations: MSQoL54* Multiple Sclerosis Quality of Life 54-item scale; *SF-36* Short-Form Health Survey, 36 items; *EQ-5D* European Quality of Life 5-Domain scale; *MSIS-29* Multiple Sclerosis Impact Scale, 29 items; *SDMT* Single-Digit Modalities Test; *PBA* pseudobulbar affect; *BDI* Beck Depression Inventory; *PHQ-2/9* Patient Health Questionnaire, 2 items/9 items; *GHQ* General Health Questionnaire

## Dysmobility: Gait and Imbalance

Dysmobility is the biggest factor affecting patient quality of life [6] and contributes to fall risk, whose assessment is a quality metric for MS care [11]. Causes include weakness, spasticity, poor vision, ataxia (sensory or cerebellar), fatigability, pain, and altered gait mechanics from causes like poor hip thrust, foot drop, decreased range of motion in the knees and ankles [23] (e.g., from heel cord tightening), agonist-antagonist mismatch, and edema [24].

Treatment strategies focus on physical rehabilitation, including conditioning programs, adaptive exercises such as transfer training, and vestibular rehabilitation [25]. Accordingly, early referral to physical therapy is an American Academy of Neurology (AAN) quality measure for MS care [11]. Assistive devices and medications can improve function and may help [26] with rehabilitation: stabilizers, braces, orthotics, canes, crutches, walkers, and functional electrical stimulators can improve gait mechanics, diminish pain, prevent contractures, and improve energy efficiency, potentially reducing fatigue [27]. Dalfampridine can improve walking speed and enable improved fitness, even in patients with severely affected gait (Expanded Disability Status Scale [EDSS] >6.0); in clinical trials, about 40% of patients were responders [28–31]. It may also improve heat sensitivity by reducing conduction block across demyelinated connections. Dalfampridine can lower the seizure threshold and worsen tonic spasms and other ephaptic electrical discharges. The risk of these side effects can be lessened by regular dosing or by eliminating the second daily dose. Since dalfampridine is renally excreted, creatinine clearance should be monitored. Discontinuing medicines that worsen dizziness or cognition can reduce the fear of falling, which itself is a source of elevated fall risk [32]. Since dysmobility tends to worsen over time [33], it should be revisited regularly.

Wheelchairs may reduce fall risk and fatigability in some circumstances, but they do not eliminate fall risk [34] and can worsen deconditioning in otherwise ambulatory patients when not used selectively. Patients in wheelchairs should be regularly evaluated for proper posture to avoid pressure ulcers.

## Spasticity

Spasticity has two overall forms: [35, 36] phasic and tonic. Both involve velocity-dependent resistance to muscle stretch. Phasic, which is more paroxysmal and dynamic, is responsive to classical anti-epileptic drugs. Tonic, which is more chronic and static, is responsive to different medicines. The advantage of tonic spasticity is that it stabilizes weakness at particular joints; the disadvantages are that it slows movement, causes painful interference with activity, and contributes to contractures and long-term joint injury. Treatment aims to soften the disadvantages of spasticity without overshooting and destabilizing weak joints. Goals of spasticity management include improving functional ability and dependence, decreasing pain, preventing and ameliorating contractures, facilitating hygiene (e.g., with a contracted upper extremity), aiding rehabilitation, and saving caregiver time.

Like dysmobility, the management of spasticity focuses on physical rehabilitation, but it is more targeted to interventions such as stretching, pilates, yoga, and functional electrical stimulation; weight-bearing exercises, pool therapy, and inhibitory casting can also be beneficial. Aside from recommending these activities, MS care providers can facilitate rehabilitation with pharmacotherapy. Medication selection should be informed not only by effectiveness but also by the treatment of comorbidities and the side-effect profile.

For phasic spasticity, MS care providers should work with patients to identify and remove noxious stimuli and prescribe medicines that break the pain-spasticity cycle, especially with anti-epileptic drugs such as gabapentin, pregabalin, levetiracetam, carbamazepine, oxcarbazepine, and lamotrigine, alone or in combination. When it is difficult to distinguish phasic spasticity from periodic limb movements [37, 38], a trial of dopamine agonists can provide both therapy and diagnosis.

For tonic spasticity, medication effectiveness depends on the affected muscles and the location of the upper motor neuron injury. With larger or more proximal muscles affected by spasticity from spinal cord lesions, baclofen, a gamma-aminobutyric acid (GABA)-B agonist, is extremely effective and can help with sleep and addiction [39], though it can worsen edema. Tizanidine, a presynaptic alpha-2 agonist, is effective for spinal and cerebral spasticity and has less of a tendency to overshoot and weaken spastic muscles or interfere with concentration; it may also lessen pain. Dantrolene, a calcium release inhibitor in muscle, is effective for spasticity of cerebral origin. All three medicines are non-habit-forming, though abrupt cessation of baclofen can precipitate a syndrome similar to benzodiazepine withdrawal including seizures [40, 41] and abrupt cessation of tizanidine can precipitate hypertension. All three require liver monitoring, and dantrolene has caused some cases of liver failure [42]. Dantrolene is less sedating than oral baclofen and tizanidine because it does not act centrally, but it can worsen weakness more than tizanidine and baclofen. Cannabinoids also show promising efficacy for spasms and associated pain, but they can also induce weakness, nausea, behavioral or mood changes, suicidal ideation and hallucinations, fatigue, and other symptoms [43]. With smaller, more distal muscles, botulinum toxin can be extremely effective with minimal systemic side effects (in addition, parotid administration also lessens sialorrhea in patients with dysphagia). Implanting an intrathecal baclofen pump can mitigate the systemic effects of orally administered agents and provide steadier dosing, though risks and benefits of implanted hardware should be considered.

Benzodiazepines are useful second-line agents, which can help with sleep and anxiety but are generally not preferred because they are habit-forming, can worsen cognition and ataxia, and increase the risk of falls; at higher doses, abrupt withdrawal can precipitate seizures. Other second-line agents include tiagabine, a GABA reuptake inhibitor similar to baclofen, and clonidine, an alpha-2 agonist similar to tizanidine. Intrathecal phenol is a second-line agent to baclofen, but it is not preferred because it causes fecal and urinary incontinence [44].

Centrally acting muscle relaxants such as cyclobenzaprine, carisoprodol, methocarbamol, metaxalone, and chlorzoxazone are helpful in patients with concomitant musculoskeletal pain [45]. Gabapentin has shown similar promise in spasticity for patients with neurogenic pain [44]. These medications can be used concurrently at low dose to treat spasticity through several mechanisms and to address comorbidities while avoiding the side effects of higher doses. Multiplying medications should proceed under strict supervision to avoid the complications of polypharmacy.

## Fatigue

Fatigue is the most commonly reported MS symptom and affects up to 75% of MS patients [46]. It is comorbid with depression and sleep disturbances (insomnia, snoring, restless legs) [9], and since these three symptoms can form a self-reinforcing triad, complaints of any one symptom should trigger screening for the others. The time course of primary MS fatigue involves a crash in energy part way through the day or in response to elevated body temperature, whereas secondary MS fatigue from sleep disturbances is usually present upon awakening; tertiary MS fatigue from depression is more pervasive and may linger long after patients report no longer feeling depressed [47]. Secondary MS fatigue is affected by spasticity, pain, deconditioning, urinary frequency (especially nocturia interfering with sleep), and medications, each of which should be addressed. In addition to treating secondary and tertiary MS fatigue, MS care providers should educate patients about strategies for dealing with primary fatigue through heat avoidance, scheduled napping, and improving strength, endurance, and cardiovascular fitness.

Patients whose MS fatigue is refractory to strategies for primary MS fatigue and the treatment of secondary and tertiary causes can be treated with stimulants, which must be managed carefully since they can interfere with sleep and thus worsen fatigue. Patients should be counseled to avoid taking stimulants close to bedtime and may need to reinforce sleep hygiene when using them. The Multiple Sclerosis Council recommends amantadine as first-line treatment for refractory fatigue; its effects are enhanced by taking a drug holiday 2 days per week [48]. Modafinil and armodafinil have shown mixed results but may benefit some patients [49–52]. Other therapies include methylphenidate, dextroamphetamine, lisdextroamphetamine, or atomoxetine, some of which have shown mixed results as augmentation therapies for depression. While some pharmacodynamic effects of these medications are understood, their complete mechanism of action is not, and MS care providers should be cautious about using them in combination, especially at higher doses. The potential for addiction and abuse must be considered, particularly with amphetamine-based agents.

## Visual Impairment

After dysmobility, MS patients report the next most bothersome symptom is visual impairment, which can contribute to falls, accidents, unemployment, and social isolation. Visual impairment in MS takes several forms including reduced visual acuity and scotomas, often due to optic neuritis; visual field deficits, which may be due to damage to the optic radiations; and eye movement dysfunction, including nystagmus, oscillopsia, and gaze palsies, which are caused by lesions to the afferents and efferents of the third, fourth, and sixth cranial nerve nuclei. Diplopia and reduced depth perception can also occur as a result of reduced eye mobility. Primary symptoms should be addressed with ophthalmology referral for prosthetics such as eye patches or prescription glasses, including prism glasses that can help diplopia in affected directions of gaze. Visual rehabilitation is available through Lighthouse for the Blind and similar local organizations.

It is imperative to check that visual impairment is not the result of medication side effects, such as macular edema from fingolimod; dry eye or blurry vision from anticholinergics used for bladder problems; open-angle glaucoma from topiramate used for spasms and headache; nystagmus from carbamazepine, lamotrigine [53], or phenytoin; or brain infection (e.g., progressive multifocal leukoencephalopathy [PML]) from immunomodulation. Eye movement dysfunction, especially myokymia and oscillopsia, can be treated with medications such as baclofen (or clonazepam), gabapentin, carbamazepine, memantine, and timolol (or other beta-blockers) [54].

## Tremor

Tremor is a hyperkinetic rhythmic oscillation of one or more body parts. In MS, it usually does not occur at rest; instead, it is usually postural or kinetic from damage to the cerebellum or its connections. Tremor is a classical feature of MS, with prevalence between 25% and 60% [55, 56], but it is not usually a presenting symptom and, on average, develops 11 years after disease onset [57]. Patients report tremor as one of the most bothersome symptoms, though its impact is highly variable, probably in proportion to its severity and location. For instance, tremor in the dominant hand, head, or trunk is typically worse than a barely perceptible palatal tremor.

Since tremor is often worse with psychological stress, excitement, and anxiety, treatment begins by encouraging coping skills for these exacerbating factors [42]. Cognitive behavioral therapy is a reasonable low-risk treatment, though its benefits are unproven in MS patients. Medications such as propranolol can treat the anxiety that worsens tremor indirectly, even when they do not treat the tremor directly [42].

There is no high-quality evidence about the effectiveness of any medications for treating cerebellar tremor in MS patients, but clinical experience suggests that they are rarely highly effective. Case series have suggested possible benefit from carbamazepine [58], topiramate [59], primidone [60], dalfampridine [61], isoniazid, ondansetron, propranolol, trihexyphenidyl, buspirone, acetazolamide, gabapentin, and hydroxyzine [61]. For rubral tremor, which is coarser and affects more proximal muscles than classical cerebellar tremor, levodopa has been effective, though it has not been tested in MS patients specifically [62]. A reasonable strategy is to use medications that treat other MS symptoms and adjust their dose to discover if they have benefit for an individual patient's tremor too.

Other potential treatments for tremor in MS treatments include botulinum toxin, deep brain stimulation (DBS), and thalamotomy. A small randomized, double-blinded, placebo-controlled, crossover study of 23 patients found that botulinum toxin improved tremor severity in the upper limb at 6 and 12 weeks, though 42% of patients reported mild to moderate weakness in affected muscles [63]. DBS and thalamotomy have proven effective in reducing limb tremor in MS patients [64], though some studies found half of patients noticed some wearing off of effectiveness [65]. In general, DBS carries a 1.7% risk of postoperative complications associated with hemiparesis or decreased consciousness, and a 5% risk of long-term hardware disruptions requiring corrective surgery [66].

## Cognitive Impairment

MS patients frequently experience cognitive decline, especially word-finding and multitasking difficulties [67]. Cognitive difficulties may present early in the disease course and worsen, eventually affecting the majority of MS patients [68–70]. In its quality metrics for MS care, the AAN recommends regular screening for cognitive impairment. Cognition screening seeks to identify affected cognitive domains in order to modify activities to reduce injury and to adjust the expectations of caregivers, employers, and acquaintances. Healthcare providers should probe secondary symptoms of cognitive decline, especially from stress, fatigue, depression, and medication side effects, such as from benzodiazepines for spasticity and anticholinergics for urinary incontinence. MS patients may benefit from cognitive rehabilitation [71], which is a low-risk emerging therapy [72].

## Emotional Disorders

Emotional disorders are divided into two categories: (1) *mood* disorders, including depression, anxiety, bipolar disorder, and adjustment disorder, and (2) *affect* disorders, including euphoria, apathy, and pseudobulbar affect (PBA). Emotional disorders are more common in MS than other neurological diseases [4], and the cause of their increased prevalence is likely multifactorial [73]. Emotional disorders are partly *primary* symptoms that result directly from the disease process [74], and they are also secondary and tertiary symptoms that can reflect reactions to the diagnosis, feelings of lost control from the unpredictability of relapses, accumulating physical disability, social isolation, and side effects of medication, especially interferons and glucocorticoids. Emotional disorders can also worsen distress from other MS symptoms, reduce quality of life [12, 75–80], and contribute to increased incidence of suicide among MS patients [81–87].

Identifying emotional disorders is challenging. Although AAN quality guidelines recommend screening for depression and assessing depression outcomes [11], AAN treatment guidelines do not recommend any particular way to do so [73]. These guidelines suggest that the Emotional Lability Scale may be useful to screen for pseudobulbar affect, the Beck Depression Inventory or Patient Health Questionnaire (PHQ)-2 to screen for depression, and the General Health Questionnaire to screen for general emotional disturbances [73]. This challenge is compounded by the fact that it can be difficult to distinguish emotional disorders from each other, and their diagnostic criteria often require that the diagnosis is "not better explained by another mental disorder," a proviso that requires the concurrent assessment of several related disorders. In complex cases, psychiatry referrals are helpful.

When an emotional disorder has been identified, it should be analyzed according to its primary, secondary, and tertiary components, and related comorbidities should be screened for, especially sleep disorders and fatigue. Once secondary and tertiary components have been addressed, residual primary components should be treated. First, MS care providers should seek to instill a sense of self-efficacy in treating MS symptoms in general, since this has shown to improve negative affect in MS patients [8]. In short, the more patients feel empowered to achieve some control over the symptoms of the disease, the more hopeful they are, and the better they feel. Second, psychotherapy should be encouraged; cognitive behavioral therapy in particular has been shown to be effective in MS patients with depression [88–90] and may be helpful in other emotional disorders as well. Local support groups [91, 92] are available through the National MS Society. Only after these important preparatory steps have been taken should MS care providers broach pharmacotherapy, whose function should be explained to patients as an important adjunct to lifestyle modifications, rather than as monotherapy [93].

Pharmacotherapy has been extensively studied for depression and PBA in MS patients, but other emotional disorders are less well studied. Dextromethorphan with quinidine is effective [94] for MS patients with PBA but may take up to 4–5 weeks to have an effect. Selective serotonin reuptake inhibitors (SSRIs) and tricyclic antidepressants (TCAs) are also effective, often at lower doses than that required for depression [95]. Refractory cases may respond to lamotrigine, venlafaxine, mirtazapine, methylphenidate, or amantadine [96].

For major depressive disorder, individual antidepressants have shown mixed results [73]. Moreover, observational studies suggest that even when patients are treated (at rates up to 85%), depressive symptoms can persist [96]. This may be due to insufficient dosing, the selection of inappropriate agents, inadequate lifestyle modifications, unsatisfactory psychotherapy, or the refractory nature of depression in MS patients. The key to effective pharmacotherapy is to select antidepressants that treat, rather than exacerbate, each individual patient's MS symptoms. SSRIs may help with depression, but they can worsen sexual dysfunction and insomnia. In particular, fluoxetine can be helpful for patients with poor medication adherence because its half-life is longer than other SSRIs and has an active metabolite, norfluoxetine, with a half-life up to 16 days [97]. Mirtazapine is effective for patients with nausea [98], insomnia [99], or anorexia [100]. Duloxetine and desvenlafaxine are in the same class of serotonin-norepinephrine reuptake inhibitors (SNRIs) as mirtazapine but, paradoxically, are associated with nausea [101], as well as insomnia, constipation, and stress incontinence [102]. Bupropion [103] and mirtazapine [104, 105] are effective for patients with sexual dysfunction and can be used in combination with SSRIs to mitigate the latter's exacerbation of sexual dysfunction. Bupropion may also hold some promise in treating fatigue [106]. However, it lowers the seizure threshold and can unmask psychosis through its dopaminergic effects [107, 108]. Tricyclics, such as nortriptyline and desipramine, are effective for migraine prophylaxis, insomnia, sialorrhea, and urinary incontinence but can worsen lethargy and urinary retention; they also can have cardiovascular effects requiring monitoring. Venlafaxine [109], nortriptyline [110], varenicline [111], and bupropion [112] (or with the latter two in combination [113, 114]) have shown some promise in smoking cessation and addiction.

## Pain

Pain is common among MS patients and significantly impacts their quality of life [9]. Complaints of pain can have multiple etiologies, and treatment should be targeted to the cause. Primary pain from axonal loss should prompt discussion about treatments for neurogenic pain, including carbamazepine, gabapentin, pregabalin, tricyclics, SNRIs (e.g., duloxetine), and lamotrigine, alone or in combination [115]. In particular, trigeminal neuralgia should prompt imaging to identify vascular loops that can be surgically addressed. Topical treatments such as lidocaine and capsaicin can be helpful. Healthcare providers should identify the causes of secondary pain, such as weakness, tonic spasticity, and altered gait mechanics, and treat those separately.

## Bladder and Bowel Dysfunction

Bladder and bowel dysfunction is a "hidden" symptom that patients and MS care providers are often reluctant to discuss. It is important to probe symptoms of storage failure—such as urgency, urge incontinence, frequency, and nocturia/enuresis—and of emptying failure, such as hesitancy, double voiding, bladder insensitivity, and poor force of stream.

Treating incontinence can improve insomnia, reduce social isolation, and prevent urinary tract infections (UTIs), which is an MS quality care metric. Easy to implement lifestyle modifications include appropriate fluid intake, bathroom mapping, scheduled voiding, double voiding (to ensure complete emptying), and avoiding stimulants. Incontinence can improve with pelvic floor training to defer urge and to enable relaxation. Intermittent catheterization or external bladder stimulators can facilitate scheduled voiding. For occasions where the risk of incontinence is greater, patients can be advised to use diapers or absorbent pads, carry a change of clothes, and wear clothes appropriate to hide wetness.

Pharmacotherapy for urinary incontinence aims to treat detrusor hyperreflexia (i.e., neurogenic or "overactive" bladder), incomplete emptying (including detrusor sphincter dyssynergia), and frequency. Distinguishing these different contributors to incontinence may require measuring the post-void residual, which, if abnormal, should prompt urology referral for urodynamic studies. Anticholinergics such as oxybutynin, tolterodine, hyoscyamine, propantheline, flavoxate, and imipramine can improve bladder spasms, but they can cause dry mouth and eyes, blurry vision, constipation, sedation, cognitive impairment, and bladder retention. Alpha-1 antagonists such as doxazosin, terazosin, and tamsulosin can improve detrusor sphincter dyssynergia as well as autonomic dysreflexia but can cause nasal congestion, orthostasis, abnormal ejaculation, and urinary incontinence; silodosin is in the same class and has lower rates of hypotension and abnormal ejaculation [116]. Tadalafil, a phosphodiesterase-5 inhibitor, may improve overactive bladder as well as improve erectile dysfunction; it can cause headaches, hypotension/orthostasis, and nausea [117]. Mirabegron, a beta-3 agonist, relaxes detrusor smooth muscle during storage without affecting voiding and can aid with overactive bladder; it is a CYP2D6 inhibitor and should be avoided in patients with severe liver or renal impairment; it can cause slight increase in blood pressure that should prompt caution in patients with uncontrolled hypertension [116]. Desmopressin, a vasopressin analogue applied intranasally, can reduce urinary frequency, especially at night, but can cause edema, hyponatremia, headache, and weight gain [118].

These medications can also be applied in combination to avoid the side effects of higher doses. Anticholinergics and alpha-1 antagonists have synergy in combination [116], and in principle, anticholinergics can also be combined with mirabegron [119].

Incontinence that is refractory to lifestyle and behavioral and pharmacological therapies can be referred to a urologist for third-line therapy, including botulinum toxin injection, which is very effective for neurogenic bladder or for a bladder stimulator, of which there are two main varieties: sacral nerve stimulation, which improves urgency, frequency, urge incontinence, nonobstructive urinary retention, and fecal incontinence [120], and posterior tibial nerve stimulation, which improves urinary urgency and urge incontinence [121]. Hardware implantation has typically been limited to advanced cases, in part because of concerns about magnetic resonance imaging (MRI) compatibility, but newer MRI-compatible models are on the horizon [122]. Cases with refractory urinary retention should be referred for the placement of suprapubic catheters. Misconceptions notwithstanding, patients with indwelling catheters often benefit from treatment with anticholinergics [116].

While urinary dysfunction in MS patients is mainly due to spinal cord disease, bowel dysfunction is less anatomically defined and may be multifactorial. One study found that 52% of MS patients with urological complaints had at least one bowel complaint, but even in those patients with both bladder and bowel dysfunction, bowel complaints did not "correlate with the patterns of urinary disturbance, the duration of MS, or the degree of disability" [123]. The connection between MS, disease-modifying therapy, and the microbiome is still under study. All types of bowel dysfunction can be improved with dietary modifications and bowel training [124]. Constipation can respond to increasing fluids to 1.5–2 quarts per day and to 20–30 g of soluble fiber, as well as including sources of insoluble fiber [124]. Laxatives, softeners, stimulants, and enemas can be beneficial when used sparingly. Diarrhea and bowel incontinence can be more difficult to treat. It can be reduced with bowel training and timed evacuations that exploit the gastrocolic reflux 20–30 minutes after eating. Bulking agents such as Citrucel or psyllium can be helpful.

## Sexual Dysfunction

Sexual dysfunction is another "hidden" symptom with widely varying estimates of its prevalence depending on how it is measured: estimates range from the same as healthy controls [125] to almost 80% of MS patients [126–129]. It has a substantial impact on patients' quality of life [9]. It typically localizes to the spinal cord, though it also results from peri-insular injury (left more commonly than right) [130, 131]. Primary sexual dysfunction includes problems with libido, sensation (numbness, paresthesias, pain), arousal, erectile dysfunction, retrograde ejaculation, lubrication, orgasm, and satisfaction [132]. Treatment is tailored to symptoms: pelvic floor exercises [133], devices (e.g., vibrators, vacuum suction) [134], and lubricants can all be helpful adjuncts. Bupropion can increase libido; sildenafil, tadalafil, vardenafil, or urethral prostaglandin can treat erectile dysfunction; topical estrogens can treat vaginal dryness and clitoral insensitivity [4]. Secondary sexual dysfunction involves different kinds of interference, such as dysmobility, pain, spasticity, fatigue, and bowel/bladder problems [132] (see other sections of this review). Tertiary sexual dysfunction involves social isolation, depression, anger, grief, guilt, spousal burden, and self-image [132], which can be successfully treated with sexual therapy [135].

## Speech and Swallowing

Dysarthria and dysphagia often co-occur but can have distinct etiologies. Dysarthria is classified as spastic, ataxic, or mixed [4] and is one of the most bothersome symptoms for MS patients [6]. Spastic dysarthria, which results from upper motor neuron injury, can respond to treatments for spasticity, while ataxic dysarthria from disrupted cerebellar connections is more difficult to treat, but can respond to carbamazepine, oxcarbazepine, and, anecdotally, topiramate. Speech and language pathologists can offer individualized therapy. Dysphagia can result from difficulty with thin liquids, reduced tongue coordination, delay in triggering swallowing, esophageal involvement, or aversion to foods because of altered sensation. It can contribute to aspiration pneumonia [136, 137], choking, malnutrition, dehydration, and decreased quality of life. Speech and language pathologists can offer individualized therapy including working on breath support, tongue and lip coordination, postural modifications (e.g., chin tuck, head turn/tilt, and Mendelsohn maneuver), and improved caregiver support. Nutritionists can suggest dietary changes, including texture restrictions and controlled eating of frequent, small meals. Advanced cases should be referred for formal swallowing studies.

## Conclusion

MS patients should have hope that their symptoms are treatable. MS care providers should help them probe neglected and hidden symptoms and, once found, unwind each symptom's primary, secondary, and tertiary contributors. Care providers must also examine the wider context, identify constellations of self-reinforcing features, and break their vicious cycle. For example, when a patient presents with depression, care providers should screen for fatigue and insomnia; they should probe for connections between depression, falls, and impaired cognition or vision. Care providers should track MS symptoms and their management over time, since a single symptom (e.g., depression) may only rise to attention when concomitant reinforcing symptoms have emerged (e.g., sexual dysfunction), and these latter symptoms may only have emerged because of chronic or secondary or tertiary exacerbations of a subclinical primary symptom (e.g., worsening incontinence). Symptom management is challenging, which can make it an especially rewarding part of high-quality MS care—for both physicians and patients alike.

## References


1. Coyle PK. Symptom management and lifestyle modifications in multiple sclerosis. Continuum. 2016;22(3):815–36.
2. Lysandropoulos AP, Havrdova E. 'Hidden' factors influencing quality of life in patients with multiple sclerosis. Eur J Neurol. 2015;22(suppl. 2):28033.
3. Tintor M, Duddy M, Jones DE, et al. The state of MS: current insight into patient-neurologist relationships, barriers to communication, and treatment satisfaction. Poster presented at: joint ACTRIMS-ECTRIMS meeting; September 10–13, 2014; Boston, MA.
4. Newsome SD, Aliotta PJ, Bainbridge J, Bennett SE, Cutter G, Fenton K, et al. A framework of care in multiple sclerosis, part 2: symptomatic care and beyond. Int J MS Care. 2017;19(1):42–56.
5. Ysrraelit MC, Fiol MP, Gaitan MI, Correale J. Quality of life assessment in multiple sclerosis: different perception between patients and neurologists. Front Neurol. 2017;8:729.
6. Barin L, Salmen A, Disanto G. The disease burden of multiple sclerosis from the individual and population perspective: which symptoms matter most? Mult Scler Relat Disord. 2018;25:112–21.
7. Heesen C, Bohm J, Reich C, Kasper J, Goebel M, Gold SM. Patient perception of bodily functions in multiple sclerosis: gait and visual function are the most valuable. Mult Scler. 2008;14(7):988–91.
8. Calandri E, Graziano F, Borghi M, Bonino S. Young adults' adjustment to recent diagnosis of multiple sclerosis: the role of identity satisfaction and self-efficacy. Disabil Health J. 2019;12(1):72–8.
9. Cioncoloni D, Innocenti I, Bartalini S, Santarnechi E, Rossi S, Rossi A, et al. Individual factors enhance health-related quality of life outcome in multiple sclerosis patients. Significance of predictive determinants. J Neurol Sci. 2014;345(1–2):213–9.
10. Ben-Zacharia AB. Therapeutics for multiple sclerosis symptoms. Mt Sinai J Med. 2011;78:176–91.
11. Rae-Grant A, Bennet A, Sanders AE, Phipps M, Cheng E, Bever C. Neurology. 2015;85(21):1904–8.
12. Vickrey BG, Hays RD, Harooni R, Myers LW, Ellison GW. A health-related quality of life measure for multiple sclerosis. Qual Life Res. 1995;4:187–206.
13. Solari A. Role of health-related quality of life measures in the routine care of people with multiple sclerosis. Health Qual Life Outcomes. 2005;3:16.
14. Hobart JC, Lamping DL, Fitzpatrick R, Riazi A, Thompson A. The Multiple Sclerosis Impact Scale (MSIS-29): a new patient-based outcome measure. Brain. 2001;124:962–73.
15. Hobart JC, Riazi A, Lamping DL, Fitzpatrick R, Thomson AJ. How responsive is the multiple sclerosis impact scale (MSIS-29)? A comparison with other self report scales. J Neurol Neurosurg Psychiatry. 2005;76:1539–43.
16. McGuigan C, Hutchinson M. The multiple sclerosis impact scale (MSIS-29) is a reliable and sensitive measure. J Neurol Neurosurg Psychiatry. 2004;75(2):266–9.
17. Van Schependom J, D'hooghe MB, Cleynhens K, D'hooge M, Haelewyck MC, DeKeyser J, et al. The symbol digit modalities test as sentinel test for cognitive impairment in multiple sclerosis. Eur J Neurol. 2014;21(9):1219–25.
18. Smith RA, Berg JE, PopE LE, Callahan JD, Wynn D, Thisted RA. Validation of the CNS emotional lability scale for pseudobulbar affect (pathological laughing and crying) in multiple sclerosis patients. Mult Scler. 2004;10(6):679–85.
19. Mohr DC, Goodkin DE, Likosky W, Beutler L, Gatto N, Langan MK. Identification of beck depression inventory items related to multiple sclerosis. J Behav Med. 1997;20:407–14.
20. Crawford P, Webster NJ. Assessment of depression in multiple sclerosis. Int J MS Care. 2009;11(4):167–73.
21. Patten SB, Burton JM, Fiest KM, Wiebe S, Bulloch AG, Koch M, Dobson KS, Metz LM, Maxwell CJ, Jette N. Validity of four screening scales for major depression in MS. Mult Scler. 2015;21(8):1064–71.
22. Gunzler DD, Perzynski A, Morris N, Bermel R, Lewis S, Miller D. Disentangling multiple sclerosis and depression: an adjusted depression screening score for patient-centered care. J Behav Med. 2015;38(2):237–50.
23. Filli L, Sutter T, Easthope CS, Killeen T, Meyer C, Reuter K, et al. Profiling walking dysfunction in multiple sclerosis : characterization, classification, and progression over time. Sci Rep. 2018;8(1):4984.
24. Pearson OR, Busse ME, van Deursen RW, Wiles CM. Quantification of walking mobility in neurological disorders. QJM. 2004;97(8):463–75.

25. Hebert JR, Corboy JR, Manago MM, Schenkman M. Effects of vestibular rehabilitation on multiple-sclerosis-related fatigue and upright postural control: a randomized controlled trial. Phys Ther. 2011;91:1166–83.
26. Severini G, Manca M, Feraresi G, Caniatii LM, Cosma M, Baldasso F, et al. Evaluation of clinical gait analysis parameters in patients affected by Multiple Sclerosis: Analysis of kinematics. Clin Biomech (Bristol, Avon). 2017;45:1–8.
27. Everaert DG, Thompson AK, Chong SL, Stein RB. Does functional electrical stimulation for foot drop strengthen corticospinal connections? Neurorehabil Neural Repair. 2010;24:168–77.
28. Hayes KC. Impact of extended-release dalfampridine on walking ability in patients with multiple sclerosis. Neuropsychiatr Dis Treat. 2011;7:229–39.
29. Goodman AD, Brown TR, Krupp LB, Schapiro RT, Schwid SR, Cohen R, et al. Fampridine MS-F203 investigators. Sustained-release oral fampridine in multiple sclerosis: a randomised, double-blind, controlled trial. Lancet. 2009;373(9665):732–8.
30. Goodman AD, Brown TR, Edwards KR, Krupp LB, Schapiro RT, Cohen R, et al. MSF204 Investigators. A phase 3 trial of extended release oral dalfampridine in multiple sclerosis. Ann Neurol. 2010;68(4):494–502.
31. Strupp M, Teufel J, Zwergal A, Schniepp R, Khodakhah K, Feil K. Aminopyridines for the treatment of neurological disorders. Neurol Clin Pract. 2017;7(1):65–76.
32. Nilsagard YE, von Koch LK, Nilsson M, Forsberg AS. Balance exercise program reduced falls in people with multiple sclerosis: a single-group pretest-posttest trial. Arch Phys Med Rehabil. 2014;95:2428–34.
33. Kister I, Bacon TE, Chamot E, Salter AR, Cutter GR, Kalina JT, et al. Natural history of multiple sclerosis symptoms. Int J MS Care. 2013;15(3):146–58.
34. Rice L, Kalron A, Berkowitz SH, Backus D, Sosnoff JJ. Fall prevalence in people with multiple sclerosis who use wheelchairs and scooters. Medicine. 2017;96(35):e7860.
35. Burke D, Knowles L, Andrews CJ, Ashby P. Spasticity decerebrate rigidity in the clasp knife phenomenon: an experimental study in the cat. Brain. 1972;95:31–4810.
36. Mukherjee A, Chakravarty A. Spasticity mechanisms – for the clinician. Front Neurol. 2010;1:149.
37. Levy J, Lansaman T, Fremondiere F, Ferrapie AL, Sher A, Dinomais M, et al. Spasticity or periodic limb movements? Lessons from a not-uncommon case report. Ann Phys Rehabil Med. 2019;62(6):453–6.
38. Levy J, Harlety S, Mauruc-Soubirac E, Leotard A, Lofaso F, Quera-Salva MA, et al. Spasticity or periodic limb movements. Eur J Phys Rehabil Med. 2018;54(5):698–704.
39. Franklin TR, Harper D, Kampman K, Kildea-McCrea S, Jens W, Lynch KG, et al. The GABA B agonist baclofen reduces cigarette consumption in a preliminary double-blind placebo-controlled smoking reduction study. Drug Alcohol Depend. 2009;103(1–2):30–6.
40. Terrence CF, Gerhard HF. Complications of baclofen withdrawal. Arch Neurol. 1931;38(9):588–9.
41. Alvis BD, Sobey CM. Oral baclofen withdrawal resulting in progressive weakness and sedation requiring intensive care admission. Neurohospitalist. 2017;7(1):39–40.
42. Schapiro RT. Managing the symptoms of MS. New York: Demos Medical Publishing, Co., Inc.; 2014.
43. Koppel BS, Brust JC, Fife T, Bronstein J, Youssof S, Gronseth G, et al. Systematic review: efficacy and safety of medical marijuana in selected neurological disorders: report of the guideline development subcommittee of the American academy of neurology. Neurology. 2014;82(17):1556–63.
44. Toosy A, Ciccarelli O, Thomson A. Symptomatic treatment and management of multiple sclerosis. Handb Clin Neurol. 2014;122:513–62.
45. Saulino M, Jacobs BW. The pharmacological management of spasticity. J Neurosci Nurs. 2006;38(6):456–9.
46. Bennett SE, Bethoud F, Finlayson M, Heyman R. Subjective symptoms breakout group discussion. Int J MS Care. 2014;16:25–32.
47. Targum SD, Fava M. Fatigue as a residual symptom of depression. Innov Clin Neurosci. 2011;8(10):40–3.
48. Multiple Sclerosis Council for Clinical Practice Guidelines. 1998. The Consortium of Multiple Sclerosis Centers website. Available at http://mscare.org/cmsc/images/pdf/fatigue.pdf. Accessed 10 Sept 2008.
49. Rammohan KW, Rosenberg JH, Lynn DJ, Pollak CP, Negaraja HN. Efficacy and safety of Modafinil (Provigil) for the treatment of fatigue in multiple sclerosis: a two-centre phase 2 study. J Neurol Neurosurg Psychiatry. 2002;72:179–83.
50. Möller F, Poettgen J, Broemel F, Neuhaus A, Daumer M, Heesen C. HAGIL (Hamburg Vigil Study): a randomized placebo-controlled double-blind study with modafinil for treatment of fatigue in patients with multiple sclerosis. Mult Scler. 2011;17:1002–9.
51. Stankoff B, Waubant E, Confavreaux C, Edan G, Debouverie M, Rumbach L, French Modafinil Study Group, et al. Modafinil for Fatigue in MS: a randomized placebo-controlled double-blind study. Neurology. 2005;64:1139–43.
52. Lapierre Y, Hum S. Treating fatigue. Int MS J. 2007;14:64–71.
53. Alkawi A, Kattah JC, Wyman K. Downbeat nystagmus as a result of lamotrigine toxicity. Epilepsy Res. 2005;63(2–3):85–8.
54. Straube A. Pharmacology of vertigo/nystagmus/oscillopsia. Curr Opin Neurol. 2005;18(1):11–4.
55. Alusi SH, Aziz TZ, Glickman S, Jahanshahi M, Stein JF, Bain PG. Stereotactic lesional surgery for the treatment of tremor in multiple sclerosis: a prospective case-controlled study. Brain. 2001;124:1576–89.
56. Koch M, Mostert J, Heersema D, De Keyser J. Tremor in multiple sclerosis. J Neurol. 2007;254:133–45.
57. Schneider SA, Deuschl G. The treatment of tremor. Neurotherapeutics. 2014;11(1):128–38.

58. Sechi GP, Zuddas M, Piredda M, Agnetti V, Sau G, Piras ML, et al. Treatment of cerebellar tremors with carbamazepine: a controlled trial with long-term follow-up. Neurology. 1989;39:1113–5.
59. Schroeder A, Linker RA, Lukas C, Kraus PH, Gold R. Successful treatment of cerebellar ataxia and tremor in multiple sclerosis with topiramate: a case report. Clin Neuropharmacol. 2010;33:317–8.
60. Naderi F, Javadi SA, Motamedi M, Sahraian MA. The efficacy of primidone in reducing severe cerebellar tremors in patients with multiple sclerosis. Clin Neuropharmacol. 2012;35:224–6.
61. Schniepp R, Jakl V, Wuehr M, Havla J, Kümpfel T, Dieterich M, et al. Treatment with 4-aminopyridine improves upper limb tremor of a patient with multiple sclerosis: a video case report. Mult Scler. 2013;19:506–8.
62. Boelmans K, Gerloff C, Munchau A. Long-lasting effect of levodopa on Holmes' tremor. Mov Disord. 2012;27:1097–8.
63. Van Der Walt A, Sung S, Spelman T, Marriott M, Kolbe S, Mitchell P, et al. A double-blind, randomized, controlled study of botulinum toxin type a in MS-related tremor. Neurology. 2012;79:92–9.
64. Mandat T, Koziara H, Tutaj M, Rola R, Bonicki W, Nauman P. Thalamic deep brain stimulation for tremor among multiple sclerosis patients. Neurol Neurochir Pol. 2010;44(6):542–5.
65. Schuurman PR, Bosch DA, Merkus MP, Speelman JD. Long-term follow-up of thalamic stimulation versus thalamotomy for tremor suppression. Mov Disord. 2008;23:1146–53.
66. Fenoy AJ, Simpson RK Jr. Risks of common complications in deep brain stimulation surgery: management and avoidance. J Neurosurg. 2014;120(1):132–9.
67. Brandstadter R, Fabian M, Krieger S, Leavitt VM, Lewis C, Pelle G, Sumowski J. Word-finding deficits in early multiple sclerosis: a behavioral and neuroimaging investigation. Neurology. 2017;27
68. Langdon D. Cognitive impairment in multiple sclerosis – recent advances and future prospects. Eur Neurol. 2010;5(1):69–72.
69. Calabrese M, Agosta F, Rinaldi F, Mattisi I, Grossi P, Favaretto A, et al. Cortical lesions and atrophy associated with cognitive impairment in relapsing- remitting multiple sclerosis. Arch Neurol. 2009;66:1144–50.
70. Penny SA, Summers MM, Swanton JK, Cipolotti L, Miller DH, Ron MA. Changing associations between cognitive impairment and imaging in multiple sclerosis as the disease progresses. J Neuropsychiatry Clin Neurosci. 2013;25(2):134–40.
71. Mani A, Chohedri E, Ravanfar P, Mowla A, Nikseresth A. Efficacy of group cognitive rehabilitation therapy in multiple sclerosis. Acta Neurol Scand. 2018;137(6):589–97.
72. Sumowski JF, Benedict R, Enzinger C, Filippi M, Geurts JJ, Hamalainen P, et al. Cognition in multiple sclerosis: state of the field and priorities for the future. Neurology. 2018;90(6):1–11.
73. Minden SL, Feinstein A, Kalb RC, Miller D, Mohr DC, Patten SB, et al. Evidence-based guideline: assessment and management of psychiatric disorders in individuals with MS. Neurology. 2014;82:174–81.
74. Haussleiter IA, Brune M, Juckel G. Psychopathology in multiple sclerosis: diagnosis, prevalence, and treatment. Ther Adv Neurol Disord. 2009;2(1):13–29.
75. Wang JL, Reimer MA, Metz LM, Patten SB. Major depression and quality of life in individuals with multiple sclerosis. Int J Psychiatry Med. 2000;30:309–17.
76. Gulick EE. Correlates of quality of life among persons with multiple sclerosis. Nurs Res. 1997;46:305–11.
77. McIvor GP, Riklan M, Reznikoff M. Depression in multiple sclerosis as a function of length and severity of illness, age, remissions, and perceived social support. J Clin Psychol. 1984;40:1028–33.
78. O'Brien MT. Multiple sclerosis: the relationship among self-esteem, social support, and coping behavior. Appl Nurs Res. 1993;6:54–63.
79. Barin L, Salmen A, Disanto G, Babacic H, Calabrese P, Chan A, Swiss Multiple Sclerosis Registry (SMSR), et al. The disease burden of multiple sclerosis from the individual and population perspective: which symptoms matter most? Mult Scler Relat Disord. 2018;25:112–21.
80. Berrigan LI, Fisk JD, Patten SB, Tremlett H, Wolfson C, Warren S, et al. CIHR Team in the Epidemiology and Impact of Comorbidity in Multiple Sclerosis (ECoMS). Health-related quality of life in multiple sclerosis: direct and indirect effects of comorbidity. Neurology. 2016;86(15):1417–24.
81. Stenager EN, Stenager E, Koch-Henriksen N, Brønnum-Hansen H, Hyllested K, Jensen K, et al. Suicide and multiple sclerosis: an epidemiological investigation. J Neurol Neurosurg Psychiatry. 1992;55:542–5.
82. Stenager EN, Stenager E. Suicide and patients with neurologic diseases: methodologic problems. Arch Neurol. 1992;49:1296–303.
83. Stenager EN, Koch-Henriksen N, Stenager E. Risk factors for suicide in multiple sclerosis. Psychother Psychosom. 1996;65:86–90.
84. Brønnum-Hansen H, Stenager E, Nylev Stenager E, Koch-Henriksen N. Suicide among Danes with multiple sclerosis. J Neurol Neurosurg Psychiatry. 2005;76:1457–9.
85. Sadovnick AD, Eisen K, Ebers GC, Paty DW. Cause of death in patients attending multiple sclerosis clinics. Neurology. 1991;41:1193–6.
86. Feinstein A. Multiple sclerosis, depression, and suicide. BMJ. 1997;315:691–2.
87. Feinstein A. An examination of suicidal intent in patients with multiple sclerosis. Neurology. 2002;59:674–8.
88. Fiest KM, Walker JR, Bernstein CN, Graff LA, Zarychanski R, Abou-Setta AM, et al. CIHR team defining the burden and managing the effects of psychiatric comorbidity in chronic immunoinflammatory disease. Systemic review and meta-analysis of interventions for depression and anxiety in persons

with multiple sclerosis. Mult Scler Relat Disord. 2016;5:12–26.
89. Cosio D, Jin L, Siddique J, Mohr DC. The effect of telephone-administered cognitive-behavioral therapy on quality of life among patients with multiple sclerosis. Ann Behav Med. 2011;41(2):227–34.
90. Moss-Morris R, Dennison L, Landau S, Yardley L, Silber E, Chalder T. A randomized controlled trial of cognitive behavioral therapy (CBT) for adjusting to multiple sclerosis (the saMS trial): does CBT work and for whom does it work? J Consult Clin Psychol. 2013;81(2):251–62.
91. Lincoln NB, Yuill F, Holmes J, Drummond AE, Constantinescu CS, Armstrong S, Phillips C. Evaluation of an adjustment group for people with multiple sclerosis and low mood: a randomized controlled trial. Mult Scler J. 2011;17(10):1250–7.
92. Graziano F, Calandri E, Borghi M, Bonino S. The effects of a group-based cognitive behavioral therapy on people with multiple sclerosis: a randomized controlled trial. Clin Rehabil. 2014;28(3):264–74.
93. Mohr DC, Boudewyn AC, Goodkin DE, Bostrom A, Epstein L. Comparative outcomes for individual cognitive-behavior therapy, supportive-expressive group psychotherapy, and sertraline for the treatment of depression in multiple sclerosis. J Consult Clin Psychol. 2001;69:942–9.
94. Garnock-Jones KP. Dextromethorphan/quinidine: in pseudobulbar affect. CNS Drugs. 2011;25:435–45.
95. Wortzel HS, Oster TJ, Anderson CA, Arciniegas DB. Pathological laughing and crying: epidemiology, pathophysiology, and treatment. CNS Drugs. 2008;22:531–45.
96. Raissi A, Bulloch AG, Fiest KM, McDonald K, Jette N, Patten SB. Exploration of undertreatment and patterns of treatment of depression in multiple sclerosis. Int J MS Care. 2015;17(6):292–300.
97. Judge R, Parry MG, Quail D, Jacobson JG. Discontinuation symptoms: comparison of brief interruption in fluoxetine and paroxetine treatment. Int Clin Psychopharmacol. 2002;17(5):217–25.
98. Watanabe N, Omori IM, Nakagawa A, Cipriani A, Barbui C, Churchill R, et al. Mirtazapine versus other antidepressive agents for depression. Cochrane Database Syst Rev. 2011;12:CD006528.
99. Winokur A, NA DM, DP MN, Gary EM, Cormier JL, Gary KA. Comparative effects of mirtazapine and fluoxetine on sleep physiology measures in patients with major depression and insomnia. J Clin Psychiatry. 2003;64(10):1224–9.
100. Domecq JP, Prutsky G, Leppin A, Sonbol MB, Altayar O, Undavalli C, et al. Clinical review: drugs commonly associated with weight change: a systematic review and meta-analysis. J Clin Endocrinol Metab. 2015;100(2):363–70.
101. Greist J, McNamara RK, Mallinckrodt CH, Rahamajhi JN, Raskin J. Incidence and duration of antidepressant-induced nausea: duloxetine compared with paroxetine and fluoxetine. Clin Ther. 2004;26(9):1446–55.
102. Li J, Yang L, Pu C, Tang Y, Yun H, Han P. The role of duloxetine in stress urinary incontinence: a systematic review and meta-analysis. Int Urol Nephrol. 2013;45(3):679–86.
103. Coleman CC, King BB, Bolden-Watson C, Book MJ, Segraves RT, Richard N, et al. A placebo-controlled comparison of the effects on sexual functioning of bupropion sustained release and fluoxetine. Clin Ther. 2001;23:1040–58.
104. Baldwin DS. Depression and sexual dysfunction. Br Med Bull. 2001;57:81–99.
105. Benkert O, Szegedi A, Kohnen R. Mirtazapine compared with paroxetine in major depression. J Clin Psychiatry. 2000;61(9):656–63.
106. Cooper JA, Tucker VI, Papakostas GI. Resolution of sleepiness and fatigue: a comparison of bupropion and selective serotonin reuptake inhibitors in subjects with major depressive disorder achieving remission at doses approved in the European Union. J Psychopharmacol. 2014;28(2):118–24.
107. Grover S, Das PP. Can bupropion unmask psychosis? Indian J Psychiatry. 2009;51(1):53–4.
108. Kumar S, Kodela S, Detweiler JG, Kim KY, Detweiler MB. Buproprion-induced psychosis: folklore or a fact? A systematic review of the literature. Gen Hosp Psychiatry. 2011;33(6):612–7.
109. Cincirpini PM, Tsoh JY, Wetter DW, Lam C, de Moor C, Cincirpini L, et al. Combined effects of venlafaxine, nicotine replacement, and brief counseling on smoking cessation. Exp Clin Psychopharmacol. 2005;13(4):282–92.
110. Hughes JR, Stead LF, Harmann-Boyce J, Cahill K, Lancaster T. Antidepressants for smoking cessation. Cochrane Database Syst Rev. 2014;8(1):CD000031.
111. Ebbert JO, Hughes JR, West RJ, Rennard SI, Russ C, McRae TD, et al. Effect of varenicline on smoking cessation through smoking reduction: a randomized clinical trial. JAMA. 2015;313(7):687–94.
112. Wilkes S. The use of Bupropion SR in cigarette smoking cessation. Int J Chron Obstruct Pulmon Dis. 2008;3(1):45–53.
113. Vogeler T, McClain C, Evoy KE. Combination bupropion SR and varenicline for smoking cessation: a systematic review. Am J Drug Alcohol Abuse. 2016;42(2):129–39.
114. Rose JE, Behm FJ. Combination varenicline/bupropion treatment benefits highly dependent smokers in adaptive smoking cessation program. Nicotine Tob Res. 2017;19(8):999–1002.
115. Solaro CM, Ferriero G. Refractory trigeminal neuralgia successfully treated by combination therapy (Pregabalin plus lamotrigine). Mult Scler Relat Disord. 2018;25:165–6.
116. Cameron AP. Medical management of neurogenic bladder with oral therapy. Transl Androl Urol. 2016;5(1):51–62.
117. Francomano D, Ilacqua A, Cortese A, Tartaglia G, Lenzi A, Inghilleri M, et al. Effects of daily Tadalafil on lower urinary tract symptoms in young men with

multiple sclerosis and erectile dysfunction: a pilot study. J Endocrinol Investig. 2017;40(3):275–9.
118. Hoverd PA, Fowler CJ. Desmopressin in the treatment of daytime urinary frequency in patients with multiple sclerosis. J Neurol Neurosurg Psychiatry. 1998;65(5):778–80.
119. Wada N, Okazaki S, Kobayashi S, Hashizume K, Kita M, Matsumoto S, et al. Efficacy of combination therapy with mirabegron for anticholinergic resistant neurogenic bladder: Videourodynamic evaluation. Hinyokika Kiyo. 2015;61(1):7–11.
120. Brazelli M, Murray A, Fraser C. Efficacy and safety of sacral nerve stimulation for urinary urge incontinence: a systematic review. J Urol. 2006;175(3, Pt 1):835–41.
121. Peters KM, Carrico DJ, Perez-Marrero RA, Khan AU, Wooldridge LS, Davis GL, et al. Randomized trial of percutaneous tibial nerve stimulation versus sham efficacy in the treatment of overactive bladder syndrome: results from the SUmiT trial. J Urol. 2010;183(4):1438–43.
122. Elkelini MS, Hassouna MM. Safety of MRI at 1.5Tesla in patients with implanted sacral nerve neurostimulator. Eur Urol. 2006;50:311–6.
123. Chia YW, Fowler CJ, Kamm MA, Henry MM, Lemieux MC, Swash M. Prevalence of bowel dysfunction in patients with multiple sclerosis and bladder dysfunction. J Neurol. 1995;242(2):105–8.
124. Namey M, Halper J. Elimination dysfunction in multiple sclerosis. Int J MS Care. 2012;14(suppl 1):1–26.
125. Nasimbera A, Rosales J, Silva B, Alonso R, Bohorquez N, Lepera S, et al. Everything you always wanted to know about sex and neurology: neurological disability and sexuality. Arq Neuropsiquiatr. 2018;76(7):430–5.
126. Bartnik P, Weilgos A, Kacperczyk J, Pisarz K, Szymusik I, Podlecka-Pietowska A, et al. Sexual dysfunction in female patients with relapsing-remitting multiple sclerosis. Brain Behav. 2017;7(6):e00699.
127. Calabro RS, Russo M, Dattola V, De Luca R, Leo A, Grisolaghi J, et al. Sexual function in young individuals with multiple sclerosis: does disability matter? J Neurosci Nurs. 2018;50(3):161–6.
128. Balsamo R, Arcaniolo D, Stizzo M, Illiano E, Autorino R, Natale F, et al. Increased risk of erectile dysfunction in men with multiple sclerosis: an Italian cross-sectional study. Cent European J Urol. 2017;70(3):289–95.
129. Masmoudi R, Aissi M, Halouani N, Fathallah S, Louribi I, Aloulou J, et al. Female sexual dysfunction and multiple sclerosis: a case-control study. Prog Urol. 2018;28(11):530–5.
130. Winder K, Linker RA, Seifert F, Deutsch M, Engelhorn T, Dorfler A, et al. Insular multiple sclerosis lesions are associated with erectile dysfunction. J Neurol. 2018;265(4):783–92.
131. Winder K, Linker RA, Seifert F, Deutsch M, Engelhorn T, Dorfler A, et al. Neuroanatomic correlates of female sexual dysfunction in multiple sclerosis. Ann Neurol. 2016;80(4):490–8.
132. Delaney KE, Donovan J. Multiple sclerosis and sexual dysfunction: a need for further education and interdisciplinary care. NeuroRehabilitation. 2017;41(2):317–29.
133. Mosalanejad F, Afrasiabifar A, Zoladi M. Investigating the combined effect of pelvic floor muscle exercise and mindfulness on sexual function in women with multiple sclerosis: a randomized controlled trial. Clin Rehabil. 2018;32(10):1340–7.
134. Alexander M, Bashir K, Alexander C, Marson L, Rosen R. Randomized trial of clitoral vacuum suction versus vibratory stimulation in neurogenic female orgasmic dysfunction. Arch Phys Med Rehabil. 2018;99(2):299–305.
135. Zamani M, Tavoli A, Khasti BY, Sedighimornani N, Zafar M. Sexual therapy for women with multiple sclerosis and its impact on quality of life. Iran J Psychiatry. 2017;12(1):58–65.
136. Kaneoka A, Pisegna JM, Inokuchi H, Ueha R, Goto T, Nito T, et al. Relationship between laryngeal sensory deficits, aspiration, and pneumonia in patients with pneumonia. Dysphagia. 2018;33(2):192–9.
137. Langmore SE, Skarupski KA, Park PS, Fries BE. Predictors of aspiration pneumonia in nursing home residents. Dysphagia. 2002;17(4):298–307.

# Index





A. L. Piquet, E. Alvarez (eds.), *Neuroimmunology*, https://doi.org/10.1007/978-3-030-61883-4

**D**

L



M

**T**

GPSR Compliance

*The European Union's (EU) General Product Safety Regulation (GPSR) is a set of rules that requires consumer products to be safe and our obligations to ensure this.*

*If you have any concerns about our products, you can contact us on ProductSafety@springernature.com*

In case Publisher is established outside the EU, the EU authorized representative is:

Springer Nature Customer Service Center GmbH
Europaplatz 3
69115 Heidelberg, Germany

**Batch number: 10732807**

Printed by Printforce, the Netherlands